T0233876

# Texts & Monographs in Symbolic Computation

A Series of the Research Institute for Symbolic Computation, Johannes Kepler University, Linz, Austria

**Founding Editor**

Bruno Buchberger

**Series Editor**

Peter Paule, Research Institute for Symbolic Computation (RISC), Johannes Kepler University, Linz, Austria

Mathematics is a key technology in modern society. Symbolic Computation is on its way to become a key technology in mathematics. "Texts and Monographs in Symbolic Computation" provides a platform devoted to reflect this evolution. In addition to reporting on developments in the field, the focus of the series also includes applications of computer algebra and symbolic methods in other subfields of mathematics and computer science, and, in particular, in the natural sciences. To provide a flexible frame, the series is open to texts of various kind, ranging from research compendia to textbooks for courses.

Indexed by zbMATH.

More information about this series at http://www.springer.com/series/3073

Wolfgang Schreiner

# Thinking Programs

## Logical Modeling and Reasoning About Languages, Data, Computations, and Executions

 Springer

Wolfgang Schreiner
RISC
Johannes Kepler University
Linz, Austria

ISSN 0943-853X                    ISSN 2197-8409   (electronic)
Texts & Monographs in Symbolic Computation
ISBN 978-3-030-80509-8           ISBN 978-3-030-80507-4   (eBook)
https://doi.org/10.1007/978-3-030-80507-4

Mathematics Subject Classification: 03B70, 03B10, 03B44, 68N15, 68N20, 68N30, 68Q55, 68Q60, 68Q65

This Springer imprint is published by the registered company Springer Nature Switzerland AG
The registered company address is: Gewerbestrasse 11, 6330 Cham, Switzerland

*Meinen Eltern*

# Foreword

When I started my study of mathematics at the University of Innsbruck in 1960, like most freshmen, I was intimidated and impressed by the apparent intelligence of the professors who gave proofs of abstract knowledge, which was far from the concrete thinking about mathematical objects which we had seen in high-school. However, after some time, in secret, I started to doubt the quality of some of the hand-waving proofs and I wanted to look behind the scene. For this, in parallel and independent of the curriculum, I started to dig through the books on logic I found in the general library of the university. (For some reason, most of them had a yellow cover—the Springer Books on Logic. Subconsciously, this may have been the reason why, many years later, I decided that the cover the Journal of Symbolic Computation should be yellow, when I founded it in 1985.)

From that time on, it became more and more clear to me that logic is the essence of mathematical thinking and, luckily, very soon after my start as a mathematics student I got the chance to become one of the first programmers on the first computer at our university and the computer appeared to me as materialized logic. Since then, for me, mathematics, logic, and computer science was just one field and I am still fascinated and convinced by the repeated algorithmic cycles through object and meta-levels to reach higher and higher states of insight and a more and more efficient grasp of the thinking process. Understanding the spiral of logic for (algorithmic) mathematics and algorithmic mathematics for logic is so important also for steering a clear course in a time of frequent new and fancy catch words that may suggest that logical clarity and brilliance is not any more relevant in a time of intelligent machines.

In 1979, in contrast to the usual analysis/linear algebra approach, I dared to give an introduction to mathematics for first semester computer science students which was nothing else than a practical introduction to predicate logic as a working language and a unified frame for proving and programming. Over the years, many of my students shared this view on the fundamental theoretical and practical importance of logic for mathematics and informatics and did remarkable work developing ideas and tools for supporting this kind of thinking. Wolfgang Schreiner embarked on this type of research, teaching, and software development already in the mid-nineties and, over the years, accumulated enormous know-how and produced impressive and extensive teaching material and software tools for the theoretical foundation and the practical application of logic in mathematics and computer science, notably in mathematics and computer science teaching.

Now, he presents this enormous amount of work in a coherent book. I think there is hardly any other book that combines the foundation of logic, the applications of logic in computer science, and software for logic in an equally rich and comprehensive way. I wish the book a wide distribution. Given the outstanding didactic qualification of Wolfgang Schreiner, I am sure that the book will be extremely helpful for students of mathematics and computer science to get a profound training of the thinking technology that is in the center of the present age and will stay and become even more important in the next turns of the spiral of innovation.

It is also a special pleasure for me that the book appears in the RISC book series on symbolic computation with the Springer Verlag whose yellow books lured me into the field of (algorithmic) logic so many years ago. When I founded the RISC book series in 1993, for some reason, we decided that the cover should be gray. However, the contents of the books in this series were very much yellow all the time. It is great to see that, since Peter Paule took over the editorship and is giving enormous drive to the series, yellow is taking over also on the covers.

Hagenberg, Austria                                                                    Bruno Buchberger
March 2021

# Preface

## Motivation

The purpose of this book is to outline some basic principles that enable developers of computer programs (computer scientists, software engineers, programmers) to more clearly *think* about the artifacts they deal with in their daily work: data types, programming languages, programs written in these languages that compute from given inputs wanted outputs, and programs for continuously executing systems. In practice, thinking about these artifacts is often muddled by not having a suitable mental framework at hand, i.e., a *language* to appropriately express this thinking.

The core message that we want to convey is that clear thinking about programs can be expressed in a single universal language, the formal language of logic. In particular, with the help of logic we can achieve the following goals:

- *Modeling*: we can unambiguously describe the meaning of syntactic entities such as the behavior of computer programs.
- *Specifying*: we can precisely formulate constraints we impose on (the meaning of) these entities such as requirements on program executions.
- *Reasoning*: we can rigorously show that the entities indeed satisfy these constraints, e.g., that programs satisfy their specifications.

However, in order to enable this clear thinking about computer programs, we also need a framework to relate the syntactic artifacts (that have a priori not any content beyond their structure) to their formal meaning (characterized by logical formulas). The description of this relationship can be generally based on three principles:

- A *grammar* that describes the basic structure of syntactic phrases.
- A *type system* that further restricts the phrases to certain well-formed ones.
- A *function* that maps every well-formed phrase to its meaning.

Thus we can give arbitrary syntactic phrases a precise meaning about which we can rigorously reason. In fact, we will use this approach of *denotational semantics* uniformly throughout this book in order to define various formal languages, starting with the language of logic itself and ending with a language of concurrent systems.

Throughout most of this book, we understand by logic the classical first-order variant of predicate logic, short *first-order logic*, the lingua franca of formal modeling and reasoning today. While some aspects of computer programming may practically profit from more general frameworks such as higher-order logic or temporal logic (which we will also discuss in this book), first-order logic is conceptually sufficient for most purposes and in any case provides a solid basis for understanding all kinds of logical extensions.

Apart from its universal elegance and expressiveness, our logical approach to the formal modeling of and reasoning about computer programs has another advantage: due to advances in computational logic (automated theorem proving, satisfiability solving, model checking), nowadays much of this process can be supported by *software*. This book therefore accompanies its theoretical elaborations by practical demonstrations of various systems and tools that are based on respectively make use of the logical underpinnings. We hope that this will convincingly demonstrate also the actual usefulness of the presented approach.

This book has been written with a broad target audience in mind that encompasses students and practitioners in computer science and computer mathematics; it therefore tries to be as self-contained as possible and not assume a particular background in logic or mathematics. However, it focuses on a "logical" perspective to the overall areas of "formal methods" and "formal semantics" which in several aspects differs from other presentations of this topic. To get a more comprehensive picture (especially on alternative approaches), the reader might want to consult additional resources; the book gives various recommendations for further reading.

## Content

To introduce the basic themes of this book and demonstrate their ultimate purpose, an introductory section Logic for Programming: A Perspective gives a short historical account on the development of logical modeling and reasoning about computer programs and presents examples of practical applications of these techniques in industrial software development today.

The main contents of this book are then organized into two parts:

Part I: The Foundations. This part introduces the basic language of logic and mathematics that is used throughout the remainder of the book. While an impatient reader (such as the author himself!) may be inclined to skip over this part, we advise to study at first reading at least Chapter 1 "Syntax and Semantics" which introduces the themes that are fundamental to this book:

- context-free grammars (inductive definitions of formal languages as sets of phrases represented by abstract syntax trees),
- type systems (logical inference systems which restrict these languages to certain well-formed phrases),
- semantics functions (inductively defined functions which give these phrases a meaning by mapping them to mathematical objects), and

- the accompanying principle of structural (more general: rule) induction which enables us to reason about these constructions.

The subsequent chapters elaborate these concepts in more detail (the impatient reader may skip over them at first reading and consult them later on demand). Chapter 2 "The Language of Logic" applies above principles to introduce the syntax and semantics of first-order logic, the core language of this book. Chapter 3 "The Art of Reasoning" continues this presentation by introducing the concept of logical proof and by introducing a variant of the sequent calculus as a formal framework for proof construction. Chapter 4 "Building Models" describes how with the language of logic models of reality (theories and data types) can be constructed in which we subsequently operate. Chapter 5 "Recursion" discusses the semantics of various forms of recursion, including inductive and coinductive definitions of functions and relations, using a restricted variant of fixed point theory. The material presented in these chapters (and much more) can be similarly found in various texts on logic and mathematics for computer science, however with non-uniform notions and notations; our goal here is to give a minimal consistent framework as a sufficient and necessary basis of Part II of this book.

Part II: The Higher Planes, this part contains the actual core contents of the book:

- Chapter 6 "Abstract Data Types" discusses the formal specification of abstract data types by logical axioms that the operations on the types must satisfy; the types may consist of finite values characterized by their constructor operations but also of potentially infinite values characterized by their observer operations. For this purpose, we gradually introduce a formal type specification language with a static type system and give it a semantics as models of first-order formulas; these models may be restricted to a particular class of candidates by special kinds of specifications (generated/free, cogenerated/cofree). We also discuss specifying in the large by various principles to compose smaller specifications to bigger ones. The chapter does not only describe the modeling of types but also the basic techniques for reasoning about them; it also discusses the refinement of more abstract types to more concrete ones.
- Chapter 7 "Programming Languages" discusses the formal semantics of programming languages. For this we extend the previously introduced type specification language to an imperative (command-based) programming language whose semantics we describe by two approaches. In denotational semantics, we first map commands to partial functions on program states; these functions are defined by logical terms. Later we generalize this classical functional style to a non-classical but much more flexible relational style where commands are mapped to state relations defined by logical formulas. In operational semantics, we give programs a semantics by mapping them to transition relations defined by logical inference systems; we then show the essential equivalence of denotational and operational semantics. We apply these techniques to model the translation of the command language to a low-level machine language and prove the correctness of the translation. Finally, we extend the command language by the abstraction mechanism of procedures and model their semantics.

- Chapter 8 "Computer Programs" discusses the formal verification of programs by various closely related calculi; the chapter thus lifts the level of reasoning from the previously discussed layer of programming languages to that of programs written in these languages. After discussing the formal specification of computational problems as the basis of program verification, we present the Hoare calculus and prove its soundness with respect to the semantics of the language. We continue with Dijkstra's predicate transformer calculus that maps commands to functions on formulas over states; further on we complement this approach by presenting a relational calculus that maps programs to formulas over state pairs. A good part of the chapter is dedicated to the pragmatics of verifying the partial correctness and termination of programs, which requires the human to devise adequate loop invariants and termination measures; here we give several concrete verification examples. Finally we discuss the abstract concept of command refinement from which we derive the concrete principles of modular reasoning about the correctness of procedure-based programs.
- Chapter 9 "Concurrent Systems" discusses the formal modeling of and reasoning about systems exhibiting nondeterministic behavior (concurrent/reactive systems). For this we extend the previously introduced command language to a language of shared systems where concurrent activities interact via a common state as well as to a language of distributed systems whose components interact by exchanging messages. We give these languages a semantics by mapping them to labeled transition systems; these are described by logical formulas that denote initial state conditions and transition relations. For specifying properties of such systems we extend first-order logic to linear temporal logic whose formulas are interpreted over system runs. The proof-based verification of such properties requires the human to devise adequate system invariants; we discuss the verification of such invariants expressing safety properties of the system and also the verification of a particular class of liveness properties which ensure the progress of the system execution. Finally we investigate the refinement of more abstract systems (models) to more concrete systems (implementations).

Altogether these chapters thus present the syntax and semantics of a language that encompasses abstract type declarations, imperative programs whose behaviors are specified by formulas in first-order logic, and concurrent systems whose behaviors are specified in linear temporal logic; they discuss corresponding calculi for the verification of the programs with respect to specifications and for the refinement of more abstract programs to more concrete ones.

## Software

Each chapter is accompanied by a section that illustrates the practical relevance of the presented theoretical material by some software system or programming language that is based on the presented concepts respectively makes use of them:

- The functional programming language OCaml.
- The proof assistant RISC ProofNavigator.
- The interactive theorem prover Isabelle/HOL.
- The algebraic specification language CafeOBJ.
- The common algebraic specification language CASL.
- The executable semantics framework K.
- The program verification environment RISC ProgramExplorer.
- The algorithm language and model checker RISCAL.
- The TLA$^+$ toolbox for modeling and checking concurrent systems.

Most of these tools have been developed by other researchers; the RISC ProofNavigator, the RISC ProgramExplorer, and RISCAL are the results of the author's own work. All the presented software is freely available; the reader may download the examples presented in each section (see below) and run them on his/her own.

## Teaching and Further Study

While of course the various chapters of this book in general linearly depend on each other in that each chapter may refer to some material presented earlier, Fig. 1 tries to outline the main dependencies by solid arrows (the dotted arrows represent weak dependencies); their consideration may be indeed useful in selecting material from this book for university courses that teach some specific topics such as

- Logic, Formal Modeling: Chaps. 1–3.
- Set Theory, Fixed Point Theory: Chaps. 4–5.
- Formal Specification of Abstract Data Types: Chap. 6.
- Formal Semantics of Programming Languages: Chap. 7 (with elements of Chap. 5).
- Formal Methods in Software Development: Chap. 8.
- Formal Models of Parallel and Distributed Systems: Chap. 9.

**Fig. 1** Chapter dependencies

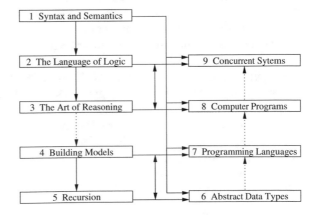

Indeed the author has taught courses (respectively participated in teaching) on most of these topics using material from this book (partially complemented by other more in-depth material).

In fact, since this book presents an integrated view on various (related but not identical) subjects, we had to choose from each field those aspects that we considered essential for conveying our core message: how by some universal principles of logical modeling and reasoning a software developer can better understand the artifacts that he/she is dealing with. This necessarily comes at the price that many other (also important) elements had to be neglected. To partially compensate for these gaps, each chapter concludes with a section Further Reading that suggests specific literature (mainly textbooks) that cover the presented topic in more depth or breadth, but typically from a different perspective. Lecturers, students, and readers may complement our presentation with additional material from these references.

## Web Page and Exercises

This book is accompanied by electronic material available from the following URL:

    https://www.risc.jku.at/people/schreine/TP

In particular, this web page contains all software examples presented in this book and exercises for the topics of the various book chapters.

## Acknowledgments

The author has written this book as an associate professor at the Research Institute of Symbolic Computation (RISC), a stimulating Austrian institution for research, education, and application of algebraic and logic methods founded by Bruno Buchberger, formerly directed by Franz Winkler, and currently directed by Peter Paule; RISC is located in the Softwarepark Hagenberg and organizationally part of the Johannes Kepler University Linz. The book is based on the author's experience in teaching corresponding courses at this university, feedback received from students, and discussions with colleagues. In particular, Patrick Riegler has helped to improve the chapters of Part II of this book; Chaps. 2 "The Language of Logic" and 3 "The Art of Reasoning" have been influenced by the course module First-Order Logic jointly taught by the author and Wolfgang Windsteiger. We are also grateful to the anonymous reviewers whose suggestions helped to improve the manuscript (in particular by adding the introductory section with examples of industrial applications of formal methods). All errors, however, are the author's own.

Hagenberg, Austria                                              Wolfgang Schreiner
March 2021

# Contents

# Logic for Programming: A Perspective

> Wer nicht von dreitausend Jahren sich weiß Rechenschaft zu
> geben, bleib im Dunkeln unerfahren, mag von Tag zu Tage
> leben. (Who cannot draw on three thousand years may stay in
> the dark, inexperienced, living from day to day.)
>
> —Johann Wolfgang von Goethe (West-östlicher Divan)

Reading this book may be conceived as traveling through a landscape of unfamiliar domains. Even if the relevance of these domains to computer programming will be continuously emphasized and also demonstrated by various software presentations, the reader may be more motivated to undertake this voyage, if she has some perspective on where the journey is heading. Therefore this section will demonstrate some real-life applications of logic to modeling and reasoning about computer programs, in particular also examples of industrial relevance. However, of equal importance is the dual understanding of where the journey has started; we will therefore begin with a short historical account on the creation of logic and its evolution to a tool of computer science. While this presentation is certainly incomplete and also influenced by the goals of this book and the personal preferences of its author, it may convey to the reader a first big picture of the landscape through which we are walking.

## Logic and Language

Logic (from the Greek word logos which may be translated as reason) emerged in antiquity from the human desire to distinguish a valid argument from an invalid one. Initially such arguments were expressed in natural language and were intended for everyday discourse, but also for discussing philosophical and scientific questions (philosophical logic); the development of formal (symbolic or mathematical) logic occurred at much later times.

The roots of logic in the Western world (there have been alternative Eastern traditions in India and China) go back to the ancient Greece of the 4th century BCE. At that time, the Greek philosopher Aristotle developed in his writings later called

Organon a theory of syllogisms; a syllogism is a particular form of logical argument that derives from two sentences, the premises, a third sentence, the conclusion. These sentences have the form of particular subject-predicate statements, namely, every $A$ is $B$ or some $A$ is $B$, respectively their negations; here $A$ and $B$ are terms that represent some basic concept like human or mortal, thus every human is mortal would be such a sentence. Although the expressiveness of this logic was quite limited, its main realization was that the validity of a logical argument (i.e., the correctness of a proof of its conclusion) only depends on the *form* of its sentences, not the interpretation of the terms in them. Indeed, Aristotle's logic (also called term logic) represented the essence of logic for more than two millennia, with only minor developments in medieval times.

A much more ambitious role for logic was envisioned in the 17th century by the German polymath Gottfried Wilhelm Leibniz (who also invented the binary numbers, the basis of digital computers today). He desired a characteristica universalis (a universal formal language), in which every mathematical, scientific, and metaphysical sentence could be expressed in such a way that a calculus ratiocinator (a logical calculation framework) could decide its truth by mechanical computation: thus, in order to solve disputes among persons with opposing opinions, Leibniz proposed calculemus! (Latin for let us calculate!). Indeed, Leibniz's writings sketched early forms of some concepts that would later appear in propositional logic and set theory. A resonating view was expressed in the 19th century by the English mathematician Ada Lovelace who wrote the world's first computer programs for Charles Babbage's analytical engine; she realized that this machine might act upon other things besides number, were objects found whose mutual fundamental relations could be expressed by those of the abstract science of operations. Thus in the future computing machines might indeed be capable of fulfilling Leibniz's dream (Fig. 2).

https://commons.wikimedia.org

**Fig. 2** The Pioneers of Logic: Aristotle, Gottlob Frege, Kurt Gödel, Alfred Tarski

## Logic and Mathematics

A first concrete step toward a more expressive kind of logic with also a more rigorous mathematical treatment was taken in the middle of the 19th century by the English mathematician and logician George Boole; he developed the principles of what was later called Boolean algebra (the formal basis of modern digital circuits), also known as propositional logic. The British mathematician Augustus De Morgane further refined Boole's system and extended it to a logic of binary relations; he also introduced the modern use of quantifiers. Based on these results, the American mathematician and logician Charles Peirce extended Boole's system to a logic of relations (the formal basis of relational databases today).

Independently of these developments in the English-speaking world, the modern kind of logic we use today was mainly developed by the German philosopher, logician, and mathematician Gottlob Frege; in his 1879 published Begriffsschrift (German for concept writing) he introduced what is today called predicate logic (later, distinctions would be made between first-order and second-order respectively higher-order predicate logic). While Frege's work was mainly ignored during his life time, it was later brought to attention by other scientists, such that Frege would be ultimately considered as the greatest logician since Aristotle. In this respect very influential was the Principia Mathematica of the English mathematician and philosopher Alfred North Whitehead and the British polymath Bertrand Russell; published in 1910, this work formulated mathematics on the basis of a variant of first-order logic and the theory of sets. This set theory had been informally developed by the German mathematician Georg Cantor in 1874 and was in 1908 formalized as an axiomatic theory in first-order logic by the German mathematician Ernst Zermelo; in 1922, this axiomatization was further refined by the German-born Isreali mathematician Abraham Fraenkel.

However, as Russel had already discovered in 1901, particular formalizations of set theory may lead to contradictions which are easy to overlook (Russel's paradox); thus the question remained open, whether mathematics was really based on a solid foundation. To settle this question, in 1920 the German mathematician David Hilbert proposed a research endeavor (Hilbert's program). The goal was to show that all of mathematics can be expressed in some formal logic as an axiomatic theory (i.e., by a list of logical formulas called *axioms*, this list may be infinite but must be enumerable) that is consistent (from the axioms no contradictory sentences can be proved) and complete (from the axioms every sentence or its negation can be proved). In 1928, this program was extended to the Entscheidungsproblem (German for decision problem): given any formula, a mechanical procedure should be able to decide in a finite amount of time whether this formula is provable or not.

Indeed Hilbert's program seemed well underway, when the Hungarian-American polymath John von Neumann proved in 1927 the consistency of a first-order axiomatization of a fragment of arithmetic and the Austrian logician Kurt Gödel proved in 1929 the completeness of first-order logic. However, in 1931 the program was utterly smashed by Gödel: he showed in his incompleteness theorems that in any logic every consistent axiomatization of arithmetic is necessarily incomplete, i.e., there are mathematical statements that can neither be proved nor disproved (this implies that second-order logic itself is incomplete, since arithmetic can be axiomatized by a single second-order formula). Even more, Gödel showed that no logical system can prove its own consistency. By these fundamental results, Gödel joined the ranks of Aristotle and Frege as one of the most significant logicians in history.

Therefore, while the formalization of mathematics today indeed rests upon first-order logic and Zermelo Fraenkel set theory, Gödel's incompleteness theorems set clear limits to this approach: neither can we be sure that the formalization is consistent (but no inconsistencies could ever be found), nor can every mathematical question be settled. For instance, in 1963 the American mathematician Paul Cohen showed that the "continuum hypothesis" (there is no set whose size is strictly between that of the integers and that of the real numbers) can neither be proved nor disproved from the first-order axioms of Zermelo-Fraenkel set theory. Thus there exist mathematical statements that are (in a certain sense) "neither true nor false".

But what if we contend ourselves with the more modest goal of deciding formulas in simpler theories which can be axiomatized in first-order logic? Indeed one aspect of the Entscheidungsproblem of first-order logic can be achieved (i.e., the problem is "semi-decidable"): given a provable formula, it is indeed possible by a mechanical procedure to find this proof in a finite amount of time. However, in 1936 the American mathematician Alonzo Church and the English polymath Alan Turing independently showed that the overall goal is impossible to reach (i.e., the Entscheidungsproblem is "undecidable"): if a first-order formula is unprovable, any mechanical procedure that attempts to prove/disprove it may run forever. Thus first-order logic can be only semi-automated: given a formula to be decided, we can never be sure whether the procedure just needs more time to find its proof or whether it runs forever in a doomed attempt to prove an unprovable formula (to show these results, Church developed the "$\lambda$-calculus" and Turing the "Turing machine", the first formal models of full-fledged "Turing-complete" programming languages).

A completely new approach to logic was established by the Polish-born American logician Alfred Tarski. From Aristotle to Gödel, logic had been treated as a purely syntactical game that investigated how formulas could be proved; thus (apart from an intuitive interpretation) a formula actually had no inherent meaning that was independent of its provability. In 1936, however, Tarski gave logic a formal "semantics" by the interpretation of formulas in some "model"; he thus created the field of "model theory" that investigates the interplay between "proof and truth". The original notion of the "consistency" of a calculus was thus widely replaced by the notion of its "soundness" (every proved formula is true) and the original notion of "completeness" was correspondingly redefined (every true formula can be proved). Nowadays, logic is mainly presented within the model-theoretic framework established by Tarski.

## Logic with Computers

After the previous decades had clarified the capabilities and limitations of formal logic, since the 1960s more and more work was dedicated to turn the theoretical foundations into practical results by applying computer programs to logical reasoning. Here essentially three strands have emerged:

- *Automated Reasoning Systems:* these are systems that prove formulas in first-order logic, typically by applying some sound and complete proving calculus of which various have been developed since the 1930s. Since the "search space" for finding a proof is of infinite size, the main challenge is to define an efficient search strategy, Here a major step forward was the "resolution algorithm" invented by Alan Robinson [151] which considerably limits the search space and is applied in many modern first-order provers, e.g., the "Vampire" prover developed by Andrei Voronkov and colleagues at the University of Manchester [176]. However, systems may be also based on more "human-oriented" strategies such as the "Theorema" system developed by Bruno Buchberger and colleagues at the RISC institute of the Johannes Kepler University Linz [175].
- *Interactive Proving Assistants:* fully automated reasoning systems reach their practical limits when dealing with complex mathematical theories. Therefore, since the 1970s research has been pursued on interactive proving assistants where a proof is developed by a collaboration of human and machine: the human generally directs the proof construction but lets the computer elaborates the tedious details. Here the "Logic for Computable Functions" (LCF) developed by Michael Gordon, Robin Milner, and colleagues at the universities of Edinburgh and Stanford was very influential [59]. LCF introduced the idea of a "trusted core" that implements a small set of logical rules (in the functional programming language "ML" which was developed for this purpose); this core is represented by an abstract datatype whose constructors are the logical rules. The core can be arbitrarily extended by derived rules in the form of proof construction procedures that may be invoked by humans or by automated proof search procedures; still the correctness of a proof only depends on the soundness of the trusted core. Since these systems do not require a complete inference systems, they typically implement a higher-order logic; notable examples are HOL family of provers [79] and the Isabelle/HOL proving assistant [132] developed at the University of Cambridge and the Technische Universität München by Lawrence Paulson, Tobias Nipkow, and colleagues.
- *Satisfiability Solvers:* the problem of deciding the validity of a formula can be reduced to deciding the satisfiability of its negation. This "SAT problem" for propositional logic is NP-complete such that we cannot expect to solve it generally faster than in exponential time. Nevertheless, since the late 1990s heuristically fast SAT solvers have been developed that are effectively able to solve problems with tens of thousands of Boolean variables [21]; these solvers have found wide application in hardware design, planning, scheduling, and

optimization. Furthermore, also for the quantifier-free fragment of first-order logic the satisfiability problem is decidable in certain (combinations of) mathematical theories such as "linear arithmetic" or "uninterpreted functions with equality". Corresponding "Satisfiability Modulo Theories" (SMT) solvers are nowadays used as central components in the verification of computer programs [10]; a well-known representative of this category is the Z3 solver developed by Leonardo De Moura and Nikolaj Bjørner at Microsoft [183].

In the area of mathematics, interactive proving assistants have been used to establish new results such as the "four-color theorem", the "Kepler conjecture", or the "Boolean Pythagorean" triples problem; however, these are structurally simple "proofs by exhaustion" where an overwhelmingly large number of cases is checked by arithmetic calculation respectively satisfiability solving. Conceptually more interesting is the use of such assistants to formally verify versions of (perhaps previously disputed) proofs such as that of the Jordan curve theorem [64]. Fully automated reasoning systems have been mainly used to confirm the validity of already known theorems; occasionally these systems also found more elegant proofs than were previously known (e.g., that of a theorem in the Principia Mathematica); a really new mathematical result was established by the automated prover EQP that found a proof of the "Robbins conjecture" [112] by equational reasoning. Generally, most success has been achieved by procedures targeted to special problems such as proofs of algebraic identities which occur in combinatorics and in the theory of special functions, such as implemented by the summation package "Sigma" developed by Carsten Schneider at the RISC institute of the Johannes Kepler University Linz [162].

Summarizing, new mathematical results were established with the help of automated or interactive reasoning systems mainly if they involved activities for which computers are more suitable than humans, such as checking many cases or establishing long chains of identities. More pragmatic success was achieved in verifying mathematical truths by formalizing existing proofs (respectively in detecting errors and gaps in these proofs); in particular, the Mizar project established by Andrzej Trybulec at the University of BiaÅ‚ystok has developed a large library of strictly formalized mathematics on the basis of the Mizar proof assistant [120].

Be that as it may, while computers have become a *tool* for logic, they have simultaneously also opened a new and very fruitful *domain* for logic, that of computer systems and computer programs themselves.

## Logic for Computer Science

With the advent of electronic computers in the 1940s, more and more complex problems were solved by computer programs; thus, however, it became more and more difficult to write computer programs that are indeed correct, i.e., that for all given inputs deliver the expected outputs respectively produce the expected effects.

Fortunately, it also became clear that it is not really necessary to run a computer program to predict its behavior. Only poor programmers write their programs by "trial and error"; good programmers apply some form of rational reasoning to deduce the behavior of their programs from the texts of the programs alone. It therefore should be possible to formalize this reasoning process in the form of a logical theory and predict the behavior of the program by logical deduction in that theory.

Indeed, based upon the pioneering works of the American computer scientists John McCarthy and Robert Floyd, the British computer scientist Tony Hoare developed in 1969 a logical inference system later called "Hoare calculus" which provides a suitable framework for this kind of reasoning [73]. Given a "precondition", a first-order formula that describes the possible inputs of a program, and a postcondition, a first-order formula that describes the desired outputs, a valid deduction in the Hoare calculus ensures that every execution of the program with inputs that satisfy the "precondition" yields outputs that satisfy the "postcondition". This deduction proceeds via the derivation of "verification conditions", first-order formulas whose truth implies the correctness of the program; thus the Hoare calculus reduces the problem to reasoning about the correctness of computer programs to the problem of proving formulas in first-order logic. However, the calculus itself does not give a strategy to build valid deductions. This was amended in 1975, when the Dutch computer scientist Edsger Dijkstra developed in his "predicate transformer semantics" an effective algorithm for deriving suitable verification conditions via the computation of "weakest preconditions" or (dually) "strongest postconditions"; it is this algorithm that was subsequently to be applied in most systems for program verification [41].

The Hoare calculus only defines the meaning of programs implicitly via rules that allow us to prove the correctness of programs with respect to specifications; in this sense it represents an "axiomatic semantics" of programs. However, in 1971 the American logician Dana Scott and the British computer scientist Christopher Strachey gave (on the basis of the "domain theory" developed by Scott) recursive functions and thus also iterative computer programs a "denotational semantics", i.e., they assigned an explicit meaning to programs [171]; this allows (in the spirit of Tarski's work on the semantics of first-order logic) to prove the soundness and completeness of the Hoare calculus with respect to the semantics of programs. Another approach was a new form of "operational semantics" of programs developed in the 1980s by the British computer scientist Gordon Plotkin ("structural operational semantics" [144]) and the French computer scientist Gilles Kahn ("natural semantics" [85]). Here the semantics of a program is defined by a logical deduction system whose inference rules mimic the execution steps of the program; thus it becomes possible to formally relate the mathematical denotation of a program to its operational interpretation. All in all, by these various approaches to formalizing the semantics of programs, computer science has established a firm basis for the execution and translation of computer programs (processors, interpreters, compilers).

So far, the main consideration was programs that transform given inputs to expected outputs; here it is only necessary to deal with *two* states, the input state and the output state of a program. However, this does not really address systems that repeatedly interact with their environment, such as the components of concurrent programs or computing systems which run through (potentially even infinite) sequences of states, each of which represents a possible point of interaction with other components. In 1977, the Israeli computer scientist Amir Pnueli proposed to apply a logic originally introduced by the New Zealand-born logician and philosopher Arthur Prior as a "temporal logic" to specify and reason about the behavior of such systems [145]. Temporal logic can be considered as an extension of first-order logic; while a first-order formula talks about a *single* program state, a temporal formula talks about *arbitrarily many* such states. This approach has become in the following decades the basis of numerous systems for concurrent system modeling and verification.

In the 1970s and 1980s, also another general approach to dealing with the problem of developing correct computer programs was widely pursued: the attempt to abandon conventional programming languages in favor of using logic itself as a much more "high-level" programming language; thus the gap between the specification of a problem and the implementation of a program solving this problem would be closed. In a certain sense, Gödel's proof of the completeness of first-order logic shows that first-order logic itself is a Turing-complete programming language: every computation can be expressed as the proof of a first-order formula. However, finding this proof is overwhelmingly more costly than performing the computation in a conventional language; therefore research was pursued on developing efficient execution mechanisms for fragments of first-order logic. Instances of this idea are "abstract data type languages" based on equational logic (in particular the "OBJ" language family initiated by the American computer scientist Joseph Goguen [58]), and "logic programming languages" based on "Horn clause" logic with applies a special form of resolution (in particular the language "Prolog" developed by the French computer scientists Alain Colmerauer and Philippe Roussel [38]). The Japanese "Fifth Generation Computer" Systems initiative even pursued the development of massively parallel computer systems based on concurrent logic programming languages [173]. While these approaches of a "direct" use of logic did ultimately not supplant conventional programming languages or computing systems, many principles developed in these endeavors formed the basis of the "indirect" use of logic for programming; in particular the type systems of modern programming languages and the theory of data types specification was substantially influenced by abstract data type languages, as well as automated reasoning over equational theories.

Returning to the topic of the verification of computer programs and systems, a core problem with corresponding logical calculi (Hoare calculus, predicate transformers, temporal logic) is that they crucially depend on additional information that adequately characterizes the behavior of iterative computations by logical formulas. These "invariants" are not inherent in the programs themselves but have to be provided by some external source, in practice by the human programmer. If the invariants are not adequate, the derived verification conditions do not hold and their proofs fail. Thus in practice program verification has to struggle with two kinds of uncertainties: the fundamental uncertainty in the correctness of the program and the additional uncertainty in the adequacy of the invariants; errors on both levels lead to unprovable verification conditions. This was especially problematic in the 1970s and 1980s, when automated reasoning systems and interactive proving assistants were only able to provide adequate reasoning support for verification conditions arising from simple "toy programs"; thus, as a third uncertainty, the failure of proving a verification condition could also be due to the inadequacy of the proof automation. For all these reasons, at that time the technique of program verification by logical deduction was generally not considered of much practical relevance.

However, in the 1980s and 1990s with the technique of "model checking" an alternative approach emerged [37]. Rather than investigating logical theories with arbitrarily many interpretations ("models") of generally infinite size, model checking focuses on a single model of finite size and analyzes its properties. This approach evolved in the area of hardware verification, because a digital circuit can be described by such a finite model (the finitely many combinations of the states of logic gates). Given a temporal logic formula that describes an expected property of such a model, sophisticated encoding and analysis techniques are able to fully automatically decide whether the model satisfies this property. Similar techniques can be also applied to computer programs if we consider the domains of all program variables as finite bit vectors and assume a finite bound for the number of execution steps ("bounded model checking" [20]); furthermore, the domain of an infinite-state system or program can be abstracted to a finite domain that is amenable to "abstraction model checking" [36]. While these approaches are typically not able to verify the general correctness of a program, they may still detect typical programming errors that lead to the abortion of programs (division by zero, null pointer dereferences, out-of-bound array indices).

In the 2000s, however, interest in the more general approach of program verification by logical deduction revived. By that time, substantial advances had been made in automated reasoning, especially fueled by the already discussed emerge of practical SMT solvers [10] which are able to decide the satisfiability of quantifier-free formulas in combinations of theories which are quite relevant in computer programming, such as linear integer arithmetic or the theory of arrays. Such "SMT solvers" have become the building blocks of many program verification environments which combine a program reasoning calculus (such as Dijstra's weakest preconditions) that automatically derives verification conditions with an automated reasoning system or interactive proving assistant that deals with the quantification structure of these conditions; finally SMT solvers can be applied to handle the resulting quantifier-free fragment. In this way, many interesting program verification problems can be nowadays successfully solved.

## Logic and Software Development

After this historical excursion, we will fulfill our initial promise of discussing some concrete examples of non-trivial software whose correctness has been actually established by logical modeling and reasoning. Here we will consider only the verification of software and altogether omit the topic of hardware verification. Furthermore, we will focus entirely on the verification of the "functional" correctness of programs; there are also approaches to establish non-functional requirements, such as "security" guarantees or "real-time" constraints. Furthermore, we will not discuss the also important topic of the verification of "cyber-physical" systems which combine digital controllers with physical sensors respectively actuators and are therefore governed not only by the laws of computing but also by the laws of nature.

Some of the following examples do indeed describe "industrial" applications, others represent major research activities with a mid-term perspective of industrial impact, some are listed because they may give a glimpse into the long term future of industrial software development. While it is in a short space not possible to describe these examples in great detail, we roughly sketch their underlying logical approaches and relate these to the topics presented in this book.

**Verified System Designs and Implementations** Engineers at Amazon Web Services (AWS) have since 2012 used the temporal logic modeling language TLA$^+$ (described on page 591 of this book) to model and verify critical components of the AWS infrastructure [128]. This work started in the context of Amazon's S3 Simple Storage Service, when the designer of the replication and fault-tolerance mechanisms of the DynamoDB data store component of that service wrote a detailed logical model of these mechanisms and applied the TLC model checker to verify its expected properties. Thus a subtle bug in the design of the fault-tolerant algorithm was detected that could lead to losing data if a particular sequence of failures and recovery steps would be interleaved with other processing; this bug had previously passed unnoticed through extensive design reviews, code reviews, and testing. By more formal modeling and verification, later two more bugs were detected in other algorithms, both serious and subtle. After these initial successes, TLA$^+$ was presented to a broader engineering community at AWS who applied it to a new fault-tolerant network algorithm (revealing two bugs and some more bugs in extended and optimized versions of the algorithm) and another critical algorithm (revealing that a proposed fix to a bug previously detected in testing actually had not removed that bug); similar experiences were reported by other engineers. Subsequent new algorithms and protocols would be routinely modeled and verified with TLA$^+$ before translating the designs to actual production code.

While AWS only verified system *designs*, Microsoft Research extended its ambition toward verified system *implementations*. In its framework "IronFleet", this organization applied a TLA-like approach to build two complex distributed systems, IronRSL (a replicated state machine library) and IronKV (a key-value store), whose correctness (with respect to safety and liveness properties, see Sects. 9.5 and 9.6) was formally proved [67]. This verification proceeded in three layers and utilized various techniques presented in this book: On the highest layer, the system is described in Microsoft's modeling and verification language "Dafny" as a state machine in a logical form (similar to the transition systems described in Sect. 9.1). On the middle layer, in Dafny a distributed state machine is defined, again in a logical form (similar to the semantics of distributed systems described in Sect. 9.3). On the lowest layer, imperative Dafny code is written for each component of the distributed state machine; this code is then automatically translated to C# code and compiled to the executable code of the system. To semantically connect the layers, it is proved that each layer is refined by the next lower one. using an abstraction function that relates the states of both layers (see the refinement techniques presented in Sects. 7.4 and 9.7). To verify the correctness of the protocol layer, TLA-style property specifications are translated into corresponding Dafny predicates over state sequences, i.e., temporal logic reasoning is reduced to first-order reasoning (see the semantics of temporal operators specified in Sect. 9.4); for the verification of the lowest layer, Hoare-style reasoning is applied (see Sect. 8.2). The actual proofs are performed within the Dafny framework with the help of Microsoft's SMT solver Z3 and human-provided annotations to guide the prover through the quantifier instantiations.

**Verified Program Components and Libraries** Rather than attempting to verify whole systems, more often efforts concentrate on modeling and verifying individual critical program components. In modern object-oriented programming, these components are mainly represented by "classes" that encapsulate data and the methods operating on these data. For various object-oriented programming languages, "behavioral interface specification languages" have been developed [66] that describe not only the syntactic interfaces of classes but their semantic behavior via logical preconditions and postconditions of their methods and logical invariants of their objects (the basics reasoning about the correctness of methods respectively procedures are discussed in Chap. 8). Furthermore, to specify the external behavior of classes without exposing their internal representation, it is usually necessary to relate these classes to "models" (examples of such models are axiomatically specified abstract data types as discussed in Chap. 6). In the ecosystem of the object-oriented programming language Java, the "Java Modeling Language" (JML) has emerged as the de facto standard behavioral interface specification language which is supported by a variety of tools [34, 136].

A prominent representative of such a tool is "KeY", a formal verifier for Java that has since the late 1990s been jointly developed by the Karlsruhe Institute of Technology, the Chalmers University of Technology, and the Technische Universität Darmstadt [3, 148]. KeY is internally based upon "dynamic logic", a logic that combines classical first-order logic with a program logic (the principles of program reasoning are essentially the same as the calculi presented in Chap. 8). KeY has been used to verify non-trivial Java code respectively detect errors in such code. The original target was Java Card, a subset of Java for smartcards and embedded devices. Until 2005, various real-life industrial examples of Java Card applications were formally specified and verified with KeY; subsequently, KeY was extended to support many features of the full Java language. In 2015, an attempt to use KeY to verify the sorting algorithm of the Java library (a hybrid combination of merge sort and insertion sort called "Timsort") revealed that this widely used Java method actually had a bug; a corrected version could be successfully verified. In 2017, the correctness of another sorting algorithm available in the Java library (Dual Pivot Quicksort) was shown. In 2018, KeY was used to verify the correctness of Hyperledger Fabric Chaincode, a protocol for smart contracts built upon blockchain technology. In 2020, core components of EVA (the Java-based main support system for elections in municipalities and counties in Norway) were formally specified in JML and verified with KeY.

Beyond individual classes, an interesting target of verification is whole class libraries such as the "container" libraries available in many programming languages; here the main challenge is to show the correctness of a concrete (optimized) internal representation with respect to an abstract mathematical specification. For instance, in 2015 researchers from MIT and ETH Zürich verified the full functional correctness of Eiffel-Base2, a container library (with more than 130 public methods and 8400 lines of code) for the object-oriented programming language Eiffel that offers all the features customary in modern language frameworks [146]. The proof was performed with the help of the automated deductive verifier AutoProof that translated Eiffel code annotated with logical specifications and invariants into Microsoft's intermediate verification language "Boogie"; the Boogie verifier then generated verification conditions that were ultimately discharged by the Microsoft's SMT solver Z3. The library specification relied heavily on mathematical model types (essentially abstract data types as discussed in Chap. 6), the relationship between the actual representation and the model type was provided by abstraction functions (as described in Sect. 6.8).

**Verified Compilers** The verification of computer programs is usually based on a logical model of the high-level language in which the source code of the program is written. However, since this source code itself is not executable, it is typically compiled into an actually executable form, a program in the machine language of the underlying computer processor. This translation process itself is a potential source of errors, i.e., the generated machine program might behave differently than expected from the logical analysis of the original source code; thus it is a worth-wile goal to verify the compiler itself.

This problem of compiler verification has been addressed by the "CompCert" project initiated in 2007 by the French computer scientist Xavier Leroy; its main outcome is a verified compiler for a large subset of the C99 programming language which generates efficient code for the PowerPC, ARM, RISC-V, x86, and x86-64 processors [82, 101]. The compiler is written in the specification language of the proof assistant Coq in purely functional style; from this Coq specification, executable code in the functional language Caml is generated. While the Caml implementation itself is unverified, the Coq specification ensures the correctness of the translation of the C99 code to machine code via a sequence of transformations through various intermediate languages. For each of the source, intermediate, and target languages, a formal operational semantics is defined (see Sect. 7.3) and for each transformation it is proved that the generated code preserves the semantics by showing that each step of the original program is simulated by a sequence of steps of the transformed program (see the techniques presented in Sect. 7.4). The whole Coq specification consists of 42000 lines of which approximately 14 represent the compilation algorithms, 10 defined the formal semantics of the various languages, and 76 represent the correctness proofs themselves. These proofs were performed in Coq by a combination of user interaction to guide the proof and the application of automated decision procedures to discharge proof obligations; the proofs are recorded in the form of proof terms that apply a small set of logical rules; their correctness can be independently verified by a trusted proof checker. Since 2015, the CompCert compiler has been commercially available; in 2017 it was used to certify a highly safety-critical industrial application (a digital engine unit that controls the backup diesel engines of nuclear power plants) according to the IEC 60880 standard for nuclear power plant control systems.

While CompCert still relies on an unverified implementation of Caml, the "CakeML" project initiated in 2012 has produced a verified compiler that is written itself in the language that it verifies, a subset of Standard ML [32, 94]. Thus the compiler can compile itself and so produce a verified executable program that provably implements the compiler itself; this "bootstrapping" process started (along the lines of the techniques used in CompCert) with a specification of the compiler in the language of the interactive theorem prover HOL4, which was also used to perform all proofs. CakeML was applied to the end-to-end verification of various Unix-like command line tools, a proof checker for the OpenTheory standard, and a certificate checker for floating-point error bounds.

**Verified Operating System Kernels** The most complex program running on a computer is usually not an application executed on behalf of some user but the "operating system", i.e., software which extends the basic capabilities of the computer processor by an additional set of services. Errors in operating systems may have devastating consequences: apart from the danger of computer crashes and data losses, they also represent security holes which malicious attackers may exploit, not only locally, but remotely over the Internet. Since an operating system is run on millions of computers worldwide, it pays off to invest some efforts to ensure its correct behavior by formal verification. Indeed, this already happens regularly, albeit in a limited form: the Microsoft Windows device driver frameworks (WDF) incorporate a "static driver verifier" (SDV), an abstract model checker that was developed by Thomas Ball and Sriram Rajamani at Microsoft Research since 1999 and originally called SLAM (the most recent version SLAM2 was released in 2010) [9, 114]. This model checker verifies that a device driver interacts correctly with the Windows operating system, e.g., by detecting illegal function calls or actions that may cause system corruption. The technique applied by SLAM is "counter-example guided abstraction refinement" which is based on the principle of "symbolic execution", essentially a form of Dijkstra's predicate transformer calculus applied to abstractions of program states; such an abstraction is represented by the values of a set of predicates on the state. If an error is detected in the abstract model of the program, Microsoft's SMT solver Z3 is applied to determine whether its abstraction corresponds to an error in the real program; if not, the model is refined by extending the predicate set. Based upon similar principles (but with many improvements) are the C language model checkers "BLAST" [17] and "CPAchecker" [16, 18]; these have been used to find bugs in kernel drivers of the GNU/Linux operating system.

A much more ambitious goal was followed by the "seL4" project pursued at the Australian NICTA Research Center under the direction of Gernot Heiser. In 2009, this project developed a formally verified microkernel and hypervisor with completely proved functional correctness and security guarantees; in 2014 the seL4 microkernel was released as open source [87, 172]. The main domain of application of seL4 are safety-critical systems such as industrial control systems, medical devices, and autonomous vehicles; for instance, it was utilized in Boeing's Unmanned Little Bird (ULB) helicopter prototype [86]. The development of seL4 started with an abstract formal specification expressed in the language of the proof assistant Isabelle/HOL (described on pages 141 and 191 of this book). This specification was essentially in the form of an abstract state machine (similar to the transition systems described in Sect. 9.1) for which certain correctness properties were proved on the basis of appropriate system invariants (see Sect. 9.5). Then a prototype of the microkernel was developed in the purely functional programming language Haskell; this prototype represented an executable form of the state machine with concrete data structures implementing the abstract types of the specification. After appropriate testing, the Haskell code was automatically translated into the language of Isabelle/HOL and it was proved that this translated form of the executable specification "refined" the previously specified abstract machine (see the refinement techniques presented in Sects. 7.4 and 9.1); thus the executable specification preserved the correctness properties of the abstract specification. Then the Haskell prototype was manually re-implemented in a subset of the C99 programming language where the high-level functional constructions of Haskell were now expressed in low-level imperative code. Based on a formal definition of the semantics of C99 in Isabelle/HOL (essentially a small step operational semantics as described in Sect. 7.3), this C99 code was automatically translated into its formal semantics in Isabelle/HOL. It was then proved that this semantics refined the executable specification; thus the C99 implementation of the seL4 microkernel preserved the correctness properties of its abstract specification. The seL4 project was one of the largest formal verification efforts so far; the various specification and coding efforts took about 2.5 person years, while the proving efforts required about 11 person years (including the setup of the proving infrastructure and learning curve); it was estimated that a subsequent effort would reduce the later figure to 6 person years, essentially twice as much as required by traditional quality assurance methods which do not provide formal correctness guarantees.

From the above descriptions, one can see that it is comparatively rarely the case that already existing ("legacy") software is verified; generally more promising it is to design new software already with the goal of verifying its correctness in mind. Often this "codesign" of a software and the proof of its correctness proceeds in multiple layers where first an abstract design is specified and verified and then gradually refined to actual code; if we can prove that each layer is properly "refined" by the next layer, the lowest layer has inherited the correctness properties from the highest one.

## Further Reading

As for the history of logic till the 20th century, good general resources are the Wikipedia pages on the corresponding persons and topics; more in-depth articles can be found in the online Stanford Encyclopedia of Philosophy [174].

For surveys on techniques and tools for the automation of logical reasoning, see for instance the various chapters of the handbook [152] edited by Robinson and Voronkov, the chapters of the book [179] edited by Wiedijk, or Harrison's handbook [65].

A historical account of the application of formal methods to computer science is given in the article [23] by Bjørner and Havelund. For surveys on applications of formal methods in industrial practice, see the paper [182] of Woodcock and others and the more recent article [57] of Gleirscher and others. The VSTTE conference series [35] provides numerous examples of actual software verifications.

# Part I
# The Foundations

# Syntax and Semantics **1**

> *Wer die Form zerstört, beschädigt auch den Inhalt. (Who destroys the form, also damages the content.)*
> — Herbert von Karajan

In this chapter, we discuss the (abstract) syntax and semantics of formal languages, how to operate on the resulting syntactic phrases, and how to reason about the properties of these languages. We will in the later chapters apply these ideas to the language of first-order logic as the basis of mathematics, as well as to formal specification languages, and ultimately to programming languages.

## 1.1 Abstract Syntax

As a first step towards the definition of formal languages, we will introduce a simple language of arithmetic expressions with binary numerals, i.e., sequences of digits 0 and 1; examples of such expressions are $1 + (10 \times 11)$ and $(1 + 10) \times 11$. Actually, we will ultimately consider these expressions just as abbreviations for the structures depicted in Fig. 1.1: e.g., every expression of form $x + y$ denotes a tree whose root is labeled with symbol $+$ and whose children represent the subexpressions $x$ and $y$. More details of such 'abstract syntax trees' will be given later.

To define a formal language, we use a variant of the well-known Backus-Naur Form (BNF) for defining context-free grammars. As a starting point, take the following grammar (whose relevance to above language will become clear soon):

$$E \in Expression$$
$$N \in Numeral$$
$$E ::= \mathsf{n}(N) \mid \mathsf{s}(E_1, E_2) \mid \mathsf{p}(E_1, E_2)$$
$$N ::= \mathsf{z}() \mid \mathsf{o}() \mid \mathsf{nz}(N) \mid \mathsf{no}(N)$$

© The Author(s), under exclusive license to Springer Nature Switzerland AG 2021
W. Schreiner, *Thinking Programs*, Texts & Monographs in Symbolic Computation,
https://doi.org/10.1007/978-3-030-80507-4_1

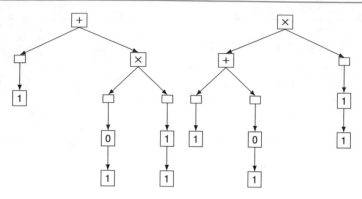

**Fig. 1.1**  The Abstract Syntax Trees for $1 + (10 \times 11)$ and $(1 + 10) \times 11$

This grammar introduces two syntactic domains *Expression* and *Numeral* with typed variables $E$ and $N$ denoting elements of these domains. The rule $E :: = \ldots$ for domain *Expression* has three alternatives that start with symbols n, s, and p; the rule $N :: = \ldots$ for domain *Numeral* has four alternatives starting with z, o, nz, and no.

The elements of the domains *Expression* and *Numeral* are syntactic phrases which we will call 'expressions' and 'numerals'. These phrases are constructed in a step-wise fashion by 'rewriting' every variable in the phrase according to the rules of the grammar, starting with a phrase that just consists of the typed variable of the corresponding domain, until no more rewriting steps are possible.

For instance, according to the rules for domain *Numeral* given above, we have the following steps (here $N \rightarrow N'$ can be read as '$N$ is rewritten to $N'$ in one step'):

$$N \rightarrow \mathsf{o}()$$
$$N \rightarrow \mathsf{nz}(N) \rightarrow \mathsf{nz}(\mathsf{o}())$$
$$N \rightarrow \mathsf{no}(N) \rightarrow \mathsf{no}(\mathsf{o}())$$

Therefore the phrases $\mathsf{o}()$, $\mathsf{nz}(\mathsf{o}())$, and $\mathsf{no}(\mathsf{o}())$ are elements of domain *Numeral*.

Consequently, we also have (considering the rules for domain *Expression* above) the following rewriting steps (here $E \rightarrow^* E'$ can be read as '$E$ is rewritten to $E'$ in multiple steps'):

$$E \rightarrow \mathsf{s}(E, E) \rightarrow \mathsf{s}(E, \mathsf{p}(E, E)) \rightarrow^* \mathsf{s}(\mathsf{n}(N), \mathsf{p}(\mathsf{n}(N), \mathsf{n}(N)))$$
$$\rightarrow^* \mathsf{s}(\mathsf{n}(\mathsf{o}()), \mathsf{p}(\mathsf{n}(\mathsf{nz}(\mathsf{o}())), \mathsf{n}(\mathsf{no}(\mathsf{o}()))))$$

Thus the phrase $\mathsf{s}(\mathsf{n}(\mathsf{o}()), \mathsf{p}(\mathsf{n}(\mathsf{nz}(\mathsf{o}())), \mathsf{n}(\mathsf{no}(\mathsf{o}()))))$ is an element of domain *Expression*. The left part of Fig. 1.2 depicts this phrase as an abstract syntax tree: every node represents the root of a phrase and is labeled with the symbol associated to the phrase.

The formal justification of above examples is given by the following definitions.

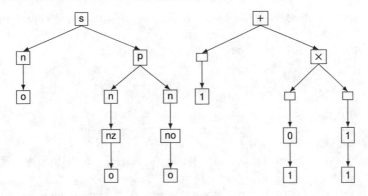

**Fig. 1.2** Abstract Syntax Trees

**Definition 1.1** (*Abstract Syntax: Grammar*) The grammar of an *abstract syntax* contains a sequence of $n \geq 1$ declarations of form

$$Var_i \in Domain_i$$

Each declaration $i$ introduces a unique name $Domain_i$ for a new syntactic domain and a unique name $Var_i$ of a variable that subsequently denotes elements of this domain; we call these variables also *nonterminals*.
Furthermore, for each declaration $i$, the grammar contains a rule of form

$$Var_i ::= Alternative_{i,1} \mid \ldots \mid Alternative_{i,n_i}$$

with $n_i \geq 1$ alternatives. Each $Alternative_{i,j}$ is a syntactic term

$$Symbol_{i,j}(V_{i,j,1}, \ldots, V_{i,j,m_{i,j}})$$

The term starts with a symbol $Symbol_{i,j}$ that is different from the symbol associated with any other alternative of the same rule; we call these symbols also *terminals*. Furthermore, the alternative contains $m_{i,j} \geq 0$ occurrences $V_{i,j,k}$ each of which denotes one of the nonterminals $Var_{i'}$. Multiple occurrences of the same nonterminal in an alternative may receive different subscripts for easier reference.

**Definition 1.2** (*Abstract Syntax: Language*) Every syntactic domain $Domain_i$ introduced by the grammar of an abstract syntax denotes a set of syntactic phrases. The first domain introduced by the grammar is considered as the *language* of the grammar. Each $Domain_i$ is defined as that set of phrases such that $p$ is in $Domain_i$ if and only if

- $p$ can be derived from $Var_i$ by a sequence of substitutions of nonterminals according to the rules of the grammar, but
- $p$ itself does not contain a nonterminal any more.

We write the first requirement as $Var_i \to^* p$, and formalize it as follows:

- Let $p \to_i p'$ denote that phrase $p'$ is identical to phrase $p$ except that some occurrence of nonterminal $Var_i$ in $p$ has been substituted in $p'$ by one of the alternatives $Alternative_{i,j}$ from the grammar rule for $Var_i$.
- Let $p \to p'$ denote that $p \to_i p'$ holds for some $0 \le i < n$, i.e., $p'$ equals $p$ except that some nonterminal in $p$ has been substituted.
- Let $p \to^* p'$ denote that there exists a sequence of $m \ge 0$ phrases $p_0, \ldots, p_m$ such that $p = p_0$ and $p_m = p'$ and $p_0 \to p_1,\ p_1 \to p_2,\ \ldots,\ p_{m-1} \to p_m$. In other words, $p'$ is derived from $p$ by a sequence of substitutions of nonterminals.

Thus a phrase $p$ without nonterminals is in $Domain_i$ if and only if $Var_i \to^* p$ holds.

In practice, our abstract syntax definitions will not rigorously stick to the standard notation where in each alternative a terminal always precedes the subphrases; it may also occur among or after the subphrases and multiple nonterminals may be used to separate the subphrases. Furthermore, terminals need not be unique or may be dropped at all, if the number/types of the subphrases uniquely determine the corresponding alternative. The grammar on Sect. 1.1 is therefore typically written in this more readable form:

$E \in Expression$
$N \in Numeral$
$E ::= N \mid E_1 + E_2 \mid E_1 \times E_2$
$N ::= 0 \mid 1 \mid N0 \mid N1$

The expression $1 + (10 \times 11)$ of the new language (we use parentheses to clarify its structure) matches the expression $\mathsf{s(n(o()), p(n(nz(o())), n(no(o())))))}$ of the original one. As depicted in Fig. 1.2, despite of the different linear representations of both expressions, their syntax trees correspond to each other node by node.

The syntax described by these grammars is 'abstract' rather than 'concrete' because it describes trees rather than strings of symbols. We are not concerned that a string like $1 + 10 \times 11$ can be parsed in two different ways (giving rise to the two different trees depicted in Fig. 1.1) because we will in the following only deal with abstract syntax trees. We will use linear notation only to describe such trees in a convenient form; if necessary, we will use parentheses as in $1 + (10 \times 11)$ or $(1 + 10) \times 11$ to make the intended tree clear.

## 1.2   **Structural Induction**

We may prove properties of syntactic domains by a particular proof principle.

> **Definition 1.3** (*Structural Induction*) Let $F[p]$ be a statement about syntactic phrase $p$ and assume that we want to prove a statement of form
>
> For every phrase $p$ in domain *Domain*, $F[p]$ holds.
>
> Then it suffices to perform, for every alternative *Alternative* in the grammar rule for *Domain*, a separate subproof where
>
> - under the assumption that $F[Var]$ holds for every occurrence of a nonterminal *Var* from *Domain* in the alternative,
> - it is proved that $F[Alternative]$ holds.

**Example 1.1** We want to prove for the language on Sect. 1.2 the (trivial) statement

For every numeral $n$ in *Numeral*, $n$ does not contain the symbol 2.

According to the four alternatives in the grammar rule for domain *Numeral*, we have to prove

1. 0 does not contain a 2.
2. 1 does not contain a 2.
3. If $N$ does not contain a 2, then $N0$ does not contain a 2.
4. If $N$ does not contain a 2, then $N1$ does not contain a 2.

which are all clearly true.                                                     □

For justifying the proof principle of structural induction, we consider that the sets of phrases defined by a grammar can be generated systematically in stages:

- In stage 0, we set $Domain_i^0$, for every $i$, to the empty set.
- In state $k + 1$, we extend $Domain_i^k$ to $Domain_i^{k+1}$: we add all phrases that we can derive from some alternative of the grammar rule for $Domain_i$ by substituting every nonterminal $Var_j$ in the alternative by some element of $Domain_j^k$.

Then each $Domain_i$ is the union of all $Domain_i^k$, i.e., it contains all phrases that can be generated from a finite number of substitutions of nonterminals. We also say that $Domain_i$ is *inductively generated*.

**Example 1.2** For the language on Sect. 1.2, we have

$$Expression^0 := \varnothing$$

$$Numeral^0 := \varnothing$$

$$Expression^{k+1} := Expression^k$$

$$\cup \left\{ \mathsf{n}(N) \mid N \in Numeral^k \right\}$$

$$\cup \left\{ E_1 + E_2 \mid E_1, E_2 \in Expression^k \right\}$$

$$\cup \left\{ E_1 \times E_2 \mid E_1, E_2 \in Expression^k \right\}$$

$$Numeral^{k+1} := Numeral^k$$

$$\cup \{0\} \cup \{1\}$$

$$\cup \left\{ N0 \mid N \in Numeral^k \right\} \cup \left\{ N1 \mid N \in Numeral^k \right\}$$

Then *Expression* is the union of all domains *Expression$^k$* and *Numeral* is the union of all domains *Numeral$^k$*. □

When applying the proof principle of structural induction, we essentially assume that all the phrases constructed in stage $k$ already have property $F$ and then show that the new phrases constructed in stage $k + 1$ preserve this property. Since we start the construction process with the empty set of phrases, all generated phrases have this property; since the syntactic domain consists only of these phrases, all phrases of the domain have this property.

Apart from *proving*, structural induction can be also applied for *defining* functions on syntactic domains. When defining such functions, we usually apply double square brackets $[\![\, \cdot \,]\!]$ in order to mark syntactic phrases and clearly separate them from the surrounding context: apart from variables, every symbol within these brackets is to be interpreted as a terminal of the underlying grammar. For instance, $[\![\, E_1 + E_2 \,]\!]$ represents an abstract syntax tree with root $+$ and two subtrees denoted by the variables $E_1$ and $E_2$. Consequently $mult[\![\, E_1 + E_2 \,]\!]$ denotes the application of function *mult* to this tree. However, in the expression $mult[\![\, E_1 \,]\!] + mult[\![\, E_2 \,]\!]$, the symbol $+$ occurs outside these brackets and thus denotes as usual the addition of two numbers.

**Example 1.3** For the language on Sect. 1.2, we define by structural induction a function *mult* such that $mult(E)$ denotes the number of occurrences of the symbol $\times$ in expression $E$:

$$mult: Expression \rightarrow \mathbb{N}$$

$$mult[\![\, N \,]\!] := 0$$

$$mult[\![\, E_1 + E_2 \,]\!] := mult[\![\, E_1 \,]\!] + mult[\![\, E_2 \,]\!]$$

$$mult[\![\, E_1 \times E_2 \,]\!] := 1 + mult[\![\, E_1 \,]\!] + mult[\![\, E_2 \,]\!]$$

By this definition, we have e.g.

$$mult[\![\,(N_1 \times N_2) + N_3\,]\!] = mult[\![\,N_1 \times N_2\,]\!] + mult[\![\,N_3\,]\!]$$
$$= (1 + mult[\![\,N_1\,]\!] + mult[\![\,N_2\,]\!]) + 0$$
$$= (1 + 0 + 0) + 0 = 1$$

i.e., in phrase $(N_1 \times N_2) + N_3$ the symbol $\times$ occurs once. $\qquad\square$

In general, to define a function $f$ on a syntax domain *Domain*, we give one equation for every alternative *Alternative* of the corresponding grammar rule where

- the left side of the equation is of form $f[\![\,Alternative\,]\!]$, and
- the right side of the equation may contain a recursive function application $f[\![\,Var\,]\!]$ for every occurrence of a variable $[\![\,Var\,]\!]$ that appears on the left side and is a nonterminal of *Domain*.

To see that such a set of equations indeed defines a function, consider this: replacing a function application that matches the left side of an equation by the term determined by the right side yields only function applications whose arguments are subphrases of the original argument; since every phrase has only a finite number of subphrases, this process can be only repeated a finite number of times before a replacement does not yield further function applications. Furthermore, the equations describe the function uniquely, because each phrase 'matches' by construction exactly one alternative such that one and only one equation is applicable. The match then uniquely determines the subphrases assigned to the variables such that all occurrences of the variables on the right side have definite values.

Structural induction as a proof principle is typically applied to proving properties of functions defined by structural induction.

**Example 1.4** Continuing Example 1.3, we also define a function *swap* such that $swap(E)$ is an expression that is identical to $E$ except that the order of arguments for the operation $+$ has been exchanged:

$swap:\ Expression \rightarrow Expression$

$$swap[\![\,N\,]\!] := [\![\,N\,]\!]$$
$$swap[\![\,E_1 + E_2\,]\!] := \text{let } E_a = swap[\![\,E_1\,]\!],\ E_b = swap[\![\,E_2\,]\!] \text{ in } [\![\,E_b + E_a\,]\!]$$
$$swap[\![\,E_1 \times E_2\,]\!] := \text{let } E_a = swap[\![\,E_1\,]\!],\ E_b = swap[\![\,E_2\,]\!] \text{ in } [\![\,E_a \times E_b\,]\!]$$

We now claim that, for every $E$ in *Expression*, we have

$$mult(swap[\![\,E\,]\!]) = mult[\![\,E\,]\!]$$

We prove this claim by structural induction which leads to three subproofs, one for each alternative of the grammar rule for domain *Expression*:

1. $mult(swap[\![\, N\, ]\!]) = 0 = mult[\![\, N\, ]\!]$.
2. We assume

$$mult(swap[\![\, E_1\, ]\!]) = mult[\![\, E_1\, ]\!] \qquad (1)$$
$$mult(swap[\![\, E_2\, ]\!]) = mult[\![\, E_2\, ]\!] \qquad (2)$$

and show

$$mult(swap[\![\, E_1 + E_2\, ]\!]) = mult[\![\, E_1 + E_2\, ]\!].$$

We have

$$
\begin{aligned}
mult(swap[\![\, E_1 + E_2\, ]\!]) ={}& \mathsf{let}\ E_a = swap[\![\, E_1\, ]\!],\ E_b = swap[\![\, E_2\, ]\!] \\
& \mathsf{in}\ mult[\![\, E_b + E_a\, ]\!] \\
={}& \mathsf{let}\ E_a = swap[\![\, E_1\, ]\!],\ E_b = swap[\![\, E_2\, ]\!] \\
& \mathsf{in}\ mult[\![\, E_b\, ]\!] + mult[\![\, E_a\, ]\!] \\
={}& mult(swap[\![\, E_2\, ]\!]) + mult(swap[\![\, E_1\, ]\!]) \\
={}& mult[\![\, E_2\, ]\!] + mult[\![\, E_1\, ]\!] \\
={}& mult[\![\, E_1\, ]\!] + mult[\![\, E_2\, ]\!] = mult[\![\, E_1 + E_2\, ]\!]. \quad (by\,1, 2)
\end{aligned}
$$

3. We assume

$$mult(swap[\![\, E_1\, ]\!]) = mult[\![\, E_1\, ]\!] \qquad (1)$$
$$mult(swap[\![\, E_2\, ]\!]) = mult[\![\, E_2\, ]\!] \qquad (2)$$

and show

$$mult(swap[\![\, E_1 \times E_2\, ]\!]) = mult[\![\, E_1 \times E_2\, ]\!].$$

We have

$$
\begin{aligned}
mult(swap[\![\, E_1 \times E_2\, ]\!]) ={}& \mathsf{let}\ E_a = swap[\![\, E_1\, ]\!],\ E_b = swap[\![\, E_2\, ]\!] \\
& \mathsf{in}\ mult[\![\, E_a \times E_b\, ]\!] \\
={}& \mathsf{let}\ E_a = swap[\![\, E_1\, ]\!],\ E_b = swap[\![\, E_2\, ]\!] \\
& \mathsf{in}\ 1 + mult[\![\, E_a\, ]\!] \times mult[\![\, E_b\, ]\!] \\
={}& 1 + mult(swap[\![\, E_1\, ]\!]) + mult(swap[\![\, E_2\, ]\!]) \\
={}& 1 + mult[\![\, E_1\, ]\!] + mult[\![\, E_2\, ]\!] \\
={}& mult[\![\, E_1 \times E_2\, ]\!]. \quad (by\,1, 2)
\end{aligned}
$$

□

## 1.3    Semantics

The 'semantics' (meaning) of the phrases of a syntactic domain can be defined by a valuation function that maps every phrase to some value of a *semantic domain*. This approach is also called *denotational semantics*; the values of the semantic domain are called *denotations*. Valuation functions are typically defined by structural induction.

Valuation functions often remain anonymous such that $[\![ D ]\!] = d$ is interpreted as 'the denotation of the syntactic phrase $D$ is the semantic value $d$'. We may thus also think of the brackets $[\![ \cdot ]\!]$ themselves as the valuation function; they are therefore also called *semantic brackets*. Which valuation function is applied in the presence of multiple syntactic domains can be easily deduced from the type of its argument.

**Example 1.5** We give the language on Sect. 1.2 a semantics by defining by structural induction the following two valuation functions that map expressions and numerals to natural numbers:

$$[\![ \cdot ]\!] : Expression \to \mathbb{N}$$
$$[\![ N ]\!] := [\![ N ]\!]$$
$$[\![ E_1 + E_2 ]\!] := [\![ E_1 ]\!] + [\![ E_2 ]\!]$$
$$[\![ E_1 \times E_2 ]\!] := [\![ E_1 ]\!] \cdot [\![ E_2 ]\!]$$

$$[\![ \cdot ]\!] : Numeral \to \mathbb{N}$$
$$[\![ 0 ]\!] := 0$$
$$[\![ 1 ]\!] := 1$$
$$[\![ N0 ]\!] := 2 \cdot [\![ N ]\!]$$
$$[\![ N1 ]\!] := 2 \cdot [\![ N ]\!] + 1$$

The first equation $[\![ N ]\!] := [\![ N ]\!]$ seems circular, but it indicates that the valuation function on domain *Expression* is defined by application of the valuation function on domain *Numeral*: the denotation of an expression that happens to be a numeral is the denotation of the numeral itself. If we give the anonymous valuation functions the names **E** and **N** and introduce the terminal $n$ for the first alternative of the grammar rule for *Expression*, the same equation is written in a more transparent way as

$$\mathbf{E}[\![ n(N) ]\!] := \mathbf{N}[\![ N ]\!].$$

We can now determine the denotation of the expression $10 \times (11 + 1)$ as

$$
\begin{aligned}
[\![ 10 \times (11 + 1) ]\!] &= [\![ 10 ]\!] \cdot [\![ 11 + 1 ]\!] \\
&= [\![ 10 ]\!] \cdot ([\![ 11 ]\!] + [\![ 1 ]\!]) \\
&= (2 \cdot [\![ 1 ]\!]) \cdot ((2 \cdot [\![ 1 ]\!] + 1) + [\![ 1 ]\!]) \\
&= (2 \cdot 1) \cdot ((2 \cdot 1 + 1) + 1) = 2 \cdot (3 + 1) = 2 \cdot 4 = 8.
\end{aligned}
$$

This semantics thus considers numerals as bit strings that denote natural numbers on which addition and multiplication are performed.                                              □

A core feature of denotational semantics is its *compositionality*, i.e., the denotation of a compound phrase is determined from the denotation of its subphrases; this simplifies the understanding of phrases and the reasoning about them considerably.
Denotations may be more complex than just numbers, as demonstrated below.

**Example 1.6** Take the grammar

$E \in Expression$
$E ::= 1 \mid \$ \mid E_1 + E_2 \mid E_1 \times E_2$

that defines a domain of arithmetic expressions in which a symbol $\$$ may occur. The informal interpretation is that the domain denotes functions on numbers where the symbol $\$$ indicates the argument of the function. This idea is formalized by the following semantics:

$$\llbracket \cdot \rrbracket : Expression \to (\mathbb{N} \to \mathbb{N})$$
$$\llbracket 1 \rrbracket := \lambda x.\, 1$$
$$\llbracket \$ \rrbracket := \lambda x.\, x$$
$$\llbracket E_1 + E_2 \rrbracket := \lambda x.\, \left( \llbracket E_1 \rrbracket(x) + \llbracket E_2 \rrbracket(x) \right)$$
$$\llbracket E_1 \times E_2 \rrbracket := \lambda x.\, \left( \llbracket E_1 \rrbracket(x) \cdot \llbracket E_2 \rrbracket(x) \right)$$

In this definition, we use the notation $\lambda x.\, T$ to denote a function with parameter $x$ and result $T$. Using a 'pattern-matching' style we may write this definition also as:

$$\llbracket \cdot \rrbracket : Expression \to (\mathbb{N} \to \mathbb{N})$$
$$\llbracket 1 \rrbracket(x) := 1$$
$$\llbracket \$ \rrbracket(x) := x$$
$$\llbracket E_1 + E_2 \rrbracket(x) := \llbracket E_1 \rrbracket(x) + \llbracket E_2 \rrbracket(x)$$
$$\llbracket E_1 \times E_2 \rrbracket(x) := \llbracket E_1 \rrbracket(x) \cdot \llbracket E_2 \rrbracket(x)$$

In any case, we get e.g. $\llbracket (\$ + 1) \times \$ \rrbracket(3) = \llbracket (\$ + 1) \rrbracket(3) \cdot \llbracket \$ \rrbracket(3) = (\llbracket \$ \rrbracket(3) + \llbracket 1 \rrbracket(3)) \cdot \llbracket \$ \rrbracket(3) = (3 + 1) \cdot 3 = 12.$                                              □

## 1.4    Type Systems

The language of the grammar

$E \in Expression$
$E ::= 1 \mid E_1 + E_2 \mid E_1 = E_2 \mid E_1 \text{ and } E_2 \mid \text{if } E \text{ then } E_1 \text{ else } E_2$

consists of a single domain of expressions; however this domain involves arithmetic expressions that denote numbers $(1, E + E)$ as well as Boolean expressions that denote truth values $(E = E, E$ and $E)$; some expressions expect their subexpressions to be numbers $(E + E, E = E)$, other expressions expect their subexpressions to be truth values $(E$ and $E)$. The expression if $E$ then $E$ else $E$ expects its first subexpression to be a Boolean expression and the other two expressions to be arithmetic expressions; as a whole it denotes an arithmetic expression. The grammar therefore also allows 'meaningless' expressions: while the expression $1 + 1 = $ if $1 = 1$ then $1 + 1$ else $1$ makes sense, the expression $(1$ and $1) + 1$ does not.

For this particular language, we can also give a grammar that captures exactly the 'meaningful' expressions by two separate domains for arithmetic and Boolean expressions:

> $A \in ArithmeticExpression$
> $B \in BooleanExpression$
> $A ::= 1 \mid A_1 + A_2 \mid$ if $B$ then $A_1$ else $A_2$
> $B ::= A_1 = A_2 \mid B_1$ and $B_2$

However, as we will see later, it is not always possible to define the intended language by grammars alone. A more general approach is to impose upon a grammar a *type system* that filters from the syntactic domains phrases that are not considered as worth of further consideration.

---

**Definition 1.4** (*Type System*) A type system is a logical inference system that introduces

- *judgements* that state that a phrase of this domain is 'well-formed' (possibly exposing some further information on that phrase), and
- for every judgement a set of *inference rules* that describe how valid judgements can be derived.

---

Figure 1.3 describes such a type system for the language on Sect. 1.4. It consists of two judgements $E$: aexp and $E$: bexp which can be read as '$E$ denotes a well-formed arithmetic expression' and '$E$ denotes a well-formed Boolean expression', respectively. Each rule is either an *axiom* like $1$: aexp that unconditionally ensures a judgement or a general rule like

$$\frac{E_1 : \text{aexp} \quad E_2 : \text{aexp}}{E_1 = E_2 : \text{bexp}}$$

which derives its *conclusion* $E_1 = E_2$: bexp under the assumption that its *premises* $E_1$: aexp and $E_2$: aexp can be derived. This rule states that, for every $E_1 \in$ *Expression* and $E_2 \in$ *Expression*,

**Rules for judgement $E$: aexp:**

$$1: \text{aexp} \qquad \frac{E_1: \text{aexp} \quad E_2: \text{aexp}}{E_1 + E_2: \text{aexp}} \qquad \frac{E_1: \text{bexp} \quad E_2: \text{aexp} \quad E_3: \text{aexp}}{\text{if } E_1 \text{ then } E_2 \text{ else } E_3: \text{aexp}}$$

**Rules for judgement $E$: bexp:**

$$\frac{E_1: \text{aexp} \quad E_2: \text{aexp}}{E_1 = E_2: \text{bexp}} \qquad \frac{E_1: \text{bexp} \quad E_2: \text{bexp}}{E_1 \text{ and } E_2: \text{bexp}}$$

**Fig. 1.3** A Type System for Arithmetic and Boolean Expressions

- if $E_1$ and $E_2$ are well-formed arithmetic expressions,
- then $E_1 = E_2$ is a well-formed Boolean expression.

According to these rules, the phrase $1 + 1 = \text{if } 1 = 1 \text{ then } 1 + 1 \text{ else } 1$ is a well-formed Boolean expression, because we can derive the judgement

$$1 + 1 = \text{if } 1 = 1 \text{ then } 1 + 1 \text{ else } 1: \text{bexp}$$

by constructing the following *inference tree*:

$$\frac{\dfrac{1: \text{aexp} \quad 1: \text{aexp}}{1 + 1: \text{aexp}} \qquad \dfrac{\dfrac{1: \text{aexp} \quad 1: \text{aexp}}{1 = 1: \text{bexp}} \quad \dfrac{1: \text{aexp} \quad 1: \text{aexp}}{1 + 1: \text{aexp}} \quad 1: \text{aexp}}{\text{if } 1 = 1 \text{ then } 1 + 1 \text{ else } 1: \text{aexp}}}{1 + 1 = \text{if } 1 = 1 \text{ then } 1 + 1 \text{ else } 1: \text{bexp}}$$

The leaves of the tree represent axioms of the typing system, the inner nodes of the tree are instances of the general rules (where variables have been replaced by concrete phrases); the root of this tree is the judgement that is ultimately derived by the inference tree. If we flip the tree upside down, we see that the inference tree resembles the abstract syntax tree of the phrase with some extra annotations describing the judgements. A type system thus can be also considered as a procedure that annotates syntax trees with extra typing information; if no such annotation can be constructed, the tree is not well-formed.

Actually, a grammar is just as a special case of a type system, as can be seen by comparing the grammar on Sect. 1.4 with the type system of Fig. 1.3. The converse, however, is generally not true: type systems are more expressive and therefore more discriminating than grammars. Typically a grammar is therefore only used to describe a 'first approximation' of the language of interest; the real language is actually defined by a type system that selects from all the phrases allowed by the grammar those that are indeed considered as 'meaningful'. This point is demonstrated below.

**Example 1.7** Consider the following language of arithmetic expressions that may also contain identifiers:

$E \in Expression$
$I \in Identifier$
$E := 1 \mid I \mid E_1 + E_2 \mid \text{var } I = E_1; E_2$
$I := \ldots$

Only those expressions are well-formed whose identifiers have been introduced by var definitions; therefore the expression

$$\text{var } x = 1; \text{var } y = x + 1; y + x$$

with identifiers $x$ and $y$ is well-formed, but this is not the case for the expression

$$\text{var } x = 1; \text{var } y = x + 1; y + z$$

with undeclared identifier $z$. The restriction that every occurrence of an identifier must occur in the context of a corresponding declaration is not expressible by a grammar.

We may, however, give the following type system that captures exactly the well-formed expressions. The judgements of this type system are of form

$$D \vdash E : \text{exp}$$

where $D$ is the set of those identifiers that have been introduced by those definitions that establish the context of expression $E$. The rules of this type system are as follows:

$$D \vdash 1 : \text{exp} \qquad \frac{I \in D}{D \vdash I : \text{exp}} \qquad \frac{D \vdash E_1 : \text{exp} \quad D \vdash E_2 : \text{exp}}{D \vdash E_1 + E_2 : \text{exp}}$$

$$\frac{D \vdash E_1 : \text{exp} \quad D \cup \{I\} \vdash E_2 : \text{exp}}{D \vdash \text{var } I = E_1; E_2 : \text{exp}}$$

The core of this type system is the rule for expression $I$ which requires in its premise that $I$ must be contained in the set $D$ of declared identifiers; the rule for var $I = E_1; E_2$ adds the declared variable to this set for type-checking $E_2$.

Then an expression $E$ is well-formed if and only if the judgement $\varnothing \vdash E : \text{exp}$ can be derived. For instance, the inference tree

$$\frac{\varnothing \vdash 1 : \text{exp} \quad \dfrac{\dfrac{x \in \{x\}}{\{x\} \vdash x : \text{exp}} \quad \{x\} \vdash 1 : \text{exp}}{\dfrac{\{x\} \vdash x + 1 : \text{exp}}{\{x\} \vdash \text{var } y = x + 1; y + x : \text{exp}} \quad \dfrac{\dfrac{y \in \{x, y\}}{\{x, y\} \vdash y : \text{exp}} \quad \dfrac{x \in \{x, y\}}{\{x, y\} \vdash x : \text{exp}}}{\{x, y\} \vdash y + x : \text{exp}}}}{\varnothing \vdash \text{var } x = 1; \text{var } y = x + 1; y + x : \text{exp}}$$

demonstrates that the expression var $x = 1$; var $y = x + 1$; $y + x$ is well-formed. □

## 1.5    The Semantics of Typed Languages

Since it is actually the typing rules, not the grammar, that define a language, we may consider a language as the set of all abstract syntax trees which the typing rules have annotated with type information. These annotation thus becomes an integral part of the tree and may also influence the semantics of the tree, as demonstrated below.

**Example 1.8** Consider the language of arithmetic expressions with integer-typed and string-typed constants defined by the grammar

$$E \in Expression$$
$$I \in Integer$$
$$S \in String$$
$$E := I \mid S \mid E_1 + E_2$$
$$I := \ldots \qquad S := \ldots$$

We impose on this language a type system with judgements $E:$ int and $E:$ str read as 'E is a well-formed expression denoting an integer' and 'E is a well-formed expression denoting a string', respectively. Their derivation rules are as fol-

$$I: \text{int} \qquad S: \text{str} \qquad \frac{E_1: \text{int} \quad E_2: \text{int}}{E_1 + E_2: \text{int}}$$

lows: $\dfrac{E_1: \text{str} \quad E_2: \text{str}}{E_1 + E_2: \text{str}} \quad \dfrac{E_1: \text{str} \quad E_2: \text{int}}{E_1 + E_2: \text{str}} \quad \dfrac{E_1: \text{int} \quad E_2: \text{str}}{E_1 + E_2: \text{str}}$ A sum therefore only denotes an integer if both arguments denote integers; it denotes a string, if at least one of the arguments denotes a string.

We are only interested in the denotation of well-formed syntax trees; the valuation function is thus not defined by induction on the grammar but by induction on the rules of the typing system. We give one equation for every rule which assigns to the abstract syntax tree annotated by the rule a denotation computed from the denotations of the annotated trees that were derived to satisfy the assumptions of the rule:

$$[\![ \cdot ]\!]: Expression \rightarrow \mathbb{Z} \cup Char^*$$
$$[\![ I: \text{int} ]\!] := [\![ I ]\!]$$
$$[\![ S: \text{str} ]\!] := [\![ S ]\!]$$
$$[\![ E_1 + E_2: \text{int} ]\!] := [\![ E_1: \text{int} ]\!] + [\![ E_2: \text{int} ]\!]$$
$$[\![ E_1 + E_2: \text{str} ]\!] := [\![ E_1: \text{str} ]\!] \circ [\![ E_2: \text{str} ]\!]$$
$$[\![ E_1 + E_2: \text{str} ]\!] := [\![ E_1: \text{str} ]\!] \circ str[\![ E_2: \text{int} ]\!]$$
$$[\![ E_1 + E_2: \text{str} ]\!] := str[\![ E_1: \text{int} ]\!] \circ [\![ E_2: \text{str} ]\!]$$

As usual, the denotations of literals are computed by applying the corresponding valuations $[\![ \cdot ]\!]: Integer \rightarrow \mathbb{Z}$ and $[\![ \cdot ]\!]: String \rightarrow Char^*$ (where $Char^*$ denotes the domain of finite character sequences with concatenation operator $\circ$). However, there are four possible denotations of the expression $E_1 + E_2$ depending on the type with which the expression was annotated: in case of annotation int, the result is computed by numerical addition; in case of str, the result is computed by application of the string concatenation operator ∘. Furthermore, if one of the arguments is a number and the other one is a string, a conversion operation $str: \mathbb{Z} \to Char^*$ is applied.  □

However, it is not a priori clear that the valuation function of above example is indeed well-defined; the operations + and ∘ may be only applied to integers and strings, respectively, but they are applied to the outcome of the valuation function which may be an integer or a string. To show the that the definition is well-formed, we apply the proof principle of *rule induction*: for every rule of the type system, we assume as the induction hypothesis that the property holds for the premises of the rule and show that under this assumption the property also holds for its conclusion. Thus it is shown that the property holds for every phrase that is well-formed according to the rules of the type system.

**Example 1.9** We claim that for every well-formed expression $E$ in the language of Example 1.8 the following holds:

- If $E$ : int can be derived, then $[\![ E ]\!]$ is an integer.
- If $E$ : str can be derived, then $[\![ E ]\!]$ is a string.

Consequently, the semantics of the language is well-defined.
   We prove this claim by induction on the typing rules of the language:

1. Case $I$: Only $I$ : int can be derived, and $[\![ I ]\!]$ is an integer.
2. Case $S$: Only $S$ : str can be derived, and $[\![ S ]\!]$ is a string.
3. Case $E_1 + E_2$: First assume that $E_1 + E_2$ : int can be derived. From the typing rules, this is only possible, if $E_1$ : int can be derived and $E_2$ : int can be derived. From the induction hypothesis, therefore $[\![ E_1 ]\!]$ is an integer and $[\![ E_2 ]\!]$ is an integer. Thus $[\![ E_1 ]\!] + [\![ E_2 ]\!]$ and consequently $[\![ E_1 + E_2 ]\!]$ is an integer.
   Now assume that $E_1 + E_2$ : str can be derived for which we have three possibilities:

   a. $E_1$ : str can be derived and $E_2$ : str can be derived. From the induction hypothesis, therefore $[\![ E_1 ]\!]$ is a string and $[\![ E_2 ]\!]$ is a string. Thus $[\![ E_1 ]\!] \circ [\![ E_2 ]\!]$ is a string and consequently $[\![ E_1 + E_2 ]\!]$ is a string.
   b. $E_1$ : str can be derived and $E_2$ : int can be derived. From the induction hypothesis, therefore $[\![ E_1 ]\!]$ is a string and $[\![ E_2 ]\!]$ is an integer. Thus $str[\![ E_2 ]\!]$ is a string, consequently $[\![ E_1 ]\!] \circ str[\![ E_2 ]\!]$ is a string, and $[\![ E_1 + E_2 ]\!]$ is a string.
   c. $E_1$ : int can be derived and $E_2$ : str can be derived. From the induction hypothesis, therefore $[\![ E_1 ]\!]$ is an integer and $[\![ E_2 ]\!]$ is a string. Thus $str[\![ E_1 ]\!]$ is a string, consequently $str[\![ E_1 ]\!] \circ [\![ E_2 ]\!]$ is a string, and $[\![ E_1 + E_2 ]\!]$ is a string.

□

## Exercises

Download from the following URL:
https://www.risc.jku.at/people/schreine/TP/exercises/ex-syntax.pdf

## Further Reading

The concept of abstract syntax in contrast to concrete syntax is most clearly explained in various texts on the denotational semantics of programming languages, e.g., in Chap. 1 of Schmidt's book [160], in Chap. 1 of Mitchell's book [119], in Chap. 1 of Fernández's book [48], in Chap. 3 of Nipkow's and Klein's book [133], or in Mosses's Chap. 11 of the handbook [177]. Also some of the more computer-science oriented presentations of logic make this distinction clear, in particular Chap. 1 of Harrison's handbook [65]; most classical logical texts, however, do not clearly separate between the two different layers.

The work of the logician Haskell Curry has since the 1930s a close relationship (the 'Curry-Howard correspondence') between types and formulas paving the way to type theory, see e.g. Mitchell's book [119], Mitchell's Chap. 8 of the handbook [177], or Gunter's book [63]. In Cardelli's and Wegner's paper [33], modern type concepts are described by logical rules. The view that ultimately not a BNF-like grammar but typing rules define a language is most clearly described in Chap. 1 of Schmidt's book [161] which has influenced our own presentation.

First-order logic was the first formal language that was given a precise semantics by defining by structural induction a function that maps syntactic structures to corresponding semantic values, see the 'Further Reading' section of Chap. 2. However, the clearest presentations of the general idea can be again found in texts on denotational semantics, e.g. Chap. 4 of [160], Chap. 1 of [119], Chaps. 3 and 11 of [133], Chaps. 1 and 4 of Nielson's and Nielson's book [131], or Chap. 5 of Winskel's book [180].

The principle of structural induction is explained in many texts on programming language semantics, e.g., in Chap. 1 of [160], Chap. 3 of [180], Chap. 1 of [131], Chap. 1 of [119], Chap. 2 of [48], Chap. 2 of [133], or Section 3 of Manna's paper [107]. Also rule induction is covered in Chap. 4 of [180], Chaps. 2 and 6 of [131], Chap. 1 of [119], Chap. 2 of [48], Chap. 4 of [133]. Also logic texts, especially if they are computer science-oriented, cover these topic as special cases of induction, see the 'Further Reading' section of Chap. 3.

## Abstract Syntax Trees in OCaml

We are now going to complement the material of the preceding chapter by some software that demonstrates the computational interpretation of the mathematical concepts encountered so far. This software is written in the programming language *OCaml* [134] that supports functional, imperative, and object-oriented programming styles. The source code of the examples presented in this section can be downloaded from the URL

  https://www.risc.jku.at/people/schreine/TP/software/syntax/syntax.ml

and can be run by executing from the command line the following command:

```
ocaml syntax.ml
```

### An Untyped Language

We start by defining the syntax domains

$E \in Expression$
$N \in Numeral$
$E ::= n(N) \mid s(E_1, E_2) \mid p(E_1, E_2)$
$N ::= z() \mid o() \mid nz(N) \mid no(N)$

as corresponding OCaml data types:

```
type expression =
| N of numeral
| S of expression * expression
| P of expression * expression
and numeral =
| Z
| O
| NZ of numeral
| NO of numeral
;;
```

We can then define the phrase respectively abstract syntax tree

  $s(n(o()), p(n(nz(o())), n(no(o()))))$

as the value

```
let e = S(N(O),P(N(NZ(O)),N(NO(O)))) ;;
```

We define by structural induction functions to convert the phrases to strings:

```
let rec estr (e:expression): string =
  match e with
  | N(n) -> nstr n
  | S(e1,e2) -> "(" ^ (estr e1) ^ "+" ^ (estr e2) ^ ")"
  | P(e1,e2) -> "(" ^ (estr e1) ^ "*" ^ (estr e2) ^ ")"
and nstr (n:numeral): string =
  match n with
  | Z -> "0"
  | O -> "1"
  | NZ(n) -> (nstr n) ^ "0"
  | NO(n) -> (nstr n) ^ "1"
;;
```

Executing the command

```
Printf.printf "%s\n" (estr e) ;;
```

then prints the linear representation of the abstract syntax tree:

```
(1+(10*11))
```

Likewise we can implement the semantics of the language defined by mappings of expressions and numerals to integer numbers

$$[\![ \cdot ]\!] : \textit{Expression} \to \mathbb{N}$$
$$[\![ N ]\!] := [\![ N ]\!]$$
$$[\![ E_1 + E_2 ]\!] := [\![ E_1 ]\!] + [\![ E_2 ]\!]$$
$$[\![ E_1 \times E_2 ]\!] := [\![ E_1 ]\!] \cdot [\![ E_2 ]\!]$$

$$[\![ \cdot ]\!] : \textit{Numeral} \to \mathbb{N}$$
$$[\![ 0 ]\!] := 0$$
$$[\![ 1 ]\!] := 1$$
$$[\![ N0 ]\!] := 2 \cdot [\![ N ]\!]$$
$$[\![ N1 ]\!] := 2 \cdot [\![ N ]\!] + 1$$

by corresponding OCaml functions:

```
let rec eval (e:expression): int =
  match e with
  | N(n) -> nval n
  | S(e1,e2) -> (eval e1) + (eval e2)
```

```
  | P(e1,e2) -> (eval e1) * (eval e2)
and nval (n:numeral): int =
  match n with
  | Z -> 0
  | O -> 1
  | NZ(n) -> 2*(nval n)
  | NO(n) -> 2*(nval n)+1
;;
```

We can then compute the semantics of the expression by executing the command

```
Printf.printf "%d\n" (eval e) ;;
```

which prints the result

7

## A Typed Language

We continue by investigating the computational aspects of type systems defined by logical inference rules. For this we implement the language

$E \in Expression$

$E ::= 1 \mid E_1 + E_2 \mid E_1 = E_2 \mid E_1 \text{ and } E_2 \mid \text{if } E \text{ then } E_1 \text{ else } E_2$

which mixes in a single domain both arithmetic and Boolean expressions by the following data types:

```
type tag = None | Aexp | Bexp ;;
type texp = Texp of tag ref * exp
and exp =
| One
| Plus   of texp * texp
| Equals of texp * texp
| And    of texp * texp
| If     of texp * texp * texp
;;
```

Type texp represents typed expressions that contain, in additional to the core expression of type exp, a mutable variable of type tag that indicates the type of the expression. The auxiliary function

```
let uexp (e:exp): texp = Texp(ref None, e);;
```

creates an 'untyped' expression by attaching the tag None to a core expression. Thus we implement the untyped abstract syntax tree

$$1 + 1 = \text{if } 1 = 1 \text{ then } 1 + 1 \text{ else } 1$$

by the expression

```
let e2 =
  uexp (Equals(
    uexp (Plus(uexp One, uexp One)),
    uexp (If(
      uexp (Equals(uexp One,uexp One)),
      uexp (Plus(uexp One,uexp One)),
      uexp One))))
;;
```

The functions

```
let rec tstr(t: tag ref): string =
  match !t with
  | None -> ""
  | Aexp -> ":aexp"
  | Bexp -> ":bexp"
;;
let rec str(e: texp): string =
  match e with
  | Texp(t,One) -> "1" ^ (tstr t)
  | Texp(t,Plus(e1,e2))    ->
      "(" ^ (str e1) ^ "+" ^ (str e2) ^ ")" ^ tstr(t)
  | Texp(t,Equals(e1,e2)) ->
      "(" ^ (str e1) ^ "=" ^ (str e2) ^ ")" ^ tstr(t)
  | Texp(t,And(e1,e2))     ->
      "(" ^ (str e1) ^ "&" ^ (str e2) ^ ")" ^ tstr(t)
  | Texp(t,If(e1,e2,e3))  ->
      "(if " ^ (str e1) ^ " then " ^ (str e2) ^
                      " else " ^ (str e3) ^ ")" ^ tstr(t)
;;
```

convert typed expressions to strings such that executing the program

```
Printf.printf "
```

prints the representation

```
((1+1)=(if (1=1) then (1+1) else 1))
```

Now we are going to implement the type system established by the judgements $E$: aexp and $E$: bexp with the following rules:

$$\frac{}{1: \text{aexp}} \qquad \frac{E_1: \text{aexp} \quad E_2: \text{aexp}}{E_1 + E_2: \text{aexp}} \qquad \frac{E_1: \text{bexp} \quad E_2: \text{aexp} \quad E_3: \text{aexp}}{\text{if } E_1 \text{ then } E_2 \text{ else } E_3: \text{aexp}}$$

$$\frac{E_1: \text{aexp} \quad E_2: \text{aexp}}{E_1 = E_2: \text{bexp}} \qquad \frac{E_1: \text{aexp} \quad E_2: \text{bexp}}{E_1 \text{ and } E_2: \text{bexp}}$$

We first define an auxiliary function

```
let check (e: texp) (t: tag): unit =
  match e with Texp(t0,_) ->
     if (!t0 <> t) then raise (Failure "type error") else ()
;;
```

that raises an exception (and thus aborts the type checking process) if a given expression $e$ does not have the type tag $t$. Based on this function, we define the function

```
let rec tset(e: texp): unit =
  match e with
  | Texp(t,One) ->
      t := Aexp
  | Texp(t,Plus(e1,e2))    ->
      tset e1; check e1 Aexp;
      tset e2; check e2 Aexp;
      t := Aexp
  | Texp(t,Equals(e1,e2)) ->
      tset e1; check e1 Aexp;
      tset e2; check e2 Aexp;
      t := Bexp
  | Texp(t,And(e1,e2))    ->
      tset e1; check e1 Bexp;
      tset e2; check e2 Bexp;
      t := Bexp
  | Texp(t,If(e1,e2,e3))  ->
      tset e1; check e1 Bexp;
      tset e2; check e2 Aexp;
      tset e3; check e3 Aexp;
      t := Aexp
;;
```

where each inference rule is translated to one case of the function. If we then execute the command

```
tset e2;;
```

the abstract syntax tree is annotated with the deduced type information. Now executing the command

```
Printf.printf "%s\n" (str e2) ;;
```

prints the linear representation of the annotated syntax tree

```
((1:aexp+1:aexp):aexp=
 (if (1:aexp=1:aexp):bexp
    then (1:aexp+1:aexp):aexp else 1:aexp):aexp):bexp
```

which corresponds to the following inference tree:

$$
\cfrac{\cfrac{}{1+1:\text{aexp}}\ \ 1:\text{aexp}\quad \cfrac{\cfrac{1:\text{aexp}\quad 1:\text{aexp}}{1=1:\text{bexp}}\quad \cfrac{\cfrac{1:\text{aexp}\quad 1:\text{aexp}}{1+1:\text{aexp}}\quad 1:\text{aexp}}{\text{if } 1=1 \text{ then } 1+1 \text{ else } 1:\text{aexp}}}{1+1=\text{if } 1=1 \text{ then } 1+1 \text{ else } 1:\text{bexp}}
$$

Type systems defined as logical rules thus can give directly rise to corresponding implementations of type checkers.

# The Language of Logic

2

*Die Grenzen meiner Sprache bedeuten die Grenzen meiner Welt.*
*(The limits of my language mean the limits of my world.)*
— Ludwig Wittgenstein (Tractatus Logico-Philosophicus)

In this chapter, we introduce the language of first-order logic, on which we may base precise thinking, speaking, and writing, in computer science and all other kinds of scientific disciplines. We present first-order logic as a formal language in the spirit of the previous chapter by an abstract syntax based on which the intuitive interpretation of the various phrases is explained. We will then equip the language with some sort of "type system" to identify the well-formed phrases, and give these phrases a formal semantics, first for the subset of "propositional logic" and then for the full language. Based on this semantics, we investigate which syntactic transformations preserve the meanings of formulas and which formulas always denote "true" statements.

## 2.1 First-Order Logic

We now introduce the language that represents the foundation of our further discourse.

**Definition 2.1** (*Syntax of First-Order Logic*) The language of *first-order logic* (*predicate logic*) consists of *terms*, *term sequences*, (first-order) *formulas*, *variables*, *constant symbols*, *function symbols*, and *predicate symbols*, which are formed according to the following grammar:

$T \in Term$
$Ts \in Terms$
$F \in Formula$
$V \in Variable$
$CS \in ConstantSymbol$
$FS \in FunctionSymbol$
$PS \in PredicateSymbol$

$T ::= V \mid CS \mid FS(Ts) \mid \text{let } V = T_1 \text{ in } T_2 \mid \text{if } F \text{ then } T_1 \text{ else } T_2$
$Ts ::= T \mid T,Ts$

$F ::= \text{true} \mid \text{false} \mid PS(Ts) \mid T_1 = T_2$
$\phantom{F ::=} \mid \neg F \mid F_1 \wedge F_2 \mid F_1 \vee F_2 \mid F_1 \Rightarrow F_2 \mid F_1 \Leftrightarrow F_2$
$\phantom{F ::=} \mid \forall V . F \mid \exists V . F \mid \text{let } V = T \text{ in } F \mid \text{if } F \text{ then } F_1 \text{ else } F_2$

$V ::= \ldots; \quad CS ::= \ldots; \quad FS ::= \ldots; \quad PS ::= \ldots$

We call the symbols $\neg$, $\wedge$, $\vee$, $\Rightarrow$, and $\Leftrightarrow$ *(logical) connectives*; they compose formulas to more complex formulas. The symbols $\forall$, $\exists$ and let ... in ... are called *quantifiers*; they introduce variables into formulas respectively terms (we also say that the quantifiers *bind* the variables).

This version of first-order logic is a bit richer than usual. Typically the phrases let ... in ... (a *variable binder*) and if ... then ... else ... (a *conditional*) are not considered, because they are theoretically not essential: they can be also introduced later as syntactic abbreviations of more complex phrases that only involve the other constructs. However, these phrases are so useful in the practice of specifying complex properties, that we make them already part of our basic framework and investigate their behavior.

Not every term or formula allowed by this grammar, however, is actually well-formed, i.e. the grammar only describes a superset of the set of well-formed terms and formulas. The reason is that every function respectively predicate may be only applied to a certain number of arguments, its *arity*. Informally, a term $FS(T_1, \ldots, T_n)$ and a formula $PS(T_1, \ldots, T_n)$ is well-formed, if and only if the arity of $FS$ respectively of $PS$ is $n$. A formal type system that captures this intuition will be introduced later.

The sets of variables and symbols are usually not explicitly defined, although often identifiers like

- $x, y, z, \ldots$ denote variables
- $a, b, c, \ldots$ denote constants,
- $f, g, h, \ldots$ denote functions, and
- $p, q, r, \ldots$ denote predicates.

Typically it becomes clear from the context or from its application whether an identifier denotes a variable or a particular kind of symbol; likewise the arities of function and predicate symbols are implicitly defined by their application. For instance, in the formula

$$\forall x.\ p_1(x) \vee p_2(x) \Rightarrow \exists y.\ q(y) \wedge r(f(x,a), y)$$

the identifiers $x$ and $y$ represent variables, because they are introduced by $\forall$ and $\exists$; we may reasonably assume that $a$ is a constant symbol because it is not introduced in this way and its name is distinctly different from the names used for the variables. Furthermore, $p_1$, $p_2$, and $q$ represent unary (1-ary) predicate symbols, $r$ represents a binary (2-ary) predicate symbol, and $f$ denotes a binary function symbol.

Above linear notation of a formula does not uniquely define its abstract syntax, e.g. it could denote

$$\forall x.\ \Big( \big( p_1(x) \vee p_2(x) \big) \Rightarrow \exists y.\ \big( q(y) \wedge r(f(x,a), y) \big) \Big)$$

but also

$$\forall x.\ \Big( p_1(x) \vee \big( p_2(x) \Rightarrow \exists y.\ q(y) \big) \wedge r(f(x,a), y) \Big)$$

as well as other choices. To make the abstract syntax of a formula clear without an excessive use of parentheses, we apply the following conventions:

- the logical connectives $\neg$, $\wedge$, $\vee$, $\Rightarrow$, $\Leftrightarrow$, and if ... then ... else ... have strictly decreasing binding power: $\neg$ binds strongest and if ... then ...binds weakest;
- the quantifiers $\forall$, $\exists$ and let ... in ... have weakest binding power; thus the body $F$ in the formulas $\forall V.\ F$, $\exists V.\ F$, and let $V = T$ in $F$ reaches as far as possible.

By these conventions, the first choice above is the intended one. This formula is also depicted by the left abstract syntax tree in Fig. 2.1. Every rectangular node in this tree represents (the root of) a formula; the leaves represent the atomic (structurally simplest) formulas. The right tree depicts the inner structure of such an atomic formula, namely that of $r(f(x,a), y)$; every circular node in that tree represents (the root of) a term while the rounded boxes represent (the roots of) term sequences.

Many binary function or predicate symbols are typically written in infix form (between the arguments) rather than in the standard prefix form (before the arguments). For instance, the function application $x + 1$ corresponds to the standard form $+(x, 1)$ with binary function symbol $+$ while the atomic formula $x \leq 1$ corresponds to the standard form $\leq(x, 1)$ with binary predicate symbol $\leq$. The atomic formula $x \cdot y \leq x \cdot (y + 1)$ is thus to be interpreted as $\leq(\cdot(x, y), \cdot(x, +(y, 1)))$. Furthermore,

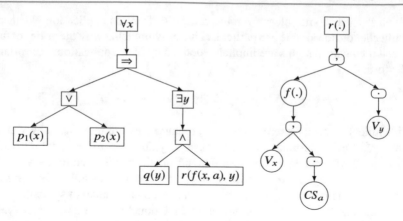

**Fig. 2.1** The abstract syntax of a formula and a term

for infix predicate symbols, a stroked version symbol indicates the negation of the predicate, in particular, $x \neq y$ and $x \notin y$ mean $\neg x = y$ and $\neg x \in y$, respectively.

Finally, if $F[x]$ denotes a formula with variable $x$, then we may use the syntactic abbreviations

$$\forall F[x].\ F \rightsquigarrow \forall x.\ (F[x] \Rightarrow F)$$
$$\exists F[x].\ F \rightsquigarrow \exists x.\ (F[x] \wedge F)$$
$$\forall x_1, \ldots, x_n.\ F \rightsquigarrow \forall x_1.\ \ldots \forall x_n.\ F$$
$$\exists x_1, \ldots, x_n.\ F \rightsquigarrow \exists x_1.\ \ldots \exists x_n.\ F$$

where $P \rightsquigarrow Q$ is to be read as "phrase $P$ expands to phrase $Q$".

Please note the connective $\Rightarrow$ in the expansion of $\forall F[x].\ F$, which corresponds to the usual interpretation of a sentence like

Every natural number $x$ is greater equal zero.

as the statement

For every $x$, if $x$ is a natural number, then $x$ is greater equal zero.

and *not* as the statement

For every $x$, $x$ is a natural number and $x$ is greater equal zero.

On the other hand, in the expansion of $\exists F[x]$. $F$ the connective $\wedge$ is used, which corresponds to the usual interpretation of a sentence like

There is a natural number $x$ greater than zero.

as the following statement:

There is some $x$, such that $x$ is a natural number and $x$ is greater than zero.

**Example 2.1** Let $0$ and $\mathbb{N}$ denote constant symbols, $+$ denote a binary function symbol and $\in$ and $\geq$ denote binary predicate symbols.

• The formula $\forall x \in \mathbb{N}. \ x \geq 0$ expands to

$$\forall x. \ x \in \mathbb{N} \Rightarrow x \geq 0$$

• The formula $\exists x, y, z. \ x + y \geq z$ expands to

$$\exists x. \ \exists y. \ \exists z. \ x + y \geq z$$

• The formula $\forall x \in \mathbb{N}, y \in \mathbb{N}. \ x + y \geq 0$ expands to

$$\forall x. \ x \in \mathbb{N} \Rightarrow (\forall y. \ y \in \mathbb{N} \Rightarrow x + y \geq 0)$$

• The formula $\exists x \in \mathbb{N}, y \in \mathbb{N}. \ x + y \geq 0$ expands to

$$\exists x. \ x \in \mathbb{N} \wedge (\exists y. \ y \in \mathbb{N} \wedge x + y \geq 0) \qquad \qquad \square$$

## 2.2    Informal Interpretation

The language of first-order logic essentially defines two kinds of syntactic phrases, *terms* and *formulas*.

**Definition 2.2** (*Informal Term Semantics*) A term denotes an "object" or "value" of some given domain. The informal interpretation of the various kinds of terms is as follows:

• The value of a *variable V* depends on the context in which it is evaluated; this context is typically set up by a quantifier that introduces the variable.
• A *constant CS* denotes a fixed value.

- A *function application* $FS(Ts)$ denotes the result of a function applied to the values denoted by the terms $Ts$.
- A *variable binder* let $V = T_1$ in $T_2$ denotes the value of term $T_2$ where every occurrence of the variable $V$ denotes the value of term $T_1$.
- A *conditional* if $F$ then $T_1$ else $T_2$ denotes the value of $T_1$, if the formula $F$ denotes "true", and the value of $T_2$, otherwise.

In natural language, the role of terms is taken by nouns and noun phrases:

- the nouns "Mary" and "John" correspond to the constants *Mary* and *John*,
- the phrase "the mother of Mary" corresponds to the application *mother*(*Mary*) of a unary function *mother*,
- the noun "she" corresponds to the variable *she* (the denoted person depends on the context),
- the phrase "the children of John and her" corresponds to the function application *children*(*John*, *she*) of a binary function *children*;
- the phrase "the children of John and her, Mary" corresponds to the let term let *she* = *Mary* in *children*(*John*, *she*).
- the phrase "if she is at least 18, then she herself, else her mother" corresponds to the conditional term if $age(she) \geq 18$ then *she* else *mother*(*she*).

**Definition 2.3** (*Informal Formula Semantics*) A formula denotes the truth value ("true" or "false") of a statement about objects. The various kinds of formulas may be understood as follows:

- The *logical constants* true and false denote the truth values "true" and "false".
- An *atomic predicate* $PS(Ts)$ denotes the truth value of a predicate for the values denoted by the terms $Ts$.
- An *equality* $T_1 = T_2$ is a special atomic predicate that is true if and only if the terms $T_1$ and $T_2$ denote the same values.
- A *negation* $\neg F$ (read "not $F$") is true if and only if the formula $F$ is false.
- A *conjunction* $F_1 \wedge F_2$ (read "$F_1$ and $F_2$") is true if and only if both formulas $F_1$ and $F_2$ are true.
- A *disjunction* $F_1 \vee F_2$ (read "$F_1$ or $F_2$") is true if at least one of the formulas $F_1$ or $F_2$ is true.
- An *implication* $F_1 \Rightarrow F_2$ (read "$F_1$ implies $F_2$" or "if $F_1$, then $F_2$") is only false if the *antecedent* $F_1$ is true and the *consequent* $F_2$ is false.
- An *equivalence* $F_1 \Leftrightarrow F_2$ (read "$F_1$ is equivalent to $F_2$" or "if $F_1$, then $F_2$, and vice versa") is true if and only if both $F_1$ and $F_2$ have the same truth value.

- A *universal quantification* $\forall V . F$ (read "for all $V$, $F$" or "for every $V$, $F$") is true if formula $F$ is true for every possible value of variable $V$.
- An *existential quantification* $\exists V . F$ (read "for some $V$, $F$" or "there exists some $V$ such that $F$") is true if formula $F$ is true for at least one value of variable $V$.
- A *variable binder* let $V = T$ in $F$ denotes the truth value of formula $F$ where every occurrence of the variable $V$ denotes the value of term $T$.
- A *conditional* if $F$ then $F_1$ else $F_2$ denotes the truth value of $F_1$, if the formula $F$ denotes "true", and the truth value of $F_2$, otherwise.

The interpretation of the logical connectives is also summarized in the tables of Fig. 2.2 and visualized by corresponding diagrams: the interior of each circle represents the set of points where the corresponding formula is true; the shaded part of the diagram represents the set of points where the compound formula is true.

It should be particularly noted that an implication $F_1 \Rightarrow F_2$ is *true*, if the antecedent $F_1$ is *false*. The rationale for this can be seen by a statement like "if it rains, then I stay at home" which is contradicted only by the fact that it rains and I do not stay at home; if it does not rain, I can stay at home or go away, as I like.

In natural language, the role of formulas is taken by sentences. For instance, the sentence

> For all pairs of persons it is true that, if the first person is the sister of the second one, then the second one is the sister or the brother of the first one.

corresponds to the formula

$$\forall x. \, \forall y. \, isSister(x, y) \Rightarrow isSister(y, x) \vee isBrother(y, x)$$

where the atomic formulas *isSister*$(x, y)$ respectively *isBrother*$(x, y)$ are to be interpreted as "$x$ is the sister of $y$" respectively "$x$ is the brother of $y$". Likewise, the sentence

> Every person has a mother and a father.

corresponds to the formula

$$\forall x. \, \bigl(\exists y. \, isMother(y, x)\bigr) \wedge \bigl(\exists y. \, isFather(y, x)\bigr)$$

where the atomic formulas *isMother*$(y, x)$ respectively *isFather*$(y, x)$ are to be interpreted as "$y$ is the mother of $x$" respectively "$y$ is the father of $x$". Finally, the sentence

> For all persons $x$ and $y$ that have the same mother it is true that, if $y$ is the brother of $x$, then $y$ is male, otherwise $y$ is female.

| $F$ | $\neg F$ |   | $F_1$ | $F_2$ | $F_1 \wedge F_2$ | $F_1 \vee F_2$ | $F_1 \Rightarrow F_2$ | $F_1 \Leftrightarrow F_2$ |
|------|-------|---|-------|-------|-----------|-----------|-------------|--------------|
| false | true |   | false | false | false | false | true | true |
| true | false |   | false | true | false | true | true | false |
|  |  |   | true | false | false | true | false | false |
|  |  |   | true | true | true | true | true | true |

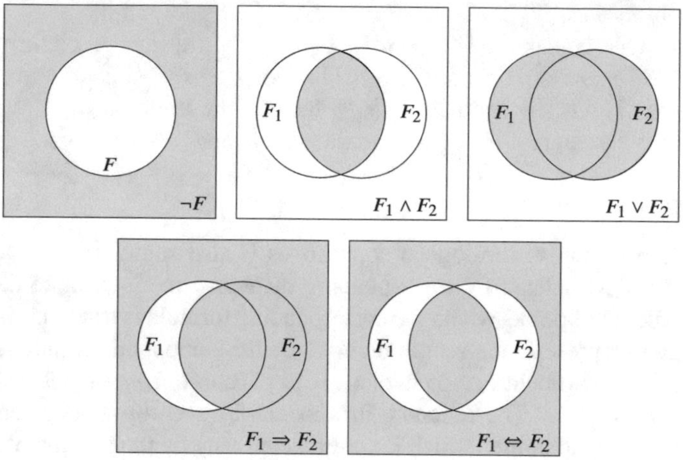

| $F$ | $F_1$ | $F_2$ | if $F$ then $F_1$ else $F_2$ |
|------|-------|-------|------------------------|
| false | false | false | false |
| false | false | true | true |
| false | true | false | false |
| false | true | true | true |
| true | false | false | false |
| true | false | true | false |
| true | true | false | true |
| true | true | true | true |

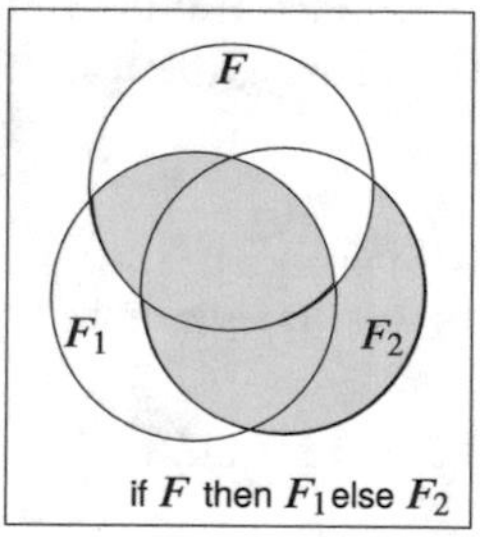

**Fig. 2.2** The logical connectives

corresponds to the formula

$$\forall x. \, \forall y.$$
$$\big(\exists z. \, isMother(z, x) \wedge isMother(z, y)\big) \Rightarrow$$
$$\text{if } isBrother(y, x) \text{ then } isMale(y) \text{ else } isFemale(y)$$

with *isMale*(*y*) and *isFemale*(*y*) interpreted as "*y* is male" and "*y* is female", respectively. It is important to note, that variables bound by quantifiers just serve as "placeholders"; their names do not influence the meaning of the formula; bound variables may be thus renamed at any time. For instance, the formula above could be also

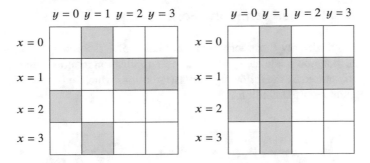

**Fig. 2.3** ($\forall x.\ \exists y.\ loves(x, y)$) versus ($\exists y.\ \forall x.\ loves(x, y)$)

written as

$$\forall u.\ \forall v.$$
$$\big(\exists w.\ isMother(w, u) \wedge isMother(w, v)\big) \Rightarrow$$
$$\text{if } isBrother(v, u) \text{ then } isMale(v) \text{ else } isFemale(v)$$

without changing its meaning.

One must be careful about the order of quantifiers, e.g., the formulas $\forall x.\ \exists y.\ F$ and $\exists y.\ \forall x.\ F$ may have different truth values. This can be also observed in natural language as demonstrated by the following example.

**Example 2.2** The two statements

- "everybody loves somebody"

$$\forall x.\ \exists y.\ loves(x, y)$$

- "somebody is loved by everybody"

$$\exists y.\ \forall x.\ loves(x, y)$$

have distinctly different meanings as demonstrated by Fig. 2.3. Here two matrices depict the love relationships of two groups of four individuals 0, 1, 2, 3: a shaded entry in row $x$ and column $y$ indicates "$x$ loves $y$".

In the left matrix "everybody loves somebody", because in every row $x$ we can find a shaded entry, for $x = 0$ this is $y = 1$, for $x = 1$ we may choose $y = 2$, for $x = 2$ we have $y = 0$, and for $x = 3$, $y = 1$ can be selected. However, here not "somebody is loved by everybody" because we cannot find a column $y$ that is completely shaded.

In the right matrix "somebody is loved by everybody", because there is such a shaded column $y = 2$. It is easy to see that this also implies that "everybody loves somebody", because for every $x$, we may here choose $y = 2$. □

We will later by a formal semantics define precisely the truth value of a first-order formula with arbitrary combinations of connectives and quantifiers.

## 2.3    Well-Formed Terms and Formulas

As already stated, only those term and formulas are well-formed that respect the
arity of function and predicate symbols in their application to term sequences. This
intuition is captured by the following formal definitions which impose a type-system
upon the grammar of Definition 2.1.

**Definition 2.4** (*Arity*) An *arity ar* is a function

$$ar\colon FunctionSymbol \cup PredicateSymbol \to \mathbb{N}_{\geq 1}$$

that maps function and predicate symbols to natural numbers greater equal
one, i.e., for every function or predicate symbol $S$, $ar(S) = n$ for some $n \in \mathbb{N}$
with $n \geq 1$.

**Definition 2.5** (*Well-formed Terms and Formulas*) A term $T$ is well-formed
with respect to arity $ar$ if and only if from the type system of Fig. 2.4 a
judgement

$$ar \vdash T \colon \mathsf{term}$$

can be derived. Likewise, a formula $F$ is well-formed with respect to $ar$ if and
only a judgement

$$ar \vdash F \colon \mathsf{formula}$$

can be derived. The rules depend on an auxiliary judgement

$$ar \vdash Ts \colon \mathsf{terms}(n)$$

that establishes that the sequence $Ts$ consists of $n$ well-formed terms.

**Example 2.3** For arity $ar = [p_1 \mapsto 1, p_2 \mapsto 1, q \mapsto 1, r \mapsto 2, f \mapsto 2]$ and formula

$$\forall x.\ p_1(x) \vee p_2(x) \Rightarrow \exists y.\ q(y) \wedge r(f(x,a), y)$$

**Rules for $ar \vdash T$ : term:**

$$\frac{ar \vdash Ts: \text{terms}(n) \quad ar(FS) = n}{ar \vdash FS(Ts): \text{term}}$$

$$ar \vdash V: \text{term} \qquad ar \vdash CS: \text{term}$$

$$\frac{ar \vdash T_1: \text{term} \quad ar \vdash T_2: \text{term}}{ar \vdash \text{let } V = T_1 \text{ in } T_2: \text{term}} \qquad \frac{ar \vdash F: \text{formula} \quad ar \vdash T_1: \text{term} \quad ar \vdash T_2: \text{term}}{ar \vdash \text{if } F \text{ then } T_1 \text{ else } T_2: \text{term}}$$

**Rules for $ar \vdash Ts$ : terms$(n)$:**

$$\frac{ar \vdash T: \text{term}}{ar \vdash T: \text{terms}(1)} \qquad \frac{ar \vdash T: \text{term} \quad ar \vdash Ts: \text{terms}(n)}{ar \vdash T, Ts: \text{terms}(n+1)}$$

**Rules for $ar \vdash F$ : formula:**

$$\frac{ar \vdash Ts: \text{terms}(n) \quad ar(PS) = n}{ar \vdash PS(Ts): \text{formula}}$$

$$ar \vdash \text{true}: \text{formula} \qquad ar \vdash \text{false}: \text{formula}$$

$$\frac{ar \vdash T_1: \text{term} \quad ar \vdash T_2: \text{term}}{ar \vdash T_1 = T_2: \text{formula}} \qquad \frac{ar \vdash F: \text{formula}}{ar \vdash \neg F: \text{formula}}$$

$$\frac{ar \vdash F_1: \text{formula} \quad ar \vdash F_2: \text{formula}}{ar \vdash F_1 \wedge F_2: \text{formula}} \qquad \frac{ar \vdash F_1: \text{formula} \quad ar \vdash F_2: \text{formula}}{ar \vdash F_1 \vee F_2: \text{formula}}$$

$$\frac{ar \vdash F_1: \text{formula} \quad ar \vdash F_2: \text{formula}}{ar \vdash F_1 \Rightarrow F_2: \text{formula}} \qquad \frac{ar \vdash F_1: \text{formula} \quad ar \vdash F_2: \text{formula}}{ar \vdash F_1 \Leftrightarrow F_2: \text{formula}}$$

$$\frac{ar \vdash F: \text{formula}}{ar \vdash \forall V. \, F: \text{formula}} \qquad \frac{ar \vdash F: \text{formula}}{ar \vdash \exists V. \, F: \text{formula}}$$

$$\frac{ar \vdash T: \text{term} \quad ar \vdash F: \text{formula}}{ar \vdash \text{let } V = T \text{ in } F: \text{formula}} \qquad \frac{ar \vdash F: \text{formula} \quad ar \vdash F_1: \text{formula} \quad ar \vdash F_2: \text{formula}}{ar \vdash \text{if } F \text{ then } F_1 \text{ else } F_2: \text{formula}}$$

**Fig. 2.4** The well-formed terms and formulas

we can derive the judgement

$$\frac{\dfrac{\cdots}{\dfrac{ar \vdash p_1(x): \text{formula} \quad ar \vdash p_2(x): \text{formula}}{ar \vdash p_1(x) \vee p_2(x): \text{formula}}} \quad \dfrac{\dfrac{\cdots}{ar \vdash q(y): \text{formula}} \quad \dfrac{\cdots}{ar \vdash r(f(x,a),y): \text{formula}}}{\dfrac{ar \vdash q(y) \wedge r(f(x,a),y): \text{formula}}{ar \vdash \exists y. \, q(y) \wedge r(f(x,a),y): \text{formula}}}}{\dfrac{ar \vdash p_1(x) \vee p_2(x) \Rightarrow \exists y. \, q(y) \wedge r(f(x,a),y): \text{formula}}{ar \vdash (\forall x. \, p_1(x) \vee p_2(x) \Rightarrow \exists y. \, q(y) \wedge r(f(x,a),y)): \text{formula}}}$$

where the inferences for the atomic formulas are given below:

$$\frac{\dfrac{ar \vdash x: \text{term}}{ar \vdash x: \text{terms}(1)} \quad ar(p_1) = 1}{ar \vdash p_1(x): \text{formula}}$$

$$\frac{\dfrac{ar \vdash x: \text{term}}{ar \vdash x: \text{terms}(1)} \quad ar(p_2) = 1}{ar \vdash p_2(x): \text{formula}}$$

$$\frac{\dfrac{ar \vdash y: \text{term}}{ar \vdash y: \text{terms}(1)} \quad ar(q) = 1}{ar \vdash q(y): \text{formula}}$$

$$\cfrac{ar \vdash x: \text{term} \quad \cfrac{ar \vdash a: \text{term}}{ar \vdash a: \text{terms}(1)}}{\cfrac{\cfrac{ar \vdash x, a: \text{terms}(2) \quad ar(f) = 2 \quad ar \vdash y: \text{term}}{ar \vdash f(x, a): \text{term} \quad \cfrac{ar \vdash y: \text{term}}{ar \vdash y: \text{terms}(1)}}}{\cfrac{ar \vdash f(x, a), y: \text{terms}(2) \quad ar(r) = 2}{ar \vdash r(f(x, a), y): \text{formula}}}}$$

Consequently, the formula is well-formed with respect to the given arity.                □

## 2.4    Propositional Logic

Having fixed the syntax of first-order logic, the rest of this chapter is dedicated to the investigation of the properties of this language. We will start, however, by considering only a simple subset of first-order logic called "propositional logic".

---

**Definition 2.6** (*Propositional Logic*) *Propositional logic* consists of the set of (propositional) *formulas* defined by the following grammar:

$F \in Formula$
$F ::= \text{true} \mid \text{false} \mid$
$\quad \mid \neg F \mid F_1 \wedge F_2 \mid F_1 \vee F_2 \mid F_1 \Rightarrow F_2 \mid F_1 \Leftrightarrow F_2$
$\quad \mid \text{if } F \text{ then } F_1 \text{ else } F_2$

---

Propositional logic is not able to formulate statements about the values of a domain; it is only concerned how the truth values of compound formulas are determined from the truth values of the basic formulas (one can imagine that every non-propositional formula of first-order logic has been abstracted to one of the logical constants true or false). Our goal is to investigate the laws that already emerge from the semantics of this fragment of first-order logic, which is is uniquely determined from the tables given in Fig. 2.2; these laws will then also hold in the full logic. Our investigations will be based on the following notions.

---

**Definition 2.7** (*Propositional Schema*) A (propositional) *schema* $F[A, B, \ldots]$ is a propositional formula $F$ which may contain *propositional variables* $A, B, \ldots$ as subformulas. If we *instantiate* the schema, i.e., we substitute its propositional variables by formulas, we get a formula which is an *instance* of the schema.

---

**Example 2.4** The propositional schema $A \wedge B$ has two propositional variables $A$ and $B$. By substitution of variable $A$ with the propositional formula false and $B$ with the propositional formula false $\vee$ true we yield the propositional formula false $\wedge$ (false $\vee$ true) as an instance of the schema. ☐

**Definition 2.8** (*Propositional Tautology*) A propositional schema $F[A, B, \ldots]$ is a (propositional) *tautology*, if it yields the truth value "true" for every instantiation of its propositional variables $A, B, \ldots$.

**Example 2.5** We claim that the schema $\neg(A \wedge B) \Leftrightarrow (\neg A) \vee (\neg B)$ is a tautology. To show this, we need not investigate the particular structure of $A$ and $B$ but only consider all possible combinations of their truth values as depicted by the following truth table:

| $A$ | $B$ | $A \wedge B$ | $\neg(A \wedge B)$ | $\neg A$ | $\neg B$ | $(\neg A) \vee (\neg B)$ | $\neg(A \wedge B) \Leftrightarrow (\neg A) \vee (\neg B)$ |
|---|---|---|---|---|---|---|---|
| False | False | False | True | True | True | True | True |
| False | True | False | True | True | False | True | True |
| True | False | false | True | false | True | True | True |
| True | True | True | False | False | False | False | True |

Since the last column always yields "true", the corresponding schema is a tautology. ☐

We will now investigate when two propositional schemas "mean the same".

**Definition 2.9** (*Logical Consequence and Equivalence*) A propositional schema $F_2[A, B, \ldots]$ is a (logical) *consequence* of $F_1[A, B, \ldots]$, written as $F_1[A, B, \ldots] \models F_2[A, B, \ldots]$, if for arbitrary substitutions of their propositional variables $A, B, \ldots$, whenever $F_1[A, B, \ldots]$ yields truth value "true", also $F_2[A, B, \ldots]$ yields "true". The two schemas are (logically) *equivalent*, written as $F_1[A, B, \ldots] \equiv F_2[A, B, \ldots]$, if $F_1[A, B, \ldots] \models F_2[A, B, \ldots]$ and $F_2[A, B, \ldots] \models F_1[A, B, \ldots]$, i.e., they yield the same truth values for arbitrary substitutions of their propositional variables.

Any formula resulting from an instantiation of a schema $F_1[\cdot]$ can be thus replaced by the formula resulting from the corresponding instantiation of an equivalent schema $F_2[\cdot]$ without changing the meaning of the formula; e.g., formula $F_1[\text{true} \wedge \text{false}]$ has the same truth value as formula $F_2[\text{true} \wedge \text{false}]$. Even more, if we instantiate the propositional variables with the same schema (rather than the same formula), we get a schema that is equivalent to the original one, e.g., $F_1[F[\cdot]] \equiv F_2[F[\cdot]]$ for every schema $F[\cdot]$. Furthermore, if in a formula schema $F[\cdot]$ a variable is instantiated once with $F_1[\cdot]$ and once with $F_2[\cdot]$, the resulting schemas are equivalent, i.e, we have $F[F_1[\cdot]] \equiv F[F_2[\cdot]]$.

**Example 2.6** We demonstrate the equivalence of

$$A \wedge (A \vee B) \equiv A$$

by the following truth table where the last and the first column are identical:

| $A$ | $B$ | $A \vee B$ | $A \wedge (A \vee B)$ |
|-------|-------|------------|------------------------|
| false | false | false | false |
| false | true | true | false |
| true | false | true | true |
| true | true | true | true |

Because of this equivalence, we know, for example, that the formula $(\text{true} \vee \text{false}) \wedge ((\text{true} \vee \text{false}) \wedge \text{false})$ has the same truth value as the formula $\text{true} \vee \text{false}$, because these formulas are instances of above schemas with propositional variable $A$ substituted by $\text{true} \vee \text{false}$ and $B$ substituted by $\text{false}$.

More generally, we know that the schema $(C \vee D) \wedge ((C \vee D) \vee D)$ is equivalent to schema $C \vee D$, because they are instances of above schemas with $A$ replaced by $C \vee D$ and $B$ replaced by $D$.

Finally, the schema $A \Rightarrow \left(B \vee \left(C \wedge (C \vee D)\right)\right)$ can be simplified to $A \Rightarrow (B \vee C)$, because both schemas are derived from instantiation of the more general schema $A \Rightarrow B \vee E$ with propositional variable $E$ once replaced by $C \wedge (C \vee D)$ and once by $C$, both of which are in turn trivial instances (variable renamings) of the schemas whose equivalence was shown above.                                                    □

As stated below, equivalences and tautologies are closely related.

**Proposition 2.1** (Equivalences as Tautologies) *If the schemas $F_1[A, B, \ldots]$ and $F_1[A, B, \ldots]$ are equivalent, i.e., $F_1[A, B, \ldots] \equiv F_2[A, B, \ldots]$, then the formula $F_1[A, B, \ldots] \Leftrightarrow F_2[A, B, \ldots]$ is a tautology, and vice versa.*

**Example 2.7** As demonstrated in Example 2.6, $A \wedge (A \vee B)$ and $A$ are equivalent and thus $A \wedge (A \vee B) \Leftrightarrow A$ is a tautology.                                         □

**Negation**

$$\neg\text{true} \equiv \text{false}$$
$$\neg\text{false} \equiv \text{true}$$
$$\neg\neg A \equiv A$$
$$\neg(A \wedge B) \equiv (\neg A) \vee (\neg B)$$
$$\neg(A \vee B) \equiv (\neg A) \wedge (\neg B)$$
$$\neg(A \Rightarrow B) \equiv A \wedge \neg B$$
$$\neg(A \Leftrightarrow B) \equiv (A \wedge \neg B) \vee (\neg A \wedge B)$$

**Conjunction and Disjunction**

$$A \wedge A \equiv A \qquad\qquad\qquad A \vee A \equiv A$$
$$A \wedge B \equiv B \wedge A \qquad\qquad A \vee B \equiv B \vee A$$
$$A \wedge \text{true} \equiv A \qquad\qquad\quad A \vee \text{true} \equiv \text{true}$$
$$A \wedge \text{false} \equiv \text{false} \qquad\qquad A \vee \text{false} \equiv A$$
$$A \wedge (\neg A) \equiv \text{false} \qquad\qquad A \vee (\neg A) \equiv \text{true}$$
$$A \wedge (B \wedge C) \equiv (A \wedge B) \wedge C \qquad A \vee (B \vee C) \equiv (A \vee B) \vee C$$
$$A \wedge (B \vee C) \equiv (A \wedge B) \vee (A \wedge C) \qquad A \vee (B \wedge C) \equiv (A \vee B) \wedge (A \vee C)$$
$$A \wedge (A \vee B) \equiv A \qquad\qquad A \vee (A \wedge B) \equiv A$$

**Implication and Equivalence**

$$A \Rightarrow B \equiv \neg A \vee B \qquad\qquad A \Leftrightarrow B \equiv (A \Rightarrow B) \wedge (B \Rightarrow A)$$
$$A \Leftrightarrow B \equiv (A \wedge B) \vee (\neg A \wedge \neg B)$$
$$A \Rightarrow A \equiv \text{true} \qquad\qquad A \Leftrightarrow A \equiv \text{true}$$
$$A \Rightarrow \text{true} \equiv \text{true} \qquad\qquad A \Leftrightarrow \text{true} \equiv A$$
$$A \Rightarrow \text{false} \equiv \neg A \qquad\qquad A \Leftrightarrow \text{false} \equiv \neg A$$
$$\text{true} \Rightarrow A \equiv A \qquad\qquad\quad \text{true} \Leftrightarrow A \equiv A$$
$$\text{false} \Rightarrow A \equiv \text{true} \qquad\qquad \text{false} \Leftrightarrow A \equiv \neg A$$
$$A \Rightarrow B \equiv (\neg B) \Rightarrow (\neg A) \qquad\qquad A \Leftrightarrow B \equiv B \Leftrightarrow A$$
$$A \Rightarrow (B \Rightarrow C) \equiv (A \wedge B) \Rightarrow C \qquad\qquad A \Leftrightarrow B \equiv \neg A \Leftrightarrow \neg B$$
$$A \Rightarrow (B \wedge C) \equiv (A \Rightarrow B) \wedge (A \Rightarrow C)$$
$$A \Rightarrow (B \vee C) \equiv (A \Rightarrow B) \vee (A \Rightarrow C)$$
$$(A \wedge B) \Rightarrow C \equiv (A \Rightarrow C) \vee (B \Rightarrow C)$$
$$(A \vee B) \Rightarrow C \equiv (A \Rightarrow C) \wedge (B \Rightarrow C)$$

**Conditional**

$$\text{if } A \text{ then } B \text{ then } C \equiv (A \Rightarrow B) \wedge (\neg A \Rightarrow C)$$

**Fig. 2.5** Logical equivalences for propositional logic

By the kind of reasoning demonstrated in the examples above, a large number of propositional equivalences (and the corresponding tautologies) can be justified; see Fig. 2.5 for a collection of some of the most important ones. For instance, "De Morgan's laws" $\neg(A \wedge B) \equiv (\neg A) \vee (\neg B)$ and $\neg(A \vee B) \equiv (\neg A) \wedge (\neg B)$ tell us that by pushing a negation into a conjunction, it becomes a disjunction, and vice versa. By the "commutativity laws" $A \wedge B \equiv B \wedge A$ and $A \vee B \equiv B \vee A$ we can freely exchange the order of arguments of $\wedge$ and $\vee$ without changing the truth value of the formula. Since we have the "associativity laws" $A \wedge (B \wedge C) \equiv (A \wedge B) \wedge C$ and $A \vee (B \vee C) \equiv (A \vee B) \vee C$, we can and usually will write $A \wedge B \wedge C$ and $A \vee B \vee C$ without parentheses (the truth value is the same however the parentheses are placed). The "distributivity laws" $A \wedge (B \vee C) \equiv (A \wedge B) \vee (A \wedge C)$ and $A \vee (B \wedge C) \equiv (A \vee B) \wedge (A \vee C)$ allow to "multiply" an $\wedge$ into a disjunction and an $\vee$ into a conjunction.

We thus get a rich system for the meaning-preserving transformation of logic formulas like algebra is a system for the meaning-preserving transformation ("calculation") of arithmetic formulas.

**Example 2.8** We transform the propositional schema if $A$ then $B$ then $C$ into various equivalent forms:

$$
\begin{aligned}
\text{if } A \text{ then } B \text{ then } C &\equiv (A \Rightarrow B) \wedge (\neg A \Rightarrow C) \\
&\equiv (\neg A \vee B) \wedge (\neg(\neg A) \vee C) \\
&\equiv \underline{(\neg A \vee B) \wedge (A \vee C)} \\
&\equiv \big((\neg A \vee B) \wedge A\big) \vee \big((\neg A \vee B) \wedge C\big) \\
&\equiv \big((\neg A \wedge A) \vee (B \wedge A)\big) \vee \big((\neg A \wedge C) \vee (B \wedge C)\big) \\
&\equiv \big(\text{false} \vee (A \wedge B)\big) \vee \big((\neg A \wedge C) \vee (B \wedge C)\big) \\
&\equiv \underline{(A \wedge B) \vee (\neg A \wedge C) \vee (B \wedge C)}.
\end{aligned}
$$

The two underlined lines show that a conditional formula can thus be written also as a conjunction of disjunctions or as a disjunction of conjunctions.

Actually the last disjunction can be simplified further. By the sequence of transformations

$$ A \equiv A \wedge \text{true} \equiv A \wedge (B \vee \neg B) \equiv (A \wedge B) \vee (A \wedge \neg B) $$

we can justify the equivalence

$$ A \equiv (A \wedge B) \vee (A \wedge \neg B). $$

By this equivalence, we can expand the three conjunctions in the underlined disjunction to contain literals for all variables $A, B, C$, then drop duplicate conjunctions, and finally contract the conjunctions by above equivalence again:

$$
\begin{aligned}
&(A \wedge B) \vee (\neg A \wedge C) \vee (B \wedge C) \\
&\equiv (A \wedge B \wedge C) \vee (A \wedge B \wedge \neg C) \vee (\neg A \wedge B \wedge C) \vee (\neg A \wedge \neg B \wedge C) \vee \\
&\quad (A \wedge B \wedge C) \vee (\neg A \wedge B \wedge C) \\
&\equiv (A \wedge B \wedge C) \vee (A \wedge B \wedge \neg C) \vee (\neg A \wedge B \wedge C) \vee (\neg A \wedge \neg B \wedge C) \\
&\equiv \underline{(A \wedge B) \vee (\neg A \wedge C)}
\end{aligned}
$$

Thus the last conjunction can be dropped without changing the truth value of the scheme; the resulting disjunction has only two conjunctions.  □

The formulas resulting from the transformations in above example are special instances of certain "normal forms".

**Definition 2.10** (*Conjunctive and Disjunctive Normal Form*) Let a *literal* be either a propositional variable or the negation of a propositional variable:

$L \in Literal$
$V \in Variable$
$L ::= V \mid \neg V$

Then a propositional schema is in *disjunctive normal form* if and only if it is an element of the syntactic domain *Disjunction* defined by the grammar

$D \in Disjunction$
$C \in Conjunction$
$D ::= C \mid D \vee C$
$C ::= L \mid L \wedge C$

i.e., it is a disjunction of conjunctions of literals.

Furthermore, a propositional schema is in *conjunctive normal form* if and only if it is an element of the syntactic domain *Conjunction* defined by the grammar

$C \in Conjunction$
$D \in Disjunction$
$C ::= D \mid C \wedge D$
$D ::= L \mid L \vee D$

i.e., it is a conjunction of disjunctions of literals.

**Proposition 2.2** (Existence of Normal Forms) *For every propositional schema that is not logically equivalent to* true *or* false, *there exists an equivalent schema in disjunctive normal form and an equivalent schema in conjunctive normal form.*

The existence of an equivalent disjunctive normal form can be easily deduced from the truth table of a propositional schema: each row that yields truth value "true" is translated into a conjunction by combining the propositional variables for that row in plain form (if its value is "true") or in negated form (if its value is "false"); the resulting conjunctions are combined by disjunction.

**Example 2.9**  Consider the following formula schema:

$$A \wedge (A \vee B)$$

From its truth table

| $A$ | $B$ | $A \vee B$ | $A \wedge (A \vee B)$ |
|-------|-------|------------|------------------------|
| false | false | false | false |
| false | true | true | false |
| true | false | true | true |
| true | true | true | true |

we derive from the last two rows of the table (which yield truth value "true") the following disjunctive normal form:

$$(A \wedge \neg B) \vee (A \wedge B)$$

Here every occurrence of "true" in column $A$ respectively $B$ becomes an occurrence of literal $A$ respectively $B$ in the formula and every occurrence of "false" becomes $\neg A$ respectively $\neg B$.                                                                    □

But also existence of an equivalent conjunctive normal form can be deduced from the truth table by considering the equivalence

$$\neg\big((A \wedge B) \vee (C \wedge D)\big) \equiv (\neg A \vee \neg B) \wedge (\neg C \vee \neg D)$$

which we can easily justify from the rules of Fig. 2.5. This shows us how we can directly translate the disjunctive normal form of a schema in the conjunctive form of its negation, respectively the disjunctive normal form of the negation of a schema into the conjunctive normal form of the original. Furthermore, from the truth table of a schema we cannot only (as shown above) determine the disjunctive normal form of the schema itself but also the disjunctive normal form of its negation: we just have to consider all rows with true value "false". Thus we combine for each such row the propositional variables in their negated form to a disjunction; the resulting disjunctions are combined by conjunction.

**Example 2.10** From the truth table of formula schema $A \wedge (A \vee B)$ presented in Example 2.9, we consider the first two rows (which yield truth value "false") from which we derive the following conjunctive normal form:

$$(A \vee B) \wedge (A \vee \neg B).$$

Here every occurrence of "false" in column $A$ respectively $B$ becomes an occurrence of literal $A$ respectively $B$ in the formula and every occurrence of "true" becomes $\neg A$ respectively $\neg B$. □

Before generalizing the laws of propositional logic to full first-order logic, we will define the semantics of first-order logic more precisely and, for this purpose, introduce some important notions.

## 2.5 Free and Bound Variables

The truth value of a formula in first-order logic depends on the values that the environment assigns to certain of its variables; e.g. the formula $\forall x \in \mathbb{N}.\ y \leq x$ is true for $y = 0$ but false for $y = 1$ (assuming that $x \in \mathbb{N}$ denotes "$x$ is one of the natural numbers $0, 1, 2, \ldots$" and $y \leq x$ denotes "$y$ is less than or equal $x$"). However, the truth value is independent of any value that the environment may assign to variable $x$, because $x$ is locally "bound" in the formula by the quantifier $\forall$. The variable $y$ is not bound by any quantifier and thus "free" in the formula. These notions are formalized by the following definition.

**Definition 2.11** (*Free and Bound Variables*) Let

$$fv\colon\ Term \cup Terms \cup Formula \to \mathsf{Set}(Variable)$$

be the function defined in Fig. 2.6 which assigns to every phrase (term, term sequence, and formula) a set of variables. We call a variable $V$ *free* in a phrase $P$, if and only if $V \in fv[\![\, P\, ]\!]$, otherwise it is *bound* in $P$. If $P$ has no free variables, i.e., $fv[\![\, P\, ]\!] = \varnothing$, we call $P$ *closed*. A closed formula is also called a *sentence*.

The definition of $fv$ recursively collects all occurrences of variables in the various parts of a phrase ($A \cup B$ denotes the set of all elements that are in $A$ or in $B$), but it also removes those occurrences that are bound by the quantifiers $\forall$, $\exists$, and let …in … ($A \setminus B$ denotes the set of all elements that are in $A$ but not in $B$).

$$fv\colon Term \cup Terms \cup Formula \to \mathrm{Set}(Variable)$$
$$fv[\![\, V \,]\!] := \{V\}$$
$$fv[\![\, CS \,]\!] := \varnothing$$
$$fv[\![\, FS(Ts) \,]\!] := fv[\![\, Ts \,]\!]$$
$$fv[\![\, \mathsf{let}\ V = T_1\ \mathsf{in}\ T_2 \,]\!] := fv[\![\, T_1 \,]\!] \cup (fv[\![\, T_2 \,]\!] \setminus \{V\})$$
$$fv[\![\, \mathsf{if}\ F\ \mathsf{then}\ T_1\ \mathsf{else}\ T_2 \,]\!] := fv[\![\, F \,]\!] \cup fv[\![\, T_1 \,]\!] \cup fv[\![\, T_2 \,]\!]$$
$$fv[\![\, T \,]\!] := fv[\![\, T \,]\!]$$
$$fv[\![\, T, Ts \,]\!] := fv[\![\, T \,]\!] \cup fv[\![\, Ts \,]\!]$$
$$fv[\![\, \mathsf{true} \,]\!] := \varnothing$$
$$fv[\![\, \mathsf{false} \,]\!] := \varnothing$$
$$fv[\![\, PS(Ts) \,]\!] := fv[\![\, Ts \,]\!]$$
$$fv[\![\, T_1 = T_2 \,]\!] := fv[\![\, V_1 \,]\!] \cup fv[\![\, T_2 \,]\!]$$
$$fv[\![\, \neg F \,]\!] := fv[\![\, F \,]\!]$$
$$fv[\![\, F_1 \wedge F_2 \,]\!] := fv[\![\, F_1 \,]\!] \cup fv[\![\, F_2 \,]\!]$$
$$fv[\![\, F_1 \vee F_2 \,]\!] := fv[\![\, F_1 \,]\!] \cup fv[\![\, F_2 \,]\!]$$
$$fv[\![\, F_1 \Rightarrow F_2 \,]\!] := fv[\![\, F_1 \,]\!] \cup fv[\![\, F_2 \,]\!]$$
$$fv[\![\, F_1 \Leftrightarrow F_2 \,]\!] := fv[\![\, F_1 \,]\!] \cup fv[\![\, F_2 \,]\!]$$
$$fv[\![\, \forall V.\ F \,]\!] := fv[\![\, F \,]\!] \setminus \{V\}$$
$$fv[\![\, \exists V.\ F \,]\!] := fv[\![\, F \,]\!] \setminus \{V\}$$
$$fv[\![\, \mathsf{let}\ V = T\ \mathsf{in}\ F \,]\!] := fv[\![\, T \,]\!] \cup (fv[\![\, F \,]\!] \setminus \{V\})$$
$$fv[\![\, \mathsf{if}\ F\ \mathsf{then}\ F_1\ \mathsf{else}\ F_2 \,]\!] := fv[\![\, F \,]\!] \cup fv[\![\, F_1 \,]\!] \cup fv[\![\, F_2 \,]\!]$$

**Fig. 2.6** The free variables of a phrase

**Example 2.11** We compute the free variables of formula $\forall x \in \mathbb{N}.\ y \leq x$:

$$
\begin{aligned}
fv[\![\, \forall x \in \mathbb{N}.\ y \leq x \,]\!] &= fv[\![\, \forall x.\ x \in \mathbb{N} \Rightarrow y \leq x \,]\!] \\
&= fv[\![\, x \in \mathbb{N} \Rightarrow y \leq x \,]\!] \setminus \{x\} \\
&= (fv[\![\, x \in \mathbb{N} \,]\!] \cup [\![\, y \leq x \,]\!]) \setminus \{x\} \\
&= (\{x\} \cup \{x, y\}) \setminus \{x\} = \{x, y\} \setminus \{x\} = \{y\} \qquad \square
\end{aligned}
$$

It should be noted that, while the names of the bound variables of a formula may be changed without affecting its semantics, the names of free variables must not be changed, since they are "visible" from the outside.

**Example 2.12** The meaning of the formula

$$\forall x \in \mathbb{N}.\ y \leq x$$

is identical to that of the formula

$$\forall z \in \mathbb{N}.\ y \leq z$$

where the bound variable $x$ is renamed to $z$. However, it is different from that of the formula

$$\forall x \in \mathbb{N}.\ z \leq x$$

because the first formula has free variable $y$ while the last has free variable $z$.   □

## 2.6   Formal Semantics

A formula is interpreted over some "domain", a non-empty collection of values; values from this domain are assigned to the free variables of the formula. The truth value of the formula depends on such an assignment as well as on the interpretation of the constant symbols as values, of the function symbols as functions, and of the predicate symbols as predicates of the domain. As an example, take the formula

$$\forall y.\ x \leq y$$

The truth value of the formula varies with the domain over which the formula is interpreted, the assignment of a value to the free variable $x$, and the interpretation of the predicate symbol $\leq$:

- In the domain of natural numbers 0, 1, 2, ..., if the free variable $x$ is assigned the number 0, and the predicate symbol $\leq$ is interpreted as the predicate "less than or equal", the formula is true; if $x$ is assigned any other number, the formula is false.
- In the domain of sets of natural numbers, if $x$ is assigned the empty set and $\leq$ is interpreted as the predicate "is subset of", the formula is true; if $x$ is assigned any other set, the formula is false.
- In the domain of integer numbers ..., $-1, 0, 1, \ldots$, if $\leq$ is interpreted as the predicate "less than or equal", the formula is false for every assignment to $x$.

These notions become more concrete by the following definitions.

**Definition 2.12** (*Domain*) A *domain D* is a non-empty (finite or infinite) collection of values.

**Definition 2.13** (*Interpretation*) An *interpretation I* over domain $D$ with arity *ar* is a mapping such that

- $I$ maps constant symbols to values of $D$: for every $CS \in ConstantSymbol$, $I(CS)$ is a value in $D$.

- *I* maps function symbols to functions on *D*: for every *FS* ∈ *FunctionSymbol*, if $ar(FS) = n$, then $I(FS)$ is a *n*-ary function on *D*, i.e. $I(FS)(d_1, \ldots, d_n)$ is a value of *D*, for all $d_1, \ldots, d_n$ in *D*.
- *I* maps predicate symbols to predicates on *D*: for every *PS* ∈ *PredicateSymbol*, if $ar(PS) = n$, then $I(PS)$ is a *n*-ary predicate on *D*, i.e. $I(PS)(d_1, \ldots, d_n)$ is a truth value, for all $d_1, \ldots, d_n$ in *D*.

**Example 2.13** We define

$$ConstantSymbol := \{0\}$$
$$FunctionSymbol := \{+\}$$
$$PredicateSymbol := \{\leq\}$$

Let $D := 0, 1, 2, \ldots$ be the domain of natural numbers and *ar* be an arity with $ar(\leq) = 2$ and $ar(+) = 2$. Then the mapping *I* defined as

$$I(0) := \text{the number "zero"}$$
$$I(+) := \lambda x, y. \text{ the sum of } x \text{ and } y$$
$$I(\leq) := \pi x, y. \ x \text{ is less than or equal to } y$$

is an interpretation over *D* with arity *ar*. Here we use the notation $\lambda x, y. T$ with term *T* to denote a function with parameters *x*, *y* and value *T*; correspondingly, the notation $\pi x, y. F$ with formula *F* denotes a predicate with parameters *x*, *y* and truth value *F*.  □

**Definition 2.14** (*Assignment*) An *assignment a* over a domain *D* is a (partial) mapping of variables to values of *D*, i.e., if *a* is defined on variable *V*, then $a(V)$ is a value of *D*.

**Definition 2.15** (*Assignment Update*) For assignment *a* on *D*, variable *V* and value *d* in *D*, the *updated assignment* $a[V \mapsto d]$ is the mapping

$$a[V \mapsto d](V') := \begin{cases} d & \text{if } V = V' \\ a(V') & \text{otherwise} \end{cases}$$

**Example 2.14** Let $D := 0, 1, 2, \ldots$ be the domain of natural numbers. Then $a := [x \mapsto 1, y \mapsto 2]$ is an assignment over *D* defined on variables *x* and *y*; in particular we have $a(x) = 1$ and $a(y) = 2$. For the updated assignment $a' := a[x \mapsto 3]$, we then have $a'(x) = 3$ and $a'(y) = a(y) = 2$.                                          □

**Definition 2.16** (*Semantics of First-Order Logic*)  Let $ar$ be an arity, $D$ a domain, $I$ an interpretation over $D$ with arity $ar$, and $a$ an assignment on $D$. Then the rules in Fig. 2.7 define for every term $T$, term sequence $Ts$, and formula $F$ its *semantics* $[\![\, T \,]\!]_a^I$, $[\![\, Ts \,]\!]_a^I$, and $[\![\, F \,]\!]_a^I$ (in the definition of $[\![\, Ts \,]\!]_a^I$, $[a]$ denotes a sequence with single element $a$, and $s_1 \circ s_2$ denotes the concatenation of sequences $s_1$ and $s_2$).

This definition implies, that, if a term $T$, term sequence $Ts$, or formula $F$ is well-formed with respect to arity $ar$, and assignment $a$ is defined on the free variables of the phrase, then

- $[\![\, T \,]\!]_a^I$ denotes a value in $D$;
- $[\![\, Ts \,]\!]_a^I$ denotes a sequence of values in $D$;
- $[\![\, F \,]\!]_a^I$ denotes a truth value.

The proof of this claim (which implies that the definition in Fig. 2.7 is well-formed) has to proceed by induction on the typing rules of first-order-logic, in analogy to the proof presented in Example 1.9; we omit the details.

The semantics of the various phrases propagates the interpretation $I$ without change recursively through the semantics of the subphrases. Also the assignment $a$ remains mostly unchanged, except for the quantifiers $\forall$, $\exists$, and let …in …that modify $a$ to the assignment $a'$ for the evaluation of their bodies. In a phrase let $V = T$ in …, assignment $a'$ maps $V$ to the value of $T$; to make formulas $\forall V.\ F$ and $\exists V.\ F$ true, the body $F$ must be true for every/some assignment $a'$ that maps $V$ to a value $d$ of $D$.

**Example 2.15**  Let domain $D$, arity $ar$, and interpretation $I$ be as in Example 2.13, and let $a$ be an arbitrary assignment on $D$. We consider the truth values $[\![\, F \,]\!]_a^I$ of the following formulas $F$ which are well-formed with respect to $ar$; since these formulas are closed, the concrete definition of $a$ does not matter.

- $[\![\, \forall x.\ \exists y.\ x \leq y \,]\!]_a^I$ : for every number $n$ in $D$, consider the truth value of $[\![\, \exists y.\ x \leq y \,]\!]_{a'}^I$ with assignment $a' = a[x \mapsto n]$. For each such $a'$ there is an assignment $a'' = a'[y \mapsto n]$ with $[\![\, x \leq y \,]\!]_{a''}^I = I(\leq)([\![\, x \,]\!]_{a''}^I, [\![\, y \,]\!]_{a''}^I) = I(\leq)$ $(n, n) = $ "$n$ is less than or equal to $n$" = true. Therefore $[\![\, \forall x.\ \exists y.\ x \leq y \,]\!]_a^I = $ true.

**Semantics** $[\![\, T \,]\!]_a^I$ **of term** $T$:

$[\![\, V \,]\!]_a^I := a(V)$    $[\![\, CS \,]\!]_a^I := I(CS)$

$[\![\, FS(Ts) \,]\!]_a^I := I(FS)([\![\, Ts \,]\!]_a^I)$

$[\![\, \text{let } V = T_1 \text{ in } T_2 \,]\!]_a^I := [\![\, T_2 \,]\!]_{a'}^I,$ where $a' = a[V \mapsto [\![\, T_1 \,]\!]_a^I]$

$[\![\, \text{if } F \text{ then } T_1 \text{ else } T_2 \,]\!]_a^I := \begin{cases} [\![\, T_1 \,]\!]_a^I & \text{if } [\![\, F \,]\!]_a^I = \text{true} \\ [\![\, T_2 \,]\!]_a^I & \text{otherwise} \end{cases}$

**Semantics** $[\![\, Ts \,]\!]_a^I$ **of term sequence** $Ts$:

$[\![\, T \,]\!]_a^I := [[\![\, T \,]\!]_a^I]$    $[\![\, T, Ts \,]\!]_a^I := [[\![\, T \,]\!]_a^I] \circ [\![\, Ts \,]\!]_a^I$

**Semantics** $[\![\, F \,]\!]_a^I$ **of formula** $F$:

$[\![\, \text{true} \,]\!]_a^I := \text{true}$    $[\![\, \text{false} \,]\!]_a^I := \text{false}$

$[\![\, PS(Ts) \,]\!]_a^I := \begin{cases} \text{true} & \text{if } I(PS)([\![\, Ts \,]\!]_a^I) = \text{true} \\ \text{false} & \text{otherwise} \end{cases}$

$[\![\, T_1 = T_2 \,]\!]_a^I := \begin{cases} \text{true} & \text{if } [\![\, T_1 \,]\!]_a^I = [\![\, T_2 \,]\!]_a^I \\ \text{false} & \text{otherwise} \end{cases}$

$[\![\, \neg F \,]\!]_a^I := \begin{cases} \text{true} & \text{if } [\![\, F \,]\!]_a^I = \text{false} \\ \text{false} & \text{otherwise} \end{cases}$

$[\![\, F_1 \wedge F_2 \,]\!]_a^I := \begin{cases} \text{true} & \text{if } [\![\, F_1 \,]\!]_a^I = \text{true and } [\![\, F_2 \,]\!]_a^I = \text{true} \\ \text{false} & \text{otherwise} \end{cases}$

$[\![\, F_1 \vee F_2 \,]\!]_a^I := \begin{cases} \text{false} & \text{if } [\![\, F_1 \,]\!]_a^I = \text{false and } [\![\, F_2 \,]\!]_a^I = \text{false} \\ \text{true} & \text{otherwise} \end{cases}$

$[\![\, F_1 \Rightarrow F_2 \,]\!]_a^I := \begin{cases} \text{false} & \text{if } [\![\, F_1 \,]\!]_a^I = \text{true and } [\![\, F_2 \,]\!]_a^I = \text{false} \\ \text{true} & \text{otherwise} \end{cases}$

$[\![\, F_1 \Leftrightarrow F_2 \,]\!]_a^I := \begin{cases} \text{true} & \text{if } [\![\, F_1 \,]\!]_a^I = [\![\, F_2 \,]\!]_a^I \\ \text{false} & \text{otherwise} \end{cases}$

$[\![\, \forall V.\, F \,]\!]_a^I := \begin{cases} \text{true} & \text{if, for every } d \text{ in } D,\ [\![\, F \,]\!]_{a'}^I = \text{true, where } a' = a[V \mapsto d] \\ \text{false} & \text{otherwise} \end{cases}$

$[\![\, \exists V.\, F \,]\!]_a^I := \begin{cases} \text{true} & \text{if, for some } d \text{ in } D,\ [\![\, F \,]\!]_{a'}^I = \text{true, where } a' = a[V \mapsto d] \\ \text{false} & \text{otherwise} \end{cases}$

$[\![\, \text{let } V = T \text{ in } F \,]\!]_a^I := \begin{cases} \text{true} & \text{if } [\![\, F \,]\!]_{a'}^I = \text{true where } a' = a[V \mapsto [\![\, T \,]\!]_a^I] \\ \text{false} & \text{otherwise} \end{cases}$

$[\![\, \text{if } F \text{ then } F_1 \text{ else } F_2 \,]\!]_a^I := \begin{cases} \text{true} & \text{if } [\![\, F \,]\!]_a^I = \text{true and } [\![\, F_1 \,]\!]_a^I = \text{true} \\ \text{true} & \text{if } [\![\, F \,]\!]_a^I = \text{false and } [\![\, F_2 \,]\!]_a^I = \text{true} \\ \text{false} & \text{otherwise} \end{cases}$

**Fig. 2.7** The semantics of first-order logic

In short: *for every number* n *assigned to* $x$, *we can assign to* $y$ *a number, namely* n, *such that* $x \leq y$. *Thus* $\forall x. \exists y. x \leq y$ *is true*.

- $[\![\exists x. \forall y. x \leq y]\!]_a^I$ : to make the formula true, there must exist a number $n$ in $D$ such that the truth value of $[\![\forall y. x \leq y]\!]_{a'}^I$ with assignment $a' = a[x \mapsto n]$ is "true". We try $n :=$ "zero", i.e., $a' = a[x \mapsto$ "zero"] and determine for every number $m$ in $D$ the truth value of $[\![x \leq y]\!]_{a''}^I$ with assignment $a'' = a'[y \mapsto m]$. Indeed we have $[\![x \leq y]\!]_{a''}^I = I(\leq)([\![x]\!]_{a''}^I, [\![y]\!]_{a''}^I) = I(\leq)(\text{"zero"}, n) =$ "zero is less than or equal to $n$" $=$ true; thus $[\![\exists x. \forall y. x \leq y]\!]_a^I = $ true.

  In short: *there exists a number* n *assigned to* $x$, *namely* 0, *such that for every number* n *assigned to* $y$, $x \leq y$ *is true. Thus* $\exists x. \forall y. x \leq y$ *is true.*

- $[\![\forall y. \exists x. x \leq y]\!]_a^I$ : for every number $y$ in $D$, consider the truth value of $[\![\exists x. x \leq y]\!]_{a'}^I$ with assignment $a' = a[x \mapsto n]$. For each such $a'$ there is an assignment $a'' = a'[y \mapsto n]$ such that $[\![x \leq y]\!]_{a''}^I = I(\leq)([\![x]\!]_{a''}^I, [\![y]\!]_{a''}^I) = I(\leq)(n, n) = $ "$n$ is less than or equal to $n$" $=$ true. Therefore $[\![\forall y. \exists x. x \leq y]\!]_a^I = $ true.

  In short: *for every number* n *assigned to* $y$, *we can assign to* $x$ *a number, namely* n, *such that* $x \leq y$. *Thus* $\forall y. \exists y. x \leq x$ *is true.*

- $[\![\exists y. \forall x. x \leq y]\!]_a^I$ : to make the formula true, there must exist a number $n$ in $D$, such that the truth value of $[\![\forall x. x \leq y]\!]_{a'}^I$ with assignment $a' = a[x \mapsto n]$ becomes "true". Let us assume for the sake of the argument that there exists such a number $n$. Then for every number $m$ in $D$ the truth value of $[\![x \leq y]\!]_{a''}^I$ with assignment $a'' = a'[x \mapsto m]$ must be "true". But if we choose $m :=$ "$n$ plus one", we have $[\![x \leq y]\!]_{a''}^I = I(\leq)([\![x]\!]_{a''}^I, [\![y]\!]_{a''}^I) = I(\leq)(\text{"}n\text{plus one"}, n) = $ "$n$ plus one is less than or equal to $n$" $=$ false, which is a contradiction. Thus our assumption about the existence of a suitable $n$ was wrong, therefore $[\![\exists x. \forall y. x \leq y]\!]_a^I = $ false.

  In short: *for every number* n *we assign to* $y$, *we may assign to* $x$ *the number* $n + 1$, *which makes* $x \leq y$ *false; thus* $\exists y. \forall x. x \leq y$ *is false.*

□

Similar to Example 2.2, above example again demonstrates that the formulas $\forall x. \exists y. F$ and $\exists y. \forall x. F$ may have different truth values.

## 2.7 Validity, Logical Consequence, and Logical Equivalence

Now, having a precise semantics of first-order logic, we will investigate the properties of its formulas, as we did for propositional logic.

First, we generalize the notion of a propositional "tautology" to the notion of "validity" in first-order logic.

**Definition 2.17** (*Validity*) Let $F$ be a first-order formula that is well-formed with respect to arity $ar$. Then $F$ is *valid*, written as $\models F$, if it always denotes "true", i.e., if for every domain $D$, interpretation $I$ over $D$ with arity $ar$, and assignment $a$ on $D$ over the free variables of $F$, we have $[\![\, F \,]\!]_a^I =$ true.

**Example 2.16**  We claim that

$$(\forall V.\, F) \Rightarrow \neg(\exists V.\, \neg F)$$

is a tautology. Take arbitrary domain $D$, interpretation $I$ over $D$ with arity $ar$, and assignment $a$ on $D$ over the free variables of $F$. We assume

$$[\![\, (\forall V.\, F) \Rightarrow \neg(\exists V.\, \neg F) \,]\!]_a^I = \text{false}$$

and show a contradiction. From $[\![ \Rightarrow ]\!]$, we know

$$[\![\, \forall V.\, F \,]\!]_a^I = \text{true} \tag{2.1}$$

$$[\![\, \neg \exists V.\, \neg F \,]\!]_a^I = \text{false} \tag{2.2}$$

From (2.16) and $[\![\, \neg \,]\!]$, we know

$$[\![\, \exists V.\, \neg F \,]\!]_a^I = \text{true} \tag{2.3}$$

From (2.3) and $[\![ \exists ]\!]$, we have a $d$ in $D$ and an $a' = a[V \mapsto d]$ such that $[\![\, \neg F \,]\!]_{a'}^I =$ true and therefore $[\![\, F \,]\!]_{a'}^I =$ false. But this contradicts (1) and $[\![ \forall ]\!]$ which states that for every $d$ in $D$ and $a' = a[V \mapsto d]$ we have $[\![\, F \,]\!]_{a'}^I =$ true.  □

We are now going to relate the results derived for propositional logic to first-order logic based on the following result.

**Proposition 2.3**  (Well-Formedness of Schema Instances) *Let $F[A, B, \ldots]$ be a propositional schema. Then for all first-order formulas $F_a, F_b, \ldots$ that are well-formed with respect to arity $ar$, the instance $F[F_a, F_b \ldots]$ is a first-order formula that is well-formed with respect to arity $ar$.*

**Example 2.17**  Take formulas $\forall x.\, p(x)$ and $\exists y.\, q(y, f(a))$ that are well-formed with respect to arity $ar = [p \mapsto 1, q \mapsto 2, f \mapsto 1]$. Then

$$(\forall x.\, p(x)) \wedge \big(\exists y.\, q(y, f(a))\big)$$

is an well-formed instance of the propositional schema $A \wedge B$.  □

By schema instantiation, propositional tautologies give rise to first-order validities in the following sense.

> **Proposition 2.4** (Tautologies as Validities) *Let $F[A, B, \ldots]$ be a propositional tautology. Then for all first-order formulas $F_a, F_b, \ldots$ that are well-formed with respect to arity ar, the instance $F[F_a, F_b \ldots]$ is valid.*

**Example 2.18** By the propositional equivalence $A \Rightarrow B \equiv \neg A \vee B$, the schema

$$(A \Rightarrow B) \Leftrightarrow (\neg B \Rightarrow \neg A)$$

is a tautology. Therefore the schema instance

$$\big((\forall x.\ p(a, x)) \Rightarrow q(f(a))\big) \Leftrightarrow \big(\neg q(f(a)) \Rightarrow \neg(\forall x.\ p(a, x))\big)$$

is a valid formula.                                                                 □

Thus all the propositional tautologies derived from the equivalences in Fig. 2.5 give rise to valid formulas in first-order logic.

Next, we extend the notions of "implication" and "equivalence" to first-order logic.

> **Definition 2.18** (*Logical Consequence and Equivalence*) Let $F_1$ and $F_2$ be two first-order formulas that are well-formed with respect to arity $ar$. Then $F_2$ is a *logical consequence* of $F_1$, written as $F_1 \models F_2$, if if for every domain $D$, interpretation $I$ over $D$ with arity $ar$, and assignment $a$ on $D$ over the free variables of $F_1$ and $F_2$, the following holds: if $[\![ F_1 ]\!]_a^I = \text{true}$, then also $[\![ F_2 ]\!]_a^I = \text{true}$. Furthermore, $F_1$ and $F_2$ are *logically equivalent*, written as $F_1 \equiv F_2$, if $F_1 \models F_2$ and $F_2 \models F_1$, i.e., $[\![ F_1 ]\!] = [\![ F_2 ]\!]$.

**Example 2.19** We claim that for every variable $V$ and formula $F$, we have

$$\neg(\forall V.\ F) \equiv \exists V.\ (\neg F)$$

Take arbitrary domain $D$, interpretation $I$ over $D$ with arity $ar$, and assignment $a$ on $D$ over the free variables of $F_1$ and $F_2$. We show

$$[\![ \neg(\forall V.\ F) ]\!]_a^I = [\![ \exists V.\ (\neg F) ]\!]_a^I$$

We proceed by case distinction:

**Universal and Existential Quantification**

$$\forall V. \, \forall V. \, F \equiv \forall V. \, F \qquad\qquad \exists V. \, \exists V. \, F \equiv \exists V. \, F$$

$$\forall V_1. \, \forall V_2. \, F \equiv \forall V_2. \, \forall V_1. \, F \qquad\qquad \exists V_1. \, \exists V_2. \, F \equiv \exists V_2. \, \exists V_1. \, F$$

$$\neg(\forall V. \, F) \equiv \exists V. \, (\neg F) \qquad\qquad \neg(\exists V. \, F) \equiv \forall V. \, (\neg F)$$

$$\forall V. \, (F_1 \wedge F_2) \equiv (\forall V. \, F_1) \wedge (\forall V. \, F_2) \qquad\qquad \exists V. \, (F_1 \vee F_2) \equiv (\exists V. \, F_1) \vee (\exists V. \, F_2)$$

$$\forall V. \, F \equiv F, \text{ if } V \notin \mathit{fv}(F) \qquad\qquad \exists V. \, F \equiv F, \text{ if } V \notin \mathit{fv}(F)$$

$$\forall V. \, (F_1 \wedge F_2) \equiv F_1 \wedge (\forall V. \, F_2) \qquad\qquad \exists V. \, (F_1 \vee F_2) \equiv F_1 \vee (\exists V. \, F_2)$$
$$\text{if } V \notin \mathit{fv}(F_1) \qquad\qquad\qquad\qquad \text{if } V \notin \mathit{fv}(F_1)$$

$$\forall V. \, (F_1 \vee F_2) \equiv F_1 \vee (\forall V. \, F_2) \qquad\qquad \exists V. \, (F_1 \wedge F_2) \equiv F_1 \wedge (\exists V. \, F_2)$$
$$\text{if } V \notin \mathit{fv}(F_1) \qquad\qquad\qquad\qquad \text{if } V \notin \mathit{fv}(F_1)$$

$$\forall V. \, (F_1 \Rightarrow F_2) \equiv F_1 \Rightarrow (\forall V. \, F_2) \qquad\qquad \exists V. \, (F_1 \Rightarrow F_2) \equiv F_1 \Rightarrow (\exists V. \, F_2)$$
$$\text{if } V \notin \mathit{fv}(F_1) \qquad\qquad\qquad\qquad \text{if } V \notin \mathit{fv}(F_1)$$

$$\forall V. \, (F_1 \Rightarrow F_2) \equiv (\exists V. \, F_1) \Rightarrow F_2 \qquad\qquad \exists V. \, (F_1 \Rightarrow F_2) \equiv (\forall V. \, F_1) \Rightarrow F_2$$
$$\text{if } V \notin \mathit{fv}(F_2) \qquad\qquad\qquad\qquad \text{if } V \notin \mathit{fv}(F_2)$$

**Let Formulas**

$$\text{let } V = T \text{ in } F \equiv \exists V. \, (V = T \wedge F)$$

**Fig. 2.8** Logical equivalences in first-order logic

- Case $[\![ \neg(\forall V. \, F) ]\!]^I_a = \text{true}$. We show $[\![ \exists V. \, (\neg F) ]\!]^I_a = \text{true}$. From (the definition of) $[\![ \neg ]\!]$, we know $[\![ \forall V. \, F ]\!]^I_a = \text{false}$. From $[\![ \forall ]\!]$, there exists some $d$ in $D$ and $a' = a[V \mapsto d]$ such that $[\![ F ]\!]^I_{a'} \neq \text{true}$. From $[\![ \cdot ]\!]$, we have $[\![ F ]\!]^I_{a'} = \text{false}$. From $[\![ \neg ]\!]$, we know $[\![ \neg F ]\!]^I_{a'} = \text{true}$. From $[\![ \exists ]\!]$, we know $[\![ \exists V. \, (\neg F) ]\!]^I_a = \text{true}$.
- Case $[\![ \neg(\forall V. \, F) ]\!]^I_a = \text{false}$. We show $[\![ \exists V. \, (\neg F) ]\!]^I_a = \text{false}$. We assume $[\![ \exists V. \, (\neg F) ]\!]^I_a = \text{true}$ and show a contradiction. From $[\![ \exists ]\!]$, we have some $d$ in $D$ and $a' = a[V \mapsto d]$ such that $[\![ \neg F ]\!]^I_{a'} = \text{true}$. From $[\![ \neg ]\!]$, we have $[\![ F ]\!]^I_{a'} = \text{false}$. But this contradicts the case assumption which states that for every $d$ in $D$ and $a' = a[V \mapsto d]$, we have $[\![ F ]\!]^I_{a'} = \text{true}$.

□

In a similar way, all the equivalences of Fig. 2.8 can be justified. The first two equivalences state that double occurrences of the same quantifier with the same variable can be merged and that the order of identical quantifiers can be exchanged. Then, similar to propositional logic, "De Morgan's laws" $\neg(\forall V. \, F) \equiv \exists V. \, (\neg F)$ and $\neg(\exists V. \, F) \equiv \forall V. \, (\neg F)$ tell us that by pushing negation into a universal or existential quantification the quantifier is flipped. The other laws tell us how to reduce the scope of quantifiers, in particular, how to "pull out" a formula out of the scope of a quantifier, if the quantified variable does not occur freely in the formula. The equivalence let $V = T$ in $F \equiv \exists V. \, (V = T \wedge F)$ shows that the let construct is just syntactic sugar.

$$(\forall V. \, F) \models (\exists V. \, F)$$
$$(\exists V_1. \, \forall V_2. \, F) \models (\forall V_2. \, \exists V_1. \, F)$$
$$(\forall V. \, F_1) \vee (\forall V. \, F_2) \models \forall V. \, (F_1 \vee V_2)$$
$$\exists V. \, (F_1 \wedge V_2) \models (\exists V. \, F_1) \wedge (\exists V. \, F_2)$$

**Fig. 2.9** Logical consequences in first-order logic

Furthermore, all the logical equivalences of propositional logic already established in Fig. 2.5 can be transformed to logical equivalences in first-order logic.

**Proposition 2.5** (Propositional and First-Order Equivalences) *Let $F_1[A, B, \ldots]$ and $F_2[A, B, \ldots]$ be two equivalent propositional schemes and let $F_a, F_b, \ldots$ be first-order logic formulas that are well-formed with arity ar. Then also the instances $F_1[F_a, F_b, \ldots]$ and $F_2[F_a, F_b, \ldots]$ are logically equivalent.*

Similar to propositional logic, also in first-order logic equivalences give rise to "true" formulas in the following sense.

**Proposition 2.6** (Logical Consequence and Equivalence as Valid Formulas) *Formula $F_2$ is a logical consequence of $F_1$ (i.e., $F_1 \models F_2$) if and only if the formula $F_1 \Rightarrow F_2$ is valid (i.e., $\models F_1 \Rightarrow F_2$). Formulas $F_1$ and $F_2$ are logically equivalent (i.e., $F_1 \equiv F_2$) if and only if the formula $F_1 \Leftrightarrow F_2$ is valid (i.e., $\models F_1 \Leftrightarrow F_2$).*

**Example 2.20** From the proof in Example 2.19, we can deduce that

$$\neg(\forall V. \, F) \Leftrightarrow \exists V. \, (\neg F)$$

is a valid formula. □

The logical equivalences in Figs. 2.5 and 2.8 thus give rise to valid equivalence formulas in first-order logic (and vice versa). Furthermore, Fig. 2.9 lists several logical consequences from which the following valid implications emerge:

$$(\forall V.\ F) \Rightarrow (\exists V.\ F)$$
$$(\exists V_1.\ \forall V_2.\ F) \Rightarrow (\forall V_2.\ \exists V_1.\ F)$$
$$(\forall V.\ F_1) \vee (\forall V.\ F_2) \Rightarrow \forall V.\ (F_1 \vee V_2)$$
$$\exists V.\ (F_1 \wedge V_2) \Rightarrow (\exists V.\ F_1) \wedge (\exists V.\ F_2)$$

However, these are only logical consequences, not equivalences; thus reverting the direction of the implication arrows yields formulas that are *not* valid.

By the results established so far, we have extended our calculus of meaning-preserving formula transformations to first-order logic.

**Example 2.21** We can transform the formula

$$\forall x.\ \text{if } p(x) \text{ then } q(y) \text{ else } r(y)$$

as follows:

$$\forall x.\ \text{if } p(x) \text{ then } q(y) \text{ else } r(y)$$
$$\equiv \forall x.\ \big(p(x) \Rightarrow q(y)\big) \wedge \big(\neg p(x) \Rightarrow r(y)\big)$$
$$\equiv \big(\forall x.\ p(x) \Rightarrow q(y)\big) \wedge \big(\forall x.\ \neg p(x) \Rightarrow r(y)\big)$$
$$\equiv \Big(\big(\exists x.\ p(x)\big) \Rightarrow q(y)\Big) \wedge \Big(\big(\exists x.\ \neg p(x)\big) \Rightarrow r(y)\Big)$$
$$\equiv \Big(\big(\neg\exists x.\ p(x)\big) \vee q(y)\Big) \wedge \Big(\big(\neg\exists x.\ \neg p(x)\big) \vee r(y)\Big)$$
$$\equiv \Big(\big(\forall x.\ \neg p(x)\big) \vee q(y)\Big) \wedge \Big(\big(\forall x.\ p(x)\big) \vee r(y)\Big)$$
$$\equiv \Big(\big(\forall x.\ \neg p(x)\big) \wedge \big(\forall x.\ p(x)\big)\Big) \vee \Big(\big(\forall x.\ \neg p(x)\big) \wedge r(y)\Big) \vee$$
$$\Big(q(y) \wedge \big(\forall x.\ p(x)\big)\Big) \vee \Big(q(y) \wedge r(y)\Big)$$
$$\equiv \Big(\big(\forall x.\ p(x)\big) \wedge q(y)\Big) \vee \Big(\big(\forall x.\ \neg p(x)\big) \wedge r(y)\Big) \vee \Big(q(y) \wedge r(y)\Big)$$

For the last transformation step we use $\big(\forall x.\ \neg p(x)\big) \wedge \big(\forall x.\ p(x)\big) \equiv \text{false}$ (which is intuitively clear but has to be proved separately) and $\text{false} \wedge F \equiv F$ such that the first conjunction vanishes from the disjunction.

In other words, to make the formula true, three possibilities exist:

1. All $x$ satisfy $p$; then $y$ must satisfy $q$.
2. All $x$ do not satisfy $p$; then $y$ must satisfy $r$.
3. Some $x$ do satisfy $p$, some do not; in thic case $y$ must satisfy both $p$ and $q$.

In the third case, if $q$ and $r$ cannot hold simultaneously, the formula is false. Perhaps, however, we actually wanted to state

$$\text{if } \forall x. \ p(x) \text{ then } q(y) \text{ else } r(y)$$

which can be transformed as follows:

$$
\begin{aligned}
&\text{if } \forall x. \ p(x) \text{ then } q(y) \text{ else } r(y) \\
&\equiv \big((\forall x. \ p(x)) \Rightarrow q(y)\big) \wedge \big((\neg \forall x. \ p(x)) \Rightarrow r(y)\big) \\
&\equiv \big((\neg \forall x. \ p(x)) \vee q(y)\big) \wedge \big((\forall x. \ p(x)) \vee r(y)\big) \\
&\equiv \big((\neg \forall x. \ p(x)) \wedge (\forall x. \ p(x))\big) \vee \big((\neg \forall x. \ p(x)) \wedge r(y)\big) \vee \\
&\quad\ \big(q(y) \wedge (\forall x. \ p(x))\big) \vee \big(q(y) \wedge r(y)\big) \\
&\equiv \big((\forall x. \ p(x)) \wedge q(y)\big) \vee \big((\neg \forall x. \ p(x)) \wedge r(y)\big) \vee \big(q(y) \wedge r(y)\big)
\end{aligned}
$$

The last formula is an instance of the scheme

$$(A \wedge B) \vee (\neg A \wedge C) \vee (B \vee C)$$

which was simplified in Example 2.8 to

$$(A \wedge B) \vee (\neg A \wedge C)$$

Above formula can be thus simplified to

$$\big((\forall x. \ p(x)) \wedge q(y)\big) \vee \big((\neg \forall x. \ p(x)) \wedge r(y)\big)$$

To make this formula true, two possibilities exist:

1. All $x$ satisfy $p$; then $y$ must satisfy $q$.
2. Not all $x$ satisfy $p$; then $y$ must satisfy $r$.

This formula has a simpler semantics and thus may well be the intended one.   □

By the rules of the calculus, we can transform also first-order formulas into a normal form.

**Definition 2.19** (*Prenex Normal Form*) A first-order formula $F$ is in *prenex normal form*, if quantifiers occur only at the outermost level. In other words, $F$ does either not contain any quantifiers or is of form $Q_1 V_1. \ \ldots Q_n V_n. \ F'$ where $F'$ does not contain quantifiers and each of $Q_1, \ldots, Q_n$ is one of the quantifiers $\forall$ or $\exists$.

**Proposition 2.7** (Existence of Prenex Normal Form) *For every first-order formula F there exists a formula F' such that F ≡ F' and F' is in prenex normal form.*

**Example 2.22** The prenex normal form of

$$\forall x.\ (\exists z.\ p(x, z)) \Rightarrow \exists y.\ q(x, y)$$

can be computed "inside-out" as

$$\forall x.\ (\exists z.\ p(x, z)) \Rightarrow \exists y.\ q(x, y)$$
$$\equiv\ \forall x.\ \exists y.\ (\exists z.\ p(x, z)) \Rightarrow q(x, y)$$
$$\equiv\ \forall x.\ \exists y.\ \forall z.\ p(x, z) \Rightarrow q(x, y).$$

□

While it is thus always possible to "pull out" all quantifiers from a formula such that their scope becomes the whole formula, this is not recommendable. Quite on the contrary, we will usually pull a subformula out of the scope of a quantifier, if the free variables of the subformula do not contain the quantified variable. As above example demonstrates, by restricting the scopes of quantifiers as much as possible, formulas usually become much more understandable than the other way round: the formula

$$\forall x.\ (\exists z.\ p(x, z)) \Rightarrow \exists y.\ q(x, y)$$

can be nicely read and understood as

For every $x$, if there is some $z$ such that $p(x, z)$ holds, then there is also some $y$, such that $q(x, y)$ holds.

To get this intuition from the logically equivalent formula

$$\forall x.\ \exists y.\ \forall z.\ p(x, z) \Rightarrow q(x, y)$$

is challenging.

**Exercises**

Download from the following URL:
https://www.risc.jku.at/people/schreine/TP/exercises/ex-logic.pdf.

## Further Reading

Several semi-formal introductions to first-order logic target a computer science audience, e.g., Lover's book [104], Makinson's book [106], O'Regan's book [138], Schenke's book [159] (in German), or Buchberger's and Lichtenberger's classical book [29] (in German).

Formal treatments of first-order logic for computer scientists can be found in Huth's and Ryan's book [81], Ben-Ari's book [13], and Li's book [102]. There exist an abundancy of introductory textbook to logic in general, e.g., Ebbinghaus et al.'s book [45] to name one. Krantz's handbook [90] summarizes the essential facts.

Treatments with more focus on automated reasoning are Bradley's and Manna's book [26], Harrison's handbook [65], and Gallier's classical text [56].

## The Logic of the RISC ProofNavigator

There exist various automated theorem provers [61] and interactive proof assistants that aid formal reasoning. In this section we present the *RISC ProofNavigator* [165, 166], a proof assistant that employs the SMT (Satisfiability Modulo Theories) solver *CVC3* [11]. The RISC ProofNavigator is integrated into the *RISC ProgramExplorer*, a computer-supported program reasoning environment [167, 168] that will be explained later. Figure 2.10 depicts the graphical user interface of the RISC ProofNavigator when executed as a standalone tool. The specifications used in the following presentations can be downloaded from the URL

> https://www.risc.jku.at/people/schreine/TP/software/logic/logic.txt

and loaded by executing from the command line the following command:

```
ProofNavigator logic.txt &
```

Propositional Logic

The language of the RISC ProofNavigator is a strongly typed variant of first-order logic. When a specification is interactively typed in or read from a file, it is parsed, typed-checked, and pretty-printed as depicted in the screenshot of Fig. 2.10.

We start with a simple example that illustrates the propositional subset of the language:

```
A: BOOLEAN;
B: BOOLEAN;
C: BOOLEAN;

F1: FORMULA
      IF A THEN B ELSE C ENDIF <=> (A => B) AND (NOT A => C);
F2: FORMULA
      IF A THEN B ELSE C ENDIF <=>
      (A AND B) OR (NOT A AND C) OR (B AND C);
```

In this specification we first declare three propositional variables $A$, $B$, and $C$. Based on these variables we introduce two formulas $F_1$ and $F_2$ that involve the various logical connectives of the language. Actually, these formulas are tautologies; this can be confirmed by selecting in the graphical user interface from the button left to the formula e.g. the menu entry prove F1. The system immediately responds by printing the line

```
Formula F1 is proved. QED.
```

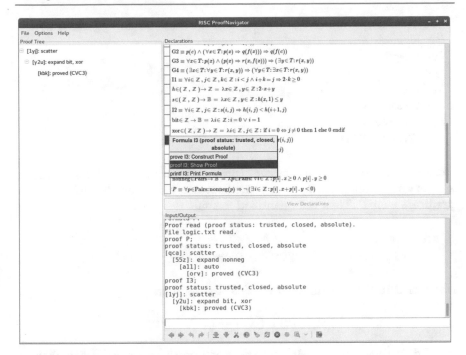

**Fig. 2.10** The RISC ProofNavigator

which shows that the formula is true for all interpretations of the propositional variables. This result has been established by the integrated SMT solver CVC3 which implements automatic decision procedures for various theories. When selecting the menu entry prove F1, indeed the "Proof State" display shows that the formula has been simplified to "true" and the "Proof Tree" window displays a line "proved (CVC3)" which indicates that the external solver has determined this result.

We may also introduce propositional variables that are just abbreviations:

```
AB: BOOLEAN = A OR B;
AC: BOOLEAN = A OR C;
BC: BOOLEAN = B AND C;


F3: FORMULA
       AB AND AC AND NOT BC => A;
```

In this specification, the propositional variables AB, AC, and BC are defined as abbreviations of certain propositional schemes; by replacing the variables with their definitions, the formula $F_3$ also becomes a tautology. This can be determined as before by selecting the menu entry prove F3.

First-Order Logic

To demonstrate full first-order language, we use the following declarations:

```
T: TYPE;
c: T;
f: T->T;
p: T->BOOLEAN;
q: T->BOOLEAN;
r: (T,T)->BOOLEAN;
```

Here $T$ is introduced as an abstract (uninterpreted) type with a constant $c$, a unary function $f$ on $T$, unary predicates $p$ and $q$ on $T$, and a binary predicate $r$ on $T$. Then the quantified formula

```
G1: FORMULA
    FORALL(x:T): p(x) AND (p(x)=>q(x)) => q(x);
```

states that in the domain denoted by $T$ the formula

$$\forall x.\ p(x) \wedge (p(x) \Rightarrow q(x)) \wedge q(x)$$

is valid; w show this by selecting the menu entry Prove G1 which changes the perspective to the one depicted in Fig. 2.11.

The formula is not immediately shown to be valid, because it involves a quantifier that the external SMT solver cannot handle. However, by pressing

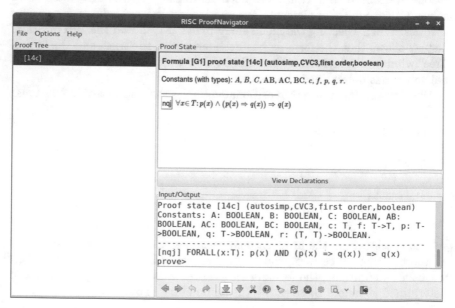

**Fig. 2.11** The RISC ProofNavigator

the button ⬇ which triggers the proof command scatter, the quantifier is appropriately discharged (as will be explained in the next chapter), such that the validity of the formula can be again shown by CVC3.

If we perform the same steps for showing the validity of the implication

G2: FORMULA
    p(c) AND (FORALL(x:T): p(x) => q(f(x))) => q(f(c));

we get the view depicted in Fig. 2.12. The system could by the application of the scatter command not yet show the validity of the formula, because for this it needs the quantified assumption $\forall x \in T.\ p(x) \Rightarrow q(f(x))$ which arises from a part of the antecedent of the implication. However, pressing the button ♻ triggers the proof command auto that appropriately uses this assumption (as will be explained in the next chapter), such that the validity of the implication can be again established.

The validity of the two formulas

G3: FORMULA
    (FORALL(x:T): p(x) AND (p(x) => r(x,f(x))) => (EXISTS(y:T): r(x,y)));
G4: FORMULA
    (EXISTS(x:T): (FORALL(y:T): r(x,y))) =>

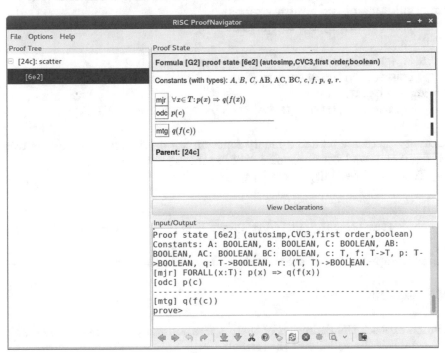

**Fig. 2.12** The RISC ProofNavigator

```
(FORALL(y:T): (EXISTS(x:T): r(x,y)));
```

can be shown by the same sequence of applications of the proof commands
scatter and auto as shown before.

Special Domains

The RISC ProofNavigator does not only deal with abstract domains; it
also has builtin-knowledge of special domains such as the integer numbers
(actually it is the SMT Solver CVC3 that provides this knowledge). This
domain is denoted by the type INT and supports the usual arithmetic operations.

For instance, the validity of the formula

```
I1: FORMULA
      FORALL(i:INT,j:INT,k:INT): i < j AND i+k = j => 2*k >= 0;
```

can be shown by a single application of the scatter command which exposes
the quantifier-free body of the formula to CVC3. In general, the validity of
any integer formula in the theory of "linear arithmetic" can be automatically decided by CVC3; this theory allows addition, subtraction, equality,
inequalities, and multiplication of a term with an integer constant (but *not*
multiplication of two arbitrary terms).

Next we define by the specification

```
bit: INT->BOOLEAN =
  LAMBDA(i:INT): i=0 OR i=1;
xor: (INT,INT)->INT =
  LAMBDA(i:INT,j:INT): IF i=0 <=> j /= 0 THEN 1 ELSE 0 ENDIF;
```

a unary predicate bit and a binary function xor on the integers; bit($i$) is true
if $i$ is 0 or 1, and xor implements the "exclusive or" operation.

To show the validity of the formula

```
I3: FORMULA
      FORALL(i:INT,j:INT): bit(i) AND bit(j) => bit(xor(i,j));
```

after the application of scatter we also execute the proof command

```
    expand bit, xor;
```

which exposes the definition of these operations to CVC3 that is then able
to establish the validity of the formula.

However, if we try the same with the formula

```
I4: FORMULA
      FORALL(i:INT,j:INT): bit(i) AND bit(j) => bit(i+j);
```

CVC3 fails, because the formula is actually *not valid.* To see why, we may
press the button ❷ which triggers the command counterexample that asks
CVC3 for a (potential) counterexample of the formula whose validity it cannot show. From the resulting message

```
This may be a counterexample:
j_0+i_0 /= 0 AND 1 /= j_0+i_0 AND i_0 /= 0 AND
1 = i_0 AND j_0 /= 0 AND 1 = j_0
```

we may deduce that the body of the universally quantified formula is not
true for i=1 and j=1.

User-Defined Domains

The language of the RISC ProofNavigator also allows users to define their
own domains. For instance, the declarations

```
Pair:  TYPE = [# x:INT, y:INT #];
Pairs: TYPE = ARRAY NAT OF Pair;
```

introduce a type Pair of "records" with labeled components $x$ and $y$ of integers and a type Pairs of arrays, i.e., mappings from natural number indices to
Pair values (since the domain NAT is unbounded, these "arrays" are actually
sequences of inifinite lengths).
    Then the declaration

```
nonneg: Pairs->BOOLEAN =
  LAMBDA(p:Pairs): FORALL(i:NAT): p[i].x >= 0 AND p[i].y >= 0;
```

introduces a unary predicate nonneg on domain Pairs where nonneg($p$) states
that all elements of array $p$ have non-negative values in both components $x$
and $y$.
    The validity of the formula

```
P: FORMULA
    FORALL(p:Pairs):
      nonneg(p) => NOT (EXISTS(i:NAT): p[i].x+p[i].y < 0);
```

can then be shown by the sequence of commands scatter, expand nonneg,
and auto.

# The Art of Reasoning

**3**

> *Logik ist die Feindin der Kunst. Aber Kunst darf nicht die Feindin der Logik sein. (Logic is the enemy of art. But art must not be the enemy of logic).*
> — *Karl Kraus*

In this chapter, we elaborate the art of formal reasoning as the basis of precise argumentation. We start by discussing the general concept of a 'proof' as an irrefutable argument why a particular statement is true. We then discuss the formalization of this concept by 'proof trees', i.e., by derivations in a system of inference rules. The focus of this chapter is on a proof system for first-order logic, which is presented both in a formal style (as is required for computer-supported proving) and in an informal style (as is typical for mathematical texts). Finally we summarize various forms of the proof principle of 'induction', which is of fundamental importance in computer science where inductive structures play a major role.

## 3.1    Reasoning and Proofs

Formal reasoning is centered around the notion of *proof* as a convincing argument why a particular statement is true. However, since some arguments may be less convincing the others, mathematical logic considers as a correct proof only an argument that cannot be questioned at all. Of crucial importance is here the existence of a calculus of rules which allow to infer correct arguments by deducing further conclusions from previously established facts. Assuming that the soundness of these rules has been established once and for all, the results derived from the application of these rules need and must not be questioned any more. A proof is therefore considered as correct only, if it is constructed by the application of these rules.

A key issue here is that these rules must operate on the syntactic structure of statements, *not* on their semantic contents. Whether the application of a rule is correct or not, does therefore not depend on a subjective interpretation: the correctness of the application can be objectively verified by simple syntactic 'pattern matching', which does not require semantic understanding and can be even mechanized, i.e., implemented by an automatic procedure. All in all, a proof is therefore not correct, because the resulting statement is true; it is correct, because the argument that leads to this statement is well-formed according to certain syntactic rules of reasoning whose soundness has been previously established.

Truly, in practice arguments are often not given in a completely formalized style that allows the immediate mechanized checking of its correctness; however, the structure of such high-level (informal or semi-formal) arguments must nevertheless match the overall structure of a formal proof: it must be possible to justify every high-level deduction step by one or more inference steps according to the rules of the formal calculus. A typical 'proof' as it can be found in a book (also in the present one) is therefore just a reasonably detailed proof sketch that allows the experienced reader to re-construct the actual formal proof. In this sense, the formal reasoning rules therefore also govern the structure of correct informal arguments. An important practical competence is therefore to 'compress' formal proofs into compact proof sketches that can be easily communicated to other people and, conversely, to 'uncompress' informal proof sketches into formal proofs, whenever the correctness of a particular argument needs to be checked.

Nevertheless, in various areas of computer-supported mathematics and logic, in particular in *program verification*, really completely formal proofs are constructed, either fully automatically by the machine alone or semi-automatically with some helpful guidance by a human. Especially for the later it is important to understand the principles of formal proof construction in order to unquestionably demonstrate that a particular piece of software satisfies its specification.

## 3.2    Inference Rules and Proof Trees

In the following, we will present a calculus of reasoning rules which belongs to the general class of *sequent calculi* (introduced by Gerhard Gentzen). Each judgement in this calculus is a (formula) *sequent* of form

$$Fs \vdash G$$

where $G$ is a first-order formula and $Fs$ is a (possibly empty) sequence $F_1, \ldots, F_n$ of such formulas. The sequent can be read as 'if $F_1$ and …and $F_n$ are true, then $G$ is true' and can be interpreted as the claim of the validity of the formula

$$F_1 \wedge \ldots \wedge F_n \Rightarrow G.$$

The formulas $F_1, \ldots, F_n$ are also collectively called *knowledge*; they are (in the current context) assumed to be true; the formula $G$ is called the *goal* whose truth follows from these assumptions. As a special case, the sequent

$$\vdash G$$

establishes the truth of goal $G$ without any knowledge.

The *(inference) rules* of the calculus have form

$$\frac{Fs_1 \vdash G_1 \quad \ldots \quad Fs_m \vdash G_m}{Fs \vdash G}$$

with $m$ *premises* $Fs_1 \vdash G_1, \ldots, Fs_m \vdash G_m$ and the *conclusion* $Fs \vdash G$. We call a rule without premises an *axiom* and write it as

$$Fs \vdash G.$$

We can read such a rule in two directions:

- *Forward:* if we have proved the premises, then we have proved the conclusion.
- *Backward:* in order to prove the conclusion, we need to prove the premises.

During the construction of a proof, the *backward* reading is the desired one: given a sequent $S$ to be proved, we match $S$ against the conclusion of the rule. The $m$ premises of the rules thus determine the sequences $S_1, \ldots, S_m$ that have to be subsequently proved. The rule therefore has reduced the (complex) problem of proving $S$ to the $m$ (simpler) subproblems of proving $S_1, \ldots, S_m$.

To let the conclusion of a rule match different sequents, the elements of the rule are generally not concrete formulas but abstract *formula patterns* that involve variables that stand for arbitrary formulas; e.g. the pattern $F_1 \wedge F_2$ with variables $F_1$ and $F_2$ describes a formula that is a conjunction of two subformulas which are denoted by the formula variables $F_1$ and $F_2$. A concrete formula that matches an abstract pattern is called an *instance of the pattern*.

An *instance of a rule*

$$\frac{Fs_1 \vdash G_1 \quad \ldots \quad Fs_m \vdash G_m}{Fs \vdash G}$$

with $m$ premises is a corresponding collection of $m + 1$ sequents

$$\frac{S_1 \quad \ldots \quad S_m}{S}$$

such that every sequent $S_i$ matches premise $Fs_i \vdash G_i$ and sequent $S$ matches conclusion $F \vdash G$. In the matching of all sequents, all occurrences of a formula variable with the same name represent the same concrete formula.

Sometimes the applicability of a rule cannot be described by simple pattern-matching; in this case the rule may be annotated as in

$$\frac{Fs_1 \vdash G_1 \quad \dots \quad Fs_m \vdash G_m \quad (\dots)}{Fs \vdash G}$$

by an additional side-condition $(\dots)$ on the variables of the rule that has to be satisfied to make the rule applicable; while this condition cannot be expressed by pattern matching, it must be nevertheless mechanically checkable.

Now, given a collection of inference rules, a *proof* of a sequent • is a tree

with root • (which is depicted at the bottom). Each node • or ∘ represents a formula sequent $S$ that, together with its parent nodes $S_1, \dots, S_n$, represents an instance of some rule. Consequently, the leaves of the tree are instances of some axioms. This proof tree establishes the truth of the sequent • at the root of the tree.

The construction of a proof typically starts in a 'goal-oriented' fashion with the sequent whose truth is to be established as the root node

•

The proof then proceeds in the direction indicated by the arrow ↑ by applying some suitable rule in the 'backward' fashion such that the proof tree can be expanded to

If a newly generated node is an instances of an axiom, it need not be considered further; the corresponding proof branch is 'closed'. However, every other node still represents an 'open' branch of the proof to which again a suitable rule has to be applied, which expands the tree further to, e.g.,

The process of proving consists of expanding the tree further and further, until all open branches can be closed, i.e., all leaves of the tree represent instances of axioms. The art of proving is to suitably select and apply the rules of the calculus such that this goal can be ultimately achieved.

While proofs are practically always constructed in this fashion by, starting with the root, applying the rules *backwards* and thus continually expanding the proof tree towards the leaves, proof sketches in books unfortunately rather often give a different impression. They start with the leaves of the tree (which represent as instances of axioms unquestioned knowledge) and then demonstrate how by the *forward* application of inference rules more and more knowledge is generated and combined until 'magically' the desired sequent is established. In this book, we will never proceed in this way but always present an argument in the backwards direction in which it was originally constructed.

## 3.3 Reasoning in First Order Logic

Figures 3.1 and 3.2 introduce a system of inference rules for first-order logic. In these rules, we use the pattern

$$Fs, F$$

**Structural Rules:**

$$\frac{Fs \vdash G}{Fs, F \vdash G} \text{ (contract)} \qquad \frac{Fs, F, F \vdash G}{Fs, F \vdash G} \text{ (expand)} \qquad \frac{Fs_1, Fs_2, F_2, F_1 \vdash G}{Fs_1, F_1, Fs_2, F_2 \vdash G} \text{ (front)}$$

**Basic Rules:**

$$Fs, G \vdash G \text{ (direct)} \qquad \frac{Fs, \neg G \vdash \text{false}}{Fs \vdash G} \text{ (indirect)} \qquad \frac{Fs \vdash F \quad Fs, F \vdash G}{Fs \vdash G} \text{ (cut)}$$

**Propositional Rules:**

$$Fs \vdash \text{true (true-G)} \qquad \frac{Fs \vdash G}{Fs, \text{true} \vdash G} \text{ (true-K)}$$

$$\frac{Fs \vdash \neg F}{Fs, F \vdash \text{false}} \text{ (false-G)} \qquad Fs, \text{false} \vdash G \text{ (false-K)}$$

$$\frac{Fs, G \vdash \text{false}}{Fs \vdash \neg G} \text{ (}\neg\text{-G)} \qquad \frac{Fs, \neg G \vdash F}{Fs, \neg F \vdash G} \text{ (}\neg\text{-K)}$$

$$\frac{Fs \vdash G_1 \quad Fs \vdash G_2}{Fs \vdash G_1 \wedge G_2} \text{ (}\wedge\text{-G)} \qquad \frac{Fs, F_1, F_2 \vdash G}{Fs, F_1 \wedge F_2 \vdash G} \text{ (}\wedge\text{-K)}$$

$$\frac{Fs, \neg G_2 \vdash G_1}{Fs \vdash G_1 \vee G_2} \text{ (}\vee\text{-G1)} \qquad \frac{Fs, \neg G_1 \vdash G_2}{Fs \vdash G_1 \vee G_2} \text{ (}\vee\text{-G2)}$$

$$\frac{Fs, F_1 \vdash G \quad Fs, F_2 \vdash G}{Fs, F_1 \vee F_2 \vdash G} \text{ (}\vee\text{-K)}$$

$$\frac{Fs, G_1 \vdash G_2}{Fs \vdash G_1 \Rightarrow G_2} \text{ (}\Rightarrow\text{-G)} \qquad \frac{Fs \vdash F_1 \quad Fs, F_2 \vdash G}{Fs, F_1 \Rightarrow F_2 \vdash G} \text{ (}\Rightarrow\text{-K)}$$

$$\frac{Fs \vdash G_1 \Rightarrow G_2 \quad Fs \vdash G_2 \Rightarrow G_1}{Fs \vdash G_1 \Leftrightarrow G_2} \text{ (}\Leftrightarrow\text{-G)} \qquad \frac{Fs, F_1 \Rightarrow F_2, F_2 \Rightarrow F_1 \vdash G}{Fs, F_1 \Leftrightarrow F_2 \vdash G} \text{ (}\Leftrightarrow\text{-K)}$$

$$\frac{Fs \vdash G \Rightarrow G_1 \quad Fs \vdash \neg G \Rightarrow G_2}{Fs \vdash \text{if } G \text{ then } G_1 \text{ else } G_2} \text{ (if-G)} \qquad \frac{Fs, F \Rightarrow F_1, \neg F \Rightarrow F_2 \vdash G}{Fs, \text{if } F \text{ then } F_1 \text{ else } F_2 \vdash G} \text{ (if-K)}$$

**Fig. 3.1** First-order logic rules (part 1)

**Quantifier Rules:**

$$(V_0 \notin fv(Fs), V_0 \notin fv(G)\setminus\{V\}, V_0 \text{ free for } V \text{ in } G)$$
$$\frac{Fs \vdash G[V_0/V]}{Fs \vdash \forall V.\, G} \quad (\forall\text{-G})$$

$$(T \text{ free for } V \text{ in } F)$$
$$\frac{Fs, F[T/V] \vdash G}{Fs, \forall V.\, F \vdash G} \quad (\forall\text{-K})$$

$$(T \text{ free for } V \text{ in } F)$$
$$\frac{Fs \vdash F[T/V]}{Fs \vdash \exists V.\, F} \quad (\exists\text{-G})$$

$$(V_0 \notin fv(Fs), V_0 \notin fv(F)\setminus\{V\}, V_0 \text{ free for } V \text{ in } F)$$
$$\frac{Fs, F[V_0/V] \vdash G}{Fs, \exists V.\, F \vdash G} \quad (\exists\text{-K})$$

**Equality Rules:**

$$(T_1 \text{ free for } V \text{ in } G, T_2 \text{ free for } V \text{ in } G)$$
$$Fs \vdash T = T \text{ (reflexivity)} \qquad \frac{Fs \vdash G[T_1/V]}{Fs, T_1 = T_2 \vdash G[T_2/V]} \quad \text{(substitution)}$$

**Let Rules:**

$$(V_0 \notin fv(Fs) \cup fv(T) \cup fv(T_1), V_0 \notin fv(T_2)\setminus\{V\}, V_0 \text{ free for } V \text{ in } T_2)$$
$$\frac{Fs, V_0 = T_1 \vdash T = T_2[V_0/V]}{Fs \vdash T = \text{let } V = T_1 \text{ in } T_2} \quad \text{(let-TG)}$$

$$(V_0 \notin fv(Fs) \cup fv(T) \cup fv(T_1) \cup fv(G), V_0 \notin fv(T_2)\setminus\{V\}, V_0 \text{ free for } V \text{ in } T_2)$$
$$\frac{Fs, V_0 = T_1, T = T_2[V_0/V] \vdash G}{Fs, T = \text{let } V = T_1 \text{ in } T_2 \vdash G} \quad \text{(let-TK)}$$

$$(V_0 \notin fv(Fs), V_0 \notin fv(G)\setminus\{V\}, V_0 \text{ free for } V \text{ in } F)$$
$$\frac{Fs, V_0 = T \vdash G[V_0/V]}{Fs \vdash \text{let } V = T \text{ in } G} \quad \text{(let-FG)}$$

$$(V_0 \notin fv(Fs) \cup fv(G), V_0 \notin fv(F)\setminus\{V\}, V_0 \text{ free for } V \text{ in } F)$$
$$\frac{Fs, V_0 = T, F[V_0/V] \vdash G}{Fs, \text{let } V = T \text{ in } F \vdash G} \quad \text{(let-FK)}$$

**Fig. 3.2** First-order logic rules (part 2)

to indicate a non-empty sequence of knowledge formulas that starts with the (possibly empty) subsequence $Fs$ and ends with the formula $F$.

We start with a notion that is central for this section.

**Definition 3.1** (*Derivability*) A sequent $F_1, \ldots, F_n \vdash G$ is *derivable*, if it is the root of a proof tree where every node of the tree is an instance of one of the rules of Figs. 3.1 and 3.2.

Now the calculus established by these rules enjoys the following core properties.

**Theorem 3.1** (Soundness of first-order logic with equality) *If the sequent $F_1, \ldots, F_n \vdash G$ is derivable, then we have the logical consequence $F_1 \wedge \ldots \wedge F_n \models G$. As a special case, if the sequent $\vdash G$ is derivable, then we have $\models G$, i.e., $G$ is valid.*

In a nutshell, the soundness of the proving calculus implies that only valid formulas can be proved.

**Theorem 3.2** (Completeness of first-order logic with equality) *If we have the logical consequence $F_1 \wedge \ldots \wedge F_n \models G$, then the sequent $F_1, \ldots, F_n \vdash G$ is derivable. As a special case, if we have $\models G$, i.e., $G$ is valid, then $\vdash G$ is derivable.*

In a nutshell, the completeness of the proving calculus implies that all valid formulas can indeed be proved.

We are now going to explain the rules in turn. To relate these formal rules to informal proofs, it is important to know that in an informal proof, a sequent

$$F_1, \ldots, F_n \vdash G$$

is usually not explicitly stated. Rather the informal proof gradually builds a sequence of formulas

$$F_1 \quad (1)$$
$$\ldots$$
$$F_n \quad (n)$$

which represents the knowledge $F_1, \ldots, F_n$ that is available in the current proof situation and can be referenced in the proof of $G$ by the tags $(1), \ldots, (n)$. Likewise, the current goal is stated by a statement like

We are going to prove $G$ (a).

where a tag like (a) may be used to indicate the current goal situation.

**Structural Rules**

These rules allow us to manipulate the knowledge formulas. They are only used in formal proofs; in semi-formal proofs they play no role, as is explained below.

$$\frac{Fs \vdash G}{Fs, F \vdash G} \text{ (contract)}$$

This rule simply tells us that we can drop an unnecessary knowledge formula $F$ in the course of a proof. Technically, this rule is not needed at all; it may however help to 'compress' formal proofs. In an informal proof, it plays no role, because we don't explicitly list the current knowledge in every proof step.

$$\frac{Fs, F, F \vdash G}{Fs, F \vdash G} \text{ (expand)}$$

This rule allows to duplicate the last knowledge formula. The rule is necessary, because some rules 'destroy' the last formula (e.g., by breaking it into subformulas); to preserve the original formula, the rule (expand) must be applied before. In an informal proof, the rule plays no role, because previously established knowledge is never destroyed.

$$\frac{Fs_1, Fs_2, F_2, F_1 \vdash G}{Fs_1, F_1, Fs_2, F_2 \vdash G} \text{ (front)}$$

This rule allows to bring any knowledge formula to the 'front', i.e, to make it the last formula in the sequence. It is needed in a formal proof, because most proof rules operate on this particular knowledge formula. In an informal proof, it plays no role, because there the rules may be applied to arbitrary knowledge formulas by referring to every formula via its tag.

**Basic Rules**

These rules establish some basic logical facts that are independent of the particular structure of the conclusion.

$$Fs, G \vdash G$$

In *direct proofs*, we proceed with the intention to ultimately apply this axiom, which allows to close a proof branch whose goal formula also appears in the knowledge. In an informal proof, the application of this rule may be indicated as follows:

We are going to prove $G$ (a). ...We therefore know $G$ and thus (a).

$$\frac{Fs, \neg G \vdash \mathsf{false}}{Fs \vdash G} \text{ (indirect)}$$

In *indirect proofs*, this rule is applied which provides an alternative proving strategy. Rather than proving goal $G$ directly, we assume $\neg G$ and derive a *contradiction*, i.e., we prove goal $\mathsf{false}$. In an informal proof, the application of this rule may be indicated as follows:

We are going to prove $G$ (a). We assume $\neg G$ and show a contradiction.

$$\frac{Fs \vdash F \quad Fs, F \vdash G}{Fs \vdash G} \text{ (cut)}$$

This rule allows to introduce an additional hypothesis $F$ for the proof of goal $G$, provided that this hypothesis is separately proved. In an informal proof, the application of this *cut rule* may be indicated as follows:

> We are going to prove $G$ (a). Let us assume $F$. ...(proof of $G$ with the additional knowledge $F$). We are now going to show our hypothesis $F$. ...(proof of $F$ with the original knowledge).

**Propositional Rules**
In the following, we introduce rules that are connected to logical connectives. The rules come in pairs: one rule can be applied, if the corresponding connective appears in a goal formula, the other rule can be applied, if the connective appears in a knowledge formula.

- true:

$$Fs \vdash \text{true (true-G)}$$

If the goal is true, the proof branch can be apparently immediately closed. In an informal proof, an application of this rule may be indicated as follows:

> We have to show true, and are therefore done.

$$\frac{Fs \vdash G}{Fs, \text{true} \vdash G} \text{ (true-K)}$$

This rule is just a special case of the structural rule (expand). If a knowledge formula is true, it can (and should) be dropped from a formal proof; in an informal proof, this rule plays no role.

- false:

$$\frac{Fs \vdash \neg F}{Fs, F \vdash \text{false}} \text{ (false-G)}$$

If the goal false (i.e., a contradiction) is to be derived, we may instead derive the negation $\neg F$ of some knowledge $F$. In an informal proof, this may be indicated as follows:

> We derive a contradiction. ...We know $F$ (1). ...We know $\neg F$ which contradicts (1).

$$Fs, \text{false} \vdash G \text{ (false-K)}$$

This rule gives another way of closing (typically an indirect) proof by deriving a contradiction (i.e., formula false) as knowledge. In an informal proof, this may be indicated as follows:

> We are going to show $G$ (a). …We know false and are therefore done.

- ¬:

$$\frac{Fs, G \vdash \mathsf{false}}{Fs \vdash \neg G} \ (\neg\text{-G})$$

This rule is just a specialization of rule (indirect) for the goal $\neg G$. In an informal proof, its application may be stated as follows.

> We have to show $\neg G$ (a). We assume $G$ and derive a contradiction. …

$$\frac{Fs, \neg G \vdash F}{Fs, \neg F \vdash G} \ (\neg\text{-K})$$

A negated assumption $\neg F$ may exchanged with the current goal $G$ by switching the negation symbols. In an informal proof, this may be stated as follows:

> We have to show $G$ (a). We know $\neg F$ (1). We assume $\neg G$ (2) and show $F$ (b). …

- ∧:

$$\frac{Fs \vdash G_1 \quad Fs \vdash G_2}{Fs \vdash G_1 \wedge G_2} \ (\wedge\text{-G})$$

To prove a conjunction $G_1 \wedge G_2$, both $G_1$ and $G_2$ have to be proved with the original knowledge $G$. In an informal proof, this may be indicated as follows:

> We have to show $G_1 \wedge G_2$ (a).
> - We prove $G_1$ (a.1). …
> - We prove $G_2$ (a.2). …

$$\frac{Fs, F_1, F_2 \vdash G}{Fs, F_1 \wedge F_2 \vdash G} \ (\wedge\text{-K})$$

The knowledge $F_1 \wedge F_2$ gives rise to the knowledge $F_1$ and $F_2$. In an informal proof, this may be indicated as follows:

> We know $F_1 \wedge F_2$ (1). We thus know $F_1$ (2) and $F_2$ (3).

- ∨:

$$\frac{Fs, \neg G_2 \vdash G_1}{Fs \vdash G_1 \vee G_2} \ (\vee\text{-G1}) \qquad \frac{Fs, \neg G_1 \vdash G_2}{Fs \vdash G_1 \vee G_2} \ (\vee\text{-G2})$$

To prove a disjunction $G_1 \vee G_2$, it suffices to prove one of the subformulas $G_1$ or $G_2$ under the additional assumption of the negation of the other subformula. In an informal proof, this may be stated as follows:

We have to prove $G_1 \vee G_2$ (a). We assume $\neg G_2$ (1) and prove $G_1$ (b). ...

$$\frac{Fs, F_1 \vdash G \quad Fs, F_2 \vdash G}{Fs, F_1 \vee F_2 \vdash G} \ (\vee\text{-K})$$

The knowledge $F_1 \vee F_2$ gives rise to a proof by case distinction, where the goal $G$ is proved once under the assumption $F_1$ and once under the assumption $F_2$. In an informal proof, this may be stated as follows:

We know $F_1 \vee F_2$ and proceed by case distinction:
- Case $F_1$: ...(proof of $G$ under assumption $F_1$).
- Case $F_2$: ...(proof of $G$ under assumption $F_2$).

- $\Rightarrow$:

$$\frac{Fs, G_1 \vdash G_2}{Fs \vdash G_1 \Rightarrow G_2} \ (\Rightarrow\text{-G})$$

To prove an implication $G_1 \Rightarrow G_2$, we assume $G_1$ and show $G_2$. In an informal proof, this may be stated as follows:

We have to prove $G_1 \Rightarrow G_2$ (a). We assume $G_1$ (1) and show $G_2$ (b). ...(proof of $G_2$ under the additional assumption $G_1$).

$$\frac{Fs \vdash F_1 \quad Fs, F_2 \vdash G}{Fs, F_1 \Rightarrow F_2 \vdash G} \ (\Rightarrow\text{-K})$$

To use knowledge $F_1 \Rightarrow F_2$, we may prove the goal $G$ under the assumption $F_2$ provided that we can show $F_1$. In an informal proof, this may be stated as follows:

We have to prove $G$. We know $F_1 \Rightarrow F_2$ (1). From ..., we know $F_1$ (2) and, from (1), therefore $F_2$ (2). ...(proof of $G$ under the additional assumption $F_2$).

- $\Leftrightarrow$:

$$\frac{Fs \vdash G_1 \Rightarrow G_2 \quad Fs \vdash G_2 \Rightarrow G_1}{Fs \vdash G_1 \Leftrightarrow G_2} \ (\Leftrightarrow\text{-G})$$

To show an equivalence $G_1 \Leftrightarrow G_2$, the implications $G_1 \Rightarrow G_2$ and $G_2 \Rightarrow G_1$ in both directions are shown. In an informal proof, this may be stated as follows:

We have to show $G_1 \Leftrightarrow G_2$ (a).
- We show $G_1 \Rightarrow G_2$ (a.1). ...
- We show $G_2 \Rightarrow G_1$ (a.2). ...

$$\frac{Fs, F_1 \Rightarrow F_2, F_2 \Rightarrow F_1 \vdash G}{Fs, F_1 \Leftrightarrow F_2 \vdash G} \quad (\Leftrightarrow\text{-K})$$

Correspondingly, the knowledge $F_1 \Leftrightarrow F_2$ gives rise to the knowledge $F_1 \Rightarrow F_2$ and $F_2 \Rightarrow F_1$. In an informal proof, this may be stated as follows:

> We know $F_1 \Rightarrow F_2$ (1). We thus know $F_1 \Rightarrow F_2$ (2) and $F_2 \Rightarrow F_1$ (3). ...

- if ... then ... else ...:

$$\frac{Fs \vdash G \Rightarrow G_1 \quad Fs \vdash \neg G \Rightarrow G_2}{Fs \vdash \text{if } G \text{ then } G_1 \text{ else } G_2} \quad (\text{if-G})$$

To show if $G$ then $G_1$ else $G_2$, we show the corresponding implications $F \Rightarrow F_1$ and $\neg F \Rightarrow F_2$. In an informal proof, this may be stated as follows:

> We have to show if $G$ then $G_1'$ else $G_2$ (a).
> - We show $G \Rightarrow G_1$ (a.1). ...
> - We show $\neg G \Rightarrow G_2$ (a.2). ...

$$\frac{Fs, F \Rightarrow F_1, \neg F \Rightarrow F_2 \vdash G}{Fs, \text{if } F \text{ then } F_1 \text{ else } F_2 \vdash G} \quad (\text{if-K})$$

Correspondingly, the knowledge if $F$ then $F_1$ else $F_2$ gives rise to the knowledge of the implications $F \Rightarrow F_1$ and $\neg F \Rightarrow F_2$. In an informal proof, this may be stated as follows:

> We know if $F$ then $F_1$ else $F_2$ (1). We thus know $F \Rightarrow F_1$ (2) and $\neg F \Rightarrow F_2$ (3). ...

**Quantifier Rules**

For dealing with quantifiers, we introduce the notation

$$F[T/V]$$

to denote a copy of formula $F$ where every free occurrence of variable $V$ is replaced by term $T$ (it is not difficult to formalize this notion by a primitively recursive function definition). However, not every such substitution is legal. Take for example the formula $F$ as

$$x \in \mathbb{N} \wedge \exists y \in \mathbb{N}. \, y + 1 = x.$$

which can be informally interpreted as '$x$ is a natural number different from zero'. Then the substitution $F[z/x]$ denotes the formula

$$z \in \mathbb{N} \wedge \exists y \in \mathbb{N}. \, y + 1 = z$$

which states the expected '$z$ is a natural number different from zero'. However, the substitution $F[y/y]$ denotes the formula

$$y \in \mathbb{N} \wedge \exists y \in \mathbb{N}.\, y + 1 = y$$

which is equivalent to false. The problem is that by the substitution the variable $y$ has been captured by the quantifier $\exists y$. In general, no free variable that occurs in $T$ may be captured by a quantifier in $F$. We formalize this constraint below.

**Definition 3.2** (*Free For*) A term $T$ is *free for* variable $V$ in formula $F$ if from the rules of Fig. 3.3 the judgement $T, V \vdash F$ : formula can be derived.

The rules for the quantifiers $\forall$, $\exists$, and let in Fig. 3.3 check that either the bound variable is $V$ (then $V$ does not occur freely in the phrase and no substitution occurs therein) or the locally bound variable does not occur among the free variables of $T$ (then the substitution is safe). Therefore, if $T$ is free for $V$ in $F$, a substitution $F[T/V]$ is safe. Please note that by renaming the locally bound variables of $F$, it is always possible to derive an equivalent formula $F'$ such that $T$ is free for $V$ in $F'$.

With this auxiliary definition, we may now introduce the rules for quantification.

- $\forall$:

$$\frac{(V_0 \notin \mathit{fv}(Fs),\ V_0 \notin \mathit{fv}(G) \backslash \{V\},\ V_0 \text{ free for } V \text{ in } G)}{Fs \vdash \forall V.\, G} \quad (\forall\text{-G})$$
$$Fs \vdash G[V_0/V]$$

To prove an universal goal $\forall V.\, G$, we introduce a 'fresh' variable $V_0$ and prove $G[V_0/V]$. This variable represents an arbitrary but fixed value from the domain of interpretation; if we can prove $G[V_0/V]$, $G$ thus holds for an arbitrary assignment to variable $V$. Technically, $V_0$ must not appear among the free variables of the knowledge, and (unless $V_0$ is $V$ itself) not among the free variables of the goal; furthermore $V_0$ must be free for $V$ in $G$ to allow the substitution $G[V_0/V]$. In an informal proof, this rule may be applied as follows:

We have to prove $\forall V.\, G$. We take arbitrary $V_0$ and prove $G[V_0/V]$. ...

$$\frac{(T \text{ free for } V \text{ in } F)}{Fs,\, \forall V.\, F \vdash G} \quad (\forall\text{-K})$$
$$Fs,\, F[T/V] \vdash G$$

A universal knowledge $\forall V.\, F$ allows us, for every term $T$ (that is free for $V$ in $F$) to generate the new knowledge formula $F[T/V]$. An universally quantified formula in the knowledge is thus like an 'engine' that may arbitrarily often be

**Rules for** $T, V \vdash T$ : **term:**

$$
T, V \vdash V : \text{term} \qquad T, V \vdash CS : \text{term} \qquad \frac{T, V \vdash Ts : \text{terms}}{T, V \vdash FS(Ts) : \text{term}}
$$

$$
\frac{(V = X \text{ or } X \notin fv(T))}{T, V \vdash T_1 : \text{term} \quad T, V \vdash T_2 : \text{term}}{T, V \vdash \text{let } X = T_1 \text{ in } T_2 : \text{term}}
$$

$$
\frac{T, V \vdash F : \text{formula} \quad T, V \vdash T_1 : \text{term} \quad T, V \vdash T_2 : \text{term}}{T, V \vdash \text{if } F \text{ then } T_1 \text{ else } T_2 : \text{term}}
$$

**Rules for** $T, V \vdash Ts$ : **terms:**

$$
\frac{T, V \vdash T : \text{term}}{T, V \vdash T : \text{terms}} \qquad \frac{T, V \vdash T : \text{term} \quad T, V \vdash Ts : \text{terms}}{T, V \vdash T, Ts : \text{terms}}
$$

**Rules for** $T, V \vdash F$ : **formula:**

$$
T, V \vdash \text{true} : \text{formula} \qquad T, V \vdash \text{false} : \text{formula} \qquad \frac{T, V \vdash Ts : \text{terms}}{T, V \vdash PS(Ts) : \text{formula}}
$$

$$
\frac{T, V \vdash T_1 : \text{term} \quad T, V \vdash T_2 : \text{term}}{T, V \vdash T_1 = T_2 : \text{formula}} \qquad \frac{T, V \vdash F : \text{formula}}{T, V \vdash \neg F : \text{formula}}
$$

$$
\frac{T, V \vdash F_1 : \text{formula} \quad T, V \vdash F_2 : \text{formula}}{T, V \vdash F_1 \wedge F_2 : \text{formula}} \qquad \frac{T, V \vdash F_1 : \text{formula} \quad T, V \vdash F_2 : \text{formula}}{T, V \vdash F_1 \vee F_2 : \text{formula}}
$$

$$
\frac{T, V \vdash F_1 : \text{formula} \quad T, V \vdash F_2 : \text{formula}}{T, V \vdash F_1 \Rightarrow F_2 : \text{formula}} \qquad \frac{T, V \vdash F_1 : \text{formula} \quad T, V \vdash F_2 : \text{formula}}{T, V \vdash F_1 \Leftrightarrow F_2 : \text{formula}}
$$

$$
\frac{(V = X \text{ or } X \notin fv(T))}{T, V \vdash F : \text{formula}}{T, V \vdash \forall X. \, F : \text{formula}} \qquad \frac{(V = X \text{ or } X \notin fv(T))}{T, V \vdash F : \text{formula}}{T, V \vdash \exists X. \, F : \text{formula}}
$$

$$
\frac{(V = X \text{ or } X \notin fv(T))}{T, V \vdash T : \text{term} \quad T, V \vdash F : \text{formula}}{T, V \vdash \text{let } X = T \text{ in } F : \text{formula}}
$$

$$
\frac{T, V \vdash F : \text{formula} \quad T, V \vdash F_1 : \text{formula} \quad T, V \vdash F_2 : \text{formula}}{T, V \vdash \text{if } F \text{ then } F_1 \text{ else } F_2 : \text{formula}}
$$

**Fig. 3.3** Term $T$ is free for variable $V$

invoked to generate new knowledge. In an informal proof, this rule may be applied as follows:

We know $\forall V. \, F$ (1). ... From (1), we know $F[T/V]$ (2). ...

- $\exists$:

$$
\frac{(T \text{ free for } V \text{ in } F)}{Fs \vdash F[T/V]}{Fs \vdash \exists V. \, F} \quad (\exists\text{-G})
$$

To prove an existential goal $\exists V.\ F$, it is sufficient to find a *witness term* $T$ that is free for $V$ in $F$ such that the goal $F[T/V]$ can be proved. In an informal proof, this rule may be applied as follows:

We have to prove $\forall V.\ F$ (a). We take $V := T$ and prove $F[T/V]$ (b). ...

$$\frac{Fs,\ F[V_0/V] \vdash G}{Fs,\ \exists V.\ F \vdash G} \ (\exists\text{-K})$$

$(V_0 \notin \mathit{fv}(Fs),\ V_0 \notin \mathit{fv}(F)\backslash\{V\},\ V_0$ free for $V$ in $F)$

An existential knowledge $\exists V.\ F$ allows us to introduce a 'fresh' variable $V_0$ for which $F[V_0/V]$ can be assumed. An existentially quantified formula is thus like an 'engine' that may be invoked once to generate new knowledge (it does not pay off to invoke the engine more than once, since it would just give us new variables; since all variables might denote the same value, this would not allow us to deduce anything we could not deduce from the first variable alone). Technically, $V_0$ must not appear among the free variables of the goal and the other knowledge formulas, and (unless $V_0$ is $V$ itself) not among the free variables of the existential knowledge; furthermore $V_0$ must be free for $V$ in $F$ to allow the substitution $F[V_0/V]$. In an informal proof, this rule may be applied as follows:

We know $\exists V.\ F$ (1). Let $V_0$ be such that $F[V_0/V]$ (2). ...

**Equality Rules**
These rules deal with knowledge about the equality predicate $=$.

$$Fs \vdash T = T \ (\text{reflexivity})$$

This rule states that every value is identical to itself, i.e., that equality is reflexive.

$$\frac{Fs \vdash G[T_1/V]}{Fs,\ T_1 = T_2 \vdash G[T_2/V]} \ (\text{substitution})$$

$(T_1$ free for $V$ in $G$, $T_2$ free for $V$ in $G)$

This rule allows us to substitute 'equals for equals'; if value $T_1$ equals value $T_2$, then in order to show goal $G$ for $T_2$, it suffices to show $G$ for $T_1$.

**Let Rules**
The remaining rules allow to reason about the $\mathsf{let}\ V = T\ \mathsf{in}\ \ldots$ construct.

$$\frac{Fs,\ V_0 = T_1 \vdash T = T_2[V_0/V]}{Fs \vdash T = \mathsf{let}\ V = T_1\ \mathsf{in}\ T_2} \ (\text{let-TG})$$

$(V_0 \notin \mathit{fv}(Fs) \cup \mathit{fv}(T) \cup \mathit{fv}(T_1),\ V_0 \notin \mathit{fv}(T_2)\backslash\{V\},\ V_0$ free for $V$ in $T_2)$

This rule allows us to prove a goal $T = \mathsf{let}\ V = T_1\ \mathsf{in}\ T_2$; in essence, we introduce a 'fresh' variable $V_0$ for which we assume $V_0 = T_1$ and then show $T = T_2[V_0/V]$.

This variable must not appear in the rest of the sequent and (unless it is $V$ itself) not in $T_2$; furthermore it must be free for $V$ in $T_2$ such that the substitution $T_2[V_0/V]$ is legal.

$$\frac{(V_0 \notin \mathit{fv}(Fs) \cup \mathit{fv}(T) \cup \mathit{fv}(T_1) \cup \mathit{fv}(G),\ V_0 \notin \mathit{fv}(T_2)\backslash\{V\},\ V_0 \text{ free for } V \text{ in } T_2)}{Fs,\ V_0 = T_1,\ T = T_2[V_0/V] \vdash G}$$
$$\overline{Fs,\ T = \text{let } V = T_1 \text{ in } T_2 \vdash G} \quad \text{(let-TK)}$$

This rule tells us that the knowledge $T = \text{let } V = T_1$ in $T_2$ gives rise to a 'fresh' variable $V_0$ for which we assume $V_0 = T_1$ and $T = T_2[V_0/V]$. This variable must not appear in the rest of the sequent and (unless it is $V$ itself) not in $T_2$; furthermore it must be free for $V$ in $T_2$ such that the substitution $T_2[V_0/V]$ is legal.

$$\frac{(V_0 \notin \mathit{fv}(Fs),\ V_0 \notin \mathit{fv}(G)\backslash\{V\},\ V_0 \text{ free for } V \text{ in } G)}{Fs,\ V_0 = T \vdash G[V_0/V]}$$
$$\overline{Fs \vdash \text{let } V = T \text{ in } G} \quad \text{(let-FG)}$$

This rule allows us to prove a goal $\text{let } V = T$ in $G$; in essence, we introduce a 'fresh' variable $V_0$ for which we assume $V_0 = T$ and then show $G[V_0/V]$. This variable must not appear in the rest of the sequent and (unless it is $V$ itself) not in $T$; furthermore it must be free for $V$ in $G$ such that the substitution $G[V_0/V]$ is legal.

$$\frac{(V_0 \notin \mathit{fv}(Fs) \cup \mathit{fv}(G),\ V_0 \notin \mathit{fv}(F)\backslash\{V\},\ V_0 \text{ free for } V \text{ in } F)}{Fs,\ V_0 = T,\ F[V_0/V] \vdash G}$$
$$\overline{Fs,\ \text{let } V = T \text{ in } F \vdash G} \quad \text{(let-FK)}$$

This rule tells us that the knowledge $\text{let } V = T$ in $F$ gives rise to a 'fresh' variable $V_0$ for which we assume $V_0 = T_1$ and $F[V_0/V]$. This variable must not appear in the rest of the sequent and (unless it is $V$ itself) not in $T$; furthermore it must be free for $V$ in $F$ such that the substitution $F[V_0/V]$ is legal.

**Derived Rules**

The rules presented in Figs. 3.1 and 3.2 are complete but they are not minimal. Several of the basic, structural, and propositional rules could be removed without affecting the completeness of the calculus; however, the calculus would be more cumbersome to use because a single inference step in the presented calculus would require several inference steps in the reduced calculus.

Furthermore, the rules presented in the calculus are not the only ones that are used in practice. For instance, a calculus might also include a variant of the classical 'modus ponens' rule

$$\frac{Fs,\ F_1,\ F_2 \vdash G}{Fs,\ F_1,\ F_1 \Rightarrow F_2 \vdash G} \quad \text{(mp)}$$

which combines two knowledge formulas $F_1$ and $F_1 \Rightarrow F_2$ to the knowledge formula $F_2$. While this rule is not included in our basic calculus, it is justified by the schematic proof tree

$$\frac{\overline{Fs,\ F_1 \vdash F_1} \ \text{(direct)} \quad Fs,\ F_1,\ F_2 \vdash G}{Fs,\ F_1,\ F_1 \Rightarrow F_2 \vdash G} \quad (\Rightarrow\text{-K})$$

which applies the rules (direct) and ($\Rightarrow$-K) of the basic calculus. This schematic tree yields a concrete partial proof tree for every instantiation of $Fs$, $F$, and $G$. By 'forgetting' the inner structure of this tree and removing its closed leaf, we derive the new rule (mp) that has the same root and the same open leaves. In practice, a lot of additional rules are in this way derived from the basic rules and added to the reasoning calculus. Another rule is derived in the following example.

**Example 3.1** The axiom

$$Fs, F, \neg F \vdash G \text{ (contradict)}$$

closes a proof if both a formula $F$ and its negation $\neg F$ is in the knowledge; it is an alternative to the rule (false-K) which closes a proof, if the knowledge false can be derived. This rule can be justified by following schematic proof tree

$$\cfrac{\cfrac{\cfrac{\cfrac{\overline{Fs, \neg F \vdash \neg F} \text{ (direct)}}{Fs, \neg F, F \vdash \text{false}} \text{ (false-G)}}{Fs, F, \neg F \vdash \text{false}} \text{ (front)}}{Fs, F, \neg F, \neg G \vdash \text{false}} \text{ (contract)}}{Fs, F, \neg F \vdash G} \text{ (indirect)}$$

that has the same root and no open leaf.                    □

**Constructing Proofs**

The proving calculus itself does not exhibit a strategy of how to successfully construct a proof for a true statement. Typically, however, a proof starts by applying the 'goal-oriented rules' named (_-G) according to the structure (the outermost quantifier respectively connective) of the goal formula; the goal is thus by a 'top-down' process decomposed into simpler subgoals. This process stops when the subgoals become atomic or when an existential quantifier is encountered (because the rule ($\exists$-G) demands the construction of a witness term that makes the body of the existential formula true).

After the top-down phase, typically a 'bottom-up' process starts where the 'knowledge-oriented rules' named (_-K) are applied to yield more and more knowledge, until the goal can be solved. Alternatively, at any step the rule (indirect) may be applied to start an indirect proof where, rather than showing the goal, it is shown that the assumption of the negation of the goal leads to a contradiction.

The following example illustrates these points by a small example proof that is given both as a formal proof tree and in the typical informal book style; both should be compared step by step.

Proof:

$$
\cfrac{
\cfrac{
\text{(Case 1)} \quad \cfrac{\text{(Cut 1)} \quad \text{(Cut 2)}}{(1),(2),(3), r(x_0) \vdash \exists y.\, q(x_0, y)}\ \text{(cut)}
}{(1),(2),(3), p(x_0) \vee r(x_0) \vdash \exists y.\, q(x_0, y)}\ \text{($\vee$-K)}
}{
\cfrac{(1),(2),(3) \vdash (p(x_0) \vee r(x_0)) \Rightarrow \exists y.\, q(x_0, y)}{(1),(2),(3) \vdash \text{(a)}}\ \text{($\forall$-G)}
}\ \text{($\Rightarrow$-G)}
$$

Case 1:

$$
\cfrac{
\cfrac{
\cfrac{
\cfrac{(2),(3), p(x_0), q(x_0, f(x_0)) \vdash q(x_0, f(x_0))}{(2),(3), p(x_0), q(x_0, f(x_0)) \vdash \exists y.\, q(x_0, y)}\ \text{(direct)}
}{(2),(3), p(x_0), p(x_0) \Rightarrow q(x_0, f(x_0)) \vdash \exists y.\, q(x_0, y)}\ \text{($\exists$-G)}
}{(2),(3), p(x_0), (1) \vdash \exists y.\, q(x_0, y)}\ \text{(mp)}
}{(1),(2),(3), p(x_0) \vdash \exists y.\, q(x_0, y)}\ \text{(front)}
$$

Cut 1:

$$
\cfrac{
\cfrac{
\cfrac{
\cfrac{
\cfrac{(1),(3), s(x_0), r(x_0), \neg r(x_0) \vdash \text{false}}{(1),(3), s(x_0), \neg r(x_0), r(x_0) \vdash \text{false}}\ \text{(contradict)}
}{(1),(3), r(x_0), s(x_0), \neg r(x_0) \vdash \text{false}}\ \text{(front)}
}{(1),(3), r(x_0), s(x_0), s(x_0) \Rightarrow \neg r(x_0) \vdash \text{false}}\ \text{(front)}
}{(1),(3), r(x_0), s(x_0), (2) \vdash \text{false}}\ \text{(mp)}
}{
\cfrac{(1),(2),(3), r(x_0), s(x_0) \vdash \text{false}}{(1),(2),(3), r(x_0) \vdash \neg s(x_0)}\ \text{(front)}
}\ \text{($\vee$-K)}
\quad\text{($\neg$-G)}
$$

Cut 2:

$$
\cfrac{
\cfrac{
\cfrac{
\cfrac{(1),(2), r(x_0), \neg s(x_0), q(x_0, a) \vdash q(x_0, a)}{(1),(2), r(x_0), \neg s(x_0), q(x_0, a) \vdash \exists y.\, q(x_0, y)}\ \text{(direct)}
}{(1),(2), r(x_0), \neg s(x_0), \neg s(x_0) \Rightarrow q(x_0, a) \vdash \exists y.\, q(x_0, y)}\ \text{($\exists$-G)}
}{(1),(2), r(x_0), \neg s(x_0), (3) \vdash \exists y.\, q(x_0, y)}\ \text{(mp)}
}{(1),(2),(3), r(x_0), \neg s(x_0) \vdash \exists y.\, q(x_0, y)}\ \text{(front)}
$$

**Fig. 3.4** Formal proof of Example 3.2

**Example 3.2** We prove that the knowledge

$$\forall x.\ p(x) \Rightarrow q(x, f(x)) \tag{3.1}$$

$$\forall x.\ s(x) \Rightarrow \neg r(x) \tag{3.2}$$

$$\forall x.\ \neg s(x) \Rightarrow q(x, a) \tag{3.3}$$

implies the goal

$$\forall x.\ ((p(x) \vee r(x)) \Rightarrow \exists y.\, q(x, y)). \tag{a}$$

The formal proof (which uses the additional rules (mp) and (contradict) derived above) is depicted in Fig. 3.4. Informally, this proof can be stated as follows: In order to prove (a), we take an arbitrary value $x_0$ and assume

$$p(x_0) \vee r(x_0) \tag{3.4}$$

to show

$$\exists y. \, q(x_0, y). \qquad \text{(b)}$$

From (3.4), we have two cases:

- Case $p(x_0)$: From this and knowledge (3.1), we know $q(x_0, f(x_0))$. Thus goal (b) holds for $y := f(x_0)$.
- Case $r(x_0)$: First we show the additional knowledge

$$\neg s(x_0) \qquad \text{(3.5)}$$

by assuming $s(x_0)$ and deriving a contradiction. From (3.2), we know $s(x_0) \Rightarrow \neg r(x_0)$ and thus $\neg r(x_0)$ which contradicts the case assumption.
We thus know (3.5) and with (3.3) also $q(x_0, a)$. Therefore (b) holds for $y := a$.

$$\square$$

Given the difficulty of constructing proofs for complex propositions, we may desire to delegate the task to an automatic procedure. However, this approach is limited by the following theorem which is due to Alonzo Church and Alan Turing.

**Theorem 3.3** (Undecidability of first-order logic) *There is no algorithm that, given a first-order formula as an input, always terminates with an output such that the output is 'valid' if and only if the formula is valid.*

While it is therefore not possible to automatically decide whether a first-order formula is valid or not, we have the following result.

**Theorem 3.4** (Semi-decidability of first-order logic) *There is an algorithm that, given a first-order formula as an input, always terminates with output 'valid' if the formula is valid. However, if the formula is not valid, the algorithm may not terminate.*

This theorem is based on the completeness of first-order logic, i.e., every valid formula has a proof, and the fact that all possible proofs can be enumerated, i.e., there exists a procedure that runs forever and generates every possible proof. The semi-decision algorithm takes every generated proof and checks whether it is a proof of the formula that it was given as input. Given a valid formula as input, the algorithm will therefore eventually encounter its proof and terminate with output 'valid'. However, if the formula is not valid, the algorithm has no way of saying when to stop its fruitless search and will therefore run forever.

As one can imagine, proving by enumerating proofs and checking them is much too inefficient in practice. Automated theorem provers therefore operate in a much more 'goal-directed' fashion by investigating the structure of the formulas to be proved and by applying sophisticated heuristics in order to find proofs in a limited amount of time with a reasonable chance of success; this search may be interrupted by a human or by a timer when a predetermined time bound has been reached. In contrast to fully automated provers, interactive proving assistants operate in close cooperation with a human; the human decides which proof strategy to apply, which additional knowledge (previously proved propositions) to use, and also provides creative insight at critical points of the proof (e.g., by determining the witness terms of existential goals); those steps of the proof that do not require much creativity are left to the machine. Whether generated by automatic provers or by a collaboration of human and machine: the proofs constructed with computer support do not suffer from human errors or laziness but are guaranteed to be correct and complete.

Another famous result (due to Kurt Gödel) fundamentally limits the expressiveness of first-order logic (indeed of every logic):

> **Theorem 3.5** (Incompleteness theorem) *It is in no sound logic possible to prove all true arithmetic statements, i.e., all statements about natural numbers with addition and multiplication.*

Consequently, since first-order logic *is* complete, this logic is not sufficiently rich to adequately describe and reason about mathematical domains such as the natural numbers. Therefore the inference rules of first-order logic are augmented by additional proof principles that are nevertheless sound over these domains. We are now going to discuss the most important such principle.

## 3.4    Reasoning by Induction

The proof rules of first-order logic represent the basis of precise reasoning. However, in various domains which play a major role in computer science, the additional proof principle of *induction* is required. Depending on the domain of interest, the principle may appear in various concrete forms; some of the most important forms are described below.

### Mathematical Induction
The domain of natural numbers $0, 1, 2, \ldots$ cannot be precisely characterized by first-order logic: for every sequence $F_1, \ldots, F_n$ of first-order formulas that are true when

interpreted over the natural numbers, there are infinitely many formulas $G$ that are also true for the natural numbers but for which the sequent $F_1, \ldots, F_n \vdash G$ is not derivable. Roughly speaking, the reason is that for every such sequent there also exists another domain $D$ which is 'bigger' than the natural numbers and in which also the formulas $F_1, \ldots, F_n$ are true but formula $G$ is not. By the soundness of the first-order calculus, sequent $F_1, \ldots, F_n \vdash G$ is only derivable, if formula $F_1 \wedge \ldots \wedge F_n \Rightarrow G$ is true in every domain; since this is not the case for $D$, the sequent cannot be derived. Indeed most truths of the natural numbers cannot be proved by the inference rules of first-order logic.

Nevertheless, for reasoning in the domain of natural numbers, we may introduce *(mathematical) induction* as an additional rule which is sound and (while it does not make the calculus complete) suffices for mathematical practice:

$$\frac{Fs \vdash G[0/n] \quad Fs \vdash \forall n.\,(G \Rightarrow G[n+1/n])}{Fs \vdash \forall n.\,G} \quad \text{(mathematical induction)}$$

Actually, (induction) is not a rule but a *rule schema* with infinitely many instantiations of concrete formulas for the schematic variable $G$. This schema states that, in order to prove that $G$ holds for all natural numbers, it is sufficient to show that

- $G$ holds for 0, and
- if $G$ holds for arbitrary $n$, then it also holds for $n+1$.

Here we assume that

- the domain of interpretation are the natural numbers,
- the constant symbol 0 is interpreted as the first natural number 'zero', and
- the unary function symbol $+1$ is interpreted as the 'successor' function that maps every natural number to the next bigger number.

More generally, if the domain includes also other elements than the natural numbers, we use the alternative schema

$$\frac{Fs \vdash G[0/n] \quad Fs \vdash \forall n \in \mathbb{N}.\,(G \Rightarrow G[n+1/n])}{Fs \vdash \forall n \in \mathbb{N}.\,G} \quad \text{(mathematical induction)}$$

where we assume that the unary predicate symbol $\in \mathbb{N}$ is interpreted as the 'is a natural number' predicate.

The practical application of induction in the informal proof style is demonstrated by the following example.

**Example 3.3** We prove the formula due to Carl Friedrich Gauß

$$\forall n \in \mathbb{N}. \ \sum_{i=1}^{n} i = n \cdot (n+1)/2$$

where $\sum_{i=1}^{n} T$ denotes the sum $T[1/i] + T[2/i] + \ldots + T[n/i]$, e.g. $\sum_{i=1}^{3} i = 1 + 2 + 3$. We proceed by induction on $n$.

- *Base case:* we prove

$$\sum_{i=1}^{0} i = 0 \cdot (0+1)/2$$

which is true because both sides of the equation yield 0.
- *Induction step:* we assume for arbitrary $n \in \mathbb{N}$ the *induction hypothesis*

$$\sum_{i=1}^{n} i = n \cdot (n+1)/2$$

and show

$$\sum_{i=1}^{n+1} i = (n+1) \cdot (n+2)/2.$$

This goal is true because of the derivation

$$\sum_{i=1}^{n+1} i = (n+1) + \sum_{i=1}^{n} i \overset{(*)}{=} (n+1) + n \cdot (n+1)/2$$
$$= (2n+2+n^2+n))/2 = (n^2+3n+2)/2$$
$$= (n+1) \cdot (n+2)/2$$

where (*) holds because of the induction hypothesis.                                         □

Mathematical induction can be further generalized to *complete induction* (also called *strong induction* or *course of values induction*):

$$\frac{Fs \vdash \forall n \in \mathbb{N}. \ \big((\forall m \in \mathbb{N}. \ m < n \Rightarrow G[m/n]) \Rightarrow G\big)}{Fs \vdash \forall n \in \mathbb{N}. \ G} \quad \text{(complete induction)}$$

This principle states that in order to prove that $G$ holds for every natural number $n$, it suffices to prove that $G$ holds for arbitrary $n$ under the assumption that it holds for every smaller number $m < n$ (since there is no natural number less than 0, the instance $n = 0$ implicitly represents the induction base).

**Example 3.4**  We prove the goal

$$\forall n \in \mathbb{N}. \ 1 < n \Rightarrow \exists a \in \mathbb{N}, b \in \mathbb{N}. \ prime(a) \wedge n = a \cdot b$$

where $prime(n)$ is logically equivalent to the formula

$$\neg \exists n_1 \in \mathbb{N}, n_2 \in \mathbb{N}. \ 1 < n_1 \wedge n_1 < n \wedge n = n_1 \cdot n_2$$

which can be interpreted as '$n$ is a prime number'. The goal thus claims that every natural number greater than one has a prime number as a factor.

We proceed by complete induction: we take arbitrary $n \in \mathbb{N}$ and assume as the induction hypothesis

$$\forall m \in \mathbb{N}. \ m < n \Rightarrow \left(1 < m \Rightarrow \exists a \in \mathbb{N}, b \in \mathbb{N}. \ prime(a) \wedge m = a \cdot b\right).$$

We have then to show

$$1 < n \Rightarrow \exists a \in \mathbb{N}, b \in \mathbb{N}. \ prime(a) \wedge n = a \cdot b.$$

We thus assume

$$1 < n$$

and show

$$\exists a \in \mathbb{N}, b \in \mathbb{N}. \ prime(a) \wedge n = a \cdot b. \tag{a}$$

Since $prime(n)$ or $\neg prime(n)$, we have two cases:

- Case $prime(n)$: we know (a) with $a := n$ and $b := 1$.
- Case $\neg prime(n)$: we have from the knowledge about $prime(n)$ some $n_1 \in \mathbb{N}$ and $n_2 \in \mathbb{N}$ with $1 < n_1$ and $n_1 < n$ and $n = n_1 \cdot n_2$. From the induction hypothesis (by substitution of $n_1$ for $m$), we then know

$$n_1 < n \Rightarrow (1 < n_1 \Rightarrow \exists a \in \mathbb{N}, b \in \mathbb{N}. \ prime(a) \wedge n_1 = a \cdot b).$$

Since $1 < n_1$ and $n_1 < n$, we thus have some $a_1 \in \mathbb{N}$ and $b_1 \in \mathbb{N}$ with

$$prime(a_1) \wedge n_1 = a_1 \cdot b_1.$$

We thus know $n = n_1 \cdot n_2 = a_1 \cdot b_1 \cdot n_2$. From this and $prime \ (a_1)$ we know (a) with $a := a_1$ and $b := b_1 \cdot n_2$. $\qquad\qquad\square$

**Structural Induction**

Also other domains than the natural numbers support corresponding induction principles. An important example are syntactical expressions respectively abstract syntax trees for which we have already introduced in Sect. 1.2 the principle of *structural induction* which we will now rephrase and generalize.

Assume we are given a grammar as described in Definition 1.1 where we would like to prove a formula of form

$$(\forall x_1 \in Domain_1. \ F_1) \wedge \ldots \wedge (\forall x_n \in Domain_n. \ F_n)$$

i.e. a conjunction of universal quantifications, one for each syntactic domain defined by the grammar. We can achieve this by *simultaneous* induction over all rules of the grammar.

Since the formulation of the general rule is quite complicated, we resort to a concrete example. Consider the grammar

$$A \in DA$$
$$B \in DB$$
$$A ::= \text{a} \ | \ \text{a1}(A) \ | \ \text{a2}(A, B)$$
$$B ::= \text{b} \ | \ \text{b1}(A) \ | \ \text{b2}(B_1, B_2)$$

where both syntax domains are defined by rules that mutually recursively refer to the other domain. This syntax domain gives rise to the following induction scheme.

$$
\begin{array}{c}
Fs \vdash G_1[\text{a}/A] \\
Fs \vdash \forall A \in DA. \ G_1 \Rightarrow G_1[\text{a1}(A)/A] \\
Fs \vdash \forall A \in DA, B \in DB. \ G_1 \wedge G_2 \Rightarrow G_1[\text{a2}(A, B)/A] \\
Fs \vdash G_2[\text{b}/B] \\
Fs \vdash \forall A \in DA. \ G_1 \Rightarrow G_2[\text{b1}(A)/B] \\
Fs \vdash \forall B_1 \in DB, B_2 \in DB. \\
G_2[B_1/B] \wedge G_2[B_2/B] \Rightarrow G_2[\text{b2}(B_1, B_2)/B] \\
\hline
Fs \vdash (\forall A \in DA. \ G_1) \wedge (\forall B \in DB. \ G_2)
\end{array}
\qquad \left( \begin{array}{l} \text{structural} \\ \text{induction} \end{array} \right)
$$

In other words, we want to simultaneously prove that each term of domain $DA$ satisfies property $G_1$ and each term of domain $DB$ satisfies property $G_2$. For this purpose, we prove each of the goals $G_1[\text{a}/A]$, $G_1[\text{a1}(A)/A]$, $G_1[\text{a2}(A, B)/A]$, $G_2[\text{b}/B]$ (one for each alternative of the first rule) and the goals $G_2[\text{b1}(A)/B]$ and $G_2[\text{b2}(B_1, B_2)/B]$ (one for each alternative of the second rule); the proofs use the induction hypothesis that property $G_1$ already holds for each subterm from domain $DA$ and that property $G_2$ holds for each subterm from domain $DB$.

**Example 3.5** Consider the grammar

$A \in DA$
$B \in DB$
$A ::= 0 \ \mid \ 0(A_1, A_2) \ \mid \ 0(B)$
$B ::= 00 \ \mid \ 1(B_1, B_2) \ \mid \ 0(A)$

and the function

$$c: DA \cup DB \to \mathbb{N}$$
$$c(0) := 1$$
$$c(0(A_1, A_2)) := 1 + c(A_1) + c(A_2)$$
$$c(0(B)) := 1 + c(B)$$
$$c(00) := 2$$
$$c(1(B_1, B_2)) := c(B_1) + c(B_2)$$
$$c(0(A)) := 1 + c(A)$$

which counts the number of occurrences of symbol 0 in a phrase. We claim

$$(\forall A \in DA.\ 2 \nmid c(A)) \wedge (\forall B \in DB.\ 2 \mid c(B))$$

where | denotes the 'divides' relation. In other words, we want to prove that every term of $DA$ has an odd number of occurrences of symbol 0 and every element of domain $DB$ has an even number of occurrences.

The proof of this claim is based on the central properties

$$\forall n \in \mathbb{N}.\ 2 \nmid n \Leftrightarrow 2 \mid (n + 1)$$
$$\forall n_1 \in \mathbb{N}, n_2 \in \mathbb{N}.\ 2 \mid n_1 \wedge 2 \mid n_2 \Rightarrow 2 \mid (n_1 + n_2)$$

and proceeds by structural induction:

- Case 0: $c(0) = 1$ and $2 \nmid 1$.
- Case $0(A_1, A_2)$: we assume $2 \nmid c(A_1)$ and $2 \nmid c(A_2)$ which is equivalent to

$$2 \mid c(A_1) + 1 \tag{3.6}$$
$$2 \mid c(A_2) + 1 \tag{3.7}$$

and show $2 \nmid c(0(A_1, A_2))$, i.e., $2 \nmid 1 + c(A_1) + c(A_2)$ which is equivalent to

$$2 \mid 2 + c(A_1) + c(A_2)$$

Since $2 + c(A_1) + c(A_2) = (1 + c(A_1)) + (1 + c(A_2))$, this goal is implied by (3.6) and (3.7).
- Case $0(B)$: we assume $2 \mid c(B)$ and show $2 \nmid c(0(B))$, i.e., $2 \nmid 1 + c(B)$ which is implied by the induction hypothesis.

- Case 00: $c(00) = 2$ and $2 \mid 2$.
- Case $1(B_1, B_2)$: we assume $2 \mid c(B_1)$ and $2 \mid c(B_2)$ and show $2 \mid c(1(B_1, B_2))$, i.e., $2 \mid c(B_1) + c(B_2)$ which is implied by the induction hypothesis.
- Case $0(A)$: we assume $2 \nmid c(A)$ and show $2 \mid c(0(A))$, i.e., $2 \mid 1 + c(A)$ which is implied by the induction hypothesis.                                                  □

Like for mathematical induction, also a generalization of the principle to a form of *complete induction* is possible, we omit the details.

In fact, mathematical induction is just a special case of structural induction since the domain of natural numbers is in one-to-one correspondence to the elements of the grammar

$$N \in Nat$$
$$N ::= 0 \mid +1(N)$$

where the terms $0, +1(0), +1(+1(0)), \ldots$ correspond to the natural numbers $0, 1, 2, \ldots$.

### Rule Induction

As indicated in Sect. 1.4, it is usually not a grammar but a rule-based type system that defines the phrases of a language that we wish to reason about. In general, given a rule-based inference system, we might wish to reason about all the judgements that can be derived from the rules of the system, i.e., all the judgements that are the roots of complete inference trees in this system. For this purpose, we have already sketched in Sect. 1.5 the principle of *rule induction* which we will now formulate more precisely.

Let $R$ be a set of inference rules of form

$$\frac{x_1 \quad \ldots \quad y_n}{y}$$

and let $\vdash_R j$ state that judgement $j$ can be derived by the rules in $R$, i.e., there exists a derivation tree whose root is $j$ and in which every node is an instance of some rule in $R$. Then the principle of rule induction can be described by the following scheme.

$$\frac{\forall \frac{x_1 \ldots x_n}{y} \in R. \; (P[x_1/j] \wedge \ldots \wedge P[x_n/j] \Rightarrow P[y/j])}{\forall j. \; (\vdash_R j \Rightarrow P)} \quad \text{(rule induction)}$$

In other words, to prove that property $P$ is true for every judgement derivable from the rules of $R$, it suffices to prove for every rule in $R$ that $P$ holds for the conclusion $y$ of the rule under the assumption that $P$ already holds for its premises $x_1, \ldots, x_n$.

**Example 3.6** Consider the rule system $R$ consisting of the the two rules

$$\frac{(m = n)}{c \to \mathsf{tick}(m, n): c} \ (\mathrm{t1}) \qquad \frac{c + 1 \to \mathsf{tick}(m + 1, n): c' \quad (m \neq n)}{c \to \mathsf{tick}(m, n): c'} \ (\mathrm{t2})$$

where the judgement $c \to \mathsf{tick}(m, n): c'$ can be interpreted as

> If a timer that runs from tick $m$ to tick $n$ has already ticked $c$ times, then it rings after it has ticked in total $c'$ times.

First we prove

$$\forall c \in \mathbb{N}, m \in \mathbb{N}, n \in \mathbb{N}, c' \in \mathbb{N}. \ (\vdash_R c \to \mathsf{tick}(m, n): c') \Rightarrow c' = c + n - m.$$

i.e., if the timer rings, it has accumulated the expected total time. We proceed by rule induction.

- Rule t1: we assume

$$m = n$$

  and show

$$c = c + (n - m)$$

  which follows from the assumption.
- Rule t2: we assume

$$c' = (c + 1) + n - (m + 1) \wedge m \neq n$$

  and show

$$c' = c + n - m$$

  which follows from the assumption.

Thus the proof is completed. Next we prove

$$\forall c \in \mathbb{N}, m \in \mathbb{N}, n \in \mathbb{N}, c' \in \mathbb{N}. \ (\vdash_R c \to \mathsf{tick}(m, n): c') \Rightarrow m \leq n$$

i.e., the timer only rings if its starting time $m$ is not bigger than its end time $n$. Again we proceed by rule induction:

- Rule t1: we assume

$$m = n$$

  and show

$$m \leq n$$

  which follows from the assumption.

- Rule t2: we assume

$$m \neq n \wedge m + 1 \leq n$$

and show

$$m \leq n$$

which follows from the assumption.                                                      □

Structural induction is just a special case of rule induction, since every grammar can be also described by a rule-based inference system.

## Exercises

Download from the following URL:
https://www.risc.jku.at/people/schreine/TP/exercises/ex-reasoning.pdf.

## Further Reading

Various books listed in the 'Further Reading' section of Chap. 2 describe the proof theory of first order logic, either in a semi-formal formal style or in some specific formal calculus. Huth's and Ryan's book [81] uses for this purpose 'natural deduction' which differs from sequent calculus that a judgement is just a single formula, which may seem conceptually simpler but makes actually the formulation of those rules more complex that require reasoning from assumptions.

Most textbooks on formal logic present some sort of sequent calculus, e.g. Gallier's book [56], Li's book [102], or Ben-Ari's book [13]. Usually sequent calculus is presented in a form where a sequent $Fs \vdash Gs$ allows multiple formulas in the goal position; in our presentation we have preferred a more restricted form with a single goal which closer corresponds to mathematical practice.

Comparatively few books deal with the practice of mathematical proof, see e.g. Buchberger's and Lichtenberger's book [29] or the more recent book of Velleman [178]. The 'Further Reading' section of Chap. 1 lists several books on language semantics that deal with the principle of structural induction and rule induction.

## Reasoning with the RISC ProofNavigator

In this section, we continue or presentation of the RISC ProofNavigator started in section "The Logic of the RISC ProofNavigator", Chap. 2 with a demonstration how to direct the proof assistant to prove some non-trivial propositions. The specifications used for this purpose can be downloaded from the URL

  https://www.risc.jku.at/people/schreine/TP/software/reasoning/reasoning.txt

and loaded by executing from the command line the following command:

```
ProofNavigator reasoning.txt &
```

A Direct Proof

We start with the following declarations of an abstract type $T$ and some predicates $p$, $q$, and $r$ on $T$:

```
T: TYPE;
p: T -> BOOLEAN;
q: T -> BOOLEAN;
r: (T,T) -> BOOLEAN;
```

From these declarations, the following formula is to be proved:

```
F: FORMULA
   (FORALL(x:T): p(x) OR r(x,x)) AND
   (FORALL(x:T): p(x) => (EXISTS(y:T): q(y))) AND
   (FORALL(x,y:T): p(x) AND q(y) => r(x,y)) =>
     (FORALL(x:T): EXISTS(y:T): r(x,y));
```

After selecting prove F from the button menu and executing the command scatter, the following view is displayed in the 'Proof State' area:

| 2i| | $(\forall x \in T: p(x) \lor r(x,x)) \land (\forall x \in T: p(x) \Rightarrow (\exists y \in T: q(y)))$ |
| --- | --- |
| | $\land$ |
| | $(\forall x \in T, y \in T: p(x) \land q(y) \Rightarrow r(x,y))$ |
| | $\Rightarrow$ |
| | $(\forall x \in T: \exists y \in T: r(x,y))$ |

This view displays a proof sequent where the knowledge formulas (here none) are displayed above the line and the goal formula is displayed below the line.

As it is typical at the beginning of the proof, we would like to apply the various rules that decompose the goal formula. For this purpose, the scatter command repeatedly applies several goal-decomposition rules, which yields the following proof state (the red bars on the right indicate those formulas that did not appear in the previous state):

| eva | $\forall x \in T : p(x) \vee r(x, x)$ |
| xnf | $\forall x \in T : p(x) \Rightarrow (\exists y \in T : q(y))$ |
| 5rx | $\forall x \in T, y \in T : p(x) \wedge q(y) \Rightarrow r(x, y)$ |

| pgq | $\exists y \in T : r(x_0, y)$ |

Here the rule ($\forall$-G) has been applied to generate a new variable $x_0$ replacing the bound variable $x$. Furthermore, the rule ($\Rightarrow$-G) has been applied to decompose the implication by making the antecedent a knowledge formula and the consequent the new goal. To the new conjunction in the knowledge, scatter has also applied twice the rule ($\wedge$-K) and thus broken up the conjunction into its parts.

The goal now is to find a suitable partner $y$ for $x_0$ to make $r$ true. From the knowledge formula labeled [eva], we see that potentially $x_0$ may itself serve as $y$. We thus invoke the command instantiate x_0 in eva (the template of this command can be selected from the formula button); this applies the rule ($\forall$-K) by instantiating the quantified variable $x$ with the term $x_0$ to yield the following situation:

| eva | $\forall x \in T : p(x) \vee r(x, x)$ |
| xnf | $\forall x \in T : p(x) \Rightarrow (\exists y \in T : q(y))$ |
| 5rx | $\forall x \in T, y \in T : p(x) \wedge q(y) \Rightarrow r(x, y)$ |
| 3ym | $p(x_0) \vee r(x_0, x_0)$ |

| pgq | $\exists y \in T : r(x_0, y)$ |

Here from the button menu of formula [3ym] the command split 3ym may be selected that applies the rule ($\vee$-K) to split the proof in two cases. We may navigate to the second case by the application of the button ➡ which yields this proof state:

| eva | $\forall x \in T : p(x) \vee r(x, x)$ |
| xnf | $\forall x \in T : p(x) \Rightarrow (\exists y \in T : q(y))$ |
| 5rx | $\forall x \in T, y \in T : p(x) \wedge q(y) \Rightarrow r(x, y)$ |
| fbb | $r(x_0, x_0)$ |

| pgq | $\exists y \in T : r(x_0, y)$ |

Here we invoke the auto command which applies to the universally quantified knowledge formulas the rule ($\forall$-K) by automatically instantiating these
formulas with various terms that appear elsewhere in the proof situation.
Indeed, the 'right' instantiations are found and the case can be closed. The
other case, however, depicted as

| eva | $\forall x \in T : p(x) \lor r(x, x)$ |
|-----|------------------------------------------|
| xnf | $\forall x \in T : p(x) \Rightarrow (\exists y \in T : q(y))$ |
| 5rx | $\forall x \in T, y \in T : p(x) \land q(y) \Rightarrow r(x, y)$ |
| cl4 | $p(x_0)$ |
| pgq | $\exists y \in T : r(x_0, y)$ |

still requires manual intervention. By executing the command instantiate
x_0 in xnf, we apply the rule ($\forall$-K) to formula [xnf] in order to derive the
situation

| eva | $\forall x \in T : p(x) \lor r(x, x)$ |
|-----|------------------------------------------|
| xnf | $\forall x \in T : p(x) \Rightarrow (\exists y \in T : q(y))$ |
| 5rx | $\forall x \in T, y \in T : p(x) \land q(y) \Rightarrow r(x, y)$ |
| cl4 | $p(x_0)$ |
| bsm | $\exists y \in T : q(y)$ |
| pgq | $\exists y \in T : r(x_0, y)$ |

which exposes the existential formula in the knowledge. The command
scatter then applies rule ($\exists$-K) to derive

| eva | $\forall x \in T : p(x) \lor r(x, x)$ |
|-----|------------------------------------------|
| xnf | $\forall x \in T : p(x) \Rightarrow (\exists y \in T : q(y))$ |
| 5rx | $\forall x \in T, y \in T : p(x) \land q(y) \Rightarrow r(x, y)$ |
| cl4 | $p(x_0)$ |
| 1fo | $q(y_0)$ |
| pgq | $\exists y \in T : r(x_0, y)$ |

from where the command auto is able to complete the proof (by applying
rule ($\forall$-K) to formula [5rx] with $x := x_0$ and $y := y_0$).

During the whole proving process, the current state of the proof tree was
displayed as a nested menu in the 'Proof Tree' area; navigation to open proof
situations was possible by double-clicking the corresponding node in the tree.
Ultimately, this proof tree had shape

[gca]: scatter
   ⊟ [yxh]: instantiate x_0 in eva
      ⊟ [wcy]: split 3ym
         ⊟ [i41]: instantiate x_0 in xnf
            ⊟ [gxa]: scatter
               ⊟ [yoy]: auto
                  [w62]: proved (CVC3)
         ⊟ [j41]: auto
           [4aj]: proved (CVC3)

which displays in the inner nodes every command that was interactively applied by the user and in the leaves the actions of the automatic SMT solver CVC3.

An Indirect Proof

We now present a formalization of Euclid's indirect proof that there exist infinitely many prime numbers. In a nutshell, the argument is as follows:

Suppose there exist only finitely many prime numbers $p_1 = 2 < p_2 < \ldots < p_n$. Let $P := 1 + p_1 \cdot p_2 \cdot \ldots \cdot p_n$. Then $P$ may be prime which, since $P > p_n$, contradicts our assumption. Otherwise, there must exist another prime $p$ that divides $P$. Since none of the $p_1, \ldots, p_n$ divides $P$, we again have a contradiction.

For a formalization, we introduce the operations

```
divides:  (NAT,NAT)->BOOLEAN;
prime:    NAT->BOOLEAN;
pproduct: NAT->NAT;
```

where *divides* and *prime* denote the predicates 'divides' and 'is prime' while *pproduct* denotes the function that maps every natural number $n$ to the product of all primes less than or equal $n$. Rather than defining these operations explicitly, we describe them by some characterizing formulas:

```
P1: AXIOM
   FORALL(n:NAT, m:NAT): n > 1 AND divides(n,m) => NOT divides(n,m+1);
P2: AXIOM
   prime(2) AND (FORALL(n:NAT): n < 2 => NOT prime(n));
P3: AXIOM
   FORALL(n:NAT): n > 1 => (EXISTS(p:NAT): prime(p) AND divides(p,n));
P4: AXIOM
```

```
    pproduct(2) = 2 AND (FORALL(n:NAT): n > 2 => pproduct(n) > n);
P5: AXIOM
    FORALL(n:NAT,p:NAT): p <= n AND prime(p) => divides(p,pproduct(n));
```

These formulas state that

1. no number $n$ greater than one divides simultaneously both $m$ and $m + 1$;
2. 2 is a prime and there is no prime less than 2;
3. every number greater than one has a prime divisor;
4. the product of all primes up to 2 is itself 2 and, for every number $n$ greater than 2, the product of all primes up to $n$ is greater than $n$;
5. the product of all primes up to $n$ is divided by every prime up to $n$.

The formulas are declared by the keyword AXIOM which implies that there are automatically added as knowledge in every subsequent proof. In a more detailed proof, we would give explicit definitions of *divides*, *prime*, and *pproduct*, and use the keyword FORMULA to introduce $P_1$–$P_6$ as normal formulas that need to be proved. Afterwards the proof command lemma can be used to introduce the formulas as knowledge in any other proof.

Anyway, our goal is now to prove the formula

```
P: FORMULA NOT (EXISTS(n:NAT): (FORALL(p:NAT): prime(p) => p <= n));
```

which states that there does not exist any upper bound $n$ for all prime numbers. The structure of the proof is exhibited by the following proof tree:

```
⊟ [qca]: scatter
    ⊟ [55z]: flip s2g
        ⊟ [a11]: case prime(pproduct(n_0)+1)
            ⊟ [orv]: auto
                [q4k]: proved (CVC3)
            ⊟ [prv]: instantiate pproduct(n_0)+1 in 4t4
                ⊟ [hht]: split pwz
                    ⊟ [plj]: scatter
                        ⊟ [hnc]: instantiate n_0, p_0 in how
                            ⊟ [pv6]: instantiate p_0, pproduct(n_0) in 3ci
                                ⊟ [hdf]: auto
                                    [ppq]: proved (CVC3)
                    ⊟ [qlj]: auto
                        [53h]: proved (CVC3)
```

Initially we perform scatter and then the command flip s2g which adds the negated goal [s2g] to the knowledge; this yields the following state:

| 3ci | $\forall n \in \mathbb{N}, m \in \mathbb{N} : \text{divides}(n,m) \wedge \text{divides}(n,m+1) \Rightarrow n \leq 1$ |
| oqp | $\text{prime}(2)$ |
| qjx | $\forall n \in \mathbb{N} : \text{prime}(n) \Rightarrow n \geq 2$ |
| 4t4 | $\forall n \in \mathbb{N} : n > 1 \Rightarrow (\exists p \in \mathbb{N} : \text{prime}(p) \wedge \text{divides}(p,n))$ |
| rgt | $\forall n \in \mathbb{N} : n > 2 \Rightarrow \text{pproduct}(n) > n$ |
| ljp | $\text{pproduct}(2) = 2$ |
| how | $\forall n \in \mathbb{N}, p \in \mathbb{N} : p \leq n \wedge \text{prime}(p) \Rightarrow \text{divides}(p, \text{pproduct}(n))$ |
| hgx | $\forall p \in \mathbb{N} : \text{prime}(p) \Rightarrow p \leq n_0$ |

There is no formula below the line; this indicates that we perform an indirect proof whose goal is to show a contradiction in the knowledge. The core idea of the proof is the application of the command

```
case prime(pproduct(n_0)+1);
```

which decomposes the proof into two parts: in one part we assume the given formula $F$, in the other one we assume $\neg F$ (this rule can be justified by the application of the (cut) rule to show the tautology $F \vee \neg F$ and then to apply $(\vee\text{-K})$ to perform the case distinction on $F$). One case can be immediately closed by the application of auto.

In the other case

| 3ci | $\forall n \in \mathbb{N}, m \in \mathbb{N} : \text{divides}(n,m) \wedge \text{divides}(n,m+1) \Rightarrow n \leq 1$ |
| oqp | $\text{prime}(2)$ |
| qjx | $\forall n \in \mathbb{N} : \text{prime}(n) \Rightarrow n \geq 2$ |
| 4t4 | $\forall n \in \mathbb{N} : n > 1 \Rightarrow (\exists p \in \mathbb{N} : \text{prime}(p) \wedge \text{divides}(p,n))$ |
| rgt | $\forall n \in \mathbb{N} : n > 2 \Rightarrow \text{pproduct}(n) > n$ |
| ljp | $\text{pproduct}(2) = 2$ |
| how | $\forall n \in \mathbb{N}, p \in \mathbb{N} : p \leq n \wedge \text{prime}(p) \Rightarrow \text{divides}(p, \text{pproduct}(n))$ |
| hgx | $\forall p \in \mathbb{N} : \text{prime}(p) \Rightarrow p \leq n_0$ |
| m5k | $\neg \text{prime}(1 + \text{pproduct}(n_0))$ |

we perform the command instantiate pproduct(n_0)+1 in 4t4 to exhibit by application of the rule $(\forall\text{-K})$ the existential knowledge formula for the instantiation $n := 1 + pprimes(n_0)$; however this formula is only part of a disjunction with second part $pproduct(n_0) \leq 0$. We can get rid of this part by the command split 6c3 which applies rule $(\vee\text{-K})$ to split the proof into two cases; the second one can be immediately closed by application of auto.

In the other case

| 3ci | $\forall n \in \mathbb{N}, m \in \mathbb{N} : \mathrm{divides}(n,m) \wedge \mathrm{divides}(n,m+1) \Rightarrow n \leq 1$ |
|-----|---|
| oqp | $\mathrm{prime}(2)$ |
| qjx | $\forall n \in \mathbb{N} : \mathrm{prime}(n) \Rightarrow n \geq 2$ |
| 4t4 | $\forall n \in \mathbb{N} : n > 1 \Rightarrow (\exists p \in \mathbb{N} : \mathrm{prime}(p) \wedge \mathrm{divides}(p,n))$ |
| rgt | $\forall n \in \mathbb{N} : n > 2 \Rightarrow \mathrm{pproduct}(n) > n$ |
| ljp | $\mathrm{pproduct}(2) = 2$ |
| how | $\forall n \in \mathbb{N}, p \in \mathbb{N} : p \leq n \wedge \mathrm{prime}(p) \Rightarrow \mathrm{divides}(p, \mathrm{pproduct}(n))$ |
| hgx | $\forall p \in \mathbb{N} : \mathrm{prime}(p) \Rightarrow p \leq n_0$ |
| m5k | $\neg\, \mathrm{prime}(1 + \mathrm{pproduct}(n_0))$ |
| lff | $\exists p \in \mathbb{N} : \mathrm{prime}(p) \wedge \mathrm{divides}(p, 1 + \mathrm{pproduct}(n_0))$ |

we perform first `scatter` in order to apply the rule ($\exists$-K) to the existential knowledge formula, and then two applications of `instantiate` to apply ($\forall$-K) on the universal knowledge formuals [how] and [3ci]. The remaining instantiations can be solved by `auto` and the proof is closed.

## An Induction Proof

The RISC ProofNavigator also supports mathematical induction on the natural numbers. To demonstrate this, we first introduce by the declarations

```
sum: NAT->NAT;
S1: AXIOM sum(0)=0;
S2: AXIOM FORALL(n:NAT): n>0 => sum(n)=n+sum(n-1);
```

a function `sum` that maps every natural number $n$ to the sum $1 + 2 + \ldots + n$; the function is described by two axioms that mimic a recursive 'definition' of `sum` with a base case (the sum of all natural numbers up to zero is zero) and a recursive case (for $n > 0$, the sum of all numbers up to $n$ is the sum of $n$ and of all numbers up to $n - 1$). From this, we want to prove the Gaussian summation formula:

```
S: FORMULA FORALL(n:NAT): sum(n) = (n+1)*n/2;
```

The proof starts by the command `induction n in ats` which applies the induction principle to the quantified variable $n$ of goal [ats]. This yields two proof states: the first state

| d3i | $\mathrm{sum}(0) = 0$ |
|---|---|
| lxe | $\forall n \in \mathbb{N} : n > 0 \Rightarrow \mathrm{sum}(n) = n + \mathrm{sum}(n-1)$ |
| cpy | $\mathrm{sum}(0) = \frac{(0+1)\cdot 0}{2}$ |

represents the base case of the induction and is automatically closed by CVC3. The second state

| d3i | $\mathrm{sum}(0) = 0$ |
|---|---|
| lxe | $\forall n \in \mathbb{N} : n > 0 \Rightarrow \mathrm{sum}(n) = n + \mathrm{sum}(n-1)$ |
| f2m | $2 \cdot \mathrm{sum}(n_0) \leq n_0^2 + n_0 \wedge n_0^2 + n_0 \leq 2 \cdot \mathrm{sum}(n_0)$ |
| 6ai | $2 \cdot \mathrm{sum}(1+n_0) \leq 2 + n_0^2 + 3 \cdot n_0 \wedge 2 + n_0^2 + 3 \cdot n_0 \leq 2 \cdot \mathrm{sum}(1+n_0)$ |

represents the induction step. Here the induction hypothesis [f2m] and the induction goal [6ai] have been transformed ('simplified') by the application of the CVC3 to a logically equivalent (but in this case admittedly not very intuitive) form; if we apply the command option autosimp="false"; before starting the induction, we can avoid this transformation. Anyway, the application of auto is able to find suitable instantiations of the knowledge to immediately close the proof.

# Building Models

**4**

*Wer hohe Türme bauen will, muss lange beim Fundament
verweilen. (They who want to build high towers have to give
good consideration to the foundation.)*
— Anton Bruckner

In this chapter, we use first-order logic to build by formal definitions higher-level
concepts from more fundamental ones; in particular, we present the types and operations that are essential for the construction of mathematical models. We start with a
general explanation of how we can introduce mathematical theories from scratch by
characterizing their fundamental entities via logical axioms; on top of these we can
later introduce other entities by explicit definitions. Then we present the theory of
sets as the prime example of an axiomatic theory; based on this theory we introduce
a collection of 'types' and 'type constructors' that also encompass in a certain sense
functions and predicates as objects of the theory. Finally we discuss how set-theoretic
functions can be also implicitly defined and how these definitions can be viewed as
'specifications' that can be subsequently 'refined' by other implicit definitions and
ultimately be 'implemented' by explicit ones.

## 4.1    Axioms and Definitions

Our first major use of the language of logic is that of building models of reality in
which we subsequently work and reason. In logic, such a model is characterized by
those sentences (closed formulas) that are true in that model. We will now discuss
how the process of building models is structured.

For this purpose, we first generalize our notions of first-order logic to varying sets
of constant, function, and predicate symbols.

W. Schreiner, *Thinking Programs*, Texts & Monographs in Symbolic Computation,
https://doi.org/10.1007/978-3-030-80507-4_4

**Definition 4.1** (*First-order language*) Let $C$, $F$, and $P$ be three sets of symbols. A *first-order language* $L[C, F, P]$ consists of a grammar as described by Definition 2.1 with the symbol sets

- *ConstantSymbol* defined as $C$,
- *FunctionSymbol* defined as $F$,
- *PredicateSymbol* defined as $P$,

and of an arity $ar \colon F \cup P \to \mathbb{N}_{\geq 1}$ that assigns to every symbol of $F$ and $P$ its arity.

$T$ is a *term* respectively $F$ is a *formula* of the first-order language $L[C, F, P]$ if $T$ respectively $F$ is constructed according to the grammar of $L[C, F, P]$ and is well-formed with respect to its arity.

$I$ is an *interpretation* of first-order language $L[C, F, P]$ for domain $D$ if for every $CS \in C$, $I(CS)$ is a value in $D$, for every $FS \in F$, $I(FS)$ is a function of arity $ar$ on $D$, and for every $PS \in P$, $I(PS)$ is a predicate of arity $ar$ on $D$.

**Definition 4.2** (*Theory*) A *theory* for the first-order language $L[C, F, P]$ (a *first-order theory*) is a collection of closed formulas of $L[C, F, P]$ that are called *axioms*.

Building a theory thus starts by stating a collection of formulas for a given first-order language.

**Example 4.1** We consider the language $L[C, F, P]$ with $C := \{0\}$, $F := \{s, +\}$, $P := \{\leq\}$ and $ar := [s \mapsto 1, + \mapsto 2, \leq \mapsto 2]$, i.e., with constant symbol 0, unary function symbol $s$, binary function symbol $+$, and binary predicate symbol $\leq$.

We introduce a theory *NAT* for $L[C, F, P]$ which consists of the following axioms:

$$\forall x.\, s(x) \neq 0$$
$$\forall x, y.\, s(x) = s(y) \Rightarrow x = y$$
$$\forall x.\, x + 0 = x$$
$$\forall x, y.\, x + s(y) = s(x + y)$$
$$\forall x.\, 0 \leq x$$
$$\forall x.\, s(x) \nleq 0$$
$$\forall x, y.\, s(x) \leq s(y) \Leftrightarrow x \leq y$$

Please note that the axioms are formulas of $L[C, F, P]$ and have no free variables; the theory is thus well-formed.                                                          $\square$

A theory may describe some model(s) of reality in the following sense.

**Definition 4.3** (*Model*) Let $T$ be a theory for $L[C, F, P]$. $I$ is a *model* of $T$ for domain $D$ if $I$ is an interpretation of $L[C, F, P]$ for $D$ such that for every formula $F$ in $L[C, F, P]$ and every assignment $a$ over $D$, the formula is true, i.e., $[\![ F ]\!]^I_a = \text{true}$.

Models of a theory in general have more properties than are explicitly stated as axioms of the theory.

**Definition 4.4** (*Logical consequence*) Let $T$ be a theory for $L[C, F, P]$. A formula $F$ is a logical consequence of $T$ if it is valid in every model $I$ of $T$, i.e., if $[\![ F ]\!]^I_a = \text{true}$ for every assignment $a$ over the domain of $I$.

A logical consequence of a theory is thus an additional property that holds in the model, although it is not explicitly stated in the theory. It can thus be implicitly added to the theory (respectively implicitly considered as an axiom of the theory), because this does not add any additional constraint: any model of the original theory remains a model of the extended theory. Logical consequences can be deduced by the proof calculus introduced in Chap. 3: given a theory $T$ with axioms $F_1, \ldots, F_n$, formula $G$ is a logical consequence of $T$ if and only if $F_1, \ldots, F_n \vdash G$ is derivable.

**Example 4.2** A logical consequence of the theory *NAT* of Example 4.1 is $s(0) + s(0) = s(s(0))$ because of the derivation

$$s(0) + s(0) \overset{(1)}{=} s(s(0) + 0) \overset{(2)}{=} s(s(0))$$

Equation (1) holds, because the axiom $\forall x, y.\ x + s(y) = s(x + y)$ implies $x + s(y) = s(x + y)$ for $x := s(0)$ and $y := 0$. Equation (2) holds, because the axiom $\forall x.\ x + 0 = x$ implies $x + 0 = x$ for $x := s(0)$.                                  □

**Definition 4.5** (*Consistent theory*) A theory $T$ for $L[C, F, P]$ is *consistent* if it has a model, i.e., if there exists a domain $D$ and an interpretation $I$ such that $I$ is a model of $T$ for $D$. Otherwise $T$ is *inconsistent*.

After having created a theory, we need to show that its consistency, otherwise all further considerations of the theory may be pointless (since it may have no model). However, a consistent theory needs not determine a unique model, as is demonstrated by the following example.

**Example 4.3** The theory *NAT* of Example 4.1 is consistent, because it has as a model the domain of natural numbers, with constant 0 interpreted as the natural number 'zero', function $s$ interpreted as the 'successor function' that assigns to every number $x$ its successor $x + 1$, function $+$ interpreted as addition and predicate $\leq$ interpreted as 'less than or equal'. For these interpretations and arbitrary assignments, the axioms of the theory are true.

While we might consider the natural numbers as the 'standard' model of *NAT*, there exist also other 'non-standard' models. For instance, also the extension of the natural numbers by an additional element $\infty$ ('infinity') with the following properties is a model of *NAT*: $s(\infty) = \infty$, $x + \infty = \infty + x = \infty$ for every $x$, $x \leq \infty \Leftrightarrow$ true for every $x$, and $\infty \leq x \Leftrightarrow$ false for every $x \neq \infty$.                                $\square$

Once we have a consistent theory, we may extend its language by new notions, i.e., we may introduce new constant, function, and predicate symbols. However, we are interested in preserving consistency: if the original theory has a model, also the extended theory shall have one. Therefore, rather than implicitly characterizing also the new symbols by axioms (which may yield an inconsistent theory), we usually define the new notions explicitly on top of the existing ones.

---

**Definition 4.6** (*Constant definition*) Let $T$ be a theory for $L[C, F, P]$. A *constant definition*

$$S := T$$

consists of a symbol $S$ that does not occur in $C$, $F$, $P$ and a closed term $T$ of $L[C, F, P]$ (in which consequently $S$ does not occur).
This definition introduces a new language $L[C', F, P]$ where the new set of constant symbols $C'$ contains, in addition to all the elements of $C$, also symbol $S$; it also creates a theory $T'$ over $L[C', F, P]$ that contains, in addition to all the elements of $T$, also the closed formula

$$S = T.$$

**Example 4.4** We extend the language and theory of Example 4.1 by the constant definitions

$$1 := s(0)$$
$$2 := 1 + 1$$

which introduce two new constants 1 and 2. The term $s(0)$ in the definition of 1 is closed and involves only the constant 0 and the function $s$ of the original language. The term $1 + 1$ in the definition of 2 is closed and involves only the function $+$ of the original theory and the previously defined constant 1. The definitions are therefore well-formed.

From the definition of constant 1, we know $1 = s(0)$. Furthermore, we know $2 = s(s(0))$ because of the derivation

$$2 \overset{(1)}{=} 1 + 1 \overset{(2)}{=} s(0) + s(0) \overset{(3)}{=} s(s(0)).$$

Equation (1) holds because of the definition of constant 2. Equation (2) holds because of the definition of constant 1. Equation (3) holds because of the derivation shown in Example 4.2. □

---

**Definition 4.7** (*Function definition*) Let $T$ be a theory over $L[C, F, P]$. A *function definition*

$$S(V_1, \ldots, V_n) := T$$

consists of a symbol $S$ that does not occur in $C, F, P$ and a term $T$ of $L[C, F, P]$ (in which consequently $S$ does not occur) with $fv(T) \subseteq V_1, \ldots, V_n$, i.e., the free variables of $T$ are among $V_1, \ldots, V_n$.

This definition introduces a new language $L[C, F', P]$ where the new set of function symbols $F'$ contains, in addition to all the elements of $F$, also symbol $S$; it also creates a theory $T'$ over $L[C, F', P]$ that contains, in addition to all the elements of $T$, also the closed formula

$$\forall V_1, \ldots, V_n. \, S(V_1, \ldots, V_n) = T.$$

---

**Example 4.5** We extend the language and theory of Example 4.1 by the function definitions

$$d(x) := x + x$$
$$e(x, y) := d(x) + y$$

which introduce a unary function $d$ and a binary function $e$. The term $x + x$ in the definition of $d$ has as its only free variable the parameter $x$ and only involves the function $+$ of the original theory. The term $d(x) + y$ in the definition of $e$ has as its free variables only the parameters $x$, $y$ and only involves $+$ and the previously defined function $d$. The definitions are therefore well-formed and introduce into the theory the following formulas:

$$\forall x.\, d(x) = x + x$$
$$\forall x, y.\, e(x, y) = d(x) + y$$

A logical consequence of the definition is $e(s(0), s(0)) = (s(0) + s(0)) + s(0)$, because of the derivation

$$e(s(0), s(0)) \overset{(1)}{=} d(s(0)) + s(0) \overset{(2)}{=} (s(0) + s(0)) + s(0).$$

Equation (1) holds because the second formula introduced above implies $e(x, y) = d(x) + y$ for $x := s(0)$ and $y := s(0)$; Equation (2) holds because the first formula implies $d(x) = x + x$ for $x := s(0)$.                                                    $\square$

---

**Definition 4.8** (*Predicate definition*) Let $T$ be a theory over $L[C, F, P]$. A *predicate definition*

$$S(V_1, \ldots, V_n) :\Leftrightarrow G$$

consists of a symbol $S$ that does not occur in $C, F, P$ and a formula $G$ of $L[C, F, P]$ (in which consequently $S$ does not occur) with $fv(G) \subseteq V_1, \ldots, V_n$, i.e., the free variables of $G$ are among $V_1, \ldots, V_n$.
This definition introduces a new language $L[C, F, P']$ where the new set of predicate symbols $P'$ contains, in addition to all the elements of $P$, also symbol $S$; it also creates a theory $T'$ over $L[C, F, P']$ that contains, in addition to all the elements of $T$, also the closed formula

$$\forall V_1, \ldots, V_n.\, S(V_1, \ldots, V_n) \Leftrightarrow G.$$

---

**Example 4.6** We extend the language and theory of Example 4.1 by the predicate definitions

$$x < y :\Leftrightarrow x \leq y \wedge y \neq x$$
$$x \geq y :\Leftrightarrow x \not< y$$
$$x > y :\Leftrightarrow y < x$$

which introduce the binary predicates $<, \geq$ and $>$. The formulas $x \leq y \wedge y \neq x$, $x \not< y$, and $y < x$ in the three definitions have as their only free variables the parameters

$x$ and $y$ and only depend on previously introduced predicates. The definitions are therefore well-formed and introduce into the theory the formulas

$$\forall x, y. \, x < y \Leftrightarrow x \le y \wedge y \neq x$$
$$\forall x, y. \, x \ge y \Leftrightarrow x \not< y$$
$$\forall x, y. \, x > y \Leftrightarrow y < x$$

A logical consequence of the extended theory is $s(0) > 0$ because of the derivation

$$s(0) > 0 \overset{(1)}{\Leftrightarrow} 0 < s(0) \overset{(2)}{\Leftrightarrow} 0 \le s(0) \wedge s(0) \neq 0 \overset{(3)}{\Leftrightarrow} \text{true}$$

Here equivalence (1) follows from $x > y :\Leftrightarrow y < x$ for $x := s(0)$ and $y := 0$ and equivalence (2) follows from $x < y :\Leftrightarrow x \le y \wedge y \neq x$ for $x := 0$ and $y := s(0)$. Equivalence (3) follows from the axiom $\forall x. \, 0 \le x$ for $x := s(0)$ and from the axiom $\forall x. \, s(x) \neq 0$ for $x := 0$. □

The role of extending theories by definitions is stated by the following fact.

**Proposition 4.1** (Conservative extension) *Let $T$ be a theory and let $T'$ be a theory in the language of $T$ extended by a sequence of constant, function, and predicate definitions. Then $T'$ is a conservative extension of $T$, i.e., every formula in the language of $T$ is a logical consequence of $T'$, if and only if it is already a logical consequence of $T$; consequently, if $T$ is consistent, then also $T'$ is consistent.*

**Example 4.7** By the definitions in the previous examples we have extended the original language $L[C, F, P]$ with $C := \{0\}$, $F := \{s, +\}$, $P := \{\le\}$ and arity $ar := [s \mapsto 1, + \mapsto 2, \le \mapsto 2]$ to the language $L[C', F', P']$ with $C := \{0, 1, 2\}$, $F := \{s, +, d, e\}$, $P := \{\le, <, \ge, >\}$ and arity $ar' := [s \mapsto 1, + \mapsto 2, d \mapsto 2, e \mapsto 2, \le \mapsto 2, < \mapsto 2, \ge \mapsto 2, > \mapsto 2]$.

Likewise, we have extended the original theory *NAT* to the theory *NAT'* with the formulas

$$\forall x. \, s(x) \neq 0$$
$$\forall x, y. \, s(x) = s(y) \Rightarrow x = y$$
$$\forall x. \, x + 0 = x$$
$$\forall x, y. \, x + s(y) = s(x + y)$$
$$\forall x. \, 0 \le x$$
$$\forall x. \, s(x) \not\le 0$$
$$\forall x, y. \, s(x) \le s(y) \Leftrightarrow x \le y$$

$$1 = s(0)$$
$$2 = 1 + 1$$
$$\forall x.\, d(x) = x + x$$
$$\forall x, y.\, e(x, y) = d(x) + y$$
$$\forall x, y.\, x < y \Leftrightarrow x \le y \land y \ne x$$
$$\forall x, y.\, x \ge y \Leftrightarrow x \not< y$$
$$\forall x, y.\, x > y \Leftrightarrow y < x$$

Since *NAT* was consistent with the natural numbers as a model, also *NAT'* is consistent, because it is a conservative extension of *NAT*.                                                        □

## 4.2    The Theory of Sets

As stated above, model building starts with axiomatizing a core theory which is subsequently enriched by defining new constants, functions, and predicates. As such a core theory, classical mathematics has chosen the theory of 'sets', informally unordered collections of objects (which are again sets). All other kinds of mathematical objects are by first-order definitions ultimately represented as sets. The formal axiomatization of sets has been very carefully elaborated to be concise, avoid redundancies, and not introduce inconsistencies; however, it is quite low-level and not very intuitive. We therefore present set theory on a higher level that is more adequate for practical work.

We consider a domain of values called *sets* with a binary predicate $x \in S$, read as '$x$ is in $S$' or '$x$ is an element of $S$'. Sets have the following core properties:

- *Extensionality*: Two sets $S_1$ and $S_2$ are equal if and only if they have the same elements:

$$\forall S_1, S_2.\, S_1 = S_2 \Leftrightarrow \forall x.\, x \in S_1 \Leftrightarrow x \in S_2$$

  Thus a set has no other features than the elements which it contains.
- *Regularity*: There is no infinite sequence of sets $S_0, S_1, S_2, S_3, \ldots$ such that

$$S_1 \in S_0,\, S_2 \in S_1,\, S_3 \in S_2, \ldots$$

  An element of a set is thus strictly 'simpler' than the set in which it is contained and sets cannot become simpler forever.
  In particular, there is no set $S$ such that $S \in S$ and there is also no sequence of sets $S_0, S_1, \ldots, S_k$ such that $S_1 \in S_0, S_2 \in S_1, \ldots, S_k \in S_{k-1}$ with $S_0 = S_k$. Thus no set can be (directly or indirectly) contained in itself.

More concretely, we have the following sets:

- *Empty set*: The constant $\varnothing$ denotes the set that has no elements:

$$\forall x.\, x \notin \varnothing$$

- *Infinite set*: The constant $\mathbb{N}$ denotes an infinite set of elements which we may identify with the natural numbers:

$$\forall x.\, x \in \mathbb{N} \Leftrightarrow x = 0 \vee x = 1 \vee x = 2 \vee \ldots$$

  Since in set theory every value is a set, also the natural numbers are defined as sets; however, we do not concern ourselves with their internal structure.
- *Finite set*: The term $\{x_1, \ldots, x_n\}$ denotes the set with the $n$ elements $x_1, \ldots, x_n$:

$$\forall x.\, x \in \{x_1, \ldots, x_n\} \Leftrightarrow x = x_1 \vee \ldots \vee x = x_n$$

  For instance, the set $\{2, 3, 5\}$ consists of the three elements 2, 3, and 5. By extensionality, it is identical to the set $\{3, 2, 2, 5\}$ but different from the set $\{2, 3, 4, 5\}$. As a special case, $\{\} = \varnothing$.
- *Specification*: Given a formula $F[x]$ with free variable $x$, the *set builder* term $\{x \in S \mid F[x]\}$ denotes the set of all elements in $S$ that have property $F[x]$:

$$\forall S, x.\, x \in \{x \in S \mid F[x]\} \Leftrightarrow x \in S \wedge F[x]$$

  For instance, the set $\{x \in \mathbb{N} \mid x > 2\}$ denotes the set of natural numbers $3, 4, 5, \ldots$.
- *Replacement*: Given a term $T[x]$ with free variable $x$, the set builder term $\{T[x] \mid x \in S\}$ denotes the set of all values $y$ such that $y = T[x]$ for some element $x$ of $S$:

$$\forall S, y.\, y \in \{T[x] \mid x \in S\} \Leftrightarrow \exists x \in S.\, y = T[x]$$

  For instance, the set $\{x + x \mid x \in \mathbb{N}\}$ denotes the set of all even natural numbers $0, 2, 4, 6, \ldots$.

We may also combine the two variants of set builder notation and generalize it to multiple variables: given a term $T[x_1, \ldots, x_n]$ and a formula $F[x_1, \ldots, x_n]$, both with free variables $x_1, \ldots, x_n$, the term

$$\{T[x_1, \ldots, x_n] \mid x_1 \in S_1 \wedge \ldots \wedge x_n \in S_n \wedge F[x_1, \ldots, x_n]\}$$

denotes the set of all values $y$ such that $y = T[x_1, \ldots, x_n]$ for some elements $x_1 \in S_1, \ldots, x_n \in S_n$ for which $F[x_1, \ldots, x_n]$ holds. For instance,

$$\{x + y \mid x \in \mathbb{N} \wedge y \in \mathbb{N} \wedge x + y \leq 2\}$$

denotes the set $\{0 + 0, 0 + 1, 1 + 0, 0 + 2, 2 + 0, 1 + 1\} = \{0, 1, 2\}$.

- *Power set*: Given a set $S$, the power set $\mathsf{Set}(S)$ is the set of all subsets of $S$, i.e., every element $x$ of $\mathsf{Set}(S)$ has only elements that are in $S$:

$$\forall S, x.\ x \in \mathsf{Set}(S) \Leftrightarrow \forall z \in x.\ z \in S$$

For instance, the set $\mathsf{Set}(\mathbb{N})$ is the set of all (finite or infinite) sets of natural numbers $\varnothing, \{0\}, \{0, 1\}, \{1, 2 \ldots\}, \ldots, \mathbb{N}$ (including $\mathbb{N}$ itself).

- *Union*: The union $\bigcup S$ of all elements of set $S$ contains every element $x$ which is an element *of some element* of $S$:

$$\forall S, x.\ x \in \bigcup S \Leftrightarrow \exists y \in S.\ x \in y$$

For instance, the set $\bigcup \{\{0, 1, 2\}, \{2, 3, 4\}\}$ is the set $\{0, 1, 2, 3, 4\}$.

We also assume the existence of a particular function that returns an arbitrary element from a given set:

- *Choice Function*: For every subset $x$ of set $S$, the term anyof $x$ denotes an element of $x$, provided that $x$ is not empty:

$$\forall x \in \mathsf{Set}(S).\ x \neq \varnothing \Rightarrow \mathsf{anyof}\ x \in x$$

For instance, anyof $\{x \in \mathbb{N} \mid x \leq 2\}$ is one of the elements $0, 1, 2$, but we don't know which one. Based on this function, we may define a quantified term

$$(\mathsf{choose}\ x \in S.\ F[x]) := \mathsf{anyof}\ \{x \in S \mid F[x]\}$$

which denotes for set $S$ and formula $F[x]$ with free variable $x$ an element $x$ of $S$ that satisfies property $F[x]$, provided that such an element exists. We then know choose $x \in \mathbb{N}.\ x \leq 2$ is one of the elements $0, 1, 2$, but we don't know which one. Still one can, e.g., show that for $f(x) := \mathsf{choose}\ y \in \mathbb{N}.\ y \leq x$ we have

$$f(2) = f(2)$$

because $0 = 0, 1 = 1, 2 = 2$, i.e., any choice for the value of $f(2)$ makes the formula true.

With the help of the notions axiomatized above, we can define certain well-known sets (we omit the details):

- $\mathbb{N}$, the set of the natural numbers $0, 1, 2, \ldots$, was already introduced above. Furthermore, we denote by $\mathbb{N}_{\geq 1}$ the set of positive natural numbers $1, 2, \ldots$ and by $\mathbb{N}_n$ the set of the first $n$ natural numbers $0, 1, \ldots, n - 1$.
- $\mathbb{Z}$ is the set of the integer numbers $\ldots, -2, -1, 0, 1, 2, \ldots$.
- $\mathbb{R}$ is the set of the real numbers such as $1.5, -2, 3.1415\ldots$; we denote by $\mathbb{R}_{\geq 0}$ the set of the non-negative reals and by $\mathbb{R}_{> 0}$ the set of the positive reals.
- $\mathbb{C}$ is the set of the complex numbers such as $3.5 + 1.2i$, $-i$, and $2.7$.

In more detail, we define the usual set operations.

- *Subset*: We have $A \subseteq B$, to be read as '$A$ is a subset of $B$' if and only if every element of $A$ is also an element of $B$:

$$A \subseteq B :\Leftrightarrow \forall x \in A.\, x \in B$$

For instance, we have $\{1\} \subseteq \{1, 2\}$, $\{1, 2\} \subseteq \{1, 2\}$, and $\varnothing \subseteq \{1, 2\}$. For the sets of numbers introduced above, we have

$$\mathbb{N} \subseteq \mathbb{Z} \subseteq \mathbb{R} \subseteq \mathbb{C}$$

From the definition, we have as a logical consequence that two sets are equal if and only if each is a subset of the other one:

$$\forall A, B.\, A = B \Leftrightarrow A \subseteq B \wedge B \subseteq A$$

Furthermore, $A$ is in the powerset of $B$ if and only if $A$ is a subset of $B$:

$$\forall A, B.\, A \in \mathsf{Set}(B) \Leftrightarrow A \subseteq B$$

- *Proper Subset*: We have $A \subset B$, to be read as '$A$ is a proper subset of $B$', if and only if $A$ is a subset of $B$ but not identical to $B$:

$$A \subset B :\Leftrightarrow A \subset B \wedge A \neq B$$

For instance, $\{1\} \subset \{1, 2\}$ and $\varnothing \subset \{1, 2\}$ but $\{1, 2\} \not\subset \{1, 2\}$.

From the definition, it follows that $A$ is a proper subset of $B$ if and only if it is a subset of $B$ and $B$ has an element that is not in $A$:

$$\forall A, B.\ A \subset B \Leftrightarrow A \subseteq B \wedge \exists x \in B.\ x \notin A$$

- *Intersection*: The intersection $A \cap B$ is the set of all elements that are both in $A$ and in $B$:

$$A \cap B := \{x \in A \mid x \in B\}$$

For instance, $\{1, 2, 3\} \cap \{2, 3, 4\} = \{2, 3\}$ and $\{1, 2, 3\} \cap \{4, 5\} = \varnothing$.

From the definition, the intersection of $A$ and $B$ is a subset of both $A$ and $B$:

$$\forall A, B.\ A \cap B \subseteq A \wedge A \cap B \subseteq B$$

We also introduce the function

$$\bigcap S := \left\{ x \in \bigcup S \,\middle|\, \forall y \in S.\ x \in y \right\}$$

For instance, the set $\bigcap \{\{1, 2, 3\}, \{2, 3, 4\}\}$ is the set $\{1, 2, 3\} \cap \{2, 3, 4\} = \{2, 3\}$.

Based on this function, we also introduce the quantified term

$$\bigcap_{x \in S} T[x] := \bigcap \{T[x] \mid x \in S\}$$

For instance, $\bigcap_{i \in \mathbb{N}_3} \mathbb{N}_i = \mathbb{N}_0 \cap \mathbb{N}_1 \cap \mathbb{N}_2 = \varnothing$.

- *Union*: The union $A \cup B$ is the set of all elements that are at least in one of $A$ or $B$:

$$A \cup B := \bigcup \{A, B\}$$

For instance, $\{1, 2, 3\} \cup \{2, 3, 4\} = \{1, 2, 3, 4\}$.

From the definition, both $A$ and $B$ are a subset of the union of $A$ and $B$

$$\forall A, B.\ A \subseteq A \cup B \wedge B \subseteq A \cup B$$

We also introduce the quantified term

$$\bigcup_{x \in S} T[x] := \bigcup \{T[x] \mid x \in S\}$$

For instance, $\bigcup_{i \in \mathbb{N}_3} \mathbb{N}_i = \mathbb{N}_0 \cup \mathbb{N}_1 \cup \mathbb{N}_2 = \mathbb{N}_2$.

- *Difference*: The difference $A \setminus B$ ('$A$ without $B$') is the set of all elements that are in $A$ but not in $B$:

$$A \setminus B := \{x \in A \mid x \notin B\}$$

For instance, $\{1, 2, 3\} \setminus \{2, 3, 4\} = \{1\}$.
From the definition, the difference of $A$ and $B$ is a subset of $A$:

$$\forall A, B.\ A \setminus B \subseteq A$$

## 4.3   Products and Sums

Considered as a 'data type', the notion of sets is quite low-level; not many real-world concepts can be directly represented as unordered collections of objects. The main usefulness of sets is their versatility as building blocks for introducing various data types that operate on a higher-level of abstraction. We introduce two such fundamental data types on which (together with plain sets) subsequently other data types will be based. However, rather than defining their low-level encoding as sets, we confine ourselves to describing the properties of their high-level 'constructor' and 'selector' operations.

- *Product Types*: For sets $S_1, \ldots, S_n$, the *product type* (also called *tuple type*)

$$S_1 \times \ldots \times S_n$$

  denotes a set of values which we call *products* or (more frequently) *tuples*. This type is equipped with
  - $n$ selectors $(\_).1, \ldots, (\_).n$ such that for every tuple $x \in S_1 \times \ldots \times S_n$ and every selector $(\_).i$ with $1 \le i \le n$ we have

$$x.i \in S_i;$$

  - a constructor $\langle \_, \cdots, \_ \rangle$ that maps $n$ values $x_1 \in S_1, \ldots, x_n \in S_n$ to a tuple

$$\langle x_1, \ldots, x_n \rangle \in S_1 \times \ldots \times S_n.$$

  The constructor is related to every selector $(\_).i$ with $1 \le i \le n$ such that for all values $x_1 \in S_1, \ldots, x_n \in S_n$ we have

$$\langle x_1, \ldots, x_n \rangle.i = x_i.$$

Intuitively, the constructor 'glues' together the $n$ values $x_1, \ldots, x_n$ to a composed value from which later each selector $.i$ can extract the original value $x_i$.

**Example 4.8** We define the product type $P := \mathbb{N} \times \mathbb{Z}$ and the tuple $t := \langle 2, -3 \rangle$. We then have $t \in P$, $t.1 = 2 \in \mathbb{N}$, and $t.2 = -3 \in \mathbb{Z}$.  □

- *Sum Types*: For sets $S_1, \ldots, S_n$, the *sum type* (also called *disjoint union*)

$$S_1 + \ldots + S_n$$

denotes a set of values which we call call *sums* or *tagged values*. This type is equipped with
  - $n$ constructors $1(\_), \ldots, n(\_)$ such that for all $x_1 \in S_1, \ldots, x_n \in S_n$ and every constructor $i(\_)$ with $1 \leq i \leq n$ we have

$$i(x_i) \in S_1 + \ldots + S_n;$$

  - a selector (case $\_$ of $f_1 \mid \ldots \mid f_n$) for arbitrary $n$ functions $f_1 : S_1 \to S$, $\ldots$, $f_n : S_n \to S$ and set $S$ such that, for every sum $x \in S_1 + \ldots + S_n$, we have

$$(\text{case } x \text{ of } f_1 \mid \ldots \mid f_n) \in S.$$

The selector is related to every constructor $i(\_)$ with $1 \leq i \leq n$ such that we have for every value $x_i \in S_i$

$$(\text{case } i(x_i) \text{ of } f_1 \mid \ldots \mid f_n) = f_i(x_i).$$

Intuitively, the constructor $i(\_)$ adds a 'tag' $i$ to value $x_i$ from $S_i$ which yields a tagged value $\langle i, x_i \rangle$ in $S_1 + \ldots + S_n$ that 'remembers' how it was constructed. The selector then knows from the tag $i$ associated to its argument $\langle i, x_i \rangle$, that it has to apply function $f_i$ to $x_i$.
Syntactically, each $f_i$ is usually defined 'inline' as

$$i(V) \to T[V]$$

with term $T[V]$ that has free variable $V$; its result, for some argument $a$, is the value of $T[V]$ where $a$ is assigned to $V$.

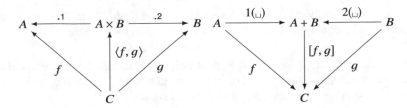

**Fig. 4.1** Products and Sums

**Example 4.9** We define the sum type $S := \{0, 2, 4\} + \{1, 3, 5\}$ with constructors $1(\_)$ from $\{0, 2, 4\}$ to $S$ and $2(\_)$ from $\{1, 3, 5\}$ to $S$. Now we consider the selector application

$$(\text{case } a \text{ of } 1(x) \rightarrow x + 1 \mid 2(x) \rightarrow x - 1)$$

If $a = 1(4)$, the result is $4 + 1 = 5$; if $a = 2(5)$, the result is $5 - 1 = 4$. Since the first function $1(x) \rightarrow x + 1$ is applied only to values in $\{0, 2, 4\}$ and the second function $2(x) \rightarrow x - 1$ is applied only to values in $\{1, 3, 5\}$, the applications maps values from $\{0, 2, 4\}$ to values in $\{1, 3, 5\}$, and vice versa.                □

Sum types are less common in classical mathematics but play an important role in computer science. Nevertheless sums are very closely related to products in the following sense:

- Consider the product type $A \times B$. For every set $C$ and all functions $f \colon C \rightarrow A$ and $g \colon C \rightarrow B$, there exists a unique function, we may call it $\langle f, g \rangle$, such that the left diagram in Fig. 4.1 'commutes', i.e., $\langle f, g \rangle \circ (.1) = f$ and $\langle f, g \rangle \circ (.2) = g$ (here $f \circ g$ denotes the composition of functions $f$ and $g$ such that $(f \circ g)(x) = g(f(x))$). This function $\langle f, g \rangle$ can be defined as

$$\langle f, g \rangle(x) := \langle f(x), g(x) \rangle.$$

Intuitively, this means that $A \times B$ is the data type that 'preserves most information' about the values of types $A$ and $B$ from which its objects were constructed. Every other type $C$ that also embeds values of these types can restore them without loss of information, rather than by directly applying its functions $f$ and $g$, by taking a small 'detour' via $A \times B$.

- Now consider the sum type $A + B$. For every set $C$ and all functions $f \colon A \rightarrow C$ and $g \colon B \rightarrow C$, there is a unique function, we may call it $[f, g]$, such that the right diagram in Fig. 4.1 'commutes', i.e., $1(\_) \circ [f, g] = f$ and $2(\_) \circ [f, g] = g$. This function $[f, g]$ can be defined as

$$[f, g](x) := \text{case } x \text{ of } 1(x) \rightarrow f(x) \mid 2(x) \rightarrow g(x).$$

Intuitively, this means that $A + B$ is the data type that 'preserves most information' about how its objects were constructed from values of either type $A$ or $B$.

Rather than directly extracting some information of type $C$ from values of type $A$ or of type $B$, we may also take the detour via $A + B$ to yield the same information.

As can be seen in Fig. 4.1, the diagrams only differ in that the directions of the the arrows are reverted, the notions of 'products' and 'sums' are therefore *dual*. For this reason, sum types are sometimes also called *coproducts*.

## 4.4   Set-Theoretic Functions and Relations

The logic introduced in Chap. 2 is called 'first-order', because variables can be only assigned values of the domain of interpretation; thus also quantification is only possible over these values. Therefore we can in first-order logic not directly formulate statements like

For every function $f$, it is true that ...

However, exactly this is necessary if one wants to discuss general properties of functions and predicates. In set theory, we can overcome this problem by defining sets that capture the essential properties of functions and predicates and in the subsequent discussion use these sets as 'proxies' for functions and predicates.

**Definition 4.9**  (*Relation*) Set $R$ is a *relation* on sets $A_1, \ldots, A_n$ if and only if

$$R \subseteq A_1 \times \ldots \times A_n$$

A relation is the set-theoretic equivalent of a logical predicate, since we can use set-builder notation to turn every predicate on sets into a relation. Usually, however, we apply a more elegant notation to denote relations.

**Definition 4.10**  (*Relation term*)  We define the *relation term* (which we also call *pi term*)

$$(\pi x_1 \in A_1, \ldots, x_n \in A_n. \ F[x_1, \ldots, x_n]) :=$$
$$\{\langle x_1, \ldots, x_n \rangle \mid x_1 \in A_1 \wedge \ldots \wedge x_n \in A_n \wedge F[x_1, \ldots, x_n]\}$$

where $A_1, \ldots, A_n$ are sets and $F[x_1, \ldots, x_n]$ is a formula whose free variables occur among the variables $x_1, \ldots, x_n$.

By definition, a term $\pi x_1 \in A_1, \ldots, x_n \in A_n.\ F[x_1, \ldots, x_n]$ denotes a relation on $A_1, \ldots, A_n$. Furthermore, by a bit of 'syntactic sugar', we can apply a relation like a logic predicate.

---

**Definition 4.11** (*Relation application*) For every $n \in \mathbb{N}$, we define the $(n+1)$-ary *relation application* $\lrcorner\langle\lrcorner, \cdots, \lrcorner\rangle$ as

$$R\langle x_1, \ldots, x_n \rangle :\Leftrightarrow \langle x_1, \ldots, x_n \rangle \in R.$$

---

**Example 4.10** Let $<$ denote the binary predicate 'less-than' on $\mathbb{N}$. Then the constant definition

$$R := \pi x \in \mathbb{N}, y \in \mathbb{N}.\ x < y$$

which means the same as

$$R := \{\langle x, y \rangle \mid x \in \mathbb{N} \wedge y \in \mathbb{N} \wedge x < y\}$$

introduces a relation $R$ on $\mathbb{N}$ and $\mathbb{N}$; we also say that $R$ is a *binary relation* on $\mathbb{N}$. By definition, we have

$$\forall x, y.\ R\langle x, y \rangle \Leftrightarrow x \in \mathbb{N} \wedge y \in \mathbb{N} \wedge x < y$$

which means the same as

$$\forall x, y.\ \langle x, y \rangle \in R \Leftrightarrow x \in \mathbb{N} \wedge y \in \mathbb{N} \wedge x < y$$

We thus have e.g. $R\langle 2, 3 \rangle$ but $\neg R\langle 3, 2 \rangle$ (since $3 \not< 2$) and $\neg R\langle -3, 2 \rangle$ (since $-3 \notin \mathbb{N}$). $\square$

Usually we will define a relation not as a constant but in a format that is closer to the definition of a predicate.

---

**Definition 4.12** (*Relation definition*) A *relation definition* has form

$$R \subseteq A_1 \times \ldots \times A_n$$
$$R\langle x_1, \ldots, x_n \rangle :\Leftrightarrow F[x_1, \ldots, x_n]$$

where $A_1, \ldots, A_n$ are sets and $F[x_1, \ldots, x_n]$ is a formula whose free variables are among $x_1, \ldots, x_n$.
This definition introduces the relation

$$R := \pi x_1 \in A_1, \ldots, x_n \in A_n.\ F[x_1, \ldots, x_n]$$

that consequently has the property

$$\forall x_1 \in A_1, \ldots, x_n \in A_n. \ R\langle x_1, \ldots, x_n \rangle \Leftrightarrow F[x_1, \ldots, x_n].$$

**Example 4.11** The definition

$$R \subseteq \mathbb{N} \times \mathbb{N}$$
$$R\langle x, y \rangle :\Leftrightarrow x < y$$

introduces the same relation as the definition in Example 4.10.                    □

The following definition will become handy in later sections.

**Definition 4.13** (*Relation updates*) Let $R \subseteq A \times B$ be a binary relation on sets $A$ and $B$ and let $x \in A$ and $y \in B$ be elements of $A$ and $B$, respectively. We define the *removal* $R[x \mapsto \bot]$ of $x$ from $R$ as

$$R[x \mapsto \bot] := R \setminus \{\langle x, y \rangle \mid x \in A\}$$

Furthermore, we define the *update* $R[x \mapsto y]$ of $R$ at $x$ by $y$ as

$$R[x \mapsto y] := R[x \mapsto \bot] \cup \{\langle x, y \rangle\}$$

Please note that for $R \subseteq A \times B$, $x \in A$, and $y \in B$, also $R[x \mapsto \bot] \subseteq A \times B$ and $R[x \mapsto y] \subseteq A \times B$.

In a similar fashion, we introduce the set-theoretic equivalent of a logical function.

**Definition 4.14** (*Function*) Set $F$ is a (set-theoretic) *function* from sets $A_1, \ldots, A_n$ (the *domain*) to set $B$ (the *range*), written as $F : A_1 \times \ldots \times A_n \to B$, if and only if

1. $F$ is a relation on $A_1, \ldots, A_n, B$:

$$F \subseteq A_1 \times \ldots \times A_n \times B$$

2. For every possible sequence of 'function arguments' $x_1, \ldots, x_n$, $F$ contains some tuple $\langle x_1, \ldots, x_n, y \rangle$ with a corresponding 'function result' $y$:

$$\forall x_1 \in A_1, \ldots, x_n \in A_n. \ \exists y \in B. \ \langle x_1, \ldots, x_n, y \rangle \in F$$

3. $F$ does not contain tuples with 'conflicting' function results:

$$\forall x_1 \in A_1, \ldots, x_n \in A_n, y \in B, y' \in B.$$
$$\langle x_1, \ldots, x_n, y \rangle \in F \wedge \langle x_1, \ldots, x_n, y' \rangle \in F \Rightarrow y = y'$$

The predicate 'is function' defined above can be lifted to a set.

**Definition 4.15** (*Set of functions*) We define by

$$A_1 \times \ldots \times A_n \to B := \{F \in \text{Set}(A_1 \times \ldots \times A_n \times B) \mid F \colon A_1 \times \ldots \times A_n \to B\}$$

the set $A_1 \times \ldots \times A_n \to B$ of a all functions from $A_1, \ldots, A_n$ to $B$.

**Example 4.12** Let $+$ denote addition on $\mathbb{N}$. Then the constant definition

$$F := \{\langle x, y, x + y \rangle \mid x \in \mathbb{N} \wedge y \in \mathbb{N}\}$$

introduces a function $F \colon \mathbb{N} \times \mathbb{N} \to \mathbb{N}$ with $\langle 2, 3, 5 \rangle \in F$ because $2 + 3 = 5$. We then naturally have $F \in \mathbb{N} \times \mathbb{N} \to \mathbb{N}$.                                    □

Functions are usually not denoted by set builders but by the following notation.

**Definition 4.16** (*Function term*) We define the *function term* (also called *lambda term* since Alonzo Church introduced it in his lambda calculus)

$$(\lambda x_1 \in A_1, \ldots, x_n \in A_n. \, T[x_1, \ldots, x_n]) :=$$
$$\{\langle x_1, \ldots, x_n, T[x_1, \ldots, x_n] \rangle \mid x_1 \in A_1 \wedge \ldots \wedge x_n \in A_n\}$$

where $A_1, \ldots, A_n$ are sets and $T[x_1, \ldots, x_n]$ is a term whose free variables occur among the variables $x_1, \ldots, x_n$.

The name 'function term' can be justified as follows: Let

$$B := \{T[x_1, \ldots, x_n] \mid x_1 \in A_1 \wedge \ldots \wedge x_n \in A_n\}$$

denote the set of all term values for variable values $x_1 \in A_1, \ldots, x_n \in A_n$. Then

$$F := \lambda x_1 \in A_1, \ldots, x_n \in A_n. \, T[x_1, \ldots, x_n]$$

denotes a function from $A_1, \ldots, A_n$ to $B$, because

1. the definition ensures $F \subseteq A_1 \times \ldots \times A_n \times B$,
2. for all $x_1 \in A_1, \ldots, x_n \in B_n$, there is some $y \in \mathbb{N}$, namely $y = T[x_1, \ldots, x_n]$, such that $\langle x_1, \ldots, x_n, y \rangle \in F$,
3. for all $x_1 \in A_1, \ldots, x_n \in B_n$, if there are some $y \in \mathbb{N}$ and $y' \in \mathbb{N}$ such that $\langle x_1, \ldots, x_n, y \rangle \in F$ and $\langle x_1, \ldots, x_n, y' \rangle \in F$, then, by definition of $F$, we have $y = T[x_1, \ldots, x_n]$ and $y' = T[x_1, \ldots, x_n]$ and thus $y = y'$.

**Example 4.13** The constant definition

$$F := \lambda x \in \mathbb{N}, y \in \mathbb{N}. \, x + y$$

introduces the same function $F \colon \mathbb{N} \times \mathbb{N} \to \mathbb{N}$ as the one defined in Example 4.12.
□

We also introduce syntactic sugar for the application of a set-theoretic functions.

**Definition 4.17** (*Function application*) For every $n \in \mathbb{N}$, we define the $(n + 1)$-ary operation $\lrcorner(\lrcorner, \cdots, \lrcorner)$ as

$$F(x_1, \ldots, x_n) := (\text{choose } t \in F. \, \exists y. \, t = \langle x_1, \ldots, x_n, y \rangle).(n + 1).$$

Thus, for $F \colon A_1 \times \ldots \times A_n \to B$ and $x_1 \in A_1, \ldots, x_n \in A_n$, the 'function application' $F(x_1, \ldots, x_n)$ denotes the unique element $y \in B$ such that $\langle x_1, \ldots, x_n, y \rangle \in F$. Any other application $F(\cdots)$ denotes an unknown value.

**Example 4.14** For the function defined in Example 4.12 respectively Example 4.13 we have $F(2, 3) = 5$.
□

Analogous to relations, we usually define a set-theoretic function not as a constant but in a format that is closer to the definition of a logical function.

**Definition 4.18** (*Function definition*) A (set-theoretic) *function definition* has the form

$$F \colon A_1 \times \ldots \times A_n \to B$$
$$F(x_1, \ldots, x_n) := T[x_1, \ldots, x_n]$$

where $A_1, \ldots, A_n$ and $B$ are sets and $T[x_1, \ldots, x_n]$ is a term whose free variables are among the variables $x_1, \ldots, x_n$.

This definition is *well-formed* if it satisfies the constraint

$$\forall x_1 \in A_1, \ldots, x_n \in A_n.\ T[x_1, \ldots, x_n] \in B$$

It then introduces a function $F : A_1 \times \ldots \times A_n \to B$ defined as

$$F := \lambda x_1 \in A_1, \ldots, x_n \in A_n.\ T[x_1, \ldots, x_n]$$

that consequently has the property

$$\forall x_1 \in A_1, \ldots, x_n \in A_n.\ F(x_1, \ldots, x_n) = T[x_1, \ldots, x_n].$$

In other words, for all arguments in the domain $A_1 \times \ldots \times A_n$, the result of function $F$ is denoted by term $T$, provided that its value is in range $B$.

A well-formed set-theoretic function definition thus automatically guarantees the existence and uniqueness of the function result. However, we are obliged to show the well-formedness of such a definition, i.e., we have to show that the value denoted by the term is indeed in the range of the function.

**Example 4.15** The definition

$$F : \mathbb{N} \times \mathbb{N} \to \mathbb{N}$$
$$F(x, y) := x + y$$

introduces the same function as the definition in Example 4.14. This definition is well-formed because

$$\forall x \in \mathbb{N}, y \in \mathbb{N}.\ x + y \in \mathbb{N}$$

□

Every function introduced by a definition is *total* in the sense that it provides a result for every argument from the domain of the function. Sometimes, however, not every possible argument of the domain 'makes sense', for example there is no reasonable result for the division $1/0$. We therefore provide an alternative definition format that allows us to constrain the domain of a function by a formula that describes those arguments for which the application of a function is legal.

**Definition 4.19** (*Function definition with input condition*) A *function defini-tion with input condition* has form

$$F : A_1 \times \ldots \times A_n \to B$$
$$F(x_1, \ldots, x_n)$$
$$\quad \text{requires } I[x_1, \ldots, x_n]$$
$$\quad \text{result } T[x_1, \ldots, x_n]$$

where $A_1, \ldots, A_n$ and $B$ are sets, $I[x_1, \ldots, x_n]$ is a formula (the *input condi-tion* or *precondition*), and $T[x_1, \ldots, x_n]$ is a term, such that the free variables of both phrases are among the variables $x_1, \ldots, x_n$.
This definition is *well-formed* if it satisfies the constraint

$$\forall x_1 \in A_1, \ldots, x_n \in A_n. \, I(x_1, \ldots, x_n) \Rightarrow T[x_1, \ldots, x_n] \in B$$

It then introduces a function $F : A_1 \times \ldots \times A_n \to B$ that has the property

$$\forall x_1 \in A_1, \ldots, x_n \in A_n. \, I(x_1, \ldots, x_n) \Rightarrow F(x_1, \ldots, x_n) = T[x_1, \ldots, x_n]$$

In other words, for all arguments in the domain $A_1 \times \ldots \times A_n$ that satisfy the input condition $I$, the result of $F$ is denoted by term $T$, provided that its value is in range $B$.

A function definition with precondition allows to deduce almost nothing about the value of a function application $F(x_1, \ldots, x_n)$ whose arguments $x_1 \in A_1, \ldots, x_n \in A_n$ violate the input condition $I[x_1, \ldots, x_n]$. We only know that this application denotes a value in $B$ (because $F : A_1 \times \ldots \times A_n \to B$), but nothing else.

By definition, a function definition in the more usual format

$$F : A_1 \times \ldots \times A_n \to B$$
$$F(x_1, \ldots, x_n) := T[x_1, \ldots, x_n]$$

is just a shortcut for

$$F : A_1 \times \ldots \times A_n \to B$$
$$F(x_1, \ldots, x_n)$$
$$\quad \text{requires true}$$
$$\quad \text{result } T[x_1, \ldots, x_n]$$

with precondition true.

**Example 4.16** The definition

$$div \colon \mathbb{R} \times \mathbb{R} \to \mathbb{R}$$

$$div(x, y)$$

$$\text{requires } x + y \neq 0$$

$$\text{result } x/(x + y)$$

is well-formed because

$$\forall x \in \mathbb{R}, y \in \mathbb{R}. \ x + y \neq 0 \Rightarrow x/(x + y) \in \mathbb{R}$$

For example, we have $div(2, 3) = 2/(2 + 3) = 2/5 = 0.4$. However, the value $div(2, -2)$ is an unknown value in $\mathbb{R}$, because $2 + (-2) = 0$ and thus the arguments $x = 2$ and $y = -2$ violate the input condition. □

By the notations introduced above, we can define set-theoretic relations and functions in a way that is very similar to the definitions of logic predicates and functions, although these definitions actually introduce constants (namely sets).

Furthermore, we can use a relation application $R\langle x_1, \ldots, x_n \rangle$ like an atomic formula and an application $F(x_1, \ldots, x_n)$ of a set-theoretic function $F$ like a logical function application. The difference between logical functions/predicates and set-theoretic functions/predicates thus gets blurred. Indeed, with the major exception of those logical functions and predicates that were introduced at the beginning of this subsection, from now on, most 'functions' and 'predicates' that we will encounter will actually be set-theoretic functions and relations, i.e., sets. We will therefore be able to write a statement

$$\forall A, B, f. \ f \colon A \to B \Rightarrow \forall x \in A. \ f(x) \in B$$

which is quantified over sets $A$ and $B$ and a (set-theoretic) function $f$ from $A$ to $B$.

From the correspondence of set-theoretic functions and relations to logical functions and predicates, there may be some confusion whether a particular symbol $f$ denotes a logical function or a set-theoretic one, respectively whether a particular symbol $r$ denotes a logic predicate or a set-theoretic relation. Actually, the answer is not so difficult: If $f$ is a set-theoretic function, there must exist some sets $A$ and $B$ with $f \in A \to B$; if $r$ is a relation, there must exist some set $S$ with $r \subseteq S$. However, some collections are 'too big' to be sets; such a collection is also called a *(proper) class*, a prominent example is the proper class $\mathcal{S}$ of all sets. Furthermore, if we have a collection $\mathcal{C}$ such that one can assign to every element of $\mathcal{S}$, i.e., to every set, a different element of $\mathcal{C}$, then also $\mathcal{C}$ is not a set but a proper class.

**Example 4.17** The symbol $\cup$ denotes a logical function, not a set-theoretical one: If $\cup$ were a function, we would have $\cup \in A \times B \to C$ for some sets $A, B, C$. However, $A$ would have to include all sets, i.e., we could assign to every element of $\mathcal{S}$

a different element of $A$ respectively $A \times B$; thus $\cup$ cannot denote a set-theoretic function. By an analogous argument, the symbol $\subseteq$ denotes a logical predicate, not a relation. $\qquad\qquad\qquad\qquad\qquad\qquad\qquad\qquad\qquad\qquad\qquad\qquad\qquad$ $\Box$

In a relation application $r\langle x \rangle$ the angle brackets were mainly chosen to differentiate its meaning $\langle x \rangle \in r$ from the meaning $r(x) = (\text{choose } t \in F.\, t.1 = x).2$ of a function application (since every function is also a relation, both could be intended). However, the syntax $r\langle x \rangle$ also makes clear that we apply a relation, not a predicate whose application is denoted by $r(x)$. For functions, we do not make this distinction, because most functions that we will encounter will be set-theoretic ones and we will apply them frequently; thus we would like to preserve for these applications the well-established notation $f(x)$ rather than inventing a new one like e.g. $f[x]$.

## 4.5   More Type Constructions

With the help of set-theoretic functions, we can define data types of finite and infinite sequences:

- *Infinite Sequences*: for every set $A$,

$$A^{\omega} := \mathbb{N} \to A$$

  is the set of infinite sequences of elements from $A$. For instance, if $s \in \mathbb{N}^{\omega}$, then $s$ consists of the natural numbers $s(0), s(1), s(2), \ldots$.
- *Sequences of Fixed Length*: for every set $A$ and natural number $n$,

$$A^{n} := \mathbb{N}_{n} \to A$$

  is the set of sequences of length $n$ with elements from $A$. For instance, if $s \in \mathbb{N}^{5}$, then $s$ consists of the five natural numbers $s(0), s(1), s(2), s(3), s(4)$.
- *Finite Sequences*: for every set $A$ and natural number $n$,

$$A^{*} := \bigcup \left\{ A^{i} \;\middle|\; i \in \mathbb{N} \right\}$$

  is the set of finite sequences with elements from $A$. We define

$$|\_| : A^{*} \to \mathbb{N}$$

$$|s| := \text{choose } i \in \mathbb{N}.\, s \in A^{i}$$

  as the length of sequence $s$. For instance, if $s \in \mathbb{N}^{*}$ and $|s| = 5$, then $s$ consists of the five natural numbers $s(0), s(1), s(2), s(3), s(4)$.

Furthermore, we extend the notions of products and sums that denote their components/alternatives by plain numbers to corresponding types that use for this purpose symbolic labels:

- *Labeled Product (Record) Type*: for pair-wise different labels $L_1, \ldots, L_n$ and sets $S_1, \ldots, S_n$, the labeled product (record) type

$$L_1{:}S_1 \times \ldots \times L_n{:}S_n := \{L_1, \ldots, L_n\} \rightarrow (S_1 \cup \ldots \cup S_n)$$

  denotes a set of values which we call *labeled products* or *records*. This type is equipped with a constructor

$$\langle L_1{:}\_, \ldots, L_n{:}\_ \rangle : (S_1 \times \ldots \times S_n) \rightarrow (L_1{:}S_1 \times \ldots \times L_n{:}S_n)$$
$$\langle L_1{:}x_1, \ldots, L_n{:}x_n \rangle := \{\langle L_1, x_1 \rangle, \ldots, \langle L_n, x_n \rangle\}.$$

  Furthermore, for record $r \in L_1{:}S_1 \times \ldots \times L_n{:}S_n$ and label $L_i$, we define

$$r.L_i := r(L_i)$$

  as the selection of the component labeled $L_i$ from $r$.

A record is thus a function that maps labels to values; by the use of the label in the application of the selector function, the role of the selected record components becomes more transparent than in the selection of a tuple component by position.

**Example 4.18** We define *Point* := x$:\mathbb{Z} \times$ y$:\mathbb{Z}$ and $p := \langle$x$:2,$ y$:3\rangle$. Then we have $p \in$ *Point* with $p.$x $= 2$ and $p.$y $= 3$.  □

- *Labeled Sum (Variant) Type*: for pair-wise different labels $L_1, \ldots, L_n$ and sets $S_1, \ldots, S_n$, the labeled sum (variant) type

$$L_1{:}S_1 + \ldots + L_n{:}S_n := \{L_1, \ldots, L_n\} \times (S_1 \cup \ldots \cup S_n)$$

  denotes a set of values which we call *labeled sums* or *variants*. This type is equipped with $n$ constructors

$$L_i(\_) : S_i \rightarrow (L_1{:}S_1 + \ldots + L_n{:}S_n)$$
$$L_i(x) := \{\langle L_1, x_1 \rangle\}.$$

Furthermore, for arbitrary set $S$ and arbitrary $n$ functions $f_1 \colon S_1 \to S, \ldots,$ $f_n \colon S_n \to S$, we define the selector

$$(\text{case } \_ \text{ of } f_1 \mid \ldots \mid f_n) \colon (L_1{:}S_1 + \ldots + L_n{:}S_n) \to S$$
$$(\text{case } x \text{ of } f_1 \mid \ldots \mid f_n) :=$$
$$\quad \text{if } x.1 = L_1 \text{ then } f_1(x.2) \text{ else}$$
$$\quad \ldots$$
$$\quad \text{if } x.1 = L_{n-1} \text{ then } f_{n-1}(x.2) \text{ else } f_n(x.2)$$

Every function $f_i$ is typically defined inline as

$$L_i(V) \to T[V]$$

with term $T[V]$ that has free variable $V$; its result, for some argument $a$, is the value of $T[a]$ where $a$ is assigned to $V$.

Labeled sums can thus be used analogous to sums; the only difference is that the numbers $1, \ldots, n$ are replaced by the labels $L_1, \ldots, L_n$.

**Example 4.19** We define type

$$Complex := \text{cart}{:}(\mathbb{R} \times \mathbb{R}) + \text{polar}{:}(\mathbb{R} \times \mathbb{R})$$

and function

$$d \colon Complex \to \mathbb{R}$$
$$d(x) := \text{case } x \text{ of } \text{cart}(c) \to \sqrt{c.1^2 + c.2^2} \mid \text{polar}(p) \to p.1$$

Then we have $d(\text{cart}(\langle 0, 1\rangle)) = d(\text{polar}(\langle 1, \pi/2\rangle)) = 1$.
Using labeled products, we may also define the type as

$$Complex := \text{cart}{:}(\text{x}{:}\mathbb{R} \times \text{y}{:}\mathbb{R}) + \text{polar}{:}(\text{r}{:}\mathbb{R} \times \alpha{:}\mathbb{R})$$

and the function as

$$d \colon Complex \to \mathbb{R}$$
$$d(x) := \text{case } x \text{ of } \text{cart}(c) \to \sqrt{c.\text{x}^2 + c.\text{y}^2} \mid \text{polar}(p) \to p.\text{r}$$

Then we have $d(\text{cart}(\langle \text{x}{:}0, \text{y}{:}1\rangle)) = d(\text{polar}(\langle \text{r}{:}1, \alpha{:}\pi/2\rangle)) = 1$.  $\qquad\square$

## 4.6   Implicit Definitions and Function Specifications

A definition of a (logic/set-theoretic) function describes explicitly how the result of the function is constructed from its arguments with the help of the previously introduced functions; explicit function definitions thus have a 'computational' flavor.

However, a function may also be described in a more abstract way, by stating the expected relationship between the function arguments and the function result without explicitly indicating how the result can be computed.

---

**Definition 4.20** (*Implicit function definition*) An *implicit function definition* has the form

$$F : A_1 \times \ldots \times A_n \to B$$
$$F(x_1, \ldots, x_n)$$
$$\text{requires } I[x_1, \ldots, x_n]$$
$$\text{ensures } O[x_1, \ldots, x_n, \text{result}]$$

where $A_1, \ldots, A_n$ and $B$ are sets, $x_1, \ldots, x_n$ are variables different from the variable result, the *input condition* or *precondition* $I[x_1, \ldots, x_n]$ is a formula whose free variables are among $x_1, \ldots, x_n$, and the *output condition* or *postcondition*. $O[x_1, \ldots, x_n, \text{result}]$ is a formula whose free variables are among $x_1, \ldots, x_n, \text{result}$.

This definition is *well-formed* if it satisfies the constraint

$$\forall x_1 \in A_1, \ldots, x_n \in A_n.$$
$$I[x_1, \ldots, x_n] \Rightarrow \exists \text{result} \in B.\ O[x_1, \ldots, x_n, \text{result}].$$

It then introduces a function $F : A_1 \times \ldots \times A_n \to B$ that has the property

$$\forall x_1 \in A_1, \ldots, x_n \in A_n.$$
$$I[x_1, \ldots, x_n] \Rightarrow$$
$$\text{let result} = F(x_1, \ldots, x_n) \text{ in } O[x_1, \ldots, x_n, \text{result}].$$

In other words, for all arguments in the domain $A_1 \times \ldots \times A_n$ that satisfy the input condition $I$, the result of $F$ is a value in range $B$ that satisfies the output condition $O$, provided that such a value exists.

---

The requires clause may be also omitted; this indicates the clause

$$\text{requires true}$$

i.e., an implicit function definition with precondition true.

**Example 4.20** We define the function

$$qr_0: \mathbb{Z} \times \mathbb{Z} \to \mathsf{q}{:}\mathbb{Z} \times \mathsf{r}{:}\mathbb{Z}$$
$$qr_0(x, y)$$
$$\text{requires } x \geq 0 \wedge y > 0$$
$$\text{ensures } x = \text{result.q} \cdot y + \text{result.r}$$

which intuitively returns 'some sort of' quotient $q$ and remainder $r$ of the division of $x$ by $y$. This definition is well-formed, i.e.,

$$\forall x \in \mathbb{Z}, y \in \mathbb{Z}.$$
$$x \geq 0 \wedge y > 0 \Rightarrow \exists \text{result} \in \mathbb{Z}. \; x = \text{result.q} \cdot y + \text{result.r}$$

because, for every $x \in \mathbb{Z}$ and $y \in \mathbb{Z}$ with $y \neq 0$, we can choose result := $\langle \mathsf{q}{:}0, \mathsf{r}{:}x \rangle$ and get result.q $\cdot y +$ result.r $= 0 \cdot y + x = x$.

The result of $qr_0$ is not uniquely determined, e.g. we have

$$qr_0(5, 2) \in \{\ldots, \langle \mathsf{q}{:}{-}1, \mathsf{r}{:}7 \rangle, \langle \mathsf{q}{:}0, \mathsf{r}{:}5 \rangle, \langle \mathsf{q}{:}1, \mathsf{r}{:}3 \rangle, \langle \mathsf{q}{:}2, \mathsf{r}{:}1 \rangle, \ldots\}$$

because $5 = \ldots = -1 \cdot 2 + 7 = 0 \cdot 2 + 5 = 1 \cdot 2 + 3 = 2 \cdot 2 + 1 = \ldots$. However, if we strengthen the output condition as in

$$qr_1: \mathbb{Z} \times \mathbb{Z} \to \mathsf{q}{:}\mathbb{Z} \times \mathsf{r}{:}\mathbb{Z}$$
$$qr_1(x, y)$$
$$\text{requires } x \geq 0 \wedge y > 0$$
$$\text{ensures } x = \text{result.q} \cdot y + \text{result.r} \wedge 0 \leq \text{result.r} \wedge \text{result.r} < y$$

we get the unique result $qr_1(5, 2) = \langle \mathsf{q}{:}2, \mathsf{r}{:}1 \rangle$ that actually denotes the quotient and remainder of the division of 5 by 2 (it remains to be shown that this definition is well-formed, which requires a longer argument). $\square$

An implicit definition may 'refine' another one, in the sense that the function introduced by the refining definition may be applied, whenever the function introduced by the refined definition is expected, without influencing the correctness of the result.

**Definition 4.21** (*Refinement of implicit definition*) Let $F_1: A_1 \times \ldots \times A_n \to B$ be a function introduced by a well-formed implicit function definition with input condition $I_1[x_1, \ldots, x_n]$ and output condition $O_1[x_1, \ldots, x_n, \text{result}]$; let $F_2: A_1 \times \ldots \times A_n \to B$ a function introduced by a well-formed implicit

function definition with input condition $I_2[x_1, \ldots, x_n]$ and output condition $O_2[x_1, \ldots, x_n, \text{result}]$.
Then $F_2$ is a *refinement* of $F_1$ if the following two conditions are satisfied:

$$\forall x_1 \in A_1, \ldots, x_n \in A_n.$$
$$I_1[x_1, \ldots, x_n] \Rightarrow I_2[x_1, \ldots, x_n];$$

$$\forall x_1 \in A_1, \ldots, x_n \in A_n, \text{result} \in B.$$
$$I_1[x_1, \ldots, x_n] \wedge O_2[x_1, \ldots, x_n, \text{result}] \Rightarrow O_1[x_1, \ldots, x_n, \text{result}].$$

One should note that while the postcondition of the refining function $F_2$ may be stronger than that of the refined function $F_1$, the relationship for the preconditions is just the inverse: while postconditions may be strengthened, preconditions may be only weakened. The justification for this particular behavior will be given below.

**Example 4.21**  Continuing Example 4.20, the function $qr_1$ is a refinement of $qr_0$, because their preconditions are identical and the precondition together with the post-condition of $qr_1$ implies the postcondition of $qr_0$:

$$x \geq 0 \wedge y > 0 \wedge x = \text{result.q} \cdot y + \text{result.r} \wedge 0 \leq \text{result.r} \wedge \text{result.r} < y \Rightarrow$$
$$x = \text{result.q} \cdot y + \text{result.r}.$$

$\square$

The notion of refinements is justified by the following property.

**Proposition 4.2**  (Refinement preserves defining property) *Let $F_1$ and $F_2$ be functions as in Definition 4.21 such that $F_2$ is a refinement of $F_1$. Then we have*

$$\forall x_1 \in A_1, \ldots, x_n \in A_n.$$
$$I_1[x_1, \ldots, x_n] \Rightarrow$$
$$\text{let result} = F_2(x_1, \ldots, x_n) \text{ in } O_1[x_1, \ldots, x_n, \text{result}]$$

*i.e., $F_2$ also satisfies the output condition of $F_1$.*

In essence, this proposition states that any application of $F_1$ may be replaced by an application of the refinement $F_2$, without affecting the correctness of its result. The proof of this statement also justifies the inverse direction of the implications in Definition 4.21.

***Proof***  We take arbitrary $x_1 \in A_1, \ldots, x_n \in A_n$ with $I_1[x_1, \ldots, x_n]$ and show

$$\text{let result} = F_2(x_1, \ldots, x_n) \text{ in } O_1[x_1, \ldots, x_n, \text{result}]. \qquad (*)$$

From $I_1[x_1, \ldots, x_n]$ and the refinement property $I_1[x_1, \ldots, x_n] \Rightarrow I_2[x_1, \ldots, x_n]$ we have $I_2[x_1, \ldots, x_n]$. From this and the property resulting from the definition of $F_2$

$$I_2[x_1, \ldots, x_n] \Rightarrow$$
$$\text{let result} = F_2(x_1, \ldots, x_n) \text{ in } O_2[x_1, \ldots, x_n, \text{result}]$$

we have let result $= F_2(x_1, \ldots, x_n)$ in $O_2[x_1, \ldots, x_n, \text{result}]$. From this, $I_1[x_1, \ldots, x_n]$, and the refinement property

$$I_1[x_1, \ldots, x_n] \wedge O_2[x_1, \ldots, x_n, \text{result}] \Rightarrow O_1[x_1, \ldots, x_n, \text{result}]$$

we have (*). $\qquad\qquad\qquad\qquad\qquad\qquad\qquad\qquad\qquad\qquad\qquad\qquad$ $\square$

The fact that preconditions may be (only) weakened is also illustrated by the following example.

**Example 4.22** We define the function

$$qr_2 \colon \mathbb{Z} \times \mathbb{Z} \to \text{q:}\mathbb{Z} \times \text{r:}\mathbb{Z}$$
$$qr_2(x, y)$$
$$\qquad \textsf{requires } y \neq 0$$
$$\qquad \textsf{ensures } x = \textsf{result.q} \cdot y + \textsf{result.r} \wedge |\textsf{result.r}| < |y| \wedge$$
$$\qquad (y \geq 0 \Leftrightarrow \textsf{result.r} \geq 0)$$

which is also defined on negative inputs. We then have for example

- $qr_2(5, 2) = \langle \text{q:2, r:1} \rangle$
- $qr_2(-5, 2) = \langle \text{q:}-3, \text{r:1} \rangle$
- $qr_2(5, -2) = \langle \text{q:2, r:}-1 \rangle$
- $qr_2(-5, -2) = \langle \text{q:2, r:}-1 \rangle$

which indicates that $qr_2$ agrees with $qr_1$ on non-negative inputs. Indeed, $qr_2$ is a refinement of $qr_1$, because

$$x \geq 0 \wedge y > 0 \Rightarrow y \neq 0$$

and in particular

$$x \geq 0 \wedge y > 0 \wedge$$
$$x = \textsf{result.q} \cdot y + \textsf{result.r} \wedge |\textsf{result.r}| < |y| \wedge$$
$$(y \geq 0 \Leftrightarrow \textsf{result.r} \geq 0) \Rightarrow$$
$$\qquad x = \textsf{result.q} \cdot y + \textsf{result.r} \wedge 0 \leq \textsf{result.r} \wedge \textsf{result.r} < y.$$

To see the truth of the second statement, assume the premises of the implication. Then the core property $x = $ result.$_q \cdot y + $ result.$_r$ holds. Furthermore, from $y > 0$ and ($y \geq 0 \Leftrightarrow$ result.$_r \geq 0$), we know result.$_r \geq 0$ and thus $0 \leq $ result.$_r$. From result.$_r \geq 0$, we know |result.$_r$| $ = $ result.$_r$. From $y > 0$, we know $|y| = y$. From |result.$_r$| $< |y|$, we thus know result.$_r < y$.

As a consequence, any application of $qr_1$ that is only defined on non-negative inputs may be replaced by an application of $qr_2$ that also allows negative inputs.  □


An implicit function definition may also be refined by an explicit definition.


**Definition 4.22** (*Implementation of an implicit definition*)   Let the function $F_1: A_1 \times \ldots \times A_n \to B$ be introduced by a well-formed implicit function definition with input condition $I_1[x_1, \ldots, x_n]$ and output condition $O_1[x_1, \ldots, x_n, \text{result}]$; let $F_2: A_1 \times \ldots \times A_n \to B$ be a function introduced by an explicit definition with input condition $I_2[x_1, \ldots, x_n]$ and term $T[x_1, \ldots, x_n]$.

Then $F_2$ is a *implementation* of $F_1$ if the following two conditions are satisfied:

$\forall x_1 \in A_1, \ldots, x_n \in A_n.$
$I_1[x_1, \ldots, x_n] \Rightarrow I_2[x_1, \ldots, x_n];$

$\forall x_1 \in A_1, \ldots, x_n \in A_n, \text{result} \in B.$
$I_1[x_1, \ldots, x_n] \wedge \text{result} = T[x_1, \ldots, x_n] \Rightarrow O_1[x_1, \ldots, x_n, \text{result}].$


**Example 4.23**   Take the implicitly defined function

$min_0: \mathbb{Z} \times \mathbb{Z} \to \mathbb{Z}$

$min_0(a, b)$

  ensures (result $= a \vee$ result $= b$) $\wedge$ result $\leq a \wedge$ result $\leq b$

and the explicitly defined function

$min_1: \mathbb{Z} \times \mathbb{Z} \to \mathbb{Z}$

$min_1(a, b) := $ if $a \leq b$ then $a$ else $b$

both with implicit precondition true. The function $min_1$ is an implementation of $min_0$ because true $\Rightarrow$ true and

  true $\wedge$ result $= $ if $a \leq b$ then $a$ else $b \Rightarrow$
    (result $= a \vee$ result $= b$) $\wedge$ result $\leq a \wedge$ result $\leq b$

which can be seen as follows:

- If $a \le b$, then result $= a$. Thus we have (result $= a \lor$ result $= b$) and result $= a \le a$ and result $= a \le b$.
- If $a \not\le b$, then result $= b$ and $b < a$ and also $b \le a$. Thus we have (result $= a \lor$ result $= b$) and result $= b \le a$ and result $= b \le b$. □

An implicit function definition can thus be also seen as a *function specification* that describes the properties of a function that is to be implemented by an explicit definition. The implementation is based on previously introduced functions which may be explicitly but also implicitly specified.

**Example 4.24** We implicitly define functions *div* and *mod* that return the integer quotient and remainder, respectively:

> $div \colon \mathbb{Z} \times \mathbb{Z} \to \mathbb{Z}$
>
> $x\ div\ y$
>
>    requires $y \ne 0$
>
>    ensures $\exists r \in \mathbb{Z}.\ x = \text{result} \cdot y + r \land |r| < |y| \land (y \ge 0 \Leftrightarrow r \ge 0)$
>
> $mod \colon \mathbb{Z} \times \mathbb{Z} \to \mathbb{Z}$
>
> $x\ mod\ y$
>
>    requires $y \ne 0$
>
>    ensures $\exists q \in \mathbb{Z}.\ x = q \cdot y + \text{result} \land |\text{result}| < |y| \land (y \ge 0 \Leftrightarrow \text{result} \ge 0)$

In each output condition, the existence of the value determined by the partner function is claimed by existential quantification. By definition (and the uniqueness of the function results, which remains to be proved), we have for all $x \in \mathbb{Z}$ and $y \in \mathbb{Z}$ with $y \ne 0$, the relationship

$$x = (x\ div\ y) \cdot y + (x\ mod\ y) \land |\text{result}| < |y| \land (y \ge 0 \Leftrightarrow \text{result} \ge 0)$$

In particular, we have $5\ div\ 2 = 2$ and $5\ mod\ 2 = 1$.

Based on these functions, we explicitly define function $qr_3$ that returns both the quotient and the remainder of two integers:

$$qr_3 \colon \mathbb{Z} \times \mathbb{Z} \to q(\mathbb{Z}) \times r(\mathbb{Z})$$
$$qr_3(x, y)$$
$$\text{requires } y \ne 0$$
$$\text{result } \langle q(x\ div\ y), r(x\ mod\ y) \rangle$$

It is easy to see that $qr_3$ is an implementation of the specification (implicitly defined function) $qr_2$ of Example 4.22. □

## Exercises

Download from the following URL:
https://www.risc.jku.at/people/schreine/TP/exercises/ex-models.pdf.

## Further Reading

There are a lot of introductory texts on set theory, e.g. Makinson's book [106], Rosen's book [155], or Buchberger's and Lichtenberger's book [29] (in German). Many logic books contain short introductions, such as Chap. 5 of Lover's book [104], the Appendix of Ben-Ari's book [13], Appendix 1 of Harrison's handbook [65], Chap. 2 of Gallier's book [56], Chap. 7 of Ebbinghaus et al.'s book [45], Chap. 12 of Schenke's book [159], or Chaps. 4 and 5 of Krantz's book [90]. Also texts on denotational semantics contain such introductions with a focus of set theory as a language for building types (semantic domains), see e.g. Chap. 2 of Schmidt's book [160], Chap. 1 of Winskel's book [180], Chap. 1 of Mitchells's book [119], or Chap. 2 of Fernandez's book [48].

Lamport's book [98] contains in Chaps. 1 and 6 set a very readable description of set theory as the basis of the TLA+ specification language, also covering some subtle points. In particular, it is one of the very view texts that clearly differentiates between logical functions (there called 'operators') and set-theoretic functions (there just called 'functions'), their relationship, and the criterion for deciding whether and operation is an operator or a function. Merz's Chapter on TLA+ in [22] gives in Section 5 a shorter account.

It is harder to find texts on the principles of building mathematical theories by axioms and definitions. The logical basis is covered in Chap. 6 and Appendix 1 of Ebbinghaus et al.'s book [45], in Chap. 1 of Kröger's and Merz's book [93], in Chap. 4 of Li's book [102], in Chap. 3 of Krantz's handbook [90], and shortly in Chap. 5 of Harrison's handbook [65]. However, as for the logic and pragmatic use of explicit and implicit definitions, still Buchberger and Lichtenberger's book [29] gives the most comprehensive presentation (in German).

Chapter 3 of Lamport's book [98] gives a shorter introduction and explains in Chap. 6 nicely the semantics of the 'choose' operator (Ernst Zermelo used a choice function for his axiomatization of set theory, David Hilbert introduced a choice operator in his epsilon calculus). The specification language VDM contains implicit function definitions, see e.g. Jones's book [84] or Chap. 4 of Alagar and Periyasamy's book [4].

## Writing Definitions in Isabelle/HOL

We demonstrate the building of models by definitions in the framework of *Isabelle/HOL* [132], a proof assistant which provides a theorem proving environment based on a higher-order logic (HOL); Fig. 4.2 shows a screenshot of the Isabelle/jEdit user interface for the system.

In contrast to the first-order logic described so far, HOL directly supports functions and predicates as values; predicates are actually just functions whose results are truth values. HOL is strongly typed based on a polymorphic type system as can be found similarly in various functional programming languages. In fact, HOL feels much like an extension of a functional programming language by logical concepts; numerous mathematical theories such as set theory are defined in this language. While first-order logic is the richest logic with a complete proving calculus, the calculus of HOL suffices to prove those propositions that occur in mathematical practice by a combination of automatic (semi-)decision procedures and human assistance. In particular, the proof language 'Isar' allows to develop scripts that resemble manually constructed proofs and can be understood by humans but simultaneously direct the system to a successful computerized proof.

The specifications used in the following presentation can be downloaded from the URL

https://www.risc.jku.at/people/schreine/TP/software/models/models.thy

and loaded by executing from the command line the following command:

```
Isabelle2021 models.thy &
```

### Defining Functions and Predicates

We start with the definition

```
type_synonym 'a seq = "nat ⇒ 'a"
```

of a polymorphic type `'a seq` which can be interpreted as the type of all infinite sequences of values from an arbitrary domain $a$; this type is actually just a shortcut for the type of all functions that map a natural number to a value from $a$. Then the definition

```
definition isconst :: "'a ⇒ 'a seq ⇒ bool" where
  "isconst c s = (∀i::nat. s i = c)"
```

introduces a binary predicate *isconst* which is true for argument value $c$ and sequence $s$, if $s$ holds $c$ at all positions. Furthermore, the definition

```
definition cseq :: "'a ⇒ 'a seq" where
  "cseq c = (λi::nat. c)"
```

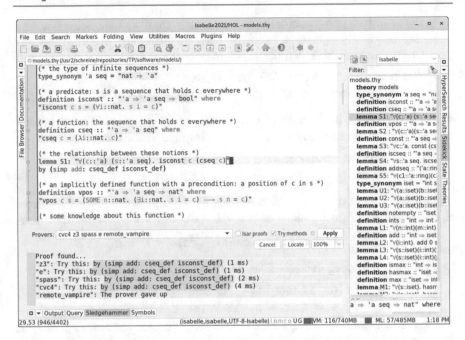

**Fig. 4.2** Isabelle/HOL

introduces a function *cseq* that maps argument *c* to exactly such a sequence, a fact which is stated by the formula

```
lemma S1: "∀(c::'a) (s::'a seq). isconst c (cseq c)" ...
```

To prove this formula, one may invoke from the Isabelle/jEdit interface the tool 'Sledgehammer' which calls several automated theorem provers and SMT solvers to find an automatic proof; if successful, it returns a proof script that allows to recreate the proof. For above formula, this script is

```
by (simp add: cseq_def isconst_def)
```

which can be substituted in the 'Ě' part of the formula; the proof is thus accepted as valid knowledge in all further proofs. All lemmas presented below are similarly 'slain' by 'Sledgehammer'; we omit the generated scripts.

The definition

```
definition vpos :: "'a ⇒ 'a seq ⇒ nat" where
"vpos c s = (SOME n::nat. (∃i::nat. s i = c) ⟶ s n = c)"
```

introduces a function *vpos* which, for argument value *c* and sequence *s*, denotes a position where *s* holds *c*. This function is defined in an implicit style with the 'choice' operator SOME $n::T.F$ that denotes a value *n* of type $T$. This value satisfies property $F$, if such a value exists; if not, the value is arbitrary. Thus in above definition the implication with the existential formula as antecedent is actually redundant; it shall just demonstrate how one may implicitly specify a function with an arbitrary precondition and postcondition. From this definition, we may prove

```
lemma S2 : "∀(c::'a)(s::'a seq).
    ¬(∀i::nat. s i ≠ c) ⟶ s(vpos c s) = c" ...
```

which claims that, if $s$ is not different from $c$ at all positions, that it holds $c$ at the position returned by *vpos*. Another implicit definition

```
definition const :: "'a seq ⟹ 'a" where
    "const s = (SOME c::'a. isconst c s)"
```

introduces a function *const* which returns for argument sequence $s$ a value $c$ that $s$ holds at all positions (if such a value exists). We consequently know

```
lemma S3: "∀c::'a. const (cseq c) = c" ...
```

i.e., that *const* indeed extracts from $cseq(c)$ the value $c$. The definition

```
definition iscseq :: "'a seq ⟹ bool" where
    "iscseq s = (∃c::'a. isconst c s)"
```

introduces a predicate *iscseq* which is true for argument sequence $s$ if it holds some value $c$ at all positions; from this, we can prove the statement

```
lemma S4: "∀s::'a seq. iscseq s ⟶ (∀i::nat. s i = const s)" ...
```

i.e., that the fact that $s$ is a constant sequence implies that it holds $const(s)$ at all positions. The definition

```
definition addseq :: "('a::ring) seq ⟹ 'a seq ⟹ 'a seq" where
    "addseq s1 s2 = (λi::nat. (s1 i) + (s2 i))"
```

introduces a binary function *addseq* that, given argument sequences $s_1$ and $s_2$ returns the sequence that results from the element-wise addition of $s_1$ and $s_2$. The annotation 'a::ring indicates that type $a$ cannot be completely arbitrary but must support the usual ring operations, in particular addition, i.e. operator $+$. Then the formula

```
lemma S5: "∀(c1::'a::ring)(c2::'a::ring).
    iscseq(addseq (cseq c1) (cseq c2))" ...
```

shows that adding two constant sequences yields another constant sequence.

**The Theory of Sets**

Isabelle/HOL provides the polymorphic datatype 'a set which denotes the type of all sets with element from arbitrary type $a$. Based on this type, we define by

```
type_synonym iset = "int set"
```

the type *iset* of integer sets (actually, the following constructions would also work with the more generic type `'a::ring set` of sets of elements from a ring $a$). This type supports the usual set operations, e.g. we can prove the properties

> lemma U1: "∀(a::iset)(b::iset)(c::iset).
>   a ∪ (b ∩ c) = (a ∪ b) ∩ (a ∪ c)" ...
> lemma U2: "∀(a::iset)(b::iset).
>   a ⊆ b ⟶ a ∪ b = b" ...
> lemma U3: "∀(a::iset)(b::iset).
>   a ∩ b ≠ {} ⟶ a - b ⊂ a" ...

We may also define our own functions, e.g.

> definition notempty :: "iset ⇒ bool" where
>   "notempty s = (∃x::int. x ∈ s)"

introduces on the basis of the set predicate ∈ the predicate 'is not empty'. The definition

> definition ints :: "int ⇒ int ⇒ iset" where
>   "ints n m = { i::int. n ≤ i ∧ i ≤ m }"

of the function *ints* uses the set builder notation to return a set of integers in a given interval. From, this, the lemma

lemma L1: "∀(n::int)(m::int). n ≤ m ⟷ notempty (ints n m)" ...

can be proved which states that the resulting set is not empty, if and only if the interval is not empty. Next, the definition

> definition add :: "int ⇒ iset ⇒ iset" where
>   "add i s = { (i+x) | (x::int). x ∈ s }"

employs another variant of the set builder notation to define a function *add* that maps a set $s$ into another set by adding to every element of the original set the value $i$. From this definition, we may prove the formulas

> lemma L2: "∀(i::int).
>   add 0 s = s" ...
> lemma L3: "∀(s::iset)(i::int)(x::int).
>   x ∈ (add i s) ⟷ x-i ∈ s" ...
> lemma L4: "∀(s::iset)(i::int)(j::int).
>   add i (add j s) = add (i+j) s" ...

which state certain properties of *add*. Formula $L_3$ states a crucial fact which is employed in the proof of formula $L_4$. While both lemmas can be automatically proved, the automatic proof of lemma $L_4$ would not succeed without the previously proved formula $L_3$: the automatic search for a proof of $L_4$ is guided by the additional knowledge $L_3$ into a direction that would not be taken without this formula. In general, the work with Isabelle/HOL encourages to prove numerous auxiliary operations and lemmas that represent a whole 'theory' of the previously introduced predicates and functions; these formulas not only serve for documentation but are crucial for further automated proofs.

For instance, assume we would like to formalize and reason about the maximum of integer sets. Rather than just introducing a corresponding function *max*, we introduce by the definitions

```
definition ismax :: "int ⇒ iset ⇒ bool" where
"ismax m s = (m ∈ s ∧ (∀x::int. x ∈ s ⟶ x ≤ m))"

definition hasmax :: "iset ⇒ bool" where
"hasmax s = (∃m::int. ismax m s)"

definition max :: "iset ⇒ int" where
"max s = (SOME (m::int). ismax m s)"
```

two predicates *ismax* and *hasmax* accompanying *max*. Predicate *ismax* is true for argument integer $m$ and set $s$, if $m$ is the maximum element of $s$. Predicate *hasmax* is true for argument set $s$, if $s$ has a maximum. Given a set $s$, the implicitly defined function *max* returns such a maximum (provided that it exists). Based on these operations, we introduce the formulas

```
lemma M1: "∀(s::iset). hasmax s ⟶
  max s ∈ s ∧ (∀x::int. x ∈ s ⟶ x ≤ max s)" ...
lemma M2: "∀(s::iset). hasmax s ⟶
  (∀x::int. x ∈ s ∧ x ≠ max s ⟶ x < max s)" ...
lemma M3: "∀(s::iset). hasmax s ∧ hasmax(s - { max s }) ⟶
  max s > max (s - { max s })" ...
```

that gradually introduce knowledge about *max*: if $s$ has a maximum, this maximum satisfies the required conditions; every other element of $s$ is smaller than the maximum; if we remove the maximum from $s$ and the result still has a maximum, this is smaller than the original maximum. All of these formulas can be automatically proved with the help of the previously proved formulas. However, without $M_1$ and $M_2$, the automatic proof of $M_3$ would fail (i.e., it would be aborted after a certain time bound).

# Recursion

**5**

*Die Lehre von der Wiederkehr ist zweifelhaften Sinns. (The doctrine of recurrence is of dubious meaning.)*

— *Wilhelm Busch*

In this chapter, we discuss the mathematical principles of recursive definitions; while such definitions are less prominent in classical mathematics, they pervade many areas of computer science and thus deserve special attention. We start with a simple logical characterization of recursion that neither determines a unique function nor guarantees its existence. We then discuss the form of "primitive recursion" that always guarantees the existence of a unique solution but is of limited expressiveness. The main part of this chapter is then dedicated to a general theory of recursion based on the concept of "least/greatest fixed points" of certain higher-order functions that arise from recursive definitions. We show how this theory gives rise to "inductive" and "coinductive" definitions of relations and functions and finally discuss the principles of proving properties about the entities defined in this way.

## 5.1 Recursive Definitions

Sometimes functions and relations are most naturally described in a "recursive" style where the value of a function/relation is determined from the result of the application of the same function/relation to another value. Take for example the following "definition" of the factorial function on the natural numbers:

$$fac(n) := \text{if } n = 0 \text{ then } 1 \text{ else } n \cdot fac(n-1)$$

If we ignore, that this definition is actually not well-formed according to our current standards (which demand that every function applied on the right side must have been defined in a *prior* definition), it implies the following equalities:

$fac(0) = 1$
$fac(1) = 1 \cdot fac(0) = 1 \cdot 1 = 1$
$fac(2) = 2 \cdot fac(1) = 2 \cdot 1 = 2$
$fac(3) = 3 \cdot fac(2) = 3 \cdot 2 = 6$

. . . .

Since the value of $fac(n)$ is therefore uniquely determined for every natural number $n$, the function seems well-defined. However, not every recursive definition has such a nice behavior. Take for instance

$f(n) :=$ if $n = 0$ then 1 else $f(n + 1)$

which allows us to deduce $f(0) = 0$ and then only

$f(1) = f(2) = f(3) = \ldots$

Therefore, for every $i \geq 1$, possibly $f(i) = 0$ might hold, but also $f(i) = 1$ or every other result is possible. A recursive function definition therefore does not necessarily denote a function uniquely. Furthermore, a recursive definition does not necessarily denote a function at all: take e.g. the definition

$g(n) :=$ if $n = 0$ then 1 else $n \cdot g(n + 1)$

which is similar to the definition of function $fac$ above, except that the recursive application has argument $n + 1$ rather than $n - 1$. One can show that there exists *no* sequence of natural numbers $n_1 = g(1)$, $n_2 = g(2)$, $n_3 = g(3)$, $\ldots$ because the constraints

$n_1 = 1 \cdot n_2, \; n_2 = 2 \cdot n_3, \; n_3 = 3 \cdot n_4, \; \ldots$

imply $n_1 = n_2 > n_3 > n_4 > \ldots$, but there exists no infinite sequence of decreasing natural numbers. Therefore there exists no function $g$ with the property required by the definition above.

Above examples demonstrate that recursive definitions may have problematic consequences. Nevertheless, from the logical perspective there is a simple way to handle such definitions: we may define $f$ (and similarly also $g$) with the help of the set-theoretic choice function as follows:

$f :=$ choose $f \in \mathbb{N} \to \mathbb{N}. \; \forall n \in \mathbb{N}. \; f(n) =$ if $n = 0$ then 1 else $f(n + 1)$

Here we treat the definition as a universally quantified equation that the function is expected to satisfy. In more detail, this definition introduces $f$ as *some* function in $\mathbb{N} \to \mathbb{N}$ for which the property

$$\forall n \in \mathbb{N}. \, f(n) := \text{if } n = 0 \text{ then } 1 \text{ else } f(n + 1)$$

holds, *provided* that such a function exists; if no such function exists, the value of $f$ is unknown.

We usually hide the use of the choice function by the following abstraction.

**Definition 5.1** *(Recursive Function Definitions)* A *recursive function definition* has form

> recursive $f: A_1 \times \ldots \times A_n \to B$
> $f(x_1, \ldots, x_n) := T_f[x_1, \ldots, x_n]$

where $f$ is a symbol that is not a constant symbol in the current first-order language, $A_1, \ldots, A_n$ and $B$ are sets, and $T_f[x_1, \ldots, x_n]$ is a term with free variables $x_1, \ldots, x_n$ in the first-order language that is derived from the current language by adding symbol $f$ as a constant symbol and (if not yet there) as a variable.

This definition introduces the function

> $f := \text{choose } f \in A_1 \times \ldots \times A_n \to B.$
> $\forall x_1 \in A_1, \ldots, x_n \in A_n. \, f(x_1, \ldots, x_n) = T_f[x_1, \ldots, x_n].$

**Example 5.1** The recursive definition

> recursive $fac: \mathbb{N} \to \mathbb{N}$
> $fac(n) := \text{if } n = 0 \text{ then } 1 \text{ else } n \cdot fac(n - 1)$

introduces the factorial function

$$fac := \text{choose } fac \in \mathbb{N} \to \mathbb{N}. \, \forall n \in \mathbb{N}. \, fac(n) = \text{if } n = 0 \text{ then } 1 \text{ else } n \cdot fac(n - 1)$$

with the property

$$fac(0) = 1, \, fac(1) = 1, \, fac(2) = 2, \, fac(3) = 6, \, \ldots$$

The function is thus uniquely defined.                                                          □

While it is tempting to think of a recursive function definition being evaluated analogously to the execution of a recursively defined computer program, this is misleading: a recursive function definition may have an unintended consequence that clearly separates it from the definition of a program.

**Example 5.2** The recursive definition

recursive $ifac\colon \mathbb{Z} \to \mathbb{Z}$
$ifac(n) :=$ if $n = 0$ then $1$ else $n \cdot ifac(n - 1)$

is similar to that of Example 5.1 except that $ifac$ is defined on $\mathbb{Z}$ rather than on $\mathbb{N}$.

Now we intuitively might expect that for non-negative arguments the results of $ifac$ coincide with those of the function $fac$ in Example 5.1, while for negative values the result of $ifac$ might be "undefined" or "arbitrary". However, this intuition is misleading: there exists no infinite sequence of integer numbers $n_1 = ifac(-1)$, $n_2 = ifac(-2)$, $n_3 = ifac(-3)$, ... with the property

$$n_1 = -1 \cdot n_2, \; n_2 = -2 \cdot n_3, \; n_3 = -3 \cdot n_4, \; \ldots$$

because this would imply the existence of an infinitely decreasing sequence of natural numbers $|n_1| = |n_2|$, $|n_2| > |n_3|$, $|n_3| > |n_4|$, ... (where $|n_i|$ denotes the absolute value of $n_i$). Therefore there also exists no function in $\mathbb{Z} \to \mathbb{Z}$ that satisfies the defining equation and consequently the value of $ifac$ is undefined: even the value of the application $ifac(0)$ is unknown.                                                    □

As demonstrated by the example above, the treatment of a recursive definition as a logical constraint does not yield a function that (corresponding to a computer program that for some inputs may not terminate) may on some arguments be "undefined". For defining such functions that behave like computer programs, the following subsection introduces the necessary mathematical machinery.

## 5.2    Primitive Recursion

While the previous section has shown us that recursive function definitions should be taken with care, there are nevertheless certain recursive function definitions such as

recursive $fac\colon \mathbb{N} \to \mathbb{N}$
$fac(n) :=$ if $n = 0$ then $1$ else $n \cdot fac(n - 1)$

that determine a unique function; the key here is that for the smallest possible argument $0$ the result of $fac(0) = 1$ is explicitly determined, while for every other argument $n$ the result of $fac(n)$ is determined by referring to the result of $fac(n - 1)$ with smaller function argument $n - 1$. Since thus for every natural number $n$ the value of $fac(n)$ can be determined by a finite number of "unfoldings" of the definition, the function is uniquely determined. The same principle can be applied to other domains than the natural numbers, provided that they satisfy a certain constraint.

**Definition 5.2**  (*Well-founded relation*) Let $S$ be a set. The binary relation $\prec \subseteq S \times S$ is *well-founded* on $S$ if and only if

$$\neg \exists s \in S^{\omega}. \, \forall i \in \mathbb{N}. \, s(i+1) \prec s(i)$$

In other words, there does not exists any infinitely descending chain $\ldots \prec s(i+1) \prec s(i) \prec \ldots \prec s(2) \prec s(1) \prec s(0)$.

The prototypical example of a well-founded relation is the "less than" relation $<$ on the natural numbers $\mathbb{N}$.

**Proposition 5.1**  (Primitive recursion) *A recursive function definition is primitively recursive, indicated as*

> primitive $f : A_1 \times \ldots \times A_n \to B$
>
> $f(x_1, \ldots, x_n) := T_f[x_1, \ldots, x_n]$

*if there exists some argument position $i$ with $1 \leq i \leq n$ and a well-founded relation $\prec$ on $A_i$, such that for every function application $f(T_1, \ldots, T_i, \ldots, T_n)$ in $T_f[x_1, \ldots, x_n]$, we have*

$$T_i \prec x_i,$$

*i.e., the function is in position $i$ only applied to an argument $T_i$ that is smaller than the original argument $x_i$ with respect to well-founded order $\prec$. The function introduces the same function as a corresponding recursive definition; by construction, however, this function is uniquely determined.*

The prototypical example of primitively recursively defined functions are functions on natural numbers.

**Example 5.3**  The function definition

> primitive $fac : \mathbb{N} \to \mathbb{N}$
>
> $fac(n) :=$ if $n = 0$ then $1$ else $n \cdot fac(n-1)$

is well-formed for argument position 1 and the well-founded relation $<$ on $\mathbb{N}$.    □

Primitive recursion is often denoted in a "rule-based" style, which is demonstrated by the following example.

**Example 5.4**  The definition of the factorial function given in Example 5.3 can also be written in a rule-based style as

primitive $fac \colon \mathbb{N} \to \mathbb{N}$

$fac(0) := 1$

$fac(m + 1) := (m + 1) \cdot fac(m)$

which can be read as the the following set of rules:

- If $n = 0$, then $fac(n) = fac(0) = 1$.
- If there exists some $m \in \mathbb{N}$ with $n = m + 1$ (i.e., $m = n - 1$), then $fac(n) = fac(m + 1) = (m + 1) \cdot fac(m) = n \cdot fac(n - 1)$.    □

In Sect. 1.2, we have introduced *structural induction* as a principle for defining functions on abstract syntax trees; actually, this principle is just another prominent application of primitive recursion in a rule-based style.

**Example 5.5**  The valuation function of Example 1.5 defined as

$[\![ \cdot ]\!] \colon Expression \to \mathbb{N}$

$$[\![ N ]\!] := [\![ N ]\!]$$
$$[\![ E_1 + E_2 ]\!] := [\![ E_1 ]\!] + [\![ E_2 ]\!]$$
$$[\![ E_1 \times E_2 ]\!] := [\![ E_1 ]\!] \cdot [\![ E_2 ]\!]$$

is primitive recursively defined with argument position 1 and the ordering "has lower height than" on the domain *Expression* of abstract syntax trees.    □

Function definition by structural induction is therefore just function definition by primitive recursion.

## 5.3    Least and Greatest Fixed Points

We will now turn our attention to recursive definitions that are not necessarily primitively recursive. The core idea for the further discussion is that we can extract from a recursive definition such as

recursive $fac \colon \mathbb{N} \to \mathbb{N}$

$fac(n) := $ if $n = 0$ then 1 else $n \cdot fac(n - 1)$.

the pattern on the right side of the equation as a non-recursive higher-order function
(also called *functional*)

$$F : (\mathbb{N} \to \mathbb{N}) \to (\mathbb{N} \to \mathbb{N})$$

$$F(f) := \lambda n \in \mathbb{N}. \text{ if } n = 0 \text{ then } 1 \text{ else } n \cdot f(n-1).$$

This functional $F$ takes a function $f : \mathbb{N} \to \mathbb{N}$ as an argument and returns another
function in $\mathbb{N} \to \mathbb{N}$ as a result. We then have

$$F(fac) = \lambda n \in \mathbb{N}. \text{ if } n = 0 \text{ then } 1 \text{ else } n \cdot fac(n-1) = fac$$

The equation $F(fac) = fac$ expresses the fact that function *fac* is a *fixed point* of
functional $F$: given *fac* as an argument, $F$ returns *fac* as a result. We are now going
to investigate a theory that ensures the existence of fixed points and that provides
means to choose among multiple possible fixed points that one which is most suitable
for a particular purpose. Since functionals like $F$ map functions to functions and
functions are sets, we will restrict our considerations to fixed points of functions on
sets.

**Definition 5.3** (*Fixed point*) Let $F : \mathsf{Set}(S) \to \mathsf{Set}(S)$ and $A \in \mathsf{Set}(S)$, i.e.,
$A \subseteq S$. Then $A$ is a *fixed point* of $F$ if

$$F(A) = A.$$

Intuitively, we will for the moment think of sets as relations and consider the
more special case of functions later. The central relationship between relations is
that of set inclusion: given a set $S$, the elements of $\mathsf{Set}(S)$, i.e., the relations on $S$,
are partially ordered by the subset relationship, with $S$ as the greatest element and
the empty set $\varnothing$ as the least one. For instance, for $S = \{1, 2, 3\}$ we have the partial
order depicted in Fig. 5.1 where $A_1 \to A_2$ indicates $A_1 \subseteq A_2$.

Given two relations $A_1$ and $A_2$ on set $S$, i.e., $A_1 \subseteq S$ and $A_2 \subseteq S$, we have
$A_1 \subseteq A_2 \Leftrightarrow \forall x \in S. \, x \in A_1 \Rightarrow x \in A_2$ and therefore

$$A_1 \subseteq A_2 \Leftrightarrow \forall x \in S. \, A_1\langle x \rangle \Rightarrow A_2\langle x \rangle.$$

In other words, if $A_1 \subseteq A_2$, then relation $A_1$ is at least as "strong" as relation $A_2$.
Figure 5.1 thus depicts the ordering of relations according to their "strength", with $\varnothing$
being the strongest relation (because for every $x \in S$, we have $x \notin \varnothing$, i.e., $\neg\varnothing\langle x \rangle$)
and $S$ being the weakest one (because for every $x \in S$, we have $x \in S$, i.e., $S\langle x \rangle$).

The ordering relation $\subseteq$ allows us to pinpoint two particular fixed points.

**Fig. 5.1** The Subset
Ordering

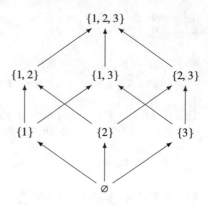

**Definition 5.4** (*Least and greatest fixed points*) Let $F : \mathsf{Set}(S) \to \mathsf{Set}(S)$ and
$A \in \mathsf{Set}(S)$, i.e., $A \subseteq S$. Then $A$ is the *least fixed point* of $F$ if

$$F(A) = A \wedge \forall A'. \, F(A') = A' \Rightarrow A \subseteq A'$$

i.e., $A$ is a fixed point of $F$ that is a subset of every other fixed pont $A'$ of $F$.
$A$ is the *greatest fixed point* of $F$ if

$$F(A) = A \wedge \forall A'. \, F(A') = A' \Rightarrow A' \subseteq A$$

i.e., $A$ is a fixed point of $F$ that is a superset of every other fixed pont $A'$ of $F$.

By definition if there exists a least/greatest fixed point, it is unique, i.e., there is
no other fixed point that is also least/greatest. Intuitively, the least fixed point is the
"strongest" one: it "implies" (is contained in) every other fixed point. the greatest
fixed point is the "weakest" one; it "is implied" by (contains) every other fixed point.
    We will now investigate under which conditions, least and greatest fixed points
exist. For this purpose, we impose a particular constraint on function $F$.

**Definition 5.5** (*Monotonicity*) A function $F : \mathsf{Set}(S) \to \mathsf{Set}(S)$ is *monotonic*
(with respect to set inclusion) if

$$\forall A \in \mathsf{Set}(S), A' \in \mathsf{Set}(S). \, A \subseteq A' \Rightarrow F(A) \subseteq F(A').$$

In other words if $F$ is monotonic, it preserves the subset order from its arguments to its results: if $A$ is at least as strong as $A'$, then $F(A)$ is at least as
strong as $F(A')$.

**Example 5.6** Function $F$ defined as

$$F: \mathsf{Set}(\mathbb{N}) \to \mathsf{Set}(\mathbb{N})$$
$$F(A) := \pi i \in \mathbb{N}. \ \neg A\langle i \rangle$$

is *not* monotonic: Define relations $A\langle i \rangle :\Leftrightarrow i < 3$ and $A'\langle i \rangle :\Leftrightarrow i < 5$ on $\mathbb{N}$. Then $A \subseteq A'$, because $\forall i \in \mathbb{N}. \ A\langle i \rangle \Rightarrow A'\langle i \rangle$. Furthermore, $F(A)\langle i \rangle \Leftrightarrow i \geq 3$ and $F(A')\langle i \rangle \Leftrightarrow i \geq 5$. Therefore $F(A)\langle 3 \rangle$ but $\neg F(A')\langle 3 \rangle$, i.e., $3 \in F(A)$ but $3 \notin F(A')$. Thus $F(A) \nsubseteq F(A')$.

Now take function $G$ defined as

$$G: \mathsf{Set}(\mathbb{N} \times \mathbb{N}) \to \mathsf{Set}(\mathbb{N} \times \mathbb{N})$$
$$G(A) := \pi x \in \mathbb{N}, y \in \mathbb{N}. \ x = 0 \vee (y \neq 0 \wedge A\langle x - 1, y - 1 \rangle)$$

This function is monotonic: Take $A$ and $A'$ with $A \subseteq A'$. We show $G(A) \subseteq G(A')$. Take arbitrary $x, y \in \mathbb{N}$ and assume $G(A)\langle x, y \rangle$. We show $G(A')\langle x, y \rangle$. If $x = 0$, we are done. If $x \neq 0$, we know $y \neq 0$ and $A\langle x - 1, y - 1 \rangle$. From this and $A \subseteq A'$, we know also $A'\langle x - 1, y - 1 \rangle$ and therefore $G(A')\langle x, y \rangle$.  □

The following theorem due to Bronislaw Knaster and Alfred Tarski establishes the existence of least and greatest fixed points of monotonic functions.

**Theorem 5.1** (Knaster–Tarski theorem) *If function $F: \mathsf{Set}(S) \to \mathsf{Set}(S)$ is monotonic, then* lfp $F$ *is the least fixed point of $F$ and* gfp $F$ *is its greatest fixed point for the following definitions of functions* lfp *and* gfp*:*

$$\textsf{lfp}: (\mathit{Set}(S) \to \mathit{Set}(S)) \to \mathit{Set}(S)$$
$$\textsf{lfp } F := \bigcap \{A \in \mathit{Set}(S) \mid F(A) \subseteq A\}$$

$$\textsf{gfp}: (\mathit{Set}(S) \to \mathit{Set}(S)) \to \mathit{Set}(S)$$
$$\textsf{gfp } F := \bigcup \{A \in \mathit{Set}(S) \mid A \subseteq F(A)\}$$

*Thus* lfp $F$ *is the least of all pre-fixed points of $F$, i.e., the least among all sets $A$ that satisfy the equation $F(A) \subseteq A$; conversely,* gfp $F$ *is the greatest among all post-fixed points of $F$, i.e., the greatest of all sets $A$ that satisfy the equation $A \subseteq F(A)$.*

***Proof*** Let $F: \mathsf{Set}(S) \to \mathsf{Set}(S)$ be monotonic. To prove that lfp $F$ is a fixed point of $F$, we first show $F(\textsf{lfp } F) \subseteq \textsf{lfp } F$ and then show lfp $F \subseteq F(\textsf{lfp } F)$.

By definition of lfp $F$, we have

$$\forall A.\ F(A) \subseteq A \Rightarrow \text{lfp } F \subseteq A.$$

Thus, since $F$ is monotonic, we also have

$$\forall A.\ F(A) \subseteq A \Rightarrow F(\text{lfp } F) \subseteq F(A).$$

and therefore

$$\forall A.\ F(A) \subseteq A \Rightarrow F(\text{lfp } F) \subseteq A.$$

and thus, by definition of lfp $F$, $F(\text{lfp } F) \subseteq \text{lfp } F$.

Furthermore, since $F(\text{lfp } F) \subseteq \text{lfp } F$, we have by the monotonicity of $F$ also $F(F(\text{lfp } F)) \subseteq F(\text{lfp } F)$ and thus, by the definition of lfp $F$, also lfp $F \subseteq F(\text{lfp } F)$.

To show that lfp $F$ is also the least fixed point of $F$, take an arbitrary fixed point $A \in \text{Set}(S)$. Since therefore $F(A) = A$, we also have $F(A) \subseteq A$ and thus, by the definition of lfp $F$, lfp $F \subseteq A$.    □

The Knaster–Tarski Theorem ensures the existence of least/greatest fixed points, which is of central importance. However, it does not help us much in finding these fixed points; we will therefore strive towards a more constructive characterization of least/greatest fixed points which is based on the following concept.

**Definition 5.6** (*Function iteration*) Let $F : \text{Set}(S) \rightarrow \text{Set}(S)$ and $i \in \mathbb{N}$. Then the function $F^i : \text{Set}(S) \rightarrow \text{Set}(S)$ defined by primitive recursion as

$$F^0 := \lambda A \in \text{Set}(A).\ A$$
$$F^{i+1} := \lambda A \in \text{Set}(A).\ F^i(F(A))$$

denotes the $i$-fold composition of $F$, i.e., $F^i(A)$ is the result of the $i$-fold iterated application of $F$ to set $A$.

We then define

$$F\!\uparrow\ := \bigcup_{i \in \mathbb{N}} F^i(\varnothing)$$

as the *infinite upward iteration* of $F$ and

$$F\!\downarrow\ := \bigcap_{i \in \mathbb{N}} F^i(S)$$

as the *infinite downward iteration* of $F$.

The upward iteration $F\uparrow$ denotes the union of the infinite sequence

$$F^0(\varnothing) = \varnothing$$
$$F^1(\varnothing) = F(\varnothing)$$
$$F^2(\varnothing) = F(F(\varnothing))$$
$$F^3(\varnothing) = F(F(F(\varnothing)))$$
$$\ldots$$

If $F$ is monotonic, then we have $\forall i \in \mathbb{N}.\ F^i(\varnothing) \subseteq F^{i+1}(\varnothing)$, i.e., every new set in the sequence may have *more* elements than the previous one and thus *add* some elements to $F\uparrow$.

**Example 5.7** Consider the monotonic function

$$E : \mathrm{Set}(\mathbb{N}) \to \mathrm{Set}(\mathbb{N})$$
$$E(Even) := \pi x \in \mathbb{N}.\ x = 0 \vee (x \neq 1 \wedge Even\langle x - 2\rangle)$$

We then have

$$E^0(\varnothing) = \varnothing$$
$$E^1(\varnothing) = E(\varnothing) = \{0\}$$
$$E^2(\varnothing) = E(\{0\}) = \{0, 2\}$$
$$E^3(\varnothing) = E(\{0, 2\}) = \{0, 2, 4\}$$
$$\ldots$$

i.e., $E^i(\varnothing)$ contains the first $i$ even natural numbers; consequently $E\uparrow$ is the set of all even natural numbers, i.e, it denotes the unary relation "is even" on $\mathbb{N}$.  □

**Example 5.8** Consider the monotonic function as

$$L : \mathrm{Set}(\mathbb{N} \times \mathbb{N}) \to \mathrm{Set}(\mathbb{N} \times \mathbb{N})$$
$$L(LessEq) := \pi x \in \mathbb{N}, y \in \mathbb{N}.\ x = 0 \vee (y \neq 0 \wedge LessEq\langle x - 1, y - 1\rangle).$$

(which is identical to the function $G$ defined in Example 5.6). We then have

$$L^0(\varnothing) = \varnothing$$
$$L^1(\varnothing) = L^0(\varnothing) \cup \{\langle 0, y\rangle \mid y \in \mathbb{N} \wedge 0 \leq y\}$$
$$L^2(\varnothing) = L^1(\varnothing) \cup \{\langle 1, y\rangle \mid y \in \mathbb{N} \wedge 1 \leq y\}$$
$$L^3(\varnothing) = L^2(\varnothing) \cup \{\langle 2, y\rangle \mid y \in \mathbb{N} \wedge 2 \leq y\}$$
$$\ldots$$

Thus $L^i(\varnothing)$ contains all pairs $x, y$ with $x < i$ and $x \le y$. Consequently $L\uparrow$ is the set of all pairs of natural numbers $x, y$ with $x \le y$, i.e., it denotes the binary relation "is less than or equal" on $\mathbb{N}$. $\qquad\qquad\qquad\qquad\qquad\qquad\qquad\qquad\qquad\qquad\qquad\qquad\qquad$ □

Correspondingly, the downward iteration $F\downarrow$ denotes the intersection of the infinite sequence

$$F^0(S) = S$$
$$F^1(S) = F(S)$$
$$F^2(S) = F(F(S))$$
$$F^3(S) = F(F(F(S)))$$
$$\cdots$$

If $F$ is monotonic, then we have $\forall i \in \mathbb{N}.\ F^{i+1}(S) \subseteq F^i(S)$, i.e., every new set in the sequence may have *less* elements than the previous one and thus *remove* some elements from $F\downarrow$.

**Example 5.9** Take the monotonic function

$$N: \mathsf{Set}(\mathbb{N}^\omega) \to \mathsf{Set}(\mathbb{N}^\omega)$$
$$N(NoZero) := \pi x \in \mathbb{N}^\omega.\ x(0) \ne 0 \wedge NoZero(tail(x))$$

which operates on relations of infinite number sequences ("streams"). The auxiliary function

$$tail: \mathbb{N}^\omega \to \mathbb{N}^\omega$$
$$tail(x) := \lambda i \in \mathbb{N}.\ x(i + 1)$$

denotes the stream that is identical to stream $x$ except that $x(0)$ has been removed, i.e. $tail(x)(0) = x(1)$, $tail(x)(1) = x(2)$, etc. We then have

$$N^0(\mathsf{Set}(\mathbb{N}^\omega)) = \mathsf{Set}(\mathbb{N}^\omega)$$
$$N^1(\mathsf{Set}(\mathbb{N}^\omega)) = N^0(\mathsf{Set}(\mathbb{N}^\omega))\setminus\{x \in \mathbb{N}^\omega \mid x(0) = 0\}$$
$$N^2(\mathsf{Set}(\mathbb{N}^\omega)) = N^1(\mathsf{Set}(\mathbb{N}^\omega))\setminus\{x \in \mathbb{N}^\omega \mid x(1) = 0\}$$
$$N^3(\mathsf{Set}(\mathbb{N}^\omega)) = N^2(\mathsf{Set}(\mathbb{N}^\omega))\setminus\{x \in \mathbb{N}^\omega \mid x(2) = 0\}$$
$$\cdots$$

Thus $N^i(\mathsf{Set}(\mathbb{N}^\omega))$ contains all streams that have no 0 in the first $i$ positions. Consequently $N\downarrow$ is the set of all streams that do not have a zero in any position, i.e., it denotes the unary relation "has no zero" on streams. $\qquad\qquad\qquad\qquad\qquad\qquad$ □

Above examples seem to suggest that the least/greatest fixed point of every mono-
tone function may be reached by infinite iteration "from below" or "from above",
but this is not not necessarily the case. The true situation is stated by the following
proposition.

**Proposition 5.2** (Least/greatest fixed points and function iteration)   *Let
$F \colon Set(S) \to Set(S)$ be monotonic. Then $F{\uparrow} \subseteq \text{lfp } F$ and $\text{gfp } F \subseteq F{\downarrow}$.*

In other words, by upward iteration we may in general only approximate the least
fixed point from below; the result is a set which is included in the least fixed point.
Conversely, by downward iteration we may in general only approximate the greatest
fixed point "from above"; the result is a set which includes the greatest fixed point.

***Proof*** We show $F{\uparrow} \subseteq \text{lfp } F$; the proof of $\text{gfp } F \subseteq F{\downarrow}$ is dual.
   By the definitions of $F{\uparrow}$ and $\text{lfp } F$, it suffices to show

$$\forall i \in \mathbb{N}, A \in Set(S). \; F(A) \subseteq A \Rightarrow F^i(\varnothing) \subseteq A.$$

We proceed by induction on $i$. In the induction base $i = 0$, we have $F^i(\varnothing) = \varnothing$ and

$$\forall A \in Set(S). \; F(A) \subseteq A \Rightarrow \varnothing \subseteq A.$$

In the induction step, we assume

$$\forall A \in Set(S). \; F(A) \subseteq A \Rightarrow F^i(\varnothing) \subseteq A$$

and show

$$\forall A \in Set(S). \; F(A) \subseteq A \Rightarrow F^{i+1}(\varnothing) \subseteq A$$

We take arbitrary $A \in Set(S)$ with $F(A) \subseteq A$ and show $F^{i+1}(\varnothing) \subseteq A$. From
$F(A) \subseteq A$ and the induction assumption, we know $F^i(\varnothing) \subseteq A$. From the monotonicity of $F$, we thus know $F^{i+1}(\varnothing) \subseteq F(A)$. From $F(A) \subseteq A$, we thus know
$F^{i+1}(\varnothing) \subseteq A$.                                                              $\square$

The results derived so far concerning the relationship between fixed points and
function iteration can be summarized as follows.

**Proposition 5.3** (Iteration and fixed points) *Let $F : Set(S) \rightarrow Set(S)$ be monotonic and $n \in \mathbb{N}$. Then the following relationships hold:*

$$\varnothing \subseteq F^n(\varnothing) \subseteq \bigcup_{i=0}^{n} F^i(\varnothing) \subseteq F\!\uparrow \;\subseteq\; \textit{lfp } F$$

$$\subseteq \textit{gfp } F \subseteq F\!\downarrow \;\subseteq\; \bigcap_{i=0}^{n} F^i(S) \subseteq F^n(S) \subseteq S$$

By iteration, the least and greatest fixed points can be thus bounded from below and from above. In order to capture by iteration the fixed points exactly, we have to demand from function $F$ a bit more than just monotonicity.

**Definition 5.7** (*Chain*) For every set $S$, we define

$$S_{\subseteq}^{\omega} := \left\{ A \in Set(S)^{\omega} \mid \forall i \in \mathbb{N}.\; A(i) \subseteq A(i+1) \right\}$$

as the set of all (infinite) *chains* in $S$.
Every element $A$ of $S_{\subseteq}^{\omega}$, i.e., every chain in $S$, is thus an infinitely increasing sequence $A(0) \subseteq A(1) \subseteq A(2) \subseteq \ldots$ of subsets of $S$.

**Definition 5.8** (*Continuity*) A monotonic function $F : Set(S) \rightarrow Set(S)$ is *upward continuous* (with respect to set inclusion) if and only if

$$\forall A \in S_{\subseteq}^{\omega}.\; F\left(\bigcup_{i\in\mathbb{N}} A(i)\right) \subseteq \bigcup_{i\in\mathbb{N}} F(A(i)).$$

$F$ is *downward continuous* (with respect to set inclusion), if and only if

$$\forall A \in S_{\subseteq}^{\omega}.\; \bigcap_{i\in\mathbb{N}} F(A(i)) \subseteq F\left(\bigcap_{i\in\mathbb{N}} A(i)\right).$$

$F$ is *continuous* (with respect to set inclusion) if and only if it is both upward and downward continuous.

Thus for every infinitely increasing chain $A(0) \subseteq A(1) \subseteq A(2) \subseteq \ldots$, the application of an upward continuous function $F$ to the limit of this sequence yields a subset of the limit of the applications of $F$ to the individual elements of the sequence. This

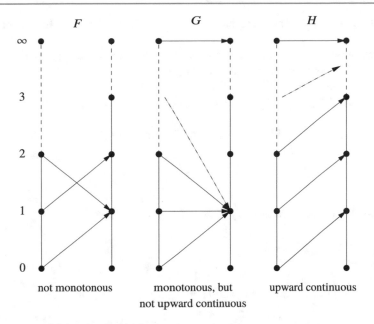

**Fig. 5.2** Monotonicity versus Continuity

relationship is illustrated in Fig. 5.2 that depicts three functions $F$, $G$, and $H$ that operate on finite sets numbered $0, 1, 2 \ldots$; we assume that set $i$ is included in set $i + 1$. Function $F$ is not monotonic, since it maps set 2 to set 1 but the smaller set 1 to the bigger set 2. Function $G$ that maps every set $i$ to set 1 is monotonic but not upward continuous, because it maps the infinite union of all sets denoted by $\infty$ to itself, while the union of all function results is the set 1. Function $H$ is upward continuous, because it maps increasing arguments to increasing results; the union of the results is $\infty$ to which also $\infty$ is mapped.

Conversely, for every infinitely decreasing chain $A(0) \supseteq A(1) \supseteq A(2) \supseteq \ldots$, the application of an downward continuous function $F$ to the limit of this sequence yields a superset of the applications of $F$ to the individual elements of the sequence.

**Example 5.10** Consider the function

$F \colon \mathsf{Set}(\mathbb{N}) \to \mathsf{Set}(\mathbb{N})$

$F(A) := \pi x \in \mathbb{N}. \ (\forall y \in \mathbb{N}. \ A\langle y\rangle)$

which, given relation $A \subseteq \mathbb{N}$, returns a relation that, for every argument $x \in \mathbb{N}$ returns "true" if $\forall y \in \mathbb{N}. \ A\langle y\rangle$, and always returns "false", otherwise. We thus deduce

$F(A) = \text{if } \forall y \in \mathbb{N}. \ A\langle y\rangle \text{ then } \mathbb{N} \text{ else } \varnothing.$

From this description, it is easy to see that $F$ is monotonic: assume $F$ is not monotonic. Then we have some $A$ and $A'$ with $A \subseteq A'$ but $F(A) \not\subseteq F(A')$. By definition of $F$, we thus know $F(A) = \mathbb{N}$ and $F(A') = \varnothing$. Then $\forall y \in \mathbb{N}. \ A\langle y\rangle$ but

$\neg \forall y \in \mathbb{N}.\ A'\langle y \rangle$. We therefore have $\neg \forall y \in \mathbb{N}.\ A\langle y \rangle \Rightarrow A'\langle y \rangle$ which contradicts $A \subseteq A'$.

However, $F$ is *not* upward continuous: assume that $F$ is upward continuous and take the infinitely increasing sequence of relations

$$\mathbb{N}_0 \subseteq \mathbb{N}_1 \subseteq \mathbb{N}_2 \subseteq \mathbb{N}_3 \subseteq \ldots$$

where $\mathbb{N}_i \langle y \rangle \Leftrightarrow y < i$. Thus for every $i \in \mathbb{N}$ we have $\neg \forall y \in \mathbb{N}.\ \mathbb{N}_i \langle y \rangle$ and therefore

$$\bigcup_{i \in \mathbb{N}} F(\mathbb{N}_i) = \bigcup_{i \in \mathbb{N}} \varnothing = \varnothing.$$

However, we clearly also have $\forall y \in \mathbb{N}.\ \mathbb{N} \langle y \rangle$ and therefore

$$F(\bigcup_{i \in \mathbb{N}} \mathbb{N}_i) = F(\mathbb{N}) = \mathbb{N}.$$

Since $\mathbb{N} \not\subseteq \varnothing$, this contradicts upward continuity. □

As demonstrated by above example, from an upward continuous function we cannot construct a relation that is true for the infinite limit of upward iteration but false for each of the finite approximations of that limit. Dually, from a downward continuous function we cannot construct a relation that is false for the limit of downward iteration but true for each of the finite approximations of that limit.

In this sense, upward/downward iteration behaves like a computer program that processes a conceptually infinite object: however many steps the program may run, it can only investigate a finite part of that object and can make decisions only on the basis of that part of the object is has seen so far. The function $F$ in Example 5.10, however, is able to distinguish between finite and infinite subsets of $\mathbb{N}$; the function is thus not continuous and a corresponding computer program cannot be implemented.

The role of continuity for fixed points is established by the following theorem which is due to Stephen Kleene.

**Theorem 5.2** (Kleene fixed point theorem) *For every function $F : Set(S) \to Set(S)$, the following holds:*

- *If $F$ is upward continuous, $F\uparrow$ is the least fixed point of $F$, i.e., $F\uparrow = lfp\ F$.*
- *If $F$ is downward continuous, $F\downarrow$ is the greatest fixed point of $F$, i.e., $F\downarrow = gfp\ F$.*

***Proof*** Let $F$ be upward continuous. We show that $F\uparrow$ is a fixed point of $F$. We have

$$F(F\uparrow) = F(\bigcup_{i\in\mathbb{N}} F^i(\varnothing)) \overset{(1)}{\subseteq} \bigcup_{i\in\mathbb{N}} F(F^i(\varnothing)) = \bigcup_{i\in\mathbb{N}} F^{i+1}(\varnothing) \overset{(2)}{=} \bigcup_{i\in\mathbb{N}} F^i(\varnothing) = F\uparrow$$

where (1) holds because $F$ is upward continuous and (2) holds because $F^0(\varnothing) = \varnothing$. Since thus $F(F\uparrow) \subseteq F\uparrow$, we have by Theorem 5.1, lfp $F \subseteq F\uparrow$. From Proposition 5.2, we also have $F\uparrow \subseteq$ lfp $F$ and thus $F\uparrow =$ lfp $F$.

The proof of the second part of the theorem is dual.                               □

From the Kleene fixed point theorem, we have the following relationship between function iteration and least respectively greatest fixed points.

> **Proposition 5.4** (Iteration and fixed points) *Let $F: Set(S) \rightarrow Set(S)$ be continuous and $n \in \mathbb{N}$. Then the following relationship hold:*
>
> $$\varnothing \subseteq F^n(\varnothing) \subseteq \bigcup_{i=0}^{n} F^i(\varnothing) \subseteq F\uparrow = \textit{lfp } F$$
>
> $$\subseteq \textit{gfp } F = F\downarrow \subseteq \bigcap_{i=0}^{n} F^i(S) \subseteq F^n(S) \subseteq S$$

This proposition establishes that for upward/downward continuous functions, the (limit of) the upward/downward iteration process described above exactly captures the least/greatest fixed point. The functions in Examples 5.7 and 5.8 are upward continuous; their upward iteration thus indeed yields the least fixed point. The function in Example 5.9 is downward continuous; its downward iteration thus indeed yields the greatest fixed point.

We will usually deal with the fixed points of functions that are continuous and establish an alternative meaning of recursive definitions which is based on the least and greatest fixed points of such functions.

## 5.4   Defining Continuous Functions

For recursively defining relations, it will be helpful to have some guidelines of how to build continuous functions on relations. The central tool is following proposition.

**Proposition 5.5** (Composition of continuous functions) *Let both*
$F \colon \mathsf{Set}(S) \to \mathsf{Set}(S)$ *and* $G \colon \mathsf{Set}(S) \to \mathsf{Set}(S)$ *be continuous. Then also*
*the composition* $F \circ G \colon \mathsf{Set}(S) \to \mathsf{Set}(S)$ *defined as* $(F \circ G)(x) := G(F(x))$
*is continuous.*

We can thus build from simpler continuous functions more complex ones by
function composition. Since we consider functions on relations which are defined in
the form

$$G \colon \mathsf{Set}(A_1 \times \ldots \times A_n) \to \mathsf{Set}(A_1 \times \ldots \times A_n)$$
$$G(A) := \pi x_1 \in A_1, \ldots, x_n \in A_n.\ F$$

we have to investigate which formula $F$ gives rise to a continuous function. As
a starting point, it is easy to see that $A\langle T_1, \ldots, T_n \rangle$ (i.e., the atomic formula
$\langle T_1, \ldots, T_n \rangle \in A$) is upward and downward continuous in argument $A$ (with the
other arguments fixed). We now consider how the continuity of atomic formulas
extends to the continuity of formulas composed from logical connectives and quantifiers.

The monotonicity of the various logical connectives can be investigated from the
truth table in Fig. 2.2: if an argument of a connective is "false" but the result is "true",
the result must not become "false" if the argument becomes "true": since formula
"false" yields the relation $\varnothing$ while "true" yields $S$, this would violate the monotonicity of the corresponding relation. Therefore both disjunction and conjunction are
monotonic (and, as can be shown, also upward and downward continuous).

However, as the equivalence of ¬false and true and the equivalence of ¬true an
false show, negation is not monotonic (this was also demonstrated by Example 5.6).
Likewise, implication is not monotonic in its first argument, because true ⇒ false is
equivalent to false, but false ⇒ false is equivalent to true; however, implication it is
monotonic in its second argument. Equivalence is not monotonic in any of its arguments, because true ⇔ false is equivalent to false, but false ⇔ false is equivalent to
true and equivalence is commutative.

Furthermore, Example 5.10 demonstrates that universal quantification is not
upward continuous; dually, existential quantification is not downward continuous.
However, existential quantification is upward continuous: if $\exists x.\ A\langle x \rangle$ is true for
every finite set $A(0), A(1), \ldots$, it is also true for their infinite union. Dually, universal quantification is downward continuous.

Taking all this into account, we can build functions that are certainly continuous
by considering simple syntactic constraints.

**Proposition 5.6** (Continuous functions on relations) *Take function $G$ on a domain $Set(S_1 \times \ldots \times S_n)$ of n-ary relations defined as*

$$G \colon Set(S_1 \times \ldots \times S_n) \to Set(S_1 \times \ldots \times S_n)$$
$$G(A) := \pi x_1 \in S_1, \ldots, x_n \in S_n.\ F[x_1, \ldots, x_n]$$

*by formula $F[x_1, \ldots, x_n]$ whose free variables are among $x_1, \ldots, x_n$ and in which parameter $A$ only occurs in the form of the relation application $A\langle \_ \rangle$, i.e., $\_ \in A$.*

*Then $G$ is upward continuous if $F$ is a formula in the first-order language where the domain Formula is constructed according to the restricted grammar*

$F ::=$
  $\mid F_1 \wedge F_2 \mid F_1 \vee F_2 \mid \exists V.\ F$
  $\mid$ *let* $V = T$ *in* $F$

*i.e., without negation, implication, equivalence, and universal quantification.*

*Then $G$ is downward continuous if $F$ is a formula in the first-order language where the domain Formula is constructed according to the restricted grammar*

$F ::=$
  $\mid F_1 \wedge F_2 \mid F_1 \vee F_2 \mid \forall V.\ F$
  $\mid$ *let* $V = T$ *in* $F$

*i.e., without negation, implication, equivalence, and existential quantification.*

**Example 5.11** The function definitions in Examples 5.7, 5.8 and Example 5.9 all satisfy the criteria for both upward and downward continuity after replacing the formulas $x \neq 1$, $y \neq 0$ and $x(0) \neq 0$ by applications $ne(x, 1)$, $ne(y, 0)$, and $ne(x(0), 0)$ of a predicate $ne \subseteq \mathbb{N} \times \mathbb{N}$, $ne(x, y) :\Leftrightarrow x \neq 0$. $\qquad\square$

However, the syntactic constraints imposed by Proposition 5.6 are in general too strict, as can be seen by the fact that, e.g., the both upward and downward continuous operation $F \vee G$ is equivalent to $\neg F \Rightarrow G$ which involves both negation and implication. The following proposition gives a more precise criterion for determining continuous functions.

**Proposition 5.7** (Continuous functions on relations) *Take function $G$ on a domain $Set(S_1 \times \ldots \times S_n)$ of n-ary relations defined as*

$$G \colon Set(S_1 \times \ldots \times S_n) \to Set(S_1 \times \ldots \times S_n)$$
$$G(A) := \pi x_1 \in S_1, \ldots, x_n \in S_n.\ F[x_1, \ldots, x_n]\ .$$

> *by formula $F[x_1, \ldots, x_n]$ whose free variables are among $x_1, \ldots, x_n$ and in which parameter A only occurs in the form of the relation application $A\langle\_\rangle$, i.e., $\_ \in A$.*
>
> *Let $F'$ be identical to $F$ except that every free occurrence of parameter A has been replaced by the new constant symbol ?. Then G is upward continuous if the judgement $F'$: up can be derived by the rules in Fig. 5.3. Likewise, G is downward continuous if the judgement $F'$: down can be derived.*

In a judgement $u, p \vdash F$: formula, $u = $ true indicates that $F$ is upward continuous and $u = $ false indicates that it is downward continuous. $p = 1$ indicates that every occurrence of ? in $F$ only occurs with positive polarity, i.e., atom ? is not negated (if $F$ is written in prenex normal form with the propositional core written in conjunctive normal form); conversely, $p = -1$ indicates the atom only occurs in form $\neg?$. When checking the body of $\neg F$ and the antecedent of $F_1 \Rightarrow F_2$, both $u$ and $p$ are inverted to indicate the change in the direction of the continuity and the change of polarity. When checking the subformulas of $F_1 \Leftrightarrow F_2$, $p = 0$ indicates that no occurrence of ? is allowed at all. When checking $\forall X. \, F$, $F$ can be checked with either $p = 0$ (i.e., $F$ does not contain an occurrence of ?) or with $u = $ false (i.e., the formula is downward-continuous); dually, we can check $\exists X. \, F$ with $p = 0$ or $u = $ true.

**Example 5.12** Take formula $\neg?\langle x \rangle \Rightarrow \exists y. \, p\langle x, y \rangle \wedge ?\langle y \rangle$ which expands to

$$\neg x \in ? \Rightarrow \exists y. \, \langle x, y \rangle \in p \wedge y \in ?$$

We can derive the judgement

$$
\cfrac{
  \cfrac{\cfrac{\cdots}{\text{true}, 1 \vdash x : \text{terms} \quad \text{true}, 1 \vdash ? : \text{formula}}}{\cfrac{\text{true}, 1 \vdash x \in ? : \text{formula}}{\text{false}, -1 \vdash \neg x \in ? : \text{formula}}}
  \quad
  \cfrac{\cdots}{\text{true}, 1 \vdash \exists y. \, \langle x, y \rangle \in p \wedge y \in ? : \text{formula}}
}{
  \text{true}, 1 \vdash \neg x \in ? \Rightarrow \exists y. \, \langle x, y \rangle \in p \wedge y \in ? : \text{formula}
}
$$

where the right branch of the derivation expands to

$$
\cfrac{
  \cfrac{
    \cfrac{\cdots}{\text{true}, 1 \vdash \langle x, y \rangle \in p : \text{formula}}
    \quad
    \cfrac{\cfrac{\cdots}{\text{true}, 1 \vdash y : \text{terms} \quad \text{true}, 1 \vdash ? : \text{formula}}}{\text{true}, 1 \vdash y \in ? : \text{formula}}
  }{\text{true}, 1 \vdash \langle x, y \rangle \in p \wedge y \in ? : \text{formula}}
}{
  \text{true}, 1 \vdash \exists y. \, \langle x, y \rangle \in p \wedge y \in ? : \text{formula}
}
$$

The formula is therefore upward continuous. □

$F$ : **up and** $F$ : **down:**

$$\frac{\text{true}, 1 \vdash F : \text{formula}}{F : \text{up}} \qquad \frac{\text{false}, 1 \vdash F : \text{formula}}{F : \text{down}}$$

$u, p \vdash T$ : **term:**

$$u, p \vdash V : \text{term} \qquad u, p \vdash CS : \text{term} \qquad u, 1 \vdash \, ? : \text{term} \qquad \frac{u, p \vdash Ts : \text{terms}}{u, p \vdash FS(Ts) : \text{term})}$$

$$\frac{u, p \vdash T_1 : \text{term} \quad u, p \vdash T_2 : \text{term}}{u, p \vdash \text{let } V = T_1 \text{ in } T_2 : \text{term}}$$

$$\frac{u, p \vdash F : \text{formula} \quad u, p \vdash T_1 : \text{term} \quad u, p \vdash T_2 : \text{term}}{u, p \vdash \text{if } F \text{ then } T_1 \text{ else } T_2 : \text{term}}$$

$u, p \vdash Ts$ : **terms:**

$$\frac{u, p \vdash T : \text{term}}{u, p \vdash T : \text{terms}} \qquad \frac{u, p \vdash T : \text{term} \quad u, p \vdash Ts : \text{terms}}{u, p \vdash T, Ts : \text{terms}}$$

$u, p \vdash F$ : **formula:**

$$u, p \vdash \text{true} : \text{formula} \qquad u, p \vdash \text{false} : \text{formula} \qquad \frac{u, p \vdash Ts : \text{terms}}{u, p \vdash PS(Ts) : \text{formula}}$$

$$\frac{u, p \vdash T_1 : \text{term} \quad u, p \vdash T_2 : \text{term}}{u, p \vdash T_1 = T_2 : \text{formula}} \qquad \frac{\neg u, -p \vdash F : \text{formula}}{u, p \vdash \neg F : \text{formula}}$$

$$\frac{u, p \vdash F_1 : \text{formula} \quad u, p \vdash F_2 : \text{formula}}{u, p \vdash F_1 \wedge F_2 : \text{formula}} \qquad \frac{u, p \vdash F_1 : \text{formula} \quad u, p \vdash F_2 : \text{formula}}{u, p \vdash F_1 \vee F_2 : \text{formula}}$$

$$\frac{\neg u, -p \vdash F_1 : \text{formula} \quad u, p \vdash F_2 : \text{formula}}{u, p \vdash F_1 \Rightarrow F_2 : \text{formula}} \qquad \frac{u, 0 \vdash F_1 : \text{formula} \quad u, 0 \vdash F_2 : \text{formula}}{u, p \vdash F_1 \Leftrightarrow F_2 : \text{formula}}$$

$$\frac{u, 0 \vdash F : \text{formula}}{u, p \vdash \forall X. \, F : \text{formula}} \qquad \frac{\text{false}, p \vdash F : \text{formula}}{\text{false}, p \vdash \forall X. \, F : \text{formula}}$$

$$\frac{u, 0 \vdash F : \text{formula}}{u, p \vdash \exists X. \, F : \text{formula}} \qquad \frac{\text{true}, p \vdash F : \text{formula}}{\text{true}, p \vdash \exists X. \, F : \text{formula}}$$

$$\frac{u, p \vdash T : \text{term} \quad u, p \vdash F : \text{formula}}{u, p \vdash \text{let } V = T \text{ in } F : \text{formula}}$$

$$\frac{u, p \vdash F : \text{formula} \quad u, p \vdash F_1 : \text{formula} \quad u, p \vdash F_2 : \text{formula}}{u, p \vdash \text{if } F \text{ then } F_1 \text{ else } F_2 : \text{formula}}$$

**Fig. 5.3** Upward and Downward Continuous Formulas

## 5.5 Inductive and Coinductive Relation Definitions

Now that we know how to define continuous functions on relations, we are ready to establish a semantics of recursive relation definitions based on fixed points.

**Definition 5.9** (*Inductive/coinductive relation definitions*) An *inductive relation definition* has form

inductive $p \subseteq A_1 \times \ldots \times A_n$
$p\langle x_1, \ldots, x_n \rangle :\Leftrightarrow F_p[x_1, \ldots, x_n]$

where $p$ be a symbol that is not a constant symbol in the current first-order language, $A_1, \ldots, A_n$ are sets, $F_p[x_1, \ldots, x_n]$ is a formula with free variables $x_1, \ldots, x_n$ in the first-order language that is derived from the current language by adding symbol $p$ as a constant symbol, $P$ is a variable that does not occur in $F_p$, and $F_P$ is identical to $F_p$ except that every occurrence of $p$ has been replaced by $P$.
This definition introduces the relation

$$p := \mathsf{lfp}\, \lambda P \in \mathsf{Set}(A_1 \times \ldots \times A_n).\, \pi x_1 \in A_1, \ldots, x_n \in A_n.\, F_P[x_1, \ldots, x_n]$$

provided that the definition is well-formed, i.e., the argument to lfp $(\cdot)$ is upward continuous.
Likewise, the corresponding *coinductive relation definition*

coinductive $p \subseteq A_1 \times \ldots \times A_n$
$p\langle x_1, \ldots, x_n \rangle :\Leftrightarrow F_p[x_1, \ldots, x_n]$

introduces the relation

$$p := \mathsf{gfp}\, \lambda P \in \mathsf{Set}(A_1 \times \ldots \times A_n).\, \pi x_1 \in A_1, \ldots, x_n \in A_n.\, F_P[x_1, \ldots, x_n]$$

provided that the definition is well-formed, i.e., the argument to gfp $(\cdot)$ is downward continuous.

Establishing the well-formedness of an inductive/coinductive definition can be achieved by demonstrating the equivalence of $F_P$ to a formula that satisfies the corresponding criteria of Propositions 5.6 or 5.7.

Since a unary relation $p \subseteq S$ denotes a subset of $S$, inductive/coinductive relation definitions can also been as particular forms of definitions of sets or *subtypes*: in an inductive definition, the subtype is constructed by, starting with the empty set $\varnothing$, gradually adding those elements that satisfy the defining formula. Dually, in a coinductive definition, the subtype is constructed by, starting with $S$, gradually removing those elements that violate the defining formula. The least/greatest fixed point denotes the limit of this iterative process and thus defines the subtype.

**Example 5.13** The definition

> inductive *even* $\subseteq \mathbb{N}$
> $even\langle x \rangle :\Leftrightarrow x = 0 \vee (x \neq 1 \wedge even\langle x - 2 \rangle)$

introduces the unary relation "is even" on $\mathbb{N}$, i.e., the subtype of the even natural numbers, with the property

> $even\langle 0 \rangle, \neg even\langle 1 \rangle, even\langle 2 \rangle, \neg even\langle 3 \rangle, \ldots.$

We have *even* = lfp $E$ where

$$E := \lambda even \in \mathsf{Set}(\mathbb{N}). \ \pi x \in \mathbb{N}. \ x = 0 \vee (x \neq 1 \wedge even\langle x - 2 \rangle)$$

which by the criterion of Proposition 5.7 ($x \neq 1$ means $\neg(x = 1)$) is upward continuous; see also Example 5.11.                                                                    □

**Example 5.14** The definition

> inductive *lesseq* $\subseteq \mathbb{N} \times \mathbb{N}$
> $lesseq\langle x, y \rangle :\Leftrightarrow x = 0 \vee (y \neq 0 \wedge lesseq\langle x - 1, y - 1 \rangle)$

introduces the binary relation "is less than or equal" on $\mathbb{N}$, i.e.,

> $lesseq\langle 0, 0 \rangle, \neg lesseq\langle 1, 0 \rangle, lesseq\langle 0, 1 \rangle, lesseq\langle 1, 1 \rangle, \ldots.$

We have *lesseq* = lfp $L$ where

$$L := \lambda lesseq \in \mathsf{Set}(\mathbb{N}). \ \pi x \in \mathbb{N}, y \in \mathbb{N}. \ x = 0 \vee (y \neq 0 \wedge lesseq\langle x - 1, y - 1 \rangle)$$

which by the criterion of Proposition 5.7 ($y \neq 0$ means $\neg y = 0$) is upward continuous; see also Example 5.11.                                                                    □

**Example 5.15** The definition

> coinductive *nozero* $\subseteq \mathbb{N}^\omega$
> $nozero\langle x \rangle :\Leftrightarrow x(0) \neq 0 \wedge nozero\langle tail(x) \rangle$

based on the auxiliary function

> $tail \colon \mathbb{N}^\omega \to \mathbb{N}^\omega$
> $tail(x) := \lambda i \in \mathbb{N}. \ x(i + 1)$

introduces the unary relation "has no zero" on infinite number sequences, i.e., the subtype of those infinite number sequences that do not contain a zero, e.g.,

$$\neg nozero\langle[1,\ldots]\rangle, \ \neg nozero\langle[0,1,\ldots]\rangle, \ \neg nozero\langle[0,0,1,\ldots]\rangle, \ldots$$
$$nozero\langle[0,0,0,0,\ldots,0,\ldots]\rangle.$$

We have $lesseq = \mathsf{gfp}\ N$ where

$$N := \lambda nozero \in \mathsf{Set}(\mathbb{N}^\omega).\ \pi x \in \mathbb{N}^\omega.\ x(0) \neq 0 \wedge nozero(tail(x))$$

which by the criterion of Proposition 5.7 ($x(0) \neq 0$ means $\neg(x(0) = 0)$) is downward continuous; see also Example 5.11. $\square$

The negation of a relation defined by the least fixed point is a relation defined by a greatest fixed point, and vice versa.

**Proposition 5.8** (Duality of least and greatest fixed point) *Let $A_1, \ldots, A_n$ be sets and $F[x_1, \ldots, x_n, P]$ be a formula whose free variables are among $x_1, \ldots, x_n, P$. We define*

$$G := \lambda P \in \mathsf{Set}(A_1 \times \ldots \times A_n).\ \pi x_1 \in A_1, \ldots, x_n \in A_n.\ F[x_1, \ldots, x_n, P]$$
$$\overline{G} := \lambda P \in \mathsf{Set}(A_1 \times \ldots \times A_n).\ \pi x_1 \in A_1, \ldots, x_n \in A_n.\ \neg F[x_1, \ldots, x_n, \overline{P}]$$

*where $\overline{P} := \mathsf{Set}(A_1 \times \ldots \times A_n) \setminus P$ denotes the negation of relation $P$. Then, if $G$ is upward continuous, then $\overline{G}$ is downward continuous and*

$$\forall x_1 \in A_1, \ldots, x_n \in A_n.\ \neg(\mathsf{lfp}\ G)\langle x_1, \ldots, x_n\rangle \Leftrightarrow (\mathsf{gfp}\ \overline{G})\langle x_1, \ldots, x_n\rangle.$$

*Dually, if $G$ is upward continuous, then $\overline{G}$ is downward continuous and*

$$\forall x_1 \in A_1, \ldots, x_n \in A_n.\ \neg(\mathsf{gfp}\ G)\langle x_1, \ldots, x_n\rangle \Leftrightarrow (\mathsf{lfp}\ \overline{G})\langle x_1, \ldots, x_n\rangle.$$

**Example 5.16** The inductive relation definition

inductive *finite* $\subseteq \mathsf{Set}(\mathbb{N})$

*finite*$\langle S\rangle :\Leftrightarrow S = \varnothing \vee finite\langle S \setminus \{\mathsf{anyof}\ S\}\rangle$

implies that *finite*$(S)$ is true if and only if $S$ is a finite subset of $S$, i.e., it denotes the subtype of all finite subsets of $S$.

Conversely, the coinductive relation definition

coinductive *infinite* $\subseteq$ Set($\mathbb{N}$)

*infinite*$\langle S \rangle$ :$\Leftrightarrow$ $S \neq \varnothing \wedge$ *infinite*$\langle S \backslash \{$anyof $S\}\rangle$

implies that *infinite*$(S)$ is true if and only if $S$ is an infinite subset of $S$, i.e., it denotes the subtype of all infinite subsets of $S$.

The definition of *finite* is inductive, because if $S$ is finite, this fact can be established in a finite number of steps; dually, the definition of *infinite* is coinductive, because if $S$ is *not* infinite, this fact can be established in a finite number of steps.

We then have $\forall S.$ *finite*$\langle S \rangle \Leftrightarrow \neg$*infinite*$\langle S \rangle$.                    $\square$

If a relation is defined by the fixed point of a function that is both upward and downward continuous, we may thus define either the relation by a least fixed point or its negation by a greatest fixed point (respectively vice versa).

## 5.6   Rule-Oriented Inductive and Coinductive Relation Definitions

Inductive respectively relation definitions are often given in a "rule-oriented style" which we illustrate by an example before giving its precise definition.

**Example 5.17** In the format that will be introduced below, we can define the function *even* from Example 5.13 as follows:

inductive *even* $\subseteq \mathbb{N}$

$\forall x \in \mathbb{N}.$

   *even*$\langle 0 \rangle \wedge$

   $($*even*$\langle x + 2 \rangle \Leftarrow$ *even*$\langle x \rangle)$

This definition can be naturally read as the following collection of rules:

- 0 is even.
- For every $x \in \mathbb{N}$, $x + 2$ is even if $x$ is even.

Furthermore, we have the implicit assumption that no other natural number is even. In other words, a natural number is not even, unless its "evenness" can be established by a repeated application of above rules.                    $\square$

This definition style is precisely introduced as follows.

**Definition 5.10** (*Rule-based inductive relation definitions*) A *rule-based inductive relation definition* is a universally quantified conjunction of right-to-left implications which we call *inductive rules* in the form

inductive $p \subseteq A_1 \times \ldots \times A_n$

$\forall x_1 \in B_1, \ldots, x_m \in B_m.$

$(p\langle T_{11}[x_1, \ldots, x_m], \ldots, T_{1n}[x_1, \ldots, x_m]\rangle \Leftarrow F_1[x_1, \ldots, x_m]) \wedge$

$\ldots$

$(p\langle T_{o1}[x_1, \ldots, x_m], \ldots, T_{on}[x_1, \ldots, x_m]\rangle \Leftarrow F_o[x_1, \ldots, x_m])$

where an inductive rule may also have the simplified form

$p\langle T_{i1}[x_1, \ldots, x_m], \ldots, T_{in}[x_1, \ldots, x_m]\rangle$

which considers the corresponding formula $F_i[x_1, \ldots, x_n]$ as true.
Here $A_1, \ldots, A_n$ and $B_1, \ldots, B_m$ are sets, for $1 \leq i \leq o$ and $1 \leq j \leq n$, $T_{i,j}[x_1, \ldots, x_n]$ are terms with free variables $x_1, \ldots, x_n$ in the current first-order language, $p$ is a symbol that is not a constant symbol in the current language and, for $1 \leq i \leq o$, $F_i[x_1, \ldots, x_n]$ are formulas with free variables $x_1, \ldots, x_n$ in the language that is derived from the current one by adding symbol $p$ as a constant symbol.

This definition has the same meaning as the following inductive relation definition:

inductive $p \subseteq A_1 \times \ldots \times A_n$

$p\langle y_1, \ldots, y_n \rangle :\Leftrightarrow$

$\exists x_1 \in B_1, \ldots, x_m \in B_m.$

$(y_1 = T_{11}[x_1, \ldots, x_m] \wedge \ldots \wedge y_n = T_{1n}[x_1, \ldots, x_m] \wedge F_1[x_1, \ldots, x_m]) \vee$

$\ldots \vee$

$(y_1 = T_{o1}[x_1, \ldots, x_m] \wedge \ldots \wedge y_n = T_{on}[x_1, \ldots, x_m] \wedge F_o[x_1, \ldots, x_m]).$

By the criterion of Proposition 5.6, the resulting inductive definition is well-formed if the formulas $F_i$ are upward continuous.

**Example 5.18** The definition given in Example 5.17 is translated to

> inductive *even* $\subseteq \mathbb{N}$
> $even\langle y \rangle :\Leftrightarrow$
>> $\exists x \in \mathbb{N}.$
>>> $y = 0 \vee$
>>> $(y = x + 2 \wedge even\langle x \rangle)$

which is logically equivalent to the original definition given in Example 5.13. The additional constraint $x \neq 1$ in the original definition was required there only to give $even\langle 1 \rangle$ truth value "false", because the value of $1 - 2$ is unknown (subtraction on $\mathbb{N}$ has a precondition that requires the first argument to be not smaller than the second one). In the new rule-based definition this constraint is not required, because for $y = 1$ there exists no $x \in \mathbb{N}$ with $y = x + 2$ and therefore $even\langle 1 \rangle$ is false. □

**Example 5.19** The definition

> inductive *lesseq* $\subseteq \mathbb{N} \times \mathbb{N}$
> $\forall x \in \mathbb{N}, y \in \mathbb{N}.$
>> $lesseq\langle 0, y \rangle \wedge$
>> $(lesseq\langle x + 1, y + 1 \rangle \Leftarrow lesseq\langle x, y \rangle)$

can be read as the rules

- 0 is less than or equal every $y \in \mathbb{N}$.
- For every $x \in \mathbb{N}$ and $y \in \mathbb{N}$, $x + 1$ is less than equal $y + 1$ if $x$ is less than equal $y$.

Again, the implicit understanding is that only those pairs $x$ and $y$ are related, for which the relationship can be established by a finite number of applications of these rules. The definition is translated to

> inductive *lesseq* $\subseteq \mathbb{N} \times \mathbb{N}$
> $lesseq\langle a, b \rangle :\Leftrightarrow$
>> $\exists x \in \mathbb{N}, y \in \mathbb{N}.$
>>> $(a = 0 \wedge b = y) \vee$
>>> $(a = x + 1 \wedge b = y + 1 \wedge lesseq\langle x, y \rangle)$

which is logically equivalent to the definition given in Example 5.14; there the constraint $y \neq 0$ is required to ensure that the truth value of $lesseq\langle 1, 0 \rangle$ becomes "false", because the value of $0 - 1$ is unknown. The same is achieved automatically in the new definition. □

As is demonstrated by above examples, in inductive relation definitions, we only have to be concerned how to make the relation "true" for certain arguments, because the default assumption is that the relation is "false" for all arguments.

For coinductive relation definitions, we have a dual format.

---

**Definition 5.11** (*Rule-based coinductive relation definitions*)  A *rule-based coinductive relation definition* is a universally quantified conjunction of left-to-right implications which we call *coinductive rules* in the form

coinductive $p \subseteq A_1 \times \ldots \times A_n$

$\forall x_1 \in B_1, \ldots, x_m \in B_m.$

$\quad (p\langle T_{11}[x_1, \ldots, x_m], \ldots, T_{1n}[x_1, \ldots, x_m]\rangle \Rightarrow F_1[x_1, \ldots, x_m]) \wedge$

$\quad \ldots$

$\quad (p\langle T_{o1}[x_1, \ldots, x_m], \ldots, T_{on}[x_1, \ldots, x_m]\rangle \Rightarrow F_o[x_1, \ldots, x_m])$

where a coinductive rule may also have the simplified form

$\quad \neg p\langle T_{i1}[x_1, \ldots, x_m], \ldots, T_{in}[x_1, \ldots, x_m]\rangle$

which considers the formula $F_i[x_1, \ldots, x_n]$ as false.

Here $A_1, \ldots, A_n$ and $B_1, \ldots, B_m$ are sets, for $1 \leq i \leq o$ and $1 \leq j \leq n$, $T_{i,j}[x_1, \ldots, x_n]$ are terms with free variables $x_1, \ldots, x_n$ in the current first-order language, $p$ is a symbol that is not a constant symbol in the current language and, for $1 \leq i \leq o$, $F_i[x_1, \ldots, x_n]$ are formulas with free variables $x_1, \ldots, x_n$ in the language that is derived from the current one by adding symbol $p$ as a constant symbol.

This definition has the same meaning as the following coinductive definition:

coinductive $p \subseteq A_1 \times \ldots \times A_n$

$p\langle y_1, \ldots, y_n\rangle :\Leftrightarrow$

$\quad \forall x_1 \in B_1, \ldots, x_m \in B_m.$

$\quad (y_1 = T_{11}[x_1, \ldots, x_m] \wedge \ldots \wedge y_n = T_{1n}[x_1, \ldots, x_m] \Rightarrow F_1[x_1, \ldots, x_m]) \wedge$

$\quad \ldots \wedge$

$\quad (y_1 = T_{o1}[x_1, \ldots, x_m] \wedge \ldots \wedge y_n = T_{on}[x_1, \ldots, x_m] \Rightarrow F_o[x_1, \ldots, x_m]).$

---

By the criterion of Proposition 5.7, the resulting coinductive definition is well-formed if the formulas $F_i$ are downward continuous ($p$ may only appear in these formulas, thus the implication is downward continuous).

**Example 5.20** The coinductive relation definition of Example 5.15 can be also written in the rule-based form

> coinductive $nozero \subseteq \mathbb{N}^\omega$
>
> $\forall x \in \mathbb{N}^\omega.$
>
> $\quad nozero\langle x \rangle \Rightarrow x(0) \neq 0 \wedge nozero\langle tail(x) \rangle$

which can be read as

> If $x$ has no zeros, then its first element is not zero and the rest has no zeros.

It is even more illustrating to revert the direction of the implication to derive the logically equivalent formula

> $x(0) = 0 \vee \neg nozero\langle tail(x) \rangle \Rightarrow \neg nozero\langle x \rangle$

which can be read as the rule

> If the head of $x$ is zero or the tail has zeros, then $x$ has zeros.

The definition is an abbreviation for

> coinductive $nozero \subseteq \mathbb{N}^\omega$
>
> $nozero\langle y \rangle :\Leftrightarrow$
>
> $\quad \forall x \in \mathbb{N}^\omega. (y = x \Rightarrow x(0) \neq 0 \wedge nozero\langle tail(x) \rangle)$

which is logically equivalent to the original definition of Example 5.15.

For a better demonstration of the rule-based style of coinductive relation definitions, we define the function

> $cons\colon \mathbb{N} \times \mathbb{N}^\omega \to \mathbb{N}^\omega$
>
> $cons(n, x) := \lambda i \in \mathbb{N}. \text{ if } i = 0 \text{ then } n \text{ else } x(i - 1)$

which prepends number $n$ to the front of infinite sequence $x$. We then may define function $nozero$ also by the definition

> coinductive $nozero \subseteq \mathbb{N}^\omega$
>
> $\forall n \in \mathbb{N}, x \in \mathbb{N}^\omega.$
>
> $\quad \neg nozero\langle cons(0, x) \rangle \wedge$
>
> $\quad (nozero\langle cons(n, x) \rangle \Rightarrow nozero\langle x \rangle)$

which can be (after reverting the implication to $\neg nozero\langle x \rangle \Rightarrow \neg nozero\langle cons(n, x) \rangle$) read as the two rules

- Every stream with head 0 has zeros.
- If $x$ has zeros, then every stream with tail $x$ has zeros.

This definition is an abbreviation for

> coinductive *nozero* $\subseteq \mathbb{N}^\omega$
>
> *nozero*$\langle y \rangle :\Leftrightarrow$
>
>    $\forall n \in \mathbb{N}, x \in \mathbb{N}^\omega.$
>
>       $(y = cons(0, x) \Rightarrow \mathsf{false}) \wedge$
>
>       $(y = cons(n, x) \Rightarrow nozero\langle x \rangle)$

which also can be shown to be logically equivalent to the definition in Example 5.15.
$\square$

**Example 5.21** The relation *finite* from Example 5.16 can be also defined in the rule-oriented style as

> inductive *finite* $\subseteq \mathsf{Set}(\mathbb{N})$
>
> $\forall S \in \mathsf{Set}(\mathbb{N}).$
>
>    *finite*$\langle S \rangle \Leftarrow S = \varnothing \vee finite\langle S \backslash \{\mathsf{anyof}\ S\}\rangle$

or as

> inductive *finite* $\subseteq \mathsf{Set}(\mathbb{N})$
>
> $\forall S \in \mathsf{Set}(\mathbb{N}).$
>
>    *finite*$\langle \varnothing \rangle \wedge$
>
>    *finite*$\langle S \rangle \Leftarrow finite\langle S \backslash \{\mathsf{anyof}\ S\}\rangle$

or also as

> inductive *finite* $\subseteq \mathsf{Set}(\mathbb{N})$
>
> $\forall S \in \mathsf{Set}(\mathbb{N}), n \in \mathbb{N}.$
>
>    *finite*$\langle \varnothing \rangle \wedge$
>
>    *finite*$\langle S \cup \{n\}\rangle \Leftarrow finite\langle S \rangle.$

These formats make it even more explicit that if $S$ is finite, this fact can be established in a finite number of steps.

Likewise, *infinite* can be defined in the rule-oriented style as

> coinductive *infinite* $\subseteq \mathsf{Set}(\mathbb{N})$
>
> $\forall S \in \mathsf{Set}(\mathbb{N}).$
>
>    *infinite*$\langle S \rangle \Rightarrow S \neq \varnothing \wedge infinite\langle S \backslash \{\mathsf{anyof}\ S\}\rangle$

or as

> coinductive *infinite* $\subseteq$ Set($\mathbb{N}$)
> $\forall S \in$ Set($\mathbb{N}$).
>   $\neg infinite\langle\varnothing\rangle \wedge$
>   $infinite\langle S\rangle \Rightarrow infinite\langle S\setminus\{\text{anyof } S\}\rangle$

or also as

> coinductive *infinite* $\subseteq$ Set($\mathbb{N}$)
> $\forall S \in$ Set($\mathbb{N}$), $n \in \mathbb{N}$.
>   $\neg infinite\langle\varnothing\rangle \wedge$
>   $infinite\langle S \cup \{n\}\rangle \Rightarrow infinite\langle S\rangle.$

By inverting the first implication to

$$S = \varnothing \vee \neg infinite\langle S\setminus\{\text{anyof } S\}\rangle \Rightarrow \neg infinite\langle S\rangle$$

this format makes it explicit that if $S$ is *not* infinite, this fact can be established in a finite number of steps.                                                              □

As is demonstrated by above examples, in coinductive relation definitions, we only have to be concerned how to make the relation "false" for certain arguments, because the default assumption is that the relation is "true" for all arguments.

## 5.7   Inductive and Coinductive Function Definitions

We will now generalize the approach presented in the previous subsection to a fixed point semantics of recursive function definitions. Here, however, we have to overcome a fundamental problem: while for a function $F$ : Set($S$) → Set($S$) the iterated applications $F^n(\varnothing)$ respectively $F^n(S)$ denote relations on $S$, they do not denote functions on $S$; therefore also the fixed points lfp $F$ and gfp $F$ in general do not denote functions.

However, consider the function $F$

> $F$ : Set($\mathbb{N} \times \mathbb{N}$) → Set($\mathbb{N} \times \mathbb{N}$)
> $F(A) := \pi n \in \mathbb{N}, r \in \mathbb{N}.$ if $n = 0$ then $r = 1$ else $\exists s \in \mathbb{N}.\ A\langle n-1, s\rangle \wedge r = n \cdot s$

where the term on the right side denotes a relation on $\mathbb{N} \times \mathbb{N}$: "argument" $n \in \mathbb{N}$ is related to "result" $r \in \mathbb{N}$ if the following holds:

- If $n = 0$, then $r = 1$.
- If $n > 0$, then $r = n \cdot s$ for some $s \in \mathbb{N}$ that $A$ relates to $n - 1$.

The upward iteration yields the following results:

$$F^0(\varnothing) = \varnothing$$
$$F^1(\varnothing) = F^0(\varnothing) \cup \{\langle 0, 1 \rangle\}$$
$$F^2(\varnothing) = F^1(\varnothing) \cup \{\langle 1, 1 \rangle\}$$
$$F^3(\varnothing) = F^2(\varnothing) \cup \{\langle 2, 2 \rangle\}$$
$$F^4(\varnothing) = F^0(\varnothing) \cup \{\langle 3, 6 \rangle\}$$

. . . .

In other words, $F^i(\varnothing)$ contains all tuples $\langle n, n! \rangle$ with $n < i$ where $n!$ denotes the $n$-th factorial of $n$. $F^i(\varnothing)$ is thus the *partial function* that maps every $n < i$ to the factorial of $n$ and is undefined for every argument $n \geq i$. The least fixed point of $F$ reached by the upward iteration $F\!\uparrow$ is thus the total factorial function, i.e., it is the (only) solution of the recursive function definition

recursive $fac \colon \mathbb{N} \to \mathbb{N}$

$fac(n) :=$ if $n = 0$ then $1$ else $n \cdot fac(n - 1)$

The least fixed point, however, may also be partial. For instance, take the function

$G \colon \mathsf{Set}(\mathbb{N} \times \mathbb{N}) \to \mathsf{Set}(\mathbb{N} \times \mathbb{N})$

$G(A) := \lambda n \in \mathbb{N}, r \in \mathbb{N}.$ if $n = 0$ then $r = 1$ else $\exists s.\ A\langle n + 1, s \rangle \wedge r = n \cdot s$

where the term $n + 1$ replaces $n - 1$ in the definition of $F$ above. We then have

$$G^0(\varnothing) = \varnothing$$
$$G^1(\varnothing) = G^0(\varnothing) \cup \{\langle 0, 1 \rangle\} = \{\langle 0, 1 \rangle\}$$
$$G^2(\varnothing) = G^1(\varnothing) = \{\langle 0, 1 \rangle\}$$
$$G^3(\varnothing) = G^2(\varnothing) = \{\langle 0, 1 \rangle\}$$
$$G^4(\varnothing) = G^3(\varnothing) = \{\langle 0, 1 \rangle\}$$

. . . .

Therefore lfp $G$ denotes the partial function that maps $0$ to $1$ and is undefined for every other argument.

It should be noted that the meaning of both lfp $F$ and lfp $G$ corresponds to a computational understanding of a recursive function definitions where the application of the function yields a result if this result can be determined by a finite number of unfoldings of the definition: We have the finite sequence of unfoldings

$$F(3) = 3 \cdot F(2) = 3 \cdot 2 \cdot F(1) = 3 \cdot 2 \cdot 1 \cdot F(0) = 3 \cdot 2 \cdot 1 \cdot 1 = 6$$

while the following sequence of unfoldings does not terminate:

$$G(3) = 3 \cdot G(4) = 3 \cdot 4 \cdot G(5) = 3 \cdot 4 \cdot 5 \cdot G(6) = 3 \cdot 4 \cdot 5 \cdot 6 \cdot G(7) = \ldots$$

The fixed point theory of recursive function definitions thus gives a formal semantics to functions understood as computer programs. Like a computer program may not terminate for certain inputs, a function may not yield a result for certain arguments. The theory is thus a theory of partial functions which we have not considered so far; we are now going to introduce the missing concepts.

An alternative (and more prominent) formal treatment of recursive function definitions within the framework of fixed point theory is based on the view that the introduced functions are total in a domain with an extra value $\bot$ ("bottom") that denotes an "undefined" result. This view generalizes the domain $Set = (\mathrm{Set}(S), \subseteq, \varnothing, S)$ considered so far to a *complete partial order* $(D, \sqsubseteq \subseteq D, \bot \in D, \top \in D)$ with certain properties that are also satisfied by *Set*. Since we prefer to stay in the more familiar domain *Set* which suffices for our purposes, we refer to the literature for presentations of the more general concept.

**Definition 5.12** (*Partial function*) Set $F$ is a *partial function* from sets $A_1, \ldots, A_n$ to $B$, written as $F \colon A_1 \times \ldots \times A_n \to_\bot B$ if there exists some sets $A_1', \ldots, A_n'$ with $A_1' \subseteq A_1, \ldots, A_n' \subseteq A_n$ such that $F \colon A_1' \times \ldots \times A_n' \to B$. Furthermore, we denote by $A_1 \times \ldots \times A_n \to_\bot B$ the set of all partial functions from $A_1, \ldots, A_n$ to $B$.

Since $F(x_1, \ldots, x_n)$ shall only denote a value if $F$ is defined on its argument tuple $\langle x_1, \ldots, x_n \rangle$, for partial functions the notion of function application as a term (which must denote a value) is problematic. Therefore we replace it by the following concept which turns partial function application into a formula.

**Definition 5.13** (*Partial function application*) Given $F \colon A_1 \times \ldots \times A_n \to_\bot B$, terms $T_1, \ldots, T_n$ and formula $G$, the formula

exists $y = F(x_1, \ldots, x_n).\ G$

is an abbreviation for the formula

$\exists y \in B.\ F\langle x_1, \ldots, x_n, y \rangle \wedge G$.

Furthermore, the formula

forall $y = F(x_1, \ldots, x_n).\ G$

is an abbreviation for the formula

$$\forall y \in B.\ F\langle x_1, \ldots, x_n, y\rangle \Rightarrow G.$$

Thus the value of formula exists $y = F(x_1, \ldots, x_n)$. $G$ is "false" if $F$ is not defined on its argument $\langle x_1, \ldots, x_n\rangle$; otherwise, it is the truth value of $F$ where variable $y$ is assigned the result of the function application. Dually, the value of formula forall $y = F(x_1, \ldots, x_n)$. $G$ is "true" if $F$ is not defined on its argument $\langle x_1, \ldots, x_n\rangle$; otherwise it is the same value as for the other formula.

Furthermore, we use formulas, not terms, to define partial functions by induction.

**Definition 5.14** (*Inductive function definition*) An *inductive function definition* has form

inductive $f : A_1 \times \ldots \times A_n \to_\perp B$
$f(x_1, \ldots, x_n) = y \Leftarrow G_f[x_1, \ldots, x_n, y]$

where $f$ is a symbol that is not a constant symbol in the current first-order language, $A_1, \ldots, A_n$ and $B$ are sets, $G_f[x_1, \ldots, x_n, y]$ is a formula with free variables $x_1, \ldots, x_n, y$ in the first-order language that is derived from the current language by adding symbol $f$ as a constant symbol.

This definition introduces the partial function

$$f := \text{lfp } \lambda F \in (A_1 \times \ldots \times A_n \to_\perp B).$$
$$\pi x_1 \in A_1, \ldots, x_n \in A_n, y \in B.\ G_F[x_1, \ldots, x_n, y]$$

where $F$ is a variable that does not occur in $G_f$, and $G_F$ is identical to $G_f$ except that every occurrence of $f$ has been replaced by $F$.

The definition is well-formed, if the argument to lfp $(\cdot)$ is an upward continuous function whose result denotes a partial function in $A_1 \times \ldots \times A_n \to_\perp B$.

Analogously to the rule-based inductive relation definitions introduced in Definition 5.10, also inductive function definitions may be introduced in a rule-based style; we omit the details but refer to the following example.

**Example 5.22** The inductive function definition

inductive $fac : \mathbb{N} \to_\perp \mathbb{N}$
$fac(n) = r \Leftarrow$ if $n = 0$ then $r = 1$ else exists $s = fac(n - 1).\ r = n \cdot s$

introduces the function

$fac :\Leftrightarrow$
  $\mathsf{lfp}\ \lambda F \in \mathbb{N} \to_\perp \mathbb{N}.$
    $\pi n \in \mathbb{N}, r \in \mathbb{N}.\ \mathsf{if}\ n = 0\ \mathsf{then}\ r = 1\ \mathsf{else\ exists}\ s = F(n-1).\ r = n \cdot s$

where the formula $\mathsf{exists}\ s = F(n-1).\ r = n \cdot s$ is syntactic sugar for

$$\exists s \in \mathbb{N}.\ F\langle n-1, s\rangle \wedge r = n \cdot s.$$

The same function may be introduced in a rule-based style as

$\mathsf{inductive}\ fac \colon \mathbb{N} \to_\perp \mathbb{N}$
$\forall n \in \mathbb{N}, r \in \mathbb{N}.$
  $fac(0) = 1\ \wedge$
  $(fac(n+1) = r \Leftarrow \mathsf{exists}\ s = fac(n-1).\ r = n \cdot s)$

It can be shown that $fac$ is actually total, i.e., we have $fac \colon \mathbb{N} \to \mathbb{N}$.                                   □

Now what about coinductive function definitions, i.e., functions defined by greatest fixed points? Here the problem arises that for $S = \mathsf{Set}(A_1 \times \ldots \times A_n \times B)$ and function $F \colon S \to S$, the downward iteration $F^n(S)$ usually does not even denote a partial function, but only a general relation. In particular $F^0(S) = S$ which relates every "argument" $\langle x_1, \ldots, x_n \rangle \in A_1 \times \ldots \times A_n$ to every "result" $y \in B$. Thus, also $\mathsf{gfp}\ F$ does not denote a function.

Nevertheless, by a certain change of perspective, we may get something similar to coinductive definitions of functions: We might argue that an inductive definition

$\mathsf{inductive}\ f \colon A_1 \times \ldots \times A_n \to_\perp B$
$f(x_1, \ldots, x_n) = y \Leftarrow G_f[x_1, \ldots, x_n, y]$

actually does not define a partial function $f$ but a way how to prove an equality $f(x_1, \ldots, x_n) = y$ for a newly introduced total function $f$. We could subsequently write a rule-based inductive relation definition

$\mathsf{inductive}\ \cdot \sim \cdot \subseteq B \times B$
$\forall x_1 \in A_1, \ldots, x_n \in A_n, y \in B.$
  $f(x_1, \ldots, x_n) \sim y \Leftarrow G_\sim[x_1, \ldots, x_n, y]$

which introduces a relation $\sim$ by a formula $G_\sim$ that is identical to $G$ except that every application of $=$ to arguments from $B$ is replaced by $\sim$. If $f(x_1, \ldots, x_n) \sim y$, thus also $f(x_1, \ldots, x_n) = y$.

This view is the key for the concept introduced below.

**Definition 5.15** (*Coinductive function definition*) A *coinductive function definition* has the form

coinductive $f[\sim]\colon A_1 \times \ldots \times A_n \to B$
$f(x_1, \ldots, x_n) = y \Rightarrow G_f[x_1, \ldots, x_n, y]$

where the symbols $f$ and $\sim$ are not constants in the current first-order language, $A_1, \ldots, A_n$ and $B$ are sets and $G_f[x_1, \ldots, x_n]$ is a formula with free variables $x_1, \ldots, x_n$ in the first-order language that is derived from the current language by adding symbol $f$ as a constant.

This definition introduces the new constant $f\colon A_1 \times \ldots \times A_n \to B$ into the current language and adds the rule-based coinductive relation definition

coinductive $\cdot \sim \cdot \subseteq B \times B$
$\forall x_1 \in A_1, \ldots, x_n \in A_n, y \in B.$
$\quad f(x_1, \ldots, x_n) \sim y \Rightarrow G_\sim[x_1, \ldots, x_n, y]$

where $G_\sim$ is identical to $G_f$ except that every application of $=$ to arguments in $B$ has been replaced by $\sim$.

The definition is well-formed if $G_\sim$ is downward continuous.

Thus a coinductive "function definition" does actually not define the function $f$ but a relation $\sim$ that is associated to $f$ which itself remains unknown. Even more, $f(x_1, \ldots, x_n) \sim y$ does not imply $f(x_1, \ldots, x_n) = y$, so one might question the overall usefulness of the concept. However, we will see later, that in certain domains $\sim$ can serve as an adequate replacement for equality such that $f$ can be considered as being "defined" by the coinductive definition; our current discussion serves as a preparation for this later presentation.

Analogously to the rule-based coinductive relation definitions introduced in Definition 5.11, also inductive function definitions may be introduced in a rule-based style; we omit the details but refer to the following example.

**Example 5.23** Based on the auxiliary functions

$head\colon \mathbb{N}^\omega \to \mathbb{N}$
$head(x) = x(0)$
$tail\colon \mathbb{N}^\omega \to \mathbb{N}^\omega$
$tail(x) = \lambda i \in \mathbb{N}.\, x(i+1)$

the coinductive function definition

coinductive $merge[\sim]: \mathbb{N}^\omega \times \mathbb{N}^\omega \to \mathbb{N}^\omega$

$merge(x_1, x_2) = y \Rightarrow head(x_1) = head(y) \wedge merge(x_2, tail(x_1)) = tail(y)$

introduces the constant $merge: \mathbb{N}^\omega \times \mathbb{N}^\omega \to \mathbb{N}^\omega$ and defines the relation

coinductive $\cdot \sim \cdot \subseteq \mathbb{N}^\omega \times \mathbb{N}^\omega$

$\forall x_1 \in \mathbb{N}^\omega, x_2 \in \mathbb{N}^\omega, y \in \mathbb{N}^\omega.$

$\quad merge(x_1, x_2) \sim y \Rightarrow head(x_1) = head(y) \wedge merge(x_2, tail(x_1)) \sim tail(y).$

We thus have for instance

$\quad merge([0, 2, 4, \ldots], [1, 3, 5, \ldots]) \sim [0, 1, 2, 3, 4, 5 \ldots]$

which indicates that *merge* creates an infinite sequence by alternatively picking values
from each of its arguments.

Defining the auxiliary function

$cons: \mathbb{N} \times \mathbb{N}^\omega \to \mathbb{N}^\omega$

$cons(h, t) := \lambda i \in \mathbb{N}.\ \text{if } i = 0 \text{ then } h \text{ else } t(i - 1)$

we may also define merge in the rule-oriented style as

coinductive $merge[\sim]: \mathbb{N}^\omega \times \mathbb{N}^\omega \to \mathbb{N}^\omega$

$merge(x_1, x_2) = cons(h, t) \Rightarrow head(x_1) = h \wedge merge(x_2, tail(x_1)) = t$

yielding the relation

coinductive $\cdot \sim \cdot \subseteq \mathbb{N}^\omega \times \mathbb{N}^\omega$

$\forall x_1 \in \mathbb{N}^\omega, x_2 \in \mathbb{N}^\omega, h \in \mathbb{N}, t \in \mathbb{N}^\omega.$

$\quad merge(x_1, x_2) \sim cons(h, t) \Rightarrow head(x_1) = h \wedge merge(x_2, tail(x_1)) \sim t.$

$\square$

As can be seen from above examples, inductive definitions introduce (in general
partial) functions denoted by least fixed points: such a function constructs its result in
a finite number of steps; the result is thus necessarily finite. Coinductive definitions,
however, introduce functions that may construct their results by an infinite number
of steps; the results may thus be infinite.

## 5.8    Inductive and Coinductive Proofs

For reasoning about fixed points, we note the following corollary of Theorem 5.1.

**Proposition 5.9** (Induction and coinduction) *For every monotonic function* $F : Set(S) \rightarrow Set(S)$, *the following holds:*

$$\forall P \in Set(S).\ F(P) \subseteq P \Rightarrow \textit{lfp}\ F \subseteq P$$
$$\forall P \in Set(S).\ P \subseteq F(P) \Rightarrow P \subseteq \textit{gfp}\ F$$

*In other words if relation* $P$ *is a "pre-fixed point" of* $F$, *it is implied by* $\textit{lfp}\ F$. *If it is a "post-fixed point" of* $F$, *it implies* $\textit{gfp}\ F$.

This proposition represents the basis of two prominent proving principles.

**Proposition 5.10**  (Induction) *Let* $P \subseteq S$ *be a property that is defined as the least fixed point of a monotonic function* $F$, *i.e.,* $P = \textit{lfp}\ F$. *We would like to show that* $P$ *implies another property* $Q \subseteq S$, *i.e.,*

$$P \subseteq Q$$

*respectively the logically equivalent proposition*

$$\forall x \in S.\ P\langle x \rangle \Rightarrow Q\langle x \rangle.$$

*For this, it suffices to show that* $Q$ *is a pre-fixed point of* $F$, *i.e.,*

$$F(Q) \subseteq Q$$

*respectively the logically equivalent proposition*

$$\forall x \in S.\ F(Q)\langle x \rangle \Rightarrow Q\langle x \rangle.$$

*More formally, we have the following inference rule:*

$$\frac{Fs \vdash \forall x \in S.\ F(Q)\langle x \rangle \Rightarrow Q\langle x \rangle \quad (P = \textit{lfp}\ F \wedge P \subseteq S \wedge Q \subseteq S)}{Fs \vdash \forall x \in S.\ P\langle x \rangle \Rightarrow Q\langle x \rangle}\ (ind)$$

This proof principle is typically applied if $P$ is considered as the definition of a "subtype" of base type $S$ about which we would like to prove some property $Q$.

---

**Proposition 5.11** (Coinduction) *Let $P \subseteq S$ be a property that is defined as the greatest fixed point of a monotonic function $F$, i.e., $P = gfp\ F$. We would like to show that another property $Q \subseteq S$ implies $P$, i.e.,*

$$Q \subseteq P$$

*respectively the logically equivalent proposition*

$$\forall x \in S.\ Q\langle x \rangle \Rightarrow P\langle x \rangle.$$

*For this, it suffices to show that $Q$ is a post-fixed point of $F$, i.e.,*

$$Q \subseteq F(Q)$$

*respectively the logically equivalent proposition*

$$\forall x \in S.\ Q\langle x \rangle \Rightarrow F(Q)\langle x \rangle.$$

*More formally, we have the following inference rule:*

$$\frac{Fs \vdash \forall x \in S.\ Q\langle x \rangle \Rightarrow F(Q)\langle x \rangle \quad (P = gfp\ F \wedge P \subseteq S \wedge Q \subseteq S)}{Fs \vdash \forall x \in S.\ Q\langle x \rangle \Rightarrow P\langle x \rangle} \ (coind)$$

---

This proof principle is typically applied to show that $P\langle x \rangle$ holds for some object $x \in S$. It then suffices to find some property $Q$ of $x$ that is a post-fixed-point of $F$, i.e., we have to show $Q\langle x \rangle$ and $\forall x \in S.\ Q\langle x \rangle \Rightarrow F(Q)\langle x \rangle$. An important special case arises if $S = D \times D$ for some domain $D$ and $P$ is a relation $\sim\ \subseteq D \times D$ (a *bisimilarity*) defined as the greatest fixed point of a monotonic function $F$ (see Definition 5.15). To show that two objects $x \in D$ and $y \in D$ are bisimilar, i.e., $x \sim y$, it suffices to find some *bisimulation* relation $Q \subseteq D \times D$ such that $Q$ relates $x$ and $y$ and $Q$ is a post-fixed point of $F$. In other words, to prove $x \sim y$, we have to show $Q\langle x, y \rangle$ and $\forall x \in D, y \in D.\ Q\langle x, y \rangle \Rightarrow F(Q)\langle x, y \rangle$.

We demonstrate these proof principles by some examples.

**Example 5.24** The inductive definition

inductive *even* $\subseteq \mathbb{N}$

$\forall x \in \mathbb{N}.\ even\langle 0 \rangle \wedge (even\langle x + 2 \rangle \Leftarrow even\langle x \rangle)$

establishes the relation "is even" on $\mathbb{N}$. We will now investigate how to prove, for an arbitrary unary relation $P$, the proposition

$$\forall x \in \mathbb{N}.\ even\langle x \rangle \Rightarrow P\langle x \rangle$$

respectively the logically equivalent

$$\forall x \in even.\ P\langle x \rangle.$$

From the inductive definition of relation *even*, we have $even = \mathsf{lfp}\ E$ where

$$E := \lambda even \in \mathsf{Set}(\mathbb{N}).\ \pi x \in \mathbb{N}.\ x = 0 \vee \exists x' \in \mathbb{N}.\ x = x' + 2 \wedge even\langle x' \rangle.$$

The proof proceeds by induction: we have to show $E(P) \subseteq P$, i.e.,

$$(\pi x \in \mathbb{N}.\ x = 0 \vee \exists x' \in \mathbb{N}.\ x = x' + 2 \wedge P\langle x' \rangle) \subseteq P$$

respectively

$$\forall x \in \mathbb{N}.\ x = 0 \vee (\exists x' \in \mathbb{N}.\ x = x' + 2 \wedge P\langle x' \rangle) \Rightarrow P\langle x \rangle.$$

This can be transformed to

$$(\forall x \in \mathbb{N}.\ x = 0 \Rightarrow P\langle x \rangle) \wedge$$
$$(\forall x \in \mathbb{N}.\ (\exists x' \in \mathbb{N}.\ x = x' + 2 \wedge P\langle x' \rangle) \Rightarrow P\langle x \rangle)$$

and further simplified to

$$P\langle 0 \rangle \wedge \forall x \in \mathbb{N}.\ P\langle x \rangle \Rightarrow P\langle x + 2 \rangle.$$

Please note how this pattern relates to the original definition that could have also be written as

inductive *even* $\subseteq \mathbb{N}$
$even\langle 0 \rangle \wedge (even\langle x \rangle \Rightarrow even\langle x + 2 \rangle)$

In other words, the proof pattern just mimics the rule-based definition. We will know apply this pattern to prove

$$\forall x \in even.\ odd\langle x + 1 \rangle$$

where relation *odd* is explicitly defined as

*odd* $\subseteq \mathbb{N}$
$odd\langle x \rangle :\Leftrightarrow \exists y \in \mathbb{N}.\ x = 2 \cdot y + 1.$

In other words, we prove that every element $x$ of "type" *even* has the "property" $odd\langle x + 1\rangle$. By instantiation of the proof pattern with $P := \pi x \in \mathbb{N}.\ odd\langle x + 1\rangle$, we derive the pr obligation

$$odd\langle 1\rangle \wedge \forall x \in \mathbb{N}.\ odd\langle x + 1\rangle \Rightarrow odd\langle x + 3\rangle.$$

We prove the two parts of the conjunction:

- We have $odd\langle 1\rangle$ because $1 = 2 \cdot 0 + 1$.
- We take arbitrary $x \in \mathbb{N}$. We assume $odd\langle x + 1\rangle$ and show $odd\langle x + 3\rangle$. From $odd\langle x + 1\rangle$ we have some $y \in \mathbb{N}$ with $x + 1 = 2 \cdot y + 1$, i.e., $x = 2 \cdot y$. Then we have $x + 3 = 2 \cdot y + 3 = 2 \cdot (y + 1) + 1$. and therefore $odd\langle x + 3\rangle$.   □

As can be seen by above example, fixed point induction generalizes the principles of mathematical induction respectively structural induction; the "induction pattern" is determined by the pattern of the inductive definition of "subtype" $T = \mathsf{lfp}\ F$ and can be directly extracted from that definition.

**Example 5.25**  Given the coinductive definition

coinductive *infinite* $\subseteq \mathsf{Set}(\mathbb{N})$
$\forall S \in \mathsf{Set}(\mathbb{N}).$
 $infinite\langle S\rangle \Rightarrow S \neq \varnothing \wedge infinite\langle S \setminus \mathsf{anyof}\ S\rangle$

of relation "is infinite" and the definition

$odd \subseteq \mathbb{N}$
$odd\langle x\rangle :\Leftrightarrow \exists y \in \mathbb{N}.\ x = 2 \cdot y + 1.$

from Example 5.24 which is logically equivalent to the constant definition

$$odd := \{x \in \mathbb{N} \mid \exists y \in \mathbb{N}.\ x = 2 \cdot y + 1\}$$

we show

$infinite\langle odd\rangle$

i.e., that there exist infinitely many odd numbers. From the coinductive definition of *infinite*, we have *infinite* $= \mathsf{gfp}\ I$ where

$$I := \lambda infinite \in \mathsf{Set}(\mathbb{N}).\ \pi S \in \mathbb{N}.\ S \neq \varnothing \wedge infinite\langle S \setminus \mathsf{anyof}\ S\rangle.$$

The proof proceeds by coinduction. For this purpose, we define relation

$R \subseteq \mathsf{Set}(\mathbb{N})$

$R\langle S \rangle :\Leftrightarrow$

  $\exists p \in S^{\omega}.$

   $(\forall x \in S.\ \exists i \in \mathbb{N}.\ p(i) = x) \wedge$

   $(\forall i \in \mathbb{N}, i' \in \mathbb{N}.\ p(i) = p(i') \Rightarrow i = i')$

such that $R\langle S \rangle$ holds if and only if there exists an infinite enumeration $p$ of all elements of $S$, i.e., a bijection (a one-to-one mapping) between $\mathbb{N}$ and $S$.

Clearly we have $R\langle odd \rangle$ for $p := \lambda i \in \mathbb{N}.\ 2 \cdot i + 1$. It remains to show

$R \subseteq I(R)$

i.e.,

$R \subseteq \pi S \in \mathbb{N}.\ S \neq \varnothing \wedge R\langle S \setminus \mathsf{anyof}\ S \rangle$

respectively

  $\forall S \in \mathbb{N}.\ R\langle S \rangle \Rightarrow S \neq \varnothing \wedge R\langle S \setminus \mathsf{anyof}\ S \rangle.$

We take arbitrary $S \in \mathbb{N}$ and assume $R\langle S \rangle$, i.e., we have some $p \in S^{\omega}$ with

  $\forall x \in S.\ \exists i \in \mathbb{N}.\ p(i) = x$

  $\forall i \in \mathbb{N}, i' \in \mathbb{N}.\ p(i) = p(i') \Rightarrow i = i'.$

Since $p \in S^{\omega}$, we have $S \neq \varnothing$. To show $R\langle S \setminus \mathsf{anyof}\ S \rangle$, we have to construct some $q \in T^{\omega}$ with

  $(\forall x \in T.\ \exists i \in \mathbb{N}.\ q(i) = x) \wedge$

  $(\forall i \in \mathbb{N}, i' \in \mathbb{N}.\ q(i) = q(i') \Rightarrow i = i')$

where $T := S \setminus \mathsf{anyof}\ S$. It is not difficult to show that this holds for

  $q := \lambda i \in \mathbb{N}.\ \text{if } (\forall j \in \mathbb{N}.\ j \leq i \Rightarrow p(j) \neq \mathsf{anyof}\ S) \text{ then } p(i) \text{ else } p(i+1).$

i.e., the enumeration that is identical to $p$ except that $\mathsf{anyof}\ S$ has been removed. $\square$

**Example 5.26** Taking from Example 5.23 the definitions

$head \colon \mathbb{N}^{\omega} \to \mathbb{N}$

$head(x) = x(0)$

$tail \colon \mathbb{N}^{\omega} \to \mathbb{N}^{\omega}$

$tail(x) = \lambda i \in \mathbb{N}.\ x(i+1)$

coinductive $merge[\sim] \colon \mathbb{N}^{\omega} \times \mathbb{N}^{\omega} \to \mathbb{N}^{\omega}$

$merge(x_1, x_2) = y \Rightarrow head(x_1) = head(y) \wedge merge(x_2, tail(x_1)) = tail(y)$

we prove by coinduction

$merge(even, odd) \sim nat$

for the streams

$even := \lambda i \in \mathbb{N}. \, 2 \cdot i$

$odd := \lambda i \in \mathbb{N}. \, 2 \cdot i + 1$

$nat := \lambda i \in \mathbb{N}. \, i$

Informally stated, we prove that

$merge([0, 2, 4, \ldots], [1, 3, 5, \ldots]) \sim [0, 1, 2, 3, 4, 5, 6, \ldots].$

By the definition of $merge$ we have $\sim = \mathsf{gfp} \, M$ where

$M := \lambda R \in \mathbb{N}^\omega \times \mathbb{N}^\omega.$
$\quad \pi x_1 \in \mathbb{N}^\omega, x_2 \in \mathbb{N}^\omega, y \in \mathbb{N}^\omega.$
$\quad head(x_1) = head(y) \wedge R\langle merge(x_2, tail(x_1)), tail(y) \rangle.$

We define the bisimulation

$R \subseteq \mathbb{N}^\omega \times \mathbb{N}^\omega \times \mathbb{N}^\omega$
$R\langle x, y \rangle :\Leftrightarrow$
$\quad \forall x_1 \in \mathbb{N}^\omega, x_2 \in \mathbb{N}^\omega.$
$\quad x = merge(x_1, x_2) \Rightarrow y = \lambda i \in \mathbb{N}. \text{ if } 2|i \text{ then } x_1(i/2) \text{ else } x_2((i-1)/2).$

It is not difficult to show $R\langle merge(even, odd), nat \rangle$, i.e.,

$nat = \lambda i \in \mathbb{N}. \text{ if } 2|i \text{ then } even(i/2) \text{ else } odd((i-1)/2).$

Subsequently we have to prove $R \subseteq F(R)$, i.e.,

$\forall x_1 \in \mathbb{N}^\omega, x_2 \in \mathbb{N}^\omega, y \in \mathbb{N}^\omega.$
$\quad R\langle merge(x_1, x_2), y \rangle \Rightarrow$
$\quad head(x_1) = head(y) \wedge R\langle merge(x_2, tail(x_1)), tail(y) \rangle.$

We take arbitrary $x_1 \in \mathbb{N}^\omega, x_2 \in \mathbb{N}^\omega, y \in \mathbb{N}^\omega$ and assume $R\langle merge(x_1, x_2), y \rangle$. Therefore we have

$$y = \lambda i \in \mathbb{N}. \text{ if } 2|i \text{ then } x_1(i/2) \text{ else } x_2((i-1)/2). \tag{1}$$

From this we have $head(y) = y(0) = x_1(0/2) = x_1(0) = head(x_1)$. It remains to be shown $R\langle merge(x_2, tail(x_1)), tail(y)\rangle$, i.e.,

$$tail(y) = \lambda i \in \mathbb{N}. \text{ if } 2|i \text{ then } x_2(i/2) \text{ else } tail(x_1)((i-1)/2).$$

From the definition of *tail*, it suffices to prove for arbitrary $i \in \mathbb{N}$

$$y(i+1) = \text{ if } 2|i \text{ then } x_2(i/2) \text{ else } tail(x_1)((i-1)/2). \tag{$*$}$$

From (1), we have

$$y(i+1) = \text{ if } 2|(i+1) \text{ then } x_1((i+1)/2) \text{ else } x_2(i/2)$$

and, since $2|i \Leftrightarrow \neg 2|(i+1)$,

$$y(i+1) = \text{ if } 2|i \text{ then } x_2(i/2) \text{ else } x_1((i+1)/2).$$

Since $tail(x_1)((i-1)/2) = x_1((i-1)/2+1) = x_1((i+1)/2)$, we have ($*$).   $\square$

The induction principle defined above allow to reason about the least fixed point of a monotonic function. However if this function is also continuous, we have a stronger proof principle for certain relations.

**Definition 5.16** (*Inclusive relation*) Relation $P \subseteq S$ is *inclusive* if and only if

$$\forall A \in S_{\subseteq}^{\omega}. \ (\forall i \in \mathbb{N}. \ P\langle A(i)\rangle) \Rightarrow P\langle \bigcup_{i \in \mathbb{N}} A(i)\rangle.$$

In other words if $P$ holds for every element of a chain, then it also holds for the union of the chain.

For inclusive relations, we have the following proof principle.

**Proposition 5.12** (Fixed point induction) *Let $F: Set(S) \to Set(S)$ be continuous and $P \subseteq S$ be inclusive. Then, in order to show*

$$P\langle lfp \ F\rangle,$$

*i.e., lfp $F \in P$, it suffices to show*

- $P\langle \emptyset \rangle$ *and*
- $\forall A \in Set(S). \ P\langle A\rangle \Rightarrow P\langle F(A)\rangle,$

*i.e.,* $\varnothing \in P$ *and* $\forall A \in Set(S). A \in P \Rightarrow F(A) \in P.$
*More formally, we have the following inference rule:*

$$\frac{Fs \vdash P\langle\varnothing\rangle \quad Fs \vdash \forall A \in Set(S). P\langle A\rangle \Rightarrow P\langle F(A)\rangle \quad (P \subseteq S \text{ is inclusive})}{Fs \vdash P\langle\mathsf{lfp}\ F\rangle}$$

The soundness of this principle is easy to see: from the two proofs, it follows by mathematical induction that $\forall i \in \mathbb{N}. P\langle F^i(\varnothing)\rangle$. Since $\lambda i \in \mathbb{N}. F^i(\varnothing)$ is a chain, we have $P(\bigcup_{i\in\mathbb{N}} F^i(\varnothing))$ and thus, since $F$ is continuous, $P\langle\mathsf{lfp}\ F\rangle$.

Not every relation is inclusive; e.g. the relation *finite* $\subseteq Set(\mathbb{N})$ that is only true for finite subsets of $\mathbb{N}$ is not: take the chain $\mathbb{N}_0, \mathbb{N}_1, \mathbb{N}_2, \ldots$. Then we have *finite*$\langle\mathbb{N}_i\rangle$ for every $i \in \mathbb{N}$, but $\neg$*finite*$\langle\bigcup_{i\in\mathbb{N}} \mathbb{N}_i\rangle$ because $\bigcup_{i\in\mathbb{N}} \mathbb{N}_i = \mathbb{N}$. Fixed point induction can therefore not be applied to every relation.

**Example 5.27** Based on the auxiliary functions

$nil := \varnothing$

$cons: \mathbb{N} \times \mathbb{N}^* \to \mathbb{N}^*$

$cons(h, t) := \lambda i \in \mathbb{N}. \text{ if } i = 0 \text{ then } h \text{ else } t(i - 1)$

we introduce by a rule-based inductive definition the concatenation *append*$(x_1, x_2)$ of two finite number sequences $x_1$ and $x_2$ as

inductive *append*: $\mathbb{N}^* \times \mathbb{N}^* \to_\perp \mathbb{N}^*$

*append*$(nil, l) = l \wedge$

$(append(cons(h, t), l) = r \Leftarrow \text{exists } s = append(t, l). r = cons(h, s)).$

We thus have, for instance, *append*$([1, 2, 3], [4, 5]) = append([1, 2, 3, 4, 5])$. Now we want to prove

$$\forall x \in \mathbb{N}^*. \text{ forall } y = append(x, nil). y = x. \tag{*}$$

Actually, one can show that *append* is total, i.e., *append*: $\mathbb{N}^* \times \mathbb{N}^* \to \mathbb{N}^*$, because the function is defined by primitive recursion in its first argument with respect to the well-founded relation "is shorter than". We could thus write (*) also as

$$\forall x \in \mathbb{N}^*. append(x, nil) = nil.$$

However, in the following proof, we stick to the notion of *append* as a partial function; it is an easy exercise to translate the proof into a form that takes into account that *append* is actually total.

Statement (*) is logically equivalent to

$$P \langle \mathsf{lfp} \ F \rangle$$

for the relation $P$ defined as

$$P \subseteq \mathbb{N} \to_\perp \mathbb{N}$$
$$P \langle a \rangle :\Leftrightarrow \forall x \in \mathbb{N}^*. \ \mathsf{forall} \ y = a(x, nil). \ y = x$$

(whose inclusiveness will be shown later) and the upward continuous function $F$ induced by the definition of *append*:

$$F : (\mathbb{N} \to_\perp \mathbb{N}) \to (\mathbb{N} \to_\perp \mathbb{N})$$
$$F(a) :=$$
$$\quad \pi x \in \mathbb{N}^*, y \in \mathbb{N}^*, z \in \mathbb{N}^*.$$
$$\quad (x = nil \wedge z = y) \vee$$
$$\quad (\exists h \in \mathbb{N}, t \in \mathbb{N}^*. \ x = cons(h, t) \wedge \mathsf{exists} \ s = a(t, y). \ z = cons(h, s)).$$

We proceed by fixed point induction:

- To show $P \langle \varnothing \rangle$, it suffices to show

$$\forall x \in \mathbb{N}^*. \ \forall y \in \mathbb{N}^*. \ \langle x, nil, y \rangle \in \varnothing \Rightarrow y = x$$

  which is trivially true.
- We assume, for arbitrary $a \in (\mathbb{N} \to_\perp \mathbb{N})$, $P \langle a \rangle$, i.e.,

$$\forall x \in \mathbb{N}^*. \ \mathsf{forall} \ y = a(x, nil). \ y = x \tag{1}$$

  and show $P \langle F(a) \rangle$, i.e.,

$$\forall x \in \mathbb{N}^*. \ \mathsf{forall} \ y = F(a)(x, nil). \ y = x.$$

  We thus take arbitrary $x \in \mathbb{N}^*$, and show

$$\forall y \in \mathbb{N}^*. \ F(a) \langle x, nil, y \rangle \Rightarrow y = x$$

  We take arbitrary $y \in \mathbb{N}^*$. We assume $F(a) \langle x, nil, y \rangle$ and show $y = x$. From $F(a) \langle x, nil, y \rangle$, we have

$$(x = nil \wedge y = nil) \vee$$
$$(\exists h \in \mathbb{N}, t \in \mathbb{N}^*. \ x = cons(h, t) \wedge \mathsf{exists} \ s = a(t, nil). \ y = cons(h, s)),$$

  and thus two cases: If $x = nil \wedge y = nil$, we immediately have $y = x$. Otherwise, there exists some $h \in \mathbb{N}, t \in \mathbb{N}^*$, and $s \in \mathbb{N}^*$ with $x = cons(h, t)$, $a \langle t, nil, s \rangle$ and $y = cons(h, s)$. From $a \langle t, nil, s \rangle$ and (1), we have $s = t$ and thus $y = x$. $\qquad \square$

Since it is not easy to see which relations are inclusive, the following syntactic criterion may be helpful.

---

**Proposition 5.13**  (Inclusive formula) *Let S denote a set and $Q_x$ be a formula with only free variable x. Then the relation*

$$\pi x \in S.\, Q_x$$

*on S is inclusive provided that the following constraints are satisfied:*

- *Formula $Q_x$ is identical to a formula $Q_?$ described below except that in $Q_?$ every free occurrence of x in $Q_x$ has been replaced by a new constant symbol ? that is not in the current first-order language; consequently $Q_?$ is closed.*

- *Formula $Q_?$ is formed according to the restricted grammar*

  $$Q ::= P \mid \forall V \in CS.\, Q$$
  $$P ::= F \mid U_1 \subseteq U_2 \mid P_1 \vee P_2 \mid P_1 \wedge P_2$$

  *where V, CS, and F are formed as in the original language but U denotes a term in an extended version of the language that also includes the constant symbol ?; consequently ? can only occur in a formula $U_1 \subseteq U_2$.*
- *Let $x_1, \ldots, x_n$ be the free variables of U (in any order) and let $CS_1, \ldots, CS_n$ be the corresponding constants in the conditions $x_1 \in CS_1, \ldots, x_n \in CS_n$ by which the variables were introduced in the universal quantifiers of Q. Then the function*

  $$\lambda x \in S.\, \text{choose } x_1 \in CS_1.\, \ldots \text{choose } x_n \in CS_n.\, U_x$$

  *is upward continuous, where x is a variable that does not occur in U and $U_x$ is identical to U except that every occurrence of constant ? has been replaced by x.*

*In other words, a relation defined by a formula $Q_x$ is inclusive if $Q_x$ is a universally quantified composition of conjunctions and disjunctions of formulas in which either x does not occur freely or which are atomic formulas $U_1 \subseteq U_2$ where the terms $U_1$ and $U_2$ denote upward continuous functions in x (with the values of the other free variables arbitrarily fixed).*

---

It should be noted that, if $U_1$ and $U_2$ satisfy the equations $U_1 = \{x \in A \mid F_1[x]\}$ and $U_2 = \{x \in A \mid F_2[x]\}$ for some set A and formulas $F_1$ and $F_2$ with free variable $x$, then in order to show $U_1 \subseteq U_2$, it suffices to prove $\forall x \in A.\, F_1[x] \Rightarrow F_2[x]$.

**Example 5.28** To justify the application of fixed point induction in Example 5.27, we have show that $P$ is inclusive: first we note that $P$ can be also written as

$$\pi a \in \mathbb{N} \rightarrow_\perp \mathbb{N}.\ \forall x \in \mathbb{N}^*, y \in \mathbb{N}^*.\ a\langle x, y\rangle \Rightarrow y = x$$

which is logically equivalent to

$$\begin{aligned}&\pi a \in \mathbb{N} \rightarrow_\perp \mathbb{N}.\\ &\quad \big\{\langle x, y\rangle \mid x \in \mathbb{N}^* \wedge y \in \mathbb{N}^* \wedge a\langle x, nil, y\rangle\big\} \subseteq\\ &\quad \big\{\langle x, y\rangle \mid x \in \mathbb{N}^* \wedge y \in \mathbb{N}^* \wedge y = x\big\}.\end{aligned}$$

Then it is not difficult to see that the two functions

$$\begin{aligned}&\lambda a \in \mathbb{N} \rightarrow_\perp \mathbb{N}.\ \mathsf{choose}\ x \in \mathbb{N}^*, y \in \mathbb{N}^*.\ \big\{\langle x \in \mathbb{N}^*, y \in \mathbb{N}^*\rangle \mid a\langle x, nil, y\rangle\big\}\\ &\lambda a \in \mathbb{N} \rightarrow_\perp \mathbb{N}.\ \mathsf{choose}\ x \in \mathbb{N}^*, y \in \mathbb{N}^*.\ \big\{\langle x, y\rangle \mid x \in \mathbb{N}^* \wedge y \in \mathbb{N}^* \wedge y = x\big\}\end{aligned}$$

are upward continuous; thus all the requirements of Proposition 5.13 are satisfied and $P$ is indeed inclusive. □

## Exercises

Download from the following URL:
https://www.risc.jku.at/people/schreine/TP/exercises/ex-recursion.pdf.

## Further Reading

The interpretation of a recursive function definition as an implicit specification is described in Chap. 6 of Lamport's book [98].

The concepts of "primitive" recursion, "structural" recursion, or, more general, "well-founded" recursion (respectively induction) are explained in many texts, e.g. Sect. 3 of Manna et al.'s paper [107], Chap. 3 of Winskel's book [180], Chap. 4 of Bruni's and Montanari's book [28], Chap. 1 of Nielson's and Nielson's book [131], Appendix B of Nielson et al.'s book [130], Chap. 1 of Mitchell's book [119], Chap. 2 of Nipkow's and Klein's book [133], Appendix 1 of Harrison's handbook [65], Chap. 4 of Makinson's book [106], Chap. 4 of Bradley's and Manna's book [26], Chap. 12 of Schenke's book [159], Chap. 2 of Gallier's book [56], Chap. 5 of Jones's book [84], Chap. 1 of Dowek's and Lévy's book [43].

Likewise the theory of least fixed points as the basis of inductive definitions is covered in many texts, but usually in the general context of complete partial orders (only Chaps. 1 and 5 of Mitchell's book [119] also briefly present fixed point theory in the domain of plain sets). See e.g. Manna et al.'s influential paper [107], Chap. 6 of Schmidt's book [160], Chaps. 5 and 8 of Winskel's book [180], Chap. 5 of Bruni's and Montanari's book [28], Chaps. 4 and 6 of Nielson's and Nielson's book [131], Chap. 4 of Gunter's book [63], Chaps. 2 and 5 of Mitchell's book [119], Chaps. 10, 11, and 13 of Nipkow's and Klein's book [133], Mosses's Chap. 11 and Gunter's and

Scott's Chap. 12 of the handbook [177], Appendix 1 of Harrison's handbook [65], Chap. 12 of Bradley's and Manna's book [26], Chap. 3 of Nebel's book [127] (in German), Chap. 1 of Dowek's and Lévy's book [43], Chap. 4 and Appendices A and B of Nielson et al.'s book [131], Appendix A of Kropf's book [91], Chap. 3 of Schneider's book [164], Möller's chapter in [27], Chap. 3 of Kröger's and Merz's book [93], or Appendix A of Roscoe's book [154].

The dual theory of greatest fixed points as the basis of coinductive definitions is not given equal treatment in textbooks: Chap. 10 of Schmidt's book [161] contains on page 339 an exercise on coinductive types and Appendix B of Nielson et al.'s book [130] contains a discussion of coinductive proofs. More material can be found in the research literature: Kozen's and Silva's paper [89] gives an accessible introduction to coinduction as well as Paulson's article [142], which describes the logic of (co)inductive definitions implemented in the proof assistant Isabelle. Jacobs's and Rutten's tutorial [83] presents the general theory in the framework of (co)algebras. Padawitz's paper [141] introduces a specification formalism for both inductively and coinductively defined types. Sangiorgi's paper [157] provides a historical account on the development of bisimulation and coinduction.

## Recursive Definitions in Isabelle/HOL

In this section, we continue our presentation of Isabelle/HOL started in section "Writing Definitions in Isabelle/HOL", Chap. 4 with a demonstration of various forms of recursion respectively induction supported by the proof assistant. The specifications used in the following presentation can be downloaded from the URL

> https://www.risc.jku.at/people/schreine/TP/software/recursion/recursion.thy

and loaded by executing from the command line the following command:

```
Isabelle2021 recursion.thy &
```

Our presentation borrows heavily from the documentation of Isabelle/HOL.

### Implicit Definitions

The approach of considering recursive function definitions as implicit definitions may be implemented in Isabelle/HOL by the following definitions:

```
definition isfactorial :: "(nat ⇒ nat) ⇒ bool" where
"isfactorial f = (∀n::nat. f n = (if n = 0 then 1 else n*(f(n-1))))"

definition factorial :: "nat ⇒ nat" where
"factorial = (SOME f::nat⇒nat. isfactorial f)"
```

We may then show that the implicitly defined function $factorial$ satisfies the expected properties:

```
lemma isfactorial:
  "(∃f::nat⇒nat. isfactorial f) ⟶ isfactorial factorial" ...
lemma factorial3:
  "(∃f::nat⇒nat. isfactorial f) ⟶ factorial 3 = 6" ...
```

However, to show that the assumption $\exists f : \mathbb{N} \to \mathbb{N}. \, isfactorial(f)$ is satisfied, indeed requires the explicit definition of such a function; this makes the implicit definition rather pointless. We therefore turn to Isabelle/HOL's mechanisms for recursive definitions of functions, predicates, and sets.

### Primitive Recursion

Isabelle/HOL supports primitive recursion the natural numbers. For instance, the definition

```
fun factorial2 ::"nat ⇒ nat" where
"factorial2 0       = 1" |
"factorial2 (Suc n) = (Suc n)*factorial2 n"
```

gives a primitive recursive definition of the factorial function. The keyword primrec enforces that this definition is given in a rule-based form that makes the well-founded relation explicit on which the definition is based. In this case, the primitive recursion uses the size of the number as the well-founded relation: the argument is matched either by the pattern 0 (zero) or by Suc n (the successor of $n$, i.e., $n + 1$); in the second case, the function is recursively applied only to $n$ which is smaller than the original argument. With this definition, we can show

```
lemma isfactorial2: "isfactorial factorial2" ...
```

i.e., the function indeed satisfies the requirements of the implicit definition given above.

As another example, the primitive recursive definition

```
primrec add :: "nat ⇒ nat ⇒ nat" where
"add 0 n = n" |
"add (Suc m) n = Suc(add m n)"
```

introduces the binary function *add* (addition) on the natural numbers. where the primitive recursion runs over the first argument. Every datatype that supports primitive recursion is equipped with a suitable induction principle that can be applied to prove properties of the functions that are defined by primitive recursion. For instance, the formula

```
lemma add_L: "∀m::nat. add m 0 = m"
```

that states that $m + 0 = 0$, for every natural number $m$, can be verified by the following proof in the Isar proof language of Isabelle/HOL:

```
proof
  fix m
  show "add m 0 = m"
  proof (induction m)
    show "add 0 0 = 0" by auto
  next
    fix m
    show "add m 0 = m ⟹ add (Suc m) 0 = Suc m" by auto
  qed
qed
```

Without going into details, one can recognize the inductive structure of the proof with the induction base and the induction step.

Primitive recursion is not limited to natural numbers; it may be applied to all (also user-defined) "inductive" data types such as finite lists. For instance, the definition

```
primrec app :: "'a list ⇒ 'a list ⇒ 'a list" where
"app Nil ys = ys" |
"app (Cons x xs) ys = Cons x (app xs ys)"
```

introduces the binary function *app* (append) on lists. Again, the primitive recursion runs on the first argument with the length of the list as the well-founded relation: the argument is matched either by the pattern Nil (the empty list) or Cons x xs (the list whose head is $x$ and tail is $xs$); in the second case, the function is recursively applied only to $xs$ which is one element shorter than the original argument.

**More General Recursion**

While the primitive recursion format supported by primrec is very restricted (exactly one rule must be provided for each constructor of the respective datatype), the keyword fun supports more general forms of recursion. This is demonstrated by the following definitions of the predicates "even" and "odd" on the natural numbers:

```
fun ev :: "nat ⇒ bool" where
"ev 0 = True" |
"ev (Suc 0) = False" |
"ev (Suc(Suc n)) = ev n"

fun od :: "nat ⇒ bool" where
"od 0 = False" |
"od (Suc 0) = True" |
"od (Suc(Suc n)) = od n"
```

Here the recursion has two base cases for 0 and 1, while the recursive case reduces $n + 2$ to $n$. Based on these definitions, we can show by an induction proof that the successor of every even number is odd and vice versa:

```
lemma ev_L2: "∀(n::nat). ev n = od (n+1) ∧ od n = ev (n+1)"
proof (rule allI)
  fix n :: "nat"
  show "ev n = od (n + 1) ∧ od n = ev (n+1)"
  proof (induction n)
    show "ev 0 = od (0 + 1) ∧ od 0 = ev (0+1)" by auto
  next
    fix n
    assume "ev n = od (n+1) ∧ od n = ev (n+1)"
    show "ev (Suc n) = od (Suc n + 1) ∧ od (Suc n) = ev (Suc n + 1)"
      using 'ev n = od (n + 1) ∧ od n = ev (n + 1)' by auto
  qed
qed
```

In general, fun allows those kinds of recursive definitions where Isabelle/ HOL can automatically find the well-founded relation of the argument domain. This is typically a lexicographic order on a certain permutation of the arguments: a prefix of the arguments in this permutation remains unchanged while the argument after the prefix gets smaller. For instance, in the following definition of "Ackermann's function"

```
fun ack :: "nat ⇒ nat ⇒ nat" where
"ack n 0 = n+1" |
"ack 0 (Suc m) = ack 1 m" |
"ack (Suc n) (Suc m) = ack (ack n (Suc m)) m"
```

in every recursive invocation the second argument may get smaller; otherwise it stays the same and the first argument gets smaller.

If Isabelle is not able to find the well-founded relation, we may prefix a recursive definition with the keyword `function`. In this case, the system generates the obligations that have to be proved to show that the definition is well-formed, i.e., the recursion terminates for every possible argument tuple. This proof typically involves the explicit definition of a natural number "measure" that gets smaller in every recursive invocation. For instance, in the definition of the function

```
function sum :: "nat ⇒ nat ⇒ nat" where
"sum i N = (if i > N then 0 else i + sum (i+1) N)" ...
```

this measure is essentially the difference between $i$ and $N + 1$. The corresponding Isar proof script looks as follows:

```
by pat_completeness auto
termination sum
proof (relation "measure (λ(i,N). N + 1 - i)")
  show "wf (measure (λ(i, N). N + 1 - i))" by simp
  show "⋀i N. ¬N<i ⟹ ((i+1,N),i,N) ∈ measure(λ(i,N). N+1-i)" by simp
qed
```

The key is the indication of the measure $\lambda i, N.\ N + 1 - i$ by the user; the rest of the termination proof proceeds automatically.

### Inductive Set and Predicate Definitions

Functions in Isabelle/HOL are always total. Since also the recursive definitions by `primrec`, `fun`, or `function` introduce only total functions, these constructions need not resort to fixed point theory. Nevertheless, one may introduce sets or predicates as least fixed points. For instance, the definition

```
definition finset :: "'a set ⇒ bool" where
"finset = lfp
  (λ(F::'a set⇒bool)(S::'a set). S = {} ∨ F(S-{SOME x::'a. x ∈ S}))"
```

defines the set *finset* of all finite sets of elements of some type $a$ as the least fixed point of the functional $\lambda F.\ \lambda s.\ s = \varnothing \lor F(s \setminus \{\text{anyof } s\})$.

However, Isabelle/HOL provides more direct ways of introducing such sets or predicates. For instance, the definitions

```
inductive_set eset :: "nat set" where
zero: "0 ∈ eset" |
step: "n ∈ eset ⟹ (Suc (Suc n)) ∈ eset"

inductive_set oset :: "nat set" where
zero: "1 ∈ oset" |
step: "n ∈ oset ⟹ (Suc (Suc n)) ∈ oset"
step: "n ∈ oset ⟹ (Suc (Suc n)) ∈ oset"
```

introduce the sets *eset* and *oset* of all even respectively odd natural numbers by rules: a natural number is in the respective set if and only if this can be deduced from these rules (the arrow ⟹ does not represent logical implication but is a symbol of the meta-logic of Isabelle/HOL that separates in an inference rule the assumptions from the conclusion).

In a similar way, corresponding predicates (i.e. functions that yield truth values) can be defined as

```
inductive even :: "nat ⟹ bool" where
even1: "even 0" |
even2: "even n ⟹ even (n + 2)"

inductive odd :: "nat ⟹ bool" where
odd1: "odd 1" |
odd2: "odd n ⟹ odd (n + 2)"
```

which are true for certain arguments if and only if this can be deduced from these rules. The sets respectively predicates are accompanied by corresponding induction principles. For instance, the fact that the successor of every even number is odd can be proved by the following Isar proof script:

```
lemma even_L2: "∀(n::nat). even n ⟶ odd (n+1)"
proof (auto)
  fix n :: "nat"
  show "recursion.even n ⟹ recursion.odd (Suc n)"
  proof (induction rule: even.induct)
    show "recursion.odd (Suc 0)" using odd1 by auto
  next
    fix n
    assume "recursion.even n"
    assume "recursion.odd (Suc n)"
    show "recursion.odd (Suc (n+2))"
      using 'recursion.odd (Suc n)' odd2 by auto
  qed
qed
```

This script refers to the induction rule even.induct that was generated from above definition of the predicate *even*.

Returning to the introductory example, the set of finite sets of elements of *a* can be defined by the rules

```
inductive_set finiteset :: "'a set set" where
finiteset1: "{} ∈ finiteset" |
finiteset2: "S-{SOME x::'a. x ∈ S} ∈ finiteset ⟹ S ∈ finiteset"
```

or by the following logically equivalent rules:

```
inductive_set finiteset2 :: "'a set set" where
finiteset1: "{} ∈ finiteset2" |
finiteset2: "S ∈ finiteset2 ⟹ S ∪ {x} ∈ finiteset2"
```

The corresponding predicate "is finite" can be defined as

```
inductive finite2 :: "'a set => bool" where
finite1: "finite2 {}" |
finite2: "finite2 S ⟹ finite2 (S ∪ {x})"
```

which describes in a very readable way when a set is finite.

## Coinductive Set and Predicate Definitions

Dually to the definition of $finset$ as a least fixed point, the definition

```
definition infset :: "'a set ⟹ bool" where
"infset = gfp
   (λ(I::'a set⟹bool)(S::'a set). S ≠ {} ∧ I(S-{SOME x::'a. x ∈ S}))"
```

introduces the set $infset$ of infinite sets of elements of type $a$ as the greatest fixed point of the functional $\lambda O.\ \lambda s.\ s \neq \varnothing \wedge I(s \backslash \{\text{anyof } s\})$. A corresponding coinductive definition has the form

```
coinductive_set infiniteset :: "'a set set" where
iset: "S ≠ {} ∧ S-{SOME x::'a. x∈S} ∈ infiniteset ⟹ S ∈ infiniteset"
```

while the corresponding predicate can be defined as

```
coinductive infinite :: "'a set ⟹ bool" where
inf: "S ≠ {} ∧ infinite (S-{SOME x::'a. x ∈ S}) ⟹ infinite (S)"
```

The definitions can be also written in the form

```
coinductive_set infiniteset2 :: "'a set set" where
iset2: "⟦ x ∉ S; S ∈ infiniteset2 ⟧ ⟹ S ∪ {x} ∈ infiniteset2"
```

```
coinductive infinite2 :: "'a set ⟹ bool" where
inf2: "⟦ x ∉ S; infinite2 S ⟧ ⟹ infinite2 (S ∪ {x})"
```

that is more idiomatic to Isabelle/HOL (here the phrase ⟦ ⋯; ⋯ ⟧ embeds two assumptions of an inference rule). An implicit constraint is that a set is only in set $infiniteset2$ respectively only represents an argument for which predicate $infinite2$ becomes true, if it matches the pattern $S \cup \{x\}$ in the conclusion of the rule; thus it is in these definitions not necessary to explicitly state that the empty set {} is not infinite.

# Part II
# The Higher Planes

# Abstract Data Types

# 6

*Der Worte sind genug gewechselt, lasst uns endlich Daten*
*sehen. (That's enough words for the moment, now let me see*
*some data!)—Gerhard Kocher (Vorsicht, Medizin!), after*
*Johannn Wolfgang von Goethe (Faust, 'Taten/actions' rather*
*than 'Daten/data'.)*

Programs operate on data. It is thus natural to start our considerations of how to think about programs by a discussion of how to think about data types. For this purpose, we do not really need to know how the objects of a type are concretely represented (such representations have been discussed in Chap. chapter:models); we may rather focus on the properties that are satisfied by the operations which have been given to us to work with these objects. This view is also in line with modern software engineering that abstracts from the implementation details of data by encapsulating them in classes that only expose a (more or less) well documented method interface to the user.

This chapter presents the core of a theory of such 'abstract data types' which is a blend of universal algebra and logic; in particular we introduce a language for specifying abstract data types and give it a formal semantics. Our presentation starts in Sect. 6.1 with some examples of this language before we elaborate in Sect. 6.2 its core of 'declarations'; this core give rise to the formal notions of 'signatures' and 'presentations' which capture the syntactic aspects of declarations. We then proceed in Sect. 6.3 to the mathematical concepts of (many-sorted) 'algebras' and 'homomorphisms' on which the notion of an 'abstract data type' is based which captures the semantics of a declaration.

In Sects. 6.4, 6.5, and 6.6, we give three classes of possible interpretations of declarations as abstract data types: the 'loose' one (which confines itself to the logical characterization of a type), the 'generated/free' one (which describes data types such

© The Author(s), under exclusive license to Springer Nature Switzerland AG 2021
W. Schreiner, *Thinking Programs*, Texts & Monographs in Symbolic Computation,
https://doi.org/10.1007/978-3-030-80507-4_6

as finite lists by the means of their construction), and the dual 'cogenerated/cofree' one (which describes data types such as infinite streams by the ways of how they can be observed). We then extend in Sect. 6.7 this language of 'specifying in the small' to a language of 'specifying in the large': that language allows to combine specifications of individual data types to compound specifications and also to develop 'generic specifications' that can be instantiated in various ways.

While the previous elaborations have given abstract data type specifications a formal semantics, it has not yet become really clear what we can practically 'do' with such specifications. We therefore conclude in Sect. 6.8 by discussing how to reason about formally specified abstract data types, for instance, in programs that operate on such types or that implement such types.

## 6.1    Introduction

We start by presenting several examples of abstract data types that shall motivate the formal concepts that will be discussed in the subsequent sections. We define abstract data types by named *specifications* such as the following one:

```
% a domain with an associative operation and a neutral element
spec MONOID :=
  {
    sort Elem
    const e: Elem
    fun op: Elem × Elem → Elem
    pred isE ⊆ Elem
    axiom ∀x:Elem. op(e,x) = x ∧ op(x,e) = x
    axiom ∀x:Elem, y:Elem, z:Elem. op(x,op(y,z)) = op(op(x,y),z)
    axiom ∀x:Elem. isE(x) ⇔ x=e
  }
```

This specification named MONOID introduces the following entities:

1. a *sort* Elem which denotes a non-empty set of elements;
2. a *constant operation* e which denotes one of these elements;
3. a *function operation* op which denotes a binary function on this set;
4. a *predicate operation* isE which denotes a unary predicate on this set.

We use the term *operations* to differentiate between a name and the entities denoted by this name; if clear from the context, we may also drop the appendix 'operation' and just speak of constants, functions, and predicates.

Not every interpretation of these names is allowed: the specification contains various *axioms*, i.e., formulas that must be true for the denoted set, constant, function, and predicate to yield a valid *implementation* of the abstract data type. Conversely, every interpretation obeying the axioms represents a valid implementation: for instance, the specification might be implemented by

- the set of character strings for Elem, the empty string for e, string concatenation for op, and the emptiness test for isE; however, it may also be implemented by
- the set of natural numbers for Elem, the number 0 for e, the addition operation for op, and the nullness test for isE.

These two implementations differ in crucial features, e.g. the axiom

**axiom** $\forall$x:Elem, y:Elem.  op(x,y) = op(y,x)

is false for the first but true for the second one. Above specification is therefore also called *loose*, because it allows implementations with observably different behaviors.

However, a specification need not be loose. Take the specification

```
% the natural numbers
spec NAT :=
  free type Nat := 0 | +1(Nat)
  then
  {
    fun +: Nat × Nat → Nat
    axiom ∀n1:Nat, n2:Nat.
      +(0, n2) = n2 ∧
      +(+1(n1), n2) = +1(+(n1, n2))
    pred is0 ⊆ Nat, is0(n) :⟺ n=0
  }
```

Here the declaration

**free type** Nat := 0 | +1(Nat)

is a shortcut for the specification

```
free {
  sort Nat
  const 0: Nat
  fun +1: Nat → Nat
}
```

which introduces a sort Nat with a constant 0 and a unary *constructor* function +1. This specification is tagged as **free**, which essentially means that every term that one can build from the constructors represents a different element and that these are the only elements of the specified sort. The set denoted by Nat thus consists of the distinct elements denoted by 0, +1(0), +1(+1(0)), ...; consequently, this set can be identified with the set of natural numbers.

Using the keyword **then**, this specification is subsequently extended by a loose specification that introduces a binary function + on Nat which is however uniquely characterized by an axiom: for every term of form $+(T_1, T_2)$ where $T_1$ and $T_2$ are only constructed by application of 0 and +1, the first argument $T_1$ 'matches' one of the two universally quantified equations in the axiom: if $T_1$ is 0, the first equation matches and determines the value of the term to be the value of $T_2$; if it is of form

$+1(U_1)$, the result is the value of $+1(+(U_1, T_2)))$. By the freedom of the specification of Nat, exactly one of the specification matches and determines unique values for the variables such that the result is uniquely determined.

Furthermore, we introduce a predicate is0 by a declaration

**pred** is0 $\subseteq$ Nat, is0(n) :$\Leftrightarrow$ n=0

which is a shortcut for

**pred** is0 $\subseteq$ Nat
**axiom** $\forall$n:Nat. is0(n) $\Leftrightarrow$ n=0

The first format, however, makes it immediately clear that the predicate is uniquely defined, because for every value of its argument, the resulting truth value is explicitly described.

A free specification introduces a sort whose values are 'finite', in the sense that they can be constructed by finitely many applications of constructors to a constant. For the following specification, this is not the case:

```
% infinite streams of natural numbers
spec NATSTREAM import NAT :=
  cofree cotype NatStream := head:Nat | tail:NatStream
  then
  {
    fun cons: Nat × NatStream → NatStream
    axiom ∀n:Nat, s:NatStream.
      head(cons(n,s)) = n ∧
      tail(cons(n,s)) = s
    fun counter: Nat → NatStream, counter(n) := cons(n, counter(n+1))
  }
```

Using the keyword **import**, this specification first 'imports' the previously written specification NAT (whose entities thus become available to the specification). It then extends it by the declaration

**cofree cotype** NatStream := head:Nat | tail:NatStream

which is a shortcut for the specification

```
cofree {
  sort NatStream
  fun head: NatStream → Nat
  fun tail: NatStream → NatStream
}
```

that introduces two *observer* functions head and tail. This specification is tagged as **cofree** which essentially means that the elements of the introduced sort are 'black boxes' which are only considered as different if they can be distinguished by the (repeated) application of observer operations; furthermore, the new sort contains elements for all possible sequences of observer operations. The sort NatStream can be thus identified with the set of infinite streams of natural numbers: given a stream $s$, head($s$) denotes the first number (the 'head') of the stream while tail($s$) denotes the remainder of $s$ (its 'tail').

This specification is subsequently extended by a loose specification

```
{
    fun cons: Nat × NatStream → NatStream
    axiom ∀n:Nat, s:NatStream.
        head(cons(n,s)) = n ∧
        tail(cons(n,s)) = s
    ...
}
```

which introduces a function cons such that cons($n,s$) denotes an infinite stream with head $n$ and tail $s$. This function is constrained by two 'pattern-matching' equations that specify for every observer of NatStream the result of its application to cons($n,s$). Because of the co-freedom of NatStream, these equations determine unique values for $n$ and $s$ such that $s$ is a proper substream of the original stream; equations of this kind can thus not introduce any inconsistencies but indeed define a function uniquely.

In the same style, we could by a specification

```
{
    fun counter: Nat → NatStream
    axiom ∀n:Nat, s:NatStream.
        head(counter(n)) = n ∧
        tail(counter(n)) = counter(n+1)
}
```

define a function counter such that that counter(n) denotes the infinite stream $[n, n + 1, n + 2, \ldots]$. However, with the help of cons, this can be much more elegantly achieved: the declaration

```
fun counter: Nat → NatStream, counter(n) := cons(n, counter(n+1))
```

determines the same function, because the equations

```
head(counter(n)) = head(cons(n, counter(n+1))) = n
tail(counter(n)) = tail(cons(n, counter(n+1))) = counter(n+1)
```

follow from the axioms of the cons operation.

The specification NATSTREAM models streams of natural numbers; however, streams behave more or less the same for all kinds of elements. We can express this by creating a generic (parameterized) specification

```
% infinite streams of elements
spec STREAM[sort Elem] :=
    cofree cotype Stream := head:Elem | tail:Stream
    then
    {
        fun cons: Elem × Stream → Stream
```

**axiom** ∀e:Elem, s:Stream.
  head(cons(e,s)) = e ∧
  tail(cons(e,s)) = s
}

from which we derive NATSTREAM as a special instance:

**spec** NATSTREAM **import** NAT :=
  STREAM[NAT **fit** Elem↦Nat] **with** Stream↦NatStream
  **then fun** counter: Nat → NatStream, counter(n) := cons(n, counter(n+1))

The specification instantiation STREAM[NAT **fit** Elem↦Nat] generates a version
of STREAM that replaces the formal parameter sort Elem by the actual argument
sort Nat; by the clause **with** Stream↦NatStream the sort Stream of the resulting
specification is then renamed to NatStream.

## 6.2   Declarations, Signatures, and Presentations

In this section, we investigate the declarations that represents the core of the specification language that we are subsequently going to discuss (the full language will be
presented in Sect. 6.7).

**Definition 6.1** (*Abstract data type declarations*) An *abstract data type declaration* is a phrase $D \in Declaration$ which is formed according to the following
grammar:

$D \in Declaration,\ F \in Formula,\ T \in Term,\ C \in Constructor,\ O \in Observer$
$Ss \in Sorts,\ Is \in Identifiers,\ S \in Sort,\ I \in Identifier$

$D ::= D_1\,D_2\ |\ \textbf{sort}\ S\ |\ \textbf{const}\ I{:}S\ |\ \textbf{const}\ I{:}S := T$
$\quad |\ \textbf{fun}\ I{:}Ss \to S\ |\ \textbf{fun}\ I_1{:}Ss \to S,\ I_2(Is) := T$
$\quad |\ \textbf{pred}\ I \subseteq Ss\ |\ \textbf{pred}\ I_1 \subseteq Ss,\ I_2(Is) :\Leftrightarrow F$
$\quad |\ \textbf{type}\ I := C\ |\ \textbf{cotype}\ I := O\ |\ \textbf{axiom}\ F$

$C ::= C_1|C_2\ |\ I\ |\ I(Ss)$
$O ::= O_1|O_2\ |\ I{:}S\ |\ I{:}Ss \to S$
$Ss ::= Ss_1 \times Ss_2\ |\ S$
$Is ::= Is_1, Is_2\ |\ I$
$S ::= I;\ I ::= \ldots$

The syntactic domains *Formula* of formulas and *Term* of terms are formed
as in Definition def:logicsyntax except for the quantified formulas which are
replaced by

$F ::= \ldots\ |\ \forall V{:}S.\ F\ |\ \exists V{:}S.\ F\ |\ \ldots$

i.e., the declaration of every variable $V$ is annotated with a sort $S$. Furthermore,
the domains *ConstantSymbol*, *FunctionSymbol*, and *PredicateSymbol* are all
replaced by the domain *Identifier* of identifiers.

In an abstract data type declaration, the order of the subdeclarations does not matter. Furthermore, some declarations are abbreviations that can be translated into other declarations. We will consequently 'distill' from a declaration its 'canonical essence', which is based upon two concepts that are subsequently introduced. We will use in our presentation the Greek capital letters $\Sigma$ (Sigma) and $\Phi$ (Phi), as is tradition in the literature of abstract data types.

The first concept is that of a 'signature' that captures the syntactic interface of a declaration.

---

**Definition 6.2** (*Signature*) A *signature*

$$\Sigma \in s{:}Set(Sort) \times c{:}Constants \times f{:}Functions \times p{:}Predicates$$

consists of a set $\Sigma.s$ of *sorts*, a set $\Sigma.c$ of *constant operations*, a set $\Sigma.f$ of *function operations*, and a set of $\Sigma.p$ of *predicate operations* (briefly called *constants*, *functions*, and *predicates*) whose elements are formed according to the following definitions:

$Constants := Identifier \rightarrow_{\perp} \Sigma.s$

$Functions := \big\{ F \subseteq Identifier \times \Sigma.s^* \times \Sigma.s \mid \exists F' \in Identifier \times \Sigma.s^* \rightarrow_{\perp} \Sigma.s.$

$\qquad\qquad \forall f \in Identifier, a \in \Sigma.s^*, s \in \Sigma.s. \ \langle f, a, s \rangle \in F \Leftrightarrow \langle \langle f, a \rangle, s \rangle \in F' \big\}$

$Predicates := \mathsf{Set}(Identifier \times \Sigma.s^*)$

For every constant $\langle c, s \rangle \in \Sigma.c$, $c$ denotes its name and $s \in \Sigma.s$ its sort. For every function $\langle f, a, s \rangle \in \Sigma.f$, $f$ denotes its name, $a \in \Sigma.s^*$ its *arity*, and $s \in \Sigma.s$ its *result sort*; for every predicate $\langle p, a \rangle \in \Sigma.p$, $p$ denotes its name and $a \in \Sigma.s^*$ its arity.

---

A signature can contain multiple functions with the same identifier and multiple predicates with the same identifier, provided that they differ in their arities; this allows the 'overloading' of operation identifiers in such a way that in a function application or atomic formula the intended operation can be identified from the sorts of the argument terms. We will use the following notions related to signatures.

**Definition 6.3** (*Signature notions*) We define the following signature notions:

- The *empty signature* $\Sigma_0$:

$$\Sigma_0 := \langle s(\varnothing), c(\varnothing), f(\varnothing), p(\varnothing) \rangle$$

- Signature $\Sigma$ is *extended* by signature $\Sigma'$, $\Sigma \subseteq \Sigma'$:

$$\Sigma \subseteq \Sigma' :\Leftrightarrow \Sigma.s \subseteq \Sigma'.s \wedge \Sigma.c \subseteq \Sigma'.c \wedge \Sigma.f \subseteq \Sigma'.f \wedge \Sigma.p \subseteq \Sigma'.p$$

- The *signature combination* $\Sigma_1 \cup \Sigma_2$ of signatures $\Sigma_1$ and $\Sigma_2$:

$$\Sigma_1 \cup \Sigma_2 := \langle s(\Sigma_1.s \cup \Sigma_2.s), c(\Sigma_1.c \cup \Sigma_2.c), f(\Sigma_1.f \cup \Sigma_2.f), p(\Sigma_1.p \cup \Sigma_2.p) \rangle$$

Please note that a combination $\Sigma_1 \cup \Sigma_2$ of signatures $\Sigma_1$ and $\Sigma_2$ only represents a signature itself, if $\Sigma_1$ and $\Sigma_2$ do not both declare a constant with the same name but different sorts and do not both declare a function with the same name and arity but different result sorts. However, multiple identical declarations do not cause harm and are therefore allowed; also the order of declarations is not relevant.

The second concept is that of a set of formulas that are well-formed according to a given signature.

**Definition 6.4** ($\Sigma$-*formulas and* $\Sigma$-*terms*) Let $\Sigma$ be a signature. We call a formula $F \in Formula$ a $\Sigma$-*formula*, if we can derive a judgement $\Sigma, \varnothing \vdash F$ : formula according to the rules of Fig. 6.1. Likewise, we call a term $T \in Term$ a $\Sigma$-*term* (of sort $S$), if we can derive a judgement $\Sigma, \varnothing \vdash T$ : term($S$). We denote by $Formula_\Sigma$ the set of all $\Sigma$-formulas and by $Term_\Sigma^S$ the set of all $\Sigma$-terms of sort $S$.

Figure 6.1 defines a type system for terms, term sequences, and formulas such that every constant, every function application, and every atomic formula may refer only to operations declared in $\Sigma$ with the correct arity and sorts; by this type system and Definition 6.4, $\Sigma$-formulas and $\Sigma$-terms are closed, i.e., do not have free variables.

In more detail, the judgements depend on the following concept.

**Rules for $\Sigma, Vt \vdash T : \mathbf{term}(S)$:**

$$\frac{\langle V, S \rangle \in Vt}{\Sigma, Vt \vdash V : \mathbf{term}(S)} \qquad \frac{\langle I, S \rangle \in \Sigma.c}{\Sigma, Vt \vdash I : \mathbf{term}(S)} \qquad \frac{\Sigma, Vt \vdash Ts : \mathbf{terms}(Ss) \quad \langle I, Ss, S \rangle \in \Sigma.t}{\Sigma, Vt \vdash I(Ts) : \mathbf{term}(S)}$$

$$\frac{\Sigma, Vt \vdash T_1 : \mathbf{term}(S_1) \quad \Sigma, Vt[V \mapsto S_1] \vdash T_2 : \mathbf{term}(S)}{\Sigma, Vt \vdash \mathsf{let}\ V = T_1\ \mathsf{in}\ T_2 : \mathbf{term}(S)}$$

$$\frac{\Sigma, Vt \vdash F : \mathbf{formula} \quad \Sigma, Vt \vdash T_1 : \mathbf{term}(S) \quad \Sigma, Vt \vdash T_2 : \mathbf{term}(S)}{\Sigma, Vt \vdash \mathsf{if}\ F\ \mathsf{then}\ T_1\ \mathsf{else}\ T_2 : \mathbf{term}(S)}$$

**Rules for $\Sigma, Vt \vdash Ts : \mathbf{terms}(Ss)$:**

$$\frac{\Sigma, Vt \vdash T : \mathbf{term}(S)}{\Sigma, Vt \vdash T : \mathbf{terms}([S])} \qquad \frac{\Sigma, Vt \vdash T : \mathbf{term}(S) \quad \Sigma, Vt \vdash Ts : \mathbf{terms}(Ss)}{\Sigma, Vt \vdash T, Ts : \mathbf{terms}([S] \circ Ss)}$$

**Rules for $\Sigma, Vt \vdash F : \mathbf{formula}$:**

$$\Sigma, Vt \vdash \mathsf{true} : \mathbf{formula} \qquad \Sigma, Vt \vdash \mathsf{false} : \mathbf{formula} \qquad \frac{\Sigma, Vt \vdash Ts : \mathbf{terms}(Ss) \quad \langle I, Ss \rangle \in \Sigma.p}{\Sigma, Vt \vdash I(Ts) : \mathbf{formula}}$$

$$\frac{\Sigma, Vt \vdash T_1 : \mathbf{term}(S) \quad \Sigma, Vt \vdash T_2 : \mathbf{term}(S)}{\Sigma, Vt \vdash T_1 = T_2 : \mathbf{formula}} \qquad \frac{\Sigma, Vt \vdash F : \mathbf{formula}}{\Sigma, Vt \vdash \neg F : \mathbf{formula}}$$

$$\frac{\Sigma, Vt \vdash F_1 : \mathbf{formula} \quad \Sigma, Vt \vdash F_2 : \mathbf{formula}}{\Sigma, Vt \vdash F_1 \wedge F_2 : \mathbf{formula}} \qquad \frac{\Sigma, Vt \vdash F_1 : \mathbf{formula} \quad \Sigma, Vt \vdash F_2 : \mathbf{formula}}{\Sigma, Vt \vdash F_1 \vee F_2 : \mathbf{formula}}$$

$$\frac{\Sigma, Vt \vdash F_1 : \mathbf{formula} \quad \Sigma, Vt \vdash F_2 : \mathbf{formula}}{\Sigma, Vt \vdash F_1 \Rightarrow F_2 : \mathbf{formula}} \qquad \frac{\Sigma, Vt \vdash F_1 : \mathbf{formula} \quad \Sigma, Vt \vdash F_2 : \mathbf{formula}}{\Sigma, Vt \vdash F_1 \Leftrightarrow F_2 : \mathbf{formula}}$$

$$\frac{S \in \Sigma.s \quad \Sigma, Vt[V \mapsto S] \vdash F : \mathbf{formula}}{\Sigma, Vt \vdash \forall V : S.\ F : \mathbf{formula}} \qquad \frac{S \in \Sigma.s \quad \Sigma, Vt[V \mapsto S] \vdash F : \mathbf{formula}}{\Sigma, Vt \vdash \exists V : S.\ F : \mathbf{formula}}$$

$$\frac{\Sigma, Vt \vdash T : \mathbf{term}(S) \quad \Sigma, Vt[V \mapsto S] \vdash F : \mathbf{formula}}{\Sigma, Vt \vdash \mathsf{let}\ V = T\ \mathsf{in}\ F : \mathbf{formula}}$$

$$\frac{\Sigma, Vt \vdash F : \mathbf{formula} \quad \Sigma, Vt \vdash F_1 : \mathbf{formula} \quad \Sigma, Vt \vdash F_2 : \mathbf{formula}}{\Sigma, Vt \vdash \mathsf{if}\ F\ \mathsf{then}\ F_1\ \mathsf{else}\ F_2 : \mathbf{formula}}$$

**Fig. 6.1** Derivation of $\Sigma$-formulas and $\Sigma$-terms

**Definition 6.5** (*Variable typings*) Let $\Sigma$ be a signature. We define

$$VarTyping^{\Sigma} := Variable \rightarrow_{\perp} \Sigma.s$$

as the set of all partial functions of variables to sorts in $\Sigma$; we call these functions *variable typings*.

Please note that, if $Vt$ is a variable typing, $V$ is a variable, and $S$ is a sort, then also the updated variant (see Definition def:relupdate) $Vt[V \mapsto S]$ is a variable typing.

The judgement $\Sigma, Vt \vdash F : \mathbf{formula}$ indicates that formula $F$ is well formed according to signature $\Sigma$ and the typing $Vt \in VarTyping^{\Sigma}$ of free variables. Correspondingly, $\Sigma, Vt \vdash T : \mathbf{term}(S)$ indicates that $T \in Term$ is a well-formed term

**Rules for** $\vdash D$ : **declaration**$(\Sigma, \Phi)$:

$$\frac{\vdash D_1 : \text{declaration}(\Sigma_1, \Phi_1) \quad \vdash D_2 : \text{declaration}(\Sigma_2, \Phi_2)}{\vdash D_1 D_2 : \text{declaration}(\Sigma_1 \cup \Sigma_2, \Phi_1 \cup \Phi_2)}$$

$$\frac{\Sigma = \langle s(\{I\}), c(\varnothing), f(\varnothing), p(\varnothing)\rangle}{\vdash \textbf{sort } I : \text{declaration}(\Sigma, \varnothing)} \qquad \frac{\Sigma = \langle s(\varnothing), c(\{\langle I, S\rangle\}), f(\varnothing), p(\varnothing)\rangle}{\vdash \textbf{const } I : S : \text{declaration}(\Sigma, \varnothing)}$$

$$\frac{I_1 = I_2 \quad \Sigma = \langle s(\varnothing), c(\{\langle I_1, S\rangle\}), f(\varnothing), p(\varnothing)\rangle \quad \Phi = \{I_2 = T\}}{\vdash \textbf{const } I_1 : S, I_2 = T : \text{declaration}(\Sigma, \Phi)}$$

$$\frac{\vdash Ss : \text{sorts}(\overline{Ss}) \quad \Sigma = \langle s(\varnothing), c(\varnothing), f(\{\langle I, \overline{Ss}, S\rangle\}), p(\varnothing)\rangle}{\vdash \textbf{fun } I : Ss \rightarrow S : \text{declaration}(\Sigma, \varnothing)}$$

$$\frac{I_1 = I_2 \quad \vdash Ss : \text{sorts}(\overline{Ss}) \quad \vdash Is : \text{parameters}(\overline{Is}) \quad |\overline{Ss}| = |\overline{Is}|}{\Sigma = \langle s(\varnothing), c(\varnothing), f(\{\langle I_1, \overline{Ss}, S\rangle\}), p(\varnothing)\rangle \quad \Phi = \{\forall Is : \overline{Ss}. \; I_2(Is) = T\}}{\vdash \textbf{fun } I_1 : Ss \rightarrow S, I_2(Is) := T : \text{declaration}(\Sigma, \Phi)}$$

$$\frac{\vdash Ss : \text{sorts}(\overline{Ss}) \quad \Sigma = \langle s(\varnothing), c(\varnothing), f(\varnothing), p(\{\langle I, \overline{Ss}\rangle\})\rangle}{\vdash \textbf{pred } I \subseteq S : \text{declaration}(\Sigma, \varnothing)}$$

$$\frac{I_1 = I_2 \quad \vdash Ss : \text{sorts}(\overline{Ss}) \quad \vdash Is : \text{parameters}(\overline{Is}) \quad |\overline{Ss}| = |\overline{Is}|}{\Sigma = \langle s(\varnothing), c(\varnothing), f(\varnothing), p(\{\langle I_1, \overline{Ss}\rangle\})\rangle \quad \Phi = \{\forall Is : \overline{Ss}. \; I_2(Is) \Leftrightarrow F\}}{\vdash \textbf{pred } I_1 \subseteq S, I_2(Is) :\Leftrightarrow F : \text{declaration}(\Sigma, \Phi)}$$

$$\overline{\vdash \textbf{axiom } F : \text{declaration}(\Sigma_0, \{F\})}$$

$$\frac{I \vdash C : \text{constructors}(Cs, Fs) \quad \Sigma = \langle s(\{I\}), c(Cs), f(Fs), p(\varnothing)\rangle}{\vdash \textbf{type } I := C : \text{declaration}(\Sigma, \varnothing)}$$

$$\frac{I \vdash O : \text{observers}(Fs) \quad \Sigma = \langle s(\{I\}), c(\varnothing), f(Fs), p(\varnothing)\rangle}{\vdash \textbf{cotype } I := O : \text{declaration}(\Sigma, \varnothing)}$$

**Fig. 6.2** Derivation of presentations

and determines its sort $S \in \Sigma.s$; finally $\Sigma, Vt \vdash Ts : \text{terms}(Ss)$ indicates that the term sequence $Ts \in Term^*$ is well-formed and determines the resulting sort sequence $Ss \in \Sigma.s^*$. In these rules, we use the notation $[e_1, \ldots, e_n]$ to denote a sequence of elements $e_1, \ldots, e_n$ while $s_1 \circ s_2$ denotes the concatenation of sequences.

In the following, we assume that by this type system in a $\Sigma$-formula every application of a function/predicate $I$ is annotated by its arity $S_1, \ldots, S_n$ as $I^{S_1, \ldots, S_n}$.

Signatures and formulas over a signature are combined to a 'presentation'.

**Definition 6.6** (*Presentation*) A *presentation* $\langle \Sigma, \Phi \rangle$ is a pair of a signature $\Sigma$ and a set $\Phi$ of $\Sigma$-formulas.

A declaration can be 'distilled' to a presentation.

**Definition 6.7** (*Presentation of a declaration*) We call $\langle \Sigma, \Phi \rangle$ the *presentation of a declaration* $D$, if from the rules in Figs. 6.2 and 6.3 a judgement $\vdash D : \text{declaration}(\Sigma, \Phi)$ can be derived such that $\langle \Sigma, \Phi \rangle$ represents a presentation.

**Rules for** $\vdash Ss$ **:** $\mathsf{sorts}(\overline{Ss})$**:**

$$\frac{\vdash Ss_1 \colon \mathsf{sorts}(\overline{Ss_1}) \quad \vdash Ss_2 \colon \mathsf{sorts}(\overline{Ss_2})}{\vdash Ss_1 \times Ss_2 \colon \mathsf{sorts}(\overline{Ss_1} \circ \overline{Ss_2})} \qquad \vdash S \colon \mathsf{sorts}([S])$$

**Rules for** $\vdash Is$ **:** $\mathsf{parameters}(\overline{Is})$**:**

$$\frac{\vdash Is_1 \colon \mathsf{parameters}(\overline{Is_1}) \quad \vdash Is_2 \colon \mathsf{parameters}(\overline{Is_2})}{\vdash Is_1, Is_2 \colon \mathsf{parameters}(\overline{Is_1} \circ \overline{Is_2})} \qquad \vdash I \colon \mathsf{parameters}([I])$$

**Rules for** $I_s \vdash C$ **:** $\mathsf{constructors}(Cs, Fs)$**:**

$$\frac{I_s \vdash C_1 \colon \mathsf{constructors}(Cs_1, Fs_1) \quad I_s \vdash C_2 \colon \mathsf{constructors}(Cs_2, Fs_2)}{I_s \vdash C_1 | C_2 \colon \mathsf{constructors}(Cs_1 \cup Cs_2, Fs_1 \cup Fs_2)}$$

$$I_s \vdash I \colon \mathsf{constructors}(\{\langle I, I_s \rangle\}, \varnothing) \qquad \frac{\vdash Ss \colon \mathsf{sorts}(\overline{Ss})}{I_s \vdash I(Ss) \colon \mathsf{constructors}(\varnothing, \{\langle I, \overline{Ss}, I_s \rangle\})}$$

**Rules for** $I_s \vdash O$ **:** $\mathsf{observers}(Fs)$**:**

$$\frac{I_s \vdash O_1 \colon \mathsf{observers}(Fs_1) \quad I_s \vdash O_2 \colon \mathsf{observers}(Fs_2)}{I_s \vdash O_1 | O_2 \colon \mathsf{observers}(Fs_1 \cup Fs_2)}$$

$$I_s \vdash I \colon S \colon \mathsf{observers}(\{\langle I, [I_s], S \rangle\}) \qquad \frac{\vdash Ss \colon \mathsf{sorts}(\overline{Ss})}{I_s \vdash I \colon Ss \to S \colon \mathsf{observers}(\{\langle I, [I_s] \circ \overline{Ss}, S \rangle\})}$$

**Fig. 6.3** Auxiliary judgements

In Fig. 6.2, the notation $\forall \overline{Is} \colon \overline{Ss}. \; F$ indicates the formula $\forall I_1 \colon S_1. \; \ldots \forall I_n \colon S_n. \; F$ for an identifier sequence $\overline{Is} = [I_1, \ldots, I_n]$ and a sort sequence $\overline{Ss} = [S_1, \ldots, S_n]$ of the same length. Figure 6.3 gives rules for several auxiliary judgements that are used in the rules of Fig. 6.2. In detail, the judgement $\vdash Ss \colon \mathsf{sorts}(\overline{Ss})$ derives from a phrase $Ss \in Sorts$ a sort sequence $\overline{Ss} \in Sort^*$; the judgement $\vdash Is \colon \mathsf{parameters}(\overline{Is})$ derives from a phrase $Is \in Identifiers$ that contains only distinct identifiers a sequence $\overline{Is} \in Identifier^*$ of these identifiers.

The judgement $I_s \vdash C \colon \mathsf{constructors}(Cs, Fs)$ translates the constructor declaration $C$ of a newly introduced type $I_s$ into a set $Cs$ of constants of sort $I_s$ and into a set $Fs$ of functions whose result is of sort $I_s$. A type declaration

**type** S := c | f(A) | g(A,B)

is thus just an abbreviation for

**sort** S
**const** c: S
**fun** f: A $\to$ S
**fun** g: A $\times$ B $\to$ S

which introduces sort $S$ with 'constructors' $c, f, g$. This kind of declaration corresponds to what in mathematics is called an 'algebra'; it also dominates the generated/free type declarations introduced in Sect. 6.5.

Finally, the judgement $I_s \vdash I : S :$ observers($Fs$) translates the observer declaration $O$ of a newly introduced cotype $I_s$ into a set $Fs$ of functions whose first argument is of sort $I_s$. A cotype declaration

**cotype** $S := c{:}A \mid f{:}A \to B \mid g{:}A \times B \to C$

is thus just an abbreviation for

**sort** $S$
**fun** $c{:}S \to A$
**fun** $f{:}S \times A \to B$
**fun** $g{:}S \times A \times B \to C$

which introduces sort $S$ with 'observers' $c$, $f$, $g$. This kind of declaration corresponds to what in mathematics is called a 'coalgebra'; it also dominates the cogenerated respectively cofree type declarations introduced in Sect. 6.6.

It should be noted that not every derivation of a judgement $\vdash D :$ declaration $(\Sigma, \Phi)$ gives rise to a presentation $\langle \Sigma, \Phi \rangle$; apart from the problem of incompatible 'clashes' of constant and function declarations by combining multiple declarations, it may be the case that an entity is referenced without declaration or applied to the wrong number/sorts of arguments. We will subsequently only consider declarations that are 'well-formed', i.e., for which indeed a presentation can be derived. Subsequently, we are not interested in the declaration any more but only use the derived presentation.

**Example 6.1** The declaration shown in the specification MONOID on Sect. 6.1 gives rise to the presentation $\langle \Sigma, \Phi \rangle$ where

$\Sigma.\mathsf{s} = \{\mathsf{Elem}\}$
$\Sigma.\mathsf{c} = \{\langle \mathsf{e}, \mathsf{Elem} \rangle\}$
$\Sigma.\mathsf{f} = \{\langle \mathsf{op}, [\mathsf{Elem}, \mathsf{Elem}], \mathsf{Elem} \rangle\}$
$\Sigma.\mathsf{p} = \{\langle \mathsf{isE}, [\mathsf{Elem}] \rangle\}$

and $\Phi = \{F_1, F_2, F_3\}$ where

$F_1 = (\forall \mathsf{x}{:}\mathsf{Elem}.\mathsf{op(e,x)} = \mathsf{x} \wedge \mathsf{op(x,e)} = \mathsf{x})$
$F_2 = (\forall \mathsf{x}{:}\mathsf{Elem}, \mathsf{y}{:}\mathsf{Elem}, \mathsf{z}{:}\mathsf{Elem}.\mathsf{op(x,op(y,z))} = \mathsf{op(op(x,y),z)})$
$F_3 = (\forall \mathsf{x}{:}\mathsf{Elem}.\mathsf{isE(x)} \Leftrightarrow \mathsf{x=e})$

$\square$

**Example 6.2** The declaration

**type** Nat $:= 0 \mid +1(\mathsf{Nat})$
**cotype** NatStream $:=$ head:Nat $\times$ tail:NatStream

induces the presentation $\langle \Sigma, \varnothing \rangle$ where

> $\Sigma.\text{s} = \{\text{Nat, NatStream}\}$
>
> $\Sigma.\text{c} = \{\langle 0, \text{Nat} \rangle\}$
>
> $\Sigma.\text{f} = \{\langle +1, [\text{Nat}], \text{Nat} \rangle, \langle \text{head}, [\text{NatStream}], \text{Nat} \rangle, \langle \text{tail}, [\text{NatStream}], \text{NatStream} \rangle\}$
>
> $\Sigma.\text{p} = \varnothing$

□

## 6.3  Algebras, Homomorphisms, and Abstract Data Types

In this section, we will introduce the 'raw-material' to give signatures (and subsequently presentations) a semantics.

---

**Definition 6.8** ($\Sigma$-*algebra*) Let $\Sigma$ be a signature. A $\Sigma$-*algebra* $A$ assigns

- to every sort $S \in \Sigma.\text{s}$ a non-empty set

  $$A(S) \neq \varnothing$$

  which is also called the *carrier* of the sort;
- to every constant $C = \langle I, S \rangle \in \Sigma.\text{c}$ a value

  $$A(C) \in A(S);$$

- to every function $F = \langle I, [S_1, \ldots, S_n], S \rangle \in \Sigma.\text{f}$ a mapping

  $$A(F) \colon A(S_1) \times \ldots \times A(S_n) \rightarrow A(S);$$

- to every predicate $P = \langle I, [S_1, \ldots, S_n] \rangle \in \Sigma.\text{p}$ a relation

  $$A(P) \subseteq A(S_1) \times \ldots \times A(S_n).$$

We denote by $Alg(\Sigma)$ the class of all $\Sigma$-algebras.

---

A $\Sigma$-algebra thus assigns to every syntactic entity declared in signature $\Sigma$ a 'semantics' that is compatible with all sort declarations. As a side note, the collection $Alg(\Sigma)$ is technically not a set but a (proper) 'class', because it is 'too big' to be a set (the domain of the sort assignment would have to be the collection of all sets which is not a set). Nevertheless we will apply the usual notation for set operations also to operations on classes; e.g. $A \in Alg(\Sigma)$ indicates that $A$ is a $\Sigma$-algebra.

**Example 6.3** We consider the signature $\Sigma$ where

$\Sigma.\text{s} = \{\text{Nat}\}$

$\Sigma.\text{c} = \{\langle 0, \text{Nat}\rangle\}$

$\Sigma.\text{f} = \{\langle +1, [\text{Nat}], \text{Nat}\rangle\}$

$\Sigma.\text{p} = \varnothing$

Then we have the $\Sigma$-algebra $A$ with

$A(\text{Nat}) = \mathbb{N}$

$A(0) = 0$

$A(+1) = \lambda x \in \mathbb{N}.\ x + 1$

but also the $\Sigma$-algebra $B$ with

$B(\text{Nat}) = \{0, 1\}$

$B(0) = 0$

$B(+1) = \lambda x \in \{0, 1\}.\ 1 - x$

and the $\Sigma$-algebra $C$ with

$C(\text{Nat}) = \{0\}$

$C(0) = 0$

$C(+1) = \lambda x \in \{0\}.\ 0$

(we write $A(0)$ rather than $A(\langle 0, \text{Nat}\rangle)$ and $A(+1)$ rather than $A(\langle +1, [\text{Nat}], \text{Nat}\rangle)$) because the operations are not overloaded, so there is no danger of confusion).

Clearly all three algebras differ in their properties, e.g. $A(+1)(A(0)) \neq A(0)$ and $B(+1)(B(0)) \neq B(0)$ but $C(+1)(C(0)) = C(0)$; furthermore we have $A(0) \neq A(+1)(A(+1)(A(0)))$ but $B(0) = B(+1)(B(+1)(B(0)))$. □

Different $\Sigma$-algebras may nevertheless be in a certain sense 'compatible' which is indicated by the existence of a 'structure-preserving map' from one algebra to another.

**Definition 6.9** ($\Sigma$-*homomorphism*) Let $A$ and $B$ be $\Sigma$-algebras. A family $h$ of functions $h_S$, for every sort $S \in \Sigma.\text{s}$, is a $\Sigma$-*homomorphism* from $A$ to $B$, which is written as $h: A \rightarrow_\Sigma B$, if for every $S \in \Sigma.\text{s}$ we have

$h_S: A(S) \rightarrow B(S)$

such that the following conditions hold:

1. for every constant $C = \langle I, S \rangle \in \Sigma.c$, we have

   $$h_S(A(C)) = B(C);$$

2. for every function $F = \langle I, [S_1, \ldots, S_n], S \rangle \in \Sigma.f$, we have

   $$\forall a_1 \in A(S_1), \ldots, a_n \in A(S_n).$$
   $$h_S(A(F)(a_1, \ldots, a_n)) = B(F)(h_{S_1}(a_1), \ldots, h_{S_n}(a_n));$$

3. for every predicate $P = \langle I, [S_1, \ldots, S_n] \rangle \in \Sigma.p$, we have

   $$\forall a_1 \in A(S_1), \ldots, a_n \in A(S_n).$$
   $$A(P)(a_1, \ldots, a_n) \Leftrightarrow B(P)(h_{S_1}(a_1), \ldots, h_{S_n}(a_n)).$$

The existence of a $\Sigma$-homomorphism $h$ from algebra $A$ to algebra $B$ expresses the fact that the interpretations that $A$ and $B$ assign to every entity in $\Sigma$ are 'compatible'. Figure 6.4 depicts this compatibility for a function $F$: on the one side, we may first apply the interpretation $A(F)$ to arguments from $A$, which yields a result in $A$ that can then be translated by $h$ to a value in $B$; on the other side, $h$ may first translate the arguments from $A$ to $B$ to which then the interpretation $B(F)$ can be applied yielding a value in $B$. The homomorphism condition ensures that going either way results in the same value in $B$.

The existence of a homomorphism can be also understood as an ordering relation among compatible structures: if $h: A \rightarrow_\Sigma B$, then $A$ has 'not less structure' than $B$. For instance, if $A$ assigns the same value to two constants $C_1$ and $C_2$, then $B$ must also assign the same value to these constants, because $A(C_1) = A(C_2)$ implies $h(A(C_1)) = h(A(C_2))$; since the homomorphism condition ensures $h(A(C_1)) = B(C_1)$ and $h(A(C_2)) = B(C_2)$, this implies $B(C_1) = B(C_2)$. Conversely, if $B$ maps two constants $C_1$ and $C_2$ two different values, then $A$ cannot map them to the same value.

Furthermore, if two applications of the interpretation $A(F)$ of a function $F$ to arguments in $A$ yield the same result, then the applications of $B(F)$ to the corresponding arguments in $B$ must also yield the same result: if $A(F)(a_1) = A(F)(a_2)$, the homomorphism condition on $F$ ensures $B(F)(h(a_1)) = h(A(F)(a_1))$ and $B(F)(h(a_2)) =$

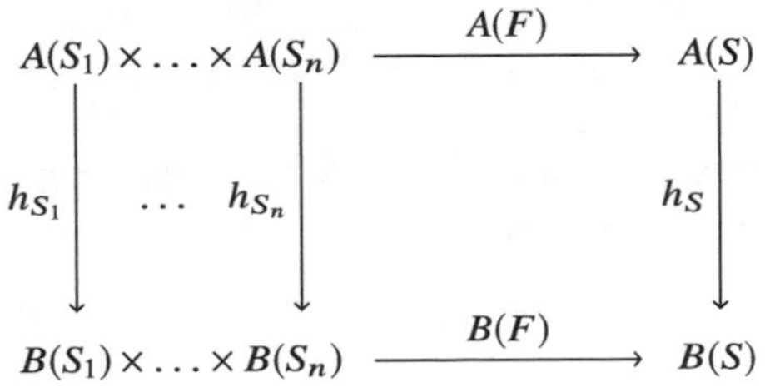

**Fig. 6.4** Homomorphism condition for function $F$

$h(A(F)(a_2))$, which implies $B(F)(h(a_1)) = B(F)(h(a_2))$; conversely, if $B(F)$ yields two different results for two arguments in $B$, then also $A(F)$ must also yield different results for the corresponding arguments in $A$.

As for atomic formulas, if the interpretation $A(P)$ of a predicate $P$ holds for arguments in $A$, then the interpretation $B(P)$ must also hold for the corresponding arguments in $B$ and vice versa; thus $A$ and $B$ must agree on their respective interpretations of all predicates.

**Example 6.4** Take the signature $\Sigma$ with

$\Sigma.s = \{Bool\}$

$\Sigma.c = \{\langle true, Bool\rangle, \langle false, Bool\rangle\}$

$\Sigma.f = \varnothing$

$\Sigma.p = \varnothing$

We consider the $\Sigma$-algebras $A, B, C$ with

$A(Bool) = \{0\}$

$A(true) = 0$

$A(false) = 0$

$B(Bool) = \{0, 1\}$

$B(true) = 1$

$B(false) = 0$

$C(Bool) = \mathbb{N}$

$C(true) = 1$

$C(false) = 0$

We then have $a \colon B \to_\Sigma A$ with

$a_{Bool}(x) := 0$

because (writing $a$ rather than $a_{Bool}$)

$a(B(true)) = 0 = A(true)$

$a(B(false)) = 0 = A(false).$

However, there exists no $\Sigma$-homomorphism from $A$ to $B$. Furthermore, we have $b \colon C \to_\Sigma B$ with

$b_{Bool}(x) := x \bmod 2$

because (writing $b$ rather than $b_{Bool}$)

$$b(C(\text{true})) = b(1) = 1 = B(\text{true})$$
$$b(C(\text{false})) = b(0) = 0 = B(\text{false}).$$

Finally, we have $c: B \to_\Sigma C$ with

$$c_{Bool}(x) := x.$$

because (writing $c$ rather than $c_{Bool}$)

$$c(B(\text{true})) = b(1) = 1 = C(\text{true})$$
$$c(B(\text{false})) = b(0) = 0 = C(\text{false}).$$

However, $c_{Bool}(b_{Bool}(2)) = c_{Bool}(0) = 0$, i.e. $c_{Bool}$ is not the inverse of $b_{Bool}$.
Let us now extend $\Sigma$ to a signature $\Sigma'$ which is identical to $\Sigma$ except for

$$\Sigma'.\mathsf{f} = \{\langle \text{not}, [\text{BOOL}], \text{BOOL} \rangle\}.$$

We correspondingly extend $A$, $B$, $C$ to $\Sigma'$-algebras by also defining

$$A(\text{not}) := \lambda x \in \{0\}.\ 0$$
$$B(\text{not}) := \lambda x \in \{0, 1\}.\ 1 - x$$
$$C(\text{not}) := \lambda x \in \mathbb{N}.\ \text{if } x \bmod 2 = 0 \text{ then } x + 1 \text{ else } x - 1$$

We are going to show that we have still $a: B \to_{\Sigma'} A$, $b: C \to_{\Sigma'} B$, and $c: B \to_{\Sigma'} C$
for the morphisms $a$, $b$, $c$ defined above. First, we have for every $x \in \{0, 1\}$

$$a(B(\text{not})(x)) = 0 = A(\text{not})(a(x)).$$

Second, we have for every $x \in \{0, 1\}$

$$c(B(\text{not})(x)) = c(1 - x) = 1 - x$$
$$\overset{(*)}{=} (\text{if } x \bmod 2 = 0 \text{ then } x + 1 \text{ else } x - 1)$$
$$= C(\text{not})(x) = C(\text{not})(c(x)).$$

where it easy to check that equality (*) holds for $x \in \{0, 1\}$.
Finally, we can show for every $x \in \mathbb{N}$

$$b(C(\text{not})(x)) = B(\text{not})(b(x))$$

by considering two cases:

1. Case $x \bmod 2 = 0$: we have

$$
\begin{aligned}
b(C(\text{not})(x)) &= b(x+1) = (x+1) \bmod 2 = 1 = 1 - 0 = 1 - (x \bmod 2) \\
&= B(\text{not})(x \bmod 2) = B(\text{not})(b(x))
\end{aligned}
$$

2. Case $x \bmod 2 = 1$: we have

$$
\begin{aligned}
b(C(\text{not})(x)) &= b(x+1) = (x+1) \bmod 2 = 0 = 1 - 1 = 1 - (x \bmod 2) \\
&= B(\text{not})(x \bmod 2) = B(\text{not})(b(x))
\end{aligned}
$$

Let us now extend $\Sigma'$ further to a signature $\Sigma''$ which is identical to $\Sigma'$ except for

$$\Sigma'.\mathsf{f} = \{\langle \text{not}, [\text{BOOL}], \text{BOOL}\rangle, \langle \text{and}, [\text{BOOL}, \text{BOOL}], \text{BOOL}\rangle\}.$$

We correspondingly extend $A, B, C$ to $\Sigma''$-algebras by also defining

$$
\begin{aligned}
A(\text{and}) &= \lambda x \in \{0\}, y \in \{0\}.\ 0 \\
B(\text{and}) &= \lambda x \in \{0, 1\}, y \in \{0, 1\}.\ x \cdot y \\
C(\text{and}) &= \lambda x \in \mathbb{N}, y \in \mathbb{N}.\ x \cdot y
\end{aligned}
$$

Then we have still $a\colon B \to_{\Sigma'} A$, $b\colon C \to_{\Sigma'} B$, and $c\colon B \to_{\Sigma'} C$ for the morphisms $a, b, c$ defined above. First, we have for all $x, y \in \{0, 1\}$

$$a(B(\text{and})(x, y)) = 0 = A(\text{and})(a(x), a(y)).$$

Second, we know for all $x, y \in \mathbb{N}$

$$b(C(\text{and})(x, y)) = x \cdot y \bmod 2 = (x \bmod 2) \cdot (y \bmod 2) = B(\text{and})(b(x), b(y))$$

from the laws of modular arithmetic. Third, we know for all $x, y \in \{0, 1\}$

$$c(B(\text{and})(x, y)) = x \cdot y = C(\text{and})(c(x), c(y)).$$

$$\square$$

Since a homomorphism corresponds to an ordering relation that compares compatible algebras for their 'amount of structure', the following concept corresponds to 'structural equality'.

**Definition 6.10** ($\Sigma$-*isomorphism*) Let $\Sigma$ be a signature and $h\colon A \to_\Sigma B$ be a homomorphism from $\Sigma$-algebra $A$ to $\Sigma$-algebra $B$. Then $h$ is a $\Sigma$-*isomorphism*, if for every sort $S \in \Sigma.s$, the function $h_S\colon A(S) \to B(S)$ is bijective, i.e.

$$\forall x_1, x_2 \in A(S).\ h_S(x_1) = h_S(x_2) \Rightarrow x_1 = x_2$$
$$\forall y \in B(S).\ \exists x \in A(S).\ h_S(x) = y.$$

We call two $\Sigma$-algebras $A$ and $B$ $\Sigma$-*isomorphism*, written as $A \simeq_\Sigma B$, if there exists some $\Sigma$-isomorphism $h\colon A \to_\Sigma B$. We denote by $[A]_\Sigma$ the class of all $\Sigma$-algebras that are isomorphic to $A$, i.e.

$$[A]_\Sigma := \{B \in Alg(\Sigma) \mid A \simeq_\Sigma B\}.$$

Above definition implies that, if $h\colon A \to_\Sigma B$ is an isomorphism, then for every element $y \in B(S)$ there exists exactly one $x \in A(S)$ with $h_S(x) = y$. Furthermore, since the homomorphism condition ensures that the algebras interpret the operations in $\Sigma$ in a compatible way, $\Sigma$-isomorphic algebras are structurally/behaviorally indistinguishably (as far as the operations in $\Sigma$ are concerned).

**Example 6.5** Consider the specification $\Sigma''$ and the $\Sigma''$-algebra $B$ defined in Example 6.4. We define the $\Sigma''$-algebra $D$ with

$$D(\mathsf{Bool}) = \mathsf{Set}(\{0\})$$
$$D(\mathsf{true}) = \{0\}$$
$$D(\mathsf{false}) = \varnothing$$
$$D(\mathsf{not}) := \lambda x \in \mathsf{Set}(\{0\}).\ \{0\}\backslash x$$
$$D(\mathsf{and}) = \lambda x \in \mathsf{Set}(\{0\}), y \in \mathsf{Set}(\{0\}).\ x \cap y$$

Then we have $B \simeq_{\Sigma''} D$ with the isomorphism $d\colon B \to_{\Sigma''} D$ defined by

$$d_{\mathsf{Bool}}(x) := \begin{cases} \{0\} & \text{if } x = 1 \\ \varnothing & \text{if } x = 0 \end{cases}$$

Clearly $d$ is bijective. We show that it is also a $\Sigma''$-homomorphism:

- We have $d(B(\mathsf{true})) = d(1) = \{0\} = D(\mathsf{true})$.
- We have $d(B(\mathsf{false})) = d(0) = \varnothing = D(\mathsf{false})$.
- We have $d(B(\mathsf{not})(0)) = d(1 - 0) = d(1) = \{0\} = \{0\}\backslash\varnothing = D(\mathsf{not})(\varnothing) = D(\mathsf{not})(d(0))$ and $d(B(\mathsf{not})(1)) = d(1 - 1) = d(0) = \varnothing = \{0\}\backslash\{0\} = D(\mathsf{not})(\{0\}) = D(\mathsf{not})(d(1))$.

- We have to show, for all $x, y \in \{0, 1\}$, $d(B(\text{and})(x, y)) = D(\text{and})(d(x), d(y))$. For the case $x = 0$, $y = 0$, we have $d(B(\text{and})(0, 0)) = d(0 \cdot 0) = d(0) = \varnothing = \varnothing \cap \varnothing = D(\text{and})(\varnothing, \varnothing) = D(\text{and})(d(0), d(0))$. The other three cases proceed analogously.                                                                                                    $\square$

We are now in the position to introduce the central notion of this section.

---

**Definition 6.11** (*Abstract data type*) An *abstract data type* $C$ for a signature $\Sigma$ is a class of $\Sigma$-algebras that is closed under isomorphism, i.e.,

$$C \subseteq Alg(\Sigma) \; \wedge$$
$$\forall A, B \in Alg(\Sigma).\; A \in C \wedge A \simeq_\Sigma B \Rightarrow B \in C.$$

$C$ is *monomorphic* if its algebras are all isomorphic to each other, i.e.

$$\forall A, B \in C.\; A \simeq_\Sigma B$$

otherwise it is called *polymorphic*.

---

The idea of an 'abstract' data type $C$ is that it collects all those algebras that represent its possible 'implementations'. However, we do not see the representation of an abstract data type but can interact with it only through the operations provided by its signature $\Sigma$; since isomorphic algebras are not distinguishable by these operations, if a $\Sigma$-algebra $A$ represents a possible implementation, then also every isomorphic 'sibling' $B$ of $A$ represents a possible implementation. Consequently we can consider for every algebra $A \in C$ the whole isomorphism class $[A]_\Sigma$ as a single 'concrete' data type that represents a possible implementation of $C$. $C$ can be thus also seen as the union of all concrete data types by which it is implemented, i.e.,

$$C = \bigcup \{[A]_\Sigma \mid A \in C\}.$$

A monomorphic abstract data type essentially allows only one implementation $[A]_\Sigma$ while a polymorphic one still leaves room for multiple implementations $[A]_\Sigma$ and $[B]_\Sigma$ with $A \not\simeq_\Sigma B$, i.e., implementations that are observationally different. Figure 6.5 illustrates the concept by an ellipse that represents a polymorphic abstract data type $C$ whose algebras are denoted by bullets; $C$ comprises three isomorphism classes $[A]_\Sigma$, $[B]_\Sigma$, and $[C]_\Sigma$ with $A \not\simeq_\Sigma B$, $A \not\simeq_\Sigma C$, and $B \not\simeq_\Sigma C$.

**Fig. 6.5** A polymorphic
abstract data type

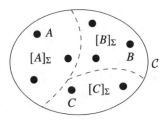

## 6.4    Loose Specifications

We may consider a presentation $\langle \Sigma, \Phi \rangle$ as the 'loose specification' of an abstract
data type $\mathcal{M}$, if we select for $\mathcal{M}$ exactly those algebras that satisfy the axioms in $\Phi$.
For this purpose, we first have to interpret formulas over algebras as domains.

---

**Definition 6.12** (*Semantics of $\Sigma$-formulas and $\Sigma$-terms*) Let $\Sigma$ be a signature, $A$ be a $\Sigma$-algebra, $F$ be a $\Sigma$-formula and $T$ be a $\Sigma$-term. We define
$[\![\,F\,]\!]^A := [\![\,F\,]\!]^A_\varnothing$ and $[\![\,T\,]\!]^A := [\![\,T\,]\!]^A_\varnothing$ as the semantics of $F$ and $T$ exactly
like in Definition def:semanticsfo for a first-order formula $F$ and term $T$ (with
the empty variable assignment) except that the $\Sigma$-algebra $A$ takes the role of
the interpretation, and the semantics of various terms and formulas is redefined as follows (in the following, $I$ denotes a constant and $I^{S_1,\ldots,S_n}$ denotes a
function/predicate with name $I$ and arity $S_1, \ldots, S_n$):

- *Terms:*

$$[\![\,I\,]\!]^A_a := A(I)$$
$$[\![\,I^{S_1,\ldots,S_n}(Ts)\,]\!]^A_a := A(\langle I, [S_1, \ldots, S_n]\rangle)([\![\,Ts\,]\!]^A_a)$$

- *Formulas:*

$$[\![\,I^{S_1,\ldots,S_n}(Ts)\,]\!]^A_a := \begin{cases} \text{true} & \text{if } A(\langle I, [S_1, \ldots, S_n]\rangle)([\![\,Ts\,]\!]^A_a) = \text{true} \\ \text{false} & \text{otherwise} \end{cases}$$

$$[\![\,\forall V\colon S.\ F\,]\!]^A_a := \begin{cases} \text{true} & \text{if, for every } d \text{ in } A(S),\ [\![\,F\,]\!]^A_{a[V\mapsto d]} = \text{true} \\ \text{false} & \text{otherwise} \end{cases}$$

$$[\![\,\exists V\colon S.\ F\,]\!]^A_a := \begin{cases} \text{true} & \text{if, for some } d \text{ in } A(S),\ [\![\,F\,]\!]^A_{a[V\mapsto d]} = \text{true} \\ \text{false} & \text{otherwise} \end{cases}$$

---

One can show that $\Sigma$-formulas cannot distinguish between isomorphic algebras.

**Proposition 6.1** (Isomorphism condition on formulas) *Let $\Sigma$ be a signature, F be a $\Sigma$-formula, and A, B be isomorphic $\Sigma$-algebras, i.e., $A \simeq_\Sigma B$. Then we have*

$$[\![\, F \,]\!]^A = [\![\, F \,]\!]^B.$$

We can select those $\Sigma$-algebras that satisfy a given set of $\Sigma$-formulas.

**Definition 6.13** ($\Sigma$-*models*) Let $\Sigma$ be a signature, $A$ be a $\Sigma$-algebra, and $\Phi$ be a set of $\Sigma$-formulas:

- $A$ is a $\Sigma$-*model* of $\Phi$, written as $A \models_\Sigma \Phi$, if $A$ makes every formula $F \in \Phi$ true:

$$A \models_\Sigma \Phi :\Leftrightarrow \forall F \in \Phi. \; [\![\, F \,]\!]^A = \text{true}$$

- We denote by $Mod_\Sigma(\Phi)$ the class of all $\Sigma$-models of $\Phi$:

$$Mod_\Sigma(\Phi) := \{A \in Alg(\Sigma) \mid A \models_\Sigma \Phi\}$$

One can show that classes of $\Sigma$-formulas cannot distinguish between isomorphic $\Sigma$-algebras.

**Proposition 6.2** (Isomorphism condition on sets formulas) *Let $\Sigma$ be a signature, $\Phi$ be a set of $\Sigma$-formulas, and A, B be isomorphic $\Sigma$-algebras, i.e., $A \simeq_\Sigma B$. Then we have*

$$A \models_\Sigma \Phi \Leftrightarrow B \models_\Sigma \Phi$$

***Proof*** A direct consequence of Proposition 6.1.                                    □

Now we are ready to give a presentation (derived from an abstract data type declaration) a loose semantics.

**Definition 6.14** (*Loose interpretation*) Let $\langle \Sigma, \Phi \rangle$ be a presentation. The *loose interpretation* of $\langle \Sigma, \Phi \rangle$ is the class of all $\Sigma$-models of $\Phi$:

$$loose(\Sigma, \Phi) := Mod_{\Sigma}(\Phi)$$

**Proposition 6.3** (Loose abstract data types) *Let $\langle \Sigma, \Phi \rangle$ be a presentation. Its loose interpretation $loose(\Sigma, \Phi)$ represents an abstract data type.*

**Proof** By definition, $loose(\Sigma, \Phi) \subseteq Alg(\Sigma)$. It remains to show that for arbitrary $A, B \in Alg(\Sigma)$ with $A \in loose(\Sigma, \Phi)$ and $A \simeq_{\Sigma} B$, we have $B \in loose(\Sigma, \Phi)$. Since $A \in loose(\Sigma, \Phi)$, we have $A \in Mod_{\Sigma}(\Phi)$ and thus $A \models_{\Sigma} \Phi$. By Proposition 6.2, we thus have $B \models_{\Sigma} \Phi$ and $B \in Mod_{\Sigma}(\Phi)$ and $B \in loose(\Sigma, \Phi)$. □

Abstract data types are only constrained by the axioms of their presentation; adding axioms makes the data type 'smaller' by excluding (isomorphism classes) of models. However, adding axioms has its perils: if an 'incompatible' axiom is added, the declaration becomes inconsistent, i.e., has no model at all.

**Example 6.6** We consider the loose interpretation $\mathcal{N}$ of the presentation derived from the declaration

**sort** Nat
**const** 0: Nat
**fun** +1: Nat → Nat

which can be also written as

**type** Nat := 0 | +1(Nat)

i.e., of a presentation $\langle \Sigma, \Phi \rangle$ where $\Sigma.s = \{\text{Nat}\}$, $\Sigma.c = \{\langle 0, \text{Nat} \rangle\}$, and $\Sigma.f = \{\langle +1, [\text{Nat}], \text{Nat} \rangle\}$. The data type $\mathcal{N}$ includes a model $N$ that represents the algebra of natural numbers, i.e., $N(\text{Nat}) = \mathbb{N}$, $N(0) = 0$, $N(+1) = \lambda n \in \mathbb{N}. \, n + 1$. However, it also includes the model $Z$ that represents the corresponding algebra of integer numbers, i.e., $Z(\text{Nat}) = \mathbb{Z}$, $Z(0) = 0$, $Z(+1) = \lambda i \in \mathbb{Z}. \, i + 1$.

We thus add the axiom

**axiom** $\forall$x:Nat. 0$\neq$+1(x)

to exclude $Z$ from $\mathcal{N}$ ($\mathcal{N}$ still includes models that are non-isomorphic to the natural numbers, but we don't care for the moment).

Now we extend the declaration by a new operation and axiom

**fun** -1: Nat ↛ Nat
**axiom** $\forall$x:Nat. +1(-1(x))=x

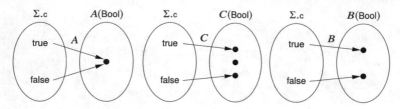

**Fig. 6.6** The loose specification of the data type of Boolean values

The result is disastrous: by instantiating the new axiom with x:=0, we derive the consequence +1(-1(0)) = 0 with contradicts the first axiom for x:=-1(0). Consequently, the declaration is inconsistent and $\mathcal{N} = \emptyset$.

The inconsistency can be actually avoided by a small change:

**fun** -1: Nat → Nat
**axiom** ∀x:Nat. -1(+1(x))=x

Now the declaration is consistent with $N \in \mathcal{N}$ and

$$N(\text{-1}) = \lambda n \in \mathbb{N}. \text{ if } n = 0 \text{ then } 0 \text{ else } n - 1.$$

This can be proved by showing that $N$ is a model of the axioms, i.e.,

$$\forall n \in \mathbb{N}. \ N(0) \neq N(+1)(n) \tag{1}$$
$$\forall n \in \mathbb{N}. \ N(\text{-1})(N(+1)(n)) = n. \tag{2}$$

To show (1), take arbitrary $n \in \mathbb{N}$. Then $N(0) = 0 \neq n + 1 = N(+1)(n)$. To show (2), take arbitrary $n \in \mathbb{N}$. Then $N(\text{-1})(N(+1)(n)) = N(\text{-1})(n + 1) = (n + 1) - 1 = n$.                                                                      □

As demonstrated by the following examples, it may be tricky to specify an abstract data type by a loose specification in such a way that all unintended models are excluded and only the intended one remains (respectively the intended ones remain).

**Example 6.7** We want to specify the abstract data type $\mathcal{B} := loose(\Sigma, \Phi)$ of Boolean values 'true' and 'false' as the loose interpretation of a presentation derived from a declaration

**sort** Bool
**const** true: Bool
**const** false: Bool

which can be also written as

**type** Bool := true | false

i.e., of a presentation $\langle \Sigma, \Phi \rangle$ where $\Sigma.s = \{\text{Bool}\}$ and $\Sigma.c = \{\langle \text{true}, \text{Bool} \rangle, \langle \text{false}, \text{Bool} \rangle\}$. The challenge is define the set of axioms, i.e., formula set $\Phi$, such that unintended models are ruled out (see also Fig. 6.6).

Without any axiom, i.e., using $\Phi = \varnothing$, the data type $\mathcal{B}$ contains a model $A$ with $|A(\mathsf{Bool})| = 1$, i.e., a carrier with a single value, and thus $A(\mathsf{true}) = A(\mathsf{false})$. To ensure that 'no confusion' of this kind may occur, we add the axiom

**axiom** true$\neq$false

i.e., define $\Phi = \{\mathsf{true}\neq\mathsf{false}\}$. However, then $\mathcal{B}$ contains a model $C$ with $|C(\mathsf{Bool})| = 3$, i.e., a carrier with three values and thus a value $b$ such that $C(\mathsf{true}) \neq b$ and $C(\mathsf{false}) \neq b$. To ensure that 'no junk' of this kind is possible, we add the axiom

**axiom** $\forall$x:Bool. x=true $\vee$ x=false

i.e., use $\Phi = \{\mathsf{true}\neq\mathsf{false}, \forall\mathsf{x}{:}\mathsf{Bool}.\mathsf{x}{=}\mathsf{true}\vee\mathsf{x}{=}\mathsf{false}\}$. Now, $\mathcal{B}$ is a monomorphic data type, namely the isomorphism class of a model $B$ with $|B(\mathsf{Bool})| = 2$, i.e., a carrier with two values $b_1$ and $b_2$ such that $B(\mathsf{true}) = b_1$ and $B(\mathsf{true}) = b_2$. Thus the declaration

**type** Bool := true | false
**axiom** true$\neq$false
**axiom** $\forall$x:Bool. x=true $\vee$ x=false

adequately specifies the expected data type.                                $\square$

Generalizing the idea presented in above example, to specify a monomorphic data type with a carrier of size $n$, we need $n \cdot (n-1)/2$ inequalities to prevent 'confusion' and a universally quantified conjunction of $n$ equations to prevent 'junk'.

In the following example, we are going to investigate how far these concepts can be also applied to data types with infinite carriers.

**Example 6.8** We want to specify the abstract data type $\mathcal{N} := loose(\Sigma, \Phi)$ of the natural numbers $0, 1 = +1(0), 2 = +1(+1(0)), \ldots$ as the loose interpretation of a presentation derived from a declaration

**sort** Nat
**const** 0: Nat
**fun** +1: Nat $\rightarrow$ Nat

which can be also written as

**type** Nat := 0 | +1(Nat)

i.e., of a presentation $\langle\Sigma, \Phi\rangle$ where $\Sigma.\mathsf{s} = \{\mathsf{Nat}\}$, $\Sigma.\mathsf{c} = \{\langle 0, \mathsf{Nat}\rangle\}$, and $\Sigma.\mathsf{f} = \{\langle +1, [\mathsf{Nat}], \mathsf{Nat}\rangle\}$ (see Fig. 6.7 for the subsequent discussion).

Without any axiom, our data type $\mathcal{N}$ includes a model $A$ whose carrier has a single value $n$ such that $n = [\![\, 0 \,]\!]^A = [\![\, +1(0) \,]\!]^A = [\![\, +1(+1(0)) \,]\!]^A = \ldots$, i.e. $n = [\![\, +1^i(0) \,]\!]$ for every $i \geq 0$. However, having learned our lesson from Example 6.7, we add the quantified inequality

**axiom** $\forall$x:Nat. 0$\neq$+1(x)

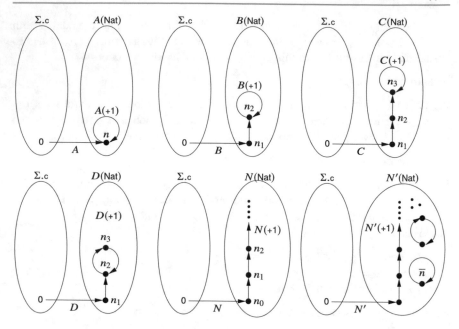

**Fig. 6.7** The loose specification of the data type of natural numbers

which rules this model out.

Nevertheless, we still have a model $B$ whose carrier has only two values $n_1$ and $n_2$, such that $n_1 = [\![\, 0 \,]\!]^B$ and $n_2 = [\![\, +1^i(s(0)) \,]\!]^B$ for every $i \geq 0$. We also have a model $C$ whose carrier has three values $n_1, n_2, n_3$, such that $n_1 = [\![\, 0 \,]\!]^C$, $n_2 = [\![\, +1(0) \,]\!]^C$, and $n_3 = [\![\, +1^i(+1(+1(0))) \,]\!]^C$, for $i \geq 0$. Furthermore, we also have another model $D$ with three values $n_1, n_2, n_3$ such that $n_1 = [\![\, 0 \,]\!]^D$, $n_2 = [\![\, +1^{2i}(+1(0)) \,]\!]^D$, and $n_3 = [\![\, +1^{2i}(+1(+1(0))) \,]\!]^D$, for every $i \geq 0$.

To rule these models (and all other models with finite carriers) out that cause 'confusion' among the term interpretations, we add the axiom

**axiom** $\forall x{:}Nat,y{:}Nat.\, x{\neq}y \Rightarrow +1(x){\neq}+1(y)$

respectively the logically equivalent

**axiom** $\forall x{:}Nat,y{:}Nat.\, +1(x){=}+1(y) \Rightarrow x{=}y$

which make the function s injective; thus every new application of $+1$ yields a new value. In particular, there is no model $A$ such that $[\![\, +1^i(0) \,]\!]^A = [\![\, +1^j(0) \,]\!]^A$ for $i > j$ because this would imply $[\![\, +1^{i-j}(0) \,]\!]^A = [\![\, 0 \,]\!]^A$ which is ruled out by the first axiom given above. The loose interpretation $\mathcal{N}$ of the declaration

**type** $Nat := 0 \mid +1(Nat)$
**axiom** $\forall x{:}Nat.\, 0{\neq}+1(x)$
**axiom** $\forall x{:}Nat,y{:}Nat.\, +1(x){=}+1(y) \Rightarrow x{=}y$

thus prevents 'confusion': it only contains models with infinite carriers such as the model $N$ whose carrier consists of values $n_i$ for every $i \geq 0$ such that $[\![ +1^i(0) ]\!]^N = n_i$.

However, $\mathcal{N}$ still allows 'junk': in addition to $N$, the data type also contains every model $N'$ whose carriers are as big as the carrier of $N$ but hold additional values, namely at least one additional value $\bar{n}$ such that $[\![ +1^i(0) ]\!]^N \neq \bar{n}$ for every $i \geq 0$. In other words, the value $\bar{n}$ is not denoted by any $\Sigma$-term, it is not 'reachable' by the operations in the signature.

Unfortunately, there is no finite set of first-order formulas that allow us to express that the model may only contain reachable values; the data type $\mathcal{N}$ is therefore the best possible approximation to the data type of natural numbers that can be derived by a loose interpretation: it contains the $\Sigma$-algebra $N$ that models the natural numbers but also models such as $N'$ that are non-isomorphic to this algebra; the data type $\mathcal{N}$ is therefore polymorphic. □

The failure to specify a monomorphic abstract data type that includes the natural numbers as a model is not accidental: it is a consequence of the following theorem.

> **Proposition 6.4** (Löwenheim-Skolem theorem) *Let $\Sigma$ be a signature and $\Phi$ be a countable set of (first-order) $\Sigma$-formulas. If there exists some $\Sigma$-algebra $A \in Mod_\Sigma(\Phi)$ and some sort $S \in \Sigma.s$ such that the carrier $A(S)$ is infinite, then there also exists a $\Sigma$-algebra $B \in Mod_\Sigma(\Phi)$ and some value $b \in B(S)$ such that $[\![ T ]\!]^B \neq b$ for every $\Sigma$-term $T$.*

This theorem (which we have stated in a special form, the actual result is more general) casts in concrete that, if the loose interpretation of a specification allows infinite carriers, it always contains models whose carriers contain 'junk', i.e., values that cannot be reached by the operations of the signature; such specifications are necessarily polymorphic.

Above results show that loose specifications are not entirely up to the task of serving as the basis of abstract data type specifications. However, as will be shown in Sect. 6.7, they may be adequately utilized for 'extending' specifications constructed by the mechanisms that we are going to present in the following sections.

## 6.5   Generated and Free Specifications

As the previous section has demonstrated, not all properties of abstract data types can be expressed by a finite set of axioms in first-order logic. We will therefore additionally constrain on the meta-level abstract data type specifications by semantic constraints that help to exclude unintended models.

Since a major problem (that is only inadequately addressed by loose specifications) is that of ruling out values that are not denoted by terms, the following definition introduces a first constraint that forbids an algebra to contain such 'junk'.

**Definition 6.15** (*Generated algebras*) Let $\Sigma$ and $\Sigma'$ be signatures such that $\Sigma \subseteq \Sigma'$ and let $A'$ be a $\Sigma'$-algebra. Then $A'$ *is generated in* $\Sigma'$ with respect to $\Sigma$, if for every sort $S \in \Sigma'.s$ and for every value $a \in A'(S)$

$$[\![\, T \,]\!]^{A''} = a$$

for some $\Sigma''$-Term $T$ where $\Sigma''$ is a signature that is identical to $\Sigma'$ except that it includes constants $\langle I_1, S_1 \rangle, \ldots, \langle I_n, S_n \rangle$ for some sorts $S_1, \ldots, S_n \in \Sigma.s$ and $A''$ is the $\Sigma''$-algebra that is identical to $A'$ except that $A''(\langle I_1, S_1 \rangle) = a_1, \ldots, A''(\langle I_n, S_n \rangle) = a_n$ for some distinct values $a_1 \in A'(S_1), \ldots, a_n \in A'(S_n)$.

In a nutshell, $\Sigma'$-algebra $A'$ is generated in $\Sigma$, if every value in the carriers of $A'$ can be described by a term where we allow additional constants for values whose sorts are in $\Sigma$. The general definition will become relevant in Sect. 6.7; now we are going to use it in a special form that gives rise to a more restricted interpretation of presentations.

**Definition 6.16** (*Generated interpretation*) Let $\langle \Sigma, \Phi \rangle$ be a presentation. The *generated interpretation* of $\langle \Sigma, \Phi \rangle$ is the class of all $\Sigma$-models of $\Phi$ that are generated in $\Sigma$ with respect to the empty signature $\Sigma_0$:

$$generated(\Sigma, \Phi) := \{A \in Mod_\Sigma(\Phi) \mid A \text{ is generated in } \Sigma \text{ w.r.t. } \Sigma_0\}$$

In the generated interpretation of $\langle \Sigma, \Phi \rangle$, it must be possible to describe every value by a term without resorting to additional constants. Also this restricted interpretation induces an abstract data type.

**Proposition 6.5** (Generated abstract data types) *Let* $\langle \Sigma, \Phi \rangle$ *be a presentation. Its generated interpretation generated*$(\Sigma, \Phi)$ *represents an abstract data type.*

***Proof*** Take $\Sigma$-algebras $A$ and $B$ such that $A \in generated(\Sigma, \Phi)$ and $A \simeq_\Sigma B$; we show $B \in generated(\Sigma, \Phi)$. Since $generated(\Sigma, \Phi) \subseteq loose(\Sigma, \Phi)$, we have $A \in loose(\Sigma, \Phi)$, thus by Property 6.3 also $B \in loose(\Sigma, \Phi)$. It remains to show that $B$ is generated in $\Sigma$. We take $S \in \Sigma_{.s}$ and $b \in B(S)$ and show that there exists a $\Sigma$-Term $T$ with $[\![ T ]\!]^B = b$. Since $A \simeq_\Sigma B$, there exists some $\Sigma$-isomorphism $h\colon A \to_\Sigma B$. Since $A \in generated(\Sigma, \Phi)$, $A$ is generated in $\Sigma$ with respect to $\Sigma_0$. Thus there exists a $\Sigma$-Term $T$ with $[\![ T ]\!]^A = h^{-1}(b)$, therefore $h([\![ T ]\!]^A) = h(h^{-1}(b)) = b$. Since $h$ is a homomorphism, we have $h([\![ T ]\!]^A) = [\![ T ]\!]^B$ and thus $[\![ T ]\!]^B = b$.  $\square$

Given a declaration $D$, we will write **generated** $D$, if the presentation introduced by $D$ shall be interpreted as generated (rather than loose). By such declarations, we may now define those abstract data types that we also wanted to specify in the previous section.

**Example 6.9** The declaration

   **generated** {
      **type** Bool := true | false
      **axiom** true≠false
   }

defines the monomorphic data type with algebra $B$ also defined in Example 6.7; the role of the axiom $\forall x$:Bool. x=true $\vee$ x=false is now taken by the keyword **generated**.

Likewise, the specification

   **generated** {
      **type** Nat := 0 | +1(Nat)
      **axiom** $\forall x$:Nat. 0≠+1(x)
      **axiom** $\forall x$:Nat,y:Nat. +1(x)=+1(y) $\Rightarrow$ x=y
   }

defines the monomorphic data type with algebra $N$ that could be only approximated by a polymorphic data type in Example 6.8; by the keyword **generated**, the non-isomorphic algebra $N'$ is excluded.  $\square$

While generated algebras automatically get rid of carriers with 'junk', they rely on explicit axioms to also avoid 'confusion' among term interpretations. We are now going to elaborate a semantic constraint that achieves the same purpose in a simpler way by singling out a special algebra from the class of potential models of a presentation.

**Definition 6.17** (*Initial algebras*) Let $\Sigma$ be a signature, $A$ a $\Sigma$-algebra, and $\mathcal{C}$ a class of $\Sigma$-algebras. Then $A$ is *initial* in $\mathcal{C}$ (with respect to $\Sigma$), if $A \in \mathcal{C}$ and, for every algebra $B \in \mathcal{C}$, there exists exactly one homomorphism $h\colon A \to_\Sigma B$ from $A$ to $B$.

Initial algebras are unique up to isomorphism.

**Proposition 6.6** (Uniqueness of initial algebras) *Let $\Sigma$ be a signature, $A$, $B$ $\Sigma$-algebras, and $C$ a class of $\Sigma$-algebras. If $A$ and $B$ are initial in $C$ with respect to $\Sigma$ then $A$ and $B$ are isomorphic, i.e., $A \simeq_\Sigma B$.*

**Proof** If $A$ and $B$ are initial in $C$, then there exists a unique homomorphism $h: A \to_\Sigma B$ from $A$ to $B$ and a unique homomorphism $h': B \to_\Sigma A$ from $B$ to $A$. Thus $h \circ h': A \to_\Sigma A$ is the unique homomorphism from $A$ to $A$. However, also the identity $id: A \to_\Sigma A$ is a homomorphism from $A$ to $A$ which, since $h \circ h'$ is unique, implies $h \circ h' = id$. Thus $h$ is an isomorphism and $A \simeq_\Sigma B$. $\square$

Since the existence of a homomorphism $h: A \to_\Sigma B$ can be understood as the relationship '$A$ has not less structure than $B$', an initial algebra is an algebra that 'has most structure' respectively is 'most discriminating' among a given class of algebras. This characteristics is formalized below.

**Proposition 6.7** (Initial algebras are most discriminating) *Let $\Sigma$ be a signature and let $A$ be a $\Sigma$-algebra that is initial in a class $C$ of $\Sigma$-algebras. Then for every sort $S \in \Sigma.s$ and all terms $T, T' \in Term_\Sigma^S$ of sort $S$ we have*

$$\left(A \models_\Sigma T = T'\right) \Leftrightarrow \left(\forall B \in C.\ B \models_\Sigma T = T'\right).$$

*Furthermore, for every predicate $\langle I, [S_1, \ldots, S_n]\rangle \in \Sigma.p$ with arity $[S_1, \ldots, S_n]$ and all terms $T_1 \in Term_\Sigma^{S_1}, \ldots, T_n \in Term_\Sigma^{S_n}$ of sorts $S_1, \ldots, S_n$, we have*

$$(A \models_\Sigma I(T_1, \ldots, T_n)) \Leftrightarrow (\forall B \in C.\ B \models_\Sigma I(T_1, \ldots, T_n)).$$

In other words, if some algebra in a class interprets two terms differently, then also the initial algebra in that class does so. Furthermore, atomic formulas are interpreted 'minimally': if some algebra in this class interprets an atomic formula as false, then also the initial algebra does so.

Initial algebras give rise to another interpretation of presentations.

**Definition 6.18** (*Free interpretation*) Let $\langle \Sigma, \Phi \rangle$ be a presentation. The *free interpretation* of $\langle \Sigma, \Phi \rangle$ is the class of all $\Sigma$-models of $\Phi$ that are initial in $Mod_\Sigma(\Phi)$ with respect to $\Sigma$:

$$free(\Sigma, \Phi) := \{A \in Mod_\Sigma(\Phi) \mid A \text{ is initial in } Mod_\Sigma(\Phi) \text{ with respect to } \Sigma\}$$

Also the free interpretation induces an abstract data type.

**Proposition 6.8** (Free abstract data types) *Let $\langle \Sigma, \Phi \rangle$ be a presentation. Its free interpretation $free(\Sigma, \Phi)$ represents an abstract data type.*

**Proof** By Proposition 6.6, $free(\Sigma, \Phi)$ is monomorphic.                         □

Given a declaration $D$, we will write **free** $D$, if the presentation introduced by $D$ shall be interpreted as free. By Proposition 6.7, in this interpretation only those equations and atomic predicates hold that follow from the axioms of the presentation.

Now the question arises under which constraints $free(\Sigma, \Phi)$ is not empty, i.e., does indeed have an initial model and what this model concretely is. A candidate for such an initial model is given below.

**Definition 6.19** (*Quotient term algebra*) Let $\langle \Sigma, \Phi \rangle$ be a presentation. Let, for every sort $S \in \Sigma.s$ and $\Sigma$-term $T \in Term_\Sigma^S$ of sort $S$,

$$[T]_\Sigma^\Phi := \left\{ T' \in Term_\Sigma^S \,\middle|\, \forall A \in Mod_\Sigma(\Phi). \, [\![ T ]\!]^A = [\![ T' ]\!]^A \right\}$$

denote the equivalence class of all terms whose values are the same as the value of $T$ in every $\Sigma$-model of $\Phi$.

Then we define the *quotient term algebra* $Term_\Sigma^\Phi$ of $\Sigma$ with respect to $\Phi$ as the following $\Sigma$-algebra:

- for every sort $S \in \Sigma.s$:

$$Term_\Sigma^\Phi(S) := \left\{ [T]_\Sigma^\Phi \,\middle|\, T \in Term_\Sigma^S \right\}$$

- for every constant $C = \langle I, S \rangle \in \Sigma.c$:

$$Term_\Sigma^\Phi(C) := [I]_\Sigma^\Phi$$

- for every function $F = \langle I, [S_1, \ldots, S_n], S \rangle \in \Sigma.\text{f}$:

  $Term_\Sigma^\Phi(F) :=$

  $\lambda x_1 \in Term_\Sigma^\Phi(S_1), \ldots, x_n \in Term_\Sigma^\Phi(S_n). [I(\text{anyof } x_1, \ldots, \text{anyof } x_n)]_\Sigma^\Phi$

- for every predicate $P = \langle I, [S_1, \ldots, S_n] \rangle \in \Sigma.\text{p}$:

  $Term_\Sigma^\Phi(P) :=$

  $\pi x_1 \in Term_\Sigma^\Phi(S_1), \ldots, x_n \in Term_\Sigma^\Phi(S_n).$

  $\forall A \in Mod_\Sigma(\Phi). [\![ I(\text{anyof } x_1, \ldots, \text{anyof } x_n) ]\!]^A$

For every sort $S$, its carrier in a quotient term algebra is a class of terms of that sort, namely all terms whose interpretations coincide in all models of the given presentation. The functions of the quotient term algebra operate on such term classes by selecting arbitrary representatives of these classes as argument terms and constructing the equivalence classes of those terms that represent the applications of the functions to these terms. The actual choice of the argument terms does not matter, because the conditions used in the definitions of the equivalence classes $[T]_\Sigma^\Phi$ represent congruence relations, i.e., equivalent arguments yield equivalent results. Likewise, the predicates of the quotient term algebra construct atomic formulas from the class representatives as argument terms; the predicates hold, if the atomic formulas hold in all models of the presentation.

The quotient term algebra of a presentation can be thus understood as that algebra where functions by default yield new values, i.e., different terms denote different values unless they are identified by axioms. Furthermore, predicates hold minimally, i.e., an atomic formula is false unless axioms force it to become true.

The role of quotient term algebras with respect to initiality is exhibited by the following proposition.

**Proposition 6.9** (Initiality of quotient term algebra) *Let $\langle \Sigma, \Phi \rangle$ be a presentation. If $Term_\Sigma^\Phi \in Mod_\Sigma(\Phi)$, i.e., the quotient term algebra of $\Sigma$ with respect to $\Phi$ is a $\Sigma$-model of $\Phi$, then $Term_\Sigma^\Phi$ is initial in $Mod_\Sigma(\Phi)$ with respect to $\Sigma$.*

**Proof** The core of the proof is to show that, if $Term_\Sigma^\Phi$ is a $\Sigma$-model of $\Phi$, then there exists the unique *evaluation homomorphism* $h_S \colon Term_\Sigma^\Phi(S) \to_\Sigma A(S), h_S(x) := [\![ \text{anyof } x ]\!]^A$ from the term classes of sort $S$ to their values in any model $A$; we omit the details.                                                                    □

The following proposition gives a criterion to satisfy the assumption established by Proposition 6.9.

**Proposition 6.10**  (Quotient term algebra as model) *Let* $\langle \Sigma, \Phi \rangle$ *be a presentation such that every formula* $F \in \Phi$ *is formed according to the grammar*

$$F ::= \forall V : S. \, Fs$$
$$Fs ::= As \Rightarrow A \mid Fs_1 \wedge Fs_2$$
$$A ::= p(T_1, \ldots, T_n) \mid T_1 = T_2$$
$$As ::= A \mid As_1 \wedge As_2$$

*Then* $Term_{\Sigma}^{\Phi} \in Mod_{\Sigma}(\Phi)$, *i.e., the quotient term algebra of* $\Sigma$ *with respect to* $\Phi$ *is a* $\Sigma$-*model of* $\Phi$.

*Proof*  The proof proceeds by showing $Term_{\Sigma}^{\Phi} \models_{\Sigma} F$ for every axiom $F \in \Phi$; we omit the details.                                                                                □

The grammar given in Proposition 6.10 defines a fragment of first-order logic called *conditional equational logic*; its core formulas $A_1 \wedge \ldots \wedge A_n \Rightarrow A$, are also called *Horn clauses* and are typically written in a 'rule-oriented' way as $A \Leftarrow A_1 \wedge \ldots \wedge A_n$. Furthermore, since every propositional formula can be transformed into disjunctive normal form, i.e., into a disjunction $As_1 \vee \ldots \vee As_n$ of conjunctions $As_i$, and since $A \Leftarrow As_1 \vee \ldots \vee As_n$ is logically equivalent to a conjunction of Horn clauses $(A \Leftarrow As_1) \wedge \ldots \wedge (A \Leftarrow As_n)$, we will allow the hypothesis $As$ of a Horn clause to be formed according to the following extended grammar that also allows disjunction:

$$As ::= A \mid As_1 \wedge As_2 \mid As_1 \vee As_2$$

The role of this logic is established by the following proposition.

**Proposition 6.11**  (Quotient term algebra as free interpretation) *Let* $\langle \Sigma, \Phi \rangle$ *be a presentation such that every formula* $F \in \Phi$ *is formed according to the grammar in Proposition 6.10. Then* $Term_{\Sigma}^{\Phi}$ *is initial in* $Mod_{\Sigma}(\Phi)$ *with respect to* $\Sigma$ *and thus*

$$Term_{\Sigma}^{\Phi} \in free(\Sigma, \Phi)$$

*i.e., the free interpretation of* $\langle \Sigma, \Phi \rangle$ *has the quotient term algebra of* $\Sigma$ *with respect to* $\Phi$ *as a model. Furthermore, we have*

$$free(\Sigma, \Phi) \subseteq generated(\Sigma, \Phi)$$

*i.e., the initial algebras are generated.*

**Proof** The first part is a direct consequence of Propositions 6.10 and 6.6. The second part follows from the first part and the fact that $Term_\Sigma^\Phi$ is generated.                  □

By this proposition, free interpretations of presentations are in practice confined to Horn clauses as axioms, because they guarantee the consistency of a free specification by ensuring the quotient term algebra as its canonical model.

**Example 6.10** The declaration

  **free type** Bool := true | false

defines the monomorphic data type with algebra $B$ also defined in Examples 6.7 and 6.9; the role of the axiom **axiom** true$\neq$false is now taken by the keyword **free**. The carrier for sort Bool consists of the two equivalence classes {true}, and {false}.
  Likewise, the declaration

  **free type** Nat := 0 | +1(Nat)

defines the monomorphic data type with algebra $N$ that was also defined in Example 6.9; the role of the axioms **axiom** $\forall$x:Nat.$0\neq+1(x)$ and **axiom** $\forall$x:Nat,y:Nat. $+1(x)=+1(y) \Rightarrow x=y$ is taken by **free**. The carrier for sort Nat consists of the infinitely many equivalence classes {0}, {+1(0)}, {+1(+1(0))}, ...   □

As above example shows, **free type** declarations (without additional axioms) correspond to the programming language types that are usually called 'enumerated', 'algebraic', or 'recursive'.
  If additional operations are introduced that shall *not* introduce new carrier values, equations have to be added to identify the new terms with existing terms of the previous subsignature.

**Example 6.11** The declaration

  **free** {
    **type** Bool := true | false
    **fun** not: Bool $\rightarrow$ Bool
    **axiom** not(true)=false $\wedge$ not(false)=true
  }

extends the specification given in Example 6.10 by the function not. However, we still have only two carrier values {true, not(false), ...} and {false, not(true), ...}.
  Likewise the declaration

  **free** {
    **type** Nat := 0 | +1(Nat)
    **fun** +: Nat $\times$ Nat $\rightarrow$ Nat
    **axiom**
      $\forall$n1:Nat,n2:Nat.
      +(0,n2) = n2 $\wedge$

$$+(+1(n1),n2) = +1(+(n1,n2))$$
}

which introduces a new operation + but does not add new carrier values (it only adds new terms to the existing ones); for instance we have the carrier

$$[0] = \{0, +(0, 0), +(0, +(0, 0)), +(+(0, 0), 0), \ldots\}$$

with infinitely many alternatives for the term 0.                               □

We are now demonstrating by a sequence of small examples, the flavor (and also the dangers) of initial specifications.

**Example 6.12** We extend the declaration of the natural numbers given in Example 6.11 by some predicates to

> **free** {
>
> ...
>
> **pred** $>0 \subseteq$ Nat
> **pred** $\leq\ \subseteq$ Nat $\times$ Nat
> **pred** $<\ \subseteq$ Nat $\times$ Nat
> **axiom**
>   ∀n1:Nat,n2:Nat.
>     $>0(+1(n1)) \wedge$
>     $\leq(0,n2) \wedge (\leq(+1(n1),+1(n2)) \Leftarrow \leq(n1,n2)) \wedge$
>     $<(0,+1(n2)) \wedge (<(+1(n1),+1(n2)) \Leftarrow <(n1,n2))$
> }

The carrier of the quotient term algebra for Nat consists of the elements $[+1^i(0)]$, for every $i \geq 0$. From the first axiom, the interpretation of formula $>0(+1^i(0))$ is 'true' for $i > 0$; since predicates are interpreted minimally and there is no axiom to state otherwise, the interpretation of $>0(0)$ is 'false'. From the second line of axioms, $\leq (0,0)$ is interpreted as 'true'. From the third line, also $<(0,+1(0))$ is interpreted as 'true'; however, since there is no axiom to state otherwise, $<(0,0)$ is interpreted as 'false'. The axioms expressed by right-to-left implications can be read as 'rewrite rules', that reduce atomic formulas with more complex arguments to one of the base cases such that $\leq(+1^i(0),+1^j(0))$ is interpreted as 'true' if and only if $i \leq j$; likewise, $<(+1^i(0),+1^j(0))$ is interpreted as 'true' if and only if $i < j$.               □

**Example 6.13** The following declaration introduces the abstract data type 'finite sequence (list) of natural numbers':

> **free** {
>   **type** Nat := 0 | +1(Nat)
>   **type** List := [ ] | [Nat, List]
> }

In addition to the sort Nat already introduced in Example 6.10, this declaration provides a sort List with two constructors [ ] ('empty list') and [n,l] ('add number $n$ to the front of list $l$'); the carrier values for this sort are (the equivalence classes) of the terms [ ], $[n_1, [\,]]$, $[n_1, [n_2, [\,]]]$, $[n_1, [n_2, [n_3, [\,]]]]$, ..., for all terms $n_1, n_2, n_3, \ldots$ of sort Nat.

Let us now extend this declaration by an additional operation:

**free** {
   ...
   **fun** o: List $\times$ List $\to$ List
   **axiom**
     $\forall$n:Nat, l1:List, l2:List.
      [ ] o l2 = l2 $\wedge$
      [n, l1] o l2 = [n, l1 o l2]
}

The operation $l_1 \circ l_2$ denotes 'concatenation of lists $l_1$ and $l_2$'. It does not introduce new elements to the carrier, because the axioms 'reduce' all terms of shape $l_1 \circ l_2$ to terms involving only the other operations. For instance, the terms $[n_1, [\,]] \circ [n_2, [\,]]$ and $[n_1, [n_2, [\,]]]$ are both elements of the same equivalence class, i.e., they represent identical values.

Now let us add by the declaration

**free** {
   ...
   **fun** head: List $\to$ Nat
   **fun** tail: List $\to$ List
   **axiom**
     $\forall$n:Nat, l:List.
      head([n, l]) = n $\wedge$
      tail([n, l]) = l
}

two more operations head(l) and tail(l) ('head/tail of list $l$'). By the axioms, e.g., head(0, [ ]) and 0 denote the same value, as do tail(0, [ ]) and [ ].

However, there are no axioms that constrain head([ ]) and tail([ ]), so that we may wish to consider these terms as 'undefined' respectively have them denote 'error values'. Nevertheless, in the free interpretation these terms represent additional carrier values; consequently, also +1(head([ ]), head(tail([ ])) and tail(tail([ ])) represent additional carrier values. Ultimately, we get infinitely many such terms that are not related to the 'basic' constructors of sorts Nat and List and all represent different error values.

While this multitude of error values does not really hurt, it might be considered as 'ugly'; the following declaration reduces all error values of a sort to a single one:

**free** {
   ...
   **const** errorN: Nat

```
const errorL: List
fun head: List → Nat
fun tail: List → List
axiom
   ∀n:Nat, l:List.
     head([ ]) = errorN ∧ head([n, l]) = n ∧
     tail([ ]) = errorL ∧ tail([n, l]) = l ∧
     +1(errorN) = errorN ∧
     [errorN, l] = errorL ∧ [n, errorL] = errorL ∧
     errorL ∘ l = errorL ∧ l ∘ errorL = errorL ∧
     head(errorL) = errorN ∧ tail(errorL) = errorL
}
```

This declaration introduces an 'error number' and an 'error list' to which all errors
of the corresponding sorts shall be mapped. The first two axiom lines introduce these
values by illegal applications of head and tail. However, now we have to consider
the application of all operations error values; the last four axiom lines make sure that
these applications do not yield new error values.

However, adding so many axioms can be dangerous. Let us assume that we add
to above declaration the innocently looking axiom

```
free {
   ...
   ∀l:List. l = [head(l),tail(l)]
}
```

The consequences are fatal: from this axiom we can deduce

$$[ ] = [head([ ]),tail([ ])] = [errorN,errorL] = errorL$$

and consequently $[n,[ ]] = errorL$, for every $n$, and thus also $[n,l] = errorL$, for
every $n, l$. In other words, the carrier for List collapses to a singleton $\{errorL\}$.

This example demonstrates the danger of adding 'too many' axioms to a declaration with free interpretation; it may let a carrier collapse so that it has fewer values
than intended, possibly just a single one.

All in all, free interpretations do not mix well with operations that are only partially defined; either we get (usually infinitely) many error values or we have to add
a large set of equations to limit the error values to a minimal set at the risk of adding
'one axiom too much' which lets the carrier collapse. Declarations with initial interpretation should be therefore restricted to the 'core' of the data type with the minimal
set of operations needed to describe its behavior; additional operations (that may be
only partially defined) should be better added later by loose extensions, as described
in Sect. 6.7.                                                                    □

**Example 6.14** The following declaration introduces the abstract data type 'finite bag (multiset) of natural numbers', a data type that stores its elements without order:

**free** {
  **type** Nat := 0 | +1(Nat)
  **type** Bag := ∅ | ∪(Nat, Bag)
  **axiom**
    ∀n1:Nat, n2:Nat, b:Bag.
      ∪(n1,∪(n2,b)) = ∪(n2,∪(n1,b))
}

Similar to the sort List introduced in Example 6.13, sort Bag has two constructors ∅ ('empty bag') and ∪(n,b) ('add number $n$ to bag $b$'). The difference between both declarations is in the axiom that states that the order in which numbers are inserted does not matter: thus the terms ∪(0,∪(+1(0),∅)) and ∪(+1(0),∪(0,∅)) denote the same bag. Nevertheless, the number of insertion matters, thus the terms ∪(0,∪(0,∅)) and ∪(0,∅) denote different bags. We may specify by the declaration

**free** {
  ...
  **fun** count: Bag × Nat → Nat
  **axiom**
    ∀n1:Nat, n2:Nat, b:Bag.
      count(∅,n2) = 0 ∧
      count(∪(n1,b),n1) = 1+count(b,n1) ∧
      count(∪(n1,b),n2) = count(b,n2) ⟸ ≠(n1,n2)
  **pred** ≠ ⊆ Nat × Nat
  **axiom**
    ∀n1:Nat, n2:Nat.
      ≠(0,+1(n2)) ∧ ≠(+1(n1),0) ∧
      ≠(+1(n1),+1(n2)) ⟸ ≠(n1,n2)
}

an operation count(b,n) that returns the number of occurrences of number $n$ in bag $b$. The corresponding axiom uses the auxiliary predicate ≠ $(n_1, n_2)$ that determines whether numbers $n_1$ and $n_2$ are different; this predicate is needed, since negation and thus inequality is not directly available in conditional equational logic.     □

**Example 6.15** The following declaration introduces the abstract data type 'finite set of natural numbers':

**free** {
  **type** Nat := 0 | +1(Nat)
  **type** Set := ∅ | ∪(Nat, Set)
  **axiom**
    ∀n1:Nat, n2:Nat, s:Set.
      ∪(n1,∪(n2,s)) = ∪(n2,∪(n1,s)) ∧

$$\cup(n1,\cup(n1,s)) = \cup(n1,s)$$
}

This declaration is very similar to the declaration of sort Bag given in Example 6.14. The only difference is the last axiom line that merges multiple occurrences of the same element in the set to one; thus the terms $\cup(0,\cup(0,\varnothing))$ and $\cup(0,\varnothing)$ denote the same bag. We may specify by the declaration

**free** {

...

**pred** $\in \subseteq$ Nat $\times$ Set
**axiom**
   $\forall$n1:Nat, n2:Nat, s:Set.
      $\in$ (n1,$\cup$(n1,s))

}

a predicate $\in$ (n,s) that holds if number $n$ is an element of set $s$; $\in$ (n,$\varnothing$) does not hold for any $n$, since there is no axiom that claims otherwise and in the free interpretation predicates hold minimally. Please note that it suffices to cover the case of a set $\cup(n1, s)$ (where the element $n_1$ appears 'at the front of the set') because (according to the first axiom of the basic specification) the order of element additions does not matter (i.e., we can always 'reshuffle' the set to have the necessary element 'at the front').                                                                                       □

The free interpretation of specifications has very much the flavor of inductive relation and function definitions that have been presented in Sects. sect:inductrel, sect:inductrel2, and sect:inductfun on the basis of least fixed points: in the free interpretation of a specification, a formula is considered as false, unless its truth can be established from the axioms of the specification; likewise, terms denote different values unless their identity follows from the axioms. The difference, however, is that specifications construct new domains together with operations on these domains; the previously presented definitions introduced new operations on already given domains.

## 6.6   Cogenerated and Cofree Specifications

Section 6.5 has focused on abstract data types where a sort $S$ is primarily determined by its constructors of form $c: Ss \rightarrow S$, i.e., functions *into* the specified sort. Observers of form $o: S \rightarrow S'$, i.e., functions *out of* the sort have a secondary status, because they can be defined with the help of the constructors. For instance, assuming a previous declaration of a sort Elem, the declaration

**free type** List = empty | cons(Elem,List)

introduces the sort *List* with constructors *empty*: *List* and *cons*: *Elem* $\times$ *List* $\rightarrow$ *List*; we can later introduce by

**fun** head: List $\to$ Elem
**fun** tail: List $\to$ List
**axiom** $\forall$e:Elem, l:List. head(cons(e,l))=e $\land$ tail(cons(e,l))=l

the corresponding observers *head*: *List* $\to$ *Elem* and *tail*: *List* $\to$ *List*. This view is primarily applicable to abstract data types whose values are 'data' in the sense of computer science: these can be constructed by a finite number of constructor applications; their contents are thus fully determined by these applications. For instance, while we may by an infinite sequence of terms

*empty*

$cons(e_1, empty)$

$cons(e_2, cons(e_1, empty))$

$cons(e_3, cons(e_2, cons(e_1, empty)))$

. . .

construct longer and longer lists without upper bound on the length of the lists, each list itself has finite length.

In this section, we will consider the dual view of abstract data types where a sort is primarily determined by its observers. Take for instance the declaration

**cofree cotype** Stream = head:Elem | tail:Stream

(the meaning of the keyword **cofree** will be explained later) which introduces the sort *Stream* with observers *head*: *Stream* $\to$ *Elem* and *tail*: *Stream* $\to$ *Stream*. Given a value *s* of sort *Stream* as a starting point, we may by an infinite sequence of terms

$head(s)$

$head(tail(s))$

$head(tail(tail(s)))$

$head(tail(tail(tail(s))))$

. . .

observe from *s* more and more elements. While each observation determines only a finite number of elements of *s*, the number of observations itself is unbounded; the stream *s* has thus to be considered of infinite length. This view is thus primarily applicable to abstract data types whose values are 'processes' in the sense of computer science: some observers (whose result is a value of another data type, e.g. *head*, which delivers an *Elem*) produce useful information while others (whose result is a value of the specified sort, e.g., *tail*, which delivers another *Stream*) let the process proceed to a new state from which another observation can be made. For such cotypes the constructors are secondary and may be defined in terms of the observers; for instance, the declaration

**fun** cons: Elem $\times$ Stream $\to$ Stream
**axiom** $\forall$e:Elem, s:Stream. head(cons(e,s))=e $\land$ tail(cons(e,s))=s

introduces a constructor $cons: Elem \times Stream \rightarrow Stream$ which adds an element $e$ to the front of stream $s$; the axiom defining the behavior of this constructor is the same as the one defining the behavior of the observers *head* and *tail* on *List*.

Observers may also have additional arguments; take for instance the declaration

**cofree** {
  **cotype** Stream = head:Elem | tail(e:Elem):Stream
  **axiom** ∀s:Stream. head(tail(s,e))=e
}

where the observer *tail*: $Stream \times Elem \rightarrow Stream$ has an additional argument $e$ that specifies the result of the next observation of *head*; we thus have

$$e_1 = head(tail(s, e_1))$$
$$e_2 = head(tail(tail(s, e_1), e_2))$$
$$e_3 = head(tail(tail(tail(s, e_1), e_2), e_3))$$

$$\cdots$$

By this feature, we may thus model 'inputs' to processes which modify their future behavior; a more illustrative example will be given at the end of this section.

As discussed in Sect. 6.4, the loose interpretation of constructor-based specifications raises the problem of 'junk', i.e., an specified abstract data type may contain algebras that are 'too big' in the sense that they contain values that are not reachable by constructors; this problem was solved by the concept of generated specifications. An analogous problem arises with the loose interpretation of observer-based specifications: the specified abstract data type may also contain algebras that are 'too big' in the sense that they contain values that are not distinguishable by observers. This problem is demonstrated by the following example.

**Example 6.16** Take the declaration

**free type** Bool := T | F

which gives rise to an interpretation of sort Bool with distinct values $T$, $F$ denoted by constants T, F.

Now consider the declaration

**cotype** Stream := head:Bool | tail:Stream
**axiom** ∀s:Stream. head(s)=T

From the later axiom, every stream has head $T$. But then also the tail of every stream has head $T$ and thus every stream contains only $T$ values. Since there seems to exist only a single such stream $s$ of infinitely many $T$ values, we thus might expect this declaration to specify a monomorphic abstract data type $S = [S]$ with $S(Stream) = \{s\}$.

In fact, however, $S$ has infinitely many non-isomorphic models each of which may have arbitrarily many different streams such as the ones depicted in Fig. 6.8 (every

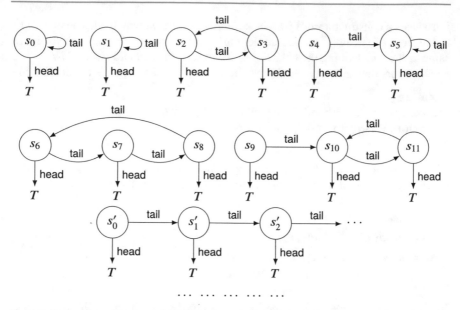

**Fig. 6.8** Various interpretations of Stream with a single visible value

stream is represented by a node whose outgoing arrows denote the results of the interpretations of operations head and tail when applied to that stream). In particular, there
are the non-isomorphic models $S_0, \ldots, S_7$ with $S_0(\text{Stream}) = \{s_0\}$, $S_1(\text{Stream}) =$
$\{s_0, s_1\}$, $S_2(\text{Stream}) = \{s_2, s_3\}$, $S_3(\text{Stream}) = \{s_4, s_5\}$, $S_4(\text{Stream}) = \{s_6, s_7, s_8\}$,
$S_5(\text{Stream}) = \{s_9, s_{10}, s_{11}\}$, $S_6(\text{Stream}) = \{s_0, \ldots, s_{11}\}$, $S_7 = \{s_0', s_1', s_2', \ldots\}$. The
only requirement is that the carrier of Stream is closed under the interpretation of
operation tail and that the interpretation of head always yields result $T$.                    □

However, the picture presented by above example changes if we consider the
Stream values themselves as non-observable, i.e., 'hidden'. In that case, all the
streams depicted in Fig. 6.8 become 'essentially the same' in that we cannot distinguish them by the observation of non-Stream values, i.e., by the observations made
through the observer head. We can thus, without changing the observations made by
head, replace every stream by the stream $s_0$, which yields the monomorphic abstract
data type $S = [S_0]$ with $S_0(\text{Stream}) = \{s_0\}$. The following definitions formalize this
intuition of 'observational equivalence' on the basis of a signature extension $\Sigma \subseteq \Sigma'$
where $\Sigma$ introduces the 'visible' sorts while $\Sigma'$ extends $\Sigma$ by the 'hidden' sorts.

---

**Definition 6.20** (*Observational equivalence*) Let $\Sigma$ and $\Sigma'$ be signatures such
that $\Sigma \subseteq \Sigma'$, $A$ a $\Sigma'$-algebra, and $S \in \Sigma'.s \setminus \Sigma.s$ a 'hidden' sort. Then two values
$d_1, d_2 \in A(S)$ are *observationally equivalent*, written as $d_1 \simeq_{\Sigma, \Sigma'}^{A, S} d_2$ (or just
$d_1 \simeq d_2$), if the following two conditions hold:

- For every visible sort $O \in \Sigma.s$ and every $\Sigma'$-term $T$ of that sort with free variable $V$ of hidden sort $S$, i.e., $\Sigma', Vt \vdash T : \mathsf{term}(O)$ holds for some variable set $Vt$ with $\langle V, S \rangle \in Vt$, we have

$$[\![\, T \,]\!]^A_{a[V \mapsto d_1]} = [\![\, T \,]\!]^A_{a[V \mapsto d_2]}$$

  for every assignment $a$ on $Vt$.

- For every $\Sigma'$-formula $F$ with free variable $V$ of hidden sort $S$, i.e. $\Sigma', Vt \vdash F : \mathsf{formula}$ holds for some variable set $Vt$ with $\langle V, S \rangle \in Vt$, we have

$$[\![\, F \,]\!]^A_{a[V \mapsto d_1]} \Leftrightarrow [\![\, F \,]\!]^A_{a[V \mapsto d_2]}$$

  for every assignment $a$ on $Vt$.

So in every term or formula with a free variable of a hidden sort, we may assign observationally equivalent values to the variable without affecting the observable outcome.

**Definition 6.21** (*Cogenerated algebra*) Let $\Sigma$ and $\Sigma'$ be signatures such that $\Sigma \subseteq \Sigma'$ and $A$ a $\Sigma'$-algebra. Then $A$ *is cogenerated in* $\Sigma'$ *with respect to* $\Sigma$, if for every 'hidden' sort $S \in \Sigma'.s \setminus \Sigma.s$ and every pair of values $d_1, d_2 \in A(S)$ of that sort, we have $d_1 \simeq d_2 \Rightarrow d_1 = d_2$.

In other words, in a cogenerated algebra the carrier of a hidden sort cannot contain distinct values that are observationally equivalent.

Our goal is now to single out from all possible cogenerated algebras those that arise from extensions of a given algebra; this notion is formalized by the following two definitions.

**Definition 6.22** (*Algebra extension and reduction*) Let $\Sigma$ and $\Sigma'$ be signatures such that $\Sigma \subseteq \Sigma'$.
We say that signature $\Sigma$-algebra $A$ is *extended* by $\Sigma'$-algebra $A'$, written $A \subseteq_{\Sigma, \Sigma'} A'$ (or just $A \subseteq A'$), if the following conditions hold:

- for every sort $S \in \Sigma.s$, we have $A(S) = A'(S)$;
- for every constant $C \in \Sigma.c$, we have $A(C) = A'(C)$;
- for every function $F \in \Sigma.f$, we have $A(F) = A'(F)$;
- for every predicate $P \in \Sigma.p$, we have $A(P) = A'(P)$.

Conversely, we denote by the *reduction* $A'|_\Sigma$ of $\Sigma'$-algebra $A'$ to signature $\Sigma$ the unique $\Sigma$-algebra $A$ for which above conditions hold.

**Definition 6.23** (*Extension class*) Let $\Sigma$, $\Sigma'$ be signatures such that $\Sigma \subseteq \Sigma'$, let $A$ be a $\Sigma$-algebra, and let $\mathcal{C}$ be a class of $\Sigma'$-algebras. We denote by $\mathcal{C}_\Sigma^A$, the *extension class* of $\mathcal{C}$ with respect to $A$, the class of all algebras in $\mathcal{C}$ that are extensions of $A$:

$$\mathcal{C}_\Sigma^A := \left\{ A' \in \mathcal{C} \mid A'|_\Sigma = A \right\}$$

We are now ready to give our intended interpretation of presentations.

**Definition 6.24** (*Cogenerated interpretation*) Let $\Sigma$ be a signature, $A$ a $\Sigma$-algebra and $\langle \Sigma', \Phi \rangle$ a presentation such that $\Sigma \subseteq \Sigma'$. The *cogenerated interpretation* of $\langle \Sigma, \Phi \rangle$ with respect to $\Sigma$ and $A$ consists of all $\Sigma'$-models of $\Phi$ that extend $A$ and that are cogenerated in $\Sigma'$ with respect to $\Sigma$:

$$cogenerated(\Sigma', \Phi)^{\Sigma,A} :=$$

$$\left\{ A' \in Mod_{\Sigma'}(\Phi)_\Sigma^A \;\middle|\; A' \text{ is cogenerated in } \Sigma' \text{ w.r.t. } \Sigma \right\}$$

The cogenerated interpretation of a presentation does not stand on its own but depends on the pair $\langle \Sigma, A \rangle$ as an additional argument. In Sect. 6.7, we will clarify where this information actually stems from. Given a declaration $D$, we will write **cogenerated** $D$, if the presentation introduced by $D$ shall be interpreted as cogenerated.

**Example 6.17** Take the declaration

**free type** Bool := T | F

of Example 6.16 and let $\langle \Sigma, \Phi \rangle$ be its presentation; then because of the free interpretation of this declaration, the quotient term algebra $Term_\Sigma^\Phi$ is its canonical model.

Let $\langle \Sigma', \Phi \rangle$ be the presentation derived from the extension of this declaration by

**cogenerated** {
  **cotype** Stream := head:Bool | tail:Stream
  **axiom** ∀s:Stream. head(s)=T
}

Then the data type denoted by this extension is $cogenerated(\Sigma', \Phi)^{\Sigma, Term_\Sigma^\Phi} = [S]_{\Sigma'}$ where $S(\mathsf{Bool}) = \{T, F\}$ for some distinct values $T$ and $F$, $S(\mathsf{Stream}) = \{s_0\}$ for some value $s_0$, $S(\mathsf{T}) = T$, $S(\mathsf{F}) = F$, $S(\mathsf{head}) = \lambda s.\,T$, and $S(\mathsf{tail}) = \lambda s.\,s$.  □

As discussed in Sect. 6.4, the loose interpretation of constructor-based specifications raises also the problem of 'confusion', i.e., an specified abstract data type may contain algebras that are 'too small' in the sense that they identify values that arise from different constructor applications; this problem was solved by the concept of free specifications. An analogous problem arises with the loose interpretation of observer-based specifications: the specified abstract data type may also contain algebras that are 'too small' in the sense that they do not include all the values that may arise from observer applications. This problem is demonstrated by the following example.

**Example 6.18** We extend the declaration

   **free type** Bool := T | F

by the declaration

   **cotype** Stream := head:Bool | tail:Stream

which is identical the one given in Example 6.16 except that it does not impose any constraint of the values observed by head; we thus might expect this declaration to specify a monomorphic abstract data type that contains all possible streams with values $T$, $F$, in all possible interleavings. Every algebra in that data type should include in particular streams that are observationally equivalent to the streams depicted in Fig. 6.9.

In fact, however, $S$ has infinitely many non-isomorphic models, each of which maps Stream to a subset of the set of possible streams. In particular, there are the non-isomorphic models $S_0, \ldots, S_4$ with $S_0(\mathsf{Stream}) = \{s_0\}$, $S_1(\mathsf{Stream}) = \{s_1\}$, $S_2(\mathsf{Stream}) = \{s_2, s_3\}$, $S_3(\mathsf{Stream}) = \{s_4, s_0\}$, $S_4(\mathsf{Stream}) = \{s_4, s_0, s_5, s_1\}$. The only requirement is that the carrier of Stream is closed under the interpretation of tail and that that it does not contain different observationally equivalent streams. □

Among all the possible models depicted in Fig. 6.8, the intended model $A$ that encompasses all possible streams distinguishes itself in the following way: it provides every model $B$ with a homomorphism $h\colon B \to_\Sigma A$ that 'embeds the visible behavior' of every stream $S'$ of $B$ into that of some stream $S$ of $A$. This characteristic feature is formalized below by a dualization of the concept of initial algebras introduced for constructor-based specifications.

**Definition 6.25** (*Final algebras*) Let $\Sigma$ be a signature, $A$ a $\Sigma$-algebra, and $C$ a class of $\Sigma$-algebras. Then $A$ is *final* in $C$ (with respect to $\Sigma$), if $A \in C$ and,

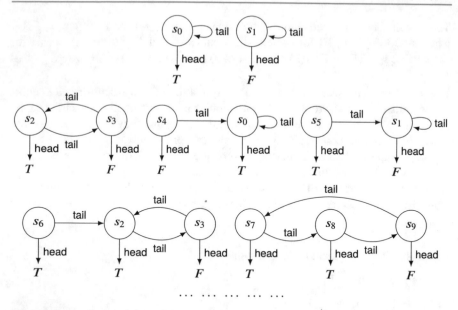

**Fig. 6.9** Various interpretations of Stream with multiple visible values

for every algebra $B \in C$, there exists exactly one homomorphism $h : B \rightarrow_\Sigma A$ from $B$ to $A$.

Final algebras are unique up to isomorphism.

**Proposition 6.12** (Uniqueness of final algebras) *Let $\Sigma$ be a signature, $A$, $B$ $\Sigma$-algebras, and $C$ a class of $\Sigma$-algebras. If $A$ and $B$ are both final in $C$ with respect to $\Sigma$, then $A$ and $B$ are isomorphic, i.e., $A \simeq_\Sigma B$.*

*Proof* Analogous to the proof of Proposition 6.6.                                        □

Since the existence of a homomorphism $h : B \rightarrow_\Sigma A$ can be understood as the relationship '$B$ has not less structure than $A$', a final algebra is an algebra that 'has least structure' respectively is 'most unifying' among a given class of algebras. This characteristics is formalized below.

**Proposition 6.13** (Final algebras are most unifying) *Let $\Sigma$ be a signature and let $A$ be a $\Sigma$-algebra that is final in a class $C$ of $\Sigma$-algebras.*

*Then for every sort $S \in \Sigma.s$ and all terms $T, T' \in Term_{\Sigma}^{S}$ of sort $S$ we have*

$$\left( A \models_{\Sigma} T = T' \right) \Leftrightarrow \left( \exists B \in C.\ B \models_{\Sigma} T = T' \right).$$

*Furthermore, for every predicate $\langle I, [S_1, \ldots, S_n] \rangle \in \Sigma.p$ with arity $[S_1, \ldots, S_n]$ and all terms $T_1 \in Term_{\Sigma}^{S_1}, \ldots, T_n \in Term_{\Sigma}^{S_n}$ of sorts $S_1, \ldots, S_n$, we have*

$$\left( A \models_{\Sigma} I(T_1, \ldots, T_n) \right) \Leftrightarrow \left( \exists B \in C.\ B \models_{\Sigma} I(T_1, \ldots, T_n) \right).$$

In other words, if some algebra in a class interprets two terms identical, then also the final algebra in that class does so. Furthermore, atomic formulas are interpreted 'maximally': if some algebra in this class interprets an atomic formula as true, then also the final algebra does so.

Final algebras give rise to another interpretation of presentations.

**Definition 6.26** (*Cofree interpretation*) Let $\Sigma$ be a signature, $A$ a $\Sigma$-algebra and $\langle \Sigma', \Phi \rangle$ a presentation such that $\Sigma \subseteq \Sigma'$. The *cofree interpretation* of $\langle \Sigma', \Phi \rangle$ with respect to $\Sigma$ and $A$ consists of all $\Sigma'$-models of $\Phi$ that extend $A$ and that are final in this class with respect to $\Sigma$:

$$cofree(\Sigma', \Phi)^{\Sigma, A} := \left\{ A' \in Mod_{\Sigma'}(\Phi)_{\Sigma}^{A} \,\middle|\, A' \text{ is final in } Mod_{\Sigma'}(\Phi)_{\Sigma}^{A} \text{ w.r.t. } \Sigma' \right\}$$

Also the cofree interpretation induces an abstract data type.

**Proposition 6.14** (Cofree abstract data type) *Let $\Sigma$ be a signature, $A$ a $\Sigma$-algebra and $\langle \Sigma', \Phi \rangle$ a presentation such that $\Sigma \subseteq \Sigma'$. Its cofree interpretation $cofree(\Sigma', \Phi)^{\Sigma, A}$ represents an abstract data type.*

**Proof** By Proposition 6.12, $cofree(\Sigma, \Phi)^{\Sigma, A}$ is monomorphic.                    □

Given a declaration $D$, we will write **cofree** $D$, if the presentation introduced by $D$ shall be interpreted as cofree. By Proposition 6.13, in this interpretation only those equations and atomic formulas do not hold that contradict the axioms of the presentation.

**Example 6.19** Given the declaration

**free type** Bool = T | F

the extension

**cofree cotype** Stream = head:Bool | tail:Stream

denotes the type of all infinite streams of values T or F.
Likewise, the extension

**cofree cotype** Tree = value:Bool | left:Tree | right:Tree

denotes the type of all infinite binary trees whose nodes carry the values T or F.   □

As for initial models, the question arises under which constraints $cofree(\Sigma, \Phi)^{\Sigma, A}$ is not empty, i.e., does indeed have a final model and what this model concretely is.

We are now going to construct a candidate for such a model; first we introduce the objects that will make up its carrier values.

**Definition 6.27** (*Trees*) A *tree t* over a set $I$ of indices and a set $L$ of labels is a partial function $t: I^* \to_\perp L$ that maps finite sequences of indices to labels such that the following conditions hold (where $def\langle t, is \rangle :\Leftrightarrow \exists l \in L. t\langle is, l \rangle$):

• $t$ has a root referenced by the empty index sequence [ ]:

  $def\langle t, [\ ]\rangle$;

• the domain of $t$ is prefix-closed:

  $\forall is_1 \in I^*, is_2 \in I^*. def\langle t, is_1 \circ is_2 \rangle \Rightarrow def\langle t, is_1 \rangle$.

**Example 6.20** Given $I := \{0, 1\}$ and $L := \{a, b, c, d, e\}$, the finite binary tree $t$ depicted as

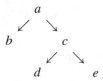

satisfies the conditions $t([]) = a, t([0]) = b, t([1]) = c, t([1, 0]) = d, t([1, 1]) = e,$ $\neg def\langle [0, i]\rangle$, and $\neg def\langle [1, i, j]\rangle$ for every $i, j \in \{0, 1\}$.   □

Please note that according to this definition a tree may be infinite; in fact, the following definition constructs infinite trees.

---

**Definition 6.28** (*Behavior algebra*)  Let $\Sigma$ and $\Sigma'$ be signatures such that $\Sigma \subseteq \Sigma'$ and the following conditions hold:

- there is no constant $C \in \Sigma'.c \setminus \Sigma.c$;
- every function $F \in \Sigma'.f \setminus \Sigma.f$ or predicate $P \in \Sigma'.p \setminus \Sigma.p$ has in its arity a sort from $\Sigma'.s$, but only as the first sort of the arity.

In other words, there are no constants of hidden sorts, the only functions operating on hidden sorts are observers, and also predicates may serve as such observers.

Then we can extend every $\Sigma$-algebra $A$ on the visible sorts to a corresponding $\Sigma'$-algebra $Beh^A_{\Sigma,\Sigma'}$, the *behavior algebra* for $A$, i.e., $A \subseteq Beh^A_{\Sigma,\Sigma'}$:

- All elements of $\Sigma$ are interpreted in $Beh^A_{\Sigma,\Sigma'}$ as in $A$, i.e., for every observable sort $S \in \Sigma.s$, constant $C = \langle I, S \rangle \in \Sigma.c$, function $F = \langle I, [S_1, \ldots, S_n], S \rangle \in \Sigma.f$, and predicate $P = \langle I, [S_1, \ldots, S_n] \rangle \in \Sigma.p$ we have:

$$Beh^A_{\Sigma,\Sigma'}(S) := A(S)$$

$$Beh^A_{\Sigma,\Sigma'}(C) := A(C)$$

$$Beh^A_{\Sigma,\Sigma'}(F) := A(F)$$

$$Beh^A_{\Sigma,\Sigma'}(P) := A(P)$$

- For every hidden sort $S \in \Sigma'.s \setminus \Sigma.s$, the carrier $Beh^A_{\Sigma,\Sigma'}(S)$ contains every *behavior tree* $t$, a tree $t$ with index set $I$ and label set $L$ where

$$I = (\Sigma'.f \setminus \Sigma.f) \times V^*$$

$$L = (\Sigma'.s \setminus \Sigma.s) \cup V$$

$$V = \bigcup_{S \in \Sigma.s} A(S)$$

  (i.e., every index in the tree is an observer on a hidden sort together with a sequence of observable values and every label is a hidden sort or an observable value)

  that satisfies the following conditions:
  - the root of $t$ is labeled with sort $S$:

$$t([\,]) = S$$

---

- every node labeled with some hidden sort $S$ has for every observer $F$ on $S$ with result sort $S'$ and every tuple of additional arguments to $F$ a child that is labeled
  - if $S'$ is hidden, with $S'$ itself, and
  - if $S'$ is visible, with a value from $A(S')$:

$$\forall is \in I^*, S, F, I', S_1, \ldots, S_n, S', a_1, \ldots, a_n.$$
$$t(is) = S \wedge S \in \Sigma'.s \backslash \Sigma.s \wedge$$
$$F \in \Sigma'.f \backslash \Sigma.f \wedge F = \langle I', [S, S_1, \ldots, S_n], S' \rangle \wedge$$
$$\langle a_1, \ldots, a_n \rangle \in A(S_1) \times \ldots \times A(S_n) \Rightarrow$$
$$\quad \text{if } S' \in \Sigma'.s \backslash \Sigma.s$$
$$\quad \quad \text{then } t(is \circ [\langle F, [a_1, \ldots, a_n] \rangle]) = S'$$
$$\quad \quad \text{else } \exists B \in A(S'). \, t(is \circ [\langle F, [A_1, \ldots, A_n] \rangle]) = B$$

- The application of observer $F = \langle I, [S, S_1, \ldots, S_n], S' \rangle \in \Sigma'.f \backslash \Sigma.f$ to a tree $t$ (with root $S$) and additional arguments $a_1, \ldots, a_n$ (from the carriers of sorts $S_1, \ldots, S_n$) is interpreted as the selection of the corresponding child (either a tree whose root is a hidden sort $S'$ or a value from the carrier of a visible sort $S'$):

$$Beh^A_{\Sigma, \Sigma'}(F) := \lambda t, a_1, \ldots, a_n.$$
$$\quad \text{if } S' \in \Sigma'.s \backslash \Sigma.s$$
$$\quad \quad \text{then } \lambda is \in I^*. \, t([\langle F, [a_1, \ldots, a_n] \rangle] \circ is)$$
$$\quad \quad \text{else } t([\langle F, [a_1, \ldots, a_n] \rangle])$$

- The application of predicate $P = \langle I, [S, S_1, \ldots, S_n] \rangle \in \Sigma'.p \backslash \Sigma.p$ to a tree $t$ (with root $S$) and additional arguments $a_1, \ldots, a_n$ (from the carriers of sorts $S_1, \ldots, S_n$) is interpreted as 'true'

$$Beh^A_{\Sigma, \Sigma'}(P) := \pi t, a_1, \ldots, a_n. \, \text{true}$$

i.e., it includes every tuple $\langle t, a_1, \ldots, a_n \rangle$.

A behavior algebra thus includes all possible 'behaviors' that may arise from the signature of an abstract data type under the cofree interpretation.

**Example 6.21** Take as $\Sigma$ the signature arising from the declaration

**free type** Bool $:= T \mid F$

and as $\Sigma'$ the signature arising from

**cofree cotype** Stream $:=$ head:Bool $\mid$ tail:Stream

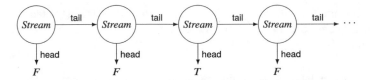

**Fig. 6.10** A behavior tree for Stream

Take as $A$ the classical $\Sigma$-algebra whose with $A(\mathsf{Bool}) = \{T, F\}$ for two distinct values $T, F$.

Then $Beh^A_{\Sigma, \Sigma'}(\mathsf{Stream})$ consists of all infinite trees of the same form as the sample tree depicted in Fig. 6.10. Each tree has a single infinite branch arising from the repeated application of the operation tail; each corresponding edge leads to a new subtree. Furthermore, from each node a path of length one emerges by the application of the operation head which exposes the observable value. Each such tree thus uniquely determines an infinite sequence of stream values. □

If the data type is further constrained by axioms, we consider only those subalgebras of the behavior algebra that conform to the axioms.

---

**Definition 6.29** (*Behavior subalgebras*) Let $\Sigma, \Sigma', A$ as in Definition 6.29. Furthermore, let $\Phi$ be such that $\langle \Sigma', \Phi \rangle$ is a presentation. We define $Beh^A_{\Sigma, \Sigma', \Phi} \subseteq Mod_{\Sigma'}(\Phi)$ as the class of $\Sigma'$-models of $\Phi$ which consists of every $\Sigma'$-algebra $A'$ that is defined exactly like $Beh^A_{\Sigma, \Sigma'}$ was defined in Definition 6.28, except that

- for every hidden sort $S \in \Sigma'.\mathsf{s} \setminus \Sigma.\mathsf{s}$, $A'(S) \subseteq Beh^A_{\Sigma, \Sigma'}(S)$, i.e., the carrier of $S$ consists of not necessarily all behavior trees,
- for every predicate $P = \langle I, [S, S_1, \ldots, S_n] \rangle \in \Sigma'.\mathsf{p} \setminus \Sigma.\mathsf{p}$ on a hidden sort $S \in \Sigma'.\mathsf{s} \setminus \Sigma.\mathsf{s}$, the interpretation $Beh^A_{\Sigma, \Sigma'}(P)$ is not necessarily 'true', i.e., it does not necessarily consist of all tuples $\langle t, a_1, \ldots, a_n \rangle$, and
- $A' \in Mod_{\Sigma'}(\Phi)$, i.e., $A'$ is a $\Sigma'$-model of $\Phi$.

---

**Example 6.22** Take as $\Sigma$ the signature arising from the declaration

**free type** Bool := T | F

and take as $\langle \Sigma', \Phi \rangle$ the presentation arising from

**cofree** {
  **cotype** Stream := head:Bool | tail:Stream
  **axiom** ∀s:Stream. head(s)=F ∧ head(tail(s))=F ⇒ head(tail(tail(s)))=T
}

Then no algebra $A' \in Beh^A_{\Sigma,\Sigma',\Phi}$ includes in $A'$ (Stream) the tree depicted in Fig. 6.10: two subsequent observations of $F$ cannot be followed by another such observation.                                                                                  □

Behavior subalgebras are candidates for final algebras as quotient term algebras were candidates for initial algebras. However, as the logic of free specifications had to be restricted to Horn clauses in order to ensure the existence of an initial model represented by the quotient algebra, also the logic of cofree specifications has to be restricted in order to ensure the existence of a final model represented by a particular behavior subalgebra.

---

**Proposition 6.15** (Maximal behavior subalgebra as cofree interpretation) *Let $\Sigma$ and $\Sigma$ be signatures such that $\Sigma \subseteq \Sigma'$ and the constraints of Definition 6.29 hold. We extend the grammar of formulas in abstract data type specifications to include also* modal formulas *of form* $[f(T_1, \ldots, T_n)]F'$:

$$F' ::= F \mid [f(T_1, \ldots, T_n)]F'$$
$$\mid \neg F' \mid F'_1 \wedge F'_2 \mid F'_1 \vee F'_2 \mid F'_1 \Rightarrow F'_2 \mid F'_1 \Leftrightarrow F'_2$$
$$\mid \forall V : S.\ F' \mid \exists V : S.\ F' \mid let\ V = T\ in\ F' \mid if\ F'\ then\ F'_1\ else\ F'_2$$

*Here $F$ is a basic formula that is formed according to Definition 6.1 but does not include the application of an observer $f$ with a hidden result sort $S \in \Sigma'.s \setminus \Sigma.s$. The application of such an observer is restricted to the modality $[f(T_1, \ldots, T_n)]$, propositional combinations of formulas with such modalities, and also quantification over such formulas provided that quantified variable $V$ has visible sort $S \in \Sigma.s$. The meaning of such a generalized formula $F'$ with respect to a variable $V$ is defined by a translation $[\![ F' ]\!]^V$ to a first-order formula:*

$$[\![ [f(T_1, \ldots, T_n)]F' ]\!]^V := \forall V' : S'.\ V' = f(V, T_1, \ldots, T_n) \Rightarrow [\![ F' ]\!]^{V'}$$
$$\text{where } f \text{ is an observer in } \Sigma'.f \setminus \Sigma.f$$
$$\text{with some hidden result sort } S'$$
$$\text{and } V' \text{ does not occur in } f(V, T_1, \ldots, T_n) \text{ or } M$$

$$[\![ f(T_1, \ldots, T_n) ]\!]^V := f(V, T_1, \ldots, T_n)$$
$$\text{where } f \text{ is an observer in } \Sigma'.f \setminus \Sigma.f$$
$$\text{with some visible result sort}$$

$$[\![ f(T_1, \ldots, T_n) ]\!]^V := f(T_1, \ldots, T_n)$$
$$\text{where } f \text{ is not an observer in } \Sigma'.f \setminus \Sigma.f$$

$$[\![ \neg M ]\!]^V := \neg [\![ M ]\!]^V$$

......

*We correspondingly extend the syntax of declarations to include apart from basic axioms of form*

   **axiom** *F*

*involving a basic formula F also* modal axioms *of form*

   **axiom**[*S*] *F'*

*where $S \in \Sigma'.s \setminus \Sigma.s$ is a hidden sort and F' is a generalized formula; such an axiom is translated to the axiom*

   **axiom** $\bigvee\!\!\bigvee \!:\! S. [\![ F' ]\!]^V$

*for some variable V that does not occur in F'.*
*Now let A be a Σ-algebra and let Φ be a set of Σ'-formulas that are either basic axioms or translations of modal axioms. Then the following holds:*

- *if there exists some $A' \in Mod_{\Sigma'}(\Phi)$ of Φ with $A \subseteq A'$, i.e., A can be extended to some Σ'-model A' of Φ,*
- *then the class $Beh^A_{\Sigma,\Sigma',\Phi}$ of behavior subalgebras contains an element A' that is* maximal *in the following sense:*

> *For every behavior subalgebra $A'' \in Beh^A_{\Sigma,\Sigma',\Phi}$, every hidden sort $S \in \Sigma'.s \setminus \Sigma.s$, and every predicate $P \in \Sigma'.p \setminus \Sigma.p$ on a hidden sort, we have $A''(S) \subseteq A'(S)$ and $A''(P) \subseteq A'(P)$.*

*Furthermore, we have*

   $A' \in cofree(\Sigma', \Phi)^{\Sigma,A}$

*i.e., the cofree interpretation of ⟨Σ, Φ⟩ has this maximal behavior algebra A' as a model. Finally, we also have*

   $cofree(\Sigma', \Phi)^{\Sigma,A} \subseteq cogenerated(\Sigma', \Phi)^{\Sigma,A}$

*i.e. the cofree algebras are also cogenerated.*

**Proof** The core of the first part of the proof is to show for every hidden sort $S$ the existence of a unique homomorphism $h_S \colon A''(S) \to A'(S)$ from every Σ'-model $A''$ of Φ that extends $A$ to the maximal behavior subalgebra $A'$. This homomorphism is the *behavior homomorphism* that maps every element $x \in A''(S)$ to its behavior tree $h_S(x) \in A(S)$ that has the same observable behavior as $x$. The second part follows from the fact that different behavior trees have different observable behaviors.    □

The modal logic presented in Proposition 6.15 is not the richest one that guarantees the existence of final models; for instance, we might also add a modality $[f_1(T_{1,1}, \ldots, T_{1,n_1}), \ldots, f_m(T_{1,1}, \ldots, T_{1,n_m})]$ which allow the next value $V'$ to be reached by the application of any of the listed observers or a modality $[f(T_1, \ldots, T_n)*]$ which allows the next value $V'$ to be reached by an arbitrary number of applications of the observer, as well as a combination of both. Furthermore, the modal formula $\langle \ldots \rangle F$ may be introduced as an abbreviation of $\neg[\ldots]\neg F$. We omit the detailed elaboration but rather present some examples on the use of modal logic below.

**Example 6.23**  Given the declaration

> **free type** Bool := T | F

the extension

> **cofree** {
>   **cotype** Stream := head:Bool | tail:Stream
>   **axiom**[Stream] head=F $\wedge$ [tail]head=F $\Rightarrow$ [tail][tail]head=T
> }

has the same meaning as

> **cofree** {
>   **cotype** Stream := head:Bool | tail:Stream
>   **axiom** $\forall$s:Stream. head(s)=F $\wedge$ head(tail(s))=F $\Rightarrow$ head(tail(tail(s)))=T
> }

It specifies the type of all infinite streams that do not contain more than two subsequent occurrences of value F.                                                                      □

**Example 6.24**  Given the declaration

> **free type** Bool := T | F

the extension

> **cofree** {
>   **cotype** Tree = value:Bool | left:Tree | right:Tree
>   **axiom**[Tree] value=F $\Rightarrow$ [left,right]value=F
> }

is equivalent to the declaration

> **cofree** {
>   **cotype** Tree = value:Bool | left:Tree | right:Tree
>   **axiom**[Tree] value=F $\Rightarrow$ [left]value=F $\wedge$ [right]value=F
> }

It specifies the type of all infinite binary trees where every subtree whose root carries value F has only F values in its nodes.                                                                      □

**Example 6.25**  Given the declaration

> **free type** Bool := T | F

the extension

> **cofree** {
>   **cotype** Stream := head:Bool | tail:Stream
>   **axiom**[Stream] ⟨tail*⟩head=T
> }

or equivalently

> **cofree** {
>   **cotype** Stream := head:Bool | tail:Stream
>   **axiom**[Stream] ¬[tail*]head=F
> }

denotes the type of all streams that eventually contain the value true. However, since also the tail of every stream is a stream with that property, this actually implies that every stream must contain T *infinitely often*.                    □

**Example 6.26**  Given the declaration

> **free type** Bool := T | F

the extension

> **cofree** {
>   **cotype** Stream := end:Bool | head:Bool | tail:Stream
>   **axiom**[Stream] end=T ⇒[tail*]head=F
> }

denotes the type of all finite or infinite streams: if the observer end returns T, the remaining stream can be considered as empty (any subsequent application of head returns F).                                                           □

**Example 6.27**  Given the declaration

> **free type** Nat := 0 | +1(Nat)

the extension

> **cofree** {
>   **cotype** Stream := head:Nat | tail:Stream
>   **axiom**[Stream] ∀e:Nat. head=e ⇒ [tail][tail*]head≠e
>   **pred** allzero⊆Stream
>   **axiom**[Stream] allzero ⇒ head=0 ∧ [tail]allzero
> }

denotes the type of all infinite streams of natural numbers that do not contain double elements: the observer head returns some element of the stream while tail proceeds to the remainder of the stream that, according to the first axiom, does not contain this element any more.

The predicate allzero states that a stream contains only zeros. Since the predicate is interpreted maximally, without further axiom, it would be 'true' for every stream. However, the second axiom ensures that the predicate becomes 'false' if the head of the stream is not zero or its tail makes the predicate false. This can be seen easier by rewriting the axiom to its dual form

**axiom**[Stream] head$\neq$0 $\vee$ $\neg$[tail]allzero $\Rightarrow$ $\neg$allzero

respectively by writing

**axiom**[Stream] head$\neq$0 $\Rightarrow$ $\neg$allzero
**axiom**[Stream] $\neg$[tail]allzero $\Rightarrow$ $\neg$allzero

which splits the axiom into two.                                                                         □

**Example 6.28**  Given the declaration

**free** {
  **type** Bool := T | F
  **type** Symbol := A | B | ... | Z
}

the extension

**cofree cotype** State := accept:Bool | read(Symbol):State

denotes the type of all automata that read arbitrarily long words of symbols from alphabet Symbol: the reading of a symbol by observer read drives the automaton from its current state to a successor state; the observer accept indicates whether the current state is accepting, i.e., whether the word read so far is accepted. For instance, for a State term $s$ denoting the initial state of the automaton and Symbol terms $a_1, a_2, a_3$, the value of the term

accept(read(read(read($s, a_1$), $a_2$), $a_3$))

indicates whether the word denoted by $a_1, a_2, a_3$ is accepted. If we extend the specification to

**cofree** {
  **cotype** State := accept:Bool | read(Symbol):State
  **axiom**[State] $\forall$a:Symbol. [read(a)]accept=T $\Rightarrow$ (a=A $\vee$ a=B)
}

the specification denotes the type of all automata that only may (but need not) accept words that end with one of the symbols A or B.                                                  □

In analogy to the relationship of the free interpretation of specifications to the inductive relation and function definitions presented in Sects. sect:inductrel, sect:inductrel2, and sect:inductfun on the basis of least fixed points, the cofree interpretation of specifications has very much the flavor of coinductive relation and function definitions that have been introduced there on the basis of greatest fixed points: in the cofree interpretation of a specification, a formula is considered as true, unless its violation can be established from the axioms of the specification; likewise, values are considered the same unless they can be distinguished by their visible behavior. Again, however, it should be noted is that specifications construct new domains together with operations on these domains; the previously presented definitions introduced new operations on already given domains.

## 6.7   Specifying in the Large

The main purpose of the language of declarations introduced in Sect. 6.2 was specifying 'in the small', i.e., to introduce individual abstract data types. We are now going to elaborate this language to a language of specifying abstract data types 'in the large' which allows to extend types and to combine multiple of them to more complex structures. Based on the algebraic and logic concepts introduced in Sects. 6.3, 6.4, 6.5, and 6.6, we will also give this language a formal semantics.

In more detail, the complete language introduces two main kinds of phrases:

- *Specification Expressions* (short *specifications*): these phrases embed declarations and extend respectively combine the denoted types to compound types.
- *Specification Definitions*: these phrases embed specification expressions and give the denoted types names for later reference; they may also make the types 'generic' by providing them with parameters that allow multiple instantiations.

The language's abstract syntax is provided by the following definition.

**Definition 6.30** (*Abstract data type specifications*) A *specification definition* is a phrase $SD \in SpecDefinition$ and a *specification (expression)* is a phrase $SE \in SpecExpression$ which are formed according to the following grammar:

$SD \in SpecDefinition, \ SE \in SpecExpression$
$EI \in ExportItemSeq, \ RI \in RenameItemSeq$
$D \in Declaration, \ Ss \in Sorts, \ I \in Identifier$

$SD ::= SD_1 \ SD_2 \ | \ \textbf{spec} \ I[SE_1] \ \textbf{import} \ SE_2 := SE_3$

$SE ::= \_ \ | \ D \ | \ \textbf{generated} \ D \ | \ \textbf{free} \ D \ | \ \textbf{cogenerated} \ D \ | \ \textbf{cofree} \ D$
$\qquad | \ SE_1 \ \textbf{then} \ SE_2 \ | \ SE_1 \ \textbf{and} \ SE2$
$\qquad | \ SE \ \textbf{export} \ EI \ | \ SE \ \textbf{with} \ RI \ | \ I[SE \ \textbf{fit} \ RI]$

$$EI ::= EI_1, EI_2 \mid \textbf{sort } I$$
$$\mid \textbf{const } I \mid \textbf{fun } I(Ss) \mid \textbf{pred } I(Ss)$$

$$RI ::= RI_1, RI_2 \mid \textbf{sort } I_1 \mapsto I_2$$
$$\mid \textbf{const } I_1 \mapsto I_2 \mid \textbf{fun } I_1(Ss) \mapsto I_2 \mid \textbf{pred } I_1(Ss) \mapsto I_2$$

The syntactic domains *Declaration* of declarations, *Sorts* of sort sequences, and *Identifier* of identifiers are formed as in Definition 6.1.

The semantics of the language will be provided in a dual form:

- *Static Semantics*: this semantics is summarized in Fig. 6.11. It is in essence the specification language's 'type system' which extends the elaboration of Sect. 6.2. There we introduced the judgement

  $\vdash D: \text{declaration}(\Sigma, \Phi)$

  which states that a declaration $D$ is well-formed and gives rise to a presentation $\langle \Sigma, \Phi \rangle$. Analogously, the extended type system introduces judgements for specification definitions and expressions that state their well-formedness and extract from them that information that is essential for their 'model semantics'.
- *Model Semantics*: this semantics is summarized in Fig. 6.12. It is given in denotational form by valuation functions that map a specification definition respectively expression to its actual 'meaning' in terms of abstract data types. We assume that a valuation function of the model semantics is only applied to a phrase that has been deduced to be well-formed according to the corresponding judgement of the static semantics and that it has been annotated with the information provided by that judgement.

In the following, we will elaborate the static and the model semantics of the various kinds of specification expressions and specification definitions.

## Specification Expressions

A specification expression $SE$ essentially denotes an abstract data type, i.e., a class $\mathcal{M}'$ of algebras with a signature $\Sigma'$. However, it does so by extending an already given class $\mathcal{M}$ of algebras with signature $\Sigma$ which must be correspondingly provided as an argument. Furthermore, it is evaluated in an environment *me* of named abstract data types whose signatures are captured by a corresponding environment *se*. These considerations give rise to the following semantics:

**Rules for** $se \vdash SD$ : **spec**$(se')$**:**

$$\frac{se \vdash SD_1 : \mathrm{spec}(se_1) \quad se_1 \vdash SD_2 : \mathrm{spec}(se')}{se \vdash SD_1 \; SD_2 : \mathrm{spec}(se')}$$

$$\frac{\begin{array}{c} \neg \exists \mathfrak{S}. \; \langle I, \mathfrak{S} \rangle \in se \\ se, \Sigma_0 \vdash SE_2 : \mathrm{specexp}(\Sigma_i) \quad se, \Sigma_i \vdash SE_1 : \mathrm{specexp}(\Sigma_p) \quad se, \Sigma_p \vdash SE_3 : \mathrm{specexp}(\Sigma_e) \\ \mathfrak{S} = \langle \mathrm{i}(\Sigma_i), \mathrm{p}(\Sigma_p), \mathrm{e}(\Sigma_e) \rangle \quad se' = se \cup \{\langle I, \mathfrak{S} \rangle\} \end{array}}{se \vdash \textbf{spec}\; I[SE_1]\; \textbf{import}\; SE_2 \; := \; SE_3 : \mathrm{spec}(se')}$$

**Rules for** $se, \Sigma \vdash SE$ : **specexp**$(\Sigma')$

$$se, \Sigma \vdash \_ : \mathrm{specexp}(\Sigma)$$

$$\frac{\begin{array}{c} \vdash D : \mathrm{declaration}(\Sigma_1, \Phi) \\ \Sigma' = \langle \Sigma.\mathrm{s} \cup \Sigma_1.\mathrm{s}, \; \langle \mathrm{c}(\Sigma.\mathrm{c} \cup \Sigma_1.\mathrm{c}), \mathrm{f}(\Sigma.\mathrm{f} \cup \Sigma_1.\mathrm{f}), \mathrm{p}(\Sigma.\mathrm{p} \cup \Sigma_1.\mathrm{p}) \rangle \rangle \\ \Sigma' \text{ is a signature} \quad \forall F \in \Phi. \; F \text{ is a } \Sigma'\text{-formula} \end{array}}{se, \Sigma \vdash D : \mathrm{specexp}(\Sigma')}$$

$$\frac{se, \Sigma \vdash D : \mathrm{specexp}(\Sigma')}{se, \Sigma \vdash \textbf{generated}\; D : \mathrm{specexp}(\Sigma')} \qquad \frac{se, \Sigma \vdash D : \mathrm{specexp}(\Sigma')}{se, \Sigma \vdash \textbf{free}\; D : \mathrm{specexp}(\Sigma')}$$

$$\frac{se, \Sigma \vdash D : \mathrm{specexp}(\Sigma')}{se, \Sigma \vdash \textbf{cogenerated}\; D : \mathrm{specexp}(\Sigma')} \qquad \frac{se, \Sigma \vdash D : \mathrm{specexp}(\Sigma')}{se, \Sigma \vdash \textbf{cofree}\; D : \mathrm{specexp}(\Sigma')}$$

$$\frac{se, \Sigma \vdash SE_1 : \mathrm{specexp}(\Sigma_1) \quad se, \Sigma_1 \vdash SE_2 : \mathrm{specexp}(\Sigma')}{se, \Sigma \vdash SE_1 \; \textbf{then}\; SE_2 : \mathrm{specexp}(\Sigma')}$$

$$\frac{se, \Sigma \vdash SE_1 : \mathrm{specexp}(\Sigma_1) \quad se, \Sigma \vdash SE_2 : \mathrm{specexp}(\Sigma_2) \quad \Sigma' = \Sigma_1 \cup \Sigma_2 \quad \Sigma' \text{ is a signature}}{se, \Sigma \vdash SE_1 \; \textbf{and}\; SE_2 : \mathrm{specexp}(\Sigma')}$$

$$\frac{se, \Sigma \vdash SE : \mathrm{specexp}(\Sigma_1) \quad \Sigma, \Sigma_1 \vdash EI : \mathrm{export}(\Sigma')}{se, \Sigma \vdash SE\; \textbf{export}\; EI : \mathrm{specexp}(\Sigma')}$$

$$\frac{se, \Sigma \vdash SE : \mathrm{specexp}(\Sigma_1) \quad \Sigma, \Sigma_1 \vdash RI : \mathrm{rename}(\Sigma', \mu : \Sigma_1 \to \Sigma')}{se, \Sigma \vdash SE\; \textbf{with}\; RI : \mathrm{specexp}(\Sigma')}$$

$$\frac{\begin{array}{c} \langle I, \mathfrak{S} \rangle \in se \\ se, \mathfrak{S}.\mathrm{i} \vdash SE : \mathrm{specexp}(\Sigma_a) \quad \mathfrak{S}.\mathrm{i}, \mathfrak{S}.\mathrm{p}, \Sigma_a \vdash RI : \mathrm{fit}(\mu : \mathfrak{S}.\mathrm{p} \to \Sigma_a) \\ \langle e' : \Sigma_a \to \Sigma_r, \mu' : \mathfrak{S}.\mathrm{e} \to \Sigma_r \rangle \text{ is the pushout of } \langle \mu : \mathfrak{S}.\mathrm{p} \to \Sigma_a, e : \mathfrak{S}.\mathrm{p} \to \mathfrak{S}.\mathrm{e} \rangle \\ \text{for the extension morphisms } e : \mathfrak{S}.\mathrm{p} \to \mathfrak{S}.\mathrm{e} \text{ and and } e' : \Sigma_a \to \Sigma_r \\ \Sigma' = \Sigma \cup \Sigma_r \quad \Sigma' \text{ is a signature} \end{array}}{se, \Sigma \vdash I[SE\; \textbf{fit}\; RI] : \mathrm{specexp}(\Sigma')}$$

**Rules for** $\Sigma, \Sigma_1 \vdash EI$ : **export**$(\Sigma')$**:**

(omitted)

**Rules for** $\Sigma, \Sigma_1 \vdash RI$ : **rename**$(\Sigma', \mu : \Sigma_1 \to \Sigma')$**:**

(omitted)

**Rules for** $\Sigma, \Sigma_1, \Sigma_2 \vdash RI$ : **fit**$(\mu : \Sigma_1 \to \Sigma_2)$**:**

(omitted)

**Fig. 6.11** The static semantics of abstract data type specifications

**Equations for** $[\![\, SD \,]\!](me) = \boldsymbol{me'}$ :

$[\![\, SD_1 \; SD_2 \,]\!](me) := [\![\, SD_2 \,]\!]([\![\, SD_1 \,]\!](me))$

$[\![\, \textbf{spec } I[SE_1^{\Sigma_i,\Sigma_p}] \textbf{ import } SE_2^{\Sigma_0,\Sigma_i} := SE_3^{\Sigma_p,\Sigma_e} \,]\!](me) :=$
  let $\mathcal{MI} = [\![\, SE_2^{\Sigma_0,\Sigma_i} \,]\!](me, \mathcal{M}_0)$
    $\mathcal{MP} = [\![\, SE_1^{\Sigma_i,\Sigma_p} \,]\!](me, \mathcal{MI})$
    $\mathcal{ME} = [\![\, SE_3^{\Sigma_p,\Sigma_e} \,]\!](me, \mathcal{MP})$
  in $me \cup \{\langle I, \mathcal{ME} \rangle\}$

**Equations for** $[\![\, SE^{\Sigma,\Sigma'} \,]\!](\boldsymbol{me}, \mathcal{M}) = \mathcal{M'}$ :

$[\![\, \_^{\Sigma,\Sigma} \,]\!](me, \mathcal{M}) := \mathcal{M}$

$[\![\, (D^{\Phi})^{\Sigma,\Sigma'} \,]\!](me, \mathcal{M}) := \bigcup_{A \in \mathcal{M}} Mod_{\Sigma'}(\Phi)^A_{\Sigma}$

$[\![\, (\textbf{generated } D^{\Phi})^{\Sigma,\Sigma'} \,]\!](me, \mathcal{M}) :=$
  $\bigcup_{A \in \mathcal{M}} \{ A' \in Mod_{\Sigma'}(\Phi)^A_{\Sigma} \mid A' \text{ is generated in } \Sigma' \text{ w.r.t. } \Sigma \}$

$[\![\, (\textbf{free } D^{\Phi})^{\Sigma,\Sigma'} \,]\!](me, \mathcal{M}) :=$
  $\bigcup_{A \in \mathcal{M}} \{ A' \in Mod_{\Sigma'}(\Phi)^A_{\Sigma} \mid A' \text{ is initial in } Mod_{\Sigma'}(\Phi)^A_{\Sigma} \}$

$[\![\, (\textbf{cogenerated } D^{\Phi})^{\Sigma,\Sigma'} \,]\!](me, \mathcal{M}) :=$
  $\bigcup_{A \in \mathcal{M}} \{ A' \in Mod_{\Sigma'}(\Phi)^A_{\Sigma} \mid A' \text{ is cogenerated in } \Sigma' \text{ w.r.t. } \Sigma \}$

$[\![\, (\textbf{cofree } D^{\Phi})^{\Sigma,\Sigma'} \,]\!](me, \mathcal{M}) :=$
  $\bigcup_{A \in \mathcal{M}} \{ A' \in Mod_{\Sigma'}(\Phi)^A_{\Sigma} \mid A' \text{ is final in } Mod_{\Sigma'}(\Phi)^A_{\Sigma} \}$

$[\![\, (SE_1^{\Sigma,\Sigma_1} \textbf{ then } SE_2^{\Sigma_1,\Sigma'})^{\Sigma,\Sigma'} \,]\!](me, \mathcal{M}) :=$
  let $\mathcal{M}_1 = [\![\, SE_1^{\Sigma,\Sigma_1} \,]\!](me, \mathcal{M})$ in $[\![\, SE_2^{\Sigma_1,\Sigma'} \,]\!](me, \mathcal{M}_1)$

$[\![\, (SE_1^{\Sigma,\Sigma_1} \textbf{ and } SE_2^{\Sigma,\Sigma_2})^{\Sigma,\Sigma'} \,]\!](me, \mathcal{M}) :=$
  $\left\{ A \in Alg(\Sigma') \,\middle|\, A|_{\Sigma_1} \in [\![\, SE_1^{\Sigma,\Sigma_1} \,]\!](me, \mathcal{M}) \wedge A|_{\Sigma_2} \in [\![\, SE_2^{\Sigma,\Sigma_2} \,]\!](me, \mathcal{M}) \right\}$

$[\![\, (SE^{\Sigma,\Sigma_1} \textbf{ export } EI)^{\Sigma,\Sigma'} \,]\!](me, \mathcal{M}) :=$
  $\left\{ A' \in Alg(\Sigma') \,\middle|\, \exists A \in [\![\, SE^{\Sigma,\Sigma_1} \,]\!](me, \mathcal{M}).\ A|_{\Sigma'} = A' \right\}$

$[\![\, (SE^{\Sigma,\Sigma_1} \textbf{ with } RI^{\,\mu\,:\,\Sigma_1 \rightarrow \Sigma'})^{\Sigma,\Sigma'} \,]\!](me, \mathcal{M}) :=$
  $\left\{ A' \in Alg(\Sigma') \,\middle|\, A'|_{\mu} \in [\![\, SE^{\Sigma,\Sigma_1} \,]\!](me, \mathcal{M}) \right\}$

$[\![\, (I[SE^{\Sigma,\Sigma_a} \textbf{ fit } RI^{\,e'\,:\,\Sigma_a \rightarrow \Sigma_r,\,\mu'\,:\,\Sigma_e \rightarrow \Sigma_r})^{\Sigma,\Sigma'} \,]\!](me, \mathcal{M}) :=$
  $\left\{ A' \in Alg(\Sigma') \,\middle|\, (A'|_{\Sigma_r})|_{e'} \in [\![\, SE^{\Sigma,\Sigma_a} \,]\!](me, \mathcal{M}) \wedge (A'|_{\Sigma_r})|_{\mu'} \in me(I) \right\}$

**Fig. 6.12** The model semantics of abstract data type specifications

- A static semantics with judgements of form

$$se, \Sigma \vdash SE \colon \mathsf{specexp}(\Sigma')$$

Such a judgement states that in an environment $se$ that maps type names to signatures, the specification $SE$ is well-formed and extends every abstract data type with signature $\Sigma$ to one with signature $\Sigma'$. We subsequently assume that $SE$ is annotated by the judgement to $SE^{\Sigma,\Sigma'}$.

- A model semantics with valuations of form

$$[\![\, SE^{\Sigma,\Sigma'} \,]\!](me, \mathcal{M}) = \mathcal{M'}$$

This valuation says that the well-formed specification *SE* that extends every abstract data type of signature $\Sigma$ to an abstract data type with signature $\Sigma'$ indeed extends, in an environment *me* that maps type names to abstract data types, the abstract data type $\mathcal{M}$ with signature $\Sigma$ to a data type $\mathcal{M}'$ with signature $\Sigma'$.

The valuation function of the model semantics could be declared as

$$[\![\,.\,]\!] : SpecExpression \rightarrow (ModelClassEnv \times ModelClass \rightarrow ModelClass)$$

where *ModelClass* denotes the class of all $\Sigma$-algebras, for some signature $\Sigma$, and *ModelClassEnv*: *Identifier* $\rightarrow_\perp$ *ModelClass* is a partial mapping of names to abstract data types. However, this would abuse set-theoretic notations, since *ModelClass* is a proper class, not a set, and thus also $[\![\,.\,]\!]$ is not a set-theoretic function. Nevertheless, for the sake of convenience and readability, we will in the following not distinguish between classes and sets and freely use set-theoretic notation in the definitions of the resulting abstract data types.

**Empty Specification**

Static Semantics:

$$se, \Sigma \vdash \_ : \mathsf{specexp}(\Sigma)$$

Model Semantics:

$$[\![\,\_^{\Sigma,\Sigma}\,]\!](me, \mathcal{M}) := \mathcal{M}$$

An empty specification _ typically serves in a specification

**spec** $I[SE_1]$ **import** $SE_2 := SE_3$

as a place-holder for the parameter $SE_1$ or the import $SE_2$ or the body $SE_3$. For instance, the specification

**spec** $I := SE$

is to be interpreted as

**spec** $I[\_]$ **import** $\_ := SE$

In the static semantics of _, the signature $\Sigma$ is not extended and the model semantics is just the identity function, i.e., the parameter $\mathcal{M}$ itself is returned as the result.

**Loose Specifications**

Static Semantics:

$$\vdash D: \text{declaration}(\Sigma_1, \Phi)$$
$$\Sigma' = \langle \Sigma.s \cup \Sigma_1.s, \langle c(\Sigma.c \cup \Sigma_1.c), f(\Sigma.f \cup \Sigma_1.f), p(\Sigma.p \cup \Sigma_1.p) \rangle \rangle$$
$$\underline{\Sigma' \text{ is a signature}\quad \forall F \in \Phi. \; F \text{ is a } \Sigma'\text{-formula}}$$
$$se, \Sigma \vdash D: \text{specexp}(\Sigma')$$

Model Semantics:

$$[\![\, (D^\Phi)^{\Sigma,\Sigma'} \,]\!](me, \mathcal{M}) := \bigcup_{A \in \mathcal{M}} Mod_{\Sigma'}(\Phi)_{\Sigma}^A$$

Given a declaration $D$, the static semantics extracts the corresponding presentation $\langle \Sigma_1, \Phi \rangle$, and ensures that by combining the current signature $\Sigma$ with $\Sigma_1$, the resulting signature $\Sigma'$ is well-formed (which rules out that both $\Sigma$ and $\Sigma_1$ contain constants with the same name but different sorts); furthermore, all formulas in $\Phi$ must be well-formed according to $\Sigma'$. It should be noted that this semantics allows an entity already declared in $\Sigma$ to show up also in $\Sigma_1$: both declarations are simply 'merged' in $\Sigma'$. Furthermore, any formula in $\Phi$ referring to an entity present in $\Sigma$ constrains the semantics of that entity.

The model semantics generalizes the semantics presented in Sect. 6.4 by an *extension class* (see Definition 6.23): the model semantics consists of every algebra $\Sigma'$-algebra $A'$ that extends some $\Sigma$-algebra $A \in \mathcal{M}$. This interpretation is still 'loose' in that its only restrictions are those imposed by the axioms in $\Phi$.

It should be noted that a loose declaration that imposes additional constraints on the entities declared in the current signature $\Sigma$, 'kills' all the models in $\mathcal{M}$ that do not conform to these constraints. For instance, in a specification

**type** Int $:= 0 \,|\, +1(\text{Int}) \,|\, -1(\text{Int})$
**then axiom** $\forall$i:Int. $i = +1(-1(i)) \wedge i = -1(+1(i))$

the model $\mathcal{M}$ set up by the declaration

**type** Int $:= 0 \,|\, +1(\text{Int}) \,|\, -1(\text{Int})$

is 'extended' by the loose declaration

**axiom** $\forall$i:Int. $i = +1(-1(i)) \wedge i = -1(+1(i))$

(see below for the semantics of 'vertical composition (extension)' denoted by the **then** clause) in such a way that all models are removed where the operations +1 and -1 are not the inverse of each other. In the declaration

**free type** Int $:= 0 \,|\, +1(\text{Int}) \,|\, -1(\text{Int})$
**then axiom** $\forall$i:Int. $i = +1(-1(i)) \wedge i = -1(+1(i))$

(see below for the semantics of the keyword **free**) this implies that the resulting data type is *empty*, because in the free interpretation of Int (represented by the set of all Int-terms) the stated equalities do not hold.

As demonstrated above, a loose declaration may thus have fewer models than the ones derived from the previous declarations and may not allow any models, i.e., make a specification inconsistent.

**(Co-)Generated and (Co-)Free Specifications**

Static Semantics:

$$\frac{se,\ \Sigma \vdash D : \mathsf{specexp}(\Sigma')}{se,\ \Sigma \vdash \textbf{generated}\ D : \mathsf{specexp}(\Sigma')} \qquad \frac{se,\ \Sigma \vdash D : \mathsf{specexp}(\Sigma')}{se,\ \Sigma \vdash \textbf{free}\ D : \mathsf{specexp}(\Sigma')}$$

$$\frac{se,\ \Sigma \vdash D : \mathsf{specexp}(\Sigma')}{se,\ \Sigma \vdash \textbf{cogenerated}\ D : \mathsf{specexp}(\Sigma')} \qquad \frac{se,\ \Sigma \vdash D : \mathsf{specexp}(\Sigma')}{se,\ \Sigma \vdash \textbf{cofree}\ D : \mathsf{specexp}(\Sigma')}$$

Model Semantics:

$$[\![\ (\textbf{generated}\ D^{\Phi})^{\Sigma,\Sigma'}\ ]\!](me,\mathcal{M}) :=$$

$$\bigcup_{A\in\mathcal{M}} \left\{ A' \in Mod_{\Sigma'}(\Phi)^A_{\Sigma} \ \middle|\ A' \text{ is generated in } \Sigma' \text{ w.r.t. } \Sigma \right\}$$

$$[\![\ (\textbf{free}\ D^{\Phi})^{\Sigma,\Sigma'}\ ]\!](me,\mathcal{M}) :=$$

$$\bigcup_{A\in\mathcal{M}} \left\{ A' \in Mod_{\Sigma'}(\Phi)^A_{\Sigma} \ \middle|\ A' \text{ is initial in } Mod_{\Sigma'}(\Phi)^A_{\Sigma} \right\}$$

$$[\![\ (\textbf{cogenerated}\ D^{\Phi})^{\Sigma,\Sigma'}\ ]\!](me,\mathcal{M}) :=$$

$$\bigcup_{A\in\mathcal{M}} \left\{ A' \in Mod_{\Sigma'}(\Phi)^A_{\Sigma} \ \middle|\ A' \text{ is cogenerated in } \Sigma' \text{ w.r.t. } \Sigma \right\}$$

$$[\![\ (\textbf{cofree}\ D^{\Phi})^{\Sigma,\Sigma'}\ ]\!](me,\mathcal{M}) :=$$

$$\bigcup_{A\in\mathcal{M}} \left\{ A' \in Mod_{\Sigma'}(\Phi)^A_{\Sigma} \ \middle|\ A' \text{ is final in } Mod_{\Sigma'}(\Phi)^A_{\Sigma} \right\}$$

The static semantics of (co)generated and (co)free specifications is as for loose ones.

The model semantics of cogenerated/cofree specifications was in essence already presented in Sect. 6.6; it is now only applied to multiple algebras $A \in \mathcal{M}$.

The model semantics of generated/free specifications generalizes the semantics presented in Sect. 6.5 in an analogous way[1]: it extracts from each extension class $Mod_{\Sigma'}(\Phi)^A_{\Sigma}$ arising from $\Sigma$-algebra $A$ in given class $\mathcal{M}$ those algebras that are generated respectively initial in that class. For instance, in

---

[1] The model semantics of free specifications is usually presented in terms of the notion of *free extensions*; see the literature references for further information.

**type** Bool := True | False
**then free type** List := empty | cons(Bool,List)

the loose declaration of Bool allows multiple non-isomorphic models, in particular one in which the equality True=False holds and one in which it does not hold; furthermore, the models need not be generated and thus may have arbitrarily many carrier values. Nevertheless, each of these models is extended by the free declaration of List in such a way that operation cons is injective and every List-value $l$ is denoted by $cons(b_1, \ldots cons(b_n, empty) \ldots)$ for some Bool-values $b_1, \ldots, b_n$ and the interpretations $cons$ and $empty$ of cons and empty.

Like loose declarations, also (co)generated and (co)free declarations may impose additional constraints on the entities in the given signature $\Sigma$ which 'kills' (some or all) models in $\mathcal{M}$ and may thus make the specification inconsistent.

Furthermore, free/cofree specifications do not necessarily induce a monomorphic abstract data type any more; for non-isomorphic algebras $A_1$ and $A_2$ in $\mathcal{M}$, in general also the initial respectively final algebras in the corresponding extension classes $Mod_{\Sigma'}(\Phi)_{\Sigma}^{A_1}$ and $Mod_{\Sigma'}(\Phi)_{\Sigma}^{A_2}$ are not isomorphic. However, provided that (such as in the example given above) the declaration does not contain any constraints or only constrains newly declared sorts as described in Sect. 6.5 respectively Sect. 6.6, the 'number' of models is not 'changed': the extension class of each given model indeed contains an initial/final algebra; the resulting data type consists of the initial/final algebras of all extension classes.

**Vertical Composition (Extension)**

Static Semantics:

$$\frac{se, \Sigma \vdash SE_1 : \mathsf{specexp}(\Sigma_1) \quad se, \Sigma_1 \vdash SE_2 : \mathsf{specexp}(\Sigma')}{se, \Sigma \vdash SE_1 \textbf{ then } SE_2 : \mathsf{specexp}(\Sigma')}$$

Model Semantics:

$$[\![ (SE_1^{\Sigma, \Sigma_1} \textbf{ then } SE_2^{\Sigma_1, \Sigma'})^{\Sigma, \Sigma'} ]\!](me, \mathcal{M}) :=$$

$$\text{let } \mathcal{M}_1 = [\![ SE_1^{\Sigma, \Sigma_1} ]\!](me, \mathcal{M}) \text{ in } [\![ SE_2^{\Sigma_1, \Sigma'} ]\!](me, \mathcal{M}_1)$$

In a *vertical composition (extension)* $SE_1$ **then** $SE_2$, the given type is first extended by $SE_1$ and the resulting type is extended by $SE_2$ to give the overall result (see the left diagram in Fig. 6.13). For instance, given the type $\mathcal{M}$ resulting from the specification expression

**sort** Elem

which introduces sort Elem, the specification

**free type** Nat := 0 | +1(Nat)
**then free** {
  **type** List := empty | cons(Elem,List)

**Fig. 6.13** Vertical
composition (left) versus
horizontal composition
(right)

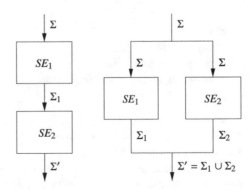

```
      fun len:List→Nat
      axiom len(empty)=0
      axiom ∀e:Elem,l:List. len(cons(e,l))=+1(len(l))
   }
```

first extends $\mathcal{M}$ by introducing sort Nat with corresponding constructors; the resulting data type is then used and extended in the second specification by a sort List with
corresponding constructors on Nat and a function len from List to Nat.

The static semantics demonstrates how the original signature $\Sigma$ is extended by $SE_1$
to $\Sigma_1$ which is used by $SE_2$ and extended to the result signature $\Sigma'$.

Correspondingly, the model semantics shows how the original class $\mathcal{M}$ of $\Sigma$-
algebras is extended by $SE_1$ to the class $\mathcal{M}_1$ of $\Sigma_1$-algebras which is then extended
by $SE_2$ to the resulting class of $\Sigma'$-algebras.

**Horizontal Composition (Union)**

Static Semantics:

$$\frac{se, \Sigma \vdash SE_1 : \mathsf{specexp}(\Sigma_1) \quad se, \Sigma \vdash SE_2 : \mathsf{specexp}(\Sigma_2)}{se, \Sigma \vdash SE_1 \textbf{ and } SE_2 : \mathsf{specexp}(\Sigma')}$$
$$\Sigma' = \Sigma_1 \cup \Sigma_2 \quad \Sigma' \text{ is a signature}$$

Model Semantics:

$$[\![ (SE_1^{\Sigma, \Sigma_1} \textbf{ and } SE_2^{\Sigma, \Sigma_2})^{\Sigma, \Sigma'} ]\!] (me, \mathcal{M}) :=$$
$$\left\{ A \in Alg(\Sigma') \;\middle|\; A|_{\Sigma_1} \in [\![ SE_1^{\Sigma, \Sigma_1} ]\!] (me, \mathcal{M}) \wedge A|_{\Sigma_2} \in [\![ SE_2^{\Sigma, \Sigma_2} ]\!] (me, \mathcal{M}) \right\}$$

In a *horizontal composition (union)* $SE_1$ **and** $SE_2$, the given type is extended 'in
parallel' by $SE_1$ and $SE_2$ (which do not 'see' each other) and the results are combined
(see the right diagram in Fig. 6.13). For instance, the specification

```
   (
      free type Bool := True | False
      and free type Nat := 0 | +1(Nat)
```

```
)
then {
  fun  bool:Nat→Bool
  axiom bool(0)=False
  axiom ∀n:Nat. bool(+1(n))=True
}
```

extends any type $\mathcal{M}$ first by introducing separate sorts Bool and Nat with corresponding constructors; the resulting types are then combined and both used in the subsequent specification.

The static semantics shows that the signatures $\Sigma_1$ and $\Sigma_2$ arising from $SE_1$ and $SE_2$ must be 'compatible' in the sense that their combination $\Sigma'$ is again a well-formed signature; the model semantics indicates that the resulting data type consists of every $\Sigma'$-specification that can be 'decomposed' in a $\Sigma_1$-algebra $A_1$ that is contained in the data type arising from $SE_1$ and a $\Sigma_2$-algebra $A_2$ that is contained in the data type arising from $SE_2$. If $\Sigma_1$ and $\Sigma_2$ are not disjoint, this implies that the entities in the overlapping parts must not have a mutually exclusive semantics, because otherwise the resulting data type is empty. For instance, the specification

```
free type Nat := 0 | +1(Nat)
then
(
  {
    const value:Nat
    axiom value=0 ∨ value=+1(0)
  }
  and
  {
    const value:Nat
    axiom value≠0
  }
)
```

specifies a non-empty data type where the equation value=+1(0) holds; however, the specification

```
free type Nat := 0 | +1(Nat)
then
(
  {
    const value:Nat
    axiom value=0
  }
  and
  {
    const value:Nat
    axiom value=+1(0)
```

```
  }
)
```

is inconsistent, i.e., it specifies an empty data type: the equation $0 = +1(0)$ cannot hold in any algebra of the resulting type because of the **free** specification of Nat with constructors 0 and 1.

### Reduction

Static Semantics:

$$\frac{se, \Sigma \vdash SE\colon \mathsf{specexp}(\Sigma_1) \quad \Sigma, \Sigma_1 \vdash EI\colon \mathsf{export}(\Sigma')}{se, \Sigma \vdash SE\ \textbf{export}\ EI\colon \mathsf{specexp}(\Sigma')}$$

Model Semantics:

$$[\![\, (SE^{\Sigma, \Sigma_1}\ \textbf{export}\ EI)^{\Sigma, \Sigma'} \,]\!](me, \mathcal{M}) :=$$

$$\left\{ A' \in Alg(\Sigma') \mid \exists A \in [\![\, SE^{\Sigma, \Sigma_1} \,]\!](me, \mathcal{M}).\ A|_{\Sigma'} = A' \right\}$$

In a *reduction SE* **export** *EI*, the signature $\Sigma_1$ resulting from the specification *SE* is restricted to a signature $\Sigma' \subseteq \Sigma_1$ that includes only the entities declared in *EI* and thus 'hides' from the outside world all entities of $\Sigma_1$ that are not listed in *EI*. For instance, the specification

```
free {
    type Nat := 0 | +1(Nat)
    type List:= empty | cons(Nat,List)
    pred notNull⊆List
    axiom notNull(empty)
    axiom ∀n:Nat,l:List. notNull(cons(+1(n),l)) ⇐ notNull(l)
}
export {
    sort Nat, const 0, fun +1(Nat)
    sort List, const empty, fun cons(Nat,List)
}
```

hides the predicate notNull which restricts the type of lists to those that do not contain value 0; the resulting signature only contains the elements listed in the **export** clause.

The static semantics assumes for this purpose a judgement

$$\Sigma, \Sigma_1 \vdash EI\colon \mathsf{export}(\Sigma')$$

which states that the exported items listed in *EI* are declared in signature $\Sigma_1$ (which extends $\Sigma$) but not in signature $\Sigma$ and that the combination of these entities together with all entities in the current signature $\Sigma$ yields a well-formed signature $\Sigma'$. While the intuition is simple, the formal details are tedious to elaborate; we thus omit the rules for this judgement.

The model semantics contains every $\Sigma'$-algebra $A'$ that results from the restriction of some $\Sigma$-algebra $A$ in the type specified by *SE*.

**Translation**

Static Semantics:

$$\frac{se,\ \Sigma \vdash SE\colon \mathsf{specexp}(\Sigma_1)\quad \Sigma,\ \Sigma_1 \vdash RI\colon \mathsf{rename}(\Sigma',\mu\colon \Sigma_1 \to \Sigma')}{se,\ \Sigma \vdash SE\ \textbf{with}\ RI\colon \mathsf{specexp}(\Sigma')}$$

Model Semantics:

$$[\![\,(SE^{\Sigma,\Sigma_1}\ \textbf{with}\ RI^{\,\mu\colon \Sigma_1 \to \Sigma'})^{\Sigma,\Sigma'}\,]\!](me,\mathcal{M}) :=$$
$$\{A' \in Alg(\Sigma')\mid A'|_\mu \in [\![\,SE^{\Sigma,\Sigma_1}\,]\!](me,\mathcal{M})\}$$

In a *translation SE* **with** *RI*, the signature $\Sigma_1$ resulting from the specification *SE* is translated to a signature $\Sigma'$ according to the renamings listed in *RI*. For instance, the specification

```
free {
  type Nat := 0 | +1(Nat)
  type List:= empty | cons(Nat,List)
}
with {
  sort Nat↦Number, const 0↦zero, fun +1(Nat)↦inc
  const empty↦nil
}
```

results in the same signature and type as the specification

```
free {
  type Number := zero | inc(Number)
  type List:= nil | cons(Number,List)
}
```

The static semantics assumes for this purpose a judgement

$$\Sigma,\ \Sigma_1 \vdash RI\colon \mathsf{rename}(\Sigma',\mu\colon \Sigma_1 \to \Sigma')$$

which states that the entities being renamed in *RI* are declared in signature $\Sigma_1$ (which extends $\Sigma$); the translation results in a new signature $\Sigma'$ by the application of a *signature morphism* $\mu\colon \Sigma_1 \to \Sigma'$, a function that maps every entity of $\Sigma_1$ to the corresponding entity of $\Sigma'$ (if the entity is not listed in *RI*, it is mapped to itself). Since the intuition about this notion is simple to understand but the details are tedious to elaborate, we omit the exact definition of a signature morphism and the rules for this judgement.

The model semantics is based on the notion $A'|_\mu$ which, for $\Sigma'$-algebra $A'$ and signature morphism $\mu\colon \Sigma_1 \to \Sigma'$, denotes that $\Sigma_1$-algebra $A$ that interprets its sorts and operations in the same way as $A'$ interprets its corresponding sorts and operations (according to the translation determined by $\mu$); again we omit the exact definition. The abstract data type denoted by the signature translation consists of all algebras $A'$ that can be derived in this way from the algebras $A$ in the type specified by *SE*.

**Specification Instantiations**

We defer the discussion of the *specification instantiation I*[*SE* **fit** *RI*] until the concept of *specification definitions* will have been introduced below.

**Specification Definitions**

A specification definition *SD* essentially denotes a set of named abstract data types, i.e., a partial mapping *me'* of identifiers to types. Actually, it is evaluated in an environment *me* that was already established by previous definitions and extends this environment by new mappings; the signatures of the types are captured by corresponding environments *se* and *se'*. These considerations give rise to the following semantics:

- A static semantics with judgements of form

$$se \vdash SD: \mathsf{spec}(se')$$

  which states that in an environment *se* that maps type names to signatures, the definition *SD* is well-formed and extends *se* to *se'*.
- A model semantics with valuations of form

$$[\![\, SD \,]\!](me) = me'$$

  which states that the well-formed definition *SD* extends the environment *me* that maps names to types to another such environment *me'*.

The valuation function of the model semantics could be declared as

$$[\![\, . \,]\!]: ModelClassEnv \to ModelClassEnv$$

i.e., as a function from type environments to type environments, but since abstract data types are in general proper classes, not sets, this would represent an abuse of set-theoretic notation.

**Definition Sequences**

Static Semantics:

$$\frac{se \vdash SD_1: \mathsf{spec}(se_1) \quad se_1 \vdash SD_2: \mathsf{spec}(se')}{se \vdash SD_1 \, SD_2: \mathsf{spec}(se')}$$

Model Semantics:

$$[\![\, SD_1 \, SD_2 \,]\!](me) := [\![\, SD_2 \,]\!]([\![\, SD_1 \,]\!](me))$$

The semantics of a sequence $SD_1 \, SD_2$ of definitions is almost trivial: in the static semantics, the definition $SD_1$ extends the given signature environment $se$ to $se_2$ which in turn is extended by $SD_2$ to the resulting environment $se'$. In the model semantics, the same is done for the corresponding type environments. For instance, given an environment that includes type identifier $A$, the definition

> **spec** $B := \ldots$
> **spec** $C :=$

first extends the environment to one that includes both $A$ and $B$; this environment is subsequently extended to one that includes $A$, $B$, and $C$.

## Generic Specifications

Static Semantics:

$$
\frac{
\begin{array}{c}
\neg \exists \mathfrak{S}. \; \langle I, \mathfrak{S} \rangle \in se \\
se, \Sigma_0 \vdash SE_2 : \mathsf{specexp}(\Sigma_i) \quad se, \Sigma_i \vdash SE_1 : \mathsf{specexp}(\Sigma_p) \\
se, \Sigma_p \vdash SE_3 : \mathsf{specexp}(\Sigma_e) \\
\mathfrak{S} = \langle \mathrm{i}(\Sigma_i), p(\Sigma_p), \mathrm{e}(\Sigma_e) \rangle \quad se' = se \cup \{\langle I, \mathfrak{S} \rangle\}
\end{array}
}{
se \vdash \textbf{spec } I[SE_1] \textbf{ import } SE_2 := SE_3 : \mathsf{spec}(se')
}
$$

Model Semantics:

$$
\begin{aligned}
&[\![\, \textbf{spec } I[SE_1^{\Sigma_i, \Sigma_p}] \textbf{ import } SE_2^{\Sigma_0, \Sigma_i} := SE_3^{\Sigma_p, \Sigma_e} \,]\!](me) := \\
&\quad \text{let } \mathcal{MI} = [\![\, SE_2^{\Sigma_0, \Sigma_i} \,]\!](me, \mathcal{M}_0) \\
&\quad \mathcal{MP} = [\![\, SE_1^{\Sigma_i, \Sigma_p} \,]\!](me, \mathcal{MI}) \\
&\quad \mathcal{ME} = [\![\, SE_3^{\Sigma_p, \Sigma_e} \,]\!](me, \mathcal{MP}) \\
&\quad \text{in } me \cup \{\langle I, \mathcal{ME} \rangle\}
\end{aligned}
$$

A *generic specification* is a definition with a 'parameter specification' that can be instantiated by an 'argument specification'; the definition also contains an 'import specification' which describes a fixed specification from which both the parameter and the argument inherit.

In its simplest (non-generic) form, such a definition can be written as

> **spec** $I := SE_3$

which is interpreted as

> **spec** $I[\_]$ **import** $\_ := SE_3$

i.e., the parameter specification and the import specification are both empty. The static semantics first ensures that $I$ is not defined in the current environment $se$. It then propagates the *empty signature* $\Sigma_0$ to become the 'import signature' $\Sigma_i = \Sigma_0$

and the 'parameter signature' $\Sigma_p = \Sigma_i = \Sigma_0$. Now it evaluates the specification body $SE_3$ to extend $\Sigma_p$ to the 'export signature' $\Sigma_e$; since $\Sigma_p = \Sigma_0$, i.e., $\Sigma_p$ is empty, the specification body $SE_3$ must be fully self-contained. Finally, the current environment $se$ is extended to the new environment $se'$ which binds type name $I$ to a triple $\mathfrak{S}$ ('German S') of signatures whose only non-empty component is $\mathfrak{S}.e = \Sigma_e$.

The model semantics correspondingly starts with the *empty data type* $\mathcal{M}_0$ whose only element is the unique 'empty' $\Sigma_0$-algebra; this data type is propagated to the 'import type' $\mathcal{MI} = \mathcal{M}_0$ of $\Sigma_i$-algebras and the 'parameter type' $\mathcal{MP} = \mathcal{MI} = \mathcal{M}_0$ of $\Sigma_p$-algebras; this empty parameter type $\mathcal{MP} = \mathcal{M}_0$ is extended by $SE_3$ to the 'export type' $\mathcal{ME}$ of $\Sigma_e$-algebras; finally the current type environment $me$ is extended by a mapping of type name $I$ to $ME$; this type thus represents the actual 'meaning' of the specification.

Now consider a definition of form

> **spec** $I[SE_1]$ **import** _ $:= SE_3$

which also includes a non-trivial parameter expression $SE_1$. A simple example of such a definition is

> **spec** LIST[**sort** Elem] :=
> **free type** List := empty | cons(Elem,List)

whose parameter loosely specifies a sort Elem while the body freely specifies a sort List of Elem values with the usual constructors. For this kind of specification, the parameter signature $\Sigma_p$ contains the sorts/operations declared in $SE_1$ which is extended by the specification body to the export signature $\Sigma_e$ that includes the entities of $SE_1$, as well as the entities of $SE_3$, i.e., in our example Elem, List, empty, and cons. The export signature $\Sigma_e$ and export type $\mathcal{ME}$ is thus the same as that of the definition

> **spec** LIST :=
> **sort** Elem
> **then free type** List := empty | cons(Elem,List)

However, as we will see in the next subsection, the application possibilities of both definitions are different; in particular, the signature environment $se'$ binds type name $I$ to a triple $\mathfrak{S}$ that records the parameter signature in $\mathfrak{S}.p = \Sigma_p$.

Finally, let us consider a general definition

> **spec** $I[SE_1]$ **import** $SE_2$ $:= SE_3$

which also includes a non-trivial import expression $SE_2$. An example of such a definition is

> **spec** LIST[**sort** Elem, **fun** nat:Elem→ Nat]
> **import**
> **free type** Nat := 0 | +1(Nat)
> **then** {

```
      fun add:Nat×Nat→Nat
      axiom ∀n1:Nat,n2:Nat. add(0,n2)=n2 ∧ add(+1(n1),n2)=+1(add(n1,n2))
   }
:=
  free {
    type List := empty | cons(Elem,List)
    fun sum:List→Nat
    axiom sum(empty)=0
    axiom ∀e:Elem,l:List. sum(cons(e,l))=add(nat(e),sum(l))
  }
```

Here the **import** clause freely specifies a sort Nat with constructors 0 and +1 extended by a loosely specified operation add. This sort is used both by the parameter (which in addition to sort Elem specifies a function nat from Elem to Nat) and the body of the specification (which introduces a function sum from List to Nat that is axiomatized with the help of add and nat).

In this kind of definition, the import signature $\Sigma_i$ contains the sorts/operations declared in $SE_2$. The static semantics first extends this signature by the parameter $SE_1$ to the parameter signature $\Sigma_p$ and then by the specification body $SE_3$ to the export signature $\Sigma_e$; in our example this signature thus contains also Nat, 0, +1, and add. Correspondingly the model semantics gradually expands the import type $\mathcal{MI}$ to the parameter type $\mathcal{MP}$ and the export type $\mathcal{ME}$. The signature environment $se'$ binds type name $I$ to a triple $\mathfrak{S}$ that records the import signature in $\mathfrak{S}.i = \Sigma_i$.

The purpose of the **import** clause and the difference to the parameter clause will become clear by the following subsection where we return to the deferred elaboration of the application of named specifications.

### Specification Instantiations: Informal Description

In a *specification instantiation* $I[SE$ **fit** $RI]$ (whose static and model semantics will be given below), the formal parameter in the specification denoted by name $I$ is replaced by the actual argument specification $SE$ according to the mapping of parameter sorts and operations to argument sorts and operations denoted by the renaming $RI$; the resulting specification extends the current signature/type correspondingly. For instance, given the generic specification

```
spec LIST[sort Elem] :=
  free type List := empty | cons(Elem,List)
```

the application

```
LIST[free type Nat := 0 | +1(Nat) fit Elem↦Nat ]
```

has the same meaning as the specification

```
free type Nat := 0 | +1(Nat)
then free type List := empty | cons(Nat,List)
```

Thus in general every specification instantiation leads to different sorts and operations; if they shall be used in the same context we thus have to rename sorts (to

avoid improper identification) and constants (to avoid conflicting declarations). Due
to overloading, functions and predicates need typically not renamed provided that
their arities are different. For instance, the specification

> LIST[**free type** Nat := 0 | +1(Nat) **fit** Elem↦Nat ]
>   **with** List↦NatList, empty↦natEmpty
> **and** LIST[**free type** Bool := True | False **fit** Elem↦Bool ]
>   **with** List↦BoolList, empty↦boolEmpty

has the same meaning as the specification

> (
>   **free type** Nat := 0 | +1(Nat)
>   **then free type** NatList := natEmpty | cons(Nat,NatList)
> )
> **and**
> (
>   **free type** Bool := True | False
>   **then free type** BoolList := boolEmpty | cons(Bool,BoolList)
> )

The parameter of a generic specification may also contain axioms that restrict the
legal instantiations. For instance, the specification

> **spec** LIST[MONOID] :=
>   **type** List := empty | cons(Elem,List)

where MONOID is the specification given in Sect. 6.1 may be only instantiated with
a type that provides operations that satisfy the corresponding axioms. Now, given
the specification NAT from the introduction, the instantiation

> **spec** LIST[NAT **fit** Elem↦Nat,op↦+,isE↦is0]

is a valid instantiation that denotes the type of lists of natural numbers. If the argument
violates the axioms satisfied by the parameters, the resulting data type is empty, i.e.,
has no models.

While the actual argument may vary with each application of a parameterized
specification, we sometimes would like to make sure that all applications share the
interpretation of some sorts and operations; the role of the **import** clause is to ensure
such a shared interpretation. Take for instance the already previously given generic
specification

> **spec** LIST[**sort** Elem, **fun** nat:Elem→Nat] **import** NAT :=
>   **free** {
>     **type** List := empty | cons(Elem,List)
>     **fun** sum:List→Nat
>     **axiom** sum(empty)=0
>     **axiom** ∀e:Elem,l:List. sum(cons(e,l))=add(nat(e),sum(l))
>   }

Here the idea is that the interpretation of sort Nat and operations 0, +1, and Nat should be shared among the different applications of LIST. Thus the specification

```
LIST[
  free type Bool := True | False
  then {
    fun nat:Bool→Nat
    axiom nat(True)=1 ∧ nat(False)=0
  }
  fit Elem↦Bool ]
  with List↦BoolList, empty↦boolEmpty
and LIST[
  free type Symbol := A | B | C
  then {
    fun nat:Symbol→Nat
    axiom nat(A)=0 ∧ nat(B)=1 ∧ nat(C)=2
  }
  fit Elem↦Symbol ]
  with List↦SymbolList, empty↦symbolEmpty
```

has the same meaning as

```
free type Nat := 0 | +1(Nat)
then {
  fun add:Nat×Nat→Nat
  axiom ∀n1:Nat,n2:Nat. add(0,n2)=n2 ∧ add(+1(n1),n2)=+1(add(n1,n2))
}
then
(
  (
    free type Bool := True | False
    then {
      fun nat:Bool→Nat
      axiom nat(True)=1 ∧ nat(False)=0
    }
    then free {
      type BoolList := boolEmpty | cons(Bool,BoolList)
      fun sum:BoolList→Nat
      axiom sum(boolEmpty)=0
      axiom ∀e:Bool,l:BoolList. sum(cons(e,l))=add(nat(e),sum(l))
    }
  )
and
  (
    free type Symbol := A | B | C
    then {
      fun nat:Symbol→Nat
```

       **axiom** nat(A)=0 ∧ nat(B)=1 ∧ nat(C)=2
   }
   **then free** {
       **type** SymbolList := symbolEmpty | cons(Symbol,SymbolList)
       **fun** sum:SymbolList→Nat
       **axiom** sum(symbolEmpty)=0
       **axiom** ∀e:Symbol,l:SymbolList. sum(cons(e,l))=add(nat(e),sum(l))
   }
 )
)

Since both versions of sum thus have a common understanding of Nat, in the context set up by above specification an application

    add(sum(cons(True,boolEmpty)), sum(cons(A,cons(B,symbolEmpty))))

is possible.

Based on this intuitive understanding of specification instantiations, we now elaborate their formal semantics.

**Specification Instantiations: Formal Semantics**

Static Semantics:

$$\langle I, \mathfrak{S}\rangle \in se$$
$$se, \mathfrak{S}.\mathsf{i} \vdash SE\colon \mathsf{specexp}(\Sigma_a) \quad \mathfrak{S}.\mathsf{i}, \mathfrak{S}.\mathsf{p}, \Sigma_a \vdash RI\colon \mathsf{fit}(\mu\colon \mathfrak{S}.\mathsf{p} \to \Sigma_a)$$
$$\langle e'\colon \Sigma_a \to \Sigma_r, \mu'\colon \mathfrak{S}.\mathsf{e} \to \Sigma_r\rangle \text{ is the pushout of}$$
$$\langle \mu\colon \mathfrak{S}.\mathsf{p} \to \Sigma_a, e\colon \mathfrak{S}.\mathsf{p} \to \mathfrak{S}.\mathsf{e}\rangle \text{ for the extension morphisms}$$
$$e\colon \mathfrak{S}.\mathsf{p} \to \mathfrak{S}.\mathsf{e} \text{ and and } e'\colon \Sigma_a \to \Sigma_r$$
$$\Sigma' = \Sigma \cup \Sigma_r \quad \Sigma' \text{ is a signature}$$

$$\overline{\qquad se, \Sigma \vdash I[SE \textbf{ fit } RI]\colon \mathsf{specexp}(\Sigma') \qquad}$$

Model Semantics:

$$[\![ (I[SE^{\Sigma,\Sigma_a} \textbf{ fit } RI^{\,e'\colon \Sigma_a \to \Sigma_r, \mu'\colon \Sigma_e \to \Sigma_r})^{\Sigma,\Sigma'} ]\!](me, \mathcal{M}) :=$$
$$\{ A' \in Alg(\Sigma') \mid (A'|_{\Sigma_r})|_{e'} \in [\![ SE^{\Sigma,\Sigma_a} ]\!](me, \mathcal{M}) \wedge (A'|_{\Sigma_r})|_{\mu'} \in me(I)\}$$

The static semantics first extracts from the current signature environment *se* the signature triple $\mathfrak{S}$ associated to specification name $I$ consisting of the import signature $\mathfrak{S}.\mathsf{i}$, the parameter signature $\mathfrak{S}.\mathsf{p}$ and the exported signature $\mathfrak{S}.\mathsf{e}$ of the generic specification. Then it evaluates the argument expression $SE$ in the context set up by the import signature $\mathfrak{S}.\mathsf{i}$ which is thus extended to an argument signature $\Sigma_a$. The judgement

$$\mathfrak{S}.\mathsf{i}, \mathfrak{S}.\mathsf{p}, \Sigma_a \vdash RI\colon \mathsf{fit}(\mu\colon \mathfrak{S}.\mathsf{p} \to \Sigma_a)$$

**Fig. 6.14** The pushout
semantics of specification
instantiations

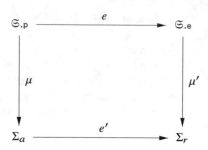

states that the entities being renamed in *RI* are declared in the parameter signa-
ture $\mathfrak{S}$.p (which also extends the import signature $\mathfrak{S}$.i) but not in $\mathfrak{S}$.i and that they
are renamed to entities declared in the argument signature $\Sigma_a$; it thus determines
the *fitting morphism* $\mu$: $\mathfrak{S}$.p $\rightarrow \Sigma_a$ that maps the parameter signature $\mathfrak{S}$.p to the
argument signature $\Sigma_a$ (if an entity is not listed in *RI*, it is mapped to itself). Again
we omit the rules for this judgement.

Furthermore, the parameter signature $\mathfrak{S}$.p and the exported signature $\mathfrak{S}$.e (which
also extends $\mathfrak{S}$.p) determine a unique *extension morphism* $e$: $\mathfrak{S}$.p $\rightarrow$ $\mathfrak{S}$.e which
'embeds' $\mathfrak{S}$.p into $\mathfrak{S}$.e. Now, under a certain constraint explained below, $e$ and $\mu$
uniquely determine a *result signature* $\Sigma_r$ that extends $\Sigma_a$ (i.e., $\Sigma_a \subseteq \Sigma_r$) together
with another extension morphism $e'$: $\Sigma_a \rightarrow \Sigma_r$ and fitting morphism $\mu'$: $\mathfrak{S}$.e $\rightarrow \Sigma_r$
whose target is $\Sigma_r$. This pair $\langle e', \mu' \rangle$, called the *pushout* of $\langle \mu, e \rangle$, lets the diagram
in Fig. 6.14 'commute': the composition $\mu \circ e'$: $\mathfrak{S}$.p $\rightarrow \Sigma_r$ and the composition
$e \circ \mu'$: $\mathfrak{S}$.p $\rightarrow \Sigma_r$ both denote identical morphisms.

Intuitively, the morphism $e'$ represents the extension of the argument signature $\Sigma_a$
into the result signature $\Sigma_r$; the morphism $\mu'$ represents the translation of the exported
signature $\mathfrak{S}$.e to $\Sigma_r$. The diagram thus states that we get the same result for the
specification instantiation, independently of whether

- we first extend (via $e$) the formal parameter to the exported specification and then
  translate (via $\mu'$) this to the result, or whether
- we first translate (via $\mu$) the formal parameter to the actual argument and then
  extend (via $e'$) the argument to the result.

The result $\Sigma'$ of the specification instantiation is then the combination of the current
signature $\Sigma$ with $\Sigma_r$, provided that $\Sigma'$ is well-formed. This last constraint may be
violated, if the specification introduces constants with names that are already in use
in the current context.

Once the pushout morphisms $\langle e', \mu' \rangle$ have been determined, the model semantics
of the specification instantiation can be easily defined: the specified type consists of
every $\Sigma'$-algebra $A'$ such that two conditions hold:

- $A'$ is the result of the application of $e'$ to some $\Sigma_a$-algebra in the type denoted by
  the argument expression *SE* (i.e., it is the extension of some argument);
- $A'$ is the result of the application of $\mu'$ to some $\Sigma_e$-algebra in the type denoted by
  the generic specification *I* (i.e., it is the translation of some result).

This model semantics thus lets only those algebras of the argument type 'pass' to become part of the instantiation result that are also contained in the parameter type; consequently, if the axioms of the argument specification are inconsistent with the axioms of the parameter specification, the instantiation results in an empty type: there is no algebra $A'$ that simultaneously represents an extension of the argument and a translation of the result, since the result contains the parameter as a submodel.[2]

The pushout described above only exists (and a specification instantiation is thus only well-formed) if the argument signature $\Sigma_a$ and the translated version of the exported signature $\mathfrak{S}$.e do not have conflicting declarations that prohibit their combination to a well-formed signature $\Sigma_r$; in our specification language this might happen, if both $\Sigma_a$ and $\mathfrak{S}$.e declare a constant with the same name but different sort. In more richer specification languages, also other problems may arise that all can be resolved by the following general constraint: if the argument signature $\Sigma_a$ shares entities with the exported signature $\mathfrak{S}$.e, these entities must stem from the parameter signature $\mathfrak{S}$.p (and thus possibly from the import signature $\mathfrak{S}$.i $\subseteq$ $\mathfrak{S}$.p) that is extended by both $\Sigma_a$ and $\mathfrak{S}$.e.

Using this constraint, given a specification

**spec** NAT :=
**free type** Nat = 0 | +1(Nat)

the generic specification

**spec** LIST[**sort** Elem] :=
NAT
**then free type** List = empty | cons(Elem,List)
**then** {
  **fun** len:List→Nat
  **axiom** len(empty)=0
  **axiom** ∀e:Elem,l:List. len(cons(e,l))=+1(len(l))
}

cannot be instantiated as

LIST[NAT **fit** Elem↦Nat]

because the sort Nat appears in the argument and in the body of the specification, but not in the parameter. However, the specification can be rewritten to

**spec** LIST[**sort** Elem] **import** NAT :=
**free type** List = empty | cons(Elem,List)
**then** {
  **fun** len:List→Nat
  **axiom** len(empty)=0

---

[2]The usual intention, however, is that the argument specification 'refines' the parameter specification in that the argument type is a subset of the parameter type and thus no argument algebra is 'filtered'; in Sect. 6.8 we will discuss the question of specification refinement.

**axiom** $\forall e{:}Elem,l{:}List.\ len(cons(e,l))=+1(len(l))$
}

which allows the instantiation

LIST[**fit** Elem$\mapsto$Nat]

(remember that the argument automatically extends the imported specification).

## 6.8    Reasoning About Specifications

While we have discussed in the previous sections the formal semantics of specifications, we now turn our attention to the precise reasoning about them. Let the data
type $\mathcal{M}$ be described by a specification $SE$, i.e.,

$$\mathcal{M} = [\![\, SE \,]\!](me, \mathcal{M}_1)$$

for some environment $me$ and data type $\mathcal{M}_1$ that is extended by $SE$. Then, among
all possible questions to be addressed about $\mathcal{M}$, the following may be of particular
interest:

- *Satisfaction:* does $\mathcal{M}$ satisfy a particular property (described by a formula $F$)?

  $$\forall A \in \mathcal{M}.\ [\![\, F \,]\!]^A = \text{true}$$

- *Consistency:* is $\mathcal{M}$ not empty (i.e., is $SE$ consistent)?

  $$\mathcal{M} \neq \varnothing$$

- *Refinement:* does $\mathcal{M}$ also represent a valid implementation of another specification $SE'$ (i.e., does $SE$ refine some $SE'$)?

  $$\mathcal{M} \subseteq [\![\, SE' \,]\!](me, \mathcal{M}_1)$$

We are now going to address these questions in turn.

### The Satisfaction of Formulas

We address the first question

$$\forall A \in \mathcal{M}.\ [\![\, F \,]\!]^A = \text{true}$$

for the various kinds of 'basic' (loose, (co)generated, and (co)free) specifications and
then for the 'structured' specifications composed by the various kinds of constructs
introduced in Sect. 6.7.

**Loose Data Types**

If an abstract data type $\mathcal{M}$ is denoted by the loose interpretation of a declaration $D$ with presentation $\langle \Sigma, \Phi \rangle$, the set $\Phi = \{F_1, \ldots, F_n\}$ of axioms already describes all knowledge that is available about $\mathcal{M}$. To show that $\mathcal{M}$ satisfies a formula $F$, it suffices to show

$$F_1, \ldots, F_n \vdash F$$

i.e., that $F$ is a logical consequence of $\Phi$.

**Example 6.29**  Given the specification

**type** Bool := True | False | Not(Bool)
**axiom** ∀b:Bool.
  True ≠ False ∧
  (b = True ∨ b = False) ∧
  Not(b) ≠ b

in order to show Not(True)=False, we have to prove that

$\forall b \in Bool.$
  $True \neq False \land$
  $(b = True \lor b = False) \land$
  $(Not(b) \neq b)$

implies

$Not(True) = False$

This can be easily shown: from the axiom, we know $Not(True) = True \lor Not(True)$ $= False$, but also $Not(True) \neq True$, which implies our goal.          □

If the specification is of one of the forms **generated** $D$, **free** $D$, **cogenerated** $D$, or **cofree** $D$, i.e., it is a declaration that is interpreted in a (co-)generated or (co-)free way, we may still attempt to prove that $F$ is a logical consequence of $\Phi$. Then, however, there is additional information available which we may help to perform these proofs.

**Generated Data Types**

In the case of a specification **generated** $D$, we may assume for every sort introduced in $D$ that every value in the carrier $S$ of that sort can be denoted by the application of some function. This gives rise to the additional axiom

$$\forall x \in S. \exists. \ldots. x = f_1(\ldots) \lor \ldots \lor x = f_n(\ldots)$$

where we denote by $f_1, \ldots, f_n$ (the interpretations of) all *constructors* of S, i.e., all functions with domain $S$ introduced by $D$.

Even more, we may actually assume that every value in $S$ can be denoted by some term. This induces the proof principle of *(structural) induction* on $S$ (compare with the induction rule introduced in Sect. sect:indproof):

$$\frac{\begin{array}{l} Fs \vdash \forall x_1, x_2, \ldots \in S. \\ \quad F[x_1/x] \wedge F[x_2/x] \wedge \ldots \Rightarrow \\ \quad\quad F[f_1(\ldots, x_i, \ldots)/x] \wedge \ldots \wedge F[f_n(\ldots, x_j, \ldots)/x] \end{array}}{Fs \vdash \forall x \in S.\ F}$$

The rule essentially states that, in order to prove that property $F$ holds for every element $x$ of $S$, it suffices to prove that $F$ holds for the result of every constructor application under the assumption that it already holds for every argument of the constructor. Since $x$ is denoted by some term which is constructed from finitely many constructor applications, this indeed demonstrates that $F$ holds for $x$.

**Example 6.30** Take the specification

**generated type** Int := 0 | 1 | Plus(Int,Int) | Minus(Int,Int)

that introduces sort Int with constructors 0, 1, Plus, and Minus. Let *Int* denote the carrier of Int and 0, 1, *Plus*, *Minus* the interpretations of the constructors. Then, in order to prove for arbitrary formula $F$ the property

$$\forall x \in Int.\ F$$

it suffices to prove

$$\begin{array}{l} \forall x_1, x_2 \in Int. \\ \quad F[x_1/x] \wedge F[x_2/x] \Rightarrow \\ \quad\quad F[0/x] \wedge \\ \quad\quad F[1/x] \wedge \\ \quad\quad F[Plus(x_1, x_2)/x] \wedge \\ \quad\quad F[Minus(x_1, x_2)/x] \end{array}$$

respectively the following four equivalent conditions:

$$\begin{array}{l} F[0/x] \\ F[1/x] \\ \forall x_1, x_2 \in Int.\ F[x_1/x] \wedge F[x_2/x] \Rightarrow F[Plus(x_1, x_2)/x] \\ \forall x_1, x_2 \in Int.\ F[x_1/x] \wedge F[x_2/x] \Rightarrow F[Minus(x_1, x_2)/x] \end{array}$$

$\square$

The induction principle can be also generalized to multiple generated sorts where the constructors for one sort may also have arguments from the other sort; for instance, for sorts $S$ and $T$ with constructors $f_1, \ldots, f_n$ for $S$ and constructors $g_1, \ldots, g_m$ for $T$, we have the following induction principle:

$$Fs \vdash \forall x_1, x_2, \ldots \in S, y_1, y_2, \ldots \in T.$$
$$F[x_1/x] \wedge F[x_2/x] \wedge \ldots \wedge G[y_1/x] \wedge G[y_2/x] \wedge \ldots \Rightarrow$$
$$F[f_1(\ldots, x_i, \ldots, y_j, \ldots)/x] \wedge \ldots \wedge F[f_n(\ldots, x_k, \ldots, y_l, \ldots)/x] \wedge$$
$$\frac{G[g_1(\ldots, x_o, \ldots, y_p, \ldots)/x] \wedge \ldots \wedge G[g_m(\ldots, x_q, \ldots, y_r, \ldots)/x]}{Fs \vdash (\forall x \in S. \ F) \wedge (\forall x \in T. \ G)}$$

**Example 6.31** Take the specification

**generated** {
    **type** Int := 0 | 1 | Plus(Int,Int) | Minus(Int,Int) | If(Bool,Int,Int)
    **type** Bool := T | Not(Bool) | And(Bool,Bool) | Eq(Int,Int)
}

introducing the generated sorts Int and Bool. In order to prove for the carriers *Int* of Int and *Bool* of Bool the property

$$(\forall i \in Int. \ F) \wedge (\forall b \in Bool. \ G)$$

it suffices to prove the following conditions:

$$F[0/i]$$
$$F[1/i]$$
$$G[T/b]$$
$$\forall i_1, i_2 \in Int, b_1, b2 \in Bool.$$
$$F[i_1/i] \wedge F[i_2/i] \wedge G[b_1/b] \wedge G[b_2/b] \Rightarrow$$
$$F[Plus(i_1, i_2)/i] \wedge F[Minus(i_1, i_2)/i] \wedge F[\mathit{If}(b_1, i_1, i_2)/i] \wedge$$
$$G[Not(b_1)/b] \wedge G[And(b_1, b_2)/b] \wedge G[Eq(i_1, i_2)/b]$$

$\square$

**Free Data Types**

For a specification **free** $D$ with a set $\Phi$ of Horn-clauses as axioms, we may assume (since every data type specified in this way is also generated) that all the axioms and inference rules induced by a specification **generated** $D$ also are applicable. To prove that the specified data type $\mathcal{M}$ satisfies a formula $F$, we may thus, as in the case of a specification **generated** $D$, try to show that $F$ is implied by the formulas in $\Phi$ with the help of the proof principle of induction on the constructors of the specified sorts.

If this does not suffice, however, we may also apply the knowledge that $\mathcal{M}$ is monomorphic, i.e., it consists of a single isomorphism class which contains the quotient term algebra $Term_\Sigma^\Phi$ as the canonical representative; it thus also suffices to show $[\![\,F\,]\!]^{Term_\Sigma^\Phi} = $ true. Since working on this 'semantic' level, however, is more involved, we prefer to remain on the 'logic' level by adding further knowledge that may be used in the proof. In fact, from the knowledge that the formulas are interpreted in the quotient term algebra, we may deduce that only those equalities and predicates hold that arise from the axioms in $D$.

In particular, if $D$ does not contain any axioms, as in a specification expression

**free type** $S := f_1(\ldots) \mid \ldots \mid f_n(\ldots)$

with $n$ constructors $f_1, \ldots, f_n$, the carriers of the quotient term algebra only contain singleton sets, i.e., every term composed from the constructors denotes a different value. The quotient term algebra is thus isomorphic to a plain term algebra that maps every term to itself, i.e., every value can be identified with a term that only consists of constructor applications.

This knowledge can be described by an additional axiom with $n + n \cdot (n-1)$ subformulas that state that the constructors are injective (i.e., two applications of a constructor yield identical results only if the arguments are identical) and that application of different constructors always yield different results:

$\forall x, y, \ldots$.
$\quad (f_1(x, \ldots) = f_1(y, \ldots) \Rightarrow x = y \land \ldots) \land$
$\quad \ldots \land$
$\quad (f_n(x, \ldots) = f_n(y, \ldots) \Rightarrow x = y \land \ldots) \land$
$\quad f_1(\ldots) \neq f_2(\ldots) \land$
$\quad \ldots \land$
$\quad f_{n-1}(\ldots) \neq f_n(\ldots)$

In order to prove $F$, we thus may make use of this additional axiom.

**Example 6.32** The specification

**free type** Nat $:= 0 \mid +1(\text{Nat})$

induces the axiom

$\forall x, y \in Nat. \ (+1(x) = +1(y) \Rightarrow x = y) \land 0 \neq +1(x)$

for every model with carrier $Nat$ for sort Nat and interpretations $0$ and $+1$ for the constructors $0$ and $+1$, respectively. From this, e.g. we can deduce

$+1(0) \neq +1(+1(0))$

If this were not the case, then from the first subaxiom we could deduce $0 = +1(0)$ which however would contradict the second subaxiom.                                 □

The fact that free type declarations without axioms represent term algebras has also an important consequence because it allows the introduction of new operations by 'recursive definitions' that apply 'pattern matching' against constructor terms; this will be further elaborated later when dealing with the consistency of specifications.

However, the situation gets considerably more complicated, if we also consider free specifications that include axioms.

**Example 6.33** Take the free specification

> **free** {
>   **type** Nat := 0 | +1(Nat)
>   **type** Int := I(Nat,Nat)
>   **axiom**
>     ∀p:Nat, m:Nat.  I(+1(p),+1(m))=I(p,m)
> }

which introduces a sort Nat with constructors 0 and +1 and a sort Int with constructor $I$ that maps a pair of Nat values to an Int value. The core idea here is that the Int value corresponds to the difference of the Nat values, i.e., the integers $-1$ is e.g. represented by $I(2, 3)$. The axiom states that the representation is not unique, e.g. $I(2, 3) = I(1, 2) = I(0, 1)$; by the axiom every Int value can be thus reduced to a value $I(p, m)$ where $p = 0$ or $m = 0$ (or both, $I(0, 0)$ represents the value zero).

The carrier of the quotient term algebra for sort Int consists only of infinite sets such as the set

$$\{I(0,0), I(+1(0),+1(0)), I(+1(+1(0)), +1(+1(0))), \ldots\}$$

which represents the integer 0. In general, every integer $i \in \mathbb{Z}$ is represented by the infinite set

$$\left\{I(+1^p(0),+1^m(0)) \mid p \in \mathbb{N} \wedge m \in \mathbb{N} \wedge p - m = i\right\}$$

where, for every $n \in \mathbb{N}$, the term $+1^n(0)$ represents the $n$-fold application of constructor +1 to constant 0. □

For reasoning about free specifications with axioms, we generally indeed have to resort to the semantic level as discussed above by choosing a representative algebra of the monomorphic data type denoted by the specification. However, to reason about quotient term algebras whose carriers hold sets of terms, as in the example above, may get complex. Fortunately, however, it is often very easily possible to extract from each such set a single element as the set's characteristic representative. We may then replace the quotient term algebra as the canonical model of a free specification by an algebra whose carriers consist of these 'characteristic terms'; thus we may subsequently reason about an algebra of terms (as in the case of a free algebra without axioms) rather than about an algebra of sets of terms. In the following, we are going to elaborate this idea.

**Definition 6.31** (*Characteristic term algebra*) Let $\langle \Sigma, \Phi \rangle$ be a presentation. A $\Sigma$-algebra $C$ is a *characteristic term algebra* for $\langle \Sigma, \Phi \rangle$ if the following conditions hold:

1. $C(S) \subseteq Term_{\Sigma}^S$ for each sort $S \in \Sigma.s$;
2. $(\wedge \Phi) \Rightarrow C(T) = T$ is valid for each term $T \in Term_{\Sigma}^S$ and sort $S \in \Sigma.s$;
3. $C \models_{\Sigma} \Phi$.

Here $(\wedge \Phi)$ denotes $F_1 \wedge \ldots \wedge F_n$ for $\Phi = \{F_1, \ldots, F_n\}$.

A characteristic term algebra $C$ for $\langle \Sigma, \Phi \rangle$ thus maps every term to a term that is equal to the argument term according to the axioms in $\Phi$; furthermore $C$ itself is a $\Sigma$-model of $\Phi$. The relevance of such an algebra is given by the following theorem.

**Proposition 6.16** (Characteristic term algebra and quotient term algebra) *Let* $\langle \Sigma, \Phi \rangle$ *be a presentation and $C$ be a characteristic term algebra for* $\langle \Sigma, \Phi \rangle$. *Then $C$ is isomorphic to the quotient term algebra of $\Sigma$ with respect to $\Phi$, i.e.,* $C \simeq_{\Sigma} Term_{\Sigma}^{\Phi}$.

**Proof** To show $C \simeq_{\Sigma} Term_{\Sigma}^{\Phi}$, it suffices to show that $C$ is generated, $C \models_{\Sigma} \Phi$, and that the evaluation homomorphism $h \colon Term_{\Sigma}^{\Phi} \to_{\Sigma} C$, $h([T]_{\Sigma}^{\Phi}) = C(T)$ is injective. The first two conditions hold by the definition of $C$. As for the last condition, we take arbitrary sort $S \in \Sigma.s$ and $T_1, T_2 \in Term_{\Sigma}^S$ with $C(T_1) = C(T_2)$. We show $[T_1]_{\Sigma}^{\Phi} = [T_2]_{\Sigma}^{\Phi}$. For this, it suffices to show that $\wedge(\Phi) \Rightarrow T_1 = T_2$ is valid, which holds, because the definition of $C$ implies $\wedge(\Phi) \Rightarrow C(T_1) = T_1$ and $\wedge(\Phi) \Rightarrow C(T_2) = T_2$ and we have assumed $C(T_1) = C(T_2)$.                                              □

A characteristic term algebra $C$ may thus serve as another canonical model of a free specification. In particular, in order to prove that the abstract data type $\mathcal{M}$ described by the specification satisfies a property $F$, we thus only have to prove $[\![ F ]\!]^C = true$. Such a proof may profit from the knowledge about the structure of the terms in the carriers of $C$.

**Example 6.34** For the specification of sorts Nat and Int given in Example 6.33, we may choose the characteristic term algebra $C$ defined as

$$C(\text{Nat}) = \{+1^n(0) \mid n \in \mathbb{N}\}$$
$$C(0) = 0$$
$$C(+1) = \lambda T \in C(\text{Nat}). +1(T)$$
$$C(\text{Int}) = \{I(0,0)\} \cup \{I(+1^n(0),0) \mid n \in \mathbb{N}_{\geq 1}\} \cup \{I(0,+1^n(0)) \mid n \in \mathbb{N}_{\geq 1}\}$$
$$C(I) = \lambda N_1 \in C(\text{Nat}), N_2 \in C(\text{Nat}).$$

> let $n_1, n_2 = \text{choose } n_1 \in \mathbb{N}, n_2 \in \mathbb{N}. N_1 = +1^{n_1}(0) \wedge N_2 = +1^{n_2}(0)$ in
>
> if $n_1 = n_2$ then
>
>> $I(0,0)$
>
> else if $n_1 > n_2$ then
>
>> $I(+1^{n_1 - n_2}(0),0)$
>
> else
>
>> $I(0,+1^{n_2 - n_1}(0))$

The carrier of $C$ for Int contains one canonical term for every integer number, $I(0,0)$ for integer 0, $I(+1^n(0),0)$ for integer $n > 0$, and $I(0,+1^n(0))$ for integer $-n < 0$. It is not difficult (but tedious) to prove that $C$ is indeed a characteristic term algebra for the specification. Its core is to prove for each Int-term $T$ that

$$A \Rightarrow C(T) = T$$

where $A$ is the single axiom given in the specification. Since there is only one constructor I, it suffices to prove for all $n_1, n_2 \in \mathbb{N}$ that

$$A \Rightarrow C(I(+1^{n_1}(0),+1^{n_2}(0))) = I(+1^{n_1}(0),+1^{n_2}(0))$$

which can be reduced to proving, for all $n, m \in \mathbb{N}$,

$$A \Rightarrow C(I(+1^{n+m}(0),+1^m(0))) = I(+1^n(0),0)$$
$$A \Rightarrow C(I(+1^m(0),+1^{n+m}(0))) = I(0,+1^n(0)).$$

This proof proceeds by induction on $n$; we omit the details.

Then, for proving that the specified data type satisfies for instance the formula

$$\forall n{:}\text{Nat},m{:}\text{Nat}. \ \exists d{:}\text{Nat}. \ I(n,m){=}I(d,0) \vee I(n,m){=}I(0,d)$$

it suffices to show, for arbitrary $n, m \in C(\text{Nat})$

$$\exists d \in C(\text{Nat}). \ C(I)(n, m) = C(I)(d, 0) \vee C(I)(n, m) = C(I)(0, d)$$

Since the range of $C(I)$ is $C(\text{Int})$, this particular result immediately follows from the definition of $C(\text{Int})$.                                                        □

**Cogenerated Data Types**

In the case of a specification **cogenerated** $D$ that introduces a single hidden sort with carrier $S$, we may assume that two values of $S$ that are observationally equivalent are indeed equal. This gives rise to the following inference rule which states that we may apply the proof principle of *coinduction* (compare with the coinduction rule introduced in Sect. sect:indproof):

$$
\begin{array}{c}
Fs \vdash T_1 \in S \wedge T_2 \in S \wedge T_1 \sim T_2 \wedge \\
\forall x, y \in S, \ldots . x \sim y \Rightarrow \\
f_1(x, \ldots) \sim f_1(y, \ldots) \wedge \ldots \wedge f_n(x, \ldots) \sim f_n(y, \ldots) \wedge \\
g_1(x, \ldots) = g_1(y, \ldots) \wedge \ldots \wedge g_m(x, \ldots) = g_m(y, \ldots) \wedge \\
p_1(x, \ldots) \Leftrightarrow p_1(y, \ldots) \wedge \ldots \wedge p_o(x, \ldots) \Leftrightarrow p_o(y, \ldots) \\
\hline
Fs \vdash T_1 = T_2
\end{array}
$$

Here the functions $f_1, \ldots, f_n$ denote all observers whose domain is $S$, the functions $g_1, \ldots, g_m$ denote all observers whose domain is the carrier of some visible sort, and $p_1, \ldots, p_o$ denote all predicates. In essence, the rule states that in order to show the equality of two $S$-values denoted by the terms $T_1$ and $T_2$, it suffices to find a *bisimulation* $\sim$ on $S$ that relates $T_1$ and $T_2$, i.e., a binary relation on $S$ such that the application of every observer with domain $S$ to arbitrary bisimilar values $x$ and $y$ yields results that are again bisimilar, that the application of every observer whose domain is the carrier of a visible sort yields equal values, and the application of every predicate yields equivalent truth values.

**Example 6.35** Take the specification

> **sort** Elem
> **then cogenerated cotype** Stream := head:Elem | tail:Stream
> **then** {
>   **fun** copy: Stream → Stream
>   **axiom** ∀s:Stream.
>     head(copy(s)) = head(s) ∧
>     tail(copy(s)) = copy(tail(s))
> }

which introduces a visible sort Elem, a cogenerated hidden sort Stream with observers head and tail, and a function copy on Stream. We would like to prove for every model of the specification with carriers *Elem* and *Stream* for the corresponding sorts and functions *head*, *tail*, and *copy* for the corresponding operations that

$$\forall s \in Stream.\ copy(s) = s.$$

We take arbitrary $s \in Stream$ and show $copy(s) = s$. Since $copy(s) \in Stream$ and $s \in Stream$, the coinduction rule can be applied. For this purpose, we define the binary relation $\sim$ on *Stream* as

$$s \sim t :\Leftrightarrow s = copy(t)$$

By definition, $copy(s) \sim s$ holds. It then remains to show that $\sim$ is indeed a bisimulation, i.e., for arbitrary $s, t \in$ *Stream* with $s \sim t$, we have

$$head(s) = head(t) \tag{a}$$
$$tail(s) \sim tail(t) \tag{b}$$

From $s \sim t$, we have

$$s = copy(t). \tag{1}$$

We know (a) from (1) and the first axiom by the derivation

$$head(s) = head(copy(t)) = head(t).$$

To show (b), we have to show by the definition of $\sim$

$$tail(s) = copy(tail(t))$$

which follows from (1) and the second axiom.                               □

### Cofree Data Types

For a specification **cofree** $D$ with a set $\Phi$ of modal formulas as axioms we may assume (since every data type specified in this way is also cogenerated) that all the axioms and inference rules induced by a specification **cogenerated** $D$ also are applicable. To prove that the specified data type $\mathcal{M}$ satisfies a formula $F$, we may thus, as in the case of a specification **generated** $D$, try to show that $F$ is implied by the formulas in $\Phi$ using the proof principle of coinduction on the observers of the specified sorts.

   If this does not suffice, however, we may also apply the knowledge that the specification extends every algebra $A$ to a monomorphic data type $\mathcal{M}$ that contains as its canonical representative the behavior subalgebra $Beh^A_{\Sigma,\Sigma',\Phi}$ whose carriers are sets of behavior trees; it thus also suffices to show $[\![\, F\, ]\!]^{Beh^A_{\Sigma,\Sigma',\Phi}} = $ true. Unfortunately the nature of this algebra cannot be adequately described by first-order formulas or simple proof principles on the logic level such that indeed one has indeed to stick to reasoning on the semantic level.

**Example 6.36** Take the specification

```
free type Nat := 0 | +1(Nat)
then cofree {
 cotype NatStream := head:Nat | tail:NatStream
 axiom[NatStream] let n=head in [tail]head=+1(n)
}
```

that first introduces a monomorphic data type that contains the algebra $N$ of natural numbers which is subsequently extended to the algebra $S = Beh^N_{\Sigma, \Sigma', \varnothing}$ of behavior trees whose leaves (reached by the observer head) are labeled with these numbers; these thus represent infinite streams of natural numbers. However, the axiom states that every value in the stream is followed by its successor, i.e., every stream is some suffix of the infinite sequence $[0, 1, 2, \ldots]$.

Suppose we wish that the specified data type satisfies the formula

$$\exists s:\text{NatStream. head}(s){=}0 \wedge \text{head}(\text{tail}(s)){\neq}0$$

i.e., there exists a stream that starts with 0 and is followed by a value different from 0. We thus have to find some $s \in S(\text{NatStream})$ such that

$$head(s) = 0 \wedge head(tail(s)) \neq 0$$

where $0$, $head$, and $tail$ are the interpretations of the corresponding operations in $S$. Indeed the behavior algebra extending $N$ contains the following behavior tree $s$

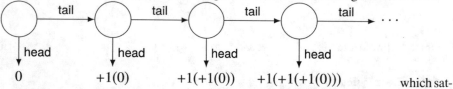

$$0 \qquad\qquad +1(0) \qquad\qquad +1(+1(0)) \qquad +1(+1(+1(0)))$$

which satisfies the axiom and thus is an element of $S$; furthermore, $s$ satisfies the stated formula such that we are done (the formalization of this proof sketch is tedious, we omit the details).                                                                                □

If the specification does not contain any axioms, i.e., $\Phi = \varnothing$, then the canonical model is the behavior algebra $Beh^A_{\Sigma, \Sigma'}$ whose carriers contain all possible behavior trees; the main benefit of such types is the possibility of introducing new operations by 'corecursive definitions' that apply 'pattern matching' against observer terms; this will be further elaborated later when dealing with the consistency of specifications.

## Structured Specifications

We now turn our attention to reasoning about a structured specification expression that arises from the composition of basic expressions by the constructs introduced in Sect. 6.7. Let us for the moment assume that all basic expressions are loose declarations, i.e., that the structured expression does not contain (co)generated or (co)free declarations.

A simple idea how to deal with this expression is to recursively 'flatten' it, i.e., to translate it into a single loose declaration that denotes the same abstract data type:

- In the case of an extension $(SE_1 \textbf{ then } SE_2)$ or union $(SE_1 \textbf{ and } SE_2)$, we first flatten $SE1_1$ and $SE_2$ to declarations $D_1$ and $D_2$ and then combine all the entities (sorts, operations, and axioms) from $D_1$ and $D_2$ to a single declaration.

- In the case of a reduction (*SE* **export** *EI*), we first flatten *SE* to a declaration *D* and then give all the entities that do *not* appear in the export clause *EI* fresh names that are different from all the fresh names that have been introduced in *D* and from all the names in the original specification.
- In the case of a translation (*SE* **with** *RI*), we first flatten *SE* to a declaration *D* and then rename the entities as described by the renaming clause *RI*.
- In the case of an instantiation *I*[*SE* **with** *RI*], we first flatten the import expression of the generic specification *I* to a declaration $D_i$, the parameter expression of that definition to a declaration $D_p$, and the body expression to a declaration $D_b$. We also flatten the argument *SE* of the instantiation to a declaration $D_a$ and then rename the entities as described by the renaming clause *RI* which yields a declaration $D_r$. Finally the entities of $D_r$, $D_i$, $D_p$ and $D_b$ are combined to a single declaration.[3]

It should be noted that the intermediate declarations constructed during the recursive flattening are not necessarily well-formed, since, e.g., in (*SE₁* **then** *SE₂*) the expression $SE_2$ may refer to entities declared in $SE_1$, and consequently also the entities in $D_2$ refer to those declared in $D_1$ (similar issues arise for *I*[*SE* **with** *RI*]). However, after the process is completed, the overall resulting declaration is well-formed. This declaration may have (since the flattening exposes under fresh names those symbols that are hidden by the **export** clause) a larger signature than the original specification expression. However, after reducing this signature to the original one, the declaration describes the same abstract data type.

Now let us consider how to deal with the non-loose declarations **generated** *D*, **free** *D*, **cogenerated** *D*, and **cofree** *D*. A simple idea is to treat such a declaration as a loose declaration *D* to which a 'constraint' is added that describes an additional semantic condition on the denoted type; since this constraint can in general not be expressed as a first-order axiom, it is just represented by a syntactic phrase of form generated$_\Sigma$(...), free$_\Sigma$(...), cogenenerated$_\Sigma$(...), or cofree$_\Sigma$(...) where $\Sigma$ is the signature that arises from *D* and '...' represents some other information that has to be preserved about *D*: thus e.g. a declaration **free** *D* is translated to a loose declaration {*D*; free$_\Sigma$(...)}. After the flattening, every constraint in the resulting declaration states a semantic condition that has to be satisfied by the reduction $A|_\Sigma$ of every algebra *A* that is a model of the denoted type. For the purpose of proving, the occurrence of such a constraint can be considered as the permission to add particular first-order axioms (such as that certain constructors are injective) or to apply particular proof rules (such as induction).

For our purpose, this presentation shall suffice; we refrain from elaborating the concepts sketched above in more detail.

---

[3]This declaration thus constrains the parameter type by both the axioms of $D_r$ and the axioms of $D_p$; if the axioms are incompatible, the declaration becomes inconsistent, i.e., it describes the empty type that satisfies every formula. The usual intention, however, is that the argument type 'refines' the parameter type, i.e., that the axioms of $D_r$ imply the axioms of $D_p$; we will discuss the question of refinement below.

## The Consistency of Specifications

We are now going to address the question whether a specification is consistent, i.e., whether the specified data type $\mathcal{M}$ is not empty:

$$\mathcal{M} \neq \varnothing$$

Rather than attempting to prove the consistency of arbitrary specifications (which is very hard), we will give simple consistency criteria for certain kinds of specification 'patterns' (which are therefore preferred in practice). These criteria are given in a 'recursive' style: first, we discuss how certain basic specifications can be demonstrated to be consistent; then we show how the consistency of compound specifications can be reduced to the consistency of the elementary specifications. The most important case is that a specification is subsequently *conservatively extended* by declarations that only introduce new constants, functions, or predicates by constructive *definitions*.

### Basic Specifications

Due to proposition 6.10, a specification **free** $D$ with presentation $\langle \Sigma, \Phi \rangle$ is consistent, if the axioms in $\Phi$ are expressed in conditional equational logic (since the specified data type contains the quotient term algebra as the canonical representative). Thus also a specification $D$ respectively **generated** $D$ is consistent, if its axioms are of this special form.

Dually, due to proposition 6.15, a specification **cofree** $D$ with presentation $\langle \Sigma, \Phi \rangle$ is consistent, if the axioms in $\Phi$ are expressed in modal logic (since the specified data type contains the behavior subalgebra as the canonical representative). Thus also a specification $D$ respectively **cogenerated** $D$ is consistent, if its axioms are of this special form.

In the general case of a loose specification $D$ with presentation $\langle \Sigma, \Phi \rangle$, the question of the consistency of $D$ can be rephrased as the question of the *satisfiability* of $\Phi$, i.e., whether there exists a model (algebra) $A$ that satisfies every formula in $\Phi$. However, the satisfiability problem of first-order logic is undecidable and thus also the consistency of all kinds of specifications whose formulas are expressed in general first-order logic.

In these cases, we thus have to explicitly prove the non-emptiness of the specified data type $\mathcal{M}$ by defining a $\Sigma$-algebra $A$ with

$$A \in \mathcal{M}.$$

which however in general requires convoluted semantic arguments (which we will not address here). Only in the case of a loosely interpreted declaration $D$, straightforward reasoning is possible: here it suffices to show

$$A \models_\Sigma \Phi$$

I.e., that $A$ satisfies the axioms of the specification.

## Conservative Extensions

Consider the specification (*SE* **then** *SE′*) of a data type $\mathcal{M}'$ where we assume that the data type $\mathcal{M}$ specified by *SE* is not empty, i.e., that the consistency of *SE* has already been shown. We will consider constraints that ensure the 'conservative extension' of $\mathcal{M}$ by the specification *SE′*, i.e., that every algebra $A \in \mathcal{M}$ is extended to at least one algebra $A' \in \mathcal{M}'$; thus also $\mathcal{M}'$ is not empty. However, we will only consider two specific cases of *SE′*.

The first case is that *SE′* is of form **generated** *D* or **free** *D*: if all axioms in *D* are expressed in conditional equational logic such that unconditional equations and the conclusions of implications only equate terms of sorts introduced in *D*, the specification is clearly consistent (since no inconsistencies can be introduced by demanding equalities for the newly introduced types).

The remainder of this section is dedicated to the second case that *SE′* is of form *D*. Clearly, the specification is consistent, if *D* does not contain any clauses of form **axiom** *F*, i.e., explicitly specified axioms. This does, however, *not* rule out definitions of form **const** $I: S := T$, **fun** $I_1: Ss \rightarrow S$, $I_2(Is) := T$, or **pred** $I_1 \subseteq Ss$, $I_2(Is) :\Leftrightarrow F$; all such definitions represent conservative extensions.

However, it is possible to express also other kinds of 'definitions' by explicit axioms. Let us assume that *SE* contains a free type declaration

**free type** $S := f\_1(...) \mid ... \mid f\_n(...)$

which we subsequently extend by the following form of a loose specification of a new function $g$ on $S$:

**fun** $g: S \times ... \rightarrow ...$
**axiom** $\forall .... \, g(f\_1(...),...) = ... \wedge ... \wedge g(f\_n(...),...) = ...$

The given axiom contains $n$ equations where

- on the left-hand side of every equation, the arguments denoted by '...' are distinct variables (i.e., no general term is allowed as an argument and every variable occurs only once as an argument),
- on the right-hand side of every equation, only those variables occur freely that also occur on the left-hand side, and
- on the right-hand side of every equation, the first argument of every application of function $g$ is some variable that occurs as an argument of the constructor application $f_i(...)$ on the left-hand side.

In other word, the function $g$ is described by a set of 'pattern-matching' equations which can be applied as 'simplification rules' to replace every application of $g$; by the axioms induced by the free data type declaration, the first argument of $g$ matches the left side of exactly one equation which determines the values of the variables on the left side; the application can be thus replaced by the term on the right side. Since this term can contain only applications of $g$ to an argument that is structurally 'simpler' than the argument on the left side, the process can be repeated only finitely often leading ultimately to a term that does not involve any application of $g$. In essence, the equations induce a definition by *structural induction* as has been described in Sect. sect:structural on abstract syntax trees respectively a *primitively recursive definition* as has been described in Sect. sect:primitive. The same constraints can be also applied to the definition of predicates (we omit the details). Then the constraints imposed by the free specification of $S$ ensure that the model of the original specification can be indeed extended by a function $g$ that satisfies these equations; furthermore there is only one such function, i.e., $g$ is even uniquely defined.

**Example 6.37**  The specification

> **free type** Nat := 0 | +1(Nat)
> **then**
> {
>     **fun** +: Nat × Nat → Nat
>     **axiom** ∀n1:Nat, n2:Nat.
>      +(0, n2) = n2 ∧
>      +(+1(n1), n2) = +1(+(n1, n2))
> }

first induces a model of the **free type** specification; this model is subsequently conservatively extended by a function + whose definition conforms to the constraints described above. By applying the equations as simplification rules, we can e.g. deduce

$$+(+1(+1(0)), +1(0)) = +1(+(+1(0), +1(0)))$$
$$= +1(+1(+(0, +1(0)))) = +1(+1(+1(0)))$$

for the interpretations $0$, $+1$, and $+$ of operations 0, +1, and +, respectively.   □

Above considerations of conservative extensions only apply to free specifications without axioms. Now let us include also axioms such as in the specification

> **free** {
>     **type** Int := 0 | +1(Int) | -1(Int)
>     **axiom** ∀i:Int.  i = +1(-1(i)) ∧ i = -1(+1(i))
> }

where different terms such as +1(-1(0)) and 0 may denote the same values. Then introducing a function by a declaration such as

> **fun** f: Int → Int
> **axiom** ∀x:Int.
>   f(0) = 0 ∧
>   f(+1(x)) = +1(0) ∧
>   f(-1(x)) = -1(0)

does not necessarily induce a conservative extension of the original model: from above declaration, we may deduce for the interpretations $0$, $+1$ and $f$ of the operations $0$, $+1$, and $f$

$$0 = f(0) = f(+1(-1(0))) = +1(0)$$

but the equation $0 = +1(0)$ does not hold in the original model.

Nevertheless, in certain cases such extensions are legit. To demonstrate this, we may simply extend a characteristic term algebra $C$ for the original specification (which according to Proposition 6.16 represents a model of this specification) by an interpretation $C(op)$ for every new operation $op$ and show that this interpretation satisfies the new axioms.

**Example 6.38** Take the specification

**free** {
   **type** Nat := 0 | +1(Nat)
   **type** Int := I(Nat,Nat)
   **axiom**
     $\forall$p:Nat, n:Nat, s:Set. I(+1(p),+1(m))=I(p,m)
}

which we extend by a binary function + as follows:

**fun** add: Int $\times$ Int $\to$ Int
**axiom**
   $\forall$p1,m1,p2,m2:Nat.
    +(I(0,0),I(p2,m2)) = I(p2,m2) $\wedge$
    +(I(+1(p1),m1),I(p2,m2)) = +(I(p1,m1),I(+1(p2),m2)) $\wedge$
    +(I(p1,+1(m1)),I(p2,m2)) = +(I(p1,m1),I(p2,+1(m2)))

The consistency of this extension can be proved by extending the characteristic term algebra $C$ defined in Example 6.34 by the interpretation

$C(+) = \lambda I_1 \in C(\text{Int}), I_2 \in C(\text{Int}).$
      let $p_1, m_1 = $ choose $p_1 \in \mathbb{N}, m_1 \in \mathbb{N}.$ $I_1 = I(+1^{p_1}(0),+1^{m_1}(0))$ in
      let $p_2, m_2 = $ choose $p_2 \in \mathbb{N}, m_2 \in \mathbb{N}.$ $I_2 = I(+1^{p_2}(0),+1^{m_2}(0))$ in
      $C(I)(+1^{p_1+p_2}(0), +1^{m_1+m_2}(0))$

and then show that, for every $p_1, m_1, p_2, m_2 \in C(\text{Nat})$, the following equations hold:

$$C(+)(C(I(0,0)), C(I)(p_2, m_2)) = C(I)(p_2, m_2) \tag{a}$$
$$C(+)(C(I)(+1(p_1), m_1), C(I)(p_2, m_2)) = C(+)(C(I)(p_1, m_1), C(I)(+1(p_2), m_2)) \tag{b}$$
$$C(+)(C(I)(p_1, +1(m_1)), C(I)(p_2, m_2)) = C(+)(C(I)(p_1, m_1), C(I)(p_2, +1(m_2))) \tag{c}$$

We sketch the proof of (c) which is a consequence of the lemma

$$\forall a_1, b_1, a_2, b_2 \in \mathbb{N}.$$
$$C(+)(C(\mathsf{l})(+1^{a_1}(0), +1^{b_1}(0)), C(\mathsf{l})(+1^{a_2}(0), +1^{b_2}(0))) =$$
$$C(\mathsf{l})(+1^{a_1+a_2}(0), +1^{b_1+b_2}(0)) \tag{1}$$

that in turn can be proved from the definitions of $C(\mathsf{l})$ and of $C(+)$, a tedious but not really difficult task.

Then, from the Definition of $C(\mathsf{Nat})$, we know $p_1 = +1^{a_1}(0), m_1 = +1^{b_1}(0), p_2 = +1^{a_2}(0), m_2 = +1^{b_2}(0)$ for some $a_1, b_1, a_2, b_2 \in \mathbb{N}$. It thus suffices to show

$$C(+)(C(\mathsf{l})(+1^{a_1}(0), +1^{b_1+1}(0)), C(\mathsf{l})(+1^{a_2}(0), +1^{b_2}(0))) =$$
$$C(+)(C(\mathsf{l})(+1^{a_1}(0), +1^{b_1}(0)), C(\mathsf{l})(+1^{a_2}(0), +1^{b_2+1}(0))).$$

By (1), it suffices to show

$$C(\mathsf{l})(+1^{a_1+a_2}(0), +1^{(b_1+1)+b_2}(0)) = C(\mathsf{l})(+1^{a_1+a_2}(0), +1^{b_1+(b_2+1)}(0))$$

which is true because $(b_1 + 1) + b_2 = b_1 + (b_2 + 1)$.                  □

In analogy to the 'pattern matching' equations on the constructors of a free type, we may also define functions by corresponding equations on the observers of a cofree type: given a cofree type declaration

**cofree type** S := f_1(...) | ... | f_n(...)

we may introduce a new function $g$ into $S$ by a loose specification of the following form:

**fun** g: ... → S
**axiom** ∀.... f_1(g(...)) = ... ∧ ... ∧ f_n(g(...)) = ...

This axiom contains $n$ equations which define the values of all observers $f_1, \ldots, f_n$ on every application $g(\ldots)$.

**Example 6.39** Example 6.35 introduced a specification similar to

**sort** Elem
**then cofree cotype** Stream := head:Elem | tail:Stream
**then** {
  **fun** copy: Stream → Stream
  **axiom** ∀s:Stream.
    head(copy(s)) = head(s) ∧
    tail(copy(s)) = copy(tail(s))
}

which uniquely defines a function *copy* on *Stream*.                  □

If the cofree specification is specified with axioms, we may apply a similar strategy as for free specifications to show that an extension of the cofree type is conservative: we define a maximal behavior subalgebra as the canonical representative of the cofree type, define an interpretation of the newly introduced operations in that algebra, and prove that their interpretations satisfy the given axioms. We leave it with this short sketch of the general strategy without going into the (complicated) details.

## Other Specifications

A union ($SE_1$ **and** $SE_2$) is consistent, if both $SE_1$ and $SE_2$ are consistent and there exist algebras $A_1$ and $A_2$ in the respective data types that 'agree on their common parts'. This is in particular true, if the signatures of both specification expressions are disjoint (then there is no common part) or if the common part just stems from the extension of the same data type such as in a specification ($I$ **then** $SE_1$) **and** ($I$ **then** $SE_2$) where $I$ is the name of a previously introduced specification.

If a specification $SE$ is consistent, then also every reduction ($SE$ **export** $EI$) is consistent and every translation ($SE$ **with** $RI$) is consistent.

Finally, take a generic specification

**spec** $I[SE_1]$ **import** $SE_2$ $:=$ $SE_3$

An instantiation $I[SE$ **fit** $RI]$ is consistent, if the specification $SE$ is consistent, the specification

($SE_2$ **then** $SE_1$) **then** $SE_3$

is consistent, and every algebra in the data type of the argument $SE$ is also an algebra in the data type of the parameter $SE_1$, i.e., $SE$ 'refines' $SE_1$. The issue of refinement will be addressed in the following subsection.

## The Refinement of Specifications

To show that a specification $SE'$ refines another specification $SE$, we have to show for every abstract data type $\mathcal{M}' = [\![ SE' ]\!](me, \mathcal{M}_1)$ and $\mathcal{M} = [\![ SE ]\!](me, \mathcal{M}_1)$

$$\mathcal{M}' \subseteq \mathcal{M}.$$

i.e., every algebra that implements $SE'$ also implements $SE$:

$$A \in \mathcal{M}' \Rightarrow A \in \mathcal{M}.$$

If also the other direction holds, i.e., $\mathcal{M}' = \mathcal{M}$, then $SE'$ and $SE$ are equivalent.

The refining specification $SE'$ thus restricts the choice of implementations, i.e., it represents a more concrete design of a data type as the original specification $SE$. As an extreme case, a specification $SE'$ that gives rise to the empty data type $\mathcal{M}' = \varnothing$

refines every specification $SE$; thus also the consistency of $SE'$ should be established as described in the previous subsection.

If $SE$ is a loose declaration $D$ with presentation $\langle \Sigma, \Phi \rangle$ (respectively can be flattened to such a declaration where $\Phi$ does not contain any non-logical constraints such as $\text{free}_\Sigma(\ldots)$), it suffices to show

$$A \in \mathcal{M}' \Rightarrow A \models_\Sigma \Phi$$

i.e., that the refining data type satisfies the axioms of $D$. Furthermore, if also $SE'$ is (respectively can be flattened to) a loose declaration $D'$ with presentation $\langle \Sigma, \Phi' \rangle$, it suffices to derive

$$\wedge(\Phi') \vdash \wedge(\Phi)$$

i.e., that from the axioms of $SE'$ the axioms of $SE$ can be proved.

In particular, such proofs often occur in the case where the consistency of an instantiation $I[SE']$ of a generic specification

**spec** $I[SE]$ **import** $SE_i := SE_b$

is to be established; in such specifications, $SE$ typically is a loose declaration $D$. The refinement proof ensures that, provided that

$$(SE_2 \textbf{ then } SE) \textbf{ then } SE_3$$

is consistent and $SE'$ is consistent, also $I[SE']$ is consistent.

If the flattening of $SE$ yields non-logical constraints and also the flattening of $SE'$ yields such constraints, any constraint in $SE$ that also occurs in $SE'$ can be trivially discharged. Furthermore, some constraints can be shown to be implied by other constraints, e.g., a constraint $\text{generated}_\Sigma(\ldots)$ that specifies that a sort $S$ is generated is implied by an analogous constraint $\text{free}_\Sigma(\ldots)$ that specifies that sort $S$ is a free type; the constraint $\text{generated}_\Sigma(\ldots)$ is also implied by another constraint $\text{generated}_{\Sigma'}(\ldots)$ or $\text{free}_{\Sigma'}(\ldots)$ that states that $S$ is generated/free in only a subset of the constructors.

In general, however, to prove the refinement of a non-loose data type requires reasoning on the semantic level, which we will not discuss further.

## Implementing Specifications

The notion of refinement presented above may be too strong when we consider a special case of refinement, the *implementation* of a (high level) abstract data type by another (lower level) representation type.

A classical example is the implementation of the abstract data type 'stack of elements' which can be specified as follows:

```
spec STACK[sort Elem] :=
  free type Stack := empty | push(Elem,Stack)
  then {
    fun top: Stack → Elem
    fun pop: Stack → Stack
    axiom ∀e:Elem,s:Stack.
      top(push(e,s)) = e ∧
      pop(push(e,s)) = s
  }
```

Here the free specification of type Stack with constructors empty and push is extended by a loose specification of selectors top and pop which can be applied to non-empty stacks (by the loose extension, the value of e.g., pop(empty) can be arbitrary; if pop were already introduced in the free specification, top(empty) would represent an additional value of the type).

Now in an imperative programming language like Java, a stack of this kind would be implemented by a class like

```
public class Stack {
    private Object[] a;
    private int n;
    public Stack() { a = new Object[100]; n = 0; }
    public void push(Object e)
    { if (n == a.length) resize(); a[n] = e; n++; }
    public Object top() { if (n > 0) return a[n-1]; else return null; }
    public void pop() { if (n > 0) n--; }
    private void resize() { ... }
}
```

where the stack is represented by a pair of an array $a$ and a counter $n$ which describes how many of the 'slots' of $a$ have been filled; popping a value from the stack then just requires decrementing this counter.

We can mimic this approach in our specification language as follows: first we introduce the following generic specification:

```
spec ARRAY[sort Elem] :=
  free type Index := 0 | +1(Index)
  then generated type Array := null | put(Array, Index, Elem)
  then {
    fun get: Array × Index → Elem
    axiom ∀a:Array, i:Index, j:Index, e:Elem.
      get(put(a,i,e),i) = e ∧
      i ≠ j ⇒ get(put(a,i,e),j) = get(a,j)
  }
```

The generic specification ARRAY introduces a generated sort Array of arrays of elements of sort Elem with constructors null and put: null represents an uninitialized array while put(a,i,e) represents the array that was derived by writing into array $a$ at index $i$ the element $e$: thus e.g., the term

put(put(put(null, 0,$e_1$), +1(0),$e_2$), 0,$e_3$)

represents the array that holds at index 0 the value $e_3$ (because the original value $e_1$ was overwritten) and at index 1 the value $e_2$. The operation get(a,i) returns the value that was written last to $a$ at position $i$: thus, if we take as $a$ the array written above, we have get($a$,0) = $e_3$ and get($a$,+1(0)) = $e_2$. Please note that the value of get(null,i) is arbitrary; furthermore, the specification leaves it open whether put(put(a,i,e),i,e)=put(a,i,e), i.e., whether writing the same value into an array twice leaves the array unchanged or not (this allows, e.g., the array to keep track of the full history of updates that have been performed on it). Arrays have in this model no upper bound, so writing to every index is legal.

With the help of ARRAY we can now develop the following generic specification:

> **spec** ARRAYSTACK[**sort** Elem] := ARRAY[**type** Elem]
>   **then free type** Stack := stack(Array,Index)
>   **then** {
>     **const** empty: Stack := stack(null,0)
>     **fun** push: Elem × Stack → Stack
>     **fun** top: Stack → Elem
>     **fun** pop: Stack → Stack
>     **axiom** ∀e:Elem,a:Array,n:Index.
>       push(e,stack(a,n)) = stack(put(a,n,e),+1(n)) ∧
>       top(stack(a,+1(n))) = get(a,n) ∧
>       top(stack(a,0)) = get(null,0) ∧
>       pop(stack(a,+1(n))) = stack(a,n) ∧
>       pop(stack(a,0)) = stack(a,0)
>   }
>   **export** { **sort** Stack, **const** empty,
>     **fun** push(Elem), **fun** top(Stack), **fun** pop(Stack) }

The generic specification ARRAYSTACK has the same export signature as the original specification ARRAY. Its free type Stack implements every 'stack' as a pair stack(a,n) of an array $a$ and an index $n$ that represents the number of elements of the stack which were written into $a$ at indices $+1^{n-1}(0), \ldots, 0$; the value written at index $+1^{n-1}(0)$ represents the top of the stack. The loosely specified operations empty, push, top and pop mimic the definitions given in the Java class above (except that resizing the array is not necessary).

Now we might naturally expect that the instantiation ARRAYSTACK[**sort** Elem] refines STACK[**sort** Elem], i.e., that the abstract data type $\mathcal{M}_{as}$ denoted by the former specification expression is a subset of the type $\mathcal{M}_s$ denoted by the later one. However, this is actually not the case: every algebra in $\mathcal{M}_s$ satisfies the equation

pop(push($e$,$s$)) = $s$.

Consequently, since every algebra in $\mathcal{M}_{as}$ satisfies the equation

pop(push(e,stack(a,n))) = pop(stack(put(a,n,e),+1(n))) = stack(put(a,n,e),n),

$\mathcal{M}_{as} \subseteq \mathcal{M}_s$ implies

stack(*a*,*n*) = stack(put(*a*,*n*,*e*),*n*)

and thus

$a = $ put(*a*,*n*,*e*)

which is generally not true. Nevertheless, while the Stack values stack(a,n) and stack(put(a,n,e),n) are not the same, they cannot be distinguished by the Elem values they contain; in particular, if $n=+1(m)$ holds, we have

top(stack(*a*,+1(*m*))) = get(*a*,*m*)

$\qquad = $ get(put(*a*,+1(*m*),*e*),*m*) = top(stack(put(*a*,+1(*m*),*e*),+1(*m*)))

where get(*a*,*m*) = get(put(*a*,+1(*m*),*e*),*m*) holds because $m \neq +1(m)$ (this is also the reason why in above Java class Stack the method pop omitted an additional assignment a[n] = null that would restore $a$ to its original state: no user of this class can recognize this omission).

In other words, both representations of sort Stack are *behaviorally equivalent* with respect to sort Elem, a notion which we are going to formalize below.

---

**Definition 6.32** (*Behavioral equivalence*) Let $\Sigma$ be a signature and $OS \subseteq \Sigma.s$ be a set of sorts in $\Sigma$ called *observable sorts*. Two $\Sigma$-algebras $A$ and $B$ are *behaviorally equivalent* on $OS$, written $A \simeq_{OS} B$, if the following conditions hold:

- For every sort $S \in OS$ and all terms $T_1, T_2 \in Term_\Sigma^S$ of this sort, we have

$$\left( [\![ T_1 ]\!]^A = [\![ T_2 ]\!]^A \right) \Leftrightarrow \left( [\![ T_1 ]\!]^B = [\![ T_2 ]\!]^B \right).$$

- For every predicate $\langle I, [S_1, \ldots, S_n] \rangle \in \Sigma.p$ with arity $S_1, \ldots, S_n$ and terms $T_1 \in Term_\Sigma^{S_1}, \ldots, T_n \in Term_\Sigma^{S_n}$, we have

$$\left( [\![ I(T_1, \ldots, T_n) ]\!]^A = \text{true} \right) \Leftrightarrow \left( [\![ I(T_1, \ldots, T_n) ]\!]^B = \text{true} \right).$$

---

Based on this concept, we can now define in what sense the the abstract data type $\mathcal{M}_{as}$ denoted by specification ARRAYSTACK[**sort** Elem] implements the type $\mathcal{M}_s$ denoted by STACK[**sort** Elem].

**Definition 6.33** (*Implementation of an abstract data type*) Let $\Sigma$ be a signature, let $OS \subseteq \Sigma.s$ be a set of observable sorts in $\Sigma$, and let $\mathcal{M}, \mathcal{M}'$ be classes of $\Sigma$-algebras. Then $\mathcal{M}'$ *implements* $\mathcal{M}$ with respect to $OS$, written as $\mathcal{M}' \subseteq_{OS} \mathcal{M}$, if the following condition holds:

$$\forall B \in \mathcal{M}'. \exists A \in \mathcal{M}. B \simeq_{OS} A$$

Please note that $\mathcal{M}' \subseteq \mathcal{M}$ implies $\mathcal{M}' \subseteq_{\Sigma.s} \mathcal{M}$, i.e., refinement by inclusion represents the special case of implementation with respect to all sorts in the signature.

Thus, although $\mathcal{M}_{as} \subseteq \mathcal{M}_s$ does not hold, we have $\mathcal{M}_{as} \subseteq_{\{Elem\}} \mathcal{M}_s$, i.e., $\mathcal{M}_{as}$ implements $\mathcal{M}_s$ with respect to observable sort Elem. However, it remains to be discussed how to actually prove $\mathcal{M}_{as} \subseteq_{\{Elem\}} \mathcal{M}_s$, i.e., the implementation of one type by another. For this, we investigate further the relationship between an algebra $B \in \mathcal{M}_{as}$ and an algebra $A \in \mathcal{M}_s$:

- While every value in carrier $A(\text{Stack})$ represents a stack, not every value of $B(\text{Stack})$ does, e.g. the value denoted by stack(empty,+1(0)) does not. More generally, stack(a,+1(n)) only denotes a stack, if array $a$ holds at every index less than $n$ some element $e$ (i.e., every application get(a,+1$^i$(0)) with $i \in \mathbb{N}_n$ must be allowed).
- While every value in carrier $A(\text{Stack})$ represents a distinct stack (uniquely denoted by a corresponding constructor term), infinitely many values of $B(\text{Stack})$ represent the same stack: e.g., the values denoted by the constructor terms

  stack(empty,0)
  stack(put(empty,$i_1$,$e_1$),0)
  stack(put(put(empty,$i_2$,$e_2$),$i_1$,$e_1$),0)
  . . .

for arbitrary terms $i_1, i_2, \ldots$ and $e_1, e_2, \ldots$ all represent the empty stack. In general, all arrays that hold the same values at indices less than $n$ (but arbitrary values at indices greater than or equal $n$) represent the same stack.

Consequently, $B \in \mathcal{M}_{as}$ is not isomorphic to any $A \in \mathcal{M}_s$. However, we may construct another algebra $B''$ that is behaviorally equivalent to $B$ and isomorphic to some $A \in \mathcal{M}_s$. Then $B$ is also behaviorally equivalent to $A$ such that $\mathcal{M}_{as}$ indeed implements $\mathcal{M}$. This construction proceeds in two steps (see Fig. 6.15):

- First we construct that subalgebra $B'$ of $B$ whose carrier $B'(S)$ contains only legal values for the non-observable sort $S$ (e.g., valid stack representations);

**Fig. 6.15** Implementing a
specification with algebra $A$
by an algebra $B$

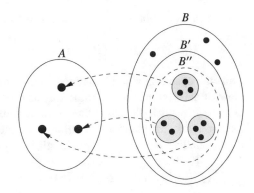

- then we construct from $B'$ that algebra $B''$ whose carrier $B''(S)$ combines to a
  single set all those 'concrete' values of $B'(S)$ that represent the same 'abstract'
  value in $A(S)$ (e.g., the same stack).

The construction of $B'$ can be based on a family $(R_S)_{S \in \Sigma.s \backslash OS} \subseteq S$ of *representation
invariants* on every non-observable sort $S \in \Sigma.s \backslash OS$; the values of $B'(S)$ are exactly
those values of $B(S)$ that satisfy this invariant, i.e., $B'(S) := \{b \in B(S) \mid R_S\langle b \rangle\}$.
However, to show that this indeed defines a well-formed $\Sigma$-algebra $B'$, one has to
show by a sort of 'induction proof' that the invariant is established by all constants
and preserved by all constructors. In detail, we have to show

- for every constant $C = \langle I, S \rangle \in \Sigma.\mathsf{c}$

  $R_S\langle B(C) \rangle$

- and for every function $F = \langle I, [S_1, \ldots, S_n], S \rangle \in \Sigma.\mathsf{f}$

  $R_S\langle B(F)(b_1, \ldots, b_n) \rangle$

  for arbitrary values $b_1 \in B(S_1), \ldots, b_n \in B(S_n)$ where we may assume $R_{S_i}(b_i)$
  for every non-observable sort $S_i \in \Sigma.s \backslash OS$.

Since all operations have in $B'$ the same interpretation as in $B$, $B'$ is behaviorally
equivalent to $B$.

The construction of $B''$ can be based on the definition of a family $(E_S)_{S \in \Sigma.s \backslash OS} \subseteq
S \times S$ of *equivalence relations* on every non-observable sort $S \in \Sigma.s \backslash OS$; each ele-
ment of $B''(S)$ is then a corresponding equivalence class, i.e.,
$B''(S) := \{[b]_{E_S} \mid b \in B'(S)\}$ where $[b]_{E_S} := \{b' \in B'(S) \mid E_S\langle b, b' \rangle\}$. We then
first have to show that $E_S$ satisfies the constraints of an equivalence relation, i.e.,
is reflexive, symmetric, and transitive. Next we have to give to every operation an
interpretation in $B''$ by defining, in the case of a non-observable argument sort and
an observable target sort,

  $B''(f)([b]_{E_S}, \ldots)) := B'(f)(b, \ldots)$

and, in case of a non-observable target sort $S$,

$$B''(f)([b]_{E_S}, \ldots)) := [B'(f)(b, \ldots)]_{E_S}$$

Thus the operation result is defined from some representative $b$ of the equivalence class of the argument in the same way as in $B'$ respectively $B$ (with an equivalence class as an result in the case of a non-observable target sort). Then, however, we also have to show that the result is uniquely defined independently of the choice of the representative: we have to prove for all $b_1, b_2 \in B'(S)$ with $E_S \langle b_1, b_2 \rangle$

$$B''(f)([b_1]_E, \ldots) = B''(f)([b_2]_E, \ldots)$$

If the target of $f$ is a non-observable sort, this amounts to proving

$$E_S \langle B'(f)(b_1, \ldots), B'(f)(b_2, \ldots) \rangle$$

i.e., that the results of the interpretations in $B'$ are equivalent with respect to $E_S$. By this kind of definition, $B''$ is behaviorally equivalent to $B'$ and thus to $B$.

   Finally, we may define the isomorphism $h \colon B'' \to_\Sigma A$ based on a family of *abstraction functions* $(a_S)_{S \in \Sigma.s \backslash OS} \colon B'(S) \to A(S)$ on every non-observable sort $S \in \Sigma.s \backslash OS$ that describe how every legal 'representation' $b \in B'(S)$ is mapped to some 'abstraction' $a(b) \in A(S)$; each value $h_S([b]_{E_S})$ on an equivalence class $[b]_{E_S}$ with representative $b \in B'(S)$ is then defined as that value to which $b$ is abstracted. To this end, we define $h_S([b]_{E_S}) := a(b)$ but then also have to show that the function result is uniquely defined independently of the choice of the representative: we have to prove for all $b_1, b_2 \in B'(S)$ with $E \langle b_1, b_2 \rangle$

$$h_S([b_1]_{E_S}) = h_S([b_2]_{E_S})$$

Finally, we then establish the homomorphism conditions of $h_S$ and its bijectivity.

**Example 6.40** Based on the principles outlined above, we sketch (without going into details) the core elements of the proof of $\mathcal{M}_{as} \subseteq_{\{Elem\}} \mathcal{M}_s$ where $\mathcal{M}_{as}$ for which we consider an arbitrary algebra $B$ in $\mathcal{M}_{as}$.

   First, we define the representation invariant $R_{Stack} \subseteq B(Stack)$. Since sorts Stack, Array, and Nat in ARRAYSTACK[**sort** Elem] are generated, every value $s \in B(Stack)$ is the value of some denotation

$$[\![ stack(a,+1^n(0)) ]\!]_x^B$$

for some term $a$, $n \in \mathbb{N}$, and assignment $x$ that assigns to variables only values from $B(Elem)$. Now we define $R_{Stack} \langle s \rangle$ to be true if and only if there also exists such a denotation where the value of $[\![\, \mathsf{get}(a, +1^i(0)) \,]\!]_x^B$ is an element of $B(\mathsf{Elem})$ for every $i \in \mathbb{N}_n$, i.e., $a$ has a well-defined element at every index position.[4]

Now we have to prove the well-formedness of the resulting sub-algebra $B'$. In particular, we have to prove that the invariant is preserved by operation push, i.e., we have to show $R_{\mathsf{Stack}}\langle B'(\mathsf{push})(e, s)\rangle$ for arbitrary $e \in B(\mathsf{Elem})$ and $s \in B(\mathsf{Stack})$ where we may assume $R_{\mathsf{Stack}}\langle s\rangle$: from the assumption, we know that $s$ is the denotation $[\![\, \mathsf{stack}(a, +1^n(0)) \,]\!]_x^B$ for some term $a$, $n \in \mathbb{N}$, and assignment $x$. From the axiom for push in ARRAYSTACK, we then know $B'(\mathsf{push})(e, s) = [\![\, \mathsf{stack}(\mathsf{put}(a, +1^n, e), +1^{n+1}(0)) \,]\!]_x^B$. To show $R_{\mathsf{Stack}}\langle B'(\mathsf{push})(e, s)\rangle$, it now suffices to show that $[\![\, \mathsf{get}(a, +1^i(0)) \,]\!]_x^B$ is an element of $B(\mathsf{Elem})$ for every $i \in \mathbb{N}_{n+1}$. For $i \in \mathbb{N}_n$, we know this from the assumption; for $i = n$, we know this for the element $e$ from the axiom for get in ARRAY.

Next we define the equivalence relation $E_{\mathsf{Stack}} \subseteq B(\mathsf{Stack}) \times B(\mathsf{Stack})$: this relation holds for two stacks $s_1$ and $s_2$ if and only if they are the values of some denotations $[\![\, \mathsf{stack}(a_1, +1^n(0)) \,]\!]_x^B$ and $[\![\, \mathsf{stack}(a_2, +1^n(0)) \,]\!]_x^B$ such that for all $i \in \mathbb{N}_n$, $[\![\, \mathsf{get}(\mathsf{stack}(a_1, +1^i(0))) \,]\!]_x^B = [\![\, \mathsf{get}(\mathsf{stack}(a_2, +1^i(0))) \,]\!]_x^B$; then $E_{\mathsf{Stack}}$ is clearly an equivalence relation.

Now we define $B''$, e.g., $B''(\mathsf{pop})([s]_{E_S}) := [B'(\mathsf{pop})(s)]_{E_S}$. This definition is consistent if we can show, for arbitrary stacks $s_1, s_2 \in B'(\mathsf{Stack})$ with $E_{\mathsf{Stack}}\langle s_1, s_2\rangle$, that we have $E_{\mathsf{Stack}}\langle B'(\mathsf{pop})(s_1), B'(\mathsf{pop})(s_2)]\rangle$ (*). From $E_{\mathsf{Stack}}\langle s_1, s_2\rangle$ we know that $s_1$ and $s_2$ are the values of some denotations $[\![\, \mathsf{stack}(a_1, +1^n(0)) \,]\!]_x^B$ and $[\![\, \mathsf{stack}(a_2, +1^n(0)) \,]\!]_x^B$. If $n = 0$, the second axiom for pop in ARRAYSTACK implies $B'(\mathsf{pop})(s_1) = [\![\, \mathsf{stack}(a_1, 0) \,]\!]_x^B$ and $B'(\mathsf{pop})(s_2) = [\![\, \mathsf{stack}(a_2, 0) \,]\!]_x^B$; this yields (*). If $n = m + 1$ for some $n \in \mathbb{N}$, the first axiom for pop in ARRAYSTACK implies $B'(\mathsf{pop})(s_1) = [\![\, \mathsf{stack}(a_1, +1^m(0)) \,]\!]_x^B$ and $B'(\mathsf{pop})(s_2) = [\![\, \mathsf{stack}(a_2, +1^m(0)) \,]\!]_x^B$. From $E_{\mathsf{Stack}}\langle s_1, s_2\rangle$, we know for all $i \in \mathbb{N}_n$ $[\![\, \mathsf{get}(\mathsf{stack}(a_1, +1^i(0))) \,]\!]_x^B = [\![\, \mathsf{get}(\mathsf{stack}(a_2, +1^i(0))) \,]\!]_x^B$, from which we know (*).

Next we define the abstraction function $a_{\mathsf{Stack}}: B'(\mathsf{Stack}) \to A(\mathsf{Stack})$ as $a_{\mathsf{Stack}}$ $([\![\, \mathsf{stack}(a, +1^n(0)) \,]\!]_x^B) := [\![\, \mathsf{push}(\mathsf{x1}, \ldots, \mathsf{push}(\mathsf{xn}, \mathsf{empty}) \ldots) \,]\!]_{[\mathsf{x1} \mapsto e_{n-1}, \ldots, \mathsf{xn} \mapsto e_0]}^A$ with $e_{n-1} = [\![\, \mathsf{get}(\mathsf{stack}(a, +1^{n-1}(0))) \,]\!]_x^{B'}$, ..., $e_0 = [\![\, \mathsf{get}(\mathsf{stack}(a, 0), 0) \,]\!]_x^{B'}$. On this basis, we define the isomorphism $h: B'' \to_\Sigma A$ as $h_{\mathsf{Elem}}(e) := e$ and $h_{\mathsf{Stack}}([s]_{E_{\mathsf{Stack}}}) := a_{\mathsf{Stack}}(s)$. It remains to be shown that this definition is indeed consistent and that the resulting function satisfies the homomorphism and bijectivity conditions. $\qquad\square$

## Partial Operations

Above strategy for proving the correctness of an implementation crucially depends on the fact that the specification of the implementing type defines the outcome of every operation such that it satisfies the representation invariant. Thus we have included in ARRAYSTACK the axiomatic equations

---

[4]In our formal framework, this is technically already a consequence of the arity of get, i.e., the relation is always 'true'; however, we glance over this point to make the following discussion also applicable to other frameworks.

top(stack(a,0)) = get(null,0) $\wedge$
pop(stack(a,0)) = stack(a,0)

Without these, it would not have been possible to show that top and pop satisfy the homomorphism conditions respectively preserve the representation invariant when being applied to the empty stack. Therefore also the methods in the Java class were correspondingly defined as

```
public Object top() { if (n > 0) return a[n-1]; else return null; }
public void pop() { if (n > 0) n--; }
```

However, we may also wish to deal with the more common implementation

```
public Object top() { return a[n-1];  }
public void pop() { n--; }
```

where an operation simply 'fails' when being applied to an object that represents the empty stack (since the documentation says that one is not allowed to use the operation in this case) and in the corresponding specification the two equations given above are omitted. For this, we may extend the semantic framework of the specification language to also consider *partial operations* which are equipped with a *precondition* that states under which assumptions on the arguments of the operation its result is well-defined. In such an extension, the precondition of the operations in the abstract type must imply the preconditions of the corresponding operations in the representation type (i.e., the representation type may only weaken a precondition, not strengthening it). Then in the implementation proof those situations where the precondition is violated need not be considered.

The question of how to adequately address the partiality of operations in abstract data type specifications goes beyond the purpose of this introductory presentation; we have therefore generally dodged it by introducing such a 'partial' operation only in loose specifications which left its outcome undefined when applied outside its natural domain of application; thus also the specification STACK did not have an axiom that defined the outcome of top(empty) or pop(empty).

## Exercises

Download from the following URL:
https://www.risc.jku.at/people/schreine/TP/exercises/ex-adt.pdf.

## Further Reading

Since the 1970s the theory of algebraic data type specifications has been elaborated in great detail and produced a large amount of research literature; still there are fewer textbooks on this subject available than one might expect. Loeckx's, Ehrich's and Wolf's book [103] achieves a very good balance between mathematical rigor and readability (by presenting many examples and relying mainly on basic notions from algebra and logic while avoiding the more advanced notions from category theory as much as possible). Wirsing's handbook chapter [181] provides a compact overview on a large amount of the theory of algebraic specifications. A most comprehensive and up-to-date resource is Sannella's and Tarlecki's monograph [158] which provides an in-depth treatment of the subject based on category theory; nevertheless, it also discusses an approach to formal program development based on algebraic specifications. The main focus of these and other resources is the semantics of abstract data types; the aspect of reasoning about abstract data type specifications is shortly sketched in [103] and treated with somewhat greater detail in [158]. Goguen's and Malcom's book [58] discusses the application of the influential algebraic specification language OBJ in software engineering by a number of case studies. Alagar's and Periyasamy's book [4] gives in Chap. 13 a short survey on algebraic specifications and OBJ and describes in Chap. 14 in more detail the specification language Larch.

The Common Algebraic Specification Language CASL designed by the Common Framework Initiative CoFI is an attempt to a comprehensive data type specification language that integrates features from many earlier languages. CASL is described from the practical point of view in the user manual [19] by Bidoit and Mossess (with chapters by Mossakowski, Sannella, and Tarlecki) and from the theoretical point of view in the reference manual [126] edited by Mossess; the later gives in particular a complete formal semantics for CASL. The specification language presented in this chapter is essentially a simplified version of CASL; also its formalization by a static semantics and a model semantics has been inspired by the corresponding CASL semantics but has been presented in a different style. The full CASL language also supports partial operations, subsorts, more expressive structuring mechanisms, and architectural specifications that have not been addressed in this chapter. Furthermore, the underlying logical system of CASL is based on the concept of institutions which allows to integrate into a common framework various sublanguages with different logics. A compact account on CASL is given in Mossakowski's, Haxthausen's, Sannella's and Tarlecki's book chapter [123].

Most texts focus on the initial semantics required for the specification of finite data types. For the dual notion of the final specification semantics required for infinite data types one mainly has to refer to the research literature. Mossakowski, Schröder, Roggenbach, and Reichl describe in [124] the language CoCASL which extends CASL by dualizing the notions of generated and free types to cogenerated and cofree types; the introduction to data types with final semantics given in this chapter is partially modeled after this presentation. The references at the end of Chap. chapter:recursion have already listed related literature on coalgebra and coinduction.

## Abstract Data Types in CafeOBJ and CASL

In this section we introduce two software systems that support algebraic specifications of abstract data types, each in its own way:

- *CafeOBJ* [40,54,55] is an algebraic specification language in the tradition of OBJ. It is based on a many-sorted equational logic extended by subsorts, unidirectional transitions, and hidden sorts with a notion of behavioral equivalence. A subset of CafeOBJ is executable: the core of the CafeOBJ software is a term rewriting system that allows to execute initial specifications with restricted forms of conditional equations as axioms. By term rewriting, also proofs by structural induction or searches for specific reduction sequences can be performed.
- The *Heterogeneous Tool Set Hets* [122] is a software framework for integrating various specification languages, most prominently *CASL* and its various extensions such as CoCASL. Hets constructs from CASL specifications 'development graphs' which structure the proofs that have to be performed to ensure various semantic constraints with which the specifications may be annotated; for proving consistency, the ideas sketched in the previous chapter have been implemented in a formal calculus [153]. Proofs of user-specified theorems are performed with the help of external automatic and interactive provers.

A comparison of the languages of CafeOBJ and CASL can be found in [125].

The specifications used in the following presentations can be downloaded from the URLs

    https://www.risc.jku.at/people/schreine/TP/software/adt/adt.cafe
    https://www.risc.jku.at/people/schreine/TP/software/adt/adt.casl

and loaded by executing from the command line the following commands:

    cafeobj adt.cafe
    hets adt.casl

### CafeOBJ

CafeOBJ is a text-only system that is operated in a terminal; see Fig. 6.16 for the startup message printed by the system. We start by writing a small specification of the abstract data type 'integer numbers':

    module! MYINTCORE {
      protecting (NAT)

```
                              Terminal                    _  □  ×
alan!65> cafeobj

-- loading standard prelude

            -- CafeOBJ system Version 1.5.9(PigNose0.99) --
                 built: 2018 Dec 21 Fri 9:06:41 GMT
                      prelude file: std.bin
                              ***
                    2021 Mar 18 Thu 12:30:12 GMT
                         Type ? for help
                              ***
                 -- Containing PigNose Extensions --
                              ---
                         built on SBCL
                         1.4.14.debian
CafeOBJ> ?
You are at top level, no context module is set.

** Here are commands for CafeOBJ online help system.
'?com [<class>]'         Shows available commands classified by <class>,
                         ommiting <classy> shows a list of <class>.
'? <name>'               Gives the reference manual description of <name>
'?ex <name>'             Similar to '? <name>', but in this case
                         shows examples if available.
'?ap <term> [<term>] ...' Searches all available online docs for the terms
                         passed. Type '? ?ap' for more detailed descriptions.
** Typing 'com' will show the list of major toplevel commands.
** URL 'http://cafeobj.org' provides anything you want to know about CafeOBJ.

CafeOBJ> ▮
```

**Fig. 6.16** CafeOBJ

```
    [ Int ]
    op int : Nat Nat -> Int

    vars N1 N2 : Nat
    ceq int(N1,N2) = int(p(N1),p(N2)) if N1 =/= 0 and N2 =/= 0.
}
```

This specification can be written either on the command-line or into a text file, e.g. adt.cafe; then the command

    input adt.

reads and processes the file. The specification introduces a module MYINTCORE which defines the core of the abstract data type; the exclamation mark in the keyword module! indicates that for the initial interpretation of the specification is desired (this is essentially just a hint for the human user, the system treats all modules alike). The module imports the abstract data type NAT which is subsequently extended by the specification; this data type is part of the system library and provides an efficient implementation of the natural

numbers (based on machine integers). The keyword protecting indicates
that the interpretation of that type shall be preserved, i.e., not modified by
the extension (again this is just a hint for the user). The module then introduces a sort Int with a constructor int that maps pairs of natural numbers to
integers; the idea is that the term int$(N_1, N_2)$ denotes the integer $N_1 - N_2$.

The vars clause introduces universally quantified variables which may be
used in subsequent axioms. The keyword ceq indicates that the given axiom is
a conditional equation; i.e., the equation on the left hand side is true, provided
that the condition on the right hand side holds. The right hand side may be
a propositional combination of equations $T_1$ == $T_2$ where $T_1$ =/= $T_2$ is a
shortcut for not $T_1$ == $T_2$. The CafeOBJ system treats axiomatic equations
as left-to-right rewrite rules; thus the given axiom says that any occurrence of
a term of form int$(N_1, N_2)$ may be rewritten to the term int$(p(N_1), p(N_2))$
provided that the stated condition holds. The operation p imported from NAT
represents the predecessor function $\lambda x.\ x - 1$; thus the conditional equation
all in all states that in an application of int$(N_1, N_2)$ to non-zero values
$N_1$ and $N_2$ both $N_1$ and $N_2$ may be replaced by their predecessors. The
predicates == respectively =/= actually represent 'reduction (in)equality'; for
determining their truth value the system reduces both argument terms as
much as possible (until no more rewriting rule can be applied); the predicates
are then considered as true if the resulting terms are identical respectively
different.

If we would have not used the builtin representation of the natural numbers but provided our own definition in a specification MYNAT, we could have
written the axiom simply as

```
eq int(s(N1),s(N2)) = int(N1,N2).
```

Here the keyword == indicates that the axiom is an unconditional equality.
The constructor s imported from MYNAT represents the successor function
$\lambda x.\ x + 1$; the constraint that the reduction rule can be only applied to non-
zero values could be then expressed by pattern-matching. In any case, the
definition is executable; by executing

```
open MYINTCORE.
```

we enter the name space of the module such that we can execute

```
reduce int(5,3).
```

which shows by the output

```
-- reduce in %MYINTCORE : (int(5,3)):Int
(int(2,0)):Int
(0.0000 sec for parse, 0.0040 sec for 26 rewrites + 36 matches)
```

that 26 rewrite rules have been applied to reduce the given term to its canonical form int(2,0). By setting the option

```
set trace on.
```

the application of all rewrite rules can be indeed monitored (we omit the verbose output). By executing

```
close.
```

we leave the name space of the module again.

We continue by extending the data type by a couple of operations:

```
module* MYINT {
  protecting (MYINTCORE)

  op 0 : -> Int
  op _ + _ : Int Int -> Int
  op _ <= _ : Int Int -> Bool

  vars N1 N2 M1 M2 : Nat
  eq 0 = int(0,0).
  eq int(N1,N2) + int(M1,M2) = int(N1 + M1,N2 + M2).
  eq int(N1,N2) <= int(M1,M2) = N1 + M2 <= M1 + N2.
}
```

Here an integer constant 0 is introduced (constants are in CafeOBJ just operations without arguments), a binary integer function + and a binary integer predicate <= (predicates are just operations into the predefined sort Bool with constants true and false); CafeOBJ allows to use infix notation for the binary operations. All three operations are uniquely defined by axiomatic equations; thus we indicate by the asterisk in the keyword module* that a loose interpretation of the extension suffices (again this is just a hint to the user). We may also compute with this specification, e.g. if we execute

```
open MYINT.
reduce int(5,3) + int(2,7).
close.
```

we get the result

```
-- reduce in %MYINT : (int(5,3) + int(2,7)):Int
(int(0,3)):Int
(0.0000 sec for parse, 0.0040 sec for 67 rewrites + 93 matches)
```

Next we are defining the core of a generic type 'list of elements':

```
module* ELEM { [ Elem ] }

module! LISTCORE[ E :: ELEM ] {
  [ List ]
  op empty : -> List
  op cons : Elem List -> List
}
```

The loosely interpreted specification ELEM introduces a sort Elem; this specification is used for the parameter of the  initially interpreted generic specification LISTCORE which introduces a sort List with constructors empty and cons. A generic module may in CafeOBJ have multiple parameters whose identities can be distinguished by the given name (E in above example); if there should be two ELEM parameters with name E1 and E2, we could distinguish by the notation Elem.E1 and Elem.E2 their respective sorts.

Furthermore, we extend the core type by the usual operations:

```
module* LIST[ E :: ELEM ] {
  protecting (LISTCORE(E))
  protecting (NAT)

  op head : List -> Elem
  op tail : List -> List
  op append : List List -> List
  op length : List -> Nat

  var E : Elem
  vars L L1 L2 : List

  eq head(cons(E, L)) = E.
  eq tail(cons(E, L)) = L.

  eq append(empty, L2) = L2.
  eq append(cons(E,L1), L2) = cons(E, append(L1, L2)).

  eq length(empty) = 0.
  eq length(cons(E,L)) = 1 + length(L).
}
```

Now we instantiate the generic type LIST with above type MYINT:

```
view INT->ELEM from ELEM to MYINT { sort Elem -> Int }
module* INTLIST { protecting (LIST(INT->ELEM)) }
```

The `view` declaration introduces a morphism `INT->ELEM` that maps the signature of `ELEM` to the signature of `MYINT`. We then define the module `INTLIST` by the application of `LIST` to this view and thus derive the type 'list of integer numbers'. By the commands

```
open INTLIST.
let L =
   append(cons(int(3,1),cons(int(5,8),empty)),cons(int(12,7),empty)).
reduce L.
reduce length(L).
close.
```

we locally define a list L; first we compute its canonical form, second its length. The resulting output is

```
-- setting let variable "L" to : append(...) : List

-- reduce in %INTLIST : (append(...)):List
(cons(int(2,0),cons(int(0,3),cons(int(5,0),empty)))):List
(0.0000 sec for parse, 0.0040 sec for 123 rewrites + 175 matches)

-- reduce in %INTLIST : (length(append(...)):Nat
(3):NzNat
(0.0000 sec for parse, 0.0000 sec for 148 rewrites + 215 matches)
```

While above examples have demonstrated the suitability of CafeOBJ for executing specifications of a certain form, the capability of the underlying term rewriting engine may be also applied to certain forms of reasoning. As an example, we demonstrate the proof of

```
length(append(L1,L2)) == length(L1)+length(L2)
```

for arbitrary integer lists L1 and L2. Since sort `List` is generated with constructors `empty` and `cons`, we may perform this proof by structural induction over L1.

First we start the proof of the base case by executing

```
open INTLIST.
op L2 : -> List.
```

Here we enter the name space of INTLIST which we extend by a new list constant L2. We then show that the equality holds for empty and L2:

```
reduce length(append(empty,L2)) == length(empty) + length(L2).
```

which is indeed confirmed:

```
-- reduce in %INTLIST : (... == ...):Bool
(true):Bool
(0.0000 sec for parse, 0.0000 sec for 4 rewrites + 13 matches)
```

Next we introduce a new list constant L1 which allows us to state the induction assumption (namely that the property holds for L1 and L2) by an additional rewrite rule:

```
op L1 : -> List.
eq length(append(L1,L2)) = length(L1) + length(L2).
```

Finally we introduce a new integer constant I which allows us to formulate the induction step (namely the claim that the property holds for cons(I,L1) and L2):

```
op I : -> Int.
reduce length(append(cons(I,L1),L2)) == length(cons(I,L1)) + length(L2).
```

Indeed the output

```
-- reduce in %INTLIST : (... == ...):Bool
(true):Bool
(0.0000 sec for parse, 0.0000 sec for 5 rewrites + 84 matches)
```

also confirms this claim. CafeOBJ may thus help to perform those kinds of proofs which can be reduced to equality reasoning (or also to a search for reduction/transition sequences, which we will not discuss further).

### CASL and Hets

The heterogeneous toolset Hets can be started from the command line with a list of CASL specification files as arguments; it then analyzes the correctness of the syntax and of the static semantics of the specifications. For instance, for the input file adt.casl whose content will be explained below, the tool produces the following output:

```
> hets adt.casl
Analyzing library adt
Downloading Basic/Numbers ...
Analyzing library Basic/Numbers version 1.0
Analyzing spec Basic/Numbers#Nat
Analyzing spec Basic/Numbers#Int
Analyzing spec Basic/Numbers#Rat
Analyzing spec Basic/Numbers#DecimalFraction
... loaded Basic/Numbers
Analyzing spec adt#MyIntCore
Analyzing spec adt#MyInt
Analyzing spec adt#Elem
Analyzing spec adt#ListCore
Analyzing spec adt#List
Analyzing spec adt#IntList
Analyzing spec adt#ListProof
```

The content of file `adt.casl` represents the CASL counterpart to the CafeOBJ specifications given in the previous section. It starts with a header

```
library adt
from Basic/Numbers get Nat
```

which ensures that the data type `Nat` from the standard library can be subsequently used. It then continues with the specification

```
spec MyIntCore = Nat then
  free {
    type Int ::= int(p:Nat;m:Nat)
    forall n1,n2:Nat
    . int(suc(n1),suc(n2)) = int(n1,n2)
  }
end
```

which defines the core of the type 'integer numbers' as an extension of the given type `Nat`: the type declaration introduces a sort `Int` with a binary constructor `int` from `Nat` to `Int` and two corresponding selectors `p` and `m`, i.e., for any `Int` value $i$, we have $i = \text{int}(\text{p}(i),\text{m}(i))$. CASL is built upon full first-order logic, thus the specification contains a quantified formula as an axiom. The free interpretation of the extension constrained by this axiom ensures that every integer has a canonical representation. The annotation %mono asserts that the extension is *monomorphic*, i.e., that every algebra $N$ of `Nat` is extended to at least one algebra $I$, and that any two extensions $I, I'$ of $N$ are isomorphic.

We continue by extending the core type by some operations:

```
spec MyInt = MyIntCore then
  op 0: Int = int(0,0)
  op  __+__(i1,i2:Int): Int = int(p(i1)+p(i2),m(i1)+m(i2))
  pred __<=__(i1,i2:Int) <=> p(i1)+m(i2) <= p(i2)+m(i1)
end
```

A constant is just a zero-ary operation, but predicates are in CASL different from operations. Above specification introduces these entities by definitions but the function and the predicate could have also been introduced in an axiomatic form:

```
op __+__: Int * Int -> Int
pred __<=__: Int * Int
forall p1, m1, p2, m2: Nat
. int(p1,m1) +  int(p2,m2) =   int(p1+p2,m1+m2)
. int(p1,m1) <= int(p2,m2) <=> p1+m2 <= p2+m1
```

The annotation %def asserts that the extension is *definitional*, i.e., that every algebra $I$ of MyIntCore is extended to exactly one algebra $I'$. The annotations %mono and %def are special cases of the annotation %cons which just states that an extension is *conservative*, i.e., that every algebra $I$ of the original type is extended to at least one algebra $I'$; as we will see below, it is easier to show that an extension is just conservative than to show that it is also monomorphic or definitional.

For specifying the type 'list of elements', we start with the specification

```
spec ListCore[sort Elem] = %mono
  free type List[Elem] ::= empty | cons(Elem,List[Elem])
end
```

where the generic specification ListCore extends by a free type declaration every argument type with a sort Elem in a monomorphic way. The specification introduces a sort with the compound name List[Elem] with constructors empty and cons; the sorts resulting from specific instantiations of the generic specification will thus receive correspondingly instantiated names.

We could have also written the type declaration as

```
free type List[Elem] ::= empty | cons(head:?Elem,tail:?List[Elem])
```

which additionally introduces two partial selectors head and tail; these operations are only defined on values constructed by application of cons. We could also introduce them in an axiomatic way

```
op head: List[Elem] ->? Elem
op tail: List[Elem] ->? List[Elem]

forall l:List[Elem]; e:Elem
. def head(l) <=> not l = empty
. head(cons(e,l)) = e
. def tail(l) <=> not l = empty
. tail(cons(e,l)) = l
```

where the arrows ->? indicate that the operations are partial and the corresponding def predicates denote by preconditions the domains of these operations. However, since the selectors are subsequently not used (and adding the additional axioms prevents a quick automatic proof given below), we do without them.

Now we equip the data type with additional operations:

```
spec List[sort Elem] given Nat = ListCore[sort Elem] then %def
  op append: List[Elem] * List[Elem] -> List[Elem]
  forall l1,l2:List[Elem]; e:Elem
  . append(empty,l2) = l2
  . append(cons(e,l1),l2) = cons(e,append(l1,l2))

  op length: List[Elem] -> Nat
  forall l:List[Elem]; e:Elem
  . length(empty) = 0
  . length(cons(e,l)) = 1+length(l)
end
```

The given clause imports the specification Nat in such a way that it also can appear as (a part of) an argument in an instantiation of the specification (the previous chapter used the keyword import for this purpose). For instance, we may now define the type 'list of integers' as

```
spec IntList = List[MyInt fit Elem |-> Int]
```

Finally, we introduce by an extension

```
spec ListProof[sort Elem] = List[sort Elem] then %implies
  forall l1,l2:List[Elem]
  . length(append(l1,l2)) = length(l1)+length(l2)
end
```

**Fig. 6.17** Hets development
graph

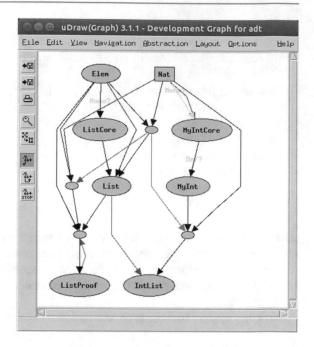

an additional axiom; the annotation %implies indicates that the extension
is *implied*, i.e., that the original type is identical to the extended type.

The annotations given in the specifications represent claims that have to
be proved; the remainder of this section demonstrates how Hets supports
these proofs. By typing

```
hets -g adt.casl
```

Hets is started in a graphical mode where the window illustrated in Fig. 6.17
is displayed. This window shows the 'development graph' of the included
specifications; in this graph the named nodes represent specifications and
the arrows represent dependencies among specifications. The black arrows
represent 'definition links' that indicate that a specification is used in the definition of another specification; the colored arrows represent 'theorem links'
that postulate relations between the theories; these links thus represent proof
obligations that have to be handled.

We start by selecting in menu Edit the entry Proofs and from the sub-
menu the Auto-DG-Prover which applies the rules of the proof calculus for
development graphs. This reduces the original proof obligations to the core
obligations that we have to deal with; the results are shown in the left diagram of Fig. 6.18. The grey labels Mono? and Def? represent the obligations
to prove that the corresponding extensions are monomorphic respectively
definitional; the red node indicates the obligation to prove the additional
axiom in the implied extension.

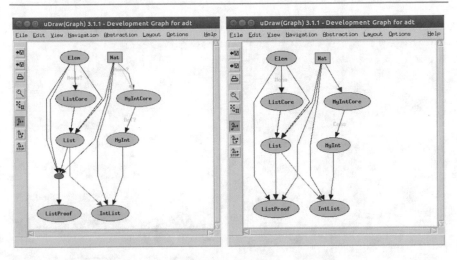

**Fig. 6.18** Hets development graph (before and after proof)

By selecting the link labeled Mono? between Elem and ListCore and right-clicking the mouse, a menu pops up from which we may select the entry Check conservativity. Indeed the builtin prover is able to deduce that the extension is monomorphic and the question mark in the label disappears. However, for the other two links labeled Mono? and Def? the resulting window shows that the prover can only deduce that the extensions are conservative, not that they are monomorphic respectively definitional. Since the first one should be actually easy to establish (only an equational axiom is provided), further investigations demonstrate that using the general kind of free { } specifications lets the proof always fail (while a corresponding proof with a free type declaration works); we thus suspect a limitation of the prover. However, that the second one could not be established, is not surprising: it demands convoluted reasoning that equations over free types with axioms (representing quotient term algebras) are indeed definitional. We thus replace the corresponding annotations by the simpler annotation %cons for which the checks succeed: the edges are subsequently labeled Cons.

It then remains to prove the formula introduced by the %implies clause. After selecting with the mouse the red node, a right-click shows a menu from which we select the Prove entry; this lets the proof management GUI pop up that is displayed as the left window in Fig. 6.19. Here we see in the list Goals the formula Ax1 to be proved; by selecting this formula and pressing the button Display, the window shown at the bottom of Fig. 6.19 pops up and displays the formula. Furthermore, we may select in the list Pick theorem prover from a choice of automatic and interactive provers the one we wish to apply for the given task.

Since the stated axiom crucially depends on equational reasoning, we choose by the entry eprover the theorem prover E which is a powerful automatic prover for first-order logic with equality. Furthermore, since the proof is

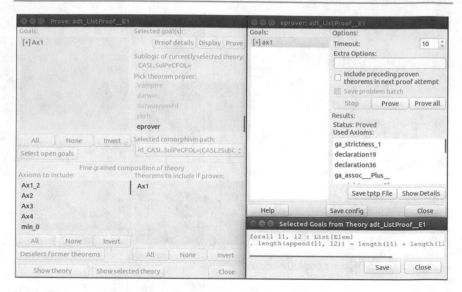

**Fig. 6.19** Hets proof management GUI

based on the principle of structural induction, we select in the list Selected
comorphism path a sequence of logic translations from CASL to E which
ends with the translation CASL2SoftFOLInduction2 that replaces goals with
induction premises. We then press the button Prove which lets the interface
to the E prover pop us that is displayed in the right window in Fig. 6.19.
Pressing the button Prove in that window lets the proof almost immediately succeed (however, if we would not have removed the partial selectors
head and tail from the specification, the proof would even after a minute
not have terminated yet). Thus also the red node in the development graph
disappears; the resulting view is depicted to the right of Fig. 6.18.

## Summary

The main benefit of CafeOBJ is that it allows to validate certain specifications by executing them and investigating the outcomes. This allows to
rapidly prototype an abstract data type by first modeling and analyzing it
in CafeOBJ; once its properties are thoroughly understood, it may be implemented in a more efficient form in a real programming language. However,
this is only possible for specifications with initial semantics whose axioms are
expressed in a restricted form of conditional equational logic, which resembles very much functional programming; the data type specifications thus
look more like concrete programs than abstract theories.

The characteristic feature of CASL is its expressiveness which allows to
write specifications on a very high-level of abstraction by leveraging the full
power of first-order logic without being restricted by considerations of exe-
cutability. Certain important aspects (such as the conservativity of exten-

sions) may be fully automatically checked, albeit only for restricted forms of specifications. Also the specifier may state general theorems which can be proved with computer assistance. Here fully automatic proving, however, is only rarely successful; typically (at least partially) interactive proofs are required. CASL/Hets thus represents a framework for building and analyzing libraries of high-level data type theories; the comprehensive CASL standard library may serve as a starting point for own developments.

# Programming Languages

7

*Alles ist eine Frage der Sprache. (Everything is a question of language.)*

— *Ingeborg Bachmann (Alles)*

In daily life, virtually all of human communication is expressed in one of the thousands of natural languages that are spoken world-wide; these languages are rich in their expressive capabilities, flexible in their applications, subtle in their nuances, and beautiful in their form. However, they are also full of gaps and ambiguities; while most of these can be usually overcome by intelligent beings that are able to deduce the intended interpretation from the context of the communication, they are from time to time are also the source of misunderstandings and disagreements, minor mishaps as well as major disasters. Thus, when communicating with ignorant partners such as computers, software developers use artificial languages that are designed in order to unambiguously express their intentions of how a computer program shall operate to solve a specific computational problem. However, even if millions of software developers use such programming languages every day, it is probably fair to say that only a minor fraction understands these languages in a sufficient depth to be able to answer subtle and critical questions about the behavior of the resulting programs. Ultimately, such an in-depth understanding requires a formal basis.

The goal of this chapter is to provide such a basis by showing how the semantics of programming languages can be precisely described in the language of logic, using the same kinds of techniques that have been introduced in the previous chapters for modeling 'mathematical' languages. For this purpose, building upon the language of data types introduced in Chap. 6, Sect. 7.1 introduces an imperative programming language, i.e., a language whose core elements are commands that operate by reading from and writing to a common store. For this language we will give a formal type system; only well-typed programs will subsequently receive a semantics. Then

Sect. 7.2 gives this language a 'denotational' semantics that interprets commands as functions on stores; these functions are partial, i.e., may not return a result, which indicates that a program aborts or loops forever. Because partial functions are comparably inconvenient to deal with, we subsequently switch from a functional semantics to a relational one that allows arbitrarily many outcomes, which will also become useful in later chapters. Based on these results, we are able to prove the correctness of program transformations such as loop unrolling.

As an alternative framework, Sect. 7.3 introduces an 'operational' semantics that models the programming language by a logical inference system whose judgements describe the transitions of a program from one state to another; here we differentiate between a 'big-step' semantics that describes transitions from the initial to the final state and a 'small-step' semantics that describes also the transitions between intermediate states. Each of these semantic forms (denotational semantics in functional and in relational flavor, operational semantics with big steps and with small steps) provides a different point of view on the same language; we formulate the precise relationships between the various forms and prove several equivalence results. As one application of semantics, Sect. 7.4 shows how to prove the correctness of a translation (compilation) of the high-level programming language considered so far to a formalized low-level machine language. Finally, Sect. 7.5 extends the programming language by procedures and discusses the formal modeling of core programming language concepts such as declaration scopes (static vs. dynamic scoping), parameter passing mechanisms (value vs. reference parameters), and recursive procedure definitions.

## 7.1    Programs and Commands

### A Program Syntax

We start by defining the syntax of a simple imperative programming language. A program in this language operates by the execution of commands that read and write the values of variables. The program communicates with its environment by the values of certain given variables, the *parameters* of the program: the environment starts the program with some initial values of the parameters and, provided that the program terminates, observes their final values. For its computation, the program may also use some *local variables* which are, however, not seen by the environment.

**Definition 7.1**  *(Program)* A *program* is a phrase $P \in Program$ which is formed according to the following grammar:

$P \in Program,\ X \in Parameters,\ C \in Command$
$F \in Formula,\ T \in Term,\ V \in Variable,\ S \in Sort,\ I \in Identifier$

$P \ ::= \ \text{program}\ I(X)\ C$

$$X ::= \_ \mid X, V : S$$
$$C ::= V := T \mid \text{var } V : S; \, C \mid C_1; C_2 \mid$$
$$\mid \text{if } F \text{ then } C_1 \text{ else } C_2 \mid \text{if } F \text{ then } C \mid \text{while } F \text{ do } C$$

Here *Parameters* denotes the syntactic domain of *parameters* while *Command* denotes the syntactic domain of *commands*. The syntactic domains *Formula* of formulas, *Term* of terms, *Variable* of variables, *Sort* of sorts, and *Identifier* of identifiers, are formed as in Definition 6.1.

For readability, we use the syntax $\{C_1; C_2; \dots; C_n\}$ to denote the command $C_1; C_2'$, where $C_2'$ denotes the command $C_2; C_3'$, where $C_3'$ denotes the command $C_3; C_4'$, ..., and where finally $C_{n-1}'$ denotes the command $C_{n-1}; C_n$. Furthermore, we use the syntax $\{\text{var } V : S; C_1; C_2; \dots; C_n\}$ to denote the command $\text{var } V : S; \{C_1; C_2; \dots; C_n\}$.

**Example 7.1** As an example, take the following program *gcd* which computes the greatest common divisor $g$ of two natural numbers $a, b$ by the Euclidean algorithm:

```
program gcd(a:Nat,b:Nat,g:Nat,n:Nat) {
  var c:Nat; c := 0;
  while a > 0 ∧ b > 0 do {
    if a ≥ b
      then a := a-b
      else b := b-a
    c := c+1
  }
  if a > 0
    then g := a
    else g := b;
  n := c
}
```

This program interacts with its environment by the parameters $a, b, g, n$. If $a$ and $b$ are initially not both zero, the program ultimately terminates with $g$ set to the greatest common divisor of $a$ and $b$; furthermore, it sets $n$ to the number of loop iterations needed to compute that value. In the course of the computation, the program modifies both $a$ and $b$ such that one becomes zero and the other one the greatest common divisor; the program also uses a local variable $c$ to keep track of the number of iterations. For instance, for the initial variable values $a = 12, b = 8$ (and $g$ and $n$ arbitrary), the program terminates with $a = 0$, $b = 4$, $g = 4$, and $n = 0$. For initial values $a = b = 0$, the program terminates with $a = b = g = n = 0$.

The program uses atomic formulas like $a > 0$ which denotes the application $gt(a, 0)$ of some predicate $gt$, terms like $a - b$ which denotes the application $minus(a, b)$ of some function $minus$, and terms like 0 which denotes the occurrence of some constant *zero*. □

**Rules for $\Sigma \vdash P$: program:**

$$\frac{\Sigma \vdash X : \text{parameters}(Vt, Ss) \quad \Sigma, Vt \vdash C : \text{command}}{\Sigma \vdash \texttt{program } I(X)\, C : \text{program}}$$

**Rules for $\Sigma \vdash X$: parameters($Vt, Ss$):**

$$\frac{Vt = \varnothing \quad Ss = [\,]}{\Sigma \vdash {}_\sqcup : \text{parameters}(Vt, Ss)}$$

$$\frac{\Sigma \vdash X : \text{parameters}(Vt', Ss') \quad S \in \Sigma.\text{s} \quad \neg \exists S' \in Sort.\ \langle V, S' \rangle \in Vt' \quad Vt = Vt' \leftarrow \{\langle V, S \rangle\} \quad Ss = Ss' \circ [S]}{\Sigma \vdash X, V : S : \text{parameters}(Vt, Ss)}$$

**Rules for $\Sigma, Vt \vdash C$: command:**

$$\frac{\langle V, S \rangle \in Vt \quad \Sigma, Vt \vdash T : \text{term}(S)}{\Sigma, Vt \vdash V := T : \text{command}} \qquad \frac{S \in \Sigma.\text{s} \quad Vt_1 = Vt \leftarrow \{\langle V, S \rangle\} \quad \Sigma, Vt_1 \vdash C : \text{command}}{\Sigma, Vt \vdash \texttt{var } V : S\,; \ C : \text{command}}$$

$$\frac{\Sigma, Vt \vdash C_1 : \text{command} \quad \Sigma, Vt \vdash C_2 : \text{command}}{\Sigma, Vt \vdash C_1 ; C_2 : \text{command}}$$

$$\frac{\Sigma, Vt \vdash F : \text{formula} \quad \Sigma, Vt \vdash C_1 : \text{command} \quad \Sigma, Vt \vdash C_2 : \text{command}}{\Sigma, Vt \vdash \texttt{if } F \texttt{ then } C_1 \texttt{ else } C_2 : \text{command}}$$

$$\frac{\Sigma, Vt \vdash F : \text{formula} \quad \Sigma, Vt \vdash C : \text{command}}{\Sigma, Vt \vdash \texttt{if } F \texttt{ then } C : \text{command}} \qquad \frac{\Sigma, Vt \vdash F : \text{formula} \quad \Sigma, Vt \vdash C : \text{command}}{\Sigma, Vt \vdash \texttt{while } F \texttt{ do } C : \text{command}}$$

**Fig. 7.1** Type checking programs

## Type Checking Programs

The well-formedness of a program is checked relative to a given signature $\Sigma$ of sorts and operations that may be used in terms and formulas (see Sect. 6.2 for the essential concepts). In particular, we recall the notion of a *variable typing* (see Definition 6.5) for which we introduce the following notion of the update of a typing by another typing (which will become handy later).

**Definition 7.2** (*Variable Typing Update*) Given variable typings $Vt_1$ and $Vt_2$ for a signature $\Sigma$, we denote by $Vt_1 \leftarrow Vt_2$ the variable typing that arises from $Vt_1$ by adding all mappings from $Vt_2$ (overriding all conflicting mappings in $Vt_1$):

$$. \leftarrow . : VarTyping^\Sigma \to VarTyping^\Sigma$$
$$Vt_1 \leftarrow Vt_2 := \left(Vt_1 \setminus \left\{\langle V, S \rangle \mid \langle V, S \rangle \in Vt_1 \wedge \exists S' \in Sort.\ \langle V, S' \rangle \in Vt_2\right\}\right) \cup Vt_2$$

Now Fig. 7.1 depicts a calculus for checking the type-correctness of a program. From this, the main judgement

$$\Sigma \vdash P : \text{program}$$

can be derived if program $P$ is well-formed with respect to $\Sigma$. The auxiliary judgement

$$\Sigma \vdash X : \mathsf{parameters}(Vt, Ss)$$

can be derived if the parameter list $X$ gives rise to the variable typing $Vt \in VarTyping^\Sigma$ and to the sequence $Ss \in Sort^*$ of parameter sorts. The judgement

$$\Sigma, Vt \vdash C : \mathsf{command}$$

can be derived if for the given variable typing $Vt$, command $C$ is well-formed. Furthermore, the typing calculus refers to judgements for checking the well-formedness of formulas and terms; these have been introduced in Sect. 6.2.

The remainder of this chapter is dedicated to giving a meaning to programs that are well-formed according to this calculus.

## 7.2 A Denotational Semantics

Our goal is to give the programming language defined in the previous section a formal semantics (as we have done for various languages in the preceding chapters) by mapping every phrase of the program to a mathematical entity. This style of program semantics is also called *denotational semantics* (we will subsequently also deal with other kinds of program semantics).

**Partial Functions**

We will essentially define the meaning of a well-formed command, the core of a program, as a function from states to states, where a state is a function from program variables to values. However, since a command does not necessarily terminate, we will actually model states as *partial functions*, i.e., as binary relations that map an argument to at most one result (see also Sect. 5.7 where such functions were introduced). For instance, if a command $C$ does not terminate on an initial state $s \in State$, the application of the partial function $[\![ C ]\!] : State \to_\perp State$ on $s$ is not defined. Given a partial function $f : A \to_\perp B$, we therefore define the *domain* of $f$

$$domain\, f := \{x \in A \mid \exists y \in B.\ \langle x, y \rangle \in f\}$$

to denote the set of all arguments $x \in A$ for which the application of $f$ yields a result $y \in B$.

However, since the domain $domain(f)$ of a partial function $f : A \to_\perp B$ is not necessarily identical to $A$, the unrestricted application of a partial function (which logically must denote a value) is problematic. We will therefore refrain from direct applications of $f$ but restrict its use to formulas of the form

$$\mathsf{exists}\ y = f(x).\ F.$$

This notation was already introduced by Definition 5.13 as an abbreviation of

$$\exists y \in B. \ \langle x, y \rangle \in f \wedge F.$$

Here $F$ is a subformula which may refer to the result $y \in B$ of the application of $f$ to argument $x \in A$. The whole formula is 'false' if there is no such value, i.e., $f$ is not defined on $x$. Dually, we have defined the formula

$$\textbf{forall } y = f(x). \ F.$$

as an abbreviation for

$$\forall y \in B. \ \langle x, y \rangle \in f \Rightarrow F.$$

This formula is 'true' if $f$ is not defined on $x$.

Furthermore, for defining a partial function $f: A \rightarrow_{\perp} B$, we will use the format

$$f(x) = y \Leftarrow F[x, y]$$

where $F$ is a formula with free variables $x$, $y$, such that for every argument $x \in A$ there is at most one result $y \in B$ such that $F[x, y]$ holds. Such a definition actually introduces a relation

$$f := \{\langle x, y \rangle \mid x \in A \wedge y \in B \wedge F[x, y]\}$$

which, by the constraint on $F$, however, denotes a partial function.

**Values and States**

Like the well-formedness of a program depends on a given signature $\Sigma$ of sorts and operations, the meaning of the program depends on a given $\Sigma$-algebra $A$ that maps the operations to constants, functions, and predicates operating on values from $A$. Likewise, the program operates on variables holding such values by reading/writing them from/to the program state. In the following, we introduce the corresponding domains relevant for the semantics of of such a program.

**Definition 7.3** (*Values and States*) Let $\Sigma$ be a signature and $A$ be a $\Sigma$-algebra. First, we define

$$Value^A := \bigcup_{S \in \Sigma.s} A(S)$$

as the set of all values from $A$, irrespective of their sort. Next, we define

$$State^A := Variable \rightarrow Value^A$$

as the set of all functions of variables to values of $A$; we call these functions
$A$-*states* (short *states*). Finally, we define

$$Default^A := \left\{ d \in \left( \Sigma.s \rightarrow Value^A \right) \, \middle| \, \forall S \in \Sigma.s. \, d(S) \in A(S) \right\}$$

as the set of functions from sorts to values in $A$ of that sort; we call these
functions *default maps*.

The role of default maps is to give newly introduced local program variables initial
values in the state (we will later also construct a semantics that copes without explicit
initial values).

For manipulating states, the notation introduced in Definition 4.13 will become
handy: given a state $s \in State^A$, a variable $V \in Variable$, and a value $v \in Value^A$, the
updated state $s[V \mapsto v] \in State^A$ maps every variable to the same value as $s$ (if any),
except that $V$ is mapped to $v$.

**Fixing the Signature and Algebra**

From now on, we will assume a fixed signature $\Sigma$ and $\Sigma$-algebra $A$ which will be
thus not explicitly introduced any more in definitions and propositions. Likewise,
the notions 'formula', 'term', and 'state' will always denote '$\Sigma$-formula', '$\Sigma$-term',
and '$A$-state'. Furthermore, we will drop from all notions depending on $\Sigma$ or $A$ this
argument, e.g., we will write $[\![ C ]\!]_s$ rather than $[\![ C ]\!]_s^A$.

**Auxiliary Notions**

We introduce the following functions for setting up the initial state of the program
from the initial values of the parameters and for extracting from the final state their
final values.

**Definition 7.4** (*Program Parameters and States*) We write $init(v)$ to denote
that state that maps every variable to value $v$:

$$init: Value \rightarrow State$$
$$init(v) := \lambda V \in Variable. \, v$$

We write $write(X, vs, s)$ to denote that state that results from state $s$ by mapping
parameters $X$ to values $vs$:

$$write: Parameters \times Value^* \times State \rightarrow State$$
$$write(\_, vs, s) := s$$
$$write((X, V{:}S), vs, s) := \text{let } n = length(vs) \text{ in}$$
$$\text{if } n = 0$$

> then $s$
>
> else $write(X, \lambda i \in \mathbb{N}_{n-1}.\ vs(i), s[V \mapsto vs(n-1)])$

We write $read(X, s)$ to denote the sequence of values of parameters $X$ in state $s$:

$$read: Parameters \times State \to Value^*$$
$$read(\_, s) := [\ ]$$
$$read((X, V:S), s) := read(X, s) \circ [s(V)]$$

Furthermore, we will use the following notions.

> **Definition 7.5** (*Functions and Relations on Value Sequences and States*) A
> *value function* is a partial function from value sequences to value sequences:
>
> $$ValueFunction := Value^* \to_\perp Value^*$$
>
> A *state function* is a partial function from states to states:
>
> $$StateFunction := State \to_\perp State$$
>
> A *value condition* is a unary condition on value sequences:
>
> $$ValueCondition := \mathsf{Set}(Value^*)$$
>
> A *state condition* is a unary relation on states:
>
> $$StateCondition := \mathsf{Set}(State)$$
>
> A *value relation* is a binary relation on value sequences:
>
> $$ValueRelation := \mathsf{Set}(Value^* \times Value^*)$$
>
> A *state relation* is a binary relation on states:
>
> $$StateRelation := \mathsf{Set}(State \times State)$$

**The Evaluation of Expressions**

Given a state $s$, we may define the semantics $[\![\, T \,]\!]_s$ of a term $T$ and the semantics $[\![\, F \,]\!]_s$ of a formula $F$ as in Definition 6.12; the role of the assignment in that definition is now taken by the state $s$ such that the meaning of $F$ and $T$ is well-defined for every state. However, this conflicts a bit with the usual interpretation of programming languages where the evaluation of expressions may fail for certain states by a premature abortion of the program.

To model also this behavior (without having to fundamentally change the semantics of terms and formulas), we will introduce conditions on states that indicate whether the state is admissible for the evaluation of a certain expression. For this purpose, we assume for every operation (function or predicate) in $\Sigma$ the existence of a *well-definedness predicate* that is true if and only if the application of the operation to given values is admissible. Based on these predicates, we may define translations of terms respectively formulas to state conditions:

$$[\![\,.\,]\!]\checkmark : Term \to StateCondition$$
$$[\![\,.\,]\!]\checkmark : Formula \to StateCondition$$

Then $[\![\,T\,]\!]\checkmark \langle s \rangle$ respectively $[\![\,F\,]\!]\checkmark \langle s \rangle$ is true if and only if term $T$ respectively formula $F$ is well-defined with respect to state $s$, i.e., if the evaluation of $T$ respectively $F$ in state $s$ is admissible. We omit the details of this simple definition.

By the appropriate application of $[\![\,T\,]\!]\checkmark \langle s \rangle$ respectively $[\![\,F\,]\!]\checkmark \langle s \rangle$ we will make sure that the semantics of a program that evaluates a term or formula in an inadmissible way does not permit any output state.

**A Functional Semantics**

We are now going to model the denotational semantics of programs and commands as partial functions. Figure 7.2 gives such a functional program semantics which consists of the following denotation functions that map syntactic phrases to their semantic values:

- $[\![\,.\,]\!] : Program \to Default \to Value \to ValueFunction$
  We write $[\![\,P\,]\!]^{d,v}(vs) = vs' \Leftarrow F[vs, vs']$ to denote that the semantics of program $P$ for default map $d$ and value $v$ is a partial function from the initial values $vs$ of the program parameters to their final values $vs'$ defined by formula $F$ with free variables $vs$ and $vs'$. The definition of the denotation function states that a program with input values $vs$ first constructs a state $s$ that maps the parameters to these values (and all other variables to the value $v$) and then invokes $C$ on $s$. If this yields a result state $s'$, then $vs'$ consists of the values of the parameters in $s'$.

$[\![\,.\,]\!]$: $\textit{Program} \rightarrow \textit{Default} \rightarrow \textit{Value} \rightarrow \textit{ValueFunction}$

$[\![\,\texttt{program } I(X)\ C\,]\!]^{d,v}(vs) = vs' \Leftarrow$
    let $s = \textit{write}(X, vs, \textit{init}(v))$ in exists $s' = [\![\,C\,]\!]^{d}(s).\ vs' = \textit{read}(X, s')$

$[\![\,.\,]\!]$: $\textit{Parameters} \rightarrow \textsf{Set}(\textit{Value}^{*})$

$[\![\,\sqcup\,]\!]\langle vs \rangle :\Leftrightarrow vs = [\,]$
$[\![\,X, V\!:\!S\,]\!]\langle vs \rangle :\Leftrightarrow \exists v \in A(S),\ vs' \in \textit{Value}^{*}.\ vs = vs' \circ [v] \wedge [\![\,X\,]\!]\langle vs' \rangle$

$[\![\,.\,]\!]$: $\textit{Command} \rightarrow \textit{Default} \rightarrow \textit{StateFunction}$

$[\![\,V\!:=\!T\,]\!]^{d}(s) = s' \Leftarrow$
    $[\![\,T\,]\!]\!\!\checkmark\langle s \rangle \wedge s' = s[V \mapsto [\![\,T\,]\!]_{s}]$
$[\![\,\texttt{var } V\!:\!S;\ C\,]\!]^{d}(s) = s' \Leftarrow$
    let $v = s(V),\ s_1 = s[V \mapsto d(S)]$ in exists $s_2 = [\![\,C\,]\!]^{d}(s_1).\ s' = s_2[V \mapsto v]$
$[\![\,C_1\,;C_2\,]\!]^{d}(s) = s' \Leftarrow$
    exists $s_1 = [\![\,C_1\,]\!]^{d}(s).$ exists $s_2 = [\![\,C_2\,]\!]^{d}(s_1).\ s' = s_2$
$[\![\,\texttt{if } F \texttt{ then } C_1 \texttt{ else } C_2\,]\!]^{d}(s) = s' \Leftarrow$
    $[\![\,F\,]\!]\!\!\checkmark\langle s \rangle \wedge$
    if $[\![\,F\,]\!]_{s}$
       then exists $s_1 = [\![\,C_1\,]\!]^{d}(s).\ s' = s_1$
       else exists $s_2 = [\![\,C_2\,]\!]^{d}(s).\ s' = s_2$
$[\![\,\texttt{if } F \texttt{ then } C\,]\!]^{d}(s) = s' \Leftarrow$
    $[\![\,F\,]\!]\!\!\checkmark\langle s \rangle \wedge$
    if $[\![\,F\,]\!]_{s}$
       then exists $s_1 = [\![\,C\,]\!]^{d}(s).\ s' = s_1$
       else $s' = s$
$[\![\,\texttt{while } F \texttt{ do } C\,]\!]^{d}(s) = s' \Leftarrow$
    let
       inductive $W: \textit{State} \rightarrow_{\perp} \textit{State}$
       $W(s) = s' \Leftarrow$
          $[\![\,F\,]\!]\!\!\checkmark\langle s \rangle \wedge$
          if $[\![\,F\,]\!]_{s}$
             then exists $s_1 = [\![\,C\,]\!]^{d}(s).$ exists $s_2 = W(s_1).\ s' = s_2$
             else $s' = s$
    in exists $s_1 = W(s).\ s' = s_1$

**Fig. 7.2** Denotational semantics of programs and commands (functional version)

- $[\![\,.\,]\!]$: $\textit{Parameters} \rightarrow \textsf{Set}(\textit{Value}^{*})$
  We write $[\![\,X\,]\!](vs) :\Leftrightarrow F[vs]$ to denote that the semantics of a parameter sequence $X$ is a unary relation on value sequences defined by formula $F$ with free variable $vs$. The definition of the denotation function constructs a relation that checks whether the sequence holds for every parameter a value of the type indicated in the parameter declaration. Actually, this denotation function is not used to determine the semantics of a program; rather it will be used later to formulate a proposition about this semantics.

- $[\![\,.\,]\!]$: *Command* $\to$ *Default* $\to$ *StateFunction*

  We write $[\![\,C\,]\!]^d(s) = s' \Leftarrow F[s, s']$ to denote that the semantics of command $C$ for given default map $d$ is a partial function from initial state $s$ to final state $s'$ defined by formula $F$ with free variables $s$ and $s'$. We discuss the semantics of the various kinds of commands in turn:

  - $[\![\,V := T\,]\!]$: here $s'$ is identical to $s$ except that the variable $V$ is mapped to the value of $T$ evaluated on $s$. As discussed in the previous subsection, the evaluation of the semantics $[\![\,T\,]\!]_s$ of term $T$ is guarded by the application $[\![\,T\,]\!]\checkmark\langle s\rangle$ of the well-definedness predicate which ensures that an inadmissible evaluation does not permit any output state $s'$.

  - $[\![\,\text{var}\ V : S;\ C\,]\!]$: from $s$, a state $s_1$ is derived that maps the locally declared variable to the default value $d(S)$ of the sort $S$ of the variable (replacing its previous value, if any). If the execution of command $C$ on $s_1$ yields a result state $s_2$, $s'$ is derived from $s_2$ by restoring its previous value $v$.

  - $[\![\,C_1 ; C_2\,]\!]$: if the execution of $C_1$ on $s$ yields a state $s_1$ and the execution of $C_2$ on $s$ yields a state $s_2$, then $s'$ equals $s_2$.

  - $[\![\,\text{if}\ F\ \text{then}\ C_1\ \text{else}\ C_2\,]\!]$: first the truth value $[\![\,F\,]\!]_s$ of $F$ is determined under the guard of the application $[\![\,F\,]\!]\checkmark\langle s\rangle$ of the well-definedness predicate, which ensures that an inadmissible evaluation does not permit any output state $s'$. In an admissible evaluation, if this value is 'true' and the execution of command $C_1$ on $s$ yields a state $s_1$, then $s'$ equals $s_1$; if the value is 'false' and the execution of command $C_2$ on $s$ yields a state $s_2$, then $s'$ equals $s_2$.

  - $[\![\,\text{if}\ F\ \text{then}\ C\,]\!]$: If this value of formula $F$ is 'true' and the execution of command $C$ on $s$ yields a state $s_1$, then $s'$ equals $s_1$; if the value is 'false', then $s'$ equals $s$, i.e., the execution of the command has no effect on the state.

  - $[\![\,\text{while}\ F\ \text{do}\ C\,]\!]$: this command is the sole reason why commands and thus programs generally denote partial functions, not total ones: if a loop does not terminate on input state $s$, there is no output state $s'$ to return. The denotation of a while loop is determined by the inductive definition of a partial function $W$, i.e., by the least fixed point interpretation of a recursive definition of $W$, see Definition 5.14 (although the formula by which $W$ is defined does not completely conform to the syntactic restrictions imposed by Proposition 5.6, it can be shown that the functional to which the fixed point operator is applied is upward-continuous and that the result indeed denotes a partial function). If the value of formula $F$ in $s$ is 'false', then $s'$ equals $s$, i.e., the execution of the loop has no effect on the state; if the value is 'true' and the execution of $C$ in $s$ (representing the first loop iteration) yields the state $s_1$ and the application of $W$ to $s_1$ (representing remaining loop iteration) yields $s_2$, then $s'$ is identical to $s_2$. If $C$ does not terminate in $s$ (such that there is no $s_1$) or $W$ is not defined on $s_1$ (such that there is no $s_2$), there is no output state $s'$.

To illustrate this semantics, we consider the following program $P$:

```
program P(a:Nat,b:Nat) {
  b := a;
  if b>0 then {
    var a:Nat;
    a := b-1;
    b := a+a
  }
  a := a+1
}
```

We now depict an execution of this program where, for simplicity, we will (ab)use the notation for total function application for the application of partial functions $[\![\,.\,]\!]^d$.

Choosing the initial parameter values $vs = [a = 3, b = 0]$, default map $d = [\text{Nat} \mapsto 0, \ldots]$ and arbitrary value $v$; we get the following evaluation

$$[\![\, \texttt{b:=a; if b>0 then \{...\}; a:=a+1}\,]\!]^d([a \mapsto 3, b \mapsto 0, \ldots])$$
$$= [\![\, \texttt{if b>0 then \{...\}; a:=a+1}\,]\!]^d([a \mapsto 3, b \mapsto 3, \ldots])$$
$$= [\![\, \texttt{a:=a+1}\,]\!]^d([a \mapsto 3, b \mapsto 4, \ldots]) \tag{*}$$
$$= [a \mapsto 4, b \mapsto 4, \ldots]$$

with final parameter values $vs' = [a = 4, b = 4]$. Here the the equality (*) can be justified by the evaluation

$$[\![\, \texttt{if b>0 then \{var a:Nat;a:=b-1;b:=a+a\}}\,]\!]^d([a \mapsto 3, b \mapsto 3, \ldots])$$
$$= [\![\, \texttt{var a:Nat;a:=b-1;b:=a+a}\,]\!]^d([a \mapsto 3, b \mapsto 3, \ldots])$$
$$= \text{let } s = [\![\, \texttt{a:=b-1;b:=a+a}\,]\!]^d([a \mapsto 0, b \mapsto 3, \ldots]) \text{ in } s[a \mapsto 3]$$
$$= \text{let } s = [\![\, \texttt{b:=a+a}\,]\!]^d([a \mapsto 2, b \mapsto 3, \ldots]) \text{ in } s[a \mapsto 3]$$
$$= \text{let } s = [a \mapsto 2, b \mapsto 4, \ldots] \text{ in } s[a \mapsto 3]$$
$$= [a \mapsto 3, b \mapsto 4, \ldots].$$

While this functional version of program semantics is essentially fine, it is a bit cumbersome to deal with because of the necessity to deal with the the application of partial functions which may not yield a result; we are therefore going to present another version that is on the one side theoretically more general and on the other side also practically easier to work with.

## A Relational Semantics

In our new variant of a denotational program semantics, rather than committing ourselves to the special case of programs that have zero or one outcome, we will handle the general case of programs that have arbitrarily many (zero, one, or more than one) outcomes. Therefore we will model programs and commands not any more as functions but rather as binary relations between value sequences respectively states; for any given input there may be arbitrarily many outputs that are in this

$[\![\,.\,]\!]$: *Program* → *ValueRelation*

$[\![\,\texttt{program } I(X)\ C\,]\!]\langle vs, vs'\rangle :\Leftrightarrow$
  $\exists v \in Value, s, s' \in State.$
    $s = write(X, vs, init(v)) \wedge [\![\,C\,]\!]\langle s, s'\rangle \wedge vs' = read(X, s')$

$[\![\,.\,]\!]$: *Parameters* → Set(*Value*\*)

$[\![\,\sqcup\,]\!]\langle vs\rangle :\Leftrightarrow vs = [\,]$
$[\![\,X, V : S\,]\!]\langle vs\rangle :\Leftrightarrow \exists v \in A(S), vs' \in Value^*.\ vs = vs' \circ [v] \wedge [\![\,X\,]\!]\langle vs'\rangle$

$[\![\,.\,]\!]$: *Command* → *StateRelation*

$[\![\,V := T\,]\!]\langle s, s'\rangle :\Leftrightarrow$
  $[\![\,T\,]\!]\!\!\surd\langle s\rangle \wedge s' = s[V \mapsto [\![\,T\,]\!]_s]$
$[\![\,\texttt{var } V : S;\ C\,]\!]\langle s, s'\rangle :\Leftrightarrow$
  $\exists v, v' \in A(S), s_1, s_2 \in State.$
    $v = s(V) \wedge s_1 = s[V \mapsto v'] \wedge [\![\,C\,]\!]\langle s_1, s_2\rangle \wedge s' = s_2[V \mapsto v]$
$[\![\,C_1; C_2\,]\!]\langle s, s'\rangle :\Leftrightarrow$
  $\exists s_1 \in State.\ [\![\,C_1\,]\!]\langle s, s_1\rangle \wedge [\![\,C_2\,]\!]\langle s_1, s'\rangle$
$[\![\,\texttt{if } F \texttt{ then } C_1 \texttt{ else } C_2\,]\!]\langle s, s'\rangle :\Leftrightarrow$
  $[\![\,F\,]\!]\!\!\surd\langle s\rangle \wedge \texttt{if } [\![\,F\,]\!]_s \texttt{ then } [\![\,C_1\,]\!]\langle s, s'\rangle \texttt{ else } [\![\,C_2\,]\!]\langle s, s'\rangle$
$[\![\,\texttt{if } F \texttt{ then } C\,]\!]\langle s, s'\rangle :\Leftrightarrow$
  $[\![\,F\,]\!]\!\!\surd\langle s\rangle \wedge \texttt{if } [\![\,F\,]\!]_s \texttt{ then } [\![\,C\,]\!]\langle s, s'\rangle \texttt{ else } s' = s$
$[\![\,\texttt{while } F \texttt{ do } C\,]\!]\langle s, s'\rangle :\Leftrightarrow$
  let
    inductive $W \subseteq State \times State$
    $W\langle s, s'\rangle :\Leftrightarrow$
      $[\![\,F\,]\!]\!\!\surd\langle s\rangle \wedge \texttt{if } [\![\,F\,]\!]_s \texttt{ then } \exists s_1 \in State.\ [\![\,C\,]\!]\langle s, s_1\rangle \wedge W\langle s_1, s'\rangle \texttt{ else } s' = s$
  in $W\langle s, s'\rangle$

**Fig. 7.3** Denotational semantics of programs and commands (relational version)

relation. If there is no such output, the program does not terminate; if there are multiple such outputs, the program is *nondeterministic*.

While nondeterminism seems to be a very special feature, it is actually a natural characteristics arising from the *underspecification* of certain aspects of programs and programming languages. For instance, some programming languages leave the initial values of locally declared variables intentionally unspecified; the semantics of such a language must therefore take all possibilities into account.

Figure 7.3 gives a relational variant of a semantics for the previously introduced programming language; here indeed the initial values of program variables are not any more determined by a default map $d$ and value $v$ but left unspecified. This semantics is based on the following denotation functions that map programs and commands to state relations:

- $[\![\,.\,]\!]$: *Program* → *ValueRelation*
  The formula $[\![\,P\,]\!]\langle vs, vs'\rangle$ can be read as 'for the initial values $vs$ of its parameters, the execution of program $P$ may terminate with $vs'$ as their final values'. The value relation denoted by $P$ is based upon the state relation denoted by the command $C$

embedded into $P$ where the initial state $s$ is constructed from the initial parameter values $vs$ and from the final state $s'$ the final parameter values $vs'$ are extracted.

- $\llbracket . \rrbracket$: *Parameters* → Set(*Value**)

  As in the previous semantics, the denotation of the parameter list is a condition that states that the values of the input parameters must be of the declared types; this condition does not contribute to the program semantics itself but will be used below for a proposition about this semantics.

- $\llbracket . \rrbracket$: *Command* → *StateRelation*

  The formula $\llbracket C \rrbracket \langle s, s' \rangle$ can be read as 'for input state $s$, the execution of command $C$ may terminate with state $s'$ as output'. The various kinds of commands basically receive the same semantics as in the previous version, now expressed in a relational form. The only exception is the semantics of variable blocks; here the initial value $v'$ assigned to a locally declared variable $V$ is left unspecified.

Comparing Figs. 7.2 and 7.3, we see that this semantics, due to the non-deterministic selection of initial variable values, copes without a default map as an additional argument.

To illustrate this semantics, we again consider the execution of program $P$

```
program P(a:Nat,b:Nat) {
  b := a;
  if b>0 then {
    var a:Nat;
    a := b-1;
    b := a+a
  }
  a := a+1
}
```

with parameter values $vs = [a = 3, b = 0]$. We then can deduce from $\llbracket P \rrbracket \langle vs, vs' \rangle$ that $vs' = [a = 3, b = 4]$ holds, because we have for every state $s'$

$$\llbracket \text{b:=a;if b>0 then } \{\ldots\};\text{a:=a+1} \rrbracket \langle [a \mapsto 3, b \mapsto 0, \ldots], s' \rangle$$
$$\Leftrightarrow \llbracket \text{if b>0 then } \{\ldots\};\text{a:=a+1} \rrbracket \langle [a \mapsto 3, b \mapsto 3, \ldots], s' \rangle$$
$$\Leftrightarrow \llbracket \text{a:=a+1} \rrbracket \langle [a \mapsto 3, b \mapsto 4, \ldots], s' \rangle$$
$$\Leftrightarrow s' = [a \mapsto 4, b \mapsto 4, \ldots]$$

where

$$\llbracket \text{if b>0 then } \{\text{var a:Nat;a:=b-1;b:=a+a}\} \rrbracket \langle [a \mapsto 3, b \mapsto 3, \ldots], s' \rangle$$
$$\Leftrightarrow \llbracket \text{var a:Nat;a:=b-1;b:=a+a} \rrbracket \langle [a \mapsto 3, b \mapsto 3, \ldots], s' \rangle$$
$$\Leftrightarrow \exists v, v', s_1, s_2.\ v = 3 \wedge s_1 = [a \mapsto v', b \mapsto 3, \ldots] \wedge$$
$$\qquad \llbracket \text{a:=b-1;b:=a+a} \rrbracket \langle s_1, s_2 \rangle \wedge s' = s_2[a \mapsto v]$$
$$\Leftrightarrow \exists v', s_2.\ \llbracket \text{a:=b-1;b:=a+a} \rrbracket \langle [a \mapsto v', b \mapsto 3, \ldots], s_2 \rangle \wedge s' = s_2[a \mapsto 3]$$
$$\Leftrightarrow \exists s_2.\ \llbracket \text{b:=a+a} \rrbracket \langle [a \mapsto 2, b \mapsto 3, \ldots], s_2 \rangle \wedge s' = s_2[a \mapsto 3]$$
$$\Leftrightarrow s' = [a \mapsto 2, b \mapsto 4, \ldots][a \mapsto 3]$$

$\Leftrightarrow s' = [a \mapsto 3, b \mapsto 4, \ldots].$

The exact relationship between the functional version and the relational version of the semantics is formulated below.

> **Proposition 7.1** (Functional   versus   Relational   Semantics)   *Let*  $P =$
> *program*  $I(X)$  $C$  *be a program for which*  $\Sigma \vdash P$ : *program can be derived.*
> *Then, if the execution of*  $P$  *with default map*  $d$  *and value*  $v$  *terminates for*
> *adequate initial parameter values*  vs  *with final values*  vs'  *according to the*
> *functional semantics,*  $P$  *may also terminate for initial parameter values*  vs
> *with final values*  vs'  *according to the relational semantics:*
>
> $$\forall vs \in Value^*, d \in Default, v \in Value.$$
>
> $$[\![ X ]\!]\langle vs \rangle \Rightarrow forall\ vs' = [\![ P ]\!]^{d,v}(vs). \ [\![ P ]\!]\langle vs, vs' \rangle$$

The relational semantics is thus 'compatible' with the functional one but potentially allows more results. For instance, consider the functional semantics with default value 0 for sort `Nat`. In this functional semantics, the program

```
program choose(x:Nat)
{
    var y:Nat; x := y
}
```

terminates for every input $x$ with output $x = 0$ (because $y$ is initialized with 0); in the relational semantics, however, every output $x = 0, 1, 2, 3, \ldots$ is possible (because local variable $y$ can be initialized with any of these values).

Since the relational semantics thus provides a more general semantic framework and, as demonstrated above, also allows conciser descriptions, we will use it as the basis of our further elaborations.

### The Soundness of Typing

We will now investigate the relationship between the type system and the denotational program semantics (in the relational form). In a nutshell, the type system determines which states may be adequately used for the evaluation of phrases such that the values of the phrases are as predicted by the type system. We start by formalizing this notion of 'adequacy'.

**Definition 7.6** (*Adequate States*) We call state $s \in State$ *adequate* with respect to variable typing $Vt \in VarTyping$ if it holds for every variable in the domain of $Vt$ a value of the corresponding sort:

$$adequate \subseteq VarTyping \times State$$
$$adequate\langle Vt, s \rangle :\Leftrightarrow \forall V \in Variable, S \in \Sigma.s. \ \langle V, S \rangle \in Vt \Rightarrow s(V) \in A(S)$$

The basis of type soundness of programs is the type soundness of term evaluation, which may then be lifted to the type soundness of command execution, and finally to the type soundness of program execution.

**Proposition 7.2** (Soundness of Term Typing) *Let* $Vt \in VarTyping$ *be a variable typing,* $S \in \Sigma.s$ *a sort, and* $T \in Term$ *a term for which* $\Sigma, Vt \vdash T : term(S)$ *can be derived. Then the evaluation of* $T$ *in a state that is adequate with respect to the typing yields a value of the sort indicated by the type system:*

$$\forall s \in State. \ adequate\langle Vt, s \rangle \Rightarrow [\![ T ]\!]_s \in A(S)$$

The proof of this statement (which we omit) proceeds by structural induction on term $T$, matching for each kind of term the applicable typing rules to the definition of the term semantics.

**Proposition 7.3** (Soundness of Command Typing) *Let* $Vt \in VarTyping$ *be a variable typing and* $C \in Command$ *a command for which* $\Sigma, Vt \vdash C : command$ *can be derived. Then, if the execution of* $C$ *on a state that is adequate with respect to the typing terminates, it yields another state that is adequate with respect to the typing:*

$$\forall s, s' \in State. \ adequate\langle Vt, s \rangle \wedge [\![ C ]\!]\langle s, s' \rangle \Rightarrow adequate\langle Vt, s' \rangle$$

The proof this statement proceeds by structural induction on command $C$ where we already may assume the soundness of term typing.

The semantics of a parameter list is the condition on the initial values of the parameters that ensures the adequacy of the initial state.

**Proposition 7.4** (Soundness of Parameter Typing) *Let $Vt \in VarTyping$ be a variable typing, $Ss \in Sort^*$ a sort list, and $X \in Parameters$ a parameter list for which $\varnothing \vdash X : parameters(Vt, Ss)$ can be derived, i.e., for which $Vt$ denotes the typing of the parameters and $Ss$ denotes the list of parameter sorts. Then the semantics of $X$ ensures that the state constructed from the parameters is adequate with respect to the parameter typing:*

$$\forall vs \in Value^*, v \in Value, s \in State.$$
$$[\![ X ]\!]\langle vs \rangle \wedge s = write(X, vs, init(v)) \Rightarrow adequate\langle Vt, s \rangle$$

The easy proof of this statement proceeds by induction on the structure of the parameter list.

Ultimately, we have reached the core claim of this section.

**Proposition 7.5** (Soundness of Program Typing) *Let $P = program\ I(X)\ C$ be a program such that $\Sigma \vdash P : program$ can be derived. Then, if the execution of $P$ on a value sequence that is adequate with respect to the parameter declaration terminates, it yields another value sequence that is also adequate with respect to the parameter declaration:*

$$\forall vs, vs' \in Value^*. [\![ X ]\!]\langle vs \rangle \wedge [\![ P ]\!]\langle vs, vs' \rangle \Rightarrow [\![ X ]\!]\langle vs' \rangle$$

This proposition is an easy consequence of the previous two propositions and the definition of the program semantics.

**Ensuring The Non-abortion of Programs**

One aspect that is not yet addressed by the relational program semantics is that a program may *abort* for certain input states due to the inadmissible evaluation of a term or formula, i.e., its evaluation in a state for which its value is not well-defined. For instance, the program

```
program mayabort(x:Nat)
{
   var y:Nat; var z:Nat; z := div(x,y);
}
```

may terminate for every input $x = n$ (only) with output $x = n$, because there is no assignment to $x$ in the program; in the relational semantics, this program is thus indistinguishable from the following one:

```
program nothing(x:Nat) { }
```

$[\![\,.\,]\!]\checkmark : \textbf{\textit{Program}} \rightarrow \textbf{\textit{ValueCondition}}$

$[\![\,\texttt{program}\ I(X)\ C\,]\!]\checkmark\langle vs\rangle :\Leftrightarrow$
  $\forall v \in \textit{Value},\ s \in \textit{State}.$
    $s = \textit{write}(X, vs, \textit{init}(v)) \Rightarrow [\![\,C\,]\!]\checkmark\langle s\rangle$

$[\![\,.\,]\!]\checkmark : \textbf{\textit{Command}} \rightarrow \textbf{\textit{StateCondition}}$

$[\![\,V := T\,]\!]\checkmark\langle s\rangle :\Leftrightarrow$
  $[\![\,T\,]\!]\checkmark\langle s\rangle$
$[\![\,\texttt{var}\ V : S;\ C\,]\!]\checkmark\langle s\rangle :\Leftrightarrow$
  $\forall v \in A(S),\ s_1 \in \textit{State}.\ s_1 = s[V \mapsto v] \Rightarrow [\![\,C\,]\!]\checkmark\langle s_1\rangle$
$[\![\,C_1 ; C_2\,]\!]\checkmark\langle s\rangle :\Leftrightarrow$
  $[\![\,C_1\,]\!]\checkmark\langle s\rangle \wedge \forall s_1 \in \textit{State}.\ [\![\,C_1\,]\!]\langle s, s_1\rangle \Rightarrow [\![\,C_2\,]\!]\checkmark\langle s_1\rangle$
$[\![\,\texttt{if}\ F\ \texttt{then}\ C_1\ \texttt{else}\ C_2\,]\!]\checkmark\langle s\rangle :\Leftrightarrow$
  $[\![\,F\,]\!]\checkmark\langle s\rangle \wedge \texttt{if}\ [\![\,F\,]\!]_s\ \texttt{then}\ [\![\,C_1\,]\!]\checkmark\langle s\rangle\ \texttt{else}\ [\![\,C_2\,]\!]\checkmark\langle s\rangle$
$[\![\,\texttt{if}\ F\ \texttt{then}\ C\,]\!]\checkmark\langle s\rangle :\Leftrightarrow$
  $[\![\,F\,]\!]\checkmark\langle s\rangle \wedge ([\![\,F\,]\!]_s \Rightarrow [\![\,C\,]\!]\checkmark\langle s\rangle)$
$[\![\,\texttt{while}\ F\ \texttt{do}\ C\,]\!]\checkmark\langle s\rangle :\Leftrightarrow$
  $\texttt{let}$
    $\texttt{inductive}\ W \subseteq \textit{State}$
    $W\langle s\rangle :\Leftrightarrow [\![\,F\,]\!]\checkmark\langle s\rangle \wedge ([\![\,F\,]\!]_s \Rightarrow [\![\,C\,]\!]\checkmark\langle s\rangle \wedge (\forall s_1 \in \textit{State}.\ [\![\,C\,]\!]\langle s, s_1\rangle \Rightarrow W\langle s_1\rangle))$
  $\texttt{in}\ W\langle s\rangle$

**Fig. 7.4** The non-abortion of programs

However, while program `nothing` can never abort, every execution of the program `mayabort` may abort, if the initial value of $y$ happens to be 0 (assuming that the term `div(x,y)` is not defined for $y = 0$).

In order to be able to differentiate between such programs, Fig. 7.4 defines a translation of programs and command to non-abortion conditions on the basis of the corresponding translations for terms and formulas: if $[\![\,P\,]\!]\checkmark\langle vs\rangle$ respectively $[\![\,C\,]\!]\checkmark\langle s\rangle$ holds, then the execution of program $P$ with parameter values $vs$ respectively the execution of command $C$ in state $s$ cannot abort. Therefore for program `nothing`, $[\![\,P\,]\!]\checkmark\langle vs\rangle$ is 'true' for every $vs$, while it is always 'false' for program `mayabort`: the definition of $[\![\,\texttt{var}\ V : S;\ C\,]\!]\checkmark\langle s\rangle$ requires, for every initial value $v \in A(S)$ and resulting state $s_1 = s[V \mapsto v]$, the condition $[\![\,C\,]\!]\checkmark\langle s_1\rangle$ to be true; but it is false for $v = 0$.

The definitions of $[\![\,C_1 ; C_2\,]\!]\checkmark\langle s\rangle$ and $[\![\,\texttt{while}\ F\ \texttt{do}\ C\,]\!]\checkmark\langle s\rangle$ make use of the relational command semantics $[\![\,C_1\,]\!]\langle s, s_1\rangle$ respectively $[\![\,C\,]\!]\langle s, s_1\rangle$ to indicate every state $s_1$ that may result from the execution of command $C_1$ respectively $C$; this state has to be admissible for the remainder of the program execution, i.e., for the execution of $C_2$ respectively of another loop iteration.

**Ensuring the Termination of Programs**

Another aspect that is not yet addressed by the relational semantics is for which input states a program command may *not terminate*. For instance, the program

```
program mayloop(x:Nat)
{
```

```
var y:Nat; x := y;
while x = 0 do
{
   var y:Nat; x := y
}
x := x-1
}
```

program may terminate for every input with some of the outputs $x = 0, 1, 2, 3, \ldots$ (after the termination of the loop, we have $x > 0$ and by the subsequent decrement operation $x \geq 0$); however, it may also not terminate at all (if in the loop body local variable $y$ is always initialized with 0). The relational semantics only captures the first fact but does not talk about the second one, which makes this program indistinguishable from the program

```
program choose(x:Nat)
{
   var y:Nat; x := y
}
```

which always terminates with with some of the outputs $x = 0, 1, 2, 3, \ldots$.

In order to rule out the possibility of non-termination, we describe the sequence of states arising from the execution of a loop.

---

**Definition 7.7** *(State Sequence of a Loop)* Let $t$ be an infinite sequence of states, $n$ a natural number, $F$ a formula, and $C$ a command. Then $t$ describes the first $n$ iterations of a loop with condition $F$ and body $C$ if, for every $i$ less than $n$, $F$ holds in state $t(i)$, and state $t(i + 1)$ is derived from $t(i)$ by the execution of $C$:

$$loop \subseteq State^{\omega} \times \mathbb{N} \times Formula \times Command$$
$$loop\langle t, n, F, C \rangle :\Leftrightarrow \forall i \in \mathbb{N}_n.\, [\![ F ]\!]\checkmark \langle t(i) \rangle \wedge [\![ F ]\!]_{t(i)} \wedge [\![ C ]\!]\langle t(i), t(i + 1) \rangle$$

---

Based on this concept, Fig. 7.5 defines a translation of programs and commands to state conditions: if $\langle\!\langle P \rangle\!\rangle \langle s \rangle$ respectively $\langle\!\langle C \rangle\!\rangle \langle s \rangle$ holds, then program $P$ respectively command $C$ terminates for every possible execution starting with input state $s$. Therefore for program `choose`, $\langle\!\langle P \rangle\!\rangle \langle s \rangle$ is 'true' for every $s$, while it is always 'false' for program `mayloop`. At the core of the translation, we have the definition of $\langle\!\langle$ while $F$ do $C \rangle\!\rangle \langle s \rangle$ which states that there is no infinite sequence of loop iterations starting with $s$, and that for every such sequence $t$ leading after $n$ iterations to a state $t(n)$ in which the loop condition $F$ is true, the execution of the loop body $C$ terminates.

$\langle\!\langle\,.\,\rangle\!\rangle : \textbf{Program} \rightarrow \textbf{ValueCondition}$

$\langle\!\langle\, \text{program } I(X)\ C \,\rangle\!\rangle\, \langle vs \rangle :\Leftrightarrow$
   $\forall v \in Value,\ s \in State.$
      $s = write(X,\ vs,\ init(v)) \Rightarrow \langle\!\langle\, C \,\rangle\!\rangle\, \langle s \rangle$

$\langle\!\langle\,.\,\rangle\!\rangle : \textbf{Command} \rightarrow \textbf{StateCondition}$

$\langle\!\langle\, V := T \,\rangle\!\rangle\, \langle s \rangle :\Leftrightarrow$
   true
$\langle\!\langle\, \text{var } V : S;\ C \,\rangle\!\rangle\, \langle s \rangle :\Leftrightarrow$
   $\forall v \in A(S),\ s_1 \in State.\ s_1 = s[V \mapsto v] \Rightarrow \langle\!\langle\, C \,\rangle\!\rangle\, \langle s_1 \rangle$
$\langle\!\langle\, C_1 ; C_2 \,\rangle\!\rangle\, \langle s \rangle :\Leftrightarrow$
   $\langle\!\langle\, C_1 \,\rangle\!\rangle\, \langle s \rangle \wedge \forall s_1 \in State.\ [\![ C_1 ]\!]\langle s, s_1 \rangle \Rightarrow \langle\!\langle\, C_2 \,\rangle\!\rangle\, \langle s_1 \rangle$
$\langle\!\langle\, \text{if } F \text{ then } C_1 \text{ else } C_2 \,\rangle\!\rangle\, \langle s \rangle :\Leftrightarrow$
   if $[\![ F ]\!]_s$ then $\langle\!\langle\, C_1 \,\rangle\!\rangle\, \langle s \rangle$ else $\langle\!\langle\, C_2 \,\rangle\!\rangle\, \langle s \rangle$
$\langle\!\langle\, \text{if } F \text{ then } C \,\rangle\!\rangle\, \langle s \rangle :\Leftrightarrow$
   $[\![ F ]\!]_s \Rightarrow \langle\!\langle\, C \,\rangle\!\rangle\, \langle s \rangle$
$\langle\!\langle\, \text{while } F \text{ do } C \,\rangle\!\rangle\, \langle s \rangle :\Leftrightarrow$
   $\forall t \in State^{\omega}.\ t(0) = s \Rightarrow$
      $\neg\,(\forall n \in \mathbb{N}.\ loop\langle t, n, F, C \rangle) \wedge$
      $(\forall n \in \mathbb{N}.\ loop\langle t, n, F, C \rangle \wedge [\![ F ]\!]_{t(n)} \Rightarrow \langle\!\langle\, C \,\rangle\!\rangle\, \langle t(n) \rangle)$

**Fig. 7.5** The termination of programs

## The Soundness of the Conditions for Non-abortion and Termination

In the following, we will re-consider the previously described translations of programs and commands to conditions, by investigating their claims to ensure the non-abortion and termination of programs with respect to the relational semantics of programs and commands.

**Proposition 7.6** (Non-abortion and Termination of Programs) *Let* $P =$ program $I(X)\ C$ *be a program such that* $\Sigma \vdash P : program$ *can be derived. If vs is a value sequence that is adequate with respect to the program parameters, that is admissible for the evaluation of the program expressions, and that leads to the termination of the program, then there exists a value sequence vs′ resulting from the execution of P:*

$$\forall vs \in Value^*.\ [\![ X ]\!]\langle vs \rangle \wedge [\![ P ]\!]\checkmark \langle vs \rangle \wedge \langle\!\langle\, P \,\rangle\!\rangle\, \langle vs \rangle \Rightarrow$$
$$\exists vs' \in Value^*.\ [\![ P ]\!]\langle vs, vs' \rangle$$

*Proof* This is a direct consequence of the following proposition.                □

**Proposition 7.7** (Non-abortion and Termination of Commands) *Let $Vt \in VarTyping$ be a variable typing and $C \in Command$ a command for which $\Sigma, Vt \vdash C$: command can be derived. If $s$ is a state that is adequate with respect to typing $Vt$, that is admissible for the evaluation of the command expressions, and that leads to the termination of the command, then there exists a post-state $s'$ resulting from the execution of $C$:*

$$\forall s \in State. \; adequate\langle Vt, s\rangle \wedge [\![ C ]\!] \checkmark \langle s \rangle \wedge \langle\!\langle C \rangle\!\rangle \langle s \rangle \Rightarrow$$
$$\exists s' \in State. \; [\![ C ]\!] \langle s, s' \rangle$$

The proof of this statement (which we omit) is based on the following alternative description of the semantics of loops that copes without fixed points.

**Proposition 7.8** (Loop Semantics) *Every loop with condition $F$ and body $C$ transforms an input state $s$ to an output state $s'$, if and only if there exist some infinite sequence $t$ of states and a natural number $n$ such that $t$ describes the first $n$ iterations of the loop, $t$ has $s$ at position $0$ and $s'$ at position $n$, and $F$ does not hold in $s'$:*

$$\forall F \in Formula, C \in Command, s, s' \in State.$$
$$[\![ \texttt{while } F \texttt{ do } C ]\!] \langle s, s' \rangle \Leftrightarrow$$
$$\exists t \in State^{\omega}, n \in \mathbb{N}.$$
$$loop\langle t, n, F, C\rangle \wedge t(0) = s \wedge t(n) = s' \wedge [\![ F ]\!] \checkmark \langle s' \rangle \wedge \neg [\![ F ]\!]_{s'}$$

*Proof* We apply a variant of *fixed point induction* (see Proposition 5.12). First we take arbitrary $F \in Formula$, $C \in Command$ and define

$$W :=$$
$$\lambda W. \pi s, s'.$$
$$[\![ F ]\!] \checkmark \langle s \rangle \wedge$$
$$\texttt{if } [\![ F ]\!]_s \texttt{ then } \exists s_1 \in State. \; [\![ C ]\!] \langle s, s_1 \rangle \wedge W \langle s_1, s' \rangle \texttt{ else } s' = s$$

such that we have from the relational semantics of commands

$$[\![ \texttt{while } F \texttt{ do } C ]\!] = \mathsf{lfp} \; W \qquad (1)$$

Then we define

$$P \langle i, W \rangle :\Leftrightarrow$$
$$\forall s, s' \in State.$$

$W \langle s, s' \rangle \Leftrightarrow$

$\exists t \in State^{\omega}, n \in \mathbb{N}_i.$

$\quad loop \langle t, n, F, C \rangle \wedge t(0) = s \wedge t(n) = s' \wedge [\![ F ]\!] \checkmark \langle s' \rangle \wedge \neg [\![ F ]\!]_{s'}$

Now suppose we are able to prove

$$\forall i \in \mathbb{N}.\ P \langle i, W^i (\varnothing) \rangle \tag{2}$$

We claim that from (2) our goal follows. To justify this claim, we take arbitrary $s, s' \in State$ and note the equivalence

$$(\mathsf{lfp}\ W) \langle s, s' \rangle \Leftrightarrow \exists i \in \mathbb{N}.\ W^i (\varnothing) \langle s, s' \rangle$$

which follows from the upward-continuity of $W$ (whose proof we omit). From (0) it thus suffices to prove

$\left( \exists i \in \mathbb{N}.\ W^i (\varnothing) \langle s, s' \rangle \right) \Leftrightarrow$

$\left( \exists t \in State^{\omega}, n \in \mathbb{N}. \right.$

$\quad loop \langle t, n, F, C \rangle \wedge t(0) = s \wedge t(n) = s' \wedge [\![ F ]\!] \checkmark \langle s' \rangle \wedge \neg [\![ F ]\!]_{s'} )$

We are now going to show both directions of this equivalence:

$\Rightarrow$    Let $i \in \mathbb{N}$ be such that $W^i (\varnothing) \langle s, s' \rangle$ holds. From (2), we then have some $t \in State^{\omega}$ and $n \in \mathbb{N}_i$ with

$$loop \langle t, n, F, C \rangle \wedge t(0) = s \wedge t(n) = s' \wedge [\![ F ]\!] \checkmark \langle s' \rangle \wedge \neg [\![ F ]\!]_{s'}$$

and thus are done.

$\Leftarrow$    Let $t \in State^{\omega}$ and $n \in \mathbb{N}$ such that

$$loop \langle t, n, F, C \rangle \wedge t(0) = s \wedge t(n) = s' \wedge [\![ F ]\!] \checkmark \langle s' \rangle \wedge \neg [\![ F ]\!]_{s'} \tag{3}$$

holds. It then suffices to show

$$W^{n+1} (\varnothing) \langle s, s' \rangle$$

From (2), we have $P \langle n + 1, W^{n+1} (\varnothing) \rangle$. It thus suffices to show $n \in \mathbb{N}_{n+1}$ and

$$loop \langle t, n, F, C \rangle \wedge t(0) = s \wedge t(n) = s' \wedge [\![ F ]\!] \checkmark \langle s' \rangle \wedge \neg [\![ F ]\!]_{s'}$$

which follows from (3).

It remains to show (2) by induction on $i \in \mathbb{N}$. The base case $P\langle 0, \varnothing \rangle$ follows from the fact that $\varnothing \langle s, s' \rangle \Leftrightarrow$ false and that there is no $n \in \mathbb{N}_0 = \varnothing$. For the induction step, we assume $P\langle i, W^i(\varnothing) \rangle$, i.e.,

$$\forall s, s' \in State. \tag{4}$$
$$W^i(\varnothing)\langle s, s' \rangle \Leftrightarrow$$
$$\exists t \in State^\omega, n \in \mathbb{N}_i.$$
$$loop\langle t, n, F, C \rangle \wedge t(0) = s \wedge t(n) = s' \wedge [\![ F ]\!] \checkmark \langle s' \rangle \wedge \neg [\![ F ]\!]_{s'}$$

and show $P\langle i + 1, W^{i+1}(\varnothing) \rangle$, i.e., for arbitrary $s, s' \in State$

$$W^{i+1}(\varnothing)\langle s, s' \rangle \Leftrightarrow$$
$$\exists t \in State^\omega, n \in \mathbb{N}_{i+1}.$$
$$loop\langle t, n, F, C \rangle \wedge t(0) = s \wedge t(n) = s' \wedge [\![ F ]\!] \checkmark \langle s' \rangle \wedge \neg [\![ F ]\!]_{s'}$$

From the definition of $W$ it thus suffices to show

$$\Big( [\![ F ]\!] \checkmark \langle s \rangle \wedge$$
$$\quad \text{if } [\![ F ]\!]_s \text{ then } \exists s_1 \in State. [\![ C ]\!]\langle s, s_1 \rangle \wedge W^i(\varnothing)\langle s_1, s' \rangle \text{ else } s' = s \Big) \Leftrightarrow$$
$$\Big( \exists t \in State^\omega, n \in \mathbb{N}_{i+1}.$$
$$\quad loop\langle t, n, F, C \rangle \wedge t(0) = s \wedge t(n) = s' \wedge [\![ F ]\!] \checkmark \langle s' \rangle \wedge \neg [\![ F ]\!]_{s'} \Big)$$

We show both directions of the equivalence:

$\Rightarrow$ We know $[\![ F ]\!] \checkmark \langle s \rangle$ and have two cases:
  - In case $[\![ F ]\!]_s$, we have some $s_1 \in State$ with $[\![ C ]\!]\langle s, s_1 \rangle$ and $W^i(\varnothing)\langle s_1, s' \rangle$. By (4), we also have $t' \in State^\omega$ and $n' \in \mathbb{N}_i$ with $loop\langle t', n', F, C \rangle$, $t'(0) = s_1$, $t'(n') = s'$, $[\![ F ]\!] \checkmark \langle s' \rangle$, and $\neg [\![ F ]\!]_{s'}$. We define $t(i) := $ if $i = 0$ then $s$ else $t'(i - 1)$ and $n := n' + 1$ and are with $[\![ F ]\!] \checkmark \langle s \rangle$ done.
  - In case $\neg [\![ F ]\!]_s$, we have $s' = s$. We thus define $t(i) := s$ and $n := 0$ and are with $[\![ F ]\!] \checkmark \langle s \rangle$ done.
$\Leftarrow$ From the assumption, we have some $t \in State^\omega$ and $n \in \mathbb{N}_{i+1}$ with $loop\langle t, n, F, C \rangle$ and $t(0) = s$, which implies $[\![ F ]\!] \checkmark \langle s \rangle$ and $[\![ F ]\!]_s$; thus it suffices to show $\exists s_1 \in State. [\![ C ]\!]\langle s, s_1 \rangle \wedge W^i(\varnothing)\langle s_1, s' \rangle$. From the assumption we also know $t(n) = s'$ and $\neg [\![ F ]\!]_{s'}$ which implies $s \neq s'$, thus $n > 0$, and ultimately $[\![ C ]\!]\langle s, t(1) \rangle$. Therefore it remains to show $W^i(\varnothing)\langle t(1), s' \rangle$, i.e., by (4),

$$\exists t' \in State^\omega, n' \in \mathbb{N}_i.$$
$$loop\langle t', n', F, C \rangle \wedge t'(0) = t(1) \wedge t'(n') = s' \wedge [\![ F ]\!] \checkmark \langle s' \rangle \wedge \neg [\![ F ]\!]_{s'}$$

This follows for $t'(i) := t(i + 1)$ and $n' := n - 1$ from the assumption.

Since Proposition 7.8 gives a more direct characterization of the behavior of loops, we will base our further considerations on this description rather than on the original semantics which relied on an inductive function definition.

**Reasoning about the Equivalence of Programs**

From the denotational semantics of programs, we may formally prove various kinds of program properties, e.g., the correctness of program transformations.

**Example 7.2 (Correctness of Loop Unrolling).** A common compiler optimization technique is to 'unroll' some iterations of a loop into conditional statements, which avoids 'back-jumps' in the generated machine code and thus improves the performance of loops that only execute very few iterations. This optimization is based on the semantic equalities

$$\llbracket \texttt{while } F \texttt{ do } C \rrbracket = \llbracket \texttt{if } F \texttt{ then } \{C; \texttt{ while } F \texttt{ do } C\} \rrbracket$$
$$\llbracket \texttt{while } F \texttt{ do } C \rrbracket \checkmark = \llbracket \texttt{if } F \texttt{ then } \{C; \texttt{ while } F \texttt{ do } C\} \rrbracket \checkmark$$
$$\langle\!\langle \texttt{while } F \texttt{ do } C \rangle\!\rangle = \langle\!\langle \texttt{if } F \texttt{ then } \{C; \texttt{ while } F \texttt{ do } C\} \rangle\!\rangle$$

We are now going to prove the first of these equalities (the other ones are comparatively simple to establish). We thus take arbitrary $s, s' \in State$ and show

$$\llbracket \texttt{while } F \texttt{ do } C \rrbracket \langle s, s' \rangle \Leftrightarrow \llbracket \texttt{if } F \texttt{ then } \{C; \texttt{ while } F \texttt{ do } C\} \rrbracket \langle s, s' \rangle$$

We show both directions of the equivalence in turn:

$\Rightarrow$ From Proposition 7.8, we have some $t \in State^\omega$ and $n \in \mathbb{N}$ with

$$loop\langle t, n, F, C \rangle \wedge t(0) = s \wedge t(n) = s' \wedge \llbracket F \rrbracket \checkmark \langle s' \rangle \wedge \neg \llbracket F \rrbracket_{s'} \qquad (1)$$

From the definition of the relational semantics, we have to show

$\llbracket F \rrbracket \checkmark \langle s \rangle \wedge$
if $\llbracket F \rrbracket_s$
  then $\exists s_1 \in State. \llbracket C \rrbracket \langle s, s_1 \rangle \wedge \llbracket \texttt{while } F \texttt{ do } C \rrbracket \langle s_1, s' \rangle$
  else $s' = s$

We have two cases:
- Case $n = 0$: from the definition of $loop$, we have $s = s'$ and thus $\llbracket F \rrbracket \checkmark \langle s \rangle$ and $\neg \llbracket F \rrbracket_s$ and are thus done.
- Case $n > 0$: from the definition of $loop$, we have $\llbracket F \rrbracket \checkmark \langle s \rangle$, $\llbracket F \rrbracket_s$, and $\llbracket C \rrbracket \langle s, t(1) \rangle$. It thus suffices to show $\llbracket \texttt{while } F \texttt{ do } C \rrbracket \langle t(1), s' \rangle$, i.e., from Proposition 7.8

$\exists t' \in State^\omega, n' \in \mathbb{N}.$
$\quad loop\langle t', n', F, C \rangle \wedge t'(0) = t(1) \wedge t'(n') = s' \wedge \llbracket F \rrbracket \checkmark \langle s' \rangle \wedge \neg \llbracket F \rrbracket_{s'}$

which follows from (1) with $t'(i) := t(i + 1)$ and $n' := n - 1$.

$\Leftarrow$ From the definition of the relational semantics, we know

> $[\![ F ]\!] \checkmark \langle s \rangle \wedge$
>
> if $[\![ F ]\!]_s$
>
>    then $\exists s_1 \in State.\ [\![ C ]\!] \langle s, s_1 \rangle \wedge [\![ \text{while } F \text{ do } C ]\!] \langle s_1, s' \rangle$
>
>    else $s' = s$                                                                            (1)

From Proposition 7.8, it suffices to show

> $\exists t \in State^{\omega}, n \in \mathbb{N}.$
>
> $loop\langle t, n, F, C \rangle \wedge t(0) = s \wedge t(n) = s' \wedge [\![ F ]\!] \checkmark \langle s' \rangle \wedge \neg [\![ F ]\!]_{s'}$

We have two cases:

- Case $[\![ F ]\!]_s$: from (1), we know $[\![ F ]\!] \checkmark \langle s \rangle$ and have some $s_1 \in State$ with $[\![ C ]\!] \langle s, s_1 \rangle$ and $[\![ \text{while } F \text{ do } C ]\!] \langle s_1, s' \rangle$. From this and Proposition 7.8, we have some $t' \in State^{\omega}$ and $n' \in \mathbb{N}$ with

$$loop\langle t', n', F, C \rangle \wedge t'(0) = s_1 \wedge t'(n') = s' \wedge [\![ F ]\!] \checkmark \langle s' \rangle \wedge \neg [\![ F ]\!]_{s'}$$

  We are thus done with $t(i) :=$ if $i = 0$ then $s$ else $t'(i - 1)$ and $n := n' + 1$.
- Case $\neg [\![ F ]\!]_s$: from (1), we know $[\![ F ]\!] \checkmark \langle s \rangle$ and $s' = s$. We are thus done with $t(i) := s$ and $n := 0$.

This completes the proof.                                                                    $\square$

## 7.3   An Operational Semantics

We are now going to present an alternative approach to formalizing the semantics of programs which is based on individual execution steps, i.e., *transitions* from one state to another one. While formally closely related to denotational semantics, this version has a much more executable flavor and is thus called *operational semantics*. We start by demonstrating this kind of semantics by a language that is simpler than a language of programs and commands.

### An Expression Language

Consider the language of expressions presented in Fig. 7.6 which consists of numerals (identified with natural numbers), variables, additions, and multiplications. The semantics $[\![ E ]\!]$ of an expression $E$ is defined in the denotational style as a function from states to natural numbers; here also a state is a function from variables

**Abstract Syntax**

$E \in Expression$
$V \in Variable$
$N \in \mathbb{N}$
$E ::= N \mid V \mid E_1{+}E_2 \mid E_1{*}E_2$

**Semantic Domain**

$State := Variable \rightarrow \mathbb{N}$

**Denotational Semantics**

$[\![\,\cdot\,]\!] : Expression \rightarrow (State \rightarrow \mathbb{N})$

$[\![\,N\,]\!] = \lambda s \in State.\ N$

$[\![\,V\,]\!] = \lambda s \in State.\ s(V)$

$[\![\,E_1{+}E_2\,]\!] = \lambda s \in State.\ [\![\,E_1\,]\!](s) + [\![\,E_2\,]\!](s)$

$[\![\,E_1{*}E_2\,]\!] = \lambda s \in State.\ [\![\,E_1\,]\!](s) \cdot [\![\,E_2\,]\!](s)$

**Fig. 7.6** An expression language and its denotational semantics

**Semantic Domains**

$ExpState := Expression \times State$

**Transitions**

$.\twoheadrightarrow. \subseteq ExpState \times \mathbb{N}$

**Inference Rules**

$$\langle N, s \rangle \twoheadrightarrow N \qquad \langle V, s \rangle \twoheadrightarrow s(V)$$

$$\frac{\langle E_1, s \rangle \twoheadrightarrow v_1 \quad \langle E_2, s \rangle \twoheadrightarrow v_2}{\langle E_1{+}E_2, s \rangle \twoheadrightarrow v_1 + v_2} \qquad \frac{\langle E_1, s \rangle \twoheadrightarrow v_1 \quad \langle E_2, s \rangle \twoheadrightarrow v_2}{\langle E_1{*}E_2, s \rangle \twoheadrightarrow v_1 \cdot v_2}$$

**Fig. 7.7** Operational semantics of the expression language (big-step version)

to natural numbers. Then the value of the expression $E := 2{*}x{+}3{*}y$ in a store $s := [x \mapsto 1, y \mapsto 2, \ldots]$ can be determined as

$$
\begin{aligned}
[\![\,E\,]\!](s) &= [\![\,2{*}x\,]\!](s) + [\![\,3{*}(y{+}1)\,]\!](s) \\
&= [\![\,2\,]\!](s) \cdot [\![\,x\,]\!](s) + [\![\,3\,]\!](s) \cdot [\![\,y{+}1\,]\!](s) \\
&= 2 \cdot s(x) + 3 \cdot ([\![\,y\,]\!](s) + [\![\,1\,]\!](s)) \\
&= 2 \cdot 1 + 3 \cdot (s(y) + 1) = 2 + 3 \cdot (2 + 1) = 11
\end{aligned}
$$

As an alternative to denotational semantics, Fig. 7.7 defines the semantics of this expression language by an inference calculus for deriving judgements of form $\langle E, s \rangle \twoheadrightarrow N$ which we call *transitions*. The left side of such a transition is a *configuration* $\langle E, s \rangle$ of an expression $E$ and a state $s$; its right side is a natural number $N$. We may then read $\langle E, s \rangle \twoheadrightarrow N$ as 'configuration $\langle E, s \rangle$ makes a transition to number $N$' respectively as 'expression $E$ in state $s$ evaluates to number $N$'.

The given inference rules describe how a transition $\langle E, s \rangle \twoheadrightarrow N$ can be derived. For instance, considering again the evaluation of the expression $E := 2*x + 3*(y + 1)$ in state $s := [x \mapsto 1, y \mapsto 2, \ldots]$, we can derive the transition $\langle E, s \rangle \twoheadrightarrow 11$ by the following inference tree:

$$\frac{\dfrac{\langle 2, s \rangle \twoheadrightarrow 2 \quad \langle x, s \rangle \twoheadrightarrow 1}{\langle 2*x, s \rangle \twoheadrightarrow 2} \quad \dfrac{\langle 3, s \rangle \twoheadrightarrow 3 \quad \dfrac{\langle y, s \rangle \twoheadrightarrow 2 \quad \langle 1, s \rangle \twoheadrightarrow 1}{\langle y+1, s \rangle \twoheadrightarrow 3}}{\langle 3*(y+1), s \rangle \twoheadrightarrow 9}}{\langle 2*x+3*(y+1), s \rangle \twoheadrightarrow 11}$$

In the following, we identify a transition $\langle E, s \rangle \twoheadrightarrow N$ with a corresponding binary relation such that a formula $\langle E, s \rangle \twoheadrightarrow N$ is true, if and only if a corresponding judgement can be derived by the stated inference rules.

Traversing above tree in a 'bottom-up' and 'left-to-right' order, we can order the individual transitions as

$$\langle 2, s \rangle \twoheadrightarrow 2$$
$$\langle x, s \rangle \twoheadrightarrow 1$$
$$\langle 2*x, s \rangle \twoheadrightarrow 2$$
$$\langle 3, s \rangle \twoheadrightarrow 3$$
$$\langle y, s \rangle \twoheadrightarrow 2$$
$$\langle 1, s \rangle \twoheadrightarrow 1$$
$$\langle y+1, s \rangle \twoheadrightarrow 3$$
$$\langle 3*(y + 1), s \rangle \twoheadrightarrow 9$$
$$\langle 2*x+3*(y+1), s \rangle \twoheadrightarrow 11$$

which resembles a sequence of *machine instructions* performed for evaluating the expression; therefore the semantics is called 'operational'. However, an individual transition $\langle 2*x+3*(y+1), s \rangle \twoheadrightarrow 11$ does not show all the intermediate evaluation steps but only the initial configuration and the final outcome; therefore this kind of semantics is also called *big-step operational semantics* (sometimes also *natural semantics*). Its close relationship to denotational semantics is stated below.

---

**Proposition 7.9** (Equivalence of Denotational and Operational Semantics) *For every expression E, store s, and natural number N, the application of the denotational semantics $[\![ E ]\!]$ to s yields N if and only if there is a big-step transition from $\langle E, s \rangle$ to N:*

$$\forall E \in \textit{Expression}, s \in \textit{State}, N \in \mathbb{N}.$$
$$[\![ E ]\!](s) = N \Leftrightarrow \langle E, s \rangle \twoheadrightarrow N$$

***Proof*** We show both directions of the equivalence in turn. First we prove

$$\forall E \in Expression, s \in State, N \in \mathbb{N}.\ [\![\,E\,]\!](s) = N \Rightarrow \langle E, s \rangle \twoheadrightarrow N$$

which is equivalent to

$$\forall E \in Expression, s \in State, N \in \mathbb{N}.\ \langle E, s \rangle \twoheadrightarrow [\![\,E\,]\!](s)$$

by structural induction on $E$:

- Case $E = N$: For arbitrary $s \in State$, we have $\langle N, s \rangle \twoheadrightarrow N$ and $N = [\![\,N\,]\!](s)$.
- Case $E = V$: For arbitrary $s \in State$, we have $\langle V, s \rangle \twoheadrightarrow s(V)$ and $s(V) = [\![\,V\,]\!](s)$.
- Case $E = E_1 + E_2$: Take arbitrary $s \in State$. From the induction assumption, we have $\langle E_1, s \rangle \twoheadrightarrow [\![\,E_1\,]\!](s)$ and $\langle E_2, s \rangle \twoheadrightarrow [\![\,E_2\,]\!](s)$. From the operational semantics, we thus have $\langle E_1 + E_2, s \rangle \twoheadrightarrow [\![\,E_1\,]\!](s) + [\![\,E_2\,]\!](s)$, while from the denotational semantics we have $[\![\,E_1\,]\!](s) + [\![\,E_2\,]\!](s) = [\![\,E_1 + E_2\,]\!](s)$.
- Case $E = E_1 * E_2$: analogous to the previous case.

Next we prove

$$\forall E \in Expression, s \in State, N \in \mathbb{N}.\ \langle E, s \rangle \twoheadrightarrow N \Rightarrow [\![\,E\,]\!](s) = N$$

by structural induction on $E$:

- Case $E = N$: Take arbitrary $s \in State$ and $N' \in \mathbb{N}$ with $\langle N, s \rangle \twoheadrightarrow N'$. From the operational semantics, we have $N' = N$; from the denotational semantics we have also $[\![\,N\,]\!](s) = N$.
- Case $E = V$: Take arbitrary $s \in State$ and $N \in \mathbb{N}$ with $\langle V, s \rangle \twoheadrightarrow N$. From the operational semantics, we have $N = s(V)$; from the denotational semantics we have also $[\![\,V\,]\!](s) = s(V)$.
- Case $E = E_1 + E_2$: Take arbitrary $s \in State$ and $N \in \mathbb{N}$ with $\langle E_1 + E_2, s \rangle \twoheadrightarrow N$. From the operational semantics, we thus know $\langle E_1, s \rangle \twoheadrightarrow N_1$ and $\langle E_2, s \rangle \twoheadrightarrow N_2$ for some $N_1$ and $N_2$ with $N = N_1 + N_2$. From the induction assumption, we thus know $[\![\,E_1\,]\!](s) = N_1$ and $[\![\,E_2\,]\!](s) = N_2$ from which we have $[\![\,E_1 + E_2\,]\!](s) = [\![\,E_1\,]\!](s) + [\![\,E_2\,]\!](s) = N_1 + N_2 = N$.
- Case $E = E_1 * E_2$: analogous to the previous case.                                    □

While in big-step operational semantics, the individual execution 'steps' are quite implicit and become only exhibited by traversing the inference tree for a transition, the steps are very explicit in the *small-step operational semantics* (sometimes also called *structural operational semantics*) defined in Fig. 7.8. Here a transition $\langle E_1, s_1 \rangle \rightarrow \langle E_2, s_2 \rangle$ can be read as 'configuration $\langle E_1, s_1 \rangle$ makes a transition to configuration $\langle E_2, s_2 \rangle$' respectively as 'expression $E_1$ in state $s_1$ is transformed to expression $E_2$ in

**Semantic Domains**

ExpState := Expression × State

**Transitions**

. → . ⊆ ExpState × ExpState

**Inference Rules**

$$\langle V, s \rangle \rightarrow \langle s(V), s \rangle$$

$$\frac{\langle E_1, s \rangle \rightarrow \langle E_1', s \rangle}{\langle E_1+E_2, s \rangle \rightarrow \langle E_1'+E_2, s \rangle} \qquad \frac{\langle E_2, s \rangle \rightarrow \langle E_2', s \rangle}{\langle E_1+E_2, s \rangle \rightarrow \langle E_1+E_2', s \rangle} \qquad \frac{N = N_1 + N_2}{\langle N_1+N_2, s \rangle \rightarrow \langle N, s \rangle}$$

$$\frac{\langle E_1, s \rangle \rightarrow \langle E_1', s \rangle}{\langle E_1*E_2, s \rangle \rightarrow \langle E_1'*E_2, s \rangle} \qquad \frac{\langle E_2, s \rangle \rightarrow \langle E_2', s \rangle}{\langle E_1*E_2, s \rangle \rightarrow \langle E_1*E_2', s \rangle} \qquad \frac{N = N_1 \cdot N_2}{\langle N_1*N_2, s \rangle \rightarrow \langle N, s \rangle}$$

**Fig. 7.8** Operational semantics of the expression language (small-step version)

state $s_2$'. The characteristic of a small-step semantics is the fact that both arguments of the transition relation $\rightarrow$ are from the same domain of configurations such that we may derive sequences of configurations $\langle E_1, s_1 \rangle \rightarrow \langle E_2, s_2 \rangle \rightarrow \ldots \rightarrow \langle E_n, s_n \rangle$ where each transition $\langle E_i, s_i \rangle \rightarrow \langle E_{i+1}, s_{i+1} \rangle$ represents one step of the evaluation.

For instance, if we again consider the evaluation of expression $E = 2*x+3*(y+1)$ in state $s = [x \mapsto 1, y \mapsto 2, \ldots]$, we derive the following execution sequence:

$$\langle E, s \rangle \rightarrow \langle 2*1+3*(y+1), s \rangle \rightarrow \langle 2+3*(y+1), s \rangle$$
$$\rightarrow \langle 2+3*(2+1), s \rangle \rightarrow \langle 2+3*3, s \rangle \rightarrow \langle 2+9, s \rangle \rightarrow \langle 11, s \rangle$$

The last configuration $\langle 11, s \rangle$ does not allow another application of the relation $\rightarrow$ and thus represents the ultimate result of the execution.

While different in style, both flavors of operational semantics represented by the big-step transition $\twoheadrightarrow$ and the small-step transition $\rightarrow$, are closely related. For formalizing this relationship, we introduce the following concept.

**Definition 7.8** *(Reflexive Transitive Closure)* Let $\rightarrow \subseteq C \times C$ be a binary relation on some domain $C$. For $n \in \mathbb{N}$, we define the $n$-fold *repetition* $\rightarrow^n \subseteq C \times C$ as:

$$c \rightarrow^0 c' :\Leftrightarrow c = c'$$

$$c \rightarrow^{n+1} c' :\Leftrightarrow \exists c_1 \in C. \, c \rightarrow c_1 \wedge c_1 \rightarrow^n c'$$

We define the *reflexive transitive closure* $\rightarrow^* \subseteq C \times C$ as:

$$c \rightarrow^* c' :\Leftrightarrow \exists n \in \mathbb{N}. \, c \rightarrow^n c'$$

In other words, $c \to^* c'$ holds, if there exist, for some $n \in \mathbb{N}$, $n$ transitions $c_0 \to c_1 \to c_2 \to \ldots \to c_{n-1} \to c_n$ with $c = c_0$ and $c' = c_n$.

**Proposition 7.10** (Induction on Reflexive Transitive Closure) *Let* $\to \subseteq C \times C$ *be a binary relation on some domain C. In order to prove for some binary relation* $P \subseteq C \times C$ *the property*

$$\forall c, c' \in C.\, c \to^* c' \Rightarrow P\langle c, c'\rangle$$

*it suffices to prove that for every* $n \in \mathbb{N}$ *the assumption*

$$\forall m \in \mathbb{N}_n.\, \forall c, c' \in C.\, c \to^m c' \Rightarrow P\langle c, c'\rangle$$

*implies the conclusion*

$$\forall c, c' \in C.\, c \to^n c' \Rightarrow P\langle c, c'\rangle$$

In other words, from the assumption that the property holds for less than $n$ transitions, it has to be shown that the property also holds for $n$ transitions; then the property holds for arbitrarily many transitions. This principle is thus just a special case of mathematical induction.

After these preliminaries, we can state the relationship between both kinds of operational semantics as follows.

**Proposition 7.11** (Equivalence of Big-Step and Small-Step Semantics) *For every expression E, store s, and natural number N, there is a big-step transition from* $\langle E, s\rangle$ *to N if and only if there is a sequence of zero or more small-step transitions from* $\langle E, s\rangle$ *to* $\langle N, s\rangle$:

$$\forall E \in Expression, s \in State, N \in \mathbb{N}.$$
$$\langle E, s\rangle \twoheadrightarrow N \Leftrightarrow \langle E, s\rangle \to^* \langle N, s\rangle$$

*In other words, the big-step relation* $\twoheadrightarrow$ *is the reflexive transitive closure of the small-step relation* $\to$.

**Proof** We show both directions of the equivalence in turn. First we show

$$\forall E \in Expression, s \in State, N \in \mathbb{N}.\, \langle E, s\rangle \twoheadrightarrow N \Rightarrow \langle E, s\rangle \to^* \langle N, s\rangle$$

by structural induction on $E$:

- Case $E = N_1$: For arbitrary $s \in State$, $\langle N_1, s \rangle \twoheadrightarrow N$ implies $N_1 = N$; since $\langle N, s \rangle \to^0 \langle N, s \rangle$, we have $\langle N, s \rangle \to^* \langle N, s \rangle$ and are done.
- Case $E = V$: For arbitrary $s \in State$, we know from $\langle V, s \rangle \twoheadrightarrow N$ that $N = s(V)$. Since we also have $\langle V, s \rangle \to \langle s(V), s \rangle$, we know $\langle V, s \rangle \to^* \langle N, s \rangle$ and are done.
- Case $E = E_1+E_2$: Take arbitrary $s \in State$ and $N \in \mathbb{N}$ with $\langle E, s \rangle \twoheadrightarrow N$. From the inference rules for $\twoheadrightarrow$, we then have $N_1, N_2 \in \mathbb{N}$ with $\langle E_1, s \rangle \twoheadrightarrow N_1$, $\langle E_2, s \rangle \twoheadrightarrow N_2$, and $N = N_1 + N_2$. From the induction assumption, we have $\langle E_1, s \rangle \to^* \langle N_1, s \rangle$ and $\langle E_2, s \rangle \to^* \langle N_2, s \rangle$. From the inference rules for $\to$, we thus have $\langle E_1+E_2, s \rangle \to^* \langle N_1+E_1, s \rangle \to^* \langle N_1+N_2, s \rangle \to \langle N, s \rangle$.
- Case $E = E_1*E_2$: analogous to the previous case.

Next we show

$$\forall E \in Expression, s \in State, N \in \mathbb{N}. \ \langle E, s \rangle \to^* \langle N, s \rangle \Rightarrow \langle E, s \rangle \twoheadrightarrow N$$

by induction on $\to^*$: we take arbitrary $n \in \mathbb{N}$ with the induction hypothesis

$$\forall m \in \mathbb{N}_n. \ \forall E \in Expression, s \in State, N \in \mathbb{N}. \ \langle E, s \rangle \to^m \langle N, s \rangle \Rightarrow \langle E, s \rangle \twoheadrightarrow N$$

and prove

$$\forall E \in Expression, s \in State, N \in \mathbb{N}. \ \langle E, s \rangle \to^n \langle N, s \rangle \Rightarrow \langle E, s \rangle \twoheadrightarrow N$$

For this, we take arbitrary $E \in Expression$, $s \in State$, and $N \in \mathbb{N}$ with $\langle E, s \rangle \to^n \langle N, s \rangle$ and prove $\langle E, s \rangle \twoheadrightarrow N$. We proceed by case distinction on $E$.

- Case $E = N_1$: From $\langle N_1, s \rangle \to^n \langle N, s \rangle$, we have $n = 0$, thus $N_1 = N$, and consequently $\langle N_1, s \rangle \twoheadrightarrow \langle N, s \rangle$.
- Case $E = V$: From $\langle V, s \rangle \to^n \langle N, s \rangle$, we have $\langle V, s \rangle \to \langle E_1, s \rangle$ and $\langle E_1, s \rangle \to^{n-1} \langle N, s \rangle$. From $\langle V, s \rangle \to \langle E_1, s \rangle$, we have $E_1 = s(V)$, thus $n = 1$ and $N = s(V)$, and consequently $\langle V, s \rangle \twoheadrightarrow N$.
- Case $E = E_1+E_2$: From $\langle E_1+E_2, s \rangle \to^n \langle N, s \rangle$, and the inference rules, we have $n > 0$ and $A_0, \ldots, A_{n-1}, B_0, \ldots, B_{n-1} \in Expression$, and $N_1, N_2 \in \mathbb{N}$ with $\langle E_1+E_2, s \rangle = \langle A_0+B_0, s \rangle \to^{n-1} \langle A_{n-1}+B_{n-1}, s \rangle = \langle N_1+N_2, s \rangle \to \langle N, s \rangle$ and $N = N_1 + N_2$; we then know for every $i \in \mathbb{N}_{n-1}$
  - $A_i \to A_{i+1}$ and $B_i = B_{i+1}$, or
  - $B_i \to B_{i+1}$ and $A_i = A_{i+1}$.
  Let $\mathcal{A}_0, \ldots, \mathcal{A}_{n_1}$ be the sequence of all $A_i$ with $A_i \to A_{i+1}$ and let $\mathcal{B}_0, \ldots, \mathcal{B}_{n_2}$ the sequence of all $B_i$ with $B_i \to B_{i+1}$, thus $n_1 + n_2 = n - 1$. We then have $\langle A_0, s \rangle = \langle \mathcal{A}_0, s \rangle \to^{n_1} \langle \mathcal{A}_{n_1}, s \rangle = \langle A_{n-1}, s \rangle = \langle N_1, s \rangle$ and $\langle B_0, s \rangle = \langle \mathcal{B}_0, s \rangle \to^{n_2} \langle \mathcal{B}_{n_2}, s \rangle = \langle B_{n-1}, s \rangle = \langle N_2, s \rangle$. By the induction hypothesis, we thus have $\langle A_0, s \rangle \twoheadrightarrow N_1$ and $\langle B_0, s \rangle \twoheadrightarrow N_2$ and consequently $\langle E_1+E_2, s \rangle = \langle A_0+B_0, s \rangle \twoheadrightarrow N_1 + N_2$.
- Case $E = E_1*E_2$: analogous to the previous case. $\qquad\qquad\qquad\square$

**Semantic Domains**

$ProgramConf := Program \times Value^*$

$CommandConf := Command \times State$

**Transitions**

$. \twoheadrightarrow . \subseteq ProgramConf \times Value^*$

$. \twoheadrightarrow . \subseteq CommandConf \times State$

**Inference Rules**

$$\frac{v \in Value \quad s = write(X, vs, init(v)) \quad \langle C, s \rangle \twoheadrightarrow s' \quad vs' = read(X, s')}{\langle \text{program } I(X)\, C, vs \rangle \twoheadrightarrow vs'}$$

$$\frac{[\![T]\!]\surd \langle s \rangle \quad s' = s[V \mapsto [\![T]\!]_s]}{\langle V := T, s \rangle \twoheadrightarrow s'}$$

$$\frac{v = s(V) \quad v' \in A(S) \quad s_1 = s[V \mapsto v'] \quad \langle C, s_1 \rangle \twoheadrightarrow s_2 \quad s' = s_2[V \mapsto v]}{\langle \text{var } V : S;\ C, s \rangle \twoheadrightarrow s'}$$

$$\frac{\langle C_1, s \rangle \twoheadrightarrow s_1 \quad \langle C_2, s_1 \rangle \twoheadrightarrow s'}{\langle C_1 ; C_2, s \rangle \twoheadrightarrow s'}$$

$$\frac{[\![F]\!]\surd \langle s \rangle \quad [\![F]\!]_s \quad \langle C_1, s \rangle \twoheadrightarrow s'}{\langle \text{if } F \text{ then } C_1 \text{ else } C_2, s \rangle \twoheadrightarrow s'} \qquad \frac{[\![F]\!]\surd \langle s \rangle \quad \neg[\![F]\!]_s \quad \langle C_2, s \rangle \twoheadrightarrow s'}{\langle \text{if } F \text{ then } C_1 \text{ else } C_2, s \rangle \twoheadrightarrow s'}$$

$$\frac{[\![F]\!]\surd \langle s \rangle \quad [\![F]\!]_s \quad \langle C, s \rangle \twoheadrightarrow s'}{\langle \text{if } F \text{ then } C, s \rangle \twoheadrightarrow s'} \qquad \frac{[\![F]\!]\surd \langle s \rangle \quad \neg[\![F]\!]_s \quad s' = s}{\langle \text{if } F \text{ then } C, s \rangle \twoheadrightarrow s'}$$

$$\frac{[\![F]\!]\surd \langle s \rangle \quad [\![F]\!]_s \quad \langle C, s \rangle \twoheadrightarrow s_1 \quad \langle \text{while } F \text{ do } C, s_1 \rangle \twoheadrightarrow s'}{\langle \text{while } F \text{ do } C, s \rangle \twoheadrightarrow s'}$$

$$\frac{[\![F]\!]\surd \langle s \rangle \quad \neg[\![F]\!]_s \quad s' = s}{\langle \text{while } F \text{ do } C, s \rangle \twoheadrightarrow s'}$$

**Fig. 7.9** A big-step operational semantics of programs

## A Big-Step Semantics of Programs

Now Fig. 7.9 introduces a big-step operational semantics of programs and commands. This semantics consists of two transitions $\langle P, vs \rangle \twoheadrightarrow vs'$ and $\langle C, s \rangle \twoheadrightarrow s'$ that are to be interpreted as 'program $P$ executed with initial parameter values $vs$ results in $vs'$ as their final parameter values' respectively 'command $C$ executed in initial state $s$ results in $s'$ as the final state'. The inference rules for most of these transitions are modeled closely after the definitions of the corresponding relations that were previously presented in Fig. 7.3 (for every transition that depends on the truth value of a formula $F$, there are two inference rules, one for each possible outcome).

To illustrate this semantics, we again consider the following program $P$:

```
program P(a:Nat,b:Nat) {
  b := a;
```

```
if b>0 then {
  var a:Nat;
  a := b-1;
  b := a+a
}
a := a+1
}
```

The big-step semantics models a possible execution of $P$ with initial parameter values $[a = 3, b = 0]$ by the following inference tree (using state $s := [a \mapsto 3, b \mapsto 0, \ldots]$):

$$\cfrac{s_1 = s[b \mapsto 3] \quad \cfrac{\cdots \quad \langle \text{if b>0 then } \{ \ \ldots\ \}, s_1 \rangle \twoheadrightarrow s_5 \quad \cfrac{s' = s_5[a \mapsto 4]}{\langle \text{a:=a+1}, s_5 \rangle \twoheadrightarrow s'}}{\langle \text{if b>0 then } \{ \ \ldots\ \}; \ \text{a:=a+1}, s_1 \rangle \twoheadrightarrow s'}}{\cfrac{\langle \text{b:=a}, s \rangle \twoheadrightarrow s_1 \qquad \langle \text{if b>0 then } \{ \ \ldots\ \}; \ \text{a:=a+1}, s \rangle \twoheadrightarrow s'}{\langle \text{b:=a; if b>0 then } \{ \ \ldots\ \}; \ \text{a:=a+1}, s \rangle \twoheadrightarrow s'}} $$
$$\cfrac{\langle \text{b:=a; if b>0 then } \{ \ \ldots\ \}; \ \text{a:=a+1}, s \rangle \twoheadrightarrow s'}{\langle P, [a = 3, b = 0] \rangle \twoheadrightarrow [a = 4, b = 4]}$$

The inference marked as '$\ldots$' is given below:

$$\cfrac{s_2 = s_1[a \mapsto 0] \quad \cfrac{\cfrac{s_3 = s_2[a \mapsto 2]}{\langle \text{a:=b-1}, s_2 \rangle \twoheadrightarrow s_3} \quad \cfrac{s_4 = s_3[b \mapsto 4]}{\langle \text{b:=a+a}, s_3 \rangle \twoheadrightarrow s_4}}{\langle \text{a:=b-1; b:=a+a}, s_2 \rangle \twoheadrightarrow s_4} \quad s_5 = s_4[a \mapsto 3]}{\cfrac{\langle \text{var a:Nat; a:=b-1; b:=a+a}, s_1 \rangle \twoheadrightarrow s_5}{\langle \text{if b>0 then } \{ \ \ldots\ \}, s_1 \rangle \twoheadrightarrow s_5}}$$

This inference tree has been simplified to only indicate the various state updates. Traversing the tree in bottom-up and left-to-right order, we note the update sequence

$$s_1 = s[b \mapsto 3]$$
$$s_2 = s_1[a \mapsto 0]$$
$$s_3 = s_2[a \mapsto 2]$$
$$s_4 = s_3[b \mapsto 4]$$
$$s_5 = s_4[a \mapsto 3]$$
$$s' = s_5[a \mapsto 4]$$

which leads from the initial state $s := [a \mapsto 3, b \mapsto 0, \ldots]$ to the final state $s' := [a \mapsto 4, b \mapsto 4, \ldots]$.

As for expressions, also the denotational semantics of programs is equivalent to the operational one.

**Proposition 7.12** (Equivalence of Denotational and Operational Semantics)
*For every program $P$ and value sequences $vs$ and $vs'$, the application of the*

*denotational semantics* $[\![\,P\,]\!]$ *to vs yields vs' if and only if there is a big-step transition from* $\langle P, vs\rangle$ *to vs':*

$$\forall P \in Program, vs, vs' \in Value^*. \; [\![\,P\,]\!]\langle vs, vs'\rangle \Leftrightarrow \langle P, vs\rangle \twoheadrightarrow vs'$$

*For every command* $C$, *and states* $s$ *and* $s'$, *the application of the denotational semantics* $[\![\,C\,]\!]$ *to* $s$ *yields* $s'$ *if and only if there is a big-step transition from* $\langle C, s\rangle$ *to* $s'$:

$$\forall C \in Command, s, s' \in State. \; [\![\,C\,]\!]\langle s, s'\rangle \Leftrightarrow \langle C, s\rangle \twoheadrightarrow s'$$

**Proof**  The equivalence for programs is a direct consequence of the equivalence for commands; we show both directions of this equivalence in turn. First we prove

$$\forall C \in Command, s, s' \in State. \; [\![\,C\,]\!]\langle s, s'\rangle \Rightarrow \langle C, s\rangle \twoheadrightarrow s'$$

by structural induction on $C$.

- Case $C = V := T$: for arbitrary $s, s' \in State$, $[\![\,V := T\,]\!]\langle s, s'\rangle$ implies $[\![\,T\,]\!]\checkmark \langle s\rangle$ and $s' = s[V \mapsto [\![\,T\,]\!]_s]$ which implies $\langle V := T, s\rangle \twoheadrightarrow s'$.
- Case $C = \mathtt{var}\ V : S;\ C_1$: for arbitrary $s, s' \in State$, from $[\![\,\mathtt{var}\ V : S;\ C_1\,]\!]\langle s, s'\rangle$ we have some $v, v' \in A(S)$ and $s_1, s_2 \in State$ with $v = s(V)$, $s_1 = s[V \mapsto v']$, $[\![\,C\,]\!]\langle s_1, s_2\rangle$, and $s' = s_2[V \mapsto v]$. From the induction hypothesis, we have $\langle C, s_1\rangle \twoheadrightarrow s_2$ and thus $\langle \mathtt{var}\ V : S;\ C_1, s\rangle \twoheadrightarrow s'$.
- Case $C = C_1;C_2$: for arbitrary $s, s' \in State$, from $[\![\,C_1;C_2\,]\!]\langle s, s'\rangle$ we have some $s_1 \in State$ with $[\![\,C_1\,]\!]\langle s, s_1\rangle$ and $[\![\,C_2\,]\!]\langle s_1, s'\rangle$. From the induction hypothesis, we have $\langle C_1, s\rangle \twoheadrightarrow s_1$ and $\langle C_2, s_1\rangle \twoheadrightarrow s'$ and thus $\langle C_1;C_2, s\rangle \twoheadrightarrow s'$.
- Case $C = \mathtt{if}\ F\ \mathtt{then}\ C_1\ \mathtt{else}\ C_2$: for arbitrary $s, s' \in State$, from the definition of $[\![\,\mathtt{if}\ F\ \mathtt{then}\ C_1\ \mathtt{else}\ C_2\,]\!]\langle s, s'\rangle$ we have $[\![\,F\,]\!]\checkmark \langle s\rangle$ and two cases:
  - Case $[\![\,F\,]\!]_s$ and $[\![\,C_1\,]\!]\langle s, s'\rangle$: from the induction hypothesis, we have $\langle C_1, s\rangle \twoheadrightarrow s'$ and thus $\langle \mathtt{if}\ F\ \mathtt{then}\ C_1\ \mathtt{else}\ C_2, s\rangle \twoheadrightarrow s'$.
  - Case $\neg[\![\,F\,]\!]_s$ and $[\![\,C_2\,]\!]\langle s, s'\rangle$: from the induction hypothesis, we have $\langle C_2, s\rangle \twoheadrightarrow s'$ and thus $\langle \mathtt{if}\ F\ \mathtt{then}\ C_1\ \mathtt{else}\ C_2, s\rangle \twoheadrightarrow s'$.
- Case $C = \mathtt{if}\ F\ \mathtt{then}\ C_1$: for arbitrary $s, s' \in State$, from $[\![\,\mathtt{if}\ F\ \mathtt{then}\ C_1\,]\!]\langle s, s'\rangle$ we have $[\![\,F\,]\!]\checkmark \langle s\rangle$ and two cases:
  - Case $[\![\,F\,]\!]_s$ and $[\![\,C_1\,]\!]\langle s, s'\rangle$: from the induction hypothesis, we have $\langle C_1, s\rangle \twoheadrightarrow s'$ and thus $\langle \mathtt{if}\ F\ \mathtt{then}\ C_1, s\rangle \twoheadrightarrow s'$.
  - Case $\neg[\![\,F\,]\!]_s$ and $s' = s$: we thus have $\langle \mathtt{if}\ F\ \mathtt{then}\ C_1, s\rangle \twoheadrightarrow s'$.
- Case $C = \mathtt{while}\ F\ \mathtt{do}\ C_1$: from Proposition 7.8, we have

$$\forall s, s' \in State. \; [\![\,\mathtt{while}\ F\ \mathtt{do}\ C_1\,]\!]\langle s, s'\rangle \Leftrightarrow \exists n \in \mathbb{N}. \; W\langle n, s, s'\rangle$$

where

$$W \langle n, s, s' \rangle :\Leftrightarrow$$
$$\exists t \in State^\omega.$$
$$loop \langle t, n, F, C \rangle \wedge t(0) = s \wedge t(n) = s' \wedge [\![ F ]\!] \checkmark \langle s' \rangle \wedge \neg [\![ F ]\!]_{s'}$$

and $n$ denotes the number of loop iterations that are required to transform $s$ into $s'$. It thus suffices to show

$$\forall n \in \mathbb{N}, s, s' \in State.\ W \langle n, s, s' \rangle \Rightarrow \langle \texttt{while } F \texttt{ do } C_1, s \rangle \twoheadrightarrow s'.$$

by mathematical induction on $n \in \mathbb{N}$:

- In the base case $n = 0$, we take arbitrary $s, s' \in State$ with $W \langle 0, s, s' \rangle$ which implies $s' = s$ and thus $[\![ F ]\!] \checkmark \langle s \rangle$ and $\neg [\![ F ]\!]_s$; we thus can conclude $\langle \texttt{while } F \texttt{ do } C_1, s \rangle \twoheadrightarrow s'$.
- In the induction step, we assume the induction hypothesis

$$\forall s, s' \in State.\ W \langle n, s, s' \rangle \Rightarrow \langle \texttt{while } F \texttt{ do } C_1, s \rangle \twoheadrightarrow s'$$

and show

$$\forall s, s' \in State.\ W \langle n + 1, s, s' \rangle \Rightarrow \langle \texttt{while } F \texttt{ do } C_1, s \rangle \twoheadrightarrow s'.$$

We take arbitrary $s, s' \in State$ with $W \langle n + 1, s, s' \rangle$. From the definition of $W$, we thus have some $t \in State^\omega$ with $loop \langle t, n + 1, F, C \rangle$ and $t(0) = s$ and $t(n + 1) = s'$ and $[\![ F ]\!] \checkmark \langle s' \rangle$ and $\neg [\![ F ]\!]_{s'}$. From the definition of $W$, we thus have $[\![ F ]\!] \checkmark \langle s \rangle$, $[\![ F ]\!]_s$, $[\![ C ]\!] \langle s, t(1) \rangle$ and $W \langle n, t(1), s' \rangle$. From $[\![ C ]\!] \langle s, t(1) \rangle$ and the hypothesis of the structural induction on commands, we know $\langle C, s \rangle \twoheadrightarrow t(1)$. From $W \langle n, t(1), s' \rangle$ and the hypothesis of the mathematical induction on $n$, we know $\langle \texttt{while } F \texttt{ do } C_1, t(1) \rangle \twoheadrightarrow s'$. We thus can conclude $\langle \texttt{while } F \texttt{ do } C_1, s \rangle \twoheadrightarrow s'$.

Next we prove

$$\forall C \in Command, s, s' \in State.\ \langle C, s \rangle \twoheadrightarrow s' \Rightarrow [\![ C ]\!] \langle s, s' \rangle$$

by rule induction on the derivation of $\langle C, s \rangle \twoheadrightarrow s'$ (as it will become clear below, here the principle of structural induction does not suffice). We therefore take arbitrary $s, s' \in State$ with $\langle C, s \rangle \twoheadrightarrow s'$ and consider all possible derivations of this transition from the rules in Fig. 7.9:

- Case $C = V := T$: From $\langle V := T, s \rangle \twoheadrightarrow s'$, we have $[\![ T ]\!] \checkmark \langle s \rangle$ and $s' = s[V \mapsto [\![ T ]\!]_s]$, which implies $[\![ V := T ]\!] \langle s, s' \rangle$.
- Case $C = \texttt{var } V : S; C_1$: From $\langle \texttt{var } V : S; C_1, s \rangle \twoheadrightarrow s'$, we have some $v = s(V)$, $v' \in A(S)$, $s_1 = s[V \mapsto v']$, and $s' = s_2[V \mapsto v]$ with $\langle C, s_1 \rangle \twoheadrightarrow s_2$. From the induction hypothesis, we thus have $[\![ C ]\!] \langle s_1, s_2 \rangle$, which implies $[\![ \texttt{var } V : S; C_1 ]\!] \langle s, s' \rangle$.

- Case $C = C_1 ; C_2$: From $\langle C_1 ; C_2, s \rangle \twoheadrightarrow s'$, we have some $s_1 \in State$ with $\langle C_1, s \rangle \twoheadrightarrow s_1$ and $\langle C_2, s_1 \rangle \twoheadrightarrow s'$. From the induction hypothesis, we thus have $[\![ C_1 ]\!] \langle s, s_1 \rangle$ and $[\![ C_2 ]\!] \langle s_1, s' \rangle$ which implies $[\![ C_1 ; C_2 ]\!] \langle s, s' \rangle$.
- Case $C = \text{if } F \text{ then } C_1 \text{ else } C_2$: From $\langle \text{if } F \text{ then } C_1 \text{ else } C_2, s \rangle \twoheadrightarrow s'$, we have two cases:
  - Case $[\![ F ]\!] \checkmark \langle s \rangle$ and $[\![ F ]\!]_s$ and $\langle C_1, s \rangle \twoheadrightarrow s'$: From the induction hypothesis, we have $[\![ C_1 ]\!] \langle s, s' \rangle$ and thus $[\![ \text{if } F \text{ then } C_1 \text{ else } C_2 ]\!] \langle s, s' \rangle$.
  - Case $[\![ F ]\!] \checkmark \langle s \rangle$ and $\neg [\![ F ]\!]_s$ and $\langle C_2, s \rangle \twoheadrightarrow s'$: From the induction hypothesis, we have $[\![ C_2 ]\!] \langle s, s' \rangle$ and thus $[\![ \text{if } F \text{ then } C_1 \text{ else } C_2 ]\!] \langle s, s' \rangle$.
- Case $C = \text{if } F \text{ then } C_1$: From $\langle \text{if } F \text{ then } C_1, s \rangle \twoheadrightarrow s'$, we have two cases:
  - We have $[\![ F ]\!] \checkmark \langle s \rangle$ and $[\![ F ]\!]_s$ and $\langle C_1, s \rangle \twoheadrightarrow s'$. From the induction hypothesis, this implies $[\![ C_1 ]\!] \langle s, s' \rangle$ and thus $[\![ \text{if } F \text{ then } C_1 \text{ else } C_2 ]\!] \langle s, s' \rangle$.
  - We have $[\![ F ]\!] \checkmark \langle s \rangle$ and $\neg [\![ F ]\!]_s$ and $s' = s$ and thus $[\![ \text{if } F \text{ then } C_1 ]\!] \langle s, s' \rangle$.
- Case $C = \text{while } F \text{ do } C_1$: From $\langle \text{while } F \text{ do } C_1, s \rangle \twoheadrightarrow s'$, we have two cases:
  - We have some $s_1 \in State$ with $[\![ F ]\!] \checkmark \langle s \rangle$ and $[\![ F ]\!]_s$ and $\langle C_1, s \rangle \twoheadrightarrow s_1$ and $\langle \text{while } F \text{ do } C_1, s_1 \rangle \twoheadrightarrow s'$. From $\langle C_1, s \rangle \twoheadrightarrow s_1$ and the induction hypothesis, we have $[\![ C_1 ]\!] \langle s, s_1 \rangle$. From $\langle \text{while } F \text{ do } C_1, s_1 \rangle \twoheadrightarrow s'$ and the induction hypothesis, we have $[\![ \text{while } F \text{ do } C_1 ]\!] \langle s_1, s' \rangle$ (here the principle of rule induction is indeed applicable because the hypothesis is applied to the inference tree for $\langle \text{while } F \text{ do } C_1, s_1 \rangle \twoheadrightarrow s'$ which is a proper subtree of the inference tree for $\langle \text{while } F \text{ do } C_1, s \rangle \twoheadrightarrow s'$; however, the principle of structural induction is here not applicable, because the hypothesis is not applied to a proper subcommand of $\text{while } F \text{ do } C_1$). From $[\![ \text{while } F \text{ do } C_1 ]\!] \langle s_1, s' \rangle$ and Proposition 7.8, we have some $t \in State^\omega$ and $n \in \mathbb{N}$ with $loop\langle t, n, F, C \rangle$, $t(0) = s_1$, $t(n) = s'$, $[\![ F ]\!] \checkmark \langle s' \rangle$, and $\neg [\![ F ]\!]_{s'}$. We thus can define $t'(i) :=$ if $i = 0$ then $s$ else $t(i - 1)$ and $n' := n + 1$ which implies $loop\langle t', n', F, C \rangle$, $t'(0) = s$, and $t'(n') = s'$. Together with $[\![ F ]\!] \checkmark \langle s' \rangle$ and $\neg [\![ F ]\!]_{s'}$, Proposition 7.8 thus implies $[\![ \text{while } F \text{ do } C_1 ]\!] \langle s, s' \rangle$.
  - We have $[\![ F ]\!] \checkmark \langle s \rangle$, $\neg [\![ F ]\!]_s$, and $s' = s$, and thus $[\![ \text{while } F \text{ do } C_1 ]\!] \langle s, s' \rangle$.

This concludes the proof of both directions of the equivalence.                                      □

### A Small-Step Semantics of Programs

When defining a small-step operational semantics of programs, every intermediate state of the program must be representable by a configuration. In the previously presented small-step semantics of expression evaluation, every intermediate state could be indeed described by an expression itself; however, for more complex languages, it is generally not the case that every execution state of a phrase can be described by a phrase itself.

Figure 7.10 gives a small-step semantics of programs where the intermediate state of the evaluation of a command is essentially represented by a sequence of instructions and a mapping of variables to values.

The first rule replaces the program by such a sequence that consists of the body of the program followed by an instruction $\text{return } X$; the initial store is constructed as in

**Semantic Domains**

$Instruction := Command$

$\cup \{\texttt{return}\ X \mid X \in Parameters\}$

$\cup \{\texttt{set}\ V := v \mid V \in Variable \wedge v \in Value\}$

$SmallStepConf := (Program \cup Instruction^*) \times (Value^* \cup State)$

**Transitions**

$. \to . \subseteq SmallStepConf \times SmallStepConf$

**Inference Rules**

$$\frac{v \in Value \quad s = write(X, vs, init(v))}{\langle \texttt{program}\ I(X)\ C, vs\rangle \to \langle [C, \texttt{return}\ X], s\rangle} \qquad \frac{vs = read(X, s)}{\langle [\texttt{return}\ X], s\rangle \to \langle [\,], vs\rangle}$$

$$\frac{[\![T]\!]\checkmark\langle s\rangle \quad s' = s[V \mapsto [\![T]\!]_s]}{\langle [V := T, \ldots], s\rangle \to \langle [\ldots], s'\rangle}$$

$$\frac{v = s(V) \quad v' \in A(S) \quad s' = s[V \mapsto v']}{\langle [\{\texttt{var}\ V : S;\ C\}, \ldots], s\rangle \to \langle [C, \texttt{set}\ V := v, \ldots], s'\rangle} \qquad \frac{s' = s[V \mapsto v]}{\langle [\texttt{set}\ V := v, \ldots], s\rangle \to \langle [\ldots], s'\rangle}$$

$$\langle [\{C_1; C_2\}, \ldots], s\rangle \to \langle [C_1, C_2, \ldots], s\rangle$$

$$\frac{[\![F]\!]\checkmark\langle s\rangle \quad [\![F]\!]_s}{\langle [\texttt{if}\ F\ \texttt{then}\ C_1\ \texttt{else}\ C_2, \ldots], s\rangle \to \langle [C_1, \ldots], s\rangle}$$

$$\frac{[\![F]\!]\checkmark\langle s\rangle \quad \neg[\![F]\!]_s}{\langle [\texttt{if}\ F\ \texttt{then}\ C_1\ \texttt{else}\ C_2, \ldots], s\rangle \to \langle [C_2, \ldots], s\rangle}$$

$$\frac{[\![F]\!]\checkmark\langle s\rangle \quad [\![F]\!]_s}{\langle [\texttt{if}\ F\ \texttt{then}\ C, \ldots], s\rangle \to \langle [C, \ldots], s\rangle} \qquad \frac{[\![F]\!]\checkmark\langle s\rangle \quad \neg[\![F]\!]_s}{\langle [\texttt{if}\ F\ \texttt{then}\ C, \ldots], s\rangle \to \langle [\ldots], s\rangle}$$

$$\frac{[\![F]\!]\checkmark\langle s\rangle \quad [\![F]\!]_s}{\langle [\texttt{while}\ F\ \texttt{do}\ C, \ldots], s\rangle \to \langle [C, \texttt{while}\ F\ \texttt{do}\ C, \ldots], s\rangle}$$

$$\frac{[\![F]\!]\checkmark\langle s\rangle \quad \neg[\![F]\!]_s}{\langle [\texttt{while}\ F\ \texttt{do}\ C, \ldots], s\rangle \to \langle [\ldots], s\rangle}$$

**Fig. 7.10** A small-step operational semantics of programs

the big-step semantics. All other rules describe the execution of the first instruction of the sequence (where '...' denotes the remainder of the sequence), removing it from the sequence, and possibly inserting other instructions instead. If the instruction sequence is empty, the program has terminated.

An instruction can be either a command or one of the following phrases:

- **return** $X$ replaces the store by the values of the parameters $X$ in that store. This instruction only appears in the last position of the sequence and is executed last; its purpose is to return after the execution of the program body the final values of the parameters.

- `set V:=v` sets the value of variable $V$ to value $v$; unlike the programming language assignment, this statement captures the value $v$ directly, not via a term $T$ denoting that value. The purpose of this instruction is to restore the value of a variable that has been overshadowed by a local variable declaration `var V:S; C` when the execution of the body $C$ of the declaration is completed.

To illustrate this semantics, we again consider the program $P$ defined as follows:

```
program P(a:Nat,b:Nat) {
  b := a;
  if b>0 then {
    var a:Nat;
    a := b-1;
    b := a+a
  }
  a := a+1
}
```

The small-step semantics models a possible execution of $P$ with initial parameter values $[a = 3, b = 0]$ by the following sequence of steps:

$\langle P, [a = 3, b = 0] \rangle$

$\rightarrow \langle [\ \{b:=a;\ if\ b>0\ \{...\}\};\ a:=a+1\},\ ret\ ],\ [a \mapsto 3, b \mapsto 0, \dots] \rangle$ (1)

$\rightarrow \langle [\ \underline{b:=a},\ \{if\ b>0\ \{...\}\};\ a:=a+1\},\ ret\ ],\ [a \mapsto 3, b \mapsto 0, \dots] \rangle$ (2)

$\rightarrow \langle [\ \{if\ b>0\ \{...\}\};\ a:=a+1\},\ ret\ ],\ [a \mapsto 3, \underline{b \mapsto 3}, \dots] \rangle$ (3)

$\rightarrow \langle [\ \underline{if\ b>0\ \{...\}},\ \underline{a:=a+1},\ ret\ ],\ [a \mapsto 3, b \mapsto 3, \dots] \rangle$ (4)

$\rightarrow \langle [\ \{\underline{var\ a};\ a:=b-1;\ b:=a+a\},\ a:=a+1,\ ret\ ],\ [a \mapsto 3, b \mapsto 3, \dots] \rangle$ (5)

$\rightarrow \langle [\ \{a:=b-1;\ b:=a+a\},\ \underline{set\ a:=3},\ a:=a+1,\ ret\ ],\ [\underline{a \mapsto 0}, b \mapsto 3, \dots] \rangle$ (6)

$\rightarrow \langle [\ \underline{a:=b-1},\ \underline{b:=a+a},\ set\ a:=3,\ a:=a+1,\ ret\ ],\ [a \mapsto 0, b \mapsto 3, \dots] \rangle$ (7)

$\rightarrow \langle [\ \underline{b:=a+a},\ set\ a:=3,\ a:=a+1,\ ret\ ],\ [\underline{a \mapsto 2}, b \mapsto 3, \dots] \rangle$ (8)

$\rightarrow \langle [\ set\ a:=3,\ a:=a+1,\ ret\ ],\ [a \mapsto 2, \underline{b \mapsto 4}, \dots] \rangle$ (9)

$\rightarrow \langle [\ a:=a+1,\ ret\ ],\ [\underline{a \mapsto 3}, b \mapsto 4, \dots] \rangle$ (10)

$\rightarrow \langle [\ ret\ ],\ [\underline{a \mapsto 4}, b \mapsto 4, \dots] \rangle$ (11)

$\rightarrow \langle [\ ],\ [\underline{a = 4, b = 4}] \rangle$ (12)

Step (1) replaces the program by the instruction sequence that contains the program body followed by instruction *ret* denoting `return a:Nat,b:Nat`; it also replaces the sequence of initial parameter values by the initial program store. Step (2) breaks up the command into two sub-commands, the assignment and the rest of the body. Step (3) executes the assignment which sets variable $b$ to 3. Step (4) breaks up the command into two sub-commands, the conditional and the final assignment. Step (5) evaluates the condition and enters the conditional branch. Step (6) enters the variable block, setting the local variable $a$ to an arbitrary value (here 0) and inserting after the body of the command block an instruction to restore after the exit from the block $a$

to its original value 3. Step (7) breaks up the body of the variable block into its sub-commands, two assignments. Step (8) executes the first assignment, setting $a$ to 2. Step (9) executes the second assignment, setting variable $b$ to 4. Step (10) executes the instruction to restore variable $a$ to its original value 3. Step (11) executes the final assignment, incrementing variable $a$ to 4. Ultimately, Step (12) replaces the final program store by the values of the parameters in that store.

We now state some fundamental properties of the small-step semantics as a basis for later proofs of higher-level properties.

**Proposition 7.13** (Replacing the First Instruction)  *A small-step transition replaces the first instruction and leaves the remaining instructions unchanged:*

$$\forall I \in Instruction, Is, Is_1 \in Instruction^*, s, s' \in State.$$

$$\langle [I] \circ Is, s \rangle \rightarrow \langle Is_1, s' \rangle \Rightarrow \exists Is_2 \in Instruction^*. Is_1 = Is_2 \circ Is$$

**Proof**  This result follows from the structure of every rule of the calculus.   □

**Proposition 7.14** (Independence from Remaining Instructions)  *The effect of a small-step transition only depends on the first instruction:*

$$\forall I \in Instruction, Is, Is_1, Is_2 \in Instruction^*, s, s' \in State.$$

$$\langle [I] \circ Is_1, s \rangle \rightarrow \langle Is \circ Is_1, s' \rangle \Rightarrow \langle [I] \circ Is_2, s \rangle \rightarrow \langle Is \circ Is_2, s' \rangle$$

**Proof**  Also this result follows from the structure of every rule of the calculus.   □

**Proposition 7.15** (Splitting Execution Sequences)  *If an instruction sequence $Is_1 \circ Is_2$ drives initial state $s$ to final state $s'$ in $n$ steps, then there exists some state $s_1$ such that $Is_1$ drives initial state $s$ to final state $s_1$ in $n_1$ steps and $Is_2$ drives initial state $s_1$ to final state $s'$ in $n_2$ steps where $n_1 + n_2 = n$:*

$$\forall n \in \mathbb{N}, Is_1, Is_2 \in Instruction^*, s, s' \in State.$$

$$\langle Is_1 \circ Is_2, s \rangle \rightarrow^n \langle [\,], s' \rangle \Rightarrow$$

$$\exists n_1, n_2 \in \mathbb{N}, s_1 \in State.$$

$$n_1 + n_2 = n \wedge \langle Is_1, s \rangle \rightarrow^{n_1} \langle [\,], s_1 \rangle \wedge \langle Is_2, s_1 \rangle \rightarrow^{n_2} \langle [\,], s' \rangle$$

***Proof*** We prove the statement by induction on $n \in \mathbb{N}$.

In the base case $n = 0$, we have $Is_1 \circ Is_2 = [\ ]$ and $s = s'$. From $s = s'$ it suffices to show $\langle Is_1, s \rangle \to^* \langle [\ ], s \rangle$ and $\langle Is_2, s \rangle \to^* \langle [\ ], s \rangle$. From $Is_1 \circ Is_2 = [\ ]$, we have $Is_1 = [\ ]$ and $Is_2 = [\ ]$ and are therefore done.

In the induction step, we take arbitrary $n \in \mathbb{N}$, assume that the statement is true for $n$, and show that it is also true for $n + 1$. For this, we take arbitrary $Is_1, Is_2 \in Instruction^*$, $s, s' \in State$ with

$$\langle Is_1 \circ Is_2, s \rangle \to^{n+1} \langle [\ ], s' \rangle \tag{0}$$

and prove

$$\exists n_1, n_2 \in \mathbb{N}, s_1 \in State.$$

$$n_1 + n_2 = n + 1 \land \langle Is_1, s \rangle \to^{n_1} \langle [\ ], s_1 \rangle \land \langle Is_2, s_1 \rangle \to^{n_2} \langle [\ ], s' \rangle \tag{a}$$

From (0) we have some $Is \in Instruction^*$ and $s_2 \in State$ with

$$\langle Is_1 \circ Is_2, s \rangle \to \langle Is, s_2 \rangle \tag{1}$$

$$\langle Is, s_2 \rangle \to^n \langle [\ ], s' \rangle \tag{2}$$

From (1) and the rules of the calculus, there exists some $I \in Instruction$ and $Is_3 \in Instruction^*$ with $Is_1 \circ Is_2 = [I] \circ Is_3$.

If $Is_1 = [\ ]$ and $Is_2 = [I] \circ Is_3$, we have $\langle Is_1, s \rangle \to^0 \langle [\ ], s \rangle$. It then suffices to show $\langle Is_2, s \rangle \to^{n+1} \langle [\ ], s' \rangle$, which follows from $Is_2 = Is_1 \circ Is_2$ and (1) and (2).

Otherwise, we have some $Is_4 \in Instruction^*$ with $Is_1 = [I] \circ Is_4$ and thus $Is_1 \circ Is_2 = [I] \circ Is_4 \circ Is_2$. Thus (1) and Proposition 7.13 give us some $Is_5 \in Instruction^*$ with $Is = Is_5 \circ Is_4 \circ Is_2$. From (1) and (2), we thus have

$$\langle [I] \circ Is_4 \circ Is_2, s \rangle \to \langle Is_5 \circ Is_4 \circ Is_2, s_2 \rangle \tag{3}$$

$$\langle Is_5 \circ Is_4 \circ Is_2, s_2 \rangle \to^n \langle [\ ], s' \rangle \tag{4}$$

From (4) and the induction hypothesis, we thus have have some $n_1, n_2 \in \mathbb{N}$ with $n_1 + n_2 = n$ and $s_1 \in State$ with

$$\langle Is_5 \circ Is_4, s_2 \rangle \to^{n_1} \langle [\ ], s_1 \rangle \tag{5}$$

$$\langle Is_2, s_1 \rangle \to^{n_2} \langle [\ ], s' \rangle \tag{6}$$

From (6), $Is_1 = [I] \circ Is_4$, and $n_1 + n_2 = n$, to prove (a), it suffices to prove

$$\langle [I] \circ Is_4, s \rangle \to^{n_1+1} \langle [\ ], s_1 \rangle \tag{b}$$

From (3) and Proposition 7.14, we know

$$\langle [I] \circ Is_4, s \rangle \to \langle Is_5 \circ Is_4, s_2 \rangle \tag{7}$$

From (7) and (5), we know (b).

Based on these results, we can formalize and prove the relationship between the big-step and the small-step semantics of programs respectively commands.

---

**Proposition 7.16** (Equivalence of Big-Step and Small-Step Semantics) *For every program $P$ and value sequences vs and vs', there is a big-step transition from configuration $\langle P, vs \rangle$ to vs' if and only if there is a sequence of small-step transitions from $\langle P, vs \rangle$ to $\langle [\,], vs' \rangle$:*

$$\forall P \in Program, vs, vs' \in Value^*.$$

$$\langle P, vs \rangle \twoheadrightarrow vs' \Leftrightarrow \langle P, vs \rangle \to^* \langle [\,], vs' \rangle$$

*For command $C$ and states $s$ and $s'$, there is a big-step transition from $\langle C, s \rangle$ to $s'$ if and only if there is a sequence of small-step transitions from $\langle [C], s \rangle$ to $\langle [\,], s' \rangle$:*

$$\forall C \in Command, s, s' \in State.$$

$$\langle C, s \rangle \twoheadrightarrow s' \Leftrightarrow \langle [C], s \rangle \to^* \langle [\,], s' \rangle$$

---

**Proof**  The equivalence of the semantics for programs is a rather direct consequence of the corresponding equivalence for commands. As for the later, we first prove

$$\forall C \in Command, s, s' \in State.\ \langle C, s \rangle \twoheadrightarrow s' \Rightarrow \langle [C], s \rangle \to^* \langle [\,], s' \rangle$$

by rule induction on the derivation of $\langle C, s \rangle \twoheadrightarrow s'$. Actually, we are going to prove the more general statement

$$\forall C \in Command, s, s' \in State.\ \langle C, s \rangle \twoheadrightarrow s' \Rightarrow \langle [C, \ldots], s \rangle \to^* \langle [\ldots], s' \rangle$$

which allows in the small-step semantics an arbitrary sequence '…' of instructions after the initial command $C$.

- Case $C = V := T$: For arbitrary $s, s' \in State$, $\langle V := T, s \rangle \twoheadrightarrow s'$ implies $[\![ T ]\!] \checkmark \langle s \rangle$ and $s' = s[V \mapsto [\![ T ]\!]_s]$, which implies $\langle V := T, \ldots \rangle, s \to \langle [\ldots], s' \rangle$.
- Case $C = \text{var } V \colon S;\ C_1$: For arbitrary $s, s' \in State$, from $\langle \text{var } V \colon S; C_1, s \rangle \twoheadrightarrow s'$, we have $v = s(V)$, $v' \in A(S)$, $s_1 = s[V \mapsto v']$, and $s' = s_2[V \mapsto v]$ with $\langle C_1, s_1 \rangle \twoheadrightarrow s_2$. The induction hypothesis implies $\langle [C_1, \text{set } V := v, \ldots], s_1 \rangle \to^* \langle [\text{set } V := v, \ldots], s_2 \rangle$. From $\langle [\text{var } V \colon S; C_1, \ldots], s \rangle \to \langle [C_1, \text{set } V := v, \ldots], s_1 \rangle$ and $\langle [\text{set } V := v, \ldots], s_2 \rangle \to \langle [\ldots], s_2[V \mapsto v] \rangle$ and $s' = s_2[V \mapsto v]$, we thus have $\langle [\text{var } V \colon S; C_1, \ldots], s \rangle \to \langle [\ldots], s' \rangle$.
- Case $C = C_1; C_2$: For arbitrary $s, s' \in State$, $\langle C_1; C_2, s \rangle \twoheadrightarrow s'$ implies the existence of some $s_1 \in State$ with $\langle C_1, s \rangle \twoheadrightarrow s_1$ and $\langle C_2, s_1 \rangle \twoheadrightarrow s'$. From the induction

hypothesis, we thus have $\langle [C_1, C_2, \ldots], s \rangle \to^* \langle [C_2, \ldots], s_1 \rangle$ and $\langle [C_2, \ldots], s_1 \rangle \to^*$ $\langle [\ldots], s' \rangle$ and thus $\langle [C_1, C_2, \ldots], s \rangle \to^* \langle [\ldots], s' \rangle$. Since we also have $\langle [\{C_1 ; C_2\}, \ldots], s \rangle \to \langle [C_1, C_2, \ldots], s \rangle$, we are done.

- Case $C = \text{if } F \text{ then } C_1 \text{ else } C_2$: For arbitrary $s, s' \in State$, from the derivation of $\langle \text{if } F \text{ then } C_1 \text{ else } C_2, s \rangle \twoheadrightarrow s'$, we have two cases:
  - Case $[\![ F ]\!] \checkmark \langle s \rangle$ and $[\![ F ]\!]_s$ and $\langle C_1, s \rangle \twoheadrightarrow s'$: From the induction hypothesis, we have $\langle [C_1, \ldots], s \rangle \to^* \langle [\ldots], s' \rangle$. Since $[\![ F ]\!] \checkmark \langle s \rangle$ and $[\![ F ]\!]_s$ implies $\langle [\text{if } F \text{ then } C_1 \text{ else } C_2, \ldots], s \rangle \to \langle [C_1, \ldots], s \rangle$, we thus also have $\langle [\text{if } F \text{ then } C_1 \text{ else } C_2, \ldots], s \rangle \to^* \langle [\ldots], s' \rangle$.
  - Case $[\![ F ]\!] \checkmark \langle s \rangle$ and $\neg [\![ F ]\!]_s$ and $\langle C_2, s \rangle \twoheadrightarrow s'$: From the induction hypothesis, we have $\langle [C_2, \ldots], s \rangle \to^* \langle [\ldots], s' \rangle$. Since $[\![ F ]\!] \checkmark \langle s \rangle$ and $\neg [\![ F ]\!]_s$ implies $\langle [\text{if } F \text{ then } C_1 \text{ else } C_2, \ldots], s \rangle \to \langle [C_2, \ldots], s \rangle$, we thus also have $\langle [\text{if } F \text{ then } C_1 \text{ else } C_2, \ldots], s \rangle \to^* \langle [\ldots], s' \rangle$.

- Case $C = \text{if } F \text{ then } C_1$: For arbitrary $s, s' \in State$, from $\langle \text{if } F \text{ then } C_1, s \rangle \twoheadrightarrow s'$, we have two cases:
  - Case $[\![ F ]\!] \checkmark \langle s \rangle$ and $[\![ F ]\!]_s$ and $\langle C_1, s \rangle \twoheadrightarrow s'$: this case is analogous to the first case of the two-sided conditional presented above.
  - Case $[\![ F ]\!] \checkmark \langle s \rangle$ and $\neg [\![ F ]\!]_s$ and $s' = s$: From $[\![ F ]\!] \checkmark \langle s \rangle$ and $\neg [\![ F ]\!]_s$, we have $\langle [\text{if } F \text{ then } C_1, \ldots], s \rangle \to \langle [\ldots], s \rangle$ and therefore as a consequence $\langle [\text{if } F \text{ then } C_1, \ldots], s \rangle \to^* \langle [\ldots], s' \rangle$. From

- Case $C = \text{while } F \text{ do } C_1$: From $\langle \text{while } F \text{ do } C_1, s \rangle \twoheadrightarrow s'$, we have two cases:
  - We have some $s_1 \in State$ with $[\![ F ]\!] \checkmark \langle s \rangle$ and $[\![ F ]\!]_s$ and $\langle C_1, s \rangle \twoheadrightarrow s_1$ and $\langle \text{while } F \text{ do } C_1, s_1 \rangle \twoheadrightarrow s'$. From the induction hypothesis, we thus have $\langle [C_1, \text{while } F \text{ do } C_1, \ldots], s \rangle \to^* \langle [\text{while } F \text{ do } C_1, \ldots], s_1 \rangle$ and furthermore $\langle [\text{while } F \text{ then } C_1, \ldots], s_1 \rangle \to^* \langle [\ldots], s' \rangle$. From $[\![ F ]\!] \checkmark \langle s \rangle$ and $[\![ F ]\!]_s$, we also have $\langle [\text{while } F \text{ do } C_1, \ldots], s \rangle \to \langle [C_1, \text{while } F \text{ do } C_1, \ldots], s \rangle$ and thus $\langle [\text{while } F \text{ do } C_1, \ldots], s \rangle \to^* \langle [\ldots], s' \rangle$.
  - We have $[\![ F ]\!] \checkmark \langle s \rangle, \neg [\![ F ]\!]_s$, and $s' = s$ and thus $\langle [\text{while } F \text{ do } C_1, \ldots], s \rangle \to^* \langle [\ldots], s' \rangle$.

Next we prove

$$\forall C \in Command, s, s' \in State. \ \langle [C], s \rangle \to^* \langle [\,], s' \rangle \Rightarrow \langle C, s \rangle \twoheadrightarrow s'$$

by induction on $\to^*$: we take arbitrary $n \in \mathbb{N}$, assume that the statement is true for every $\to^m$ with $m < n$ and show that it is true for $\to^n$.

If $n = 0$, then $[C] \neq [\,]$ makes the statement trivially true. If $n > 0$, we take $C \in Command, Is \in Instruction^*, s, s', s_1 \in State$ with

$$\langle [C], s \rangle \to \langle Is, s_1 \rangle \tag{1}$$

$$\langle Is, s_1 \rangle \to^{n-1} \langle [\,], s' \rangle \tag{2}$$

and show $\langle C, s \rangle \twoheadrightarrow s'$ by case distinction on $C$:

- Case $C = V := T$: (1) implies $[\![\, T \,]\!]\checkmark \langle s \rangle$ and $Is = [\ ]$ and $s_1 = s[V \mapsto [\![\, T \,]\!]_s]$ and thus $\langle V := T, s \rangle \twoheadrightarrow s_1$. $Is = [\ ]$ and (2) imply $s' = s_1$ and thus $\langle V := T, s \rangle \twoheadrightarrow s'$.
- Case $C = \mathtt{var}\ V : S;\ C_1$: (1) implies $Is = [C_1, \mathtt{set}\ V := v]$ and $s_1 = s[V \mapsto v']$ for $v = s(V)$ and some $v' \in A(S)$. From (2) and Proposition 7.15, we have some $n_1, n_2 \in \mathbb{N}$ with $n_1 + n_2 = n - 1$ and $s_2 \in State$ with $\langle [C], s_1 \rangle \to^{n_1} \langle [\ ], s_2 \rangle$ and $\langle [\mathtt{set}\ V := v], s_2 \rangle \to^{n_2} \langle [\ ], s' \rangle$. From $\langle [C], s_1 \rangle \to^{n_1} \langle [\ ], s_2 \rangle$ and the induction hypothesis, we have $\langle C, s_1 \rangle \twoheadrightarrow s_2$. From $\langle [\mathtt{set}\ V := v], s_2 \rangle \to^{n_2} \langle [\ ], s' \rangle$, we have $\langle [\mathtt{set}\ V := v], s_2 \rangle \to \langle [\ ], s' \rangle$, which implies $s' = s_2[V \mapsto v]$. Thus we have all prerequisites to derive $\langle \mathtt{var}\ V : S;\ C_1 \rangle \twoheadrightarrow s'$.
- Case $C = C_1; C_2$: (1) implies $Is = [C_1, C_2]$ and $s_1 = s$. Thus (2) and Proposition 7.15 give us some $n_1, n_2 \in \mathbb{N}$ with $n_1 + n_2 = n - 1$ and $s_1 \in State$ with $\langle [C_1], s \rangle \to^{n_1} \langle [\ ], s_1 \rangle$ and $\langle [C_2], s_1 \rangle \to^{n_2} \langle [\ ], s' \rangle$. The induction hypothesis then implies $\langle C_1, s \rangle \twoheadrightarrow s_1$ and $\langle C_2, s_1 \rangle \twoheadrightarrow s'$ which yields $\langle C_1; C_2, s \rangle \twoheadrightarrow s'$.
- Case $C = \mathtt{if}\ F\ \mathtt{then}\ C_1\ \mathtt{else}\ C_2$: From (1), we have two cases:
  - Case $[\![\, F \,]\!]\checkmark \langle s \rangle$ and $[\![\, F \,]\!]_s$ and $Is = [C_1]$ and $s_1 = s$: From (2), the induction hypothesis implies $\langle C_1, s \rangle \twoheadrightarrow s'$ and thus $\langle \mathtt{if}\ F\ \mathtt{then}\ C_1\ \mathtt{else}\ C_2, s \rangle \twoheadrightarrow s'$.
  - Case $[\![\, F \,]\!]\checkmark \langle s \rangle$ and $\neg [\![\, F \,]\!]_s$ and $Is = [C_2]$ and $s_1 = s$: From (2), the induction hypothesis implies $\langle C_2, s \rangle \twoheadrightarrow s'$ and thus $\langle \mathtt{if}\ F\ \mathtt{then}\ C_1\ \mathtt{else}\ C_2, s \rangle \twoheadrightarrow s'$.
- Case $C = \mathtt{if}\ F\ \mathtt{then}\ C_1$: From (1), we have two cases:
  - Case $[\![\, F \,]\!]\checkmark \langle s \rangle$ and $[\![\, F \,]\!]_s$ and $Is = [C_1]$ and $s_1 = s$: From (2), the induction hypothesis implies $\langle C_1, s \rangle \twoheadrightarrow s'$ and thus $\langle \mathtt{if}\ F\ \mathtt{then}\ C_1, s \rangle \twoheadrightarrow s'$.
  - Case $[\![\, F \,]\!]\checkmark \langle s \rangle$ and $\neg [\![\, F \,]\!]_s$ and $Is = [\ ]$ and $s_1 = s$: Thus (2) implies $s' = s$ from which we have $\langle \mathtt{if}\ F\ \mathtt{then}\ C_1, s \rangle \twoheadrightarrow s'$.
- Case $C = \mathtt{while}\ F\ \mathtt{do}\ C_1$: From (1), we have two cases:
  - Case $[\![\, F \,]\!]\checkmark \langle s \rangle$ and $[\![\, F \,]\!]_s$ and $Is = [C_1, \mathtt{while}\ F\ \mathtt{do}\ C_1]$ and $s_1 = s$: From (2) and Proposition 7.15, we have some $n_1, n_2 \in \mathbb{N}$ with $n_1 + n_2 = n - 1$ and $s_2 \in State$ with $\langle [C_1], s \rangle \to^{n_1} \langle [\ ], s_2 \rangle$ and $\langle [\mathtt{while}\ F\ \mathtt{do}\ C_1], s_2 \rangle \to^{n_2} \langle [\ ], s' \rangle$. The induction hypothesis implies $\langle C_1, s \rangle \twoheadrightarrow s_2$ and $\langle \mathtt{while}\ F\ \mathtt{do}\ C_1, s_2 \rangle \twoheadrightarrow s'$ from which we ultimately derive $\langle \mathtt{while}\ F\ \mathtt{do}\ C_1, s \rangle \twoheadrightarrow s'$.
  - Case $[\![\, F \,]\!]\checkmark \langle s \rangle$ and $\neg [\![\, F \,]\!]_s$ and $Is = [\ ]$ and $s_1 = s$: Thus (2) implies $s' = s$ from which we have $\langle \mathtt{while}\ F\ \mathtt{do}\ C_1, s \rangle \twoheadrightarrow s'$.

This completes the proof.                                                                 $\square$

## 7.4   The Correctness of Translations

In this section we are going to discuss the relationship between programs in different languages. This is of central concern in the translation of programs from one language to another one, for instance in the compilation of a high-level language (that operates according to some abstract computational model suitable for human understanding) to a low-level machine language (which reflects the concrete hardware that performs the actual execution). To be useful, the translation must guarantee a certain relationship between the abstract execution of the source program and the concrete

**Fig. 7.11** A machine model

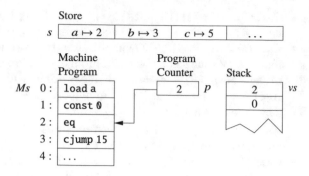

execution of the target program. If the languages of both programs are defined by a formal semantics, the expected relationship can be stated in a precise way and the soundness of the translation with respect to this expectation can be formally proved.

**A Machine Language**

We start with the definition of a low-level language of machine instructions.

**Definition 7.9** *(Machine Instruction)* A *machine instruction* is a phrase $M \in Machine$ which is formed according to the following grammar:

$M \in Machine,\ V \in Variable,\ v \in Value,\ n \in \mathbb{N}$
$M ::=$ const $v$ | add | mult | eq | not | and
    | load $V$ | store $V$ | push $V$ | pop $V$
    | jump $n$ | cjump $n$

A machine program $Ms \in Machine^*$ is simply a sequence of machine instructions. These instructions drive the execution of a machine which incorporates apart from $Ms$ the following elements (see also Fig. 7.11):

- a *program counter* $p$, the number of the execution to be performed next;
- a *state* (*store*) $s$, a mapping of variables to values;
- a *stack* $vs$, a sequence of intermediate values arising in the course of the execution.

The program is executed by iteratively executing the instruction denoted by $p$ and setting $p$ to the number of the next instruction; the execution halts, if $p$ leaves the range of instruction numbers of the given program. By default, the instruction number is incremented from $p$ to $p + 1$ by the execution of an instruction, but two instructions jump $n$ and cjump $n$ may explicitly set $p$ to the denoted number $n$.

The precise meaning of the individual kinds of instructions is defined in Fig. 7.12 by a small-step operational semantics which operates on configurations $\langle p, s, vs \rangle$ of program counter, store, and stack; a transition $\langle p, s, vs \rangle \rightarrow^{Ms} \langle p', s', vs' \rangle$ describes

**Semantic Domains**

$MachineConf := \mathbb{N} \times State \times Value^*$

**Transitions**

$\langle . \rangle \rightarrow^{\cdot} \langle . \rangle \subseteq MachineConf \times Machine^* \times MachineConf$

**Inference Rules**

$$\frac{\langle p, \text{const } v \rangle \in Ms \quad vs' = [v] \circ vs}{\langle p, s, vs \rangle \rightarrow^{Ms} \langle p + 1, s, vs' \rangle}$$

$$\frac{\langle p, \text{add} \rangle \in Ms \quad vs = [n_2, n_1] \circ vs_1 \quad n_1, n_2 \in \mathbb{N} \quad n = n_1 + n_2 \quad vs' = [n] \circ vs_1}{\langle p, s, vs \rangle \rightarrow^{Ms} \langle p + 1, s, vs' \rangle}$$

$$\frac{\langle p, \text{mult} \rangle \in Ms \quad vs = [n_2, n_1] \circ vs_1 \quad n_1, n_2 \in \mathbb{N} \quad n = n_1 \cdot n_2 \quad vs' = [n] \circ vs_1}{\langle p, s, vs \rangle \rightarrow^{Ms} \langle p + 1, s, vs' \rangle}$$

$$\frac{\langle p, \text{eq} \rangle \in Ms \quad vs = [n_2, n_1] \circ vs_1 \quad n_1, n_2 \in \mathbb{N} \quad vs' = [t] \circ vs_1}{\text{if } n_1 = n_2 \text{ then } t = \text{true else } t = \text{false}}{\langle p, s, vs \rangle \rightarrow^{Ms} \langle p + 1, s, vs' \rangle}$$

$$\frac{\langle p, \text{not} \rangle \in Ms \quad vs = [t] \circ vs_1 \quad t \in \{\text{true}, \text{false}\} \quad vs' = [t'] \circ vs_1}{\text{if } t = \text{true then } t' = \text{false else } t' = \text{true}}{\langle p, s, vs \rangle \rightarrow^{Ms} \langle p + 1, s, vs' \rangle}$$

$$\frac{\langle p, \text{and} \rangle \in Ms \quad vs = [t_2, t_1] \circ vs_1 \quad t_1, t_2 \in \{\text{true}, \text{false}\} \quad vs' = [t] \circ vs_1}{\text{if } t_1 = \text{true then } t = t_2 \text{ else } t = \text{false}}{\langle p, s, vs \rangle \rightarrow^{Ms} \langle p + 1, s, vs' \rangle}$$

$$\frac{\langle p, \text{load } V \rangle \in Ms \quad v = s(V) \quad vs' = [v] \circ vs}{\langle p, s, vs \rangle \rightarrow^{Ms} \langle p + 1, s, vs' \rangle}$$

$$\frac{\langle p, \text{store } V \rangle \in Ms \quad vs = [v] \circ vs_1 \quad s' = s[V \mapsto v] \quad vs' = vs_1}{\langle p, s, vs \rangle \rightarrow^{Ms} \langle p + 1, s', vs' \rangle}$$

$$\frac{\langle p, \text{push } V \rangle \in Ms \quad v = s(V) \quad v' \in Value \quad s' = s[V \mapsto v'] \quad vs' = [v] \circ vs}{\langle p, s, vs \rangle \rightarrow^{Ms} \langle p + 1, s', vs' \rangle}$$

$$\frac{\langle p, \text{pop } V \rangle \in Ms \quad vs = [v] \circ vs_1 \quad s' = s[V \mapsto v] \quad vs' = vs_1}{\langle p, s, vs \rangle \rightarrow^{Ms} \langle p + 1, s', vs' \rangle}$$

$$\frac{\langle p, \text{jump } n \rangle \in Ms \quad p' = n}{\langle p, s, vs \rangle \rightarrow^{Ms} \langle p', s, vs \rangle}$$

$$\frac{\langle p, \text{cjump } n \rangle \in Ms \quad vs = [t] \circ vs_1 \quad vs' = vs_1}{\text{if } t = \text{true then } p' = n \text{ else } p' = p + 1}{\langle p, s, vs \rangle \rightarrow^{Ms} \langle p', s, vs' \rangle}$$

**Fig. 7.12**  A small-step operational semantics of the machine language

how the execution of instruction $p$ of machine program $Ms$ with current store $s$ and stack $vs$ produces the updated store $s'$ and stack $vs'$ and sets the program counter to $p'$. In this semantics, we assume *Value* to be a superset of the set of natural numbers $\mathbb{N}$ and of the set of truth values {true, false}.

Each rule contains, for some machine instruction $M \in Machine$, a precondition $\langle p, M \rangle \in Ms$ which implies $p < |Ms|$ and $Ms(p) = M$, i.e., the program counter $p$ addresses machine instruction $M$ in program $Ms$; if $p \geq |Ms|$, no rule is applicable, which can be considered as indicating the termination of the program.

Informally, the various kinds of instructions behave as follows:

- **const** $v$ pushes the value $v$ to the stack.
- **add** pops the top values $n_2$ and $n_1$, two natural numbers, from the stack and replaces them by the sum $n_1 + n_2$.
- **mult** pops the top values $n_2$ and $n_1$, two natural numbers, from the stack and replaces them by the product $n_1 \cdot n_2$.
- **eq** pops the top values $n_2$ and $n_1$, two natural numbers, from the stack; if $n_1 = n_2$, it replaces them by the truth value 'true', else by 'false'.
- **not** pops the top value $t$, a truth value, from the stack; if $t$ is 'true', it is replaced by 'false', else by 'true'.
- **and** pops the top values $t_2$ and $t_1$, two truth values, from the stack; if both $t_1$ and $t_2$ are 'true', they are replaced by 'true', else by 'false'.
- **load** $V$ reads from the store the value $v$ of variable $V$ and pushes $v$ on the stack.
- **store** $V$ pops the top value $v$ from the stack and updates the store by mapping variable $V$ to $v$.
- **push** $V$ pushes the value $v$ of variable $V$ in the store to the stack and updates the store by mapping $V$ to some new value $v'$.
- **pop** $V$ pops the top value $v$ from the stack and updates the store by mapping variable $V$ to $v$.
- **jump** $n$ sets the program counter to natural number $n$.
- **cjump** $n$ pops the top value $t$, a truth value, from the stack; if $t$ is 'true', it sets the program counter to natural number $n$, else it increments it by one.

**Example 7.3** We give a machine program that produces the same effect as the following high-level program:

```
if a = 0 then
    b := 0
else {
    var c:Nat
    c := a+1
    b := a*c
}
a := b
```

The machine program called $Ms$ is defined as

```
0:   load a
1:   const 0
```

```
 2:  eq
 3:  cjump 15
 4:  push c
 5:  load a
 6:  const 1
 7:  add
 8:  store c
 9:  load a
10:  load c
11:  mult
12:  store b
13:  pop c
14:  jump 17
15:  const 0
16:  store b
17:  load b
18:  store a
```

where the numerical labels indicate the numbers of the various instructions. If we start $Ms$ with program counter $p = 0$, store $s = [a \mapsto 2, b \mapsto 3, c \mapsto 5]$, and stack $vs = [\ ]$, it executes the following sequence of transitions (the configuration with $p = 2$ is the one depicted in Fig. 7.11):

$$\langle 0, [a \mapsto 2, b \mapsto 3, c \mapsto 5], [\ ]\rangle$$
$$\to^{Ms} \langle 1, [a \mapsto 2, b \mapsto 3, c \mapsto 5], [2]\rangle$$
$$\to^{Ms} \langle 2, [a \mapsto 2, b \mapsto 3, c \mapsto 5], [0, 2]\rangle$$
$$\to^{Ms} \langle 3, [a \mapsto 2, b \mapsto 3, c \mapsto 5], [\text{false}]\rangle$$
$$\to^{Ms} \langle 4, [a \mapsto 2, b \mapsto 3, c \mapsto 5], [\ ]\rangle$$
$$\to^{Ms} \langle 5, [a \mapsto 2, b \mapsto 3, c \mapsto 0], [5]\rangle$$
$$\to^{Ms} \langle 6, [a \mapsto 2, b \mapsto 3, c \mapsto 0], [2, 5]\rangle$$
$$\to^{Ms} \langle 7, [a \mapsto 2, b \mapsto 3, c \mapsto 0], [1, 2, 5]\rangle$$
$$\to^{Ms} \langle 8, [a \mapsto 2, b \mapsto 3, c \mapsto 0], [3, 5]\rangle$$
$$\to^{Ms} \langle 9, [a \mapsto 2, b \mapsto 3, c \mapsto 3], [5]\rangle$$
$$\to^{Ms} \langle 10, [a \mapsto 2, b \mapsto 3, c \mapsto 3], [2, 5]\rangle$$
$$\to^{Ms} \langle 11, [a \mapsto 2, b \mapsto 3, c \mapsto 3], [3, 2, 5]\rangle$$
$$\to^{Ms} \langle 12, [a \mapsto 2, b \mapsto 3, c \mapsto 3], [6, 5]\rangle$$
$$\to^{Ms} \langle 13, [a \mapsto 2, b \mapsto 6, c \mapsto 2], [5]\rangle$$
$$\to^{Ms} \langle 14, [a \mapsto 2, b \mapsto 6, c \mapsto 5], [\ ]\rangle$$
$$\to^{Ms} \langle 17, [a \mapsto 2, b \mapsto 6, c \mapsto 5], [\ ]\rangle$$
$$\to^{Ms} \langle 18, [a \mapsto 2, b \mapsto 6, c \mapsto 5], [6]\rangle$$
$$\to^{Ms} \langle 19, [a \mapsto 6, b \mapsto 6, c \mapsto 5], [\ ]\rangle$$

**Abstract Syntax**

$E \in Expression$ (see Figure 7.6)
$B \in BoolExp$
$B ::= E_1 = E_2 \mid !B \mid B_1 \&\& B_2$

**Denotational Semantics**

$[\![ \cdot ]\!] : Expression \rightarrow (State \rightarrow \mathbb{N})$

(see Figure 7.6)

$[\![ \cdot ]\!] : BoolExp \rightarrow (State \rightarrow \{true, false\})$

$[\![ E_1 = E_2 ]\!] = \lambda s \in State.$ if $[\![ E_1 ]\!](s) = [\![ E_2 ]\!](s)$ then true else false
$[\![ !B ]\!] = \lambda s \in State.$ if $[\![ B ]\!](s) = true$ then false else true
$[\![ B_1 \&\& B_2 ]\!] = \lambda s \in State.$ if $[\![ B_1 ]\!](s) = false$ then false else $[\![ B_2 ]\!](s)$

**Fig. 7.13** A language of Boolean expressions

Thus the program terminates with the empty stack and new values for variables $a$ and $b$ but preserving the original value of variable $c$. The stack has been used to record the intermediate values in the computation of the arithmetic and Boolean expressions and to preserve the original value of variable $c$ which is temporarily superseded by other values but finally restored. □

**Translating Expressions**

We start by discussing the translation an expression into a machine program and the relationship between the semantics of the expression and the semantics of the program to which the expression was translated. Since we want to deal with expressions representing terms as well as with expressions representing formulas, Fig. 7.13 extends the expression language of Fig. 7.6 (which introduced a domain *Expression* of expressions denoting natural numbers) by a domain *BoolExp* of Boolean expressions, i.e., expressions denoting truth values.

Based on these domains Fig. 7.14, defines a translation $\mathbf{T}[\![ E ]\!]$ of a natural number expression $E$ and the translation $\mathbf{T}[\![ B ]\!]$ of a Boolean expression $B$ to a machine program, i.e., a sequence of machine instructions. The core idea of the translation is for both kinds of expressions the same:

- An atomic (constant or variable) expression is translated into a single instruction that pushes the corresponding value on the stack;
- A unary expression $f(E)$ is translated by appending to the translation of $E$ an instruction that pops the top element $v$ (denoting the value of $E$) from the stack and replaces it by $f(v)$.
- A binary expression $f(E_1, E_2)$ is translated by appending to the translations of $E_1$ and $E_2$ (in that order) an instruction that pops the top elements $v_2$ and $v_1$ (denoting the values of $E_2$ respectively $E_1$) from the stack and replaces them by $f(v_1, v_2)$.

$$\begin{aligned}
&\textbf{T}: \textit{Expression} \rightarrow \textit{Machine}^* \\
&\quad \textbf{T}[\![\, N \,]\!] := [\text{const } N] \\
&\quad \textbf{T}[\![\, V \,]\!] := [\text{load } V] \\
&\textbf{T}[\![\, E_1 + E_2 \,]\!] := \textbf{T}[\![\, E_1 \,]\!] \circ \textbf{T}[\![\, E_2 \,]\!] \circ [\text{add}] \\
&\textbf{T}[\![\, E_1 * E_2 \,]\!] := \textbf{T}[\![\, E_1 \,]\!] \circ \textbf{T}[\![\, E_2 \,]\!] \circ [\text{mult}] \\[6pt]
&\textbf{T}: \textit{BoolExp} \rightarrow \textit{Machine}^* \\
&\quad \textbf{T}[\![\, E_1 = E_2 \,]\!] := \textbf{T}[\![\, E_1 \,]\!] \circ \textbf{T}[\![\, E_2 \,]\!] \circ [\text{eq}] \\
&\qquad \textbf{T}[\![\, !B \,]\!] := \textbf{T}[\![\, B \,]\!] \circ [\text{not}] \\
&\quad \textbf{T}[\![\, B_1 \&\& B_2 \,]\!] := \textbf{T}[\![\, B_1 \,]\!] \circ \textbf{T}[\![\, B_2 \,]\!] \circ [\text{and}]
\end{aligned}$$

**Fig. 7.14** The translation of expressions to a machine program

We have to note that the values on the stack are popped in the reverse order in which they were pushed, thus the second argument of a binary operation is popped first. The sequence of instructions generated in this fashion can be also considered as a variant of the 'Reverse Polish Notation' for denoting expressions, where first the arguments and then the operator to be applied is written (some pocket calculators used this entry format in the past).

**Example 7.4** The expression (a+1)*b is translated to the machine program $Ms$ defined as

```
0: load a
1: const 1
2: add
3: load b
4: mult
```

When started with program counter $p = 0$, store $s = [a \mapsto 3, b \mapsto 5]$, and stack $vs = [\,]$, $Ms$ executes the following transitions:

$$\begin{aligned}
\langle 0, s, [\,] \rangle \rightarrow^{Ms} \langle 1, s, [3 = a] \rangle \rightarrow^{Ms} \langle 2, s, [1, 3] \rangle \rightarrow^{Ms} \langle 3, s, [4 = 3 + 1] \rangle \\
\rightarrow^{Ms} \langle 4, s, [5 = b, 4] \rangle \rightarrow^{Ms} \langle 5, s, [20 = 4 \cdot 5] \rangle
\end{aligned}$$

The final configuration has at the top of the stack the value 20 which denotes the result of the expression.                                                                      □

For discussing sequences of machine transitions, we introduce a notation that resembles a big-step semantics of machine execution.

**Proposition 7.17** (Reflexive Transitive Closure of Machine Transitions) *Let $Ms \in Machine^*$ be a machine program and let $mc, mc' \in MachineConf$ be machine configurations as defined in Fig. 7.11. We then write $mc \twoheadrightarrow^{Ms} mc'$ if there is a sequence of transitions from $mc$ to $mc'$:*

$$\twoheadrightarrow^{Ms} := \left(\to^{Ms}\right)^*$$

We will now formulate the core claim of the translation for natural number expressions; this claim can be correspondingly extended to Boolean expressions.

**Proposition 7.18** (Correctness of Expression Translation) *The translation of an expression yields a machine program whose execution with an initially empty stack terminates with a stack whose only element is that value that the denotational semantics assigns to the expression:*

$$\forall E \in Expression, Ms \in Machine^*, m \in \mathbb{N}, s \in State.$$
$$Ms = \mathbf{T}[\![\, E \,]\!] \wedge m = |Ms| \Rightarrow \langle 0, s, [\,] \rangle \twoheadrightarrow^{Ms} \langle m, s, [[\![\, E \,]\!](s)] \rangle$$

***Proof*** A direct consequence of the following more general proposition.                                       □

**Proposition 7.19** (Correctness of Expression Translation Generalized) *The translation of an expression yields a machine program whose execution (also when embedded in a bigger program and with a non-empty initial stack) terminates with the initial stack to which that value has been pushed that the denotational semantics assigns to the expression:*

$$\forall E \in Expression, Ms, Ms_a, Ms_b \in Machine^*, m, m_a, \in \mathbb{N}, s \in State, vs \in Value^*.$$
$$Ms = \mathbf{T}[\![\, E \,]\!] \wedge m = |Ms| \wedge m_a = |Ms_a| \Rightarrow$$
$$\langle m_a, s, vs \rangle \twoheadrightarrow^{Ms_a \circ Ms \circ Ms_b} \langle m_a + m, s, [[\![\, E \,]\!](s)] \circ vs \rangle$$

According to this proposition, the considered part of the program execution starts with program counter $m_a$ pointing to the first instruction in the translation of the expression; it 'terminates' with program counter $m_a + m$ pointing to the first instruction after the translation. The stack $vs$ with which this part of the execution starts is extended upon termination by the value of the expression.

***Proof*** We prove this statement by structural induction on $E$. For this, we take arbitrary $Ms, Ms_a, Ms_b \in Machine^*$, $m, m_a, \in \mathbb{N}$, $s \in State$, and $vs \in Value^*$ with

$$Ms = \mathbf{T}[\![\, E \,]\!] \wedge m = |Ms| \wedge m_a = |Ms_a| \tag{1}$$

We define

$$Ms' := Ms_a \circ Ms \circ Ms_b \tag{2}$$

and show

$$\langle m_a, s, vs \rangle \twoheadrightarrow^{Ms'} \langle m_a + m, s, [[\![\, E \,]\!](s)] \circ vs \rangle \tag{a}$$

by case distinction on the structure of $E$:

- If $E = N$, we have $[\![\, E \,]\!](s) = N$, $Ms = [\texttt{const } N]$, and $m = 1$. Since thus (1) and (2) imply $\langle m_a, \texttt{const } N \rangle \in Ms'$, the definition of $\rightarrow^{Ms'}$ implies

$$\langle m_a, s, vs \rangle \rightarrow^{Ms'} \langle m_a + 1, s, [N] \circ vs \rangle$$

from which $[\![\, E \,]\!](s) = N$ and $m = 1$ yield (a).
- If $E = V$, we have $[\![\, E \,]\!](s) = s(V)$, $Ms = [\texttt{load } V]$, and $m = 1$. Since thus (1) and (2) imply $\langle m_a, \texttt{load } V \rangle \in Ms'$, the definition of $\rightarrow^{Ms'}$ implies

$$\langle m_a, s, vs \rangle \rightarrow^{Ms'} \langle m_a + 1, s, [s(V)] \circ vs \rangle$$

from which $[\![\, E \,]\!](s) = s(V)$ and $m = 1$ yield (a).
- If $E = E_1 + E_2$, we have $[\![\, E \,]\!](s) = n_1 + n_2$ where $n_1 := [\![\, E_1 \,]\!](s)$ and $n_2 := [\![\, E_2 \,]\!](s)$. Furthermore, we have $Ms = \mathbf{T}[\![\, E_1 \,]\!] \circ \mathbf{T}[\![\, E_2 \,]\!] \circ [\texttt{add}]$ and $m = m_1 + m_2 + 1$ where $m_1 := |Ms_1|$ and $m_2 := |Ms_2|$. From (2), we thus have

$$Ms' = Ms_a \circ \mathbf{T}[\![\, E_1 \,]\!] \circ \mathbf{T}[\![\, E_2 \,]\!] \circ [\texttt{add}] \circ Ms_b \tag{3}$$

From the induction hypothesis for $E_1$ and from $n_1 = [\![\, E_1 \,]\!](s)$, we thus have

$$\langle m_a, s, vs \rangle \twoheadrightarrow^{Ms'} \langle m_a + m_1, s, [n_1] \circ vs \rangle \tag{4}$$

From the induction hypothesis for $E_2$ and from $n_2 = [\![\, E_2 \,]\!](s)$, we also have

$$\langle m_a + m_1, s, [n_1] \circ vs \rangle \twoheadrightarrow^{Ms'} \langle m_a + m_1 + m_2, s, [n_2] \circ [n_1] \circ vs \rangle \tag{5}$$

From (3) and $m_a = |Ms_a|$, $m_1 = |Ms_1|$, as well as $m_2 = |Ms_2|$, we know that $\langle m_a + m_1 + m_2, \text{add}\rangle \in Ms'$. The definition of $\rightarrow^{Ms'}$ and $m = m_1 + m_2 + 1$ imply

$$\langle m_a + m_1 + m_2, s, [n_2] \circ [n_1] \circ vs\rangle \rightarrow^{Ms'} \langle m_a + m, s, [n_1 + n_2] \circ vs\rangle \qquad (6)$$

From (4) to (6), and $[\![\, E \,]\!](s) = n_1 + n_2$, we have (a).
- If $E = E_1 {}^* E_2$, the proof is analogous.

This completes the proof of the proposition.                                      □

Since the full programming languages is based on the syntactic domains *Term* and *Formula* rather than the expression domains discussed above, we will (without going into details) simply stipulate the existence of corresponding translation functions that satisfy the generalized correctness condition.

---

**Proposition 7.20** (Formula and Term Translation)  *Let* $\mathbf{T}[\![\, F \,]\!]$ *respectively* $\mathbf{T}[\![\, T \,]\!]$ *denote the translation of a formula F respectively term T to a machine program. The execution of this program (also when embedded in a bigger program and with a non-empty initial stack) terminates with the semantics of the phrase on top of the initial stack (and it does not push any value to the stack, if the semantics of the phrase is not well-defined in the initial state):*

$\forall F \in Formula, Ms, Ms_a, Ms_b \in Machine^*, m, m_a, \in \mathbb{N}, s \in State, vs \in Value^*.$
$\quad Ms = \mathbf{F}[\![\, F \,]\!] \wedge m = |Ms| \wedge m_a = |Ms_a| \Rightarrow$
$\qquad if\ [\![\, F \,]\!]\checkmark\langle s\rangle$
$\qquad\quad then\ \langle m_1, s, vs\rangle \rightarrow^{Ms_a \circ Ms \circ Ms_b} \langle m_a + m, s, [\,[\![\, F \,]\!]_s] \circ vs\rangle$
$\qquad\quad else\ \neg\exists v \in Value.\ \langle m_1, s, vs\rangle \rightarrow^{Ms_a \circ Ms \circ Ms_b} \langle m_a + m, s, [v] \circ vs\rangle$

$\forall T \in Term, Ms, Ms_a, Ms_b \in Machine^*, m, m_a, \in \mathbb{N}, s \in State, vs \in Value^*.$
$\quad Ms = \mathbf{T}[\![\, T \,]\!] \wedge m = |Ms| \wedge m_a = |Ms_a| \wedge [\![\, T \,]\!]\checkmark\langle s\rangle \Rightarrow$
$\qquad if\ [\![\, T \,]\!]\checkmark\langle s\rangle$
$\qquad\quad then\ \langle m_1, s, vs\rangle \rightarrow^{Ms_a \circ Ms \circ Ms_b} \langle m_a + m, s, [\,[\![\, T \,]\!]_s] \circ vs\rangle$
$\qquad\quad else\ \neg\exists v \in Value.\ \langle m_1, s, vs\rangle \rightarrow^{Ms_a \circ Ms \circ Ms_b} \langle m_a + m, s, [v] \circ vs\rangle$

---

This proposition will be subsequently used in the proof of the correctness of the translation of programs to machine instructions.

$\mathbf{T}$: *Program* $\rightarrow$ *Machine*$^*$

$\mathbf{T}[\![\,\text{program } I(X)\ C\,]\!] :=$

  $\text{let } Ms_1 = \mathbf{T_a}[\![\,X\,]\!],\ m = |Ms_1|,\ Ms_2 = \mathbf{T_b}[\![\,X\,]\!] \text{ in } Ms_1 \circ \mathbf{T}[\![\,C\,]\!]^m \circ Ms_2$


$\mathbf{T_a}$: *Parameter* $\rightarrow$ *Machine*$^*$

$\mathbf{T_a}[\![\,\sqcup\,]\!] := [\,]$

$\mathbf{T_a}[\![\,X, V{:}S\,]\!] := \mathbf{T_a}[\![\,X\,]\!] \circ [\text{store } V]$


$\mathbf{T_b}$: *Parameter* $\rightarrow$ *Machine*$^*$

$\mathbf{T_b}[\![\,\sqcup\,]\!] := [\,]$

$\mathbf{T_b}[\![\,X, V{:}S\,]\!] := [\text{load } V] \circ \mathbf{T_b}[\![\,X\,]\!]$


$\mathbf{T}$: *Command* $\times\ \mathbb{N} \rightarrow$ *Machine*$^*$

$\mathbf{T}[\![\,V := T\,]\!]^P := \mathbf{T}[\![\,T\,]\!] \circ [\text{store } V]$

$\mathbf{T}[\![\,\text{var } V{:}S;\ C\,]\!]^P := [\text{push } V] \circ \mathbf{T}[\![\,C\,]\!]^{P+1} \circ [\text{pop } V]$

$\mathbf{T}[\![\,C_1;C_2\,]\!]^P :=$

  $\text{let } Ms_1 = \mathbf{T}[\![\,C_1\,]\!]^P,\ m_1 = |Ms_1| \text{ in } Ms_1 \circ \mathbf{T}[\![\,C_2\,]\!]^{P+m_1}$

$\mathbf{T}[\![\,\text{if } F \text{ then } C_1 \text{ else } C_2\,]\!]^P :=$

  $\text{let } Ms = \mathbf{T}[\![\,F\,]\!],\ m = |Ms|,\ p_2 = p + m + 1 \text{ in}$

  $\text{let } Ms_2 = \mathbf{T}[\![\,C_2\,]\!]^{P_2},\ m_2 = |Ms_2|,\ p_1 = p_2 + m_2 + 1 \text{ in}$

  $\text{let } Ms_1 = \mathbf{T}[\![\,C_1\,]\!]^{P_1},\ m_1 = |Ms_1| \text{ in}$

  $Ms \circ [\text{cjump } p_1] \circ Ms_2 \circ [\text{jump } p_1 + m_1] \circ Ms_1$

$\mathbf{T}[\![\,\text{if } F \text{ then } C\,]\!]^P :=$

  $\text{let } Ms = \mathbf{T}[\![\,F\,]\!],\ m = |Ms|,\ p_1 = p + m + 2 \text{ in}$

  $\text{let } Ms_1 = \mathbf{T}[\![\,C\,]\!]^{P_1},\ m_1 = |Ms_1| \text{ in}$

  $Ms \circ [\text{not, cjump } p_1 + m_1] \circ Ms_1$

$\mathbf{T}[\![\,\text{while } F \text{ then } C\,]\!]^P :=$

  $\text{let } p_1 = p + 1$

  $\text{let } Ms_1 = \mathbf{T}[\![\,C\,]\!]^{P_1},\ m_1 = |Ms_1|,\ p_f = p_1 + m_1 \text{ in}$

  $\text{let } Ms = \mathbf{T}[\![\,F\,]\!] \text{ in}$

  $[\text{jump } p_f] \circ Ms_1 \circ Ms \circ [\text{cjump } p_1]$

**Fig. 7.15**  The translation of programs to machine instructions

## Translating Programs

Based on the translation of formulas and terms stipulated above, Fig. 7.15 defines a
translation of programs, parameters, and commands to instruction sequences:

- $\mathbf{T}[\![\,\text{program } I(X)\ C\,]\!]$ first translates the $m$ program parameters $X$ to a sequence
  $Ms_1$ of $m$ instructions that pop $m$ values from the stack and write them into the
  store as the initial values of the denoted variables. To these instructions it appends
  the translation $\mathbf{T_a}[\![\,C\,]\!]^m$ of the program body $C$ whose first instruction receives

position $m$ in the machine code. The resulting machine program is extended by the instruction sequence $Ms_2$ that read the $m$ final values of the program parameters from the store and push them to the stack.

- $\mathbf{T_a}[\![\, X \,]\!]$ translates the program parameters $V_1:S_1, \ldots, V_n:S_n$ to the machine instructions store $V_1, \ldots,$ store $V_n$ that pop $n$ values $v_1, \ldots, v_n$ from the stack and store them in variables $V_1, \ldots, V_n$ respectively. Dually, $\mathbf{T_b}[\![\, X \,]\!]$ generates the machine instructions load $V_n, \ldots,$ load $V_1$ that read the values $v'_n, \ldots, v'_1$ of variables $V_n, \ldots, V_1$ from the store and push them (in that order) to the stack whose top values thus become $v'_1, \ldots, v'_n$.

- $\mathbf{T}[\![\, C \,]\!]^p$ translates a command $C$ to a machine program whose first instruction receives position $p$ in the machine code:

  - $\mathbf{T}[\![\, V := T \,]\!]^p$ first generates code for the evaluation of term $T$ which leaves its value on the top of the stack; it then adds the instruction store $V$ which pops that value from the stack and writes it into variable $V$.

  - $\mathbf{T}[\![\, \mathtt{var}\ V\!:\! S;\ C \,]\!]^p$ first generates the instruction push $V$ which saves the current value of variable $V$ on the stack. Then it generates code for the command body $C$ starting at position $p + 1$. Finally it adds the instruction pop $V$ which restores the original value of $V$ from the saved stack value.

  - $\mathbf{T}[\![\, C_1 ; C_2 \,]\!]^p$ generates first code $Ms_1$ for the execution of command $C_1$ (the first instruction receives position $p$) and then for the execution of command $C_2$ (the first instruction receives position $p + m_1$ where $m_1$ is the length of $Ms_1$.

  - $\mathbf{T}[\![\, \mathtt{if}\ F\ \mathtt{then}\ C_1\ \mathtt{else}\ C_2 \,]\!]^p$ generates the sequence of machine instructions $Ms \circ [\mathtt{cjump}\ p_1] \circ Ms_2 \circ [\mathtt{jump}\ p_1 + m_1] \circ Ms_1$ where the $m$ instructions of prefix $Ms$ evaluate formula $F$ leaving the resulting truth value on the top of the stack. Then instruction $\mathtt{cjump}\ p_1$ pops that value from the stack and examines it. If it is true, the program continues with the instruction $p_1$, the first of the $m_1$ instructions of sequence $Ms_1$, which is in charge of the execution of command branch $C_1$ and represents the suffix of the generated code (the code for the execution of $C_2$ is placed before the one for the execution of $C_1$ in order to avoid the necessity to negate the truth value of formula $F$). If the truth value is false, the program continues with the execution of $Ms_2$ which is in charge of the execution of command branch $C_2$; this execution terminates with the instruction $\mathtt{jump}\ p_1 + m_1$ which continues the execution with the first instruction after the end of the generated sequence.

  - $\mathbf{T}[\![\, \mathtt{if}\ F\ \mathtt{then}\ C \,]\!]^p$ generates code $Ms \circ [\mathtt{not}, \mathtt{cjump}\ p_1 + m_1] \circ Ms_1$ where the $m_1$ instructions of $Ms_1$ starting at position $p_1$ are in charge of the execution of conditional branch $C_1$. As for the two-sided conditional, the prefix $Ms$ evaluates formula $F$ leaving the resulting truth value on the top of the stack. However, now the instruction not negates that truth value such that the instruction $\mathtt{cjump}\ p_1 + m_1$ jumps to the end of the generated code if formula $F$ is false; otherwise the program continues with the execution of $Ms_1$.

  - $\mathbf{T}[\![\, \mathtt{while}\ F\ \mathtt{then}\ C \,]\!]^p$ generates code $[\mathtt{jump}\ p_2] \circ Ms_1 \circ Ms_2 \circ [\mathtt{cjump}\ p_1]$ whose first instruction $\mathtt{jump}\ p_2$ immediately continues with the first instruction of $Ms_2$ which is in charge of the evaluation of formula $F$ leaving its truth value on the top of the stack for inspection by instruction $\texttt{cjump}\ p_1$. If the value is false, the execution of the generated code is completed. However, if it is true, the program continues with the instruction $p_1$, the first instruction of sequence $Ms_1$ which is in charge of executing the loop body $C$; this sequence is immediately followed by the sequence $Ms_2$ for the next evaluation of formula $F$. This particular code organization ensures that very loop iteration is represented by the execution of the sequence $Ms_1 \circ Ms_2 \circ [\texttt{cjump}\ p_1]$ which only contains one instruction in addition to the instructions for the evaluation of the loop condition and the loop body, respectively.

The machine program given in Example 7.3 actually resulted from the application of this translation to the command depicted there. Other examples are given below.

**Example 7.5**  The translation of the loop $C$

```
while a ≠ b do a := a+1
```

yields the following machine code $Ms = \mathbf{T}[\![\, C \,]\!]^0$:

```
0: jump 5
1: load a
2: const 1
3: add
4: store a
5: load a
6: load b
7: eq
8: not
9: cjump 1
```

In this code, instructions 1–4 represent the execution of the loop body while instructions 5–8 represent the evaluation of the loop condition.

For initial program counter $p = 0$, store $s = [a \mapsto 2, b \mapsto 5]$, and stack $vs = [\ ]$, this program performs the following sequences of transitions:

$$\langle 0, [a \mapsto 2, b \mapsto 5], [\ ] \rangle$$
$$\to^{Ms} \langle 5, [a \mapsto 2, b \mapsto 5], [\ ] \rangle \to^{Ms} \ldots \to^{Ms} \langle 9, [a \mapsto 2, b \mapsto 5], [\mathrm{true}] \rangle$$
$$\to^{Ms} \langle 1, [a \mapsto 2, b \mapsto 5], [\ ] \rangle \to^{Ms} \ldots \to^{Ms} \langle 9, [a \mapsto 3, b \mapsto 5], [\mathrm{true}] \rangle$$
$$\to^{Ms} \langle 1, [a \mapsto 2, b \mapsto 5], [\ ] \rangle \to^{Ms} \ldots \to^{Ms} \langle 9, [a \mapsto 4, b \mapsto 5], [\mathrm{true}] \rangle$$
$$\to^{Ms} \langle 1, [a \mapsto 2, b \mapsto 5], [\ ] \rangle \to^{Ms} \ldots \to^{Ms} \langle 9, [a \mapsto 5, b \mapsto 5], [\mathrm{false}] \rangle$$
$$\to^{Ms} \langle 10, [a \mapsto 5, b \mapsto 5], [\ ] \rangle$$

The loop thus performs three iterations until termination.                 □

**Example 7.6**  The translation of the program $P$

```
program sum(a:Nat,b:Nat,c:Nat) { c := a+b }
```

yields the following machine code $Ms = \mathbf{T}[\![\, P \,]\!]$:

```
0: store a
1: store b
2: store c
3: load a
4: load b
5: add
6: store c
7: load c
8: load b
9: load a
```

In this code, instructions 0–3 respectively 7–9 handle the transfer of the parameter values from stack to store and vice versa while instructions 3–5 execute the program body. For initial program counter $p = 0$, store $s = [a \mapsto 0, c \mapsto 0, c \mapsto 0]$ and stack $vs = [2, 3, 1]$, this program performs the following sequence of transitions:

$$\langle 0, [a \mapsto 0, b \mapsto 0, c \mapsto 0], [2, 3, 1] \rangle$$

$$\to^{Ms} \langle 1, [a \mapsto 2, b \mapsto 0, c \mapsto 0], [3, 1] \rangle$$

$$\to^{Ms} \langle 2, [a \mapsto 2, b \mapsto 3, c \mapsto 0], [1] \rangle$$

$$\to^{Ms} \langle 3, [a \mapsto 2, b \mapsto 3, c \mapsto 1], [\ ] \rangle$$

$$\to^{Ms} \langle 4, [a \mapsto 2, b \mapsto 3, c \mapsto 1], [2] \rangle$$

$$\to^{Ms} \langle 5, [a \mapsto 2, b \mapsto 3, c \mapsto 1], [3, 2] \rangle$$

$$\to^{Ms} \langle 6, [a \mapsto 2, b \mapsto 3, c \mapsto 1], [5] \rangle$$

$$\to^{Ms} \langle 7, [a \mapsto 2, b \mapsto 3, c \mapsto 5], [\ ] \rangle$$

$$\to^{Ms} \langle 8, [a \mapsto 2, b \mapsto 3, c \mapsto 5], [5] \rangle$$

$$\to^{Ms} \langle 9, [a \mapsto 2, b \mapsto 3, c \mapsto 5], [3, 5] \rangle$$

$$\to^{Ms} \langle 10, [a \mapsto 2, b \mapsto 3, c \mapsto 5], [2, 3, 5] \rangle$$

By this translation, the initial stack $vs = [2, 3, 1]$ is thus transformed to the stack $vs = [2, 3, 5]$ reflecting the final values of the program parameters.                    □

We now come to the central theorem of this section.

**Proposition 7.21** (Correctness of Translations of Programs) *Let program* $P = \text{program } I(X) \ C$ *be such that* $\Sigma \vdash P : \text{program}$ *can be derived. Then the translation of P yields a machine program Ms such that for all value sequences vs, vs' that are adequate with respect to the typing of parameters X the following is true: given initial parameter values vs, P terminates with final parameter values vs' if and only if Ms started with stack vs and some store s* *terminates with stack vs′ and some store s′:*

$$\forall P \in Program, Ms \in Machine^*, m \in \mathbb{N}, vs, vs' \in Value^*.$$
$$Ms = \mathbf{T}[\![\, P\, ]\!] \wedge m = |Ms| \wedge [\![\, X\, ]\!]\langle vs \rangle \Rightarrow$$
$$\left( [\![\, P\, ]\!]\langle vs, vs' \rangle \Leftrightarrow \exists s, s' \in State.\ \langle 0, s, vs \rangle \twoheadrightarrow^{Ms} \langle m, s', vs' \rangle \right)$$

***Proof***  A straight-forward consequence of the following propositions.    □

**Proposition 7.22**  (Correctness of Parameter Translations) *Let parameters* $X$ *be such that* $\Sigma \vdash X : parameters(Vt, Vs)$ *can be derived and let* $m = |Vs|$ *be the number of the parameters. Then the translation* $\mathbf{T_a}[\![\, X\, ]\!]$ *yields a machine program that pops* $m$ *values from the stack and stores them in the variables* $Vs$:

$$\forall X \in Parameter, Ms, Ms_a, Ms_b \in Machine^*, m \in \mathbb{N}, s, s' \in State, vs, ps \in Value^*.$$
$$m_a = |Ms_a| \wedge Ms = \mathbf{T_a}[\![\, X\, ]\!] \wedge m = |Ms| \wedge m = |Vs| \wedge m = |ps| \wedge$$
$$\left( \forall V \in Variable \backslash domain(Vt).\ s'(V) = s(V) \right) \wedge \left( \forall i \in \mathbb{N}_m.\ s'(Vs(i)) = ps(i) \right) \Rightarrow$$
$$\langle m_a, s, ps \circ vs \rangle \twoheadrightarrow^{Ms_a \circ Ms \circ Ms_b} \langle m_a + m, s', vs \rangle$$

*Furthermore the translation* $\mathbf{T_b}[\![\, X\, ]\!]$ *yields a machine program that pushes the* $m$ *values stored in variables* $Vs$ *to the stack:*

$$\forall X \in Parameter, Ms, Ms_a, Ms_b \in Machine^*, m \in \mathbb{N}, s \in State, vs, ps \in Value^*.$$
$$m_a = |Ms_a| \wedge Ms = \mathbf{T_a}[\![\, X\, ]\!] \wedge m = |Ms| \wedge m = |Vs| \wedge m = |ps| \wedge$$
$$\forall i \in \mathbb{N}_m.\ s(Vs(i)) = ps(i) \Rightarrow$$
$$\langle m_a, s, vs \rangle \twoheadrightarrow^{Ms_a \circ Ms \circ Ms_b} \langle m_a + m, s, ps \circ vs \rangle$$

We omit this proof but will prove in detail the following core proposition.

**Proposition 7.23**  (Correctness of Command Translation) *Let* $Vt$ *be a variable typing and command* $C$ *be such that* $\Sigma, Vt \vdash C : command$ *can be derived. Then the translation of* $C$ *yields a machine program* $Ms$ *such that for all stores* $s, s'$ *the following is true: started in initial store* $s$, $C$ *terminates with final store* $s'$, *if and only if* $Ms$ *started with arbitrary stack* $vs$ *and store* $s$ *terminates with* *stack vs and store s′:*

$$\forall C \in Command, Ms, Ms_a, Ms_b \in Machine^*, m, \overline{m} \in \mathbb{N}, s, s' \in State, vs \in Value^*.$$
$$m_a = |Ms_a| \wedge Ms = \mathbf{T}[\![\, C \,]\!]^{m_a} \wedge m = |Ms| \Rightarrow$$
$$\left( [\![\, C \,]\!] \langle s, s' \rangle \Leftrightarrow \langle m_a, s, vs \rangle \twoheadrightarrow^{Ms_a \circ Ms \circ Ms_b} \langle m_a + m, s', vs \rangle \right)$$

***Proof*** We prove this statement by structural induction on $C$. For this, we take arbitrary $Ms, Ms_a, Ms_b \in Machine^*, m, \overline{m} \in \mathbb{N}, s, s' \in State$, and $vs \in Value^*$ with

$$m_a = |Ms_a| \wedge Ms = \mathbf{T}[\![\, C \,]\!]^{m_a} \wedge m = |Ms| \tag{1}$$

and define

$$Mt' := Ms_a \circ Ms \circ Ms_b \tag{2}$$

To prove the implication from left to right, we assume

$$[\![\, C \,]\!] \langle s, s' \rangle \tag{3}$$

and show

$$\langle m_a, s, vs \rangle \twoheadrightarrow^{Mt'} \langle m_a + m, s', vs \rangle \tag{a}$$

by case distinction on the structure of $C$:

* Case $C = V := T$: from the definition of $\mathbf{T}[\![\, C \,]\!]^{m_a}$, we know

$$Ms = \mathbf{T}[\![\, T \,]\!] \circ [\texttt{store}\ V] \wedge m = m_t + 1 \tag{4}$$

where $m_t := |\mathbf{T}[\![\, T \,]\!]|$. Furthermore, from (2), we know

$$[\![\, T \,]\!] \checkmark \langle s \rangle \wedge s' = s[V \mapsto v] \tag{5}$$

where $v := [\![\,[\![\, T \,]\!]\,]\!]_s]$. From Proposition 7.20 and (5), we know

$$\langle m_a, s, vs \rangle \twoheadrightarrow^{Ms'} \langle m_a + m_t, s, [v] \circ vs \rangle \tag{6}$$

From (1), (2), and (4), we know $\langle m_a + m_t, \texttt{store}\ V \rangle \in Ms'$. Thus the definition of $\to^{Ms'}$ implies

$$\langle m_a + m_t, s, [v] \circ vs \rangle \twoheadrightarrow^{Ms'} \langle m_a + m_t + 1, s[V \mapsto v], vs \rangle \tag{7}$$

From (6) and (7) together with (4) and (5), we know (a).

- Case $C = $ var $V: S; C_1$: from the definition of $\mathbf{T}[\![\, C \,]\!]^{m_a}$, we know

$$Ms = [\text{push } V] \circ \mathbf{T}[\![\, C \,]\!]^{m_a+1} \circ [\text{pop } V] \wedge m = m_c + 2 \qquad (4)$$

where $m_c := |\mathbf{T}[\![\, C \,]\!]^{m_a+1}|$. Furthermore, from (2), we know for some $s_1, s_2 \in State$

$$[\![\, C \,]\!]\langle s_1, s_2 \rangle \qquad (5)$$

where $s_1 = s[V \mapsto v']$ for some $v' \in Value$ and $s' = s_2[V \mapsto v]$ with $v := s(V)$. From (1), (2), and (4), we know $\langle m_a, \text{push } V \rangle \in Ms'$. Consequently the definition of $\rightarrow^{Ms'}$ implies

$$\langle m_a, s, vs \rangle \twoheadrightarrow^{Ms'} \langle m_a + 1, s_1, [v] \circ vs \rangle \qquad (6)$$

From (2), (5), $m_c = |\mathbf{T}[\![\, C \,]\!]^{m_a+1}|$, and the induction hypothesis, we know

$$\langle m_a + 1, s_1, [v] \circ vs \rangle \twoheadrightarrow^{Ms'} \langle m_a + 1 + m_c, s_2, [v] \circ vs \rangle \qquad (7)$$

From (1), (2), and (4), we know $\langle m_a + 1 + m_c, \text{pop } V \rangle \in Ms'$. The definition of $\rightarrow^{Ms'}$ with $s' = s_2[V \mapsto v]$ thus implies

$$\langle m_a + 1 + m_c, s_2, [v] \circ vs \rangle \twoheadrightarrow^{Ms'} \langle m_a + 1 + m_c + 1, s', vs \rangle \qquad (8)$$

From (6), (7), and (8) with $m = m_c + 2$ we know (a).
- Case $C = C_1; C_2$: from the definition of $\mathbf{T}[\![\, C \,]\!]^{m_a}$, we know

$$Ms = \mathbf{T}[\![\, C_1 \,]\!]^{m_a} \circ \mathbf{T}[\![\, C_2 \,]\!]^{m_a+m_1} \wedge m = m_1 + m_2 \qquad (4)$$

where $m_1 := |\mathbf{T}[\![\, C_1 \,]\!]^{m_a}|$ and $m_2 := |\mathbf{T}[\![\, C_2 \,]\!]^{m_a+m_1}|$. Furthermore, from (2) we know for some $s_1 \in State$

$$[\![\, C_1 \,]\!]\langle s, s_1 \rangle \wedge [\![\, C_2 \,]\!]\langle s_1, s' \rangle \qquad (5)$$

From (2), (5), and the induction hypothesis, we thus know

$$\langle m_a, s, vs \rangle \twoheadrightarrow^{Ms'} \langle m_a + m_1, s_1, vs \rangle \qquad (6)$$

$$\langle m_a + m_1, s_1, vs \rangle \twoheadrightarrow^{Ms'} \langle m_a + m_1 + m_2, s', vs \rangle \qquad (7)$$

From (6), (7), and $m = m_1 + m_2$, we know (a).

- Case $C = \text{if } F \text{ then } C_1 \text{ else } C_2$: from the definition of $\mathbf{T}[\![\,C\,]\!]^{m_a}$, we know

$$Ms = Ms_f \circ [\text{cjump } p_1] \circ Ms_2 \circ [\text{jump } p_1 + m_1] \circ Ms_1 \tag{4}$$
$$m = m_f + m_1 + m_2 + 2 \tag{5}$$

where $Ms_f = \mathbf{T}[\![\,F\,]\!]$, $Ms_2 := \mathbf{T}[\![\,C_2\,]\!]^{p_2}$, and $Ms_1 := \mathbf{T}[\![\,C_1\,]\!]^{p_1}$ with $m_f := |\mathbf{T}[\![\,F\,]\!]|$, $p_2 := m_a + m_f + 1$, $m_2 := |\mathbf{T}[\![\,C_1\,]\!]^{p_2}|$, $p_1 := p_2 + m_2 + 1$, and $m_1 := |\mathbf{T}[\![\,C_2\,]\!]^{p_1}|$.

From (3), we know $[\![\,F\,]\!]\checkmark\langle s\rangle$. Thus, Proposition 7.20 implies with (1), (2), and (4)

$$\langle m_a, s, vs\rangle \twoheadrightarrow^{Ms'} \langle m_a + m_f, s, [\![\![\,F\,]\!]_s] \circ vs\rangle \tag{6}$$

From (1), (2), and (4), we know $\langle m_a + m_f, \text{cjump } p_1\rangle \in Ms'$. The definition of $\twoheadrightarrow^{Ms'}$ thus implies

$$\langle m_a + m_f, s, [\![\![\,F\,]\!]_s] \circ vs\rangle \twoheadrightarrow^{Ms'} \langle p', s, vs\rangle \tag{7}$$
$$\text{if } [\![\,F\,]\!]_s = \text{true then } p' = p_1 \text{ else } p' = m_a + m_f + 1 \tag{8}$$

We have two cases.
- Case $[\![\,F\,]\!]_s$: from (3) and (8), we know

$$[\![\,C_1\,]\!]\langle s, s'\rangle \wedge p' = p_1 \tag{11}$$

From (1), (2), (4), (9), and the induction hypothesis, we know

$$\langle p_1, s, vs\rangle \twoheadrightarrow^{Ms'} \langle p_1 + m_1, s', vs\rangle \tag{10}$$

Then (a) follows from (6), (7), (9), and (10) with (7), $p_1 = p_2 + m_2 + 1$, and $p_2 = m_a + m_f + 1$.
- Case $\neg[\![\,F\,]\!]_s$: from (3) and (8), we know

$$[\![\,C_2\,]\!]\langle s, s'\rangle \wedge p' = m_a + m_f + 1 \tag{9}$$

From (1), (2), (4), (9), and the induction hypothesis, we know

$$\langle m_a + m_f + 1, s, vs\rangle \twoheadrightarrow^{Ms'} \langle m_a + m_f + 1 + m_2, s', vs\rangle \tag{10}$$

From (1), (2), and (4), we know $\langle m_a + m_f + 1 + m_2, \text{jump } p_1 + m_1\rangle \in Ms'$. The definition of $\twoheadrightarrow^{Ms'}$ thus implies

$$\langle m_a + m_f + 1 + m_2, s', vs\rangle \twoheadrightarrow^{Ms'} \langle p_1 + m_1, s', vs\rangle \tag{11}$$

Then (a) follows from (6), (7), (9), (10), and (11) with (5), $p_1 = p_2 + m_2 + 1$, and $p_2 = m_a + m_f + 1$.

- Case $C = \text{if } F \text{ then } C_1$: from the definition of $\mathbf{T}[\![\, C \,]\!]^{ma}$, we know

$$Ms = Ms_f \circ [\text{not}, \text{cjump } p_1 + m_1] \circ Ms_1 \tag{4}$$
$$m = m_f + 2 + m_1 \tag{5}$$

where $Ms_f := \mathbf{T}[\![\, F \,]\!]$ and $Ms_1 := \mathbf{T}[\![\, C \,]\!]^{p_1}$ with $m_f := |Ms_f|$, $p_1 := m_a + m_f + 2$, and $m_1 := |Ms_1|$.

From (3), we know $[\![\, F \,]\!]\checkmark \langle s \rangle$. Thus, Proposition 7.20 implies with (1), (2), and (4)

$$\langle m_a, s, vs \rangle \twoheadrightarrow^{Ms'} \langle m_a + m_f, s, [\![\, F \,]\!]_s] \circ vs \rangle \tag{6}$$

From (1), (2), and (4), we know $\langle m_a + m_f, \text{not} \rangle \in Ms'$. The definition of $\rightarrow^{Ms'}$ with $s' = s_2[V \mapsto v]$ thus implies

$$\langle m_a + m_f, s, [\![\, F \,]\!]_s] \circ vs \rangle \rightarrow^{Ms'} \langle m_a + m_f + 1, s, [v] \circ vs \rangle \tag{7}$$
$$\text{if } [\![\, F \,]\!]_s = \text{true then } v = \text{false else } v = \text{true} \tag{8}$$

From (1), (2), and (4), we know $\langle m_a + m_f + 1, \text{cjump } p_1 + m_1 \rangle \in Ms'$. The definition of $\rightarrow^{Ms'}$ with (8) thus implies

$$\langle m_a + m_f + 1, s, [v] \circ vs \rangle \rightarrow^{Ms'} \langle p', s, vs \rangle \tag{9}$$
$$\text{if } [\![\, F \,]\!]_s = \text{false then } p' = p_1 + m_1 \text{ else } p' = m_a + m_f + 2 \tag{10}$$

We have two cases.
- Case $[\![\, F \,]\!]_s$: from (3) and (10), we know

$$[\![\, C_1 \,]\!]\langle s, s' \rangle \wedge p' = m_a + m_f + 2 \tag{11}$$

From (1), (2), (4), (11), and the induction hypothesis, we know

$$\langle m_a + m_f + 2, s, vs \rangle \twoheadrightarrow^{Ms'} \langle m_a + m_f + 2 + m_1, s', vs \rangle \tag{12}$$

From (5), (6), (7), (9), (11), and (12), we know (a).
- Case $\neg[\![\, F \,]\!]_s$: from (3) and (10), we know

$$s' = s \wedge p' = p_1 + m_1 \tag{11}$$

From (5), (6), (7), (9), (11), and $p_1 := m_a + m_f + 2$, we know (a).

- Case $C = \texttt{while } F \texttt{ do } C_1$: from the definition of $\mathbf{T}[\![\, C\, ]\!]^{ma}$, we know

$$Ms = [\texttt{jump } p_f] \circ Ms_1 \circ Ms_f \circ [\texttt{cjump } p_1] \tag{4}$$
$$m = 1 + m_1 + m_f + 1 \tag{5}$$

where $Ms_1 := \mathbf{T}[\![\, C\, ]\!]^{p_1}$ and $Ms_f = \mathbf{T}[\![\, F\, ]\!]$ with $p_1 := m_a + 1$, $m_1 := |Ms_1|$, $p_f := p_1 + m_1$, and $m_f = |Ms_f|$.

From (3) and Proposition 7.8, we know for some $t \in State^\omega$ and $n \in \mathbb{N}$

$$loop\langle t, n, F, C\rangle \wedge t(0) = s \wedge t(n) = s' \wedge [\![\, F\, ]\!]\checkmark\langle s'\rangle \wedge \neg[\![\, F\, ]\!]_{s'} \tag{6}$$

We are now going to show

$$\langle m_a, s, vs\rangle \twoheadrightarrow^{Ms'} \langle p_f, t(0), vs\rangle \tag{b}$$
$$\forall i \in \mathbb{N}_{n-1}.\ \langle p_f, t(i), vs\rangle \twoheadrightarrow^{Ms'} \langle p_f, t(i+1), vs\rangle \tag{c}$$
$$\langle p_f, t(n), vs\rangle \twoheadrightarrow^{Ms'} \langle m_a + m, s', vs\rangle \tag{d}$$

which imply together (a) (this can be easily shown by induction on $n$).
- Proof of (b): from (1), (2), and (4), we know $\langle m_a, \texttt{jump } p_f\rangle \in Ms'$. The definition of $\twoheadrightarrow^{Ms'}$ thus implies

$$\langle m_a, s, vs\rangle \twoheadrightarrow^{Ms'} \langle p_f, s, vs\rangle \tag{7}$$

Together with $t(0) = s$, this implies (b).

- Proof of (d): from (6), we know $[\![\, F\, ]\!]\checkmark\langle t(n)\rangle$ and $\neg[\![\, F\, ]\!]_{t(n)}$. Thus Proposition 7.20 together with (1), (2), and (4) implies

$$\langle p_f, t(n), vs\rangle \twoheadrightarrow^{Ms'} \langle p_f + m_f, t(n), [\text{false}] \circ vs\rangle \tag{7}$$

From (1), (2), and (4), we know $\langle p_f + m_f, \texttt{cjump } p_1\rangle \in Ms'$. The definition of $\twoheadrightarrow^{Ms'}$ thus implies

$$\langle p_f + m_f, t(n), [\text{false}] \circ vs\rangle \twoheadrightarrow^{Ms'} \langle p_f + m_f + 1, t(n), vs\rangle \tag{8}$$

Together with $t(n) = s'$ and (5), (7) and (8) imply (d).

- Proof of (c): we take arbitrary $i \in \mathbb{N}_{n-1}$ and show

$$\langle p_f, t(i), vs\rangle \twoheadrightarrow^{Ms'} \langle p_f, t(i+1), vs\rangle \tag{e}$$

From (6) and the definition of *loop*, we know

$$[\![\, F \,]\!]\checkmark \langle t(i) \rangle \wedge [\![\, F \,]\!]_{t(i)} \tag{7}$$

$$[\![\, C \,]\!] \langle t(i), t(i+1) \rangle \tag{8}$$

From (7), Proposition 7.20 together with (1), (2), and (4) implies

$$\langle p_f, t(i), vs \rangle \rightarrow^{Ms'} \langle p_f + m_f, t(i), [\text{true}] \circ vs \rangle \tag{9}$$

From (1), (2), and (4), we know $\langle p_f + m_f, \texttt{cjump}\ p_1 \rangle \in Ms'$. The definition of $\rightarrow^{Ms'}$ thus implies

$$\langle p_f + m_f, t(i), [\text{true}] \circ vs \rangle \rightarrow^{Ms'} \langle p_1, t(i), vs \rangle \tag{10}$$

From (1), (2), (4), (8), and the induction hypothesis, we know

$$\langle p_1, t(i), vs \rangle \rightarrow^{Ms'} \langle p_1 + m_1, t(i+1), vs \rangle \tag{11}$$

Together with $p_f = p_1 + m_1$, (9), (10), and (11) imply (e).

This completes the proof of the implication from left to right; the proof of the implication from right to left proceeds in a very similar way. $\qquad\square$

**Relating Two Small Step Semantics**

So far we have shown the correctness of a translation by relating the denotational semantics of the source language of the translation to the small-step operational semantics of the target language. For the source language, we might have also used the big-step operational semantics; this would have yielded essentially the same kinds of propositions and proofs. However, we may be also interested in relating the semantics of two languages that are both given in a small-step operational style. Here every execution step of the program in one (higher-level) language has to be mimicked by some (one or more) execution steps of the corresponding program in the other (lower-level) language. The formal framework for this kind of 'mimicry' is given below (see also the left diagram in Fig. 7.16).

---

**Definition 7.10** (*Simulation and Bisimulation*) Let $\rightarrow_A \subseteq A \times A$ be a binary relation on domain $A$ and let $\rightarrow_C \subseteq C \times C$ be a binary relation on domain $C$. We call a binary relation $\simeq\ \subseteq A \times C$ between $A$ and $C$ a *simulation* if the following holds:

$$\forall a, a' \in A, c \in C.\ a \rightarrow_A a' \wedge a \simeq c \Rightarrow \exists c' \in C.\ c \rightarrow_C c' \wedge a' \simeq c'$$

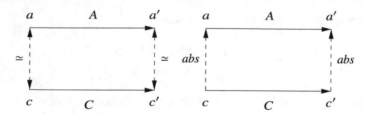

**Fig. 7.16** Bisimulation of transitions

A simulation $\simeq \subseteq A \times C$ is a *bisimulation* if additionally the following holds:

$$\forall c, c' \in C, a \in A.\ c \to_C c' \wedge a \simeq c \Rightarrow \exists a' \in A.\ a \to_C a' \wedge a' \simeq c'$$

Thus a bisimulation ensures that, if there is a transition in one domain and the source of the transition is related to a partner in the other domain, then this partner is also the source of a transition that leads to a target that is a partner of the target of the original transition. Of particular relevance are therefore bisimulations where every element in one domain has a partner in the other domain.

**Proposition 7.24** (Equivalence of Transitions) *Let* $\to_A \subseteq A \times A$ *be a binary relation on domain* $A$ *and let* $\to_C \subseteq C \times C$ *be a binary relation on domain* $C$. *Let* $\simeq \subseteq A \times C$ *be a bisimulation between* $A$ *and* $C$ *that relates every element of one domain to some element of the respective other domain:*

$$\forall a \in A.\ \exists c \in C.\ a \simeq c$$
$$\forall c \in C.\ \exists a \in A.\ a \simeq c$$

*Then* $\to_A$ *and* $\to_C$ *are* equivalent *in the sense that every transition in one domain is mimicked by some transition in the respective other domain:*

$$\forall a, a' \in A.\ a \to_A a' \Rightarrow \exists c, c' \in C.\ c \to_C c' \wedge a \simeq c \wedge a' \simeq c'$$
$$\forall c, c' \in C.\ c \to_C c' \Rightarrow \exists a, a' \in A.\ a \to_A a' \wedge a \simeq c \wedge a' \simeq c'$$

**Proof** Straight-forward from the assumptions and the definition of bisimulation. $\square$

Thus, given a suitable bisimulation between two transition systems, the set of possible transitions in both domain are in essence (apart from a translation between the domains) 'identical'.

In the following, we consider $A$ as an 'abstract' (higher-level) domain and $C$ as a 'concrete' (lower-level) domain $C$ where the bisimulation $\simeq$ is defined by a surjective *abstraction function abs*: $C \to A$, i.e., $a \simeq c :\Leftrightarrow a = abs(c)$ (see also the right diagram in Fig. 7.16). Then relation $\simeq$ satisfies the requirements of Proposition 7.24: for every $a \in A$, we have by the surjectivity of *abs* some $c \in C$ with $a = abs(c)$ and thus $a \simeq c$; conversely, for every $c \in C$, we have $a(c) \simeq c$. It should be noted that *abs* is not necessarily injective: since thus different values in $C$ may be abstracted to the same value in $A$, the inverse of *abs* is not necessarily a function; we call this inverse also the *concretization relation*.

As an example, take the previously introduced expression language where we define as the the the abstract domain $A := ExpState = Expression \times State$ and as the concrete domain $C := MachineConf = \mathbb{N} \times State \times Value^*$, i.e., the domains of configurations of the respective small step semantics. In order to determine the appropriate abstraction function, we consider the execution of expression $E = (x+1)*y$ in store $s = [x \mapsto 2, y \mapsto 5, \ldots]$ by the following 'abstract' transitions:

$$\langle (x+1)*y, s \rangle \to_A \langle (2+1)*y, s \rangle \to_A \langle 3*y, s \rangle \to_A \langle 3*5, s \rangle \to_A \langle 15, s \rangle$$

On the other side, the execution of the corresponding machine program $Ms := \mathbf{T}[\![\, E\,]\!]$

```
0: load x
1: const 1
2: add
3: load y
4: mult
```

gives rise to the following machine transitions:

$$\langle 0, s, [\,] \rangle \to^{Ms} \langle 1, s, [2] \rangle \to^{Ms} \langle 2, s, [1, 2] \rangle$$
$$\to^{Ms} \langle 3, s, [3] \rangle \to^{Ms} \langle 4, s, [5, 3] \rangle \to^{Ms} \langle 5, s, [15] \rangle$$

We notice that 4 transitions in the original small-step semantics correspond to 5 transitions in the machine semantics, i.e., there has to be one abstract transition that is in bisimulation with a sequence of two machine transitions.

A little inspection reveals that every abstract transition replaces one non-constant (variable or operator) expression by a number; this corresponds to the machine transitions executing instructions of type `load`, `add`, or `mult`; however, there is no abstract transition corresponding to an instruction of type `const`. Therefore we have to consider the execution of an instruction of type `const` as a 'non-transition' whose effect has to be integrated into the first succeeding instruction of a different type. Consequently we define the concrete transition relation $\to_C$ as follows (from now on, we consider the machine program $Ms$ as an hidden argument of all definitions):

$$\langle p, s, vs \rangle \to_C \langle p', s', vs' \rangle :=$$
$$s' = s \wedge$$
$$\exists n \in \mathbb{N}, ps \in \mathbb{N}^{n+2}, vss \in (Value^*)^{n+2}.$$

$abs: \mathbb{N} \times State \times Value^* \rightarrow_\perp Expression \times State$

$abs(p, s, vs) = a \Leftarrow$

    exists $E = abs0(Ms, p, vs)$. $a = \langle E, s \rangle$

$abs0: Machine^* \times \mathbb{N} \times Value^* \rightarrow_\perp Expression$

$abs0(Ms, p, vs) = E \Leftarrow$

    if $vs = [\,]$

        then exists $E' = abs1(Ms, p, [\,])$. $E = E'$

        else $\exists N, vs'$. $vs = [N] \circ vs' \wedge$

                     exists $E' = abs0(Ms_{<p} \circ [\text{const } N] \circ Ms_{\geq p}, p, vs')$. $E = E'$

$abs1: Machine^* \times \mathbb{N} \times Value^* \rightarrow_\perp Expression$

$abs1(Ms, p, Es) = E \Leftarrow$

    if $p \geq |Ms|$ then

        $\langle 0, E \rangle \in Es$

    else $\exists Es' \in Expression^*$.

      case $Ms(p)$ of

        $\text{const } N \rightarrow Es' = [N] \circ Es$

        $\text{load } V \rightarrow Es' = [V] \circ Es$

        $\text{add} \rightarrow \exists E_1, E_2, Es_0$. $Es = [E_1, E_2] \circ Es_0 \wedge Es' = [E_1 + E_2] \circ Es_0$

        $\text{mult} \rightarrow \exists E_1, E_2, Es_0$. $Es = [E_1, E_2] \circ Es_0 \wedge Es' = [E_1 * E_2] \circ Es_0$

      $\wedge$ exists $E' = abs1(Ms, p + 1, Es')$. $E = E'$

**Fig. 7.17** The abstraction function for expression evaluation

$$\forall i \in \mathbb{N}_{n+1}. \ \langle ps(i), s, vss(i) \rangle \rightarrow^{Ms} \langle ps(i + 1), s, vss(i + 1) \rangle \wedge$$
$$ps(0) = p \wedge vss(0) = vs \wedge ps(n + 1) = p' \wedge vss(n + 1) = vs' \wedge$$
$$(\neg const(Ms(ps(n)))) \wedge \forall i \in \mathbb{N}_n. \ const(Ms(ps(i)))$$

$const \subseteq Machine$

$const\langle M \rangle :\Leftrightarrow \exists N \in \mathbb{N}. \ M = \text{const } N$

A $\rightarrow_C$ step performs a sequence of one or more machine instructions such that the instruction that is executed last is not of type const (but all instructions executed before are). The corresponding sequence of transitions for above expression evaluation is now as follows:

$$\langle 0, s, [\,] \rangle \rightarrow_C \langle 1, s, [2] \rangle \rightarrow_C \langle 3, s, [3] \rangle \rightarrow_C \langle 4, s, [5, 3] \rangle \rightarrow_C \langle 5, s, [15] \rangle$$

We thus need to define an abstraction function *abs* with the following mappings:

$$abs(0,s,[\ ]) = \langle (x+1)^*y, s \rangle$$
$$abs(1,s,[2]) = \langle (2+1)^*y, s \rangle$$
$$abs(3,s,[3]) = \langle 3^*y, s \rangle$$
$$abs(4,s,[5, 3]) = \langle 3^*5, s \rangle$$
$$abs(5,s,[15]) = \langle 15, s \rangle$$

In these mappings, the elements on the stack represent the values of subexpressions that have been evaluated by the instructions before the current instruction counter. However, this purpose can be also achieved by inserting appropriate `const` instructions immediately before the next instructions, for instance the effect of the stack [5, 3] before the execution of instruction 4: `mult` can be also achieved by having the instructions `const 3`; `const 5` be executed before.

Based on this observation, we define *abs* as shown in Fig. 7.17 (this function and the subsequent ones are defined as partial functions, but they are total on the domain of machine programs arising from the translation of expressions): *abs* invokes the auxiliary function *abs0* with the program *Ms* as an additional argument. Function *abs0* replaces every stack element $N$ by an instruction `const N` that is inserted before the instruction with number $p$ ($Ms_{<p}$ denotes the part of the program before that instruction, $Ms_{\geq p}$ denotes the remaining part). Once the stack is empty, *abs0* invokes another auxiliary function *abs1* with an (initially empty) stack *Es* of subexpressions of the final expression to be constructed. Function *abs1* interprets each instruction of *Ms* one after each other, pushing a corresponding new expression to the stack (after having popped its subexpressions from the stack). When all instructions have been interpreted, the stack holds the complete expression denoted by the machine program.

Clearly, *abs* is not injective; for instance, considering the program

```
0: const 1
1: const 2
3: add
```

for every store $s$ we have $abs(0, s, [\ ]) = abs(1, s, [1]) = 2{+}1$, because the effect of the instruction 0: `const 1` is subsumed by the value 1 on the stack; the concretization relation thus relates 2+1 to both $\langle 0, s, [\ ] \rangle$ and $\langle 1, s, [1] \rangle$.

Now we first have to show that *abs* is surjective on the domain of expressions arising from the reduction of $E$. For this, it suffices to show that for every such expression $E'$ and every state $s$ there exists some program counter $p$ with $abs(p, s, [\ ]) = \langle E', s \rangle$. Furthermore, we have to show that the relation $\simeq$ induced by this definition of *abs* indeed represents a bisimulation. However, these proofs get quite complex and we will refrain from giving them; generally, the farther two languages are 'semantically apart', the more challenging such proofs become. Usually it is thus a less demanding strategy to define a high-level semantics of the language (equivalent to the reflexive transitive closure of the small-step transition) and prove the correctness of the translation with respect to this high-level semantics.

## 7.5   Procedures

We are now going to enrich our programming language by *procedures*, i.e., commands which are given names and which are executed by *procedure calls*, i.e., by referring to the name given to the command. To make procedures more generally usable, they may be equipped with *parameters*, i.e., variables which enable the caller of a procedure to transfer values from its own context to the context of the procedure (*value parameters*), and which also allow the procedure to update variables in the context of the caller (*reference parameters*). Furthermore, procedures may read and write *global variables* and thus cause *side effects*, i.e., changes to the store that are not restricted to the parameters. To support all these features, we introduce the concept of *environments* that assign meanings to identifiers (variables/parameters and procedures). This gives us the opportunity to formally model and discuss name binding mechanisms such as *static (lexical) scoping* and *dynamic scoping*, and investigate parameter passing strategies such as *call by value* and *call by reference*.

### Syntax

We start by extending the grammar of our programming language.

**Definition 7.11** *(Programs with Procedures)* A *program with procedures* is a phrase $P \in Program$ which is formed according to the following grammar:

$P \in Program, \; Ds \in Declarations, \; D \in Declaration$
$X \in Parameters, \; C \in Command, \; Ts \in Terms, \; Vs \in Variables$
$F \in Formula, \; T \in Term, \; V \in Variable, \; S \in Sort, \; I \in Identifier$

$P \; ::= \; Ds; \text{program } I(X) \, C$
$Ds \; ::= \; \lrcorner \mid Ds; D$
$D \; ::= \; \text{var } V : S \mid \text{procedure } I(X_1; \text{ref } X_2) \, C$
$X \; ::= \; \lrcorner \mid X, V : S$
$C \; ::= \; V := T \mid \text{var } V : S; C \mid \text{call } I(Ts; Vs) \mid \ldots$
$Ts \; ::= \; \lrcorner \mid Ts, T$
$Vs \; ::= \; \lrcorner \mid Vs, V$

The omitted parts are formed as in Definition 7.1.

In this language, the core program $Ds$; program $I(X) \, C$ is preceded by a sequence of *declarations* $Ds$ that introduce global variables and procedures; a procedure declaration procedure $I(X_1; \text{ref } X_2) \, C$ introduces two lists of parameters, a list of value parameters $X_1$ and a list of reference parameters $X_2$. Correspondingly, a procedure call call $I(Ts; Vs)$ passes two lists of arguments, a list of terms $Ts$ and a list of variables $Vs$.

**Example 7.7** As an example, take the following program *main* which operates in the context of a global variable $c$ and two procedures *div* and *gcd*:

```
var c:Nat;
procedure div(a:Nat,b:Nat; ref q:Nat,r:Nat) {
  q := 0; r := a;
  while r >= b do {
    q := q+1; r := r-b
  }
  c := c+1
}
procedure gcd(a:Nat,b:Nat; ref g:Nat,n:Nat) {
  c := 0;
  while a > 0 ∧ b > 0 do {
    var c:Nat;
    if a ≥ b
      then call div(a,b;c,a)
      else call div(b,a;c,b)
  }
  if a > 0 then g := a else g := b;
  n := c
}
program main(a:Nat,b:Nat,c:Nat,d:Nat) {
  call gcd(a*b,a+b;c,d)
}
```

Procedure *div* has two value parameters $a$ and $b$ and two reference parameters $q$ and $r$; by repeated subtraction, it computes the quotient respectively remainder of $a/b$, writing into $q$ the quotient and into $r$ the remainder; furthermore it increments the global variable $c$ by one. Thus, for instance, for parameter/variable values $a = 25$, $b = 10$, and $c = 1$ (and arbitrary values for the reference parameters $q$ and $r$), an invocation call div(a,b;q,r) terminates with $q = 2, r = 5$, and $c = 2$.

Procedure *gcd* has two value parameters $a$ and $b$ and two reference parameters $g$ and $n$; it applies the Euclidean algorithm to write into $g$ the greatest common divisor of $a$ and $b$ by repeatedly calling procedure *div*; for instance the invocation call div(a,b;c,a) sets $c$ to the quotient of $a$ divided by $b$ and $a$ to the remainder. Here $c$ denotes a local variable introduced in the body of the loop; thus in this body the declaration of the local variable overshadows the declaration of the global one. Furthermore, *gcd* initially sets the global variable $c$ to 0 and ultimately $n$ to $c$; since every call of procedure *div* increments $c$ by one, $n$ thus receives the number of invocations of *div*. Thus, for instance, for parameter values $a = 25$ and $b = 10$ (and arbitrary values for $g$, $n$, and the global variable $c$), an invocation call gcd(a,b,g,n) terminates with $g = 5$ and $n = c = 2$ (since the remainder of $25/10$ is 5 and the remainder of $10/5$ is 0, the loop is iterated twice).

Program *main* has parameters $a$, $b$, $c$, and $d$ and executes the command call gcd(a*b,a+b;c,d); here we see that the arguments for the value parameters of procedure *gcd* can be not only variables but general terms. For initial values $a = 6$, $b = 4$, and arbitrary $c$ and $d$, the program terminates with $a = 6$, $b = 4$, $c = 2$ (the greatest common divisor of $24 = 6 \cdot 4$ and $10 = 6 + 4$) and $d = 3$ (since the remainder of $24/10$ is 4, the remainder of $10/4$ is 2 and the remainder of $4/2$ is 0, three iterations are needed to compute the result).

### Type Checking

To describe the types of procedures, we introduce the following concept.

---

**Definition 7.12** *(Procedure Typing)* We define

$$ProcTyping := Identifier \rightarrow \mathsf{Set}(\Sigma.s^* \times \Sigma.s^*)$$

as the set of all functions from identifiers to sets of two finite sequences of sorts; we call these functions *procedure typings*.

---

The type of a procedure is indicated by two sequences of sorts, the sorts of the value parameters and the sorts of the reference parameters. Our model supports the *overloading* of procedures, i.e., different procedures with the same name but different sorts; thus every procedure identifier is mapped to a set of such pairs of sequences.

Figure 7.18 gives a type system for this language where $\Sigma$ denotes a signature, $Vt$, a variable typing, $Pt$ a procedure typing, and $Ss$ a sequence of sorts. This type system consists of the following kinds of judgements.

- $\Sigma \vdash P$: program: program $P$ is well-formed with respect to signature $\Sigma$.
- $\Sigma \vdash Ds$: declarations($Vt, Pt$): declarations $Ds$ are well-formed with respect to signature $\Sigma$ and introduce a (global) variable typing $Vt$ and procedure typing $Pt$.
- $\Sigma, Vt, Pt \vdash D$: declaration($Vt', Pt'$): declaration $D$ is well-formed with respect to signature $\Sigma$; it extends the typings $Vt$ and $Pt$ of variables respectively procedures to $Vt'$ respectively $Pt'$.
- $\Sigma, Vt, Pt \vdash C$: command: command $C$ is well-formed with respect to $\Sigma$ in a context described by variable typing $Vt$ and procedure typing $Pt$.
- $\Sigma, Vt \vdash Ts$: terms($Ss$): term sequence $Ts$ is well formed according to $\Sigma$ in a context where variables are typed according to $Vt$; the terms have sorts $Ss$.
- $\Sigma, Vt \vdash Vs$: variables($Ss$): variable sequence $Vs$ is well formed according to $\Sigma$ in a context where variables are typed according to $Vt$; the variables have sorts $Ss$.

As for programs without procedures, we expect that well-typed programs relate sequences of argument values to sequences of result values; however, we will refrain from formalizing this statement again.

**Rules for $\Sigma \vdash P$: program:**

$$\frac{\Sigma \vdash Ds: \text{declarations}(Vt, Pt) \quad \Sigma \vdash X: \text{parameters}(Vt', Ss) \quad Vt'' = Vt \leftarrow Vt' \quad \Sigma, Vt'', Pt \vdash C: \text{command}}{\Sigma \vdash Ds; \text{ program } I(X)\, C: \text{program}}$$

**Rules for $\Sigma \vdash Ds$: declarations($Vt, Pt$):**

$$\frac{Vt = \varnothing \quad Pt = \varnothing}{\Sigma \vdash \sqcup: \text{declarations}(Vt, Pt)} \qquad \frac{\Sigma \vdash Ds: \text{declarations}(Vt', Pt') \quad \Sigma, Vt', Pt' \vdash D: \text{declaration}(Vt, Pt)}{\Sigma \vdash Ds; D: \text{declarations}(Vt, Pt)}$$

**Rules for $\Sigma, Vt, Pt \vdash D$: declaration($Vt', Pt'$):**

$$\frac{S \in \Sigma.s \quad Vt' = Vt \leftarrow \{\langle V, S \rangle\} \quad Pt' = Pt}{\Sigma, Vt, Pt \vdash \text{var } V: S: \text{declaration}(Vt', Pt')}$$

$$\frac{\begin{array}{c} \Sigma \vdash X_1: \text{parameters}(Vt_1, Ss_1) \quad \Sigma \vdash X_2: \text{parameters}(Vt_2, Ss_2) \quad Vt_3 = Vt \leftarrow (Vt_1 \cup Vt_2) \\ \neg \exists V \in Variable, S_1, S_2 \in Sort.\ \langle V, S_1 \rangle \in Vt_1 \wedge \langle V, S_2 \rangle \in Vt_2 \\ \Sigma, Vt_3, Pt \vdash C: \text{command} \quad Vt' = Vt \quad Pt' = Pt \cup \{\langle I, \langle Ss_1, Ss_2 \rangle \rangle\} \end{array}}{\Sigma, Vt, Pt \vdash \text{procedure } I(X_1; \text{ref } X_2)\, C: \text{declaration}(Vt', Pt')}$$

**Rules for $\Sigma, Vt, Pt \vdash C$: command:**

$$\frac{\langle V, S \rangle \in Vt \quad \Sigma, Vt \vdash T: \text{term}(S)}{\Sigma, Vt, Pt \vdash V := T: \text{command}} \qquad \frac{Vt_1 = Vt \leftarrow \{\langle V, S \rangle\} \quad \Sigma, Vt_1, Pt \vdash C: \text{command}}{\Sigma, Vt, Pt \vdash \text{var } V: S; C: \text{command}}$$

$$\frac{\Sigma, Vt \vdash Ts: \text{terms}(Ss_1) \quad \Sigma, Vt \vdash Vs: \text{variables}(Ss_2) \quad \langle I, \langle Ss_1, Ss_2 \rangle \rangle \in Pt}{\Sigma, Vt, Pt \vdash \texttt{call } I(Ts; Vs): \text{command}}$$

$$\cdots$$

**Rules for $\Sigma, Vt \vdash Ts$: terms($Ss$):**

$$\frac{Ss = [\,]}{\Sigma, Vt \vdash \sqcup: \text{terms}(Ss)} \qquad \frac{\Sigma, Vt \vdash Ts: \text{terms}(Ss') \quad \Sigma, Vt \vdash T: \text{term}(S) \quad Ss = Ss' \circ [S]}{\Sigma, Vt \vdash Ts, T: \text{terms}(Ss)}$$

**Rules for $\Sigma, Vt \vdash Vs$: variables($Ss$):**

$$\frac{Ss = [\,]}{\Sigma, Vt \vdash \sqcup: \text{variables}(Ss)} \qquad \frac{\Sigma, Vt \vdash Vs: \text{variables}(Ss') \quad \langle V, S \rangle \in Vt \quad Ss = Ss' \circ [S]}{\Sigma, Vt \vdash Vs, V: \text{variables}(Ss)}$$

**Fig. 7.18** Type checking programs with procedures

In the following, we will assume that procedure parameters and the arguments of procedure calls are annotated with the sorts $Ss$ determined by the type system; the denotational semantics will use this type information to look up procedures in the environments from the sorts of the arguments passed in procedure calls.

**Environments and Stores**

The example program at the beginning of this section demonstrates that there exist different views on which variable denotes which value: for instance, program main denotes by $c$ its parameter, while gcd refers in its first assignment to the global variable $c$, but later, in the body of the loop, $c$ denotes a local variable; div refers by $c$ to to the global variable. Furthermore, the same variable may be denoted by different names: the reference parameter $g$ of div denotes the local variable $c$ in main, the reference parameter $q$ in div denotes the local variable $c$ in gcd.

**Fig. 7.19** Environments and store

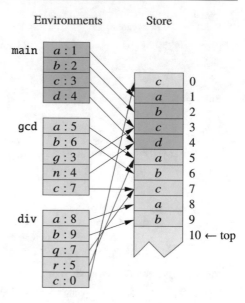

To address these issues, we will give up the idea that a program state is itself a mapping of variables to values, but split this mapping into two parts,

- an *environment* that maps variables to *addresses*, and
- a *store* (which we will also call *state*) that maps addresses to values,

where by an 'address' we understand a natural number that enumerates the values in the store. We thus also get a formal model that is closer to the actual implementation of a programming language on real computers.

Since a store now denotes a linear sequence of values indexed by addresses, we have to devise a strategy for assigning addresses to variables. As in real implementations, we will organize the store as a *stack*: there exists a particular address, the *top of stack* such that only the values at addresses less than the top are used; if new variables are to be allocated in the store, the top is correspondingly incremented; the variables allocated last can be freed by decrementing the top correspondingly.

The right half of Fig. 7.19 illustrates this model for the program given at the beginning of the section in a state where program main has invoked procedure gcd which in turn has invoked procedure div in the then branch. The stack now holds the values of all variables that are currently alive: first the global variable $c$ at address 0, then the parameters $a, b, c, d$ of main at addresses 1–4, then the value parameters $a, b$ and local variable $c$ of gcd at addresses 5–7, finally the value parameters $a, b$ of div at addresses 8–9. There are no values in the memory for the reference parameters of gcd and div, because these parameters are not variables on their own, but just synonyms for other variables.

The left half of Fig. 7.19 illustrates the various environments, i.e., the views that different parts of the program have of the store:

- The environment for program `main` maps the parameter names $a, b, c, d$ to the addresses of the corresponding local variables.
- The environment for procedure `gcd` maps the names $a, b$ of the value parameters to the addresses of the corresponding local variables, while the names $g, n$ of the reference parameters are mapped to the addresses of the local variables $c$ and $d$ of `main`; the name $c$ is mapped to the address of the local variable.
- The environment for procedure `div` maps the name $c$ to the address of the global variable, the names $a, b$ of the value parameters to addresses of the corresponding local variables, the names $q$ and $r$ of the reference parameters to the addresses of the local variables $c$ and $a$ of `gcd`.

On the basis of this model, we are now going to give our programming language a denotational semantics.

**Semantic Domains**

We begin by redefining the notion of a 'state'.

---

**Definition 7.13** (*State (Revised)*) A *address* is a natural number, a *state* is a mapping of addresses to values, and a *state relation* is a binary relation on states:

$$Address := \mathbb{N}$$
$$State := Address \rightarrow Value$$
$$StateRelation := \mathsf{Set}(State \times State)$$

---

Next we are introducing the core concept on which the new model is based.

---

**Definition 7.14** (*Environment*) A *variable environment* is a partial mapping of variables to addresses.

$$VarEnv := Variable \rightarrow_{\perp} Address$$

The *semantics of a procedure* is a function that maps two sequences of addresses (of the value parameters and of the reference parameters) and another address (the top of stack) to a state relation. A *procedure environment* is a partial function of (procedure) identifiers and their parameter sorts to their semantic values:

$$ProcSem := Address^* \times Address^* \times Address \rightarrow StateRelation$$
$$ProcEnv := Identifier \times Sort^* \times Sort^* \rightarrow_{\perp} ProcSem$$

An *environment* consists of a variable environment and a procedure environment:

$$Environment := \text{var}{:}VarEnv \times \text{proc}{:}ProcEnv$$

We will translate pairs of environments and states to variable assignments such that we can also in the new model evaluate terms and formulas without redefining their original semantics.

**Definition 7.15** *(Variable Assignment)* A *variable assignment* is a partial mapping of variables to values:

$$Assignment := Variable \rightarrow_{\perp} Value$$

We write $ass(e, s)$ to denote the variable assignment that results from an environment $e$ and state $s$ by mapping every variable $V$ to $s(e.\text{var}(V))$ (provided that $e.\text{var}$ is defined on $V$):

$$ass: Environment \times State \rightarrow Assignment$$
$$ass(e, s) = \{\langle V, s(a) \rangle \mid V \in Variable \wedge a \in Address \wedge \langle V, a \rangle \in e.\text{var}\}$$

Furthermore, we stipulate that the previously introduced 'well-definedness' predicates can be also applied to assignments in above sense, i.e.

$$[\![\,.\,]\!]\checkmark : Term \rightarrow \text{Set}(Assignment)$$
$$[\![\,.\,]\!]\checkmark : Formula \rightarrow \text{Set}(Assignment)$$

Then $[\![\, T \,]\!]\checkmark \langle a \rangle$ respectively $[\![\, F \,]\!]\checkmark \langle a \rangle$ is true if and only if term $T$ respectively formula $F$ is well-defined with respect to assignment $a$.

**Denotational Semantics**

Figure 7.20 depicts a denotational semantics for our language in the relational style. This semantics consists of the following valuation functions:

$[\![ . ]\!]$: *Program* → *ValueRelation*

$[\![ Ds;\ \text{program}\ I(X)\ C ]\!]\langle vs, vs' \rangle :\Leftrightarrow$
$\quad \exists n \in \mathbb{N},\ as \in Address^*,\ var \in VarEnv,\ e, e' \in Environment,\ a \in Address,\ s, s' \in State.$
$\quad\quad n = length(vs) \wedge n = length(vs') \wedge n = length(as) \wedge \langle e, a \rangle = [\![ Ds ]\!] \wedge$
$\quad\quad (\forall i \in \mathbb{N}_n.\ as(i) = a + i) \wedge [\![ X ]\!]^{as}\langle e.\mathsf{var}, var \rangle \wedge e' = \langle \mathsf{var}{:}var,\ \mathsf{proc}{:}e.\mathsf{proc} \rangle \wedge$
$\quad\quad (\forall i \in \mathbb{N}_n.\ s(as(i)) = vs(i)) \wedge [\![ C ]\!]^{e'}_{a+n}\langle s, s' \rangle \wedge (\forall i \in \mathbb{N}_n.\ vs'(i) = s'(as(i)))$

$[\![ . ]\!]$: *Declarations* → (*Environment* × *Address*)

$[\![ {}_\sqcup ]\!] := \langle \langle \mathsf{var}{:}\varnothing, \mathsf{proc}{:}\varnothing \rangle, 0 \rangle$
$[\![ Ds;\ D ]\!] := \mathsf{let}\ ea = [\![ Ds ]\!],\ e = ea.1,\ a = ea.2\ \mathsf{in}\ [\![ D ]\!]^e_a$

$[\![ . ]\!]$: *Declaration* → (*Environment* × *Address*) → (*Environment* × *Address*)

$[\![ \mathsf{var}\ V{:}S ]\!]^e_a := \langle \langle \mathsf{var}{:}e.\mathsf{var}[V \mapsto a], \mathsf{proc}{:}e.\mathsf{proc} \rangle, a + 1 \rangle$
$[\![ \mathsf{procedure}\ I^{Ss_1, Ss_2}(X_1;\ \mathsf{ref}\ X_2)\ C ]\!]^e_a :=$
$\quad \mathsf{let}$
$\quad\quad P^{as_1, as_2}_{a'}\langle s, s' \rangle :\Leftrightarrow$
$\quad\quad\quad \exists var_1 \in VarEnv,\ var_2 \in VarEnv.$
$\quad\quad\quad\quad [\![ X_1 ]\!]^{as_1}\langle e.\mathsf{var}, var_1 \rangle \wedge [\![ X_2 ]\!]^{as_2}\langle var_1, var_2 \rangle \wedge [\![ C ]\!]^{\langle \mathsf{var}{:}var_2, \mathsf{proc}{:}e.\mathsf{proc} \rangle}_{a'}\langle s, s' \rangle$
$\quad \mathsf{in}\ \langle \langle \mathsf{var}{:}e.\mathsf{var}, \mathsf{proc}{:}e.\mathsf{proc}[\langle I, Ss_1, Ss_2 \rangle \mapsto P] \rangle, a \rangle$

$[\![ . ]\!]$: *Parameters* → *Address*$^*$ → **Set**(*VarEnv* × *VarEnv*)

$[\![ {}_\sqcup ]\!]^{as}\langle var, var' \rangle :\Leftrightarrow var' = var$
$[\![ X, V{:}S ]\!]^{as}\langle var, var' \rangle :\Leftrightarrow$
$\quad \exists as' \in Address^*,\ a \in Address,\ var_1 \in VarEnv.$
$\quad\quad as = as' \circ [a] \wedge [\![ X ]\!]^{as'}\langle var, var_1 \rangle \wedge var' = var_1[V \mapsto a]$

$[\![ . ]\!]$: *Command* → *Environment* → *Address* → *StateRelation*

$[\![ V := T ]\!]^e_a\langle s, s' \rangle :\Leftrightarrow$
$\quad \exists a_1 \in Address.\ \langle V, a_1 \rangle \in e.\mathsf{var} \wedge [\![ T ]\!]\surd \langle ass(e, s) \rangle \wedge s' = s[a_1 \mapsto [\![ T ]\!]_{ass(e, s)}])$
$[\![ \mathsf{var}\ V{:}S;\ C ]\!]^e_a\langle s, s' \rangle :\Leftrightarrow$
$\quad \mathsf{let}\ e_1 = \langle \mathsf{var}{:}e.\mathsf{var}[V \mapsto a], \mathsf{proc}{:}e.\mathsf{proc} \rangle\ \mathsf{in}\ [\![ C ]\!]^{e_1}_{a+1}\langle s, s' \rangle$
$[\![ \mathsf{call}\ I^{Ss_1, Ss_2}(Ts; Vs) ]\!]^e_a\langle s, s' \rangle :\Leftrightarrow$
$\quad \exists vs \in Value^*,\ n \in \mathbb{N},\ as_1, as_2 \in Address^*,\ s_1 \in State,\ P \in ProcSem.$
$\quad\quad [\![ Ts ]\!]^{ass(e, s)}\langle vs \rangle \wedge n = length(vs) \wedge n = length(as_1) \wedge$
$\quad\quad (\forall i \in \mathbb{N}_n.\ as_1(i) = a + i) \wedge as_2 = [\![ Vs ]\!]^e \wedge$
$\quad\quad (\forall i \in \mathbb{N}_{a+n}.\ s_1(i) = \mathsf{if}\ i < a\ \mathsf{then}\ s(i)\ \mathsf{else}\ vs(i - a)) \wedge$
$\quad\quad \langle \langle I, Ss_1, Ss_2 \rangle, P \rangle \in e.\mathsf{proc} \wedge P^{as_1, as_2}_{a+n}\langle s_1, s' \rangle$

$\ldots$

$[\![ . ]\!]$: *Terms* → *Assignment* → **Set**(*Value*$^*$)

$[\![ {}_\sqcup ]\!]^{ass}\langle vs \rangle :\Leftrightarrow vs = [\,]$
$[\![ Ts, T ]\!]^{ass}\langle vs \rangle :\Leftrightarrow \exists vs_1 \in Value^*.\ [\![ Ts ]\!]^{ass}\langle vs_1 \rangle \wedge [\![ T ]\!]\surd \langle ass \rangle \wedge vs = vs_1 \circ [[\![ T ]\!]_{ass}]$

$[\![ . ]\!]$: *Variables* → *Environment* → *Address*$^*$

$[\![ {}_\sqcup ]\!]^e := [\,]\quad [\![ Vs, V ]\!]^e := [\![ Vs ]\!]^e \circ [e.\mathsf{var}(V)]$

**Fig. 7.20** Denotational semantics with procedures

- $[\![\,.\,]\!]$ : *Program* → *ValueRelation* maps a program to a relation between the initial values of the program parameters and their final ones. In detail, the definition of $[\![\,Ds\,;\, \texttt{program}\ I(X)\ C\,]\!]\langle vs, vs'\rangle$ describes how the program relates the initial values $vs$ of the program parameters $X$ to their final values $vs'$ where both value sequences must be of the same length called $n$: first $[\![\,Ds\,]\!]$ determines the environment $e$ established by declarations $Ds$ and the top of stack $a$ after allocation of the global variables. Then an address sequence $as$ of length $n$ is constructed that maps parameter number $i$ to address $a + i$. Now $[\![\,X\,]\!]$ uses $as$ to extend the global variable environment $e$.var to the environment $var$ which also maps the parameters to the addresses listed in $as$. Subsequently $var$ is combined with the procedure environment $e$.proc to the new environment $e'$. Furthermore, the store $s$ is initialized by setting these variables to the values provided by $vs$. Now $[\![\,C\,]\!]$ maps the initial store $s$ to the final store $s'$ using the environment $e'$ for looking up variables and procedures and using the initial top of stack $a + n$ for allocating new variables. From $s'$ the new values of the parameters are extracted to yield $vs'$.
- $[\![\,.\,]\!]$ : *Declarations* → (*Environment* × *Address*) maps a sequence of declarations to the environment arising from the declarations and to the top of stack after the allocation of the global variables.
  - $[\![\,\llcorner\,]\!]$ maps the empty sequence to the empty environment and to address 0.
  - $[\![\,Ds\,;\,D\,]\!]$ first determines the intermediate environment $e$ and top of stack $a$ arising from the declarations $Ds$; then it applies $[\![\,D\,]\!]$ to the declaration $D$ to extend $e$ and to increment $a$.
- $[\![\,.\,]\!]$ :     *Declaration* → (*Environment* × *Address*) → (*Environment* × *Address*) extends by a declaration the environment and increments the top of stack.
  - $[\![\,\texttt{var}\ V : S\,]\!]^e_a$ adds to the environment $e$ a mapping of variable $V$ to address $a$; it then increments $a$ by one.
  - $[\![\,\texttt{procedure}\ I^{Ss_1,Ss_2}(X_1\,;\ \texttt{ref}\ X_2)\ C\,]\!]^e_a$ adds to the environment $e$ a mapping of a triple $\langle I, Ss_1, Ss_2\rangle$ to the procedure semantics $P$ and leaves the address $a$ unchanged. Here $I$ denotes the identifier of the procedures, $Ss_1$ denotes the sorts of the value parameters $X_1$, and $Ss_2$ denotes the sorts of the reference parameters $X_2$; these sort annotations have been provided by the type system. The semantics $P$ is a function that takes as arguments the addresses $as_1$ of the value parameters (the addresses of new variables allocated by the caller of the procedure and initialized with the values of the terms passed as arguments), the addresses $as_2$ of the reference parameters (the addresses of the variables passed as arguments), and the current top of stack $a$ (after allocation of the value parameters); it returns a relation between the procedure's initial store $s$ (after initialization of the input parameters) and its final store $s'$. This relation extends the environment $e$ (the environment in which the procedure was declared, not the environment in which the procedure was invoked) to an environment $e_1$ that maps the input parameters $X_1$ to the addresses listed in $as_1$; it then extends $e_1$ to an environment $e_2$ that maps the reference parameters $X_2$ to the addresses listed in $as_2$. Finally, it applies $[\![\,C\,]\!]$ to determine the transition from initial state $s$ to final state $s'$ using $e_2$ to look up variables and procedures and using the region of the stack starting at $a$ to allocate new variables.

- $[\![\,.\,]\!]$ : *Parameters* → *Address** → Set(*VarEnv* × *VarEnv*) extends a variable environment by mapping the given parameters to the given addresses.
  - $[\![\,\llcorner\,]\!]^{as}\langle var, var'\rangle$ sets for the empty parameter sequence the result environment $var'$ to the input environment $var$.
  - $[\![\,X, V\!:\!S\,]\!]^{as}\langle var, var'\rangle$ decomposes the address sequence $as$ into the initial sequence $as'$ and the last address $a$. It uses $as'$ to determine the intermediate environment $var_1$ arising from the mapping of the initial parameter sequence $X$ to the addresses in $as'$. It then extends $var_1$ to the final environment $var'$ by mapping the last parameter $V$ to $a$.
- $[\![\,.\,]\!]$ : *Command* → *Environment* → *Address* → *StateRelation* maps a command for a given environment and a given top of stack address to a state relation.
  - $[\![\,V\!:=\!T\,]\!]_a^e\langle s, s'\rangle$ looks up the environment $e$ for the address $a_1$ of variable $V$. Provided that term $T$ is well-defined in initial state $s$, the final state $s'$ is determined from $s$ by mapping $a_1$ to the value of $T$ in the variable assignment arising from $e$ and $s$.
  - $[\![\,\texttt{var}\ V\!:\!S\,;\ C\,]\!]_a^e\langle s, s'\rangle$ constructs from environment $e$ a new environment $e_1$ that maps variable $V$ to the current top of stack address $a$. It then applies $[\![\,C\,]\!]$ to determine the relation between input state $s$ and final state $s'$ using $e_1$ to look up procedures and variables and using the region of the stack starting at address $a + 1$ to allocate new variables.
  - $[\![\,\texttt{call}\ I^{Ss_1, Ss_2}(Ts\,;\,Vs)\,]\!]_a^e\langle s, s'\rangle$ first determines the values $vs$ of the $n$ terms $Ts$ provided for the value parameters where the evaluation uses the assignment arising from the current environment $e$ and initial store $s$. Then it determines the sequence $as_1$ of addresses of new variables to hold these values: argument $i$ is stored at address $a + i$ where $a$ is the current top of stack. Next, it determines the addresses $as_2$ of the variables $Vs$ passed for the reference parameters. Then it constructs the intermediate store $s_1$ that is identical to $s$ at all addresses less than $a$ and that holds at the next $n$ addresses the values enumerated in $vs$. Using the procedure name $I$ and the sorts $Ss_1$ and $Ss_2$ of the arguments for the value parameters respectively reference parameters (the sort annotations have been provided by the type system), the environment $e$ is looked up for the semantics $P$ of the procedure. Given the address sequences $as_1$ and $as_2$ and the new top of stack $a + n$, $P$ finally determines the relationship between intermediate state $s_1$ and final state $s'$.
- $[\![\,.\,]\!]$ : *Terms* → *Assignment* → Set(*Value**) maps a term sequence to a condition on a value sequence that is true if the terms are all well-defined in the given variable assignment and the value sequence holds their values; the denotation applies $[\![\,T\,]\!]\!\sqrt{}\,\langle ass\rangle$ to determine whether a single term $T$ is well-defined in assignment $ass$ and it applies $[\![\,T\,]\!]_{ass}$ to determine the corresponding value.
- $[\![\,.\,]\!]$ : *Variables* → *Environment* → *Address** maps a sequence of variables to the sequence of their addresses in the given environment; it applies $e.\text{var}(V)$ to determine the address of a single variable $V$ in environment $e$.

The denotational semantics highlights the difference between *value parameters* and *reference parameters*. Value parameters are bound to the addresses of newly allocated local variables that receive the values of the argument terms; any subsequent modifications of the contents of these variables are thus hidden from the caller. On the contrary, reference parameters are bound to the addresses of the variables that are passed as arguments; modifications of the contents of these variables are immediately visible to the caller.

**Static Scoping versus Dynamic Scoping**

As highlighted above, the denotation $[\![\text{ procedure } I(X_1^{Ss_1}; \text{ ref } X_2^{Ss_2})\, C\,]\!]$ evaluates the body command $C$ in an environment that is derived from the environment $e$ active at the point of declaration; thus the meaning of any variable that does neither appear in the parameter list nor has been introduced locally in the procedure body is taken from the global environment. This name binding strategy called *static scoping* or *lexical scoping* is used in most modern programming languages. As an example, take the program

```
var c:Nat;
procedure p(ref a:Nat) {
  a := c
}
program main(a:Nat) {
  c := 1
  var c:Nat; {
    c := 2
    call p(a)
  }
}
```

where procedure $p$ sets reference parameter $a$ to the value of variable $c$. Even if at the point of the invocation of $p$ a local variable $c$ with value 2 is visible, $p$ uses the value of the global variable $c$ initialized to 1; the program therefore terminates with final parameter value $a = 1$, not $a = 2$.

An alternative to above strategy is given by the following modifications to the definitions of procedures and procedure calls:

$$[\![\text{ procedure } I^{Ss_1,Ss_2}(X_1; \text{ ref } X_2)\, C\,]\!]_a^e :=$$
$$\text{let}$$
$$\quad P_{a,e_0}^{as_1,as_2}\langle s, s'\rangle :\Leftrightarrow$$
$$\quad\quad \exists e_1 \in Environment, e_2 \in Environment.$$
$$\quad\quad\quad [\![X_1]\!]^{as_1}\langle e_0, e_1\rangle \wedge [\![X_2]\!]^{as_2}\langle e_1, e_2\rangle \wedge [\![C]\!]_a^{e_2}\langle s, s'\rangle$$
$$\text{in} \ldots$$
$$[\![\text{ call } I^{Ss_1,Ss_2}(Ts; Vs)\,]\!]_a^e\langle s, s'\rangle :\Leftrightarrow$$
$$\ldots$$
$$\langle\langle I, Ss_1, Ss_2\rangle, P\rangle \in e.\text{proc} \wedge P_{a+n,e}^{as_1,as_2}\langle s_1, s'\rangle$$
$$\ldots$$

Here the procedure semantics $P$ has an additional parameter $e_0$, the environment active at the point of the invocation of the procedure; it is this environment, rather than the declaration environment $e$, that is extended to the environment in which the command body $C$ is evaluated; correspondingly, the procedure call passes currently active environment $e$ to $P$. In this name binding strategy called *dynamic scoping* the value of any variable neither appearing in the parameter list nor having been introduced locally in the procedure body comes from the context of the caller. With dynamic scoping, thus the example program above terminates with $a = 2$.

Static scoping is slightly more complex to implement because a procedure has to be represented by a *closure* consisting of the code of the procedure and the environment active at the point of the procedure definition; in contrast, to implement dynamic scoping, only the code is required. However, dynamically scoped languages are hard to understand and do not lend themselves to static type checking. For instance, the type system provided by Fig. 7.1 assumes static scoping by using the variable typing $Vt$ at the point of the procedure definition as the basis of type-checking the definition. Therefore, while the early implementations of the programming language Lisp were dynamically scoped, today (apart from some scripting languages) virtually all programming languages are based on static scoping.

**Translating Variables to Addresses**

The denotational semantics depicted in Fig. 7.20 crucially depends on a variable environment $e$.var that maps variables to addresses; such an environment is set up in every procedure call from the declaration environment $e$ and the address lists $as_1$ and $as_2$ provided by the caller of the procedure. It thus looks as if the variable environment has to exist at runtime, during the execution of the program, exactly like the store does. However, this is actually not the case: the assignment of variables to addresses can be essentially performed by a static analysis without executing the program, e.g., by an extended version of the type checker. In detail, this static analysis assigns to every variable a tag of one of the following forms:

- global:$a$: this tag indicates that the variable is globally declared and receives absolute address $a$.
- local:$n$: this tag indicates that the variable is declared locally in a procedure, either as a value parameter, or a local variable. Here the value $n$ is not an absolute address but a relative offset to which some not yet known base address $a$ has to be added to denote the actual address $a + n$ of the variable.
- reference:$n$: this tag indicates a reference parameter; where $n$ is again a relative offset, not an absolute address.

We may thus annotate the previously given example program as shown below (we only show the annotations in the declarations; all references to the variables can be correspondingly annotated).

```
var c^global:0 :Nat;
procedure div^4(a^local:0 :Nat,b^local:1 :Nat; ref q^reference:2 :Nat,r^reference:3 :Nat) {
  q := 0; r := a;
  while r >= b do {
    q := q+1; r := r-b
  }
  c := c+1
}
```

**Fig. 7.21** An alternative store representation

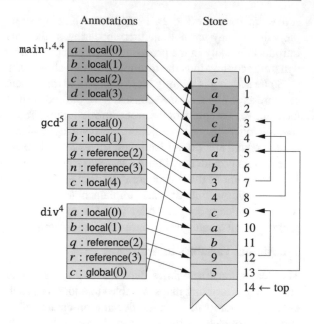

```
procedure gcd⁵(a^{local:0}:Nat,b^{local:1}:Nat; ref g^{reference:2}:Nat,n^{reference:3}:Nat) {
  c := 0;
  while a > 0 ∧ b > 0 do {
    var c^{local:4}:Nat;
    if a ≥ b
      then call div(a,b;c,a)
      else call div(b,a;c,b)
  }
  if a > 0 then g := a else g := b;
  n := c
}
program main^{1,4,4}(a^{local:0}:Nat,b^{local:1}:Nat,c^{local:2}:Nat,d^{local:3}:Nat) {
  call gcd(a*b,a+b;c,d)
}
```

Furthermore, we annotate every procedure $p$ as $p^n$ where $n$ is the number of local variables (including the value and reference parameters) declared in the procedure. The program $P$ itself is annotated as $P^{d,p,c}$ where $d$ is the number of globally declared variables, $p$ is the number of parameters of the program, and $c$ is the number of local variables (including the parameters) declared in the program. The purpose of these annotations will be explained below.

The goal of the variable annotations is to represent the assignments of variables to addresses that was originally presented in Fig. 7.19 in a modified form depicted in Fig. 7.21. All variables of a function marked as local:$n$ and reference:$n$ are represented by a contiguous portion of the stack which is called the 'frame' of the function. For a value parameter, the stack contains the argument value assigned to the parameter. For a reference parameter, the stack contains the address of the actual variable to which the parameter refers; in the diagram this is indicated by an arrow to this variable. Thus a stack contains both values and addresses.

Since the position $n$ of a variable within a frame is determined by static analysis, at runtime additionally only the initial address $a$ of the frame (the top of the stack before the function is called) has to be known to determine the actual variable address $a + n$. When a function $f$ annotated by the static analysis as $f^s$ is called, the top of the stack is incremented by the size $s$ of its frame (essentially the number of its local variables, but see below for more details). Thus, since also for every variable marked as global:$a$ its absolute address $a$ is known from the static analysis, it is not at all necessary to represent the variable environment at runtime; this environment becomes a data structure that is only used for the purpose of the static analysis.

The following semantics uses the result of the static analysis to drop the variable environment $e$.var and retain in parameter $e$ only the procedure environment which maps every procedure identifier to its semantics. Actually, also this mapping is independent of the values in the store such that in another static analysis step the abstract syntax tree of the program can be traversed and every procedure call be annotated by its semantics; then also the procedure environment $e$ can be completely dropped from the semantics, i.e., it need not be represented at runtime. However, since this additional step is very simple, we will refrain from its formalization.

The static analysis depends on an extended definition of a variable typing.

**Definition 7.16** *(Extended Typings)* Let $\Sigma$ be a signature. We define

$$Location := \text{global:}\textit{Address} + \text{local:}\mathbb{N} + \text{reference:}\mathbb{N}$$
$$VarTyping := Variable \rightarrow_\perp (\Sigma.s \times Location)$$
$$ProcTyping := Identifier \rightarrow \text{Set}(\Sigma.s^* \times \Sigma.s^* \times \mathbb{N})$$

as the sets of all *locations*, *extended variable typings*, and *extended procedure typings*.

For every variable, thus a variable typing additionally records its location; for every procedure, a procedure typing additionally records the size of its frame.

We will in the following denote by *variable typings* and *procedure typings* the extended forms and use the notion $T_1 \leftarrow T_2$ to denote the correspondingly generalized notion for updating extended variable typings.

Figure 7.22 now generalizes the type checking calculus presented in Fig. 7.1 to also assign addresses to variables. In detail, we have the following judgements (whose derivation can be considered as the execution of a procedure that annotates the program, its procedures, and variables, with the information outlined above):

**Rules for $\Sigma \vdash P$: program:**

$$\frac{\Sigma \vdash Ds: \text{declarations}(Vt, Pt, A_d) \quad \Sigma \vdash X: \text{vparams}(Vt', Ss, A_p) \quad Vt'' = Vt \leftarrow Vt' \\ \Sigma, Vt'', Pt, A_p \vdash C: \text{command}(A_c)}{\Sigma \vdash Ds;\ \text{program}\ I(X)\ C: \text{program}}$$

**Rules for $\Sigma \vdash Ds$: declarations$(Vt, Pt, A)$:**

$$\frac{Vt = \varnothing \quad Pt = [\ ] \quad A = 0}{\Sigma \vdash {}_\sqcup: \text{declarations}(Vt, Pt, A)}$$

$$\frac{\Sigma \vdash Ds: \text{declarations}(Vt', Pt', A') \quad \Sigma, Vt', Pt', A' \vdash D: \text{declaration}(Vt, Pt, A)}{\Sigma \vdash Ds;\ D: \text{declarations}(Vt, Pt, A)}$$

**Rules for $\Sigma, Vt, Pt, A \vdash D$: declaration$(Vt', Pt', A')$:**

$$\frac{S \in \Sigma.s \quad Vt' = Vt \leftarrow \{\langle V, \langle S, \text{global}:A\rangle\rangle\} \quad Pt' = Pt \quad A' = A + 1}{\Sigma, Vt, Pt, A \vdash \text{var}\ V: S: \text{declaration}(Vt', Pt', A')}$$

$$\frac{\begin{array}{c}\Sigma \vdash X_1: \text{vparams}(Vt_1, Ss_1, A_1) \quad \Sigma, A_1 \vdash X_2: \text{rparams}(Vt_2, Ss_2, A_2) \quad Vt_3 = Vt \leftarrow (Vt_1 \cup Vt_2) \\ \neg\exists V \in Variable, S_1, S_2 \in Sort, L_1, L_2 \in Location.\ \langle V, \langle S_1, L_1\rangle\rangle \in Vt_1 \wedge \langle V, \langle S_2, L_2\rangle\rangle \in Vt_2 \\ \Sigma, Vt_3, Pt, A_2 \vdash C: \text{command}(A_I) \quad Vt' = Vt \quad Pt' = Pt \cup \{\langle I, \langle Ss_1, Ss_2, A_I\rangle\rangle\} \quad A' = A\end{array}}{\Sigma, Vt, Pt, A \vdash \text{procedure}\ I(X_1;\ \text{ref}\ X_2)\ C: \text{declaration}(Vt', Pt', A')}$$

**Rules for $\Sigma \vdash X$: vparams$(Vt, Ss, A)$:**

$$\frac{Vt = \varnothing \quad Ss = [\ ] \quad A = 0}{\Sigma \vdash {}_\sqcup: \text{vparams}(Vt, Ss, A)}$$

$$\frac{\Sigma \vdash X: \text{vparams}(Vt', Ss', A') \quad S \in \Sigma.s \quad \neg\exists S' \in Sort, L_V \in Location.\ \langle V, \langle S', L_V\rangle\rangle \in Vt' \\ Vt = Vt' \leftarrow \{\langle V, \langle S, \text{local}:A'\rangle\rangle\} \quad Ss = Ss' \circ [S] \quad A = A' + 1}{\Sigma \vdash X, V: S: \text{vparams}(Vt, Ss, A)}$$

**Rules for $\Sigma, A \vdash X$: rparams$(Vt, Ss, A')$:**

$$\frac{Vt = \varnothing \quad Ss = [\ ] \quad A' = A}{\Sigma, A \vdash {}_\sqcup: \text{rparams}(Vt, Ss, A')}$$

$$\frac{\Sigma, A \vdash X: \text{rparams}(Vt', Ss', A'') \quad S \in \Sigma.s \quad \neg\exists S' \in Sort, L_V \in Location.\ \langle V, \langle S', L_V\rangle\rangle \in Vt' \\ Vt = Vt' \leftarrow \{\langle V, \langle S, \text{reference}(A'')\rangle\rangle\} \quad Ss = Ss' \circ [S] \quad A' = A'' + 1}{\Sigma, A \vdash X, V: S: \text{rparams}(Vt, Ss, A')}$$

**Rules for $\Sigma, Vt, Pt, A \vdash C$: command$(A')$:**

$$\frac{\Sigma, Vt \vdash T: \text{term}(S) \quad \langle V, \langle S, L_V\rangle\rangle \in Vt \quad A' = A}{\Sigma, Vt, Pt, A \vdash V := T: \text{command}(A')}$$

$$\frac{Vt_1 = Vt \leftarrow \{\langle V, \langle S, \text{local}:A\rangle\rangle\} \quad \Sigma, Vt_1, Pt, A + 1 \vdash C: \text{command}(A')}{\Sigma, Vt, Pt, A \vdash \text{var}\ V: S;\ C: \text{command}(A')}$$

$$\frac{\Sigma, Vt \vdash Ts: \text{terms}(Ss_1) \quad \Sigma, Vt \vdash Vt: \text{variables}(Ss_2) \quad \langle I, \langle Ss_1, Ss_2, A_I\rangle\rangle \in Pt \quad A' = A}{\Sigma, Vt, Pt, A \vdash \text{call}\ I(Ts; Vs): \text{command}(A')}$$

$$\frac{\Sigma, Vt, Pt, A \vdash C_1: \text{command}(A_1) \quad \Sigma, Vt, Pt, A \vdash C_2: \text{command}(A_2) \quad A' = \max\{A_1, A_2\}}{\Sigma, Vt, Pt, A \vdash C_1; C_2: \text{command}(A')}$$

$$\dots$$

**Fig. 7.22** Translating variables to addresses

- $\Sigma \vdash P$ : program analyzes program $P$ with respect to signature $\Sigma$. It determines the number of global variable declarations $A_d$, the number $A_p$ which adds to $A_d$ the number of program parameters, and the number $A_c$ which adds to $A_p$ the space required for the local variables of the program body. Thus $A_p - A_d$ denotes the number of program parameters and $A_c - A_d$ denotes the frame size of the program. The derivation of this judgement annotates program $P$ as $P^{A_d, A_p, A_c}$.
- $\Sigma \vdash Ds$: declarations($Vt, Pt, A$) analyzes the program declarations $Ds$ determining the variable typing $Vt$, the procedure typing $Pt$, and the number $A \in \mathbb{N}$ of addresses allocated for the global variable declarations in $Ds$.
- $\Sigma, Vt, Pt, A \vdash D$: declaration($Vt', Pt', A'$) analyzes the declaration $D$ and correspondingly extends the variable typing $Vt$, procedure typing $Pt$, and number $A$ of global variable declarations to $Vt'$, $Pt'$, and $A'$, respectively:
  - For a global variable declaration var $V: S$, the variable typing is extended by a mapping that assigns to variable $V$ the location global:$A$. Furthermore, the analysis determines $A' = A + 1$.
  - For a procedure declaration procedure $I(X_1; \text{ref } X_2)\, C$, first the analysis of the value parameters $X_1$ determines their number $A_1$, which is then extended by the analysis of the reference parameters $X_2$ to the number $A_2$ of all parameters. By analysis of the program body $C$, this number is extended to the total frame size $A_I$ of the procedure; the procedure typing is extended by a mapping that assigns to procedure identifier $I$ this value $A_I$.
- $\Sigma \vdash X$: vparams($Vt, Ss, A$) analyzes the value parameters $X$, deriving the corresponding variable typing $Vt$ and determining their number $A$. The variable typing maps each parameter to a location local:$n$ where $n$ denotes the position of the parameter in the sequence.
- $\Sigma, A \vdash X$: rparams($Vt, Ss, A'$) analyzes the reference parameters $X$, deriving the corresponding variable typing $Vt$ and extending the number $A$ of the value parameters to the total parameter number $A'$. The variable typing maps each parameter to a location reference:$n$ where $n$ denotes the position of the parameter in the whole parameter sequence.
- $\Sigma, Vt, Pt, A \vdash C$: command($A'$) analyzes the command $C$ in a context where the procedure frame already requires $A$ slots for the local variables, incrementing this value to $A'$ to also accommodate space for the local variables of $C$:
  - For an assignment $V := T$, we have $A' = A$, i.e., the frame size remains unchanged. However, the analysis determines from the variable typing the location $L_V$ of variable $V$ and annotates $V$ as $V^{L_V}$.
  - For a local variable declaration var $V: S$; $C$, the analysis extends the current variable typing by a mapping of $V$ to location local:$A$, thus allocating one cell of the frame for $A$ and analyzing the command body $C$ for frame size $A + 1$ yielding the final frame size $A'$.
  - For a procedure call call $I(Ts; Vs)$, we have $A' = A$, i.e., the frame size remains unchanged. However, the analysis determines from the procedure typing the sorts $Ss_1$ of the procedure's value parameters, the sorts $Ss_2$ of the procedure's reference parameters, and the frame size $A_I$; consequently it annotates $I$ as $I^{Ss_1, Ss_2, A_I}$.

- For a command sequence $C_1; C_2$, the analysis determines the frame size requirement $A_1$ of command $C_1$ and the requirement $A_2$ of command $C_2$. The total requirement is then $A' = \max\{A_1, A_2\}$, because after the execution of $C_1$ the space that was used by additional variables in $C_1$ can be reused for additional variables in the execution of $C_2$.
- The analysis of the other commands can be easily generalized from the corresponding judgements given in Fig. 7.1. For the two-sided conditional, as for the command sequence, the total frame size requirement is determined by the maximum of the requirements of both branches of the conditional.

As depicted in Fig. 7.21, in the revised execution model a store needs to hold not only values, but also addresses. We thus extend our notion of states and state relations correspondingly.

**Definition 7.17** *(Extended State)* An *extended state* is a mapping of addresses to values or addresses; an extended state relation is a binary relation on such states:

$$State := Address \rightarrow (Value \cup Address)$$
$$StateRelation := \mathsf{Set}(State \times State)$$

Furthermore, we will need to convert the abstract locations generated by the static analysis to actual addresses usable in the semantics. Here the problem arises that the memory layout depicted in Fig. 7.21, while being quite intuitive, can be actually not realized without some extra effort: since after the call of a procedure with frame size $s$ the top of the stack has been incremented to the next address $a$ after the frame, from the position $n$ of a variable in that frame alone we cannot determine its actual address $a - s + n$, unless either $s$ (the frame size) or $a - s$ (the base address of the last frame) are explicitly available. Rather than extending our model by such an extra value, we solve this problem by *reverting* the order of variables in the stack frame: local variable 0 of the frame is stored at address $a - 1$, local variable 1 is stored at address $a - 2$, and so on, until the last local variable $s - 1$ is stored at address $a - s$. We formalize this idea as follows.

**Definition 7.18** *(Converting Locations to Addresses)* Given a top of stack address $a$, store $s$, and location $L$, we refer by $addr_s^a(L)$ to the address denoted by location $L$ with respect to $a$ and $s$:

$$addr : Address \rightarrow State \rightarrow Location \rightarrow Address$$
$$addr_s^a(L) :=$$

case $L$ of

    global$(a) \rightarrow a$ |

    local$(n) \rightarrow a - n - 1$ |

    reference$(n) \rightarrow s(a - n - 1)$

Finally, we assume that all occurrences of program variables in terms and formulas have been appropriately annotated such that, given the current top of stack address $a$ and store $s$, it is possible to check whether a term respectively formula is well-defined with respect to $a$ and $s$ and to determine its value correspondingly. In particular $[\![\, T \,]\!]\checkmark_s^a$ shall indicate whether term $T$ is well-defined and $[\![\, T \,]\!]_s^a$ shall denote its value; we omit the details of the formalization.

Figure 7.23 now revises the denotational semantics given in Fig. 7.20; since the mapping of variables to addresses has been delegated to the type checker, this version of the semantics is considerably simpler than the original one:

- $[\![\,.\,]\!]$: *Program* $\rightarrow$ *ValueRelation* first determines from the declarations $Ds$ the environment $e$ that maps procedure identifier to their semantics. Then it assigns in the initial store $s$ to every slot that has been allocated for parameter $i$ at address $a$ (in the first frame below address $A_c$) the corresponding initial parameter value from the sequence $vs$. Now it determines the final store $s'$ arising from the execution of program body $C$ in the frame below address $A_c$ using environment $e$ for the lookup of procedures. Finally, it extracts from $s'$ the final values of the program parameters into the result sequence $vs'$.
- $[\![\,.\,]\!]$: *Declarations* $\rightarrow$ *ProcEnv* iterates over the sequence of declarations to determine the corresponding procedure environment.
- $[\![\,.\,]\!]$: *Declaration* $\rightarrow$ *ProcEnv* $\rightarrow$ *ProcEnv* extends by a declaration a procedure environment. For a global variable declaration `var` $V : S$, the environment remains unchanged. For a procedure declaration `procedure` $I^{Ss_1, Ss_2}(X_1 ; $ `ref` $ X_2)$ $C$ (annotated by the type inference), the environment is extended by a mapping of the procedure identifier $I$ and the sorts $Ss_1$ and $Ss_2$ of its parameters to a function $P$ that just denotes the semantics of the procedure body $C$ executed in declaration environment $e$; to this semantics the top of stack address $a$ provided the caller of the procedure (already appropriately incremented by the procedure's frame size, see below) is forwarded.
- $[\![\,.\,]\!]$: *Command* $\rightarrow$ *ProcEnv* $\rightarrow$ *Address* $\rightarrow$ *StateRelation* maps a command executed in a given procedure environment $e$ with a current top of stack address $a$ (indicating the limit of the current frame) to a state relation:
  - For an annotated assignment $V^{L_V} := T$, the semantics determines from the location $L_V$ of variable $V$ the address of the variable with respect to top of stack $a$ and store $s$ and updates $s$ by the value of term $T$ in that store.
  - For a local variable declaration block `var` $V : S$; $C$, the semantics just determines the state relation from the command body $C$.

$[\![\,.\,]\!]$: **Program → ValueRelation**

$[\![\,Ds;\ \text{program } I^{A_d,A_p,A_c}(X)\ C\,]\!]\langle vs, vs'\rangle :\Leftrightarrow$
$\quad \exists n \in \mathbb{N},\ e \in ProcEnv,\ s,\ s' \in State.$
$\quad\quad n = A_p - A_d \wedge n = length(vs) \wedge n = length(vs') \wedge e = [\![\,Ds\,]\!] \wedge [\![\,C\,]\!]^e_{A_c}\ \langle s, s'\rangle\ \wedge$
$\quad\quad \Big(\forall i \in \mathbb{N}_n.\ \text{let } a = addr^{A_c}_s(\text{local}:i) \text{ in } s(a) = vs(i) \wedge vs'(i) = s'(a)\Big)$

$[\![\,.\,]\!]$: **Declarations → ProcEnv**

$\quad [\![\,\sqcup\,]\!] := \emptyset$
$\quad [\![\,Ds;\ D\,]\!] := \text{let } e = [\![\,Ds\,]\!] \text{ in } [\![\,D\,]\!]^e$

$[\![\,.\,]\!]$: **Declaration → ProcEnv → ProcEnv**

$\quad [\![\,\text{var } V:S\,]\!]^e := e$
$\quad [\![\,\text{procedure } I^{Ss_1,Ss_2}(X_1;\ \text{ref } X_2)\ C\,]\!]^e :=$
$\quad\quad \text{let}$
$\quad\quad\quad P^a\langle s, s'\rangle :\Leftrightarrow [\![\,C\,]\!]^e_a\langle s, s'\rangle$
$\quad\quad \text{in } e[\langle I, Ss_1, Ss_2\rangle \mapsto P]$

$[\![\,.\,]\!]$: **Command → ProcEnv → Address → StateRelation**

$\quad [\![\,V^{Lv} := T\,]\!]^e_a\langle s, s'\rangle :\Leftrightarrow$
$\quad\quad \text{let } a' = addr^a_s(L_V) \text{ in } [\![\,T\,]\!]^a_s \wedge s' = s[a' \mapsto [\![\,T\,]\!]^a_s])$
$\quad [\![\,\text{var } V:S;\ C\,]\!]^e_a\langle s, s'\rangle :\Leftrightarrow [\![\,C\,]\!]^e_a\langle s, s'\rangle$
$\quad [\![\,\text{call } I^{Ss_1,Ss_2,A_I}(Ts;Vs)\,]\!]^e_a\langle s, s'\rangle :\Leftrightarrow$
$\quad\quad \exists vs \in Value^*,\ as \in Address^*,\ n_1, n_2 \in \mathbb{N},\ s_1 \in State,\ P \in ProcSem.$
$\quad\quad [\![\,Ts\,]\!]^a_s\langle vs\rangle \wedge n_1 = length(vs) \wedge as = [\![\,Vs\,]\!]^a_s \wedge n_2 = length(as) \wedge$
$\quad\quad (\forall i \in \mathbb{N}_a.\ s_1(i) = s(i)) \wedge$
$\quad\quad \Big(\forall i \in \mathbb{N}_{n_1}.\ \text{let } a' = addr^{a+A_I}_s(\text{local}:i) \text{ in } s_1(a') = vs(i)\Big) \wedge$
$\quad\quad \Big(\forall i \in \mathbb{N}_{n_2}.\ \text{let } a' = addr^{a+A_I}_s(\text{local}:n_1 + i) \text{ in } s_1(a') = as(i)\Big) \wedge$
$\quad\quad \langle\langle I, Ss_1, Ss_2\rangle, P\rangle \in e \wedge P^{a+A_I}\langle s_1, s'\rangle$

$\quad \dots$

$[\![\,.\,]\!]$: **Variables → Address → State → Address***

$\quad [\![\,\sqcup\,]\!]^a_s := [\,]$
$\quad [\![\,Vs, V^{Lv}\,]\!]^a_s := \text{let } a' = addr^a_s(L_V) \text{ in } [\![\,Vs\,]\!]^a_s \circ [a']$

**Fig. 7.23** Denotational semantics with variables translated to addresses

- For a procedure call call $I^{Ss_1,Ss_2,A_I}(Ts;Vs)$, the semantics determines the $n_1$ values $vs$ of the argument terms $Ts$ passed for the value parameters and the $n_2$ addresses $as$ denoted by the argument variables $Vs$ passed for the value parameters. It then constructs an intermediate store $s_1$ which is identical to the initial store $s$ below the top of the stack address $a$; however it holds in a new frame of size $A_I$ starting at $a$ the $n_1$ elements of $vs$ respectively $n_2$ elements of $as$ at those addresses that are appropriate for the first $n_1 + n_2$ local variables of the frame. Then the semantics $P$ of the procedure is looked up in the current environment and invoked with the new top of stack address $a + A_I$ and intermediate store $s_1$ to determine the final store $s'$.

- $[\![\,.\,]\!]$: *Variables* $\to$ *Address* $\to$ *State* $\to$ *Address** determines the address sequence denoted by the given variable sequence for the current top of stack $a$ and store $s$.

This denotational semantics resembles in some aspects a *calling convention* in the actual implementation of procedures for a programming language. However, while in our model all work for the procedure call is performed by the caller of a procedure, actual implementations typically let both caller and callee share this work such that, e.g., the callee is responsible for saving the values of processor registers to the stack.

## Recursive Procedures

The programming language described so far only allows every procedure to call other procedures that have been declared previously. In order to allow a procedure also to call itself recursively, we would have to define the semantics of a procedure declaration as follows:

$$[\![\text{procedure } I^{Ss_1,Ss_2}(X_1; \text{ ref } X_2) \, C \,]\!]^e :=$$
$$\text{let}$$
$$P^a\langle s, s'\rangle :\Leftrightarrow [\![\,C\,]\!]_a^{e[\langle I,Ss_1,Ss_2\rangle \mapsto P]}\langle s, s'\rangle$$
$$\text{in } e[\langle I, Ss_1, Ss_2\rangle \mapsto P]$$

Here the procedure would be allowed to already refer to the new procedure environment that is constructed by the procedure definition; however, this definition is technically not correct because the definition of $P$ now refers to $P$ itself on the right hand side. Ignoring this problem for the moment and giving the new environment the name $e'$, we can write this also as

$$[\![\text{procedure } I^{Ss_1,Ss_2}(X_1; \text{ ref } X_2) \, C \,]\!]^e :=$$
$$\text{let}$$
$$P^a\langle s, s'\rangle :\Leftrightarrow [\![\,C\,]\!]_a^{e'}\langle s, s'\rangle$$
$$e' := e[\langle I, Ss_1, Ss_2\rangle \mapsto P]$$
$$\text{in } e'$$

or more succinctly as

$$[\![\text{procedure } I^{Ss_1,Ss_2}(X_1; \text{ ref } X_2) \, C \,]\!]^e :=$$
$$\text{let}$$
$$e' := e[\langle I, Ss_1, Ss_2\rangle \mapsto \pi a \in Address, s, s' \in State. [\![\,C\,]\!]_a^{e'}\langle s, s'\rangle]$$
$$\text{in } e'$$

This highlights that by such a construction actually the procedure environment $e'$ itself is defined recursively.

Now, remembering that an environment is a partial function and expanding the definition of the function update notation $e[\langle I, Ss_1, Ss_2\rangle \mapsto P]$, we can write this definition also in a technically correct way using the inductive style introduced in Definition 5.14:

$[\![\,\text{procedure } I^{Ss_1,Ss_2}(X_1;\, \text{ref } X_2)\, C\,]\!]^e :=$
   let
      inductive $e'$: $Identifier \times Sort^* \times Sort^* \rightarrow_\perp ProcSem$
      $e'(I', Ss'_1, Ss'_2) = P' \Leftarrow$
         if $I' = I \wedge Ss'_1 = Ss_1 \wedge Ss'_2 = Ss_2$
            then $P' = \pi a \in Address,\, s, s' \in State.\ [\![\,C\,]\!]^{e'}_a \langle s, s' \rangle$
            else $\exists P = e(I, Ss_1, Ss_2).\ P' = P$
   in $e'$

This form interprets the meaning of the recursive function definition as the least fixed point of the corresponding functional, i.e.

$$e' := \mathsf{lfp}\ \lambda e' \in ProcEnv.$$
$$\pi\, I' \in Identifier,\, Ss'_1 \in Sort^*,\, Ss'_2 \in Sort,\, P' \in ProcEnv^*.$$
$$\text{if } I' = I \wedge Ss'_1 = Ss_1 \wedge Ss'_2 = Ss_2$$
$$\text{then } P' = \pi a \in Address,\, s, s' \in State.\ [\![\,C\,]\!]^{e'}_a \langle s, s' \rangle$$
$$\text{else } \exists P = e(I, Ss_1, Ss_2).\ P' = P$$

(actually, this functional has to be upward-continuous to ensure the existence of the least fixed point; we omit the corresponding proof).

Intuitively, we can thus consider $e'$ as the limit of a sequence of environments $e_0, e_1, e_2, \ldots$ where $e_0 = e$ only contains the procedures in $e$, $e_1 = e_0[\langle I, Ss_1, Ss_2 \rangle \mapsto P_0]$ additionally contains a procedure $P_0$ that can only call the procedures in $e$, $e_2 = e_1[\langle I, Ss_1, Ss_2 \rangle \mapsto P_1]$ supersedes the procedure definition of $e_1$ by a procedure $P_1$ that can call $P_0$ and the procedures in $e$, and so on: environment $e_i = e_{i-1}[\langle I, Ss_1, Ss_2 \rangle \mapsto P_{i-1}]$ thus contains a procedure $P_{i-1}$ that can call $P_{i-2}$ and the procedures in $e$. Since each $P_i$ is derived from the same functional, this amounts to saying that $P_i$ is the version of the procedure that allows at most $i$ (recursive) calls of $P$. The limit $e'$ contains a procedure $P$ that allows arbitrarily many (recursive) procedure applications.

In order to allow also mutually recursive definitions of multiple procedures $P, Q, R, \ldots$, this technique can be generalized to an inductive definition of $e'$ that updates $e$ by bindings for all procedures in the mutual recursion cycle; the fixed point semantics defines $e'$ as the limit of the sequence of environments $e_0, e_1, e_2, \ldots$ where for each $i > 0$ environment $e_i$ contains procedures $P_i, Q_i, R_i, \ldots$ that may call procedures $P_{i-1}, Q_{i-1}, R_{i-1}, \ldots$; thus we can consider $P_i, Q_i, R_i, \ldots$ as versions of $P, Q, R, \ldots$ where the total number of applications in an execution is at most $i$. The limit $e'$ contains procedures $P, Q, R, \ldots$ that allow arbitrarily many mutually recursive procedure applications.

The semantics of the programming language might be modified to consider all procedure declarations as parts of one big (potential) mutual recursion cycle allowing each procedure to call each other one (programming languages like Java have this kind of semantics); then the whole declaration environment would be defined by a single fixed point. Alternatively, we might use a syntactic construct such as recursive{...} to surround the declarations of (single or multiple) procedures that are allowed to call themselves recursively; thus we would limit the fixed point to the particular cycle and would have multiple fixed points for multiple cycles (ML-like languages have such constructs; also 'forward' declarations in languages such as C and C++ serve a similar purpose). However, having sketched the basic principles, we will leave it at that and not elaborate the formal details further.

## Exercises

Download from the following URL:
https://www.risc.jku.at/people/schreine/TP/exercises/ex-languages.pdf.

## Further Reading

A number of textbooks provide good overviews on the various approaches to the semantic modeling of programming languages.

Schmidt's book [160] discusses in depth the denotational semantics of programming languages, imperative and functional ones; as in most texts on denotation semantics, programs are modeled as partial functions (while our book focuses on the more general view of programs as relations). The book also includes an ample description of the corresponding theory of fixed points, where partial functions are described as total functions with a co-domain that is extended by a special element $\perp$ ('bottom') representing 'undefinedness'; consequently, semantic functions are defined in a style such that from $\perp$ as a partial result only $\perp$ as a total result can emerge. While the languages discussed in [160] are only dynamically typed (as part of the denotational semantics), Schmidt's successor book [161] discusses languages with a static type system where only well-typed programs are given a denotational semantics; to this resource the present book owes much of the inspiration of describing languages both by a type system in the form of an inference calculus and by a mapping of well-typed language phrases to semantic values. Also in Van Leeuwen's handbook [177], Chap. 11 by Peter Mosses presents in depth the theory of denotational semantics.

Winskel's book [180] gives a compact overview on both the denotational semantics (where the problem of describing partial functions is solved by a set-theoretic form of function definitions) and the big-step operational semantics of a simple imperative language, also proving their equivalence; Bruni's and Montanari's book [28] gives similar presentations. Our presentation of the relationship of the semantics owes to Winskel, but we model a much more expressive language. Nielson's and Nielson's book [131] contains a more in depth discussion of denotational semantics, big-step semantics, small-step semantics, and their relationships; it also proves the correctness of a translation of a high-level language to a low-level machine-oriented one and presents implementations of operational and denotational semantics in the functional programming language Miranda. While we took inspiration from this presentation, we actually use a richer higher-level language that also includes local variable declarations; this requires a more elaborate domain of configurations than the one in Nielson's and Nielson's book. Furthermore, our machine language is more expressive (by also using the stack for local variable declarations) and the execution model is closer to a real machine language (by having a single machine program in which the current instruction is referenced by a program counter).

Fernandez's book [48] gives a concise overview on operational semantics (both big and small step); while going less into the details of formal proofs, it also discusses the operational semantics of functional and logic languages. Nebel's compact book [127] contains a chapter on denotational and operational semantics. O'Regans book [138] contains a chapter on language theory and semantics. More in-depth background on the mathematics of programming language theory is given in Gunter's book [63] and in Mitchell's book [119].

Rather than presenting language semantics as an 'abstract' (paper and pencil) exercise, Nipkow's book [133] gives a 'concrete semantics': both denotational and operational programming language semantics are elaborated in the language of the proof assistant Isabelle. All proofs are formally checked and programs are executable by application of Isabelle's rewriting engine to the semantic definitions.

While most approaches to denotational language semantics focus on the model of programs as (partial) functions, Hoare's text [76] models programs as predicates (i.e., relations). Furthermore, Hoare's and Jifeng's later book [78] presents a unified view on programming language theories that embeds denotational and operational semantics in a relational framework. More references on the relational view of programs will be given in the 'Further Reading' section of the following chapter.

# Language Semantics in OCaml and the K Framework

In this section, we present two implementations of the core of the imperative programming language that was introduced in Definition 7.1; both implementations are directly based on the formal semantics of the language:

- The first implementation is derived from its denotational semantics (in the functional flavor); it is written in the programming language *OCaml* [134] (on page ?? we already demonstrated the implementation of the semantics of expressions in OCaml).
- The second implementation is rooted in its small step operational semantics; it is developed the *K framework* [135, 156, 184], a rewrite-based executable semantic framework and corresponding software for the definition of programming languages. From a single syntactic and semantic definition of the language, the software generates a parser and an interpreter for executing programs in the language, as well as program analysis tools such as a state space explorer. The K Framework has been used for formalizing the operational semantics of various real programming languages, for example C, Java, JavaScript, Scheme, and Haskell.

The definitions used in the following presentations can be downloaded from the URLs

https://www.risc.jku.at/people/schreine/TP/software/languages/imp.ml
https://www.risc.jku.at/people/schreine/TP/software/languages/imp.k
https://www.risc.jku.at/people/schreine/TP/software/languages/prog.imp

and loaded by executing from the command line the following commands:

```
ocaml imp.ml
kompile imp.k
krun prog.imp
```

### Denotational Semantics in OCaml

Our goal is to implement the (core of the) denotational semantics introduced in Fig. 7.2. We start by defining the abstract syntax of the command part of the imperative language (we omit the top-level domain of programs) as data types in OCaml. Here `expression` denotes a domain of terms denoting integer values and `boolexp` denotes a domain of formulas denoting truth values; the core type `command` denotes the type of programming language commands:

```
type numeral = int ;;
type identifier = string ;;
```

```
type expression =
| N of numeral
| I of identifier
| S of expression * expression
| P of expression * expression
;;

type boolexp =
| Eq of expression * expression
| Not of boolexp
| And of boolexp * boolexp
;;

type command =
| Assign of identifier * expression
| Var of identifier * command
| Seq of command * command
| If2 of boolexp * command * command
| If1 of boolexp * command
| While of boolexp * command
;;
```

Based on this, it is not difficult to define a function `cstr(c:command):
string` that translates the abstract syntax of a command to its concrete
textual representation (see the source code for details). We can then define
a sample command and print it as follows:

```
let b = Not(Eq(I("i"),I("n"))) ;;
let c1 = Assign("j", S(P(N(2),I("i")), N(1))) ;;
let c2 = Assign("a", S(I("a"), I("j"))) ;;
let c3 = Assign("i", S(I("i"),N(1))) ;;
let c = While(b, Var("j", Seq(c1, Seq(c2, c3)))) ;;
Printf.printf "
```

The corresponding output shows the text of the program as follows:

```
while (~i=n) do {var j; {j:=((2*i)+1); {a:=(a+j); i:=(i+1)}}}
```

After the syntactic domains, we define the semantic algebras, in our case
the algebras of stores. A store consists in our implementation of a default
value (an integer that describes the value of every variable that has no other
value) and a list of identifier/integer pairs (the values of all variables that are
different from the default; it thus represents conceptually a (total) function
from the domain of identifiers to the domain of integer values:

```
type value = int ;;
type map = (identifier*value)list ;;
type store = map * value ;;
```

For this domain, we define three functions `lookup` for reading the value of an identifier from a given store, a function `init` to construct a store where all variables hold the default value, and a function `update` that updates the value of a variable in a store (again see the source for the definitions):

```
let rec lookup ((m,d):store) (i:identifier) = ... ;;
let init(v:value): store = ([],v) ;;
let update((m,d):store)(i:identifier)(v:value):store = ... ;;
```

Now, we can give the denotational semantics of value expressions and Boolean expressions as functions from stores to (truth) values:

```
let rec eval(e:expression)(s:store): value =
  match e with
  | N(n) -> n
  | I(i) -> lookup s i
  | S(e1,e2) -> (eval e1 s) + (eval e2 s)
  | P(e1,e2) -> (eval e1 s) * (eval e2 s)
;;

let rec bval (b:boolexp)(s:store): bool =
  match b with
  | Eq(e1,e2) -> (eval e1 s) = (eval e2 s)
  | Not(b) -> not (bval b s)
  | And(b1,b2) -> (bval b1 s) && (bval b2 s)
;;
```

In OCaml function definitions are automatically 'Curried', thus for instance the function `eval(e:expression)(s:store): value` can be also applied as `eval e` to a single argument e yielding a function from stores to values. In the same way we can define the semantics of a command c as a function from stores to stores by the application `cval c` of a function `cval`:

```
let rec cval(c:command)(s:store): store =
  match c with
  | Assign(i,e) -> update s i (eval e s)
  | Var(i,c) ->
      let v = lookup s i in
      let s1 = update s i 0 in
      let s2 = cval c s1 in
      update s2 i v
  | Seq(c1,c2) -> let s1 = cval c1 s in cval c2 s1
  | If2(b,c1,c2) -> if bval b s then cval c1 s else cval c2 s
  | If1(b,c) -> if bval b s then cval c s else s
```

```
| While(b,c) ->
    let rec w(s:store): store =
      if bval b s
        then let s1 = cval c s in w s1
        else s
    in w s
;;
```

Here the semantics of a loop is defined by a recursive function; thus given non-terminating command as input the interpreter runs into a non-terminating recursion.

Based on these definitions, we can now create the initial store s that maps variable *n* to 5 (and every other variable to 0) and by application of cval  c s the final store s0 that results from the execution of command c in s. The stores can be printed with the help of an auxiliary function sstr(s:store): string that maps stores to strings (see the source file for its definition):

```
let s = update (init 0) "n" 5 ;;
Printf.printf "
let s0 = cval c s ;;
Printf.printf "
```

The corresponding output is as follows:

```
n=5 (default 0)
n=5 a=25 i=5 (default 0)
```

This result shows that the program has indeed terminated with the expected output store holding in variable a the value $a = \sum_{i=0}^{5}(2 \cdot i + 1) = 5^2 = 25$.

**Operational Semantics in the K Framework**

Our goal is to implement the (core of the) small-step operational semantics introduced in Fig. 7.10. In the K Framework, we define the imperative language by two 'modules' stored in file imp.k. The first module

```
module IMP-SYNTAX
  syntax Numeral ::= Int
  syntax Expression ::= ...
  syntax BoolExp ::= ...
  syntax Command ::= ...
endmodule
```

defines simultaneously the concrete and the abstract syntax of the language and annotates them with hints for the evaluation of the corresponding phrases by the interpreter. For instance, the definition

```
syntax Expression ::=
  Numeral
| Id
| Expression "*" Expression   [left, strict]
> Expression "+" Expression   [left, strict]
| "(" Expression ")"          [bracket]
```

introduces the domain of expressions denoting integer values. The annotation `left` indicates that the binary expressions are to be evaluated from left to right. The annotation `strict` indicates that the result value of the whole expression can be only determined after the subexpressions have been evaluated. The annotation `bracket` indicates that the abstract syntax tree need not contain a separate node for this expression, because the meaning of the whole expression is identical to the meaning of its subexpression. The separator > of grammatical alternatives indicates that the preceding alternatives have higher precedence than the later ones, thus e.g. 2+3*4 is parsed as 2+(3*4), not as (2+3)*4.

Similarly the domain of expressions denoting truth values is defined:

```
syntax BoolExp ::=
  Bool
| Expression "=" Expression   [left, strict]
| "~" BoolExp                 [strict]
> BoolExp "/\\" BoolExp        [left, strict(1)]
| "(" BoolExp ")"             [bracket]
```

Here the annotation `strict(1)` indicates that the value of the whole expression can only be determined after the value of the first subexpression has been determined (but the second subexpression need not be necessarily evaluated).

Finally, we encounter the core domain of the language:

```
syntax Command ::=
  Id ":=" Expression ";"                                [strict(2)]
| "var" Id ";" "{" Command "}"
| "if" BoolExp "then" "{" Command "}" "else" "{" Command "}"
                                                        [strict(1)]
| "if1" BoolExp "then" "{" Command "}"       [strict(1)]
| "while" BoolExp "do" "{" Command "}"
| "while" BoolExp BoolExp "do" "{" Command "}" [strict(1)]
```

```
   | "remove" Id ";"
   > Command Command                                          [left]
endmodule
```

This domain basically introduces the commands of our language (the keyword
if1 for the one-sided conditional is chosen to avoid problems with parsing
ambiguities). However, two additional commands are introduced to represent
intermediate execution states (the domain thus corresponds to the domain
*Instruction* of the small step operational semantics introduced in Fig. 7.10);
the meaning of the additional instructions (a while loop with two Boolean
expressions and a remove instruction) will be explained below.

The second module gives the semantics of the language:

```
module IMP
  imports IMP-SYNTAX
  syntax KResult ::= Int | Bool

  configuration <top> ... </top>
  rule ...
endmodule
```

Having imported the syntactic definitions, it defines by the special type
KResult the domains of those values that shall be reached by the evaluation
of phrases marked as strict in the syntax; thus phrases of type Expression
and BoolExp will be evaluated to values of type Int and Bool, respectively.

The behavior of the imperative language is defined by a small-step operational semantics where the shape of a configuration is given as follows:

```
configuration
  <top>
    <k> $PGM:K </k>
    <state> .Map </state>
  </top>
```

A configuration consists of nested labeled 'cells'. Inside the cell labeled as
top, there is a cell with special name k and another one with label state.
The initial content of the state cell is simply the empty map (a partial
function from keys to values) indicated by the constant .Map; it holds the
values of the program variables. The k cell is special in that its content is
a computation $T_1 \curvearrowright T_2 \curvearrowright \ldots \curvearrowright T_n$, i.e., a finite sequence of 'computational
tasks' $T_i$ of the special type K; initially this sequence consists of only element,
the parsed program denoted by the special variable $PGM. The rewrite rules
given below process one task after each other, removing every processed task
from the head of the list and potentially replacing it by new ones; in the

course of this processing, also the other cells of the configuration may be modified. When the list becomes empty, the computation terminates with the final configuration as its result.

The execution of computations is defined by 'rewrite rules' that describe how a configuration is modified from one step to the next. Here a rule needs only to describe those parts of the configuration that are modified; if a part is not explicitly modified, it remains unchanged. For instance, the evaluation of integer expressions is defined by the following rules:

```
rule <k>I:Id => N:Int ...</k> <state>... I |-> N ...</state>
rule N1:Int * N2:Int => N1 *Int N2
rule N1:Int + N2:Int => N1 +Int N2
```

The first rule refers both to the k cell and to the state cell; it indicates, that, if the first element of the computation is an identifier I and the state cell maps variable I to integer N, then the first element of the computation can be rewritten to N (leaving the rest of the computation and the state cell unchanged). The second rule does not involve the state cell; it states that, if a configuration contains a phrase N1*N2 with integers N1 and N2 (such a phrase can only be part of a computation), it can be rewritten to the product of the two values; similar the third rule defines that N1+N2 can be rewritten to the sum of the values.

Similarly, the evaluation of Boolean expression is defined as follows:

```
rule N1:Int = N2:Int => true requires N1 ==Int N2
rule N1:Int = N2:Int => false requires N1 =/=Int N2
rule ~ T:Bool => notBool T
rule true /\ T:Bool => T
rule false /\ _ => false
```

Here the clauses marked with the keyword requires indicate semantic side conditions that have to be satisfied to make the corresponding rule applicable.

As for commands, the semantics of the assignment command can be expressed as follows:

```
rule <k> Id := N:Int; => . ...</k> <state>S:Map => S[Id<-N]</state>
```

This rule states that if a computation starts with an assignment of an integer N to variable Id, this assignment can be removed (a rule ... => . indicates 'rewriting to nothing') and the store S can be rewritten to the store S[Id<-N] that is identical to S but maps Id to N.

Local variables are handled by the following rules:

```
rule <k>var Id; { C } =>  C Id := N; ...</k>
```

```
           <state>... Id |-> (N => 0) ...</state>
  rule <k>var Id; { C } =>  C remove Id; ...</k>
           <state>S:Map => S[Id<-0]</state> requires notBool(Id in keys(S))
  rule <k>remove Id; => . ...</k> <state>S:Map => S[Id<-undef]</state>
```

The first rule states that if a computation starts with command var Id; {C}
and the state holds a mapping Id |-> N, then the command can be rewritten
to the sequence C Id := N; that first executes C and then executes Id :=
N; to restore the original value of N (here we use for simplicity an ordinary
assignment statement rather than a special set instruction of the operational
semantics). Furthermore, the state is rewritten to map Id to 0 (different from
the original operational semantics, we thus fix here the initialization value).

However, if the state has no value for Id yet, it is not the first, but
the second rule is applicable: this rule states that, if Id does not occur
in the variables of state S, then the command may be rewritten to the
sequence C remove Id; that first executes C and then executes remove Id;
to remove the binding of Id from the state. Furthermore, S may be rewritten
to S[Id<-0] that is identical to S apart from mapping Id to 0. The third
rule specifies the effect of the instruction remove Id; by rewriting state S to
S[Id<-undef] that is identical to S apart from not mapping Id to any value.

A command sequence is handled by the following rule:

```
  rule C1:Command C2:Command => C1 ˜> C2
```

This rule simply replaces the initial element of the computation (a sequence
of two commands) by the two subcommands.

Two-sided and one-sided conditionals are handled as follows:

```
  rule if true then { C1 } else { C2 } => C1
  rule if false then { C1 } else { C2 } => C2
  rule if true then { C } => C
  rule if false then { C } => .
```

These rules very closely resemble the corresponding rules of the operational
semantics.

Loops might be simply handled by the following rule which unrolls the
first iteration of the loop:

```
  rule while B do { C } => if1 B then { C while B do { C } }
```

However, there also exists a more direct representation of the execution in
the following form:

```
  rule while B do { C } => while B B do { C }
```

```
rule while true B do { C } => C while B do { C }
rule while false B do { C } => .
```

Here the first rule doubles the loop condition such that the other two rules may become applicable; by annotating the position of the first condition as strict, it is evaluated to a truth value, while the second position preserves the original expression. Then one of the two rules is applicable which leads either to the termination of the loop or to the execution of the loop body followed by the loop again.

After compiling the semantics definition with kompile imp.k, we write the following program and store it in file program.imp:

```
n := 5;
a := 0;
i := 0;
while ˜(i = n) do
{
  var b;
  {
    b:=2*i+1;
    a:=a+b;
  }
  i:=i+1;
}
```

The execution of krun prog.imp then parses the program and executes its semantics. The output

```
<top> <k> . </k> <state> n |-> 5 i |-> 5 a |-> 25 </state> </top>
```

indicates the expected final configuration of the program with an empty computation and a state depicting the final values of the program variables.

# Computer Programs

**8**

> *Der liebe Gott gibt uns die Logik des Algorithmus. Der Teufel*
> *fügt den Programmierer hinzu und schon ist das irdische*
> *Gleichgewicht wieder hergestellt. (God gives us the logic of the*
> *algorithm. The devil adds the programmer and already is the*
> *earthly balance restored.)*
> — *Heinz Zemanek*

The semantic formalisms presented in Chap. 7 define the meaning of programming languages and thus implicitly also of computer programs written in these languages. However, they are too low-level to be of practical use when our actual goal is to verify the correctness of a program with respect to its specification. In this chapter, we are going to discuss various (closely related) calculi that allow us to reason about programs on this high level. These calculi can be subsumed under the term *axiomatic semantics*, because they are based on axioms and inference rules that describe how we can derive expected program properties. The soundness of these calculi can be shown from the underlying denotational semantics of the programming language.

Our presentation starts in Sect. 8.1 with a discussion of *problem specifications* that serve as *program contracts*; we subsequently develop programs that satisfy these contracts, i.e., solve the specified problems. Then Sect. 8.2 introduces the *Hoare Calculus*, the father of axiomatic semantics; this calculus can be viewed as an abstract procedure for the generation of *verification conditions*, i.e., logical formulas whose validity implies the correctness of a program. Section 8.3 further elaborates the theory of verification condition generation to the calculi of *weakest preconditions*, *strongest postconditions*, and *commands as relations*, which are more suitable for implementation in automated verification systems. While the previous presentation have only considered a program's *partial correctness* (the program does not produce a wrong result), Sect. 8.4 deals with its *total correctness* (the program indeed produces a result, i.e., it neither aborts prematurely nor does it fail to terminate).

© The Author(s), under exclusive license to Springer Nature Switzerland AG 2021
W. Schreiner, *Thinking Programs*, Texts & Monographs in Symbolic Computation,
https://doi.org/10.1007/978-3-030-80507-4_8

So far the chapter has focused mainly on the "mechanics" of program verification; in Sect. 8.5 we also discuss its "pragmatics". In particular, we investigate by a series of examples strategies for the elaboration of "meta-information" in the form of *loop invariants* and *termination measure* which the human verifier has to provide to make a verification succeed. Also the discussion so far has centered entirely on commands, i.e., on "verification on the small". As a preparation for "verification in the large", Sect. 8.6 addresses the *refinement* of commands. Section 8.7 builds upon the insights gained there to finally discuss the contract-based *modular verification* of programs composed of procedures.

To keep the presentation concise, we mostly omit formal proofs of the variously stated claims; the major exception is Sect. 8.2 where we prove in detail the soundness of the Hoare calculus with respect to the partial correctness of programs.

## 8.1    Specifying Problems

The goal of *software engineering* is to provide a systematic step-wise approach to programming such that correspondingly developed programs solve certain problems for their users; naturally such an approach has to start with a description of the problem to be solved. The purpose of *formal methods* is to aid software engineering by a mathematical framework that enables the rigorous reasoning about the correctness of each development step; consequently such a method has to start with a formalized description of the problem, i.e., a (formal) *problem specification*.

### Specifications as Contracts

In this chapter, we will focus our attention on programs that solve specific problems called *computational problems*. Such a program has only two points of interaction with its external environment: when it is started, the program receives from its environment some data as "inputs"; the program then runs without any influence from or effect to its environment; only when the program terminates, it delivers to the environment some data as "outputs" (in the next chapter we will address "reactive systems" that may interact with their environment also during their execution). However, the program may also prematurely abort with an error or run forever; in both cases it does not deliver any output. Our requirement to the program is that it always terminates normally and that its outputs satisfy some constraint with respect to the inputs it was given; a precise formulation of this constraint, the *output condition* or *postcondition* ("post" is Latin for "after", i.e., "condition after the program execution"), is the core element of every problem specification.

However, not all possible data are legitimate inputs for the problem to be solved; thus we specify by another constraint, the *input condition* or *precondition* ("pre" is Latin for "before", i.e., "condition before the program execution") the assumptions that the program may make about the inputs that it is given. The problem correctly solves the problem if, for all inputs that satisfies the precondition, the program normally terminates and returns some outputs that satisfy the postcondition; if the program is erroneously invoked on inputs that violate the precondition, the program has no obligation: it may produce wrong outputs, abort with an error, or run forever. Together this pre- and postcondition form the two sides of a *contract* between the user of the program and its developer: the user has to ensure that the inputs given to the program satisfy the precondition, the developer has to ensure that the program terminates normally with outputs that satisfy the postcondition. At the core of all formal methods we have specification of computational problems by such contracts consisting of pre- and postconditions.

**Formal Problem Specifications**

To formalize the ideas sketched above, we will in the following assume a fixed signature $\Sigma$ and a fixed $\Sigma$-algebra $A$ (see Definition 6.8). Furthermore we fix a sequence $IS = [I_1, \ldots, I_n] \in \Sigma.s^*$ of "input sorts" and a sequence $OS = [O_1, \ldots, I_m] \in \Sigma.s^*$ of "output sorts". Correspondingly we call values of the domains $A(I_1), \ldots, A(I_n)$ "inputs" and values of the domains $A(O_1), \ldots, A(O_m)$ "outputs".

---

**Definition 8.1** (*Problem Specification*) A *(problem) specification* $S = \langle I, O \rangle$ consists of a relation

$$I \subseteq A(I_1) \times \ldots \times A(I_n)$$

on the inputs, the *input condition* or *precondition*, and of a relation

$$O \subseteq A(I_1) \times \ldots \times A(I_n) \times A(O_1) \times \ldots \times A(O_m)$$

between inputs and outputs, the *output condition* or *postcondition*.

---

Typically we write a specification in a concrete syntax such as the following:

**Input:** $x_1 \colon I_1, \ldots, x_n \colon I_n$ **with**

$I_x$

**Output:** $y_1 \colon O_1, \ldots, y_m \colon O_m$ **with**

$O_{x,y}$

Here $I_x$ is a formula such that $(\forall x_1 \colon I_1, \ldots, x_n \colon I_n.\ I_x)$ is a $\Sigma$-formula (see Definition 6.4). Thus $I_x$ is well-typed with respect to the given variable declarations and has only free variables $x_1, \ldots, x_n$; the set of all tuples of values that assigned to these variables satisfy $I_x$ (i.e., give the formula the denotation "true") represents the input condition. Likewise, $O_{x,y}$ is a formula such that $(\forall x_1 \colon I_1, \ldots, x_n \colon I_n. \exists y_1 \colon O_1, \ldots, y_m \colon O_m.\ O_{x,y})$ is a $\Sigma$-formula, i.e., $O_{x,y}$ has only free variables $x_1, \ldots, x_n, y_1, \ldots, y_m$ and is well-typed with respect to the given variable declarations; the set of all tuples of values that assigned to these variables satisfy $O_{x,y}$ represents the output condition.

**Example 8.1** Consider the following specification of the problem of "truncated division":

> **Input:** x:Nat, y:Nat **with**
> $y \neq 0$
> **Output:** q:Nat, r:Nat **with**
> $x = y \cdot q + r \wedge r < y$

Here we assume the interpretation of sort Nat as the set $\mathbb{N}$ of the natural numbers with the usual interpretation of the various arithmetic operations. The precondition allows inputs $x \in \mathbb{N}$ and $y \in \mathbb{N}$ with non-zero $y$; the postcondition demands outputs $q \in \mathbb{N}$ and $r \in \mathbb{N}$ that represent the quotient respectively the remainder of $x$ divided by $y$.

For instance, inputs $x = 15$ and $y = 0$ do not satisfy the precondition $y \neq 0$ and are thus not allowed. However, inputs $x = 15$ and $y = 6$ are allowed; for these, the postcondition allows outputs $q = 2$ and $r = 3$ (because $15 = 6 \cdot 2 + 3 \wedge 3 < 6$); no other outputs satisfy the postcondition.                                                     □

**Example 8.2** Consider the following specification of the problem of "linear search":

> **Input:** a:Array, x:Elem, n:Int
> **Output:** i:Int **with**
> if  $\neg \exists j:Int. 0 \leq j \wedge j < n \wedge get(a,j)=x$ then
>     i = -1
> else
>     $0 \leq i \wedge i < n \wedge get(a,i) = x \wedge$
>     $\forall j:Int. 0 \leq j \wedge j < n \wedge get(a,j)= x \Rightarrow i \leq j$

Here we assume the interpretation of sort Array as the domain $E^{\omega}$ of all unbounded sequences ("arrays") of elements from some domain $E$, sort Elem as that domain $E$, and sort Int as the set $\mathbb{Z}$ of the integer numbers. We interpret get(a,i) as $a(i)$ which returns the element at index $i$ in array $a$.

Since there is no precondition specified, the default true is assumed, i.e., all inputs $a \in E^{\omega}, x \in E$, and $n \in \mathbb{Z}$ are allowed. The postcondition demands some output $i \in \mathbb{Z}$ with the following properties:

- If there is no index greater equal zero and less than $n$ at which $x$ occurs in $a$, then the output $i$ must be $-1$.
- Otherwise, $i$ must be such an index, i.e., it must be greater equal zero and less than $n$ and $a$ must hold at $i$ value $x$; moreover, $i$ must be the smallest index with this property.

For instance, for inputs $a = [2, 3, 5, 7, 5, 7, \ldots]$, $n = 3$, $x = 7$, the output condition allows as the only output $i = -1$, because $a(0) = 2 \neq 7$, $a(1) = 3 \neq 7$ and $a(2) = 5 \neq 7$. However, for inputs $a = [2, 3, 5, 7, 5, 7, \ldots]$, $n = 6$, $x = 7$, allows as the only output $i = 3$, because $a(0) = 2 \neq 7$, $a(1) = 3 \neq 7$, $a(2) = 5 \neq 7$, and $a(3) = 7$.                                                     □

**Validating Specifications**

When writing a specification, it is not guaranteed that our formulation indeed matches our intention; actually, it is very easy to make errors in the formulations of preconditions and, in particular, postconditions. To validate a specification, we should investigate in a first step, whether it allows/prohibits certain example inputs and allows/prohibits certain example outputs. For a more general investigation, we may then investigate whether a specification with precondition formula $I_x$ and postcondition formula $O_{x,y}$ does or does not enjoy the following properties.

**Satisfiability of Precondition** There exist values for the input variables that make $I_x$ true, i.e., the following formula is valid:

$$\exists x_1 : I_1, \ldots, x_n : I_n.\ I_x$$

If the precondition of a specification is not satisfiable, the problem does not allow any inputs (which certainly indicates an error in the specification).

**Non-Triviality of Precondition** There exist values for the input variables that make $I_x$ not true, i.e., the following formula is valid:

$$\exists x_1 : I_1, \ldots, x_n : I_n.\ \neg I_x$$

If the precondition of a specification is trivial, the problem allows all inputs (which should only be the case if $I_x$ is indeed syntactically the formula true; otherwise, this very likely indicates an error in the specification).

**Satisfiability of Postcondition** For all input values allowed by the precondition, there exist values for the output variables that make $O_{x,y}$ true, i.e., the following formula is valid:

$$\forall x_1 : I_1, \ldots, x_n : I_n.\ (I_x \Rightarrow \exists y_1 : O_1, \ldots, y_m : O_m.\ O_{x,y})$$

If the postcondition of a specification is not satisfiable, the problem does for some inputs not allow any outputs (which certainly indicates an error in the specification).

**Non-Triviality of Postcondition** There are some input values allowed by the precondition such that some values for the output variables make $O_{x,y}$ not true, i.e., the following formula is valid:

$$\exists x_1 : I_1, \ldots, x_n : I_n, y_1 : O_1, \ldots, y_m : O_m.\ (I_x \wedge \neg O_{x,y})$$

If the postcondition of a specification is trivial, the problem allows for all inputs all outputs (which certainly indicates an error in the specification).

**Uniqueness of Outputs**  For all inputs allowed by the precondition, there is not more than one tuple of outputs that satisfies the postcondition, i.e., the following formula is valid:

$$\forall x_1: I_1, \ldots, x_n: I_n, y_1: O_1, \ldots, y_m: O_m, z_1: O_1, \ldots, z_m: O_m.$$
$$(I_x \wedge O_{x,y} \wedge O_{x,z} \Rightarrow y_1 = z_1 \wedge \ldots \wedge y_m = z_m)$$

Here $O_{x,z}$ stands for the formula that is identical to $O_{x,y}$ except that every free occurrence of every variable $y_i$ (for $i = 1, \ldots, m$) has been replaced by variable $z_i$.

If the outputs are not unique, the same tuple of inputs allows more than one tuple of outputs (which, depending on the intentions of the problem specification, may or may not indicate an error).

**Example 8.3**  Consider the following specification of the problem of "searching for the position of the maximum":

**Input:** a:Array, n:Int **with**
  n > 0
**Output:** i:Int **with**
  $0 \leq i \wedge i < n \wedge$ get(a,i)=x $\wedge$
  $\forall$j:Int. $0 \leq j \wedge j < n \wedge$ get(a,j)$\leq$ get(a,i)

Here we interpret sort **Array** as the sort of all arrays of integers and everything else as in Example 8.2.

Thus the problem is to find among the first $n > 0$ elements of array $a$ a position $i$ at which $a$ holds the maximum of the first $n$ array elements. In this problem the input condition is satisfiable $(1 > 0)$ but not trivial $(0 \not> 0)$. For every array $a$ and $n > 0$, there exists a maximum, thus the postcondition is satisfiable; generally, not every element is such a maximum, thus the postcondition is not trivial.

However, outputs are not unique: for array $a = [2, 3, 1, 3, \ldots]$ and $n = 4$ both $i = 1$ and $i = 3$ are allowed, because $a(1) = a(3) = 3$, $a(0) = 2 \leq 3$ and $a(2) = 1 \leq 3$.                                                                 □

### Implementing Specifications

The meaning of a specification $S$ is determined by its relationship to some program $P$, which may or may not correctly implement a solution to the problem specified by $S$. In Sect. 7.2 we have modeled $P$ by

- a relation $[\![ P ]\!] \langle vs, vs' \rangle$ between the program's inputs $vs$ and its outputs $vs'$,
- conditions $[\![ P ]\!] \checkmark \langle vs \rangle$ and $\langle\!\langle P \rangle\!\rangle \checkmark \langle vs \rangle$ on the inputs that ensure that $P$ terminates normally with some outputs, i.e., $P$ does neither abort nor run forever.

Based on these notions, we are able to formally describe when a program satisfies its specification.

**Definition 8.2** (*Partial and Total Correctness*)  Let $P = $ program $I(X)\ C$ be a program for which we can derive $\Sigma \vdash P$: program and $\Sigma \vdash X$: parameters$(Vt, Ss)$ for some variable typing $Vt$ and some sort sequence $Ss = [S_1, \ldots, S_n]$ (see Fig. 7.1). Furthermore, let $S = \langle I, O \rangle$ be a specification with input sorts $IS = Ss$ and output sorts $OS = Ss$, i.e., the sorts of the program parameters correspond to the input sorts and to the output sorts of the specification.

Then $P$ is *partially correct* with respect to $S$ if and only if, for all possible input values that satisfy the precondition $I$ of $S$, all possible outputs of $P$ satisfy the postcondition $O$ of $S$:

$$\forall x_1 \in A(I_1), \ldots, x_n \in A(I_n), y_1 \in A(O_1), \ldots, y_n \in A(O_n).$$
$$I\langle x_1, \ldots, x_n \rangle \wedge [\![\, C \,]\!] \langle \langle x_1, \ldots, x_n \rangle, \langle y_1, \ldots, y_n \rangle \rangle \Rightarrow$$
$$O\langle x_1, \ldots, x_n, y_1, \ldots, y_n \rangle$$

$P$ *terminates normally* with respect to $S$ if and only if all inputs that satisfy the precondition of $S$ let $P$ terminate without an error:

$$\forall x_1 \in A(I_1), \ldots, x_n \in A(I_n).\ I\langle x_1, \ldots, x_n \rangle \Rightarrow$$
$$[\![\, P \,]\!]\checkmark \langle x_1, \ldots, x_n \rangle \wedge \langle\!\langle\, P \,\rangle\!\rangle\checkmark \langle x_1, \ldots, x_n \rangle$$

$P$ *is totally correct* with respect to $S$ if it is partially correct and also terminates normally with respect to $S$.

Therefore an implementation is already partially correct if for all allowed inputs it does not produce any incorrect outputs. However, an implementation is only totally correct, if it also produces for all allowed inputs some (correct) outputs.

Since our program model does not differentiate between input and output parameters, above formalization requires a specification with identical lists of input and output parameters. However, this is not a restriction as shown in the example below.

**Example 8.4**  Consider the following specification of truncated division, a version of the specification given in Example 8.1 where now both parameter lists encompass the inputs and the outputs:

**Input:** x:Nat, y:Nat, q:Nat, r:Nat **with**
   $y \neq 0$
**Output:** x0:Nat, y0:Nat, q0:Nat, r0:Nat **with**
   $x = y \cdot q0 + r0 \wedge r0 < y$

Here the input parameters model the values of the program variables before the execution while the output parameters model the values afterwards. This specification is implemented by the following program:

```
program div(x:Nat,y:Nat,q:Nat,r:Nat) {
  q := 0; r := x;
  while r >= y do {
    q := q+1
    r := r-y
  }
}
```

Since in the specification (and therefore also in the implementation) the initial values of the output variables and the final values of the input variables do not matter, above specification could be also more succinctly written as in Example 8.1 with the implicit understanding that it is to be interpreted as shown above.                              □

The core problem is now to devise a way of reasoning that verifies that a program correctly implements a specification; this is the problem that we are going to address in the subsequent sections.

## 8.2    Verifying Programs

As a starting point for program verification, we will in this and in the subsequent sections investigate various (closely related) formalisms to verify the correctness of the commands that make up a program.

**The Hoare Calculus**

We begin with a formalism for program reasoning that was invented in 1969 by the British computer scientists C.A.R. Hoare; this formalism was later called "Hoare Calculus" or "axiomatic program semantics" (in contrast to denotational semantics or operational semantics, axiomatic semantics describes how we can reason about the externally observable effect of a command without describing its internal mechanism). This calculus is the most popular one, when it comes to manually proving the correctness of programs (automated or semi-automated program verifiers typically implement somewhat more advanced calculi; some of these will be discussed later).

The core judgement of the Hoare calculus is the "Hoare triple"

$$\{P\}\, C\, \{Q\}$$

where $C$ is a command and $P$ and $Q$ are formulas. For the moment, our informal interpretation of this judgement is the following:

If the execution of command $C$ starts in a state in which the precondition $P$ holds, every normal termination of the command yields a state in which the postcondition $Q$ holds.

The pair $\langle P, Q \rangle$ thus represents a "specification" to be implemented by command $C$. Please note that this specification only claims "partial correctness" of the command execution; it does not rule out that the command aborts or runs forever (we will deal with these issues later).

The Hoare calculus is an inference system for deriving Hoare triples; it is sound in the sense that by its rules we can only derive Hoare triples that are true with respect to the partial correctness interpretation given above.

## Variable Substitutions

Before presenting the Hoare calculus, we introduce a technical concept that is applied in some rules of the calculus.

> **Proposition 8.1** (Variable Substitution) *Let E be a phrase (formula or term), V a variable and T a term. The $E[T/V]$ is defined as the phrase*
>
> $$E[T/V] := \text{let } V = T \text{ in } E$$
>
> *whose value is the value of E when V is interpreted as the value of T (see Definition 2.16).*

Alternatively (and more intuitively), $E[T/V]$ can be understood as phrase $E$ where every free occurrence of variable $V$ has been syntactically replaced by term $T$. For instance, for $F := x > z$, formula $F[x + 1/x]$ may be understood either as let $x = x + 1$ in $x > z$ or as $x + 1 > z$. While the alternative understanding of $E[T/V]$ by syntactic substitution seems simpler, it actually requires the appropriate renaming of all quantified variables in $E$ such that they do not coincide with any free variables in $T$. Since this makes the general formalization more complicated, we prefer the simple definition given above (but will use the easier to understand interpretation by syntactic substitution, if $E$ does not contain any quantifiers).

## The Generation of Verification Conditions

The Hoare calculus is depicted in Fig. 8.1. Before we dive into the details of the individual inference rules of the calculus, we demonstrate the big picture of their application. For instance, we may use these rules to derive the Hoare triple

$$\{\text{true}\} \text{ m:=x; if m<y then m:=y} \{Q\}$$

for postcondition $Q :\Leftrightarrow (m = x \vee m = y) \wedge (m \leq x \wedge m \leq y)$. This triple expresses the correctness of a command that computes the maximum $m$ of two numbers $x$ and $y$; it is derived by the following inference tree:

$$
\frac{
\models \boxed{\text{true} \Rightarrow x = x} \quad \{x = x\} \text{ m:=x} \{m = x\} \qquad T \quad \models \boxed{m = x \wedge \neg(m < y) \Rightarrow Q}
}{
\dfrac{\{\text{true}\} \text{ m:=x} \{m = x\} \qquad\qquad \{m = x\} \text{ if m<y then m:=y} \{Q\}}{\{\text{true}\} \text{ m:=x; if m<y then m:=y} \{Q\}}
}
$$

**Judgement**

$\{\sqcup\} \sqcup \{\sqcup\} \subseteq Formula \times Command \times Formula$

**Rules for $\{P\}\, C\, \{Q\}$:**

$$\frac{\models P \Rightarrow P' \quad \{P'\}\, C\, \{Q'\} \quad \models Q' \Rightarrow Q}{\{P\}\, C\, \{Q\}}$$

$$\{Q[T/V]\}\, V := T\, \{Q\}$$

$$\frac{\{P\}\, C\, \{Q[V_0/V]\}}{V, V_0, V_1 \text{ distinct} \quad V_0, V_1 \text{ not in } P, C, Q}{\{(\forall V_1\!:\! S.\ P[V_1/V])[V/V_0]\}\, \mathtt{var}\ V\!:\! S;\ C\, \{Q\}}$$

$$\frac{\{P\}\, C_1\, \{R\} \quad \{R\}\, C_2\, \{Q\}}{\{P\}\, C_1 ; C_2\, \{Q\}}$$

$$\frac{\{P \wedge F\}\, C_1\, \{Q\} \quad \{P \wedge \neg F\}\, C_2\, \{Q\}}{\{P\}\, \mathtt{if}\ F\ \mathtt{then}\ C_1\ \mathtt{else}\ C_2\, \{Q\}} \qquad \frac{\{P \wedge F\}\, C\, \{Q\} \quad \models P \wedge \neg F \Rightarrow Q}{\{P\}\, \mathtt{if}\ F\ \mathtt{then}\ C\, \{Q\}}$$

$$\frac{\models P \Rightarrow I \quad \{I \wedge F\}\, C\, \{I\} \quad \models I \wedge \neg F \Rightarrow Q}{\{P\}\, \mathtt{while}\ F\ \mathtt{do}\ C\, \{Q\}}$$

**Fig. 8.1** The Hoare calculus (Partial Correctness)

Here $T$ represents the following subtree:

$$\frac{\models \boxed{m = x \wedge m < y \Rightarrow Q[y/m]} \quad \{Q[y/m]\}\, \mathtt{m}:=\mathtt{y}\, \{Q\}}{\{m = x \wedge m < y\}\, \mathtt{m}:=\mathtt{y}\, \{Q\}}$$

The application of these rules is mostly (but not completely) mechanically guided by the structure of the command. There is one inference rule for every command; in above example, the rules for command sequences, assignments, and one-sided conditional were applied. The rules for the compound commands (command sequences and one-sided conditional) reduce the problem of deriving the Hoare triple for a compound command to the problem of deriving Hoare triples for its subcommands; this process terminates with the rule for the for assignment command, which is atomic, i.e., it has no subcommand. The assignment rule therefore has no premise, i.e., it is an axiom. The leaves of the tree are either Hoare triples that are instances of the assignment axiom or they are of the form $\models F$. The later represents the validity of formula $F$, which has to be established by logical reasoning; we call $F$ a *verification condition* (in above derivation the verification conditions are marked with frames). The Hoare calculus thus implicitly describes a "verification condition generator", a syntax-guided procedure that derives from a Hoare triple a set of verification conditions: these conditions are formulas in first order logic whose combined validity implies the correctness of the triple.

Figure 8.2 makes this procedure explicit by the definition of a function *vcg* that takes a Hoare triple and returns the set of verification conditions determined by the Hoare calculus; the verification condition generated for each kind of command is

$vcg: Formula \times Command \times Formula \rightarrow Set(Formula)$

$$vcg(\{P\}\ V := T\ \{Q\}) := \{P \Rightarrow Q[T/V]\}$$
$$vcg(\{P\}\ \text{var}\ V : S;\ C\ \{Q\}) := \text{let}\ P_1 = \text{choose}\ F.\ F \in Formula\ \text{in}$$
$$vcg(\{P_1\}\ C\ \{Q[V_0/V]\}) \cup$$
$$\{P \Rightarrow (\forall V_1 : S.\ P_1[V_1/V])[V/V_0]\}$$
$$vcg(\{P\}\ C_1;\ C_2\ \{Q\}) := \text{let}\ R = \text{choose}\ F.\ F \in Formula\ \text{in}$$
$$vcg(\{P\}\ C_1\ \{R\}) \cup vcg(\{R\}\ C_1\ \{Q\})$$
$$vcg(\{P\}\ \text{if}\ F\ \text{then}\ C_1\ \text{else}\ C_2\ \{Q\}) := vcg(\{P \wedge F\}\ C_1\ \{Q\}) \cup$$
$$vcg(\{P \wedge \neg F\}\ C_2\ \{Q\})$$
$$vcg(\{P\}\ \text{if}\ F\ \text{then}\ C\ \{Q\}) := vcg(\{P \wedge F\}\ C_1\ \{Q\}) \cup \{P \wedge \neg F \Rightarrow Q\}$$
$$vcg(\{P\}\ \text{while}\ F\ \text{do}\ C\ \{Q\}) := \text{let}\ I = \text{choose}\ F.\ F \in Formula\ \text{in}$$
$$\{(P \Rightarrow I), (I \wedge \neg F \Rightarrow Q)\} \cup vcg(\{I \wedge F\}\ C\ \{I\})$$

**Fig. 8.2** The Hoare calculus as a verification condition generator

directly derived from the corresponding inference rule. Since the inference rules for the assignment command and for the variable declaration command do not have arbitrary preconditions, the function *vcg* applies the knowledge of the first rule of the calculus to generate a verification condition that demands that the given precondition $P$ implies the precondition derived by the inference rule.

Considering the example elaborated above, the verification condition generator therefore generates the following result:

$$vcg(\{\text{true}\}\ \texttt{m:=x;}\ \texttt{if m<y then m:=y}\ \{Q\}) :=$$

$$\left\{ \boxed{\text{true} \Rightarrow x = x}, \boxed{m = x \wedge m < y \Rightarrow Q[y/m]}, \boxed{m = x \wedge \neg m < y \Rightarrow Q} \right\}$$

Thus, to verify the correctness of the given Hoare triple, it suffices to show the validity of the three generated verification conditions.

Program verification thus consists of two essentially independent parts:

- the syntax-guided derivation of verification conditions (e.g., by the Hoare calculus),
- the establishment of the validity of the verification conditions (e.g., by proving the formulas with the inference rules of first order logic).

Most rules of the Hoare calculus have a form that allows a mechanization of the first part, but this is not true for all rules: as the definition of the verification condition generator *vcg* makes very explicit, the three rules for a variable declaration, a command sequence, and a loop require choices of formulas that are not directly derived from the input; indeed, only for suitable choices, the derived verification conditions are valid and the verification can succeed. We will see later how other calculi replace most of these choices by systematic calculations, but nevertheless for the verification of a loop command always the choice of a *loop invariant I* is required. This choice is the main creative step that has to be performed by a clever reasoner (human or machine); we will discuss this creative aspect later. To subsequently establish the validity of the generated verification conditions is a separate process that can be also performed by a combination of human intelligence and sophisticated automation.

In a manual verification, the application of the Hoare rules to derive verification conditions is rarely given in the form of detailed inference trees but are typically indicated by a semi-formal text such as the following one:

We verify {true} m:=x; if m<y then m:=y{Q}. Clearly we have {true}m:=x{$m = x$}. Thus it suffices to show {$m = x$}if m<y then m:=y{Q}:

- We verify {$m = x \wedge m < y$} m:=y{Q}. For this we assume $m = x$ and $m < y$ and show $Q[y/m]$...
- We show $m = x \wedge \neg(m < y) \Rightarrow Q$, i.e., we assume $m = x$ and $\neg(m < y)$ and show $Q$...

Also we will use in the following such descriptions from which the reader has to deduce the corresponding inference trees.

### The Rules of the Hoare Calculus

After this first exposition of the Hoare calculus, we go into the details of its rules (which are depicted in Fig. 8.1) and illustrate them by small examples.

The first rule of the Hoare calculus is "generic" (command-independent).

- **Weakening/Strengthening:**

$$\frac{\models P \Rightarrow P'  \quad \{P'\}\, C\, \{Q'\} \quad \models Q' \Rightarrow Q}{\{P\}\, C\, \{Q\}}$$

This rule states that the problem of deriving a Hoare triple $\{P\}\, C\, \{Q\}$ can be reduced to the problem of deriving another Hoare triple $\{P'\}\, C\, \{Q'\}$ provided that
- the new precondition $P'$ is not stronger than the original precondition $P$ (i.e., $P'$ is implied by $P$), and
- the new postcondition $Q'$ is not weaker than the original postcondition $Q$ (i.e., $Q'$ implies $Q$).

In a nutshell, this rule states that, in the course of a correctness proof, we may resort to the following principle:

Preconditions may be weakened and postconditions may be strengthened.

**Example 8.5** According to the following rule instance

$$\frac{\models x > 0 \Rightarrow x \geq 0 \quad \{x \geq 0\}\, C\, \{y = 1 + x\} \quad \models y = 1 + x \Rightarrow y \neq x}{\{x > 0\}\, C\, \{y \neq x\}}$$

the problem of deriving the triple $H_1 := \{x > 0\}\, C\, \{y \neq x\}$ may be reduced to the problem of deriving $H_2 := \{x \geq 0\}\, C\, \{y = 1 + x\}$, because $x > 0$ implies $x \geq 0$ and $y = 1 + x$ implies $y \neq x$, Thus any command $C$ that satisfies the interpretation of $H_2$ also satisfies the interpretation of $H_1$. $\qquad\square$

To illustrate the soundness of the reasoning applied in above example, we may assume that $C$ satisfies the interpretation of $H_2$. To show that then $C$ also satisfies the interpretation of $H_1$, we have to show that if $C$ starts execution in a state in which $x > 0$ holds, it terminates in a state in which $y \neq x$ holds: if $C$ starts execution in a state in which $x > 0$ holds, this state also satisfies $x \geq 0$. Thus, since $C$ satisfies $H_2$, $C$ terminates normally in a state in which $y = 1 + x$ holds and therefore also in a state in which $y \neq x$ holds.

The rule does *not* hold in the other direction, e.g., we cannot reduce the problem of deriving $\{x \geq 0\}\, C\, \{y \neq x\}$ to the problem of deriving $\{x > 0\}\, C\, \{y \neq x\}$ (i.e., we must not strengthen the precondition). The second Hoare triple is, e.g., satisfied by the command `y:=1/x` which, however, does not satisfy the first triple: if started in a state $x \geq 0$, we might have $x = 0$ which would let the execution of $C$ abort.

Each other rule can be applied to one specific kind of command.

- **Command $\{V := T\}$:**

$$\{Q[T/V]\}\ \mathtt{V := T}\ \{Q\}$$

This rule states that in order to derive a postcondition $Q$ of the command, it suffices to derive the precondition $Q[T/V]$, which may be understood as the syntactic substitution of term $T$ for every free occurrence of variable $V$ in $Q$.

**Example 8.6** By above rule, we for instance derive the triple

$$\{x + y > z\}\ \mathtt{x := x + y}\ \{x > z\}$$

The precondition is a duplicate of the postcondition where variable $x$ has been replaced by term $x + y$. $\qquad\square$

The intuition for this rule is simple: if condition $Q$ holds in the poststate of the assignment, it holds then for the value of the assigned variable $V$ in that state. This value was determined in the prestate of the assignment by the evaluation of term $T$. Thus in the prestate of the assignment, $Q$ must hold, not for $V$, but for the value of $T$ in that state.

- **Command $\{\mathtt{var}\ V : S;\ C\}$:**

$$\frac{\{P\}\, C\, \{Q[V_0/V]\} \qquad V, V_0, V_1\ \text{distinct} \qquad V_0, V_1\ \text{not in}\ P, C, Q}{\{(\forall V_1 : S.\ P[V_1/V])[V/V_0]\}\ \mathtt{var}\ V : S;\ C\ \{Q\}}$$

This rule seems quite complex (it was not part of the original Hoare calculus which did not deal with declarations) but it can be simply understood as deriving, from a given postcondition $Q$, the necessary precondition of the command in two steps:

- In the first step, we replace in $Q$ the variable $V$ introduced in the declaration by a new variable $V_0$ that does not occur elsewhere; this variable represents the original value of $V$ before the execution of the command was started. Then we derive from the modified postcondition $Q[V_0/V]$ the precondition $P$ from which the execution of declaration body $C$ establishes this modified postcondition; $P$ may contain references to both $V_0$ (denoting the original value of $V$ before the declaration was entered) and $V$ (denoting the unknown initial value of the program variable after the declaration has been entered).
- In the second step, we construct from the derived $P$ the precondition of the declaration: first we replace every free occurrence of $V$ by a new variable $V_1$ that does not occur elsewhere; this variable correspondingly represents the unknown initial value of the program variable $V$ after the declaration has been entered. In the resulting formula $P[V_1/V]$ we universally quantify this variable, yielding $(\forall V\colon S.\ P[V_1/V])$ which indicates that the condition must hold for arbitrary initial values in the declaration. Finally, we replace every occurrence of $V_0$ by $V$ which in the prestate indeed refers to the original value of $V$ before the declaration command is executed; this yields the ultimate precondition $(\forall V\colon S.\ P[V_1/V])[V/V_0]$.

**Example 8.7**  We derive the following Hoare triple:

$$\{x = 3 \wedge y = 2\}\ \{\texttt{var y:nat;x:=x+y}\}\ \{x \geq 3 \wedge y = 2\}$$

By the weakening/strengthening rule, it suffices to derive a Hoare triple

$$\{P\}\ \{\texttt{var y:nat;x:=x+y}\}\ \{x \geq 3 \wedge y = 2\}$$

and then show $(x = 3 \wedge y = 2 \Rightarrow P)$. To determine a suitable formula $P$, we apply the declaration rule:

- In the first step, we replace in postcondition $(x \geq 3 \wedge y = 2)$ variable $y$ by a new variable $y_0$ which yields the postcondition $(x \geq 3 \wedge y_0 = 2)$ of the declaration body $\texttt{x:=x+y}$; from this we derive by the assignment rule the precondition $(x + y \geq 3 \wedge y_0 = 2)$ of the declaration body.
- In the second step, we replace in this precondition variable $y$ by a new variable $y_1$ which yields after universal quantification the precondition $(\forall y_1\colon \mathsf{nat}.\ (x + y_1 \geq 3 \wedge y_0 = 2))$. In this condition we replace $y_0$ by $y$ to ultimately derive precondition $P$ as

$$\forall y_1\colon \mathsf{nat}.\ (x + y_1 \geq 3 \wedge y = 2)$$

such that it remains to prove

$$x = 3 \land y = 2 \Rightarrow \forall y_1 \colon \mathsf{nat}. \ (x + y_1 \geq 3 \land y = 2)$$

This formula is equivalent to

$$x = 3 \land y = 2 \Rightarrow (\forall y_1 \colon \mathsf{nat}. \ x + y_1 \geq 3) \land y = 2$$

which is clearly true for the interpretation of the formula over the domain of natural numbers. $\qquad\square$

- **Command $\{C_1 ; C_2\}$:**

$$\frac{\{P\} \ C_1 \ \{R\} \quad \{R\} \ C_2 \ \{Q\}}{\{P\} \ C_1 ; C_2 \ \{Q\}}$$

This rule is quite intuitive: to show that from every state satisfying precondition $P$ the sequential execution of two commands $C_1$ and $C_2$ only yields states satisfying postcondition $Q$, we have to find some "intermediate" condition $R$ for which we can show that
- from every state satisfying $P$, executing $C_1$ only yields states satisfying $R$, and
- from every state satisfying $R$, executing $C_2$ only yields states satisfying $Q$.

**Example 8.8** We derive the following Hoare triple:

$$\{x = \mathsf{oldx} \land y = \mathsf{oldy}\} \ \mathtt{z:=x;x:=y;y:=z} \ \{x = \mathsf{oldy} \land y = \mathsf{oldx}\}$$

This triple specifies the correctness of a command for swapping the values of two variables $x$ and $y$. The specification is expressed with the help of two constants $\mathsf{oldx}$ and $\mathsf{oldy}$ that denote the values of $x$ and $y$ before the execution of the command; this demonstrates a general technique by which in the Hoare calculus the postcondition of a command (which has only access to the values of program variables in the poststate) can also refer to the values of these variables in the prestate.

From the rule for command sequences, it suffices to find a condition $R_1$ such that we can derive the following two Hoare triples:

$$\{x = \mathsf{oldx} \land y = \mathsf{oldy}\} \ \mathtt{z:=x;x:=y} \ \{R_1\} \quad \{R_1\} \ \mathtt{y:=z} \ \{x = \mathsf{oldy} \land y = \mathsf{oldx}\}$$

How to find a suitable $R_1$? The core idea is that the application of the assignment rule to the last assignment with its given postcondition gives us the triple

$$\{x = \mathsf{oldy} \land z = \mathsf{oldx}\} \ \mathtt{y:=z} \ \{x = \mathsf{oldy} \land y = \mathsf{oldx}\}$$

and thus the candidate $R_1 :\Leftrightarrow x = \mathsf{oldy} \wedge z = \mathsf{oldx}$. Therefore our problem is reduced to deriving the following Hoare triple:

$$\{x = \mathsf{oldx} \wedge y = \mathsf{oldy}\} \ \mathsf{z:=x;x:=y} \ \{x = \mathsf{oldy} \wedge z = \mathsf{oldx}\}$$

Now, according to the rule for command sequences, it suffices to find a condition $R_2$ such that we can derive the following two Hoare triples:

$$\{x = \mathsf{oldx} \wedge y = \mathsf{oldy}\} \ \mathsf{z:=x} \ \{R_2\} \quad \{R_2\} \ \mathsf{x:=y} \ \{x = \mathsf{oldy} \wedge z = \mathsf{oldx}\}$$

Another application of the assignment rule to the last assignment gives us

$$\{y = \mathsf{oldy} \wedge z = \mathsf{oldx}\} \ \mathsf{x:=y} \ \{x = \mathsf{oldy} \wedge z = \mathsf{oldx}\}$$

with the candidate $R_2 :\Leftrightarrow y = \mathsf{oldy} \wedge z = \mathsf{oldx}$. Therefore our problem is again reduced, now to deriving the triple

$$\{x = \mathsf{oldx} \wedge y = \mathsf{oldy}\} \ \mathsf{z:=x} \ \{y = \mathsf{oldy} \wedge z = \mathsf{oldx}\}$$

Another application of the assignment rule gives us the following triple:

$$\{y = \mathsf{oldy} \wedge x = \mathsf{oldx}\} \ \mathsf{z:=x} \ \{y = \mathsf{oldy} \wedge z = \mathsf{oldx}\}$$

Finally, by application of the weakening/strengthening rule, it suffices to prove

$$x = \mathsf{oldx} \wedge y = \mathsf{oldy} \Rightarrow y = \mathsf{oldy} \wedge x = \mathsf{oldx}$$

which is clearly true.                                                                        □

Above example demonstrates an important technique for dealing with sequences of assignments: we "propagate" by repeated application of the assignment rule the initially given postcondition "backwards" through the sequence and thus yield a suitable precondition for the sequence. Finally, by application of the weakening/strengthening rule, it suffices to show that the initially given precondition implies the one derived from this "backwards-propagation".

Unfortunately, in the Hoare calculus this propagation technique is mainly applicable only to assignment commands (actually, also to declaration commands), because only the rule for this kind of commands tell us explicitly how to "compute" from a given post-condition of the command a sufficient precondition (we will later encounter another calculus where this kind of computation is also applicable to other kinds of commands). For the moment, if the last command of a sequence is not one of this special kind, some "ingenuity" is required from the verifier to determine a suitable condition that holds before that command.

- **Commands** $\{\texttt{if } F \texttt{ then } C_1 \texttt{ else } C_2\}$ **and** $\{\texttt{if } F \texttt{ then } C\}$:

$$\frac{\{P \wedge F\}\, C_1\, \{Q\} \quad \{P \wedge \neg F\}\, C_2\, \{Q\}}{\{P\}\, \texttt{if } F \texttt{ then } C_1 \texttt{ else } C_2\, \{Q\}} \qquad \frac{\{P \wedge F\}\, C\, \{Q\} \quad \models P \wedge \neg F \Rightarrow Q}{\{P\}\, \texttt{if } F \texttt{ then } C\, \{Q\}}$$

The rules for the conditional commands are not difficult to understand: each conditional gives rise to two execution paths: the first one starts in a state in which the precondition $P$ and the command condition $F$ holds; the second one starts in a state in which also $P$ holds but $F$ does not hold, i.e., $\neg F$ holds. Both execution paths may only lead to states in which the postcondition $Q$ holds. In case of the one-sided condition statement, the second execution path is empty; here the conditions $P$ and $\neg F$ have to immediately imply $Q$.

**Example 8.9** To derive the Hoare triple

$$\{x \neq 0\}\, \texttt{if } x > 0 \texttt{ then } y := x \texttt{ else } y := -x\, \{y > 0\}$$

it suffices to derive the following two triples:

- $\{x \neq 0 \wedge x > 0\}\, y := x\, \{y > 0\}$: by applying the rule for assignments we can derive $\{x > 0\}\, y := x\, \{y > 0\}$; to derive the required triple, it suffices to apply the weakening/strengthening rule and to prove $(x \neq 0 \wedge x > 0 \Rightarrow x > 0)$, which clearly holds.
- $\{x \neq 0 \wedge \neg(x > 0)\}\, y := -x\, \{y > 0\}$ by applying the rule for assignments we can derive $\{-x > 0\}\, y := -x\, \{y > 0\}$; then, to derive the required triple, it suffices to apply the weakening/strengthening rule and to prove $(x \neq 0 \wedge \neg(x > 0) \Rightarrow -x > 0)$, which follows from basic arithmetic knowledge.     □

- **Command** $\{\texttt{while } F \texttt{ do } C\}$:

$$\frac{\models P \Rightarrow I \quad \{I \wedge F\}\, C\, \{I\} \quad \models I \wedge \neg F \Rightarrow Q}{\{P\}\, \texttt{while } F \texttt{ do } C\, \{Q\}}$$

This rule for the loop command is the main reason why program verification is complex and can be generally not fully automated. While this rule is the one that is applied in practice, it is actually a consequence of the following more basic rule:

$$\frac{\{P \wedge F\}\, C\, \{P\}}{\{P\}\, \texttt{while } F \texttt{ do } C\, \{P \wedge \neg F\}}$$

Combining this rule with the strengthening/weakening rule gives us a derivation

$$\frac{\models P \Rightarrow I \quad \dfrac{\{I \wedge F\}\, C\, \{I\}}{\{I\}\, \texttt{while } F \texttt{ do } C\, \{I \wedge \neg F\}} \quad \models I \wedge \neg F \Rightarrow Q}{\{P\}\, \texttt{while } F \texttt{ do } C\, \{Q\}}$$

which yields the more general rule.

The soundness of the basic rule itself is not difficult to understand, it can be verified by the principle of *induction*: if

- (induction base) condition $P$ holds in the initial execution state (i.e., after 0 iterations of the loop body $C$), and the assumption that
- (induction assumption) $P$ holds (after $n$ loop iterations) before the next iteration of $C$ (which only takes place if loop condition $F$ holds) lets us conclude that
- (induction step) $P$ holds also after this iteration (i.e., after $n + 1$ loop iterations),

then $P$ holds after arbitrary many iteration of the loop, in particular also in the state in which the loop terminates, i.e., in the state in which also $\neg F$ holds.

However, in practice the induction step cannot be shown for the given precondition $P$: the Hoare triple $\{P \wedge F\}\, C\, \{P\}$ can generally *not* be derived, because the induction assumption $P$ is too weak to perform the induction step (i.e., $P$ is not "inductive"). Therefore in practice the general rule is applied where $P$ and $Q$ represent the actual pre- respectively postcondition of interest, while $I$ represents a stronger *loop invariant*, i.e., the condition for which the Hoare triple $\{I \wedge F\}$ $C\,\{I\}$ *can* be derived. The problem is that it is not a priori clear how to determine a suitable loop invariant: finding a condition $I$ such that

- $I$ is implied by the precondition $P$,
- $I$ remains invariant under the execution of the loop body $C$, and
- $I$ implies (together with the negated loop condition $\neg F$) the postcondition $Q$,

is the main challenge of program verification with only limited success in automation: this is the place where usually human assistance is required (we will discuss in Sect. 8.5 techniques for the determination of suitable loop invariants).

**Example 8.10** We sketch the verification of the Hoare triple

$$\{i = 0 \wedge s = 0\}\, \texttt{while i<n do \{s:=s+i;  i:=i+1\}}\, \{s = \sum_{k=0}^{n-1} k\}$$

which represents the core of a program that computes the sum $s$ of all natural numbers from 0 up to some bound $n - 1$.

For application of the generalized invariant rule given above, we choose the invariant $I :\Leftrightarrow i \leq n \wedge s = \sum_{k=0}^{i-1} k$, such that it suffices to prove the two conditions

$$i = 0 \wedge s = 0 \Rightarrow I$$

$$I \wedge \neg(i < n) \Rightarrow s = \sum_{k=0}^{n-1} k$$

(whose truth is a consequence of simple arithmetic facts) and to derive the following Hoare triple:

$$\{I \wedge i < n\} \; \texttt{s:=s+i; \; i:=i+1} \; \{I\}$$

This triple can be derived by applying the assignment rule, the sequence rule, and the weakening/strengthening rule, and proving the following condition:

$$I \wedge i < n \Rightarrow I[i+1/i][s+i/s]$$

This condition can be expanded to the formula

$$i \le n \wedge s = \sum_{k=0}^{i-1} k \wedge i < n \Rightarrow i+1 \le n \wedge s+n = \sum_{k=0}^{i} k$$

whose truth can be established by simple arithmetic arguments.                      □

As already discussed, the Hoare calculus lends itself mostly to an automatic derivation of verification conditions except in the following aspects:

- In case of a variable declaration command, ingenuity is needed to determine a suitable condition before the execution of the body of the declaration.
- In case of a sequence of commands, if the last command is not an assignment, ingenuity is needed to determine from the postcondition of the last statement a suitable precondition for that statement.
- In case of a loop, ingenuity is required to determine a suitable loop invariant.

In Sect. 8.3 we will present other reasoning calculi (a bit more advanced than the Hoare calculus but closely related to it) that solve the first two problems and thus lend themselves more to mechanical program reasoning. In Sect. 8.5, we will discuss some heuristics that may help to address the last problem.

### Programs with Arrays

The example programs shown so far all dealt with atomic data such as plain numbers. However, the Hoare calculus can also easily deal with structured data such as arrays, provided that we make the simplifying assumption that a variable $a$ holds the complete content of an array and not just an address in a computer store (such that the same array might be referenced and updated via different variables). By this assumption, a variable $a$ of type "array of integers" holds a value from the domain $\mathbb{Z}^*$ (see Sects. 4.5 and 4.4) such that

- $|a|$ denotes the "length" of $a$,
- $a(i)$ denotes for $i \in \mathbb{N}_{|a|}$ the value of $a$ at index $i$,
- $a[i \mapsto e]$ denotes the array derived from $a$ by writing value $e$ at index $i$.

Thus the "array access" a[i] can be understood as the function application $a(i)$ and the "array assignment" a [i]:=e as the plain assignment a:=$a[i \mapsto e]$.

From the definition of the "update term" $a[i \mapsto e]$ we can derive the so-called "array axioms" due to John McCarthy (which are in our framework actually propositions that follow from the definition of finite sequences).

---

**Proposition 8.2** (Array Axioms) *Let $a \in E^*$ be an array with elements of some set E. Then we have for all indices $i \in \mathbb{N}_{|a|}$, $j \in \mathbb{N}_{|a|}$, and elements $e \in E$:*

$$|a[i \mapsto e]| = |a|$$
$$i = j \Rightarrow a[i \mapsto e](j) = e$$
$$i \neq j \Rightarrow a[i \mapsto e](j) = a(j)$$

*More generally, from the last two formulas we have*

$$F[a[i \mapsto e](j)] \Leftrightarrow (i = j \Rightarrow F[e]) \wedge (i \neq j \Rightarrow F[a(j)])$$
$$\Leftrightarrow \text{if } i = j \text{ then } F[e] \text{ else } F[a(j)]$$

*where $F[T]$ denotes a formula F with an occurrence of a term T.*

---

The array axioms play a fundamental role in the verification of programs that update arrays.

**Example 8.11** We derive the Hoare triple

$$\{P\} \text{ b:=a[i];a[i]:=a[j];a[j]:=b } \{Q\}$$

with precondition $P$ and postcondition $Q$ defined as follows:

$P :\Leftrightarrow i = \text{oldi} \wedge j = \text{oldj} \wedge a = \text{olda}$
$Q :\Leftrightarrow i = \text{oldi} \wedge j = \text{oldj} \wedge$
$\quad a(i) = \text{olda}(j) \wedge a(j) = \text{olda}(i) \wedge \forall k. \, k \neq i \wedge k \neq j \Rightarrow a(k) = \text{olda}(k)$

This triple expresses the correctness of a command for swapping in array $a$ the elements at indices $i$ and $j$; to simplify the following reasoning, we assume $i, j \in \mathbb{N}_{|a|}$.

For deriving this Hoare triple, we first apply the "back-propagation" technique for determining conditions $R_1$, $R_2$, $R_3$ with

$$\{R_3\} \text{ b:=a[i] } \{R_2\} \text{ a[i]:=a[j] } \{R_1\} \text{ a[j]:=b } \{Q\}$$

and then establish the validity of $P \Rightarrow R_3$. In the following, remember that an array assignment $a[i]:=e$ is interpreted as the plain assignment $\mathsf{a}:=a[i \mapsto e]$.

By applying the assignment rule to the last assignment we derive

$R_1 \Leftrightarrow Q[a[j \mapsto b]/a]$
$\quad \Leftrightarrow i = \mathsf{oldi} \wedge j = \mathsf{oldj} \wedge$
$\qquad a[j \mapsto b](i) = \mathsf{olda}(j) \wedge a[j \mapsto b](j) = \mathsf{olda}(i) \wedge$
$\qquad \forall k. \, k \neq i \wedge k \neq j \Rightarrow a[j \mapsto b](k) = \mathsf{olda}(k)$
$\quad \Leftrightarrow i = \mathsf{oldi} \wedge j = \mathsf{oldj} \wedge$
$\qquad (\text{if } j = i \text{ then } b = \mathsf{olda}(j) \text{ else } a(i) = \mathsf{olda}(j)) \wedge b = \mathsf{olda}(i) \wedge$
$\qquad \forall k. \, k \neq i \wedge k \neq j \Rightarrow a(k) = \mathsf{olda}(k)$

where the last equivalence is justified by the array axioms. Another application of the assignment rule gives us

$R_2 \Leftrightarrow R_1[a[i \mapsto a(j)]/a]$
$\quad \Leftrightarrow i = \mathsf{oldi} \wedge j = \mathsf{oldj} \wedge$
$\qquad (\text{if } j = i \text{ then } b = \mathsf{olda}(j) \text{ else } a[i \mapsto a(j)](i) = \mathsf{olda}(j)) \wedge b = \mathsf{olda}(i) \wedge$
$\qquad \forall k. \, k \neq i \wedge k \neq j \Rightarrow a[i \mapsto a(j)](k) = \mathsf{olda}(k)$
$\quad \Leftrightarrow i = \mathsf{oldi} \wedge j = \mathsf{oldj} \wedge$
$\qquad (\text{if } j = i \text{ then } b = \mathsf{olda}(j) \text{ else } a(j) = \mathsf{olda}(j)) \wedge b = \mathsf{olda}(i) \wedge$
$\qquad \forall k. \, k \neq i \wedge k \neq j \Rightarrow a(k) = \mathsf{olda}(k)$

A last application of the assignment rule gives us

$R_3 \Leftrightarrow R_2[a(i)/b]$
$\quad \Leftrightarrow i = \mathsf{oldi} \wedge j = \mathsf{oldj} \wedge$
$\qquad (\text{if } j = i \text{ then } a(i) = \mathsf{olda}(j) \text{ else } a(j) = \mathsf{olda}(j)) \wedge a(i) = \mathsf{olda}(i) \wedge$
$\qquad \forall k. \, k \neq i \wedge k \neq j \Rightarrow a(k) = \mathsf{olda}(k)$
$\quad \Leftrightarrow i = \mathsf{oldi} \wedge j = \mathsf{oldj} \wedge$
$\qquad a(j) = \mathsf{olda}(j) \wedge a(i) = \mathsf{olda}(i) \wedge$
$\qquad \forall k. \, k \neq i \wedge k \neq j \Rightarrow a(k) = \mathsf{olda}(k)$

Then $P \Rightarrow R_3$ is the formula

$$i = \mathsf{oldi} \wedge j = \mathsf{oldj} \wedge a = \mathsf{olda} \Rightarrow$$
$$i = \mathsf{oldi} \wedge j = \mathsf{oldj} \wedge$$
$$a(j) = \mathsf{olda}(j) \wedge a(i) = \mathsf{olda}(i) \wedge$$
$$\forall k. \, k \neq i \wedge k \neq j \Rightarrow a(k) = \mathsf{olda}(k)$$

which is clearly valid; thus our verification is completed.                    □

**The Soundness of the Hoare Calculus**

In the discussion of the various inference rules of the Hoare calculus, we gave some informal "hand-waving" arguments about the soundness of these rules. We are now going to put these arguments on a firm formal basis.

---

**Proposition 8.3** (Partial Soundness of the Hoare Calculus) *Let $Vt$ be a variable typing for signature $\Sigma$. Let $P$ and $Q$ be $\Sigma$-formulas such that judgement $\Sigma, Vt \vdash F$ : formula can be derived for $F :\Leftrightarrow P$ and $F :\Leftrightarrow Q$ (see Fig. 6.1). Let $C$ be a command such that judgement $\Sigma, Vt \vdash C$: command can be derived (see Fig. 7.1).*

*If we can according to the rules of Fig. 8.1 derive a judgement (a "Hoare triple") $\{P\} C \{Q\}$, every execution of $C$ that starts in a state satisfying $P$ can only terminate in a state satisfying $Q$:*

$$\{P\} C \{Q\} \Rightarrow \left(\forall s \in State.\ [\![ P ]\!]_s \Rightarrow (\forall s' \in State.\ [\![ C ]\!]\langle s, s'\rangle \Rightarrow [\![ Q ]\!]_{s'})\right)$$

*Thus $C$ is partially correct with respect to precondition $P$ and postcondition $Q$.*

---

As presented so far, the Hoare calculus does not yet ensure the normal termination of a command; the command may still abort or run forever. We postpone these issues to Sect. 8.4 and settle for the moment with the proof of partial soundness.

***Proof*** Take formulas $P$, $Q$ and statement $C$ with the stated assumptions such that $\{P\} C \{Q\}$ can be derived. Take arbitrary state $s$ with $[\![ P ]\!]_s$ and arbitrary state $s'$ with $[\![ C ]\!]\langle s, s'\rangle$. We show $[\![ Q ]\!]_{s'}$ by induction on the rules of the calculus:

- In case of the "weakening/strengthening" rule, from $[\![ P ]\!]_s$ and $[\![ P \Rightarrow P' ]\!]_s$, we have $[\![ P' ]\!]_s$. Thus $\{P'\} C \{Q'\}$, $[\![ C ]\!]\langle s, s'\rangle$, and the induction assumption gives us $[\![ Q' ]\!]_{s'}$. From this and $[\![ Q' \Rightarrow Q ]\!]_{s'}$, we have $[\![ Q ]\!]_{s'}$.
- Case $C = \{V := T\}$: From $[\![ C ]\!]\langle s, s'\rangle$ we know $s' = s[V \mapsto [\![ T ]\!]_s]$; thus we show $[\![ Q ]\!]_{s[V \mapsto [\![ T ]\!]_s]}$. From $\{P\} C \{Q\}$ we know $P \Leftrightarrow Q[T/V]$; from $[\![ P ]\!]_s$ we thus know $[\![ Q[T/V] ]\!]_s$, i.e., $[\![$ let $V = T$ in $Q ]\!]_s$, i.e., $[\![ Q ]\!]_{s[V \mapsto [\![ T ]\!]_s]}$.
- Case $C = \{\text{var } V: S;\ C_1\}$: From $[\![ C ]\!]\langle s, s'\rangle$ we have values $v$, $v'$ and state $s_1$ with $v = s(V)$, $s_1 = s[V \mapsto v']$, $[\![ C_1 ]\!]\langle s_1, s_2\rangle$, and $s' = s_2[V \mapsto v]$; thus it suffices to show $[\![ Q ]\!]_{s_2[V \mapsto s(V)]}$.

From $\{P\} C \{Q\}$ we know $P \Leftrightarrow (\forall V_1: S.\ P_1[V_1/V])[V/V_0]$ and $\{P_1\} C_1 \{Q[V_0/V]\}$ for some formula $P_1$ and variables $V_0$, $V_1$ with $V$, $V_0$, $V_1$ distinct and $V_0$, $V_1$ not in $P_1, C_1, Q$. From $[\![ P_1 ]\!]_s$, we know $[\![$ let $V_0 = V$ in $\forall V_1: S.\ P_1[V_1/V] ]\!]_s$ and thus $[\![ \forall V_1: S.\ P_1[V_1/V] ]\!]_{s[V_0 \mapsto s(V)]}$. From this we have $[\![ P_1[V_1/V] ]\!]_{s[V_0 \mapsto s(V)][V_1 \mapsto v']}$ and consequently also $[\![$ let $V = V_1$ in $P_1 ]\!]_{s[V_0 \mapsto s(V)][V_1 \mapsto v']}$.

Since $V, V_0, V_1$ are distinct and $V_1$ is not in $P_1$, we thus also know $[\![\, P_1\,]\!]_{s[V_0\mapsto s(V)][V\mapsto v']}$ and $[\![\, P_1\,]\!]_{s[V\mapsto v'][V_0\mapsto s(V)]}$, i.e., $[\![\, P_1\,]\!]_{s_1[V_0\mapsto s(V)]}$. From $[\![\, C_1\,]\!]\langle s_1, s_2\rangle$, since $V_0$ is not in $C_1$, we also have $[\![\, C_1\,]\!]\langle s_1[V_0 \mapsto s(V)],\rangle$ $s_2[V_0 \mapsto s(V)]$.

From these last two formulas and $\{P_1\}\, C_1\,\{Q[V_0/V]\}$, the induction assumption gives us $[\![\, Q[V_0/V]\,]\!]_{s_2[V_0\mapsto s(V)]}$, i.e., $[\![\,\text{let } V = V_0 \text{ in } Q\,]\!]_{s_2[V_0\mapsto s(V)]}$, which implies, since $V_0$ is not in $Q$, the goal $[\![\, Q\,]\!]_{s_2[V\mapsto s(V)]}$.

- Case $C = \{C_1 ; C_2\}$: From $[\![\, C\,]\!]\langle s, s'\rangle$ we have a state $s_1$ with $[\![\, C_1\,]\!]\langle s, s_1\rangle$ and $[\![\, C_2\,]\!]\langle s_1, s'\rangle$. From $\{P\}\, C\,\{Q\}$, we have some formula $R$ with $\{P\}\, C_1\,\{R\}$ and $\{R\}\, C_2\,\{Q\}$. From $[\![\, P\,]\!]_s$, $[\![\, C_1\,]\!]\langle s, s_1\rangle$, and $\{P\}\, C_1\,\{R\}$, the induction assumption gives us $[\![\, R\,]\!]_{s_1}$. From $[\![\, R\,]\!]_{s_1}$, $[\![\, C_2\,]\!]\langle s_1, s'\rangle$, and $\{R\}\, C_2\,\{Q\}$, the induction assumption gives us $[\![\, Q\,]\!]_{s'}$.

- Case $C = \{\text{if } F \text{ then } C_1 \text{ else } C_2\}$: From $[\![\, C\,]\!]\langle s, s'\rangle$, we have the knowledge if $[\![\, F\,]\!]_s$ then $[\![\, C_1\,]\!]\langle s, s'\rangle$ else $[\![\, C_2\,]\!]\langle s, s'\rangle$. From $\{P\}\, C\,\{Q\}$, we have the triples $\{P \wedge F\}\, C_1\,\{Q\}$ and $\{P \wedge \neg F\}\, C_2\,\{Q\}$.

  - In case $[\![\, F\,]\!]_s$, we have $[\![\, P \wedge F\,]\!]_s$ and $[\![\, C_1\,]\!]\langle s, s'\rangle$. From $\{P \wedge F\}\, C_1\,\{Q\}$, the induction assumption gives us $[\![\, Q\,]\!]_{s'}$.
  - In case $\neg[\![\, F\,]\!]_s$, we have $[\![\, P \wedge \neg F\,]\!]_s$ and $[\![\, C_2\,]\!]\langle s, s'\rangle$. From $\{P \wedge \neg F\}\, C_1\,\{Q\}$, the induction assumption gives us $[\![\, Q\,]\!]_{s'}$.

- Case $C = \{\text{if } F \text{ then } C_1\}$: This case proceeds very similarly to the previous one.
- Case $C = \{\text{while } F \text{ do } C_1\}$: here we show the soundness of the basic rule from which (by the soundness of the weakening/strengthening rule) the soundness of the general rule follows: from $[\![\, C\,]\!]\langle s, s'\rangle$ and Proposition 7.8, we have some $t \in State^\omega$ and $n \in \mathbb{N}$ with $loop\langle t, n, F, C_1\rangle$, $t(0) = s$, $t(n) = s'$, and $\neg[\![\, F\,]\!]_{s'}$. From $\{P\}\, C\,\{Q\}$, we have the triple $\{P \wedge F\}\, C_1\,\{P\}$ and $Q \Leftrightarrow P \wedge \neg F$. Since we thus have to show $[\![\, P \wedge \neg F\,]\!]_{s'}$ but already know $\neg[\![\, F\,]\!]_{s'}$, it remains to show $[\![\, P\,]\!]_{s'}$, i.e., $[\![\, P\,]\!]_{t(n)}$. We show this by induction on $n$:

  - In the induction base $n = 0$, with $[\![\, P\,]\!]_s$ and $t(0) = s$ we are done.
  - In the induction step $n > 0$, we assume $[\![\, P\,]\!]_{t(n-1)}$ and show $[\![\, P\,]\!]_{t(n)}$. From $loop\langle t, n, F, C_1\rangle$, we know $[\![\, F\,]\!]_{t(n-1)} \wedge [\![\, C_1\,]\!]\langle t(n-1), t(n)\rangle$. From $[\![\, P\,]\!]_{t(n-1)}$ and $[\![\, F\,]\!]_{t(n-1)}$, we know $[\![\, P \wedge F\,]\!]_{t(n-1)}$. From $[\![\, P \wedge F\,]\!]_{t(n-1)}$, $[\![\, C_1\,]\!]\langle t(n-1), t(n)\rangle$, and $\{P \wedge F\}\, C_1\,\{P\}$, the hypothesis of the structural induction gives us $[\![\, P\,]\!]_{t(n)}$.

This completes the soundness proof.                                          □

## 8.3    Predicate Transformers and Commands as Relations

The Hoare calculus is a good basis for manual program verifications, but it has some stumbling blocks for automation; in particular, it is deplorable that the rule for command sequences depends on the choice of an intermediate predicate, which is

**wp**: *Command* × *Formula* → *Formula*

$$\mathsf{wp}(V := T, Q) := Q[T/V]$$
$$\mathsf{wp}(\{\texttt{var } V : S;\ C\}, Q) := \forall V_1 : S.\ \mathsf{wp}(C, Q[V_0/V])[V_1/V][V/V_0]$$
$$\mathsf{wp}(\{C_1;\ C_2\}, Q) := \mathsf{wp}(C_1, \mathsf{wp}(C_2, Q))$$
$$\mathsf{wp}(\{\texttt{if } F \texttt{ then } C_1 \texttt{ else } C_2\}, Q) := \text{if } F \text{ then } \mathsf{wp}(C_1, Q) \text{ else } \mathsf{wp}(C_2, Q)$$
$$\mathsf{wp}(\{\texttt{if } F \texttt{ then } C\}, Q) := \text{if } F \text{ then } \mathsf{wp}(C_1, Q) \text{ else } Q$$
$$\mathsf{wp}(\{\texttt{while } F \texttt{ do } C\}, Q) := \text{let } I = \text{choose } F.\ F \in \textit{Formula} \text{ in}$$
$$I \wedge (\forall Vs : Ss.\ I \wedge F \Rightarrow \mathsf{wp}(C, I)) \wedge$$
$$(\forall Vs : Ss.\ I \wedge \neg F \Rightarrow Q)$$

**Fig. 8.3** Weakest preconditions (Partial Correctness)

(unless the last command of that sequence is an assignment statement) not obvious. In this section, we present several calculi that overcome this problem. These calculi are closely related to the Hoare calculus and can be also used in combination with the Hoare calculus for the manual reasoning about programs.

**Weakest Preconditions**

As shown in Sect. 8.2, the Hoare calculus naturally leads to a particular technique when reasoning about sequences of assignments: we "backward-propagate" the desired postcondition through the assignments and thus derive a precondition for the sequence that establishes the postcondition. Then it is only necessary to show that the actually desired precondition implies the derived one.

The Dutch computer scientist Edsger W. Dijkstra generalized in 1975 this technique to a calculus whose basis is a function $\mathsf{wp}(C, Q)$ that, given an arbitrary command $C$ and postcondition $Q$ calculates a precondition $P$ such that $\{P\}\ C\ \{Q\}$ holds. This function thus transforms predicates to predicates, i.e., it is a "predicate transformer"; the calculus is therefore also called called "predicate transformer semantics". The term "weakest" indicates that the calculated precondition $P$ imposes as few constraints as possible, which rules out, e.g., that we always get the trivial result $P \Leftrightarrow \mathsf{false}$ (since no prestate satisfies $\mathsf{false}$, every Hoare triple with precondition $\mathsf{false}$ is trivially true).

Figure 8.3 gives a definition of $\mathsf{wp}(C, Q)$ (derived from the original one which used other kinds of commands) by structural induction over command $C$:

- The definition for command $\{V := T\}$ is directly derived from the corresponding rule of the Hoare calculus.
- Also the definition for $\{\texttt{var } V : S;\ C\}$ is derived from the Hoare calculus: the formula $\mathsf{wp}(C_1, Q[V_0/V])[V_1/V]$ represents the precondition $P$ in the corresponding Hoare calculus rule; the variables $V_0$, $V_1$ must also satisfy the constraints that are listed in that rule (but are omitted here for brevity).
- The definition for $\{C_1;\ C_2\}$ encodes the "back-propagation" technique, now not only for assignments but for arbitrary commands $C_1$ and $C_2$: from $C_2$ and $Q$

we establish the precondition $R :\Leftrightarrow \mathsf{wp}(C_2, Q)$ of $C_2$ from which we derive the precondition $P :\Leftrightarrow \mathsf{wp}(C_1, R)$ of $C_1$ and thus of the whole command.

- The definitions for $\{\mathtt{if}\ F\ \mathtt{then}\ C_1\ \mathtt{else}\ C_2\}$ and $\{\mathtt{if}\ F\ \mathtt{then}\ C\}$ can be directly read from the corresponding Hoare rules.
- Dijkstra's original definition for $\{\mathtt{while}\ F\ \mathtt{do}\ C\}$ calculated the semantics of the precondition of a loop as the limit of an infinite sequence of relations derived from the desired postcondition. While such a definition indeed gives the "weakest" precondition, it is not practical for actual program verifications.

  The definition presented here instead applies (as in the generalized form of the Hoare rule) the concept of a loop invariant $I$ that is independent of the desired postcondition (i.e., it must be separately provided): the derived precondition is essentially $I$ (the invariant must hold in the prestate of the loop) but also adds two formulas that correspond to the premises of the Hoare rule: the first formula shows that $I$ is preserved by the loop body $C$, the second one shows that $I$ and the negation $\neg F$ of the loop condition establish postcondition $Q$. Both formulas are universally quantified by all variables $Vs$ of sorts $Ss$ used in the program, thus they become independent of any particular program state; as in the Hoare calculus they can be considered as separate "verification conditions" that are generated by the application of $\mathsf{wp}(C, Q)$ and can be proved independently.

  The precondition returned by our definition is therefore actually (depending on the choice of $I$) not the "weakest" one that establishes $Q$; it is, however, one that suffices to perform the verification.

We now formulate correctness claim of the weakest precondition calculation.

**Proposition 8.4** (Partial Soundness of Weakest Preconditions) *Let $Q$ and $C$ be as in Proposition 8.3. Then every execution of $C$ that starts in a state satisfying* $\mathsf{wp}(C, Q)$ *(where the "predicate transformer" $\mathsf{wp}$ is defined in Fig. 8.3) can only terminate in a state satisfying $Q$:*

$$\forall s \in State.\ [\![ \mathsf{wp}(C, Q) ]\!]_s \Rightarrow (\forall s' \in State.\ [\![ C ]\!]\langle s, s' \rangle \Rightarrow [\![ Q ]\!]_{s'})$$

*Consequently, to verify the partial correctness of any Hoare triple $\{P\}\ C\ \{Q\}$ it suffices to show the validity of the formula $(P \Rightarrow \mathsf{wp}(C, Q))$.*

Rather than proving the soundness of weakest precondition calculation, we demonstrate its use by some examples.

**Example 8.12** We verify the Hoare triple

$$\{m = a\}\ \mathtt{if}\ \mathtt{b{<}m}\ \mathtt{then}\ \mathtt{m{:}{=}b};\ \mathtt{if}\ \mathtt{c{<}m}\ \mathtt{then}\ \mathtt{m{:}{=}c}\ \{Q\}$$

for postcondition $Q :\Leftrightarrow (m = a \lor m = b \lor m = c) \land (m \leq a \land m \leq b \land m \leq c)$; this represents a fragment of the verification of a program for computing the maximum $m$ of three number $a, b, c$ (where $m$ has been already set to $a$ and it remains to be seen whether the value of $m$ has to be overwritten).

For this verification it suffices to calculate $P :\Leftrightarrow \mathsf{wp}(\{C_1 ; C_2\}, Q)$ (where $C_1, C_2$ are the two conditional commands given above) and to establish $(m = a \Rightarrow P)$. The calculation of $P$ proceeds as follows:

$$
\begin{aligned}
P &\Leftrightarrow \mathsf{wp}(\{C_1 ; C_2\}, Q) \\
&\Leftrightarrow \mathsf{wp}(C_1, \mathsf{wp}(C_2, Q)) \\
&\Leftrightarrow \mathsf{wp}(C_1, \text{if } c < m \text{ then } \mathsf{wp}(\mathtt{m:=c}, Q) \text{ else } Q) \\
&\Leftrightarrow \mathsf{wp}(C_1, \text{if } c < m \text{ then } Q[c/m] \text{ else } Q) \\
&\Leftrightarrow \text{if } b < m \text{ then } \mathsf{wp}(\mathtt{m:=b}, \text{if } c < m \text{ then } Q[c/m] \text{ else } Q) \\
&\qquad \text{else } (\text{if } c < m \text{ then } Q[c/m] \text{ else } Q) \\
&\Leftrightarrow \text{if } b < m \text{ then } (\text{if } c < m \text{ then } Q[c/m] \text{ else } Q)[b/m] \\
&\qquad \text{else } (\text{if } c < m \text{ then } Q[c/m][b/m] \text{ else } Q) \\
&\Leftrightarrow \text{if } b < m \text{ then } (\text{if } c < m \text{ then } Q[c/m] \text{ else } Q[b/m]) \\
&\qquad \text{else } (\text{if } c < m \text{ then } Q[c/m] \text{ else } Q) \\
&\Leftrightarrow \text{if } b < m \text{ then } (\text{if } c < m \text{ then } Q[c/m] \text{ else } Q[b/m]) \\
&\qquad \text{else } (\text{if } c < m \text{ then } Q[c/m] \text{ else } Q)
\end{aligned}
$$

Then $(m = a \Rightarrow P)$ is equivalent to

$$
\begin{aligned}
&\text{if } b < a \text{ then } (\text{if } c < a \text{ then } Q[c/m] \text{ else } Q[b/m]) \\
&\qquad \text{else } (\text{if } c < a \text{ then } Q[c/m] \text{ else } Q[a/m])
\end{aligned}
$$

which is in turn equivalent to the conjunction of the following four formulas:

$$
\begin{aligned}
b < a \land c < a &\Rightarrow Q[c/m] \\
b < a \land \neg(c < a) &\Rightarrow Q[b/m] \\
\neg(b < a) \land c < a &\Rightarrow Q[c/m] \\
\neg(b < a) \land \neg(c < a) &\Rightarrow Q[a/m]
\end{aligned}
$$

Each of these formulas establishes the postcondition for one of the four possible execution branches of the program and the last assigned value to variable $m$. Here, e.g., $Q[a/m]$ can be understood as a duplicate of the formula $Q$ where every occurrence of variable $m$ has been syntactically replaced by $a$:

$$
\begin{aligned}
Q[a/m] &\Leftrightarrow (a = a \lor a = b \lor a = c) \land (a \leq a \land a \leq b \land a \leq c) \\
Q[b/m] &\Leftrightarrow (b = a \lor b = b \lor b = c) \land (b \leq a \land b \leq b \land b \leq c) \\
Q[c/m] &\Leftrightarrow (c = a \lor c = b \lor c = c) \land (c \leq a \land c \leq b \land c \leq c)
\end{aligned}
$$

After these substitutions, it is easy to check the validity of above four formulas. $\square$

As shown in the previous section, we can deal with programs that update arrays by utilizing "update terms" of the form $a[i \mapsto e]$ and applying the "array axioms".

**Example 8.13** We calculate the following weakest precondition:

$$\text{wp}(\{\texttt{if a[i]>0 then a[i]:=a[i]-1}\}, a(k) = 0)$$
$$\Leftrightarrow (a(i) > 0 \Rightarrow \text{wp}(\texttt{a[i]:=a[i]-1}, a[k] = 0)) \wedge (\neg(a(i) > 0) \Rightarrow a(k) = 0)$$
$$\Leftrightarrow (a(i) > 0 \Rightarrow \underline{a[i \mapsto a(i) - 1](k) = 0}) \wedge (\neg(a(i) > 0) \Rightarrow a(k) = 0)$$
$$\Leftrightarrow (a(i) > 0 \Rightarrow \underline{\text{if } i = k \text{ then } a(i) - 1 = 0 \text{ else } a(k) = 0})$$
$$\wedge (\neg(a(i) > 0) \Rightarrow a(k) = 0)$$

In this calculation, the array axioms imply the underlined equivalence

$$a[i \mapsto a(i) - 1](k) = 0 \Leftrightarrow \text{if } i = k \text{ then } a(i) - 1 = 0 \text{ else } a(k) = 0. \qquad \square$$

**Example 8.14** We calculate the following weakest precondition:

$$\text{wp}(\{\{\texttt{if a[i]<b then i:=i-1;a[i]:=a[i]+1}\}; \texttt{i:=i+1}\}, a(i) = c)$$
$$\Leftrightarrow \text{wp}(\{\texttt{if a[i]<b then i:=i-1;a[i]:=a[i]+1}\}, \text{wp}(\texttt{i:=i+1}, a(i) = c))$$
$$\Leftrightarrow \text{wp}(\{\texttt{if a[i]<b then i:=i-1;a[i]:=a[i]+1}\}, a(i + 1) = c)$$
$$\Leftrightarrow \text{if } a(i) < b \text{ then wp}(\{\texttt{i:=i-1;a[i]:=a[i]+1}\}, a(i + 1) = c) \text{ else } a(i + 1) = c$$
$$\Leftrightarrow \text{if } a(i) < b \text{ then wp}(\texttt{i:=i-1}, \text{wp}(\texttt{a[i]:=a[i]+1}, a(i + 1) = c)) \text{ else } a(i+1) = c$$
$$\Leftrightarrow \text{if } a(i) < b \text{ then wp}(\texttt{i:=i-1}, \underline{a[i \mapsto a(i) + 1](i + 1) = c}) \text{ else } a(i + 1) = c$$
$$\Leftrightarrow \text{if } a(i) < b \text{ then wp}(\texttt{i:=i-1}, a(i + 1) = c) \text{ else } a(i + 1) = c$$
$$\Leftrightarrow \text{if } a(i) < b \text{ then } a(i) = c \text{ else } a(i + 1) = c$$

In this calculation, the array axioms imply from $i \neq i + 1$ the equality

$$a[i \mapsto a(i) + 1](i + 1) = a(i + 1)$$

and thus the underlined equivalence

$$a[i \mapsto a(i) + 1](i + 1) = c \Leftrightarrow a(i + 1) = c.$$

$\square$

## Strongest Postconditions

As we have seen, the weakest precondition calculus naturally leads to a "backwards" style of reasoning, from a given postcondition to a precondition sufficient to establish that postcondition. Actually, also the reverse is possible: from a given command and precondition, we may reason "forwards" to determine the postcondition that is necessarily established from the precondition. Figure 8.4 depicts the definition of a corresponding function $\mathsf{sp}(C, P)$, another predicate transformer that, given arbitrary command $C$ and precondition $P$, calculates a postcondition $Q$ such that $\{P\}\,C\,\{Q\}$ holds. The term "strongest" indicates that $Q$ contains as much information as possible, which rules, e.g., out that we always get the trivial result $Q \Leftrightarrow \mathsf{true}$ (since every poststate satisfies true, every Hoare triple with postcondition true is trivially true). As usual, the function is defined by structural induction over command $C$:

- The definition for command $\{V := T\}$ requires an existentially quantified variable $V_0$ that has the same sort $S$ as the program variable $V$ and does neither occur in $T$ nor in $P$; $V_0$ denotes the value of the program variable $V$ before the assignment. In the prestate we know that $P$ holds for the value of $V$ in that state; in the poststate we only know that $P$ previously held in the prestate for the now unknown prestate value denoted by $V_0$; furthermore we know that the new value of $V$ is the value of term $T$ evaluated in the prestate when $V$ still had value $V_0$. However, if neither the precondition $P$ nor the term $T$ refers to the updated variable $V$, we can drop the quantified variable (see the examples given below).
- Also the definition for the command $\{\mathsf{var}\ V\colon S;\ C\}$ requires existentially quantified variables, a variable $V_0$ denoting the value of $V$ in the prestate of the command and a variable $V_1$ denoting the value of $V$ immediately after the execution of the command body $C$. Since $P$ holds in the prestate of the new command, $P[V_0/V]$ holds immediately before the execution of $C$ (when $V$ has received an arbitrary new value); immediately after the execution of $C$, we therefore have $\mathsf{sp}(C, P[V_0/V])$. However, in the poststate of the whole command, $V$ is restored to its original value $V = V_0$, thus we know in that poststate only $\mathsf{sp}(C, P[V_0/V])[V_1/V]$ for the value $V_1$ that $V$ had immediately after the execution of $C$.
- The definition for $\{C_1;\ C_2\}$ encodes the technique for the "forward-propagation" of preconditions: from $C_1$ and $P$ we establish the postcondition $R :\Leftrightarrow \mathsf{sp}(C_1, P)$ of $C_1$ from which we derive the postcondition $Q :\Leftrightarrow \mathsf{wp}(C_2, R)$ of $C_2$ and thus of the whole command.
- The definition for $\{\mathsf{if}\ F\ \mathsf{then}\ C_1\ \mathsf{else}\ C_2\}$ follow the intuition that the poststate of the command has been reached from its prestate via one of two execution branches: either the prestate satisfied $P \wedge F$, from which the execution of branch $C_1$ established the postcondition $\mathsf{sp}(C_1, P \wedge F)$, or the prestate satisfied $P \wedge \neg F$, from which the execution of branch $C_1$ established the postcondition $\mathsf{sp}(C_1, P \wedge \neg F)$. The definition for $\{\mathsf{if}\ F\ \mathsf{then}\ C\}$ can be explained analogously.
- The definition for $\{\mathsf{while}\ F\ \mathsf{do}\ C\}$ is again based on a loop invariant $I$; essentially the definition says that the poststate of the loop satisfies $I$ and the negation $\neg F$ of the loop condition. However, this is only true, under two assumptions:

  - The prestate of the loop initially satisfied $I$; since we know in the poststate about the prestate only that $P$ held, $P \Rightarrow I$ must hold.
  - The loop body $C$ preserves $I$. Since the execution of the loop starts in a state in which both $I$ and $F$ hold, the execution of $C$ in that state must establish a poststate in which $I$ holds, i.e., $\mathsf{sp}(C, I \wedge F) \Rightarrow I$ must hold.

---

**sp:** *Command* × *Formula* → *Formula*

$$\text{sp}(V := T, P) := \exists V_0 \colon S.\ P[V_0/V] \wedge V = T[V_0/V]$$
$$\text{sp}(\{\text{var } V \colon S;\ C\}, P) := \exists V_0 \colon S, V_1 \colon S.\ \text{sp}(C, P[V_0/V])[V_1/V] \wedge V = V_0$$
$$\text{sp}(\{C_1;\ C_2\}, P) := \text{sp}(C_2, \text{sp}(C_1, P))$$
$$\text{sp}(\{\text{if } F \text{ then } C_1 \text{ else } C_2\}, P) := \text{sp}(C_1, P \wedge F) \vee \text{sp}(C_2, P \wedge \neg F)$$
$$\text{sp}(\{\text{if } F \text{ then } C\}, P) := \text{sp}(C_1, P \wedge F) \vee (P \wedge \neg F)$$
$$\text{sp}(\{\text{while } F \text{ do } C\}, P) := \text{let } I = \text{choose } F.\ F \in \textit{Formula} \text{ in}$$
$$(\forall Vs \colon Ss.\ P \Rightarrow I) \wedge (\forall Vs \colon Ss.\ \text{sp}(C, I \wedge F) \Rightarrow I) \Rightarrow$$
$$I \wedge \neg F$$

**Fig. 8.4** Strongest postconditions (Partial Correctness)

Since these two assumptions talk about the values of the program variables in other states than the poststate of the loop, the corresponding formulas must be universally quantified by variables $Vs$ whose sorts $Ss$ are the sorts of these variables. These assumptions are therefore independent of any particular states; the calculation of $\text{sp}(C, P)$ can be considered to produce these as additional "verification conditions" whose validity has to be established separately.

As for weakest preconditions, the use of an invariant actually implies that the calculated postcondition is generally not the "strongest" one; however, this form of the definition is the one necessary and sufficient for practical verification.

It is mainly the necessity of a quantifier in the definition of the strongest postcondition of the assignment statement, which has lead to the preference of the backwards reasoning style based on weakest preconditions over the forwards reasoning style based on strongest postconditions.

**Example 8.15** For a program with variables $i$ and $x$ of type nat we have:

$$\text{sp}(\{\texttt{i:=n;x:=-1}\}, n > 0)$$
$$\Leftrightarrow \text{sp}(\texttt{x:=-1}, \text{sp}(\texttt{i:=n}, n > 0))$$
$$\Leftrightarrow \text{sp}(\texttt{x:=-1}, \exists i_0 \colon \text{nat}.\ n > 0 \wedge i = n)$$
$$\Leftrightarrow \text{sp}(\texttt{x:=-1}, n > 0 \wedge i = n)$$
$$\Leftrightarrow \exists x_0 \colon \text{nat}.\ n > 0 \wedge i = n \wedge x = -1$$
$$\Leftrightarrow n > 0 \wedge i = n \wedge x = -1 \qquad \qquad \square$$

As usual array updates can be handled by utilizing update terms and applying the array axioms.

**Example 8.16** We calculate the following strongest postcondition:

$$\mathsf{sp}(\{\text{if } \texttt{i<n then } \{\texttt{a[i]:=a[i]+1;i:=i+1}\}\}, a(i) = 1)$$
$$\Leftrightarrow \mathsf{sp}(\{\texttt{a[i]:=a[i]+1;i:=i+1}\}, a(i) = 1 \wedge i < n) \vee (a(i) = 1 \wedge \neg(i < n))$$
$$\Leftrightarrow \mathsf{sp}(\texttt{i:=i+1}, \mathsf{sp}(\texttt{a[i]:=a[i]+1}, a(i) = 1 \wedge i < n)) \vee (a(i) = 1 \wedge \neg(i < n))$$
$$\Leftrightarrow \mathsf{sp}(\texttt{i:=i+1}, \underline{a[i \mapsto a(i) + 1](i) = 1} \wedge i < n) \vee (a(i) = 1 \wedge \neg(i < n))$$
$$\Leftrightarrow \mathsf{sp}(\texttt{i:=i+1}, \underline{a(i) + 1 = 1} \wedge i < n) \vee (a(i) = 1 \wedge \neg(i < n))$$
$$\Leftrightarrow \mathsf{sp}(\texttt{i:=i+1}, a(i) = 0 \wedge i < n) \vee (a(i) = 1 \wedge \neg(i < n))$$
$$\Leftrightarrow (a(i + 1) = 0 \wedge i + 1 < n) \vee (a(i) = 1 \wedge \neg(i < n))$$

In this calculation, the array axioms imply from $i = i$ the equality

$$a[i \mapsto a(i) + 1](i) = a(i) + 1$$

and thus the underlined equivalence

$$a[i \mapsto a(i) + 1](i) = 1 \Leftrightarrow a(i) + 1 = 1.$$
□

The soundness claim for strongest postconditions is dual to that of weakest preconditions (Proposition 8.4).

---

**Proposition 8.5** (Partial Soundness of Strongest Postconditions) *Let P and C
be as in Proposition 8.3. Then every execution of C that starts in a state satisfying P can only terminate in a state satisfying* $\mathsf{sp}(C, P)$ *(where the "predicate
transformer"* $\mathsf{sp}$ *is defined in Fig. 8.4):*

$$\forall s \in State.\ [\![ P ]\!]_s \Rightarrow (\forall s' \in State.\ [\![ C ]\!]\langle s, s' \rangle \Rightarrow [\![ \mathsf{sp}(C, P) ]\!]_{s'})$$

*Consequently, to verify the partial correctness of any Hoare triple* $\{P\}\ C\ \{Q\}$
*it suffices to show the validity of the formula* $(\mathsf{sp}(C, P) \Rightarrow Q)$.

---

### Predicate Transformers and the Hoare Calculus

The calculation of weakest preconditions and strongest postconditions can also
become handy in manual verifications performed in the Hoare calculus. For instance,
the verification of a Hoare triple

$$\{P\}\ \{C;\ \text{if } F \text{ then } C_1 \text{ else } C_2\}\ \{Q\}$$

that embeds a command sequence whose last command is *not* an assignment can be reduced to verifying the following triple

$$\{P\}\, C\, \{\text{wp}(\{\texttt{if } F \texttt{ then } C_1 \texttt{ else } C_2\}, Q)\}$$

which is equivalent to the following triple:

$$\{P\}\, C\, \{\text{if } F \text{ then wp}(C_1, Q) \text{ else wp}(C_2, Q)\}$$

Provided that $C_1$ and $C_2$ do not contain any loops, we can thus calculate a sufficient condition for the intermediate state of the execution sequence in a purely mechanical way, i.e., without requiring deep insight into the program.

As another example, the verification of a Hoare triple

$$\{P\}\, C_1; \{\texttt{while } F \texttt{ do } C\}; C_2\, \{Q\}$$

that embeds a loop within two program fragments $C_1$ and $C_2$ can be reduced to the verification of the triple

$$\{\text{sp}(C_1, P)\}\, \texttt{while } F \texttt{ do } C\, \{\text{wp}(C_2, Q)\}$$

that only embeds the loop. Thus, from the given precondition $P$ and postcondition $Q$ of the whole command we can derive a sufficient precondition $\text{sp}(C_1, P)$ and postcondition $\text{wp}(C_2, Q)$ of the loop. The remaining verification can be then again performed according to the rules of the Hoare calculus, based on a suitable loop invariant. This reduction is useful if the commands $C_1$ and $C_2$ themselves do not contain any further loops such that $\text{sp}(C_1, P)$ and $\text{wp}(C_2, Q)$ indeed can be mechanically calculated. Thus the problem is immediately reduced to the core problem, the determination of the loop invariant, which may require insight and creativity.

**Pre- and Poststate Relationships**

Both the Hoare calculus and the predicate transformer calculus suffer in some respect from the fact that they are based on *conditions* about individual program states; this makes it somehow awkward to describe *relations* between a prestate and a poststate. As already sketched before, the usual solution is to introduce mathematical constants that are fixed in the precondition to the values of the program variables in the prestate and to refer to these constants in the poststate. For instance, the Hoare triple

$$\{x = \text{old\_x} \wedge y = \text{old\_y}\}\ \texttt{x:=x+1}\ \{x = \text{old\_x} + 1 \wedge y = \text{old\_y}\}$$

uses constants $\text{old\_x}$ and $\text{old\_y}$ to state that the assignment increments variable $x$ by 1 and leaves variable $y$ unchanged. Similarly we can express this fact by the strongest postcondition

$$\text{sp}(\texttt{x:=x+1}, x = \text{old\_x} \wedge y = \text{old\_y}) \Leftrightarrow x = \text{old\_x} + 1 \wedge y = \text{old\_y}$$

that also uses such constants. The same problem also arises when for the verification of a loop the loop invariant has to relate the current state of the program variables to the values of these variables before the loop was entered. Take for example the triple

$$\{i = \mathsf{old\_i} \wedge s = \mathsf{old\_s}\}$$

$$\texttt{while i<n do \{s:=s+i; i:=i+1\}}$$

$$\{s = \mathsf{old\_s} + \sum_{k=\mathsf{old\_i}}^{n} k\}$$

which expresses the correctness of a summation for arbitrary start values $\mathsf{old\_i}$ and $\mathsf{old\_s}$; its verification requires the invariant $I :\Leftrightarrow i \le n \wedge s = \mathsf{old\_s} + \sum_{k=\mathsf{old\_i}}^{i-1} k$.

However, it is now not clear where these mathematical constants are actually declared such that type-checking the corresponding formulas succeeds. To solve this problem, verification systems automatically introduce such constants for the initial values of procedure parameters or for initial values of variables before a loop is executed. By this they effectively extend the corresponding verification rules such as the Hoare calculus rule to

$$\frac{\ldots}{\{P \wedge \mathsf{old\_} Vs = Vs\}\,\texttt{while F do C}\,\{Q\}}$$

or the definition of the weakest precondition to

$$\mathsf{wp}(\{\texttt{while } F \texttt{ do } C\}, Q) := \mathsf{let}\ \mathsf{old\_} Vs = Vs\ \mathsf{in}\ \ldots$$

where $\mathsf{old\_} Vs$ denote constants for the initial values of all variables $Vs$ occurring in the program; the human then does not have to introduce these constants explicitly any more but can assume them as implicitly given. We will not formalize these extensions in detail but turn our attention to another calculus whose fundamental design already solves the problem of relating prestates to poststates.

## Commands as Relations

We now introduce a "relational" verification calculus which is based on judgements of the following form:

$$C : [\,F\,]^{Vs}$$

Such a judgement states that formula $F$ describes the effect of command $C$ that only changes variables from the set $Vs$. Formula $F$ here describes a *relation* between the prestate of $C$ and its poststate; in this description it uses the syntax $\mathsf{old\_} V$ to refer to the prestate value of program variable $V$ and $\mathsf{new\_} V$ to refer to its poststate value (we assume that $C$ only refers to variables whose names do not have these particular forms). Variable set $Vs$ is the *frame* of the command that bounds its effect; for every variable $V$ which is not in the frame, we automatically know $\mathsf{new\_} V = \mathsf{old\_} V$.

For example, by this calculus we are able to derive the judgement

$$\texttt{x:=x+1}: [\, \mathsf{new\_}x = \mathsf{old\_}x + 1\,]^{\{x\}}$$

which can be literally read as "command x:=x+1 increments the value of program variable $x$ by 1 and leaves every other program variable unchanged".

Therefore, while the Hoare calculus and the predicate transformer semantics are "axiomatic" in the sense that they allow to verify the correctness of commands, this relational calculus is "denotational" in the sense that it gives a command $C$ an explicit meaning $F$ (which, however, can be also applied to the verification of $C$). Thus this calculus can be also considered as the translation of an "imperative" description $C$ of a command (the sequence of steps executed) to a "declarative" form $F$ (the effect achieved by the execution).

The formulation of this calculus that translates commands to state relations requires in several places the translation of every variable reference $V$ to the form old_$V$ respectively new_$V$.

> **Definition 8.3**  (*Variable Translation*) Let $E$ be a syntactic phrase (term or formula). We denote by $E'$ a duplicate of $E$ where every free variable occurrence $V$ has been replaced by old_$V$. Furthermore, we denote by $E''$ a duplicate of $E$ where every free variable occurrence $V$ has been replaced by new_$V$. Finally, we denote by $E^0$ a duplicate of $E$ where every free variable occurrence old_$V$ and every free variable occurrence new_$V$ has been replaced by $V$.

In the following, we identify a set $Vs$ of variables with a corresponding sequence in arbitrary order. Furthermore, we use the notations $Vs_0$, old_$Vs$, respectively new_$Vs$ to denote the variables derived from the variables in $Vs$ by replacing every variable $V \in Vs$ by a new variable $V_0$, respectively by old_$V$ or new_$V$. We write $Vs: Ss$ to indicate a sequence of typed variables $V: S$ that contains every variable $V$ in $Vs$ whose sort is $S$ (to simplify the presentation, we refrain from formally describing the derivation of these sorts). Finally, we will generalize the notation of a variable substitution $E[T/V]$ to the form $E[Ts/Vs]$ which replaces in phrase (term or formula) $E$ every reference to a variable in set $Vs$ by the corresponding term in $Ts$.

Figure 8.5 gives the rules of this calculus, two generic rules, and one rule for every kind of command:

- The first generic rule is a logical "weakening" rule that allows to replace a derived state relation $F_1$ by a weaker (respectively equivalent) form $F$.

**Example 8.17** By this rule the judgement

$$\texttt{x:=x+1}: [\, \mathsf{new\_}x = \mathsf{old\_}x + 1\,]^{\{x\}}$$

leads to the following judgement:

**Judgement**

⊔: [ ⊔ ]<sup>⊔</sup> ⊆ *Command* × *Formula* × Set(*Variable*)

**Rules for** $C: [F]^{Vs}$

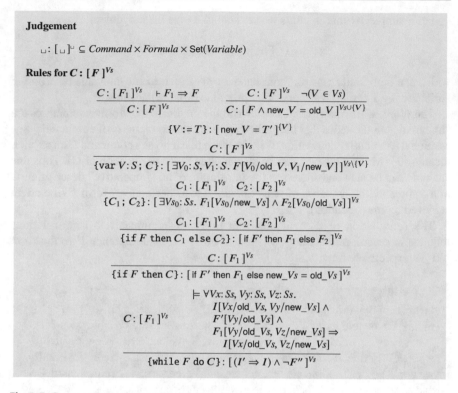

**Fig. 8.5** Commands as relations (Partial Correctness)

$$\texttt{x:=x+1}: [\, \mathsf{old\_}x = \mathsf{new\_}x - 1\,]^{\{x\}}$$

□

- The more interesting "frame extension" rule describes how a judgement with frame $Vs$ can be extended to a frame $Vs \cup V$ for an arbitrary variable $V$ that is not in $Vs$: we simply add the constraint $\mathsf{new\_}V = \mathsf{old\_}V$ to the derived formula.

**Example 8.18** By this rule the judgement

$$\texttt{x:=x+1}: [\, \mathsf{new\_}x = \mathsf{old\_}x + 1\,]^{\{x\}}$$

can be extended to the following judgement:

$$\texttt{x:=x+1}: [\, \mathsf{new\_}x = \mathsf{old\_}x + 1 \land \mathsf{new\_}y = \mathsf{old\_}y\,]^{\{x,y\}}$$

□

- The axiom for the assignment $\{V := T\}$ translates this assignment to an equality $\mathsf{new\_}V = T'$ with frame $\{V\}$. Here $T'$ is a duplicate of term $T$ where every occurrence of every program variable $X$ has been replaced by $\mathsf{old\_}X$.

**Example 8.19** By this rule we can derive the following judgement:

$$\texttt{x:=x+y}: [\, \mathsf{new\_}x = \mathsf{old\_}x + \mathsf{old\_}y \,]^{\{x\}}$$

□

- The rule for the declaration command {var $V$: $S$; $C$} states how the judgement $F$ derived for the declaration body $C$ can be extended to a judgement for the whole command: first, we introduce two existentially quantified variables $V_0$ and $V_1$ denoting the values of variable $V$ immediately before the execution of $C$ respectively immediately afterwards; then we replace every occurrence of $\mathsf{old\_}V$ in $F$ by $V_0$ and every occurrence of $\mathsf{new\_}V$ by $V_1$. Finally, since the execution of the declaration command leaves the value of $V$ unchanged, we compute the frame of the declaration command by removing $V$ from the frame $Vs$ of $F$.

**Example 8.20** Assume we have derived the following judgement (this derivation will be explained in Example 8.21 below):

$$\{\texttt{y:=x+1;x:=y+x}\}: [\, \mathsf{new\_}y = \mathsf{old\_}x + 1 \wedge \mathsf{new\_}x = 2 \cdot \mathsf{old\_}x + 1 \,]^{\{x,y\}}$$

By above rule, we can then derive the judgement

$$\{\texttt{var y:nat};\{\texttt{y:=x+1;x:=y+x}\}\}: [\, F \,]^{\{x\}}$$

where

$$
\begin{aligned}
F &\Leftrightarrow \exists y_0\colon \mathsf{nat},\, y_1\colon \mathsf{nat}.\ y_1 = \mathsf{old\_}x + 1 \wedge \mathsf{new\_}x = 2 \cdot \mathsf{old\_}x + 1 \\
&\Leftrightarrow \exists y_1\colon \mathsf{nat}.\ y_1 = \mathsf{old\_}x + 1 \wedge \mathsf{new\_}x = 2 \cdot \mathsf{old\_}x + 1 \\
&\Leftrightarrow (\exists y_1\colon \mathsf{nat}.\ y_1 = \mathsf{old\_}x + 1) \wedge \mathsf{new\_}x = 2 \cdot \mathsf{old\_}x + 1 \\
&\Leftrightarrow \mathsf{true} \wedge \mathsf{new\_}x = 2 \cdot \mathsf{old\_}x + 1 \\
&\Leftrightarrow \mathsf{new\_}x = 2 \cdot \mathsf{old\_}x + 1
\end{aligned}
$$

Thus we have derived

$$\{\texttt{var y:nat};\{\texttt{y:=x+1;x:=y+x}\}\}: [\, \mathsf{new\_}x = 2 \cdot \mathsf{old\_}x + 1 \,]^{\{x\}}$$

□

- The rule for the command sequence {$C_1$; $C_2$} states that we have first to derive the formulas $F_1$ and $F_2$ describing the effects of command $C_1$ and $C_2$, respectively, for the *same* frame $Vs$ (in practice, this means that we have to apply the extension rule to adapt the originally derived formulas with generally different frames $Vs_1$ and $Vs_2$ to the common frame $Vs := Vs_1 \cup Vs_2$). Then for all program variables $Vs$ in the frame we introduce existentially quantified variables $Vs_0$ denoting the values of the program variables after the execution of $C_1$, i.e., before the execution of $C_2$. In $F_1$ we therefore replace all occurrences of variables $\mathsf{new\_}Vs$ by $Vs_0$ while in $F_2$ we replace all occurrences of variables $\mathsf{old\_}Vs$ by $Vs_0$.

**Example 8.21** By the axiom for assignments, we can derive the following judgements:

$$y\text{:=}x\text{+}1\text{: } [\text{ new\_}y = \text{old\_}x + 1\ ]^{\{y\}}$$
$$x\text{:=}x\text{+}y\text{: } [\text{ new\_}x = \text{old\_}x + \text{old\_}y\ ]^{\{x\}}$$

From these, we can derive by the extension rule the following judgements with common frame $\{x, y\}$:

$$y\text{:=}x\text{+}1\text{: } [\text{ new\_}y = \text{old\_}x + 1 \wedge \text{new\_}x = \text{old\_}x\ ]^{\{x,y\}}$$
$$x\text{:=}x\text{+}y\text{: } [\text{ new\_}x = \text{old\_}x + \text{old\_}y \wedge \text{new\_}y = \text{old\_}y\ ]^{\{x,y\}}$$

From these, we can derive by the rule for command sequences the judgement

$$\{y\text{:=}x\text{+}1\text{;}x\text{:=}y\text{+}x\}\text{: } [\ F\ ]^{\{x,y\}}$$

where

$$F \Leftrightarrow \exists x_0\text{: nat}, y_0\text{: nat}.$$
$$\qquad y_0 = \text{old\_}x + 1 \wedge x_0 = \text{old\_}x \wedge$$
$$\qquad \text{new\_}x = x_0 + y_0 \wedge \text{new\_}y = y_0$$
$$\Leftrightarrow \exists x_0\text{: nat}, y_0\text{: nat}.$$
$$\qquad y_0 = \text{old\_}x + 1 \wedge x_0 = \text{old\_}x \wedge$$
$$\qquad \text{new\_}x = \text{old\_}x + \text{old\_}x + 1 \wedge \text{new\_}y = \text{old\_}x + 1$$
$$\Leftrightarrow \text{new\_}x = \text{old\_}x + \text{old\_}x + 1 \wedge \text{new\_}y = \text{old\_}x + 1$$
$$\Leftrightarrow \text{new\_}y = \text{old\_}x + 1 \wedge \text{new\_}x = 2 \cdot \text{old\_}x + 1$$

Thus we have derived the following judgement:

$$\{y\text{:=}x\text{+}1\text{;}x\text{:=}y\text{+}x\}\text{: } [\text{ new\_}y = \text{old\_}x + 1 \wedge \text{new\_}x = 2 \cdot \text{old\_}x + 1\ ]^{\{x,y\}} \qquad \square$$

- The rule for the conditional command $\{\text{if } F \text{ then } C_1 \text{ else } C_2\}$ also requires to derive from commands $C_1$ and $C_2$ formulas $F_1$ and $F_2$ with a common frame $Vs$. From this the conditional command with condition $F$ can be translated to a corresponding conditional formula with condition $F'$, a duplicate of $F$ where every occurrence of every program variable $V$ has been replaced by $\text{old\_}V$.

**Example 8.22** As shown in Example 8.21, we can derive the following judgements with common frame $\{x, y\}$:

$$y\text{:=}x\text{+}1\text{: } [\text{ new\_}y = \text{old\_}x + 1 \wedge \text{new\_}x = \text{old\_}x\ ]^{\{x,y\}}$$
$$x\text{:=}x\text{+}y\text{: } [\text{ new\_}x = \text{old\_}x + \text{old\_}y \wedge \text{new\_}y = \text{old\_}y\ ]^{\{x,y\}}$$

From this, we can derive

$$\{\text{if } x>z \text{ then } y:=x+1 \text{ else } x:=x+y\}: [\,F\,]^{\{x,y\}}$$

where

$$F \Leftrightarrow \text{if old\_}x > \text{old\_}z \text{ then}$$
$$\text{new\_}y = \text{old\_}x + 1 \wedge \text{new\_}x = \text{old\_}x$$
$$\text{else}$$
$$\text{new\_}x = \text{old\_}x + \text{old\_}y \wedge \text{new\_}y = \text{old\_}y \qquad \square$$

- The rule for the one-sided conditional command $\{\text{if } F \text{ then } C\}$ is analogous to the two sided conditional command except that the formula for the second branch states that all variables in the frame are unchanged.

**Example 8.23** As shown in Example 8.21, we can derive

$$\{y:=x+1; x:=y+x\}: [\,\text{new\_}y = \text{old\_}x + 1 \wedge \text{new\_}x = 2 \cdot \text{old\_}x + 1\,]^{\{x,y\}}$$

Thus we can derive

$$\{\text{if } x>z \text{ then } \{y:=x+1; x:=y+x\}\}: [\,F\,]^{\{x,y\}}$$

where

$$F \Leftrightarrow \text{if old\_}x > \text{old\_}z \text{ then}$$
$$\text{new\_}y = \text{old\_}x + 1 \wedge \text{new\_}x = 2 \cdot \text{old\_}x + 1$$
$$\text{else}$$
$$\text{new\_}x = \text{old\_}x \wedge \text{new\_}y = \text{old\_}y \qquad \square$$

- The rule for the loop command $\{\text{while } F \text{ do } C\}$ again depends on a suitable invariant $I$: however, in the contrast to the previously presented calculi, now $I$ is not any more just a state condition but a state relation, referring to the values of a program variable $V$ only by old\_$V$ respectively new\_$V$. Here old\_$V$ denotes the value of $V$ in the prestate of the loop, i.e., immediately before the first loop iteration, while new\_$V$ denotes the value of $V$ immediately before/after an arbitrary number of loop iterations.
  The rule derives for the loop command the state relation $((I' \Rightarrow I) \wedge \neg F'')$. In a nutshell, this condition states that both the invariant $I$ and the negation $\neg F$ of the loop condition holds in the poststate of the loop, the later under the assumption that the invariant also holds in the prestate. Here $I'$ is a duplicate of $I$ where, for every variable $V$, every occurrence new\_$V$ has been replaced by old\_$V$ (thus stating the assumption that the identity relationship on states satisfies $I$); $F''$ is

a duplicate of the loop condition $F$ where every occurrence of every variable $V$ has been replaced by new_$V$ (thus stating the conclusion that the negation of this condition is satisfied in the poststate).

However, his is only true if the verification condition stated in the premise of the rule holds: this condition claims that the invariance of $I$ is preserved by the execution of the loop body $C$ (which is described by the state relation $F_1$ derived from $C$) and does not change any other variable than those in the frame $Vs$ of $C$ (thus $Vs$ is also the frame of the loop command). This verification condition can be visualized as follows:

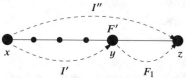

Here label $x$ marks the prestate of the loop (immediately before its first iteration), label $y$ marks the state before the current iteration, and label $z$ marks the state after that iteration. Correspondingly the verification condition uses universally quantified variables $Vx$, $Vy$, and $Vs$ with the sorts $Ss$ of the program variables to refer to the values of the program variables in the respective states. In a nutshell the condition states that, if the invariant $I$ holds between arbitrary initial state $x$ and arbitrary current state $y$, the loop condition $F$ holds in $y$ (here $F'$ refers to a duplicate of $F$ where every occurrence $V$ of every variable $V$ has been replaced by old_$V$), the state relation $F_1$ of command $C$ holds between $y$ and arbitrary state $z$ (i.e., $z$ is the state after another iteration of the loop body), then $I$ also holds between $x$ and $z$. The verification condition thus represents the induction step in the proof of the invariance of $I$ (while the assumption $I'$ in the derived state relation represents the induction base).

**Example 8.24** Consider the following loop command:

$$C := \text{while } \text{i<n do } \{\text{s:=s+i; } \text{i:=i+1}\}$$

We will now derive the judgement $C \colon [\, F \,]^{\{s,i\}}$ for the following definitions of the state relation

$$F :\Leftrightarrow (I' \Rightarrow I) \wedge \neg(\text{new\_}i < \text{new\_}n)$$

which is based on the following loop invariant:

$$I :\Leftrightarrow \text{new\_}i \leq \text{old\_}n \wedge \text{new\_}s = \text{old\_}s + \sum_{k=\text{old\_}i}^{\text{new\_}i-1} k$$

It is easy to see that we can derive the following judgement for the loop body (we omit the details):

$$\{\text{s:=s+i; } \text{i:=i+1}\} \colon [\, \text{new\_}s = \text{old\_}s + \text{old\_}i \wedge \text{new\_}i = \text{old\_}i + 1 \,]^{\{s,i\}}$$

So, according to the rule for the loop command, it remains to show, for arbitrary but fixed $s_x, s_y, s_z, i_x, i_y, i_z$, the validity of the following verification condition:

$$i_y \leq \mathsf{new\_}n \wedge s_y = s_x + \sum_{k=i_x}^{i_y-1} k \; \wedge$$

$$i_y < \mathsf{new\_}n \; \wedge$$
$$s_z = s_y + i_y \wedge i_z = i_y + 1 \Rightarrow$$
$$i_z \leq \mathsf{new\_}n \wedge s_z = s_x + \sum_{k=i_x}^{i_z-1} k$$

From the assumptions $i_y < \mathsf{new\_}n$ and $i_z = i_y + 1$, we already know $i_z \leq \mathsf{new\_}n$. From $s_y = s_x + \sum_{k=i_x}^{i_y-1} k$, $s_z = s_y + i_y$, and $i_z = i_y + 1$, we also know

$$s_z = s_y + i_y = s_x + \left(\sum_{k=i_x}^{i_y-1} k\right) + i_y = s_x + \sum_{k=i_x}^{i_y} k = s_x + \sum_{k=i_x}^{i_z-1} k.$$

which completes the derivation.

From the judgement $C : [\, F \,]^{\{s,i\}}$ we can actually derive another judgement with a more comprehensible state relation. First, we can simplify $I'$ as follows:

$$I' \Leftrightarrow \mathsf{old\_}i \leq \mathsf{old\_}n \wedge \mathsf{old\_}s = \mathsf{old\_}s + \sum_{k=\mathsf{old\_}i}^{\mathsf{old\_}i-1} k$$
$$\Leftrightarrow \mathsf{old\_}i \leq \mathsf{old\_}n \wedge \mathsf{old\_}s = \mathsf{old\_}s + 0$$
$$\Leftrightarrow \mathsf{old\_}i \leq \mathsf{old\_}n$$

Then we can derive

$$C : [\, \mathsf{old\_}i \leq \mathsf{old\_}n \Rightarrow \mathsf{new\_}i = \mathsf{old\_}n \wedge \mathsf{new\_}s = \mathsf{old\_}s + \sum_{k=\mathsf{old\_}i}^{\mathsf{old\_}n-1} k \,]^{\{s,i\}}$$

because from $C : [\, F \,]^{\{s,i\}}$ and $n \notin \{s, i\}$, we have $C : [\, F \wedge \mathsf{new\_}n = \mathsf{old\_}n \,]^{\{s,i\}}$, and $\mathsf{new\_}n = \mathsf{old\_}n$ together with $\mathsf{new\_}i \leq \mathsf{old\_}n$ and $\neg(\mathsf{new\_}i < \mathsf{new\_}n)$ implies $\mathsf{new\_}i = \mathsf{old\_}n$.

This form of the judgement expresses succinctly the behavior of the code fragment which can be spelled out as follows:

> The command changes only the values of variables $s$ and $i$ (i.e., the value of variable $n$ remains unchanged). If, before the execution of the command, the old value of $i$ is not bigger than $n$, then, after the execution, the new value of $i$ is $n$ and the new value of $s$ is the old value of $s$ plus the sum of all the values from the old value of $i$ up to $n - 1$. □

As the previous examples have demonstrated, by the presented calculus we can translate individual commands into formulas that (after some simplification) clearly describe the effects of the commands. This is also true for bigger program fragments.

**Example 8.25** We derive a judgement $P: [\, F \,]^{\{s,i\}}$ for the program fragment $P$ defined as follows:

```
if n < 0 then
  s := -1
else
  s := 0; i := 0;
  while i < n do
  {
     s := s+i; i := i+1
  }
```

As shown in Example 8.24, we can derive the judgement

$$C: [\, \text{old\_}i \le \text{old\_}n \Rightarrow \text{new\_}i = \text{old\_}n \land \text{new\_}s = \text{old\_}s + \sum_{k=\text{old\_}i}^{\text{old\_}n-1} k \,]^{\{s,i\}}$$

for the loop command $C$. Clearly we can also derive the following judgement for the assignments before $C$:

$$\{s:=0;\ i:=0\}: [\, s = 0 \land i = 0 \,]^{\{s,i\}}$$

From this we can derive the judgement $E: [\, F_e \,]^{\{s,i\}}$ where $E$ is the "else" branch in above command and $F_e$ is the following transition relation:

$$F_e :\Leftrightarrow \exists s_0\colon \text{int},\ i_0\colon \text{int}.$$
$$s_0 = 0 \land i_0 = 0 \land$$
$$(i_0 \le \text{old\_}n \Rightarrow \text{new\_}i = \text{old\_}n \land \text{new\_}s = s_0 + \sum_{k=i_0}^{\text{old\_}n-1} k)$$
$$\Leftrightarrow 0 \le \text{old\_}n \Rightarrow \text{new\_}i = \text{old\_}n \land \text{new\_}s = \sum_{k=0}^{\text{old\_}n-1} k$$

As for the "then" branch, we clearly have the judgement $\text{s}:=\text{-1}: [\, \text{new\_}s = -1 \,]^{\{s\}}$ and thus also the judgement $\text{s}:=\text{-1}: [\, F_t \,]^{\{s,i\}}$ with

$$F_t :\Leftrightarrow \text{new\_}s = -1 \land \text{new\_}i = \text{old\_}i$$

We thus can derive $P : [\, F \,]^{\{s,i\}}$ with the transition relation $F$ defined as follows:

$$F :\Leftrightarrow \text{if } \text{old\_}n < 0 \text{ then } F_t \text{ else } F_e$$

$\Leftrightarrow$ if $\text{old\_}n < 0$ then

$\qquad \text{new\_}s = -1 \wedge \text{new\_}i = \text{old\_}i$

else

$$0 \leq \text{old\_}n \Rightarrow \text{new\_}i = \text{old\_}n \wedge \text{new\_}s = \sum_{k=0}^{\text{old\_}n-1} k$$

$\Leftrightarrow$ if $\text{old\_}n < 0$ then

$\qquad \text{new\_}s = -1 \wedge \text{new\_}i = \text{old\_}i$

else

$$\text{new\_}i = \text{old\_}n \wedge \text{new\_}s = \sum_{k=0}^{\text{old\_}n-1} k$$

The judgement thus describes succinctly the behavior of the program fragment:

The program fragment changes only the values of variables $s$ and $i$ (i.e., the value of variable $n$ remains unchanged). If, before the execution of the program fragment, $n$ is negative, then variable $s$ becomes $-1$ and variable $i$ remains unchanged; otherwise, $i$ becomes $n$ and $s$ becomes the sum of all natural numbers from 0 up to $n - 1$. $\qquad\qquad\qquad\qquad\qquad\qquad\qquad\qquad\qquad\quad$ □

The formal interpretation of the judgements derived by the relational calculus is given by the following proposition.

**Proposition 8.6** (Partial Soundness of State Relations) *Let C be as in Proposition 8.3 and assume that $C : [\, F \,]^{Vs}$ can be derived for a formula F and variable set Vs. Then every pair of pre- and post-state arising from an execution of C satisfies F and differs only in variables contained in Vs:*

$$C : [\, F \,]^{Vs} \Rightarrow$$
$$\forall s, s' \in State. \, [\![ C ]\!] \langle s, s' \rangle \Rightarrow$$
$$[\![ F ]\!]_{s,s'} \wedge \forall V \in Variable. \, V \notin Vs \Rightarrow s(V) = s'(V)$$

*Here $[\![ F ]\!]_{s,s'}$ denotes a variant of the semantics of F where every occurrence old\_V is evaluated as an occurrence V in state s and every occurrence of new\_V is evaluated as an occurrence V in state s'.*

> *Consequently, to verify the partial correctness of any Hoare triple $\{P\}\,C\,\{Q\}$ it suffices to show the validity of the formula $(F \Rightarrow (P' \Rightarrow Q''))$; here $P'$ respectively $Q''$ denote variants of $P$ respectively $Q$ where all variable references have been translated as indicated in Definition 8.3.*

The last statement in above proposition indicates the greater expressiveness of the relational calculus compared to the Hoare calculus. The relational calculus is able to capture the behavior of a command in a single formula $F$ referring to both the prestate $s$ and the poststate $s'$ of the command; this formula is at least as strong as the implication $(P' \Rightarrow Q'')$ derived from the precondition $P$ and postcondition $Q$ of the original Hoare triple. From the forms $P'$ and $Q''$ it becomes however clear that $P$ only refers to $s$ while $Q$ only refers to $s'$; thus the Hoare calculus requires free mathematical variables shared between $P$ and $Q$ to express relationships between $s$ and $s'$.

The relational calculus also encompasses the predicate transformer semantics.

> **Proposition 8.7** (Predicate Transformers from State Relations) *Let $C, P, Q$ be as in Proposition 8.3 and assume that $C : [\,F\,]^{Vs}$ can be derived for a formula $F$ and set $Vs$ of variables to type $Ss$. Then we have the following equivalences:*
>
> $$\mathsf{wp}(C, Q) \Leftrightarrow \forall Vs_0 \colon Ss.\ F[Vs_0/\mathsf{new\_}Vs]^0 \Rightarrow Q[Vs_0/Vs]$$
> $$\mathsf{sp}(C, P) \Leftrightarrow \exists Vs_0 \colon Ss.\ P[Vs_0/Vs] \wedge F[Vs_0/\mathsf{old\_}Vs]^0$$
>
> *Here $F[\ldots]^0$ denotes the variant of $F[\ldots]$ where variable references have been translated as indicated in Definition 8.3.*

**Example 8.26** Consider the judgement

$$\mathtt{x:=x+y}\colon [\,\mathsf{new\_}x = \mathsf{old\_}x + \mathsf{old\_}y\,]^{\{x\}}$$

Then we can derive the weakest precondition of the given assignment command for arbitrary postcondition $Q$ as follows:

$$\mathsf{wp}(\mathtt{x:=x+y}, Q) \Leftrightarrow \forall x_0 \colon \mathsf{int}.\ x_0 = x + y \Rightarrow Q[x_0/x]$$
$$\Leftrightarrow Q[x + y/x]$$

Likewise we can derive the strongest postcondition of the given assignment command for arbitrary precondition $P$ as follows:

$$\mathsf{sp}(\mathtt{x:=x+y}, P) \Leftrightarrow \exists x_0 \colon \mathsf{int}.\ Q[x_0/x] \wedge x = x_0 + y$$

Thus the derivations confirm the definitions of the predicate transformers.                 □

## 8.4     Non-abortion and Termination

So far we have dealt only with the *partial correctness* of a program which ensures that a program does not produce a wrong result: if the program terminates normally, the resulting poststate is in a certain relationship to the prestate; however, it is still possible that the program aborts with an error or runs forever. In this section, we will extend our considerations to a program's *total correctness* which also ensures *non-abortion* and *termination*.

To define a formal semantics of programming languages that deals with the issue of abortion, we have assumed in Sect. 7.2 functions $[\![\,.\,]\!]\checkmark : Term \to StateCondition$ and $[\![\,.\,]\!]\checkmark : Formula \to StateCondition$ that map a phrase (term respectively formula) to a semantic condition on states that ensures that the evaluation of the phrase in that state does not abort. Analogously, we will assume in this section functions

$$[\![\,.\,]\!]\checkmark : Term \to Formula$$
$$[\![\,.\,]\!]\checkmark : Formula \to Formula$$

that map a phrase to a syntactic *formula* that is satisfied by a state if the evaluation of the phrase does not abort (the semantics of this formula is thus the corresponding state condition).

**Total Correctness of the Hoare Calculus**

Figure 8.6 gives now an extended form of the Hoare calculus presented in Fig. 8.1 that ensures the total correctness of a program; for greater clarity the extensions have been marked by frames.

The rule for the assignment command has been extended to include as an additional precondition the well-definedness formula for the term that is evaluated. Analogously, the rules for the two-sided and the one-sided conditional statement have been extended such that the given precondition ensures the well-definedness of the condition evaluated by the corresponding statement.

While in the other commands we thus only have to ensure non-abortion, the rule for the loop command also has to ensure termination (which we will discuss in more detail below). To ensure non-abortion, this rule has an additional precondition that demands that the given invariant $I$ must imply the well-definedness formula for the loop condition. Since the invariant must hold before and after every loop iteration, this ensures that it is indeed always safe to evaluate the loop condition.

The major part of the extensions, however, deals with the problem of ensuring the termination of the loop. This problem is solved by assuming an extra element that (analogously to the loop invariant $I$) is not part of the original Hoare triple but has to be provided separately: a term $T$, called the *termination measure* or *termination term* or variously *loop variant*. This term must satisfy the following properties:

**Judgement**

$\{\sqcup\} \sqcup \{\sqcup\} \subseteq Formula \times Command \times Formula$

**Rules for $\{P\} C \{Q\}$:**

$$\frac{\models P \Rightarrow P' \quad \{P'\} C \{Q'\} \quad \models Q' \Rightarrow Q}{\{P\} C \{Q\}}$$

$$\{\boxed{[\![T]\!]\checkmark} \wedge Q[T/V]\} V := T \{Q\}$$

$$\frac{\{P\} C \{Q[V_0/V]\}}{V, V_0, V_1 \text{ distinct} \quad V_0, V_1 \text{ not in } P, C, Q}{\{(\forall V_1 : S.\ P[V_1/V])[V/V_0]\} \text{ var } V : S;\ C \{Q\}}$$

$$\frac{\{P\} C_1 \{R\} \quad \{R\} C_2 \{Q\}}{\{P\} C_1 ; C_2 \{Q\}}$$

$$\frac{\boxed{\models P \Rightarrow [\![F]\!]\checkmark} \quad \{P \wedge F\} C_1 \{Q\} \quad \{P \wedge \neg F\} C_2 \{Q\}}{\{P\} \text{ if } F \text{ then } C_1 \text{ else } C_2 \{Q\}}$$

$$\frac{\boxed{\models P \Rightarrow [\![F]\!]\checkmark} \quad \{P \wedge F\} C \{Q\} \quad \models P \wedge \neg F \Rightarrow Q}{\{P\} \text{ if } F \text{ then } C \{Q\}}$$

$$\frac{\models P \Rightarrow I \quad \boxed{\models I \Rightarrow [\![F]\!]\checkmark \wedge T \geq 0} \quad \models I \wedge \neg F \Rightarrow Q}{\{I \wedge F \wedge \boxed{T = N}\} C \{I \wedge \boxed{T < N}\}}{\{P\} \text{ while } F \text{ do } C \{Q\}}$$

**Fig. 8.6** The Hoare calculus (Total Correctness)

- The value of the term must be an integer number; this is usually ensured by the corresponding type-checking mechanism and therefore not explicitly addressed in the given rule.
- The value of the term must not be negative in the context of the loop, i.e., it must actually denote a natural number. This is ensured by requiring that the invariant $I$ implies $T \geq 0$; thus $T$ is non-negative immediately before the first loop iteration and after every subsequent iteration.
- the value of the term must be decreased by every loop iteration. In the context of the Hoare calculus, this is achieved by introducing a fresh constant $N$ and showing as part of the corresponding Hoare triple that, if $T = N$ holds before the execution of the loop body, then $T < N$ holds afterwards.

Since thus $T$ denotes a natural number that is decreased by every iteration of the loop, the number of loop iterations is bounded by the initial value of $T$. Thus the loop is only executed finitely often, i.e., the loop eventually terminates.

**Example 8.27** We derive the Hoare triple

$$\{0 \leq i \wedge i \leq n\} \text{ while } i{<}n \text{ do } \{s{:}{=}s{+}i;\ i{:}{=}i{+}1\} \{true\}$$

whose total correctness only implies that the given loop terminates. For this we choose invariant $I$ and termination measure $T$ defined as

$$I :\Leftrightarrow 0 \le i \wedge i \le n$$
$$T := n - i$$

We then have to show the following verification conditions:

$$0 \le i \wedge i \le n \Rightarrow 0 \le i \wedge i \le n$$
$$0 \le i \wedge i \le n \Rightarrow n - i \ge 0$$
$$0 \le i \wedge i \le n \wedge \neg(i < n) \Rightarrow \mathsf{true}$$

These are all clearly valid. Furthermore we have to derive the following Hoare triple:

$$\{0 \le i \wedge i \le n \wedge i < n \wedge n - i = N\}\ \mathsf{s{:}{=}s{+}i;i{:}{=}i{+}1}\ \{0 \le i \wedge i \le n \wedge n - i < N\}$$

To derive this triple, it suffices to show

$$0 \le i \wedge i \le n \wedge i < n \wedge n - i = N \Rightarrow$$
$$0 \le i + 1 \wedge i + 1 \le n \wedge n - (i + 1) < N$$

which is clearly valid.                                                                               □

   Above version of the rule for the loop command demands that the termination measure $T$ denotes a natural number that bounds the number of loop iterations. While this is indeed very often possible, the rule can be actually generalized such that the domain of $T$ is any "well-founded" domain $D$, a set with a *well-founded relation* $\_ \prec \_ \subseteq D \times D$, i.e., a relation that does not allow infinitely descending chains $x_0 \succ x_1 \succ x_2 \succ \ldots$ (where $x_i \succ x_{i+1} :\Leftrightarrow x_{i+1} \prec x_i$). The generalized version of the rule for the loop command then replaces the formula $T \ge 0$ by the formula $T \in D$ and $T < N$ by $T \prec N$.

   The set $\mathbb{N}$ with the "less-than" relation $\_ < \_ \subseteq \mathbb{N} \times \mathbb{N}$ is that example of a well-founded domain that is most frequently used in program verifications. However, another occasionally useful well-founded domain is the set $\mathbb{N}^n$ of sequences of natural numbers of some fixed length $n$ where the well-founded ordering is the *lexicographical ordering* defined as follows:

$$\_ \prec \_ \subseteq \mathbb{N}^n \times \mathbb{N}^n$$
$$x \prec y :\Leftrightarrow \exists i \in \mathbb{N}_n.\, x(i) < y(i) \wedge \forall j \in \mathbb{N}_i.\, x(j) = y(j)$$

This domain becomes particularly useful if it is not always the same value that is decreased in every loop iteration.

**Example 8.28** Consider the loop $C$ defined as follows:

```
while a > 0 ∨ b > 0 do
{
  if b > 0 then
    b := b-1
  else
  {
    b := a+1;
    a := a-1
  }
}
```

We want to verify the termination of this loop for $a$ and $b$ initialized with arbitrary natural numbers, i.e., we want to show the total correctness of the following triple:

$$\{a \geq 0 \wedge b \geq 0\} \; C \; \{\text{true}\}$$

For this the following invariant $I$ should suffice:

$$I :\Leftrightarrow a \geq 0 \wedge b \geq 0$$

As for the termination measure $T$, we investigate an example: if initially $\langle a, b \rangle = \langle 2, 1 \rangle$, the sequence of values for $\langle a, b \rangle$ after every iteration of the loop is $\langle 2, 1 \rangle \rightsquigarrow \langle 2, 0 \rangle \rightsquigarrow \langle 1, 3 \rangle \rightsquigarrow \langle 1, 2 \rangle \rightsquigarrow \langle 1, 1 \rangle \rightsquigarrow \langle 1, 0 \rangle \rightsquigarrow \langle 0, 2 \rangle \rightsquigarrow \langle 0, 1 \rangle \rightsquigarrow \langle 0, 1 \rangle \rightsquigarrow \langle 0, 0 \rangle$ until the loop terminates with $\langle a, b \rangle = \langle 0, 0 \rangle$. Thus in every iteration either variable $b$ is decremented by one while $a$ stays the same or $b$ is incremented while $a$ is decremented by 1. We correspondingly choose $T$ as follows:

$$T :\Leftrightarrow \langle a, b \rangle$$

The domain of $T$ is $\mathbb{N}^2$, i.e., $\langle a, b \rangle \in \mathbb{N}^2 \Leftrightarrow a \geq 0 \wedge b \geq 0$; the well-founded ordering $\prec$ on $\mathbb{N}^2$ is the lexicographic order on pairs of numbers:

$$[a', b'] \prec [a, b] :\Leftrightarrow a' < a \vee (a' = a \wedge b' < b)$$

To verify above triple, we have to show the following verification conditions:

$a \geq 0 \wedge b \geq 0 \Rightarrow a \geq 0 \wedge b \geq 0$

$a \geq 0 \wedge b \geq 0 \Rightarrow a \geq 0 \wedge b \geq 0$

$a \geq 0 \wedge b \geq 0 \Rightarrow (a \geq 0 \wedge b \geq 0) \wedge a \geq 0 \wedge b \geq 0 \wedge \neg(a > 0 \vee b > 0) \Rightarrow \text{true}$

These conditions are clearly valid. Furthermore, we have to show the correctness of the following Hoare triple where $B$ represents the body of the loop:

$$\{a \geq 0 \wedge b \geq 0 \wedge (a > 0 \vee b > 0) \wedge \langle a, b \rangle = N\} \; B \; \{a \geq 0 \wedge b \geq 0 \wedge \langle a, b \rangle \prec N\}$$

For this, it suffices to show the following two verification conditions

$$a \geq 0 \wedge b \geq 0 \wedge (a > 0 \vee b > 0) \wedge \langle a, b \rangle = N \wedge b > 0 \Rightarrow$$
$$a \geq 0 \wedge b - 1 \geq 0 \wedge \langle a, b - 1 \rangle \prec N$$
$$a \geq 0 \wedge b \geq 0 \wedge (a > 0 \vee b > 0) \wedge \langle a, b \rangle = N \wedge \neg(b > 0) \Rightarrow$$
$$a - 1 \geq 0 \wedge a + 1 \geq 0 \wedge \langle a - 1, a + 1 \rangle \prec N$$

From the definition of $\prec$, we have $\langle a, b - 1 \rangle \prec \langle a, b \rangle$ and $\langle a - 1, a + 1 \rangle \prec \langle a, b \rangle$; from this we can show that indeed both conditions are valid. $\qquad\square$

We now state the correctness claim of the extended version of the Hoare calculus.

---

**Proposition 8.8** (Total Soundness of the Hoare Calculus) *Let command $C$ and formulas $P$, $Q$ be as stated in Proposition 8.3. If we can according to the rules of Fig. 8.6 derive the Hoare triple $\{P\} C \{Q\}$, every execution of $C$ that starts in a state satisfying $P$ terminates normally in a state satisfying $Q$:*

$$\{P\} C \{Q\} \Rightarrow$$
$$\forall s \in State. \llbracket P \rrbracket_s \Rightarrow$$
$$\llbracket C \rrbracket \checkmark \langle s \rangle \wedge \langle\!\langle C \rangle\!\rangle \langle s \rangle \wedge (\forall s' \in State. \llbracket C \rrbracket \langle s, s' \rangle \Rightarrow \llbracket Q \rrbracket_{s'})$$

*Here $\llbracket C \rrbracket \checkmark \langle s \rangle$ respectively $\langle\!\langle C \rangle\!\rangle \langle s \rangle$ denote the well-definedness respectively termination condition of command $C$ as defined in Sect. 7.2.*
*Thus $C$ is* totally correct *with respect to precondition $P$ and postcondition $Q$.*

---

In the remainder of this section we focus on the total correctness of the other calculi presented so far.

**Total Correctness of Weakest Preconditions**

Figure 8.7 presents some extensions to the weakest precondition calculus that ensure total correctness. For assignments and conditionals, these extensions (marked in frames) are minor: we just have to add the well-definedness formula for the evaluated term respectively formula to the generated precondition.

In the case of the loop command, a universally quantified condition is added that ensures that (for all values of the program variables) the invariant implies the well-definedness of the loop condition and the non-negativity of the termination measure $T$. Furthermore, the condition to ensure the invariance of formula $I$ is extended to show that $T$ is decreased, with the help of a universally quantified variable $N$ denoting the value of $T$ in the prestate of the loop iteration.

**wp:** *Command × Formula → Formula*

$$\mathrm{wp}(V := T, Q) := \boxed{[\![T]\!]\checkmark} \wedge Q[T/V]$$

$$\mathrm{wp}(\{\mathtt{var}\ V\!:\!S;\ C\}, Q) := \forall V_1\!:\!S.\ \mathrm{wp}(C, Q[V_0/V])[V_1/V][V/V_0]$$

$$\mathrm{wp}(\{C_1;\ C_2\}, Q) := \mathrm{wp}(C_1, \mathrm{wp}(C_2, Q))$$

$$\mathrm{wp}(\{\mathtt{if}\ F\ \mathtt{then}\ C_1\ \mathtt{else}\ C_2\}, Q) := \boxed{[\![F]\!]\checkmark} \wedge \mathtt{if}\ F\ \mathtt{then}\ \mathrm{wp}(C_1, Q)\ \mathtt{else}\ \mathrm{wp}(C_2, Q)$$

$$\mathrm{wp}(\{\mathtt{if}\ F\ \mathtt{then}\ C\}, Q) := \boxed{[\![F]\!]\checkmark} \wedge \mathtt{if}\ F\ \mathtt{then}\ \mathrm{wp}(C_1, Q)\ \mathtt{else}\ Q$$

$$\mathrm{wp}(\{\mathtt{while}\ F\ \mathtt{do}\ C\}, Q) := \mathtt{let}\ I = \mathtt{choose}\ F.\ F \in Formula\ \mathtt{in}$$

$$I \wedge \boxed{(\forall Vs\!:\!Ss.\ I \Rightarrow [\![F]\!]\checkmark \wedge T \geq 0)} \wedge$$

$$(\forall Vs\!:\!Ss, \boxed{N\!:\!\mathtt{nat}}.\ I \wedge F \wedge \boxed{T = N} \Rightarrow$$

$$\mathrm{wp}(C, I \wedge \boxed{T < N})) \wedge$$

$$(\forall Vs\!:\!Ss.\ I \wedge \neg F \Rightarrow Q)$$

**Fig. 8.7** Weakest preconditions (Total Correctness)

**Proposition 8.9** (Total Soundness of Weakest Preconditions) *Let Q and C be as in Proposition 8.4. Then every execution of C that starts in a state satisfying* $\mathrm{wp}(C, Q)$ *(where the "predicate transformer"* $\mathrm{wp}$ *is defined in Fig. 8.7) terminates normally in a state satisfying Q:*

$$\forall s \in State.\ [\![\mathrm{wp}(C, Q)]\!]_s \Rightarrow$$
$$[\![C]\!]\checkmark \langle s \rangle \wedge \langle\!\langle C \rangle\!\rangle \langle s \rangle \wedge (\forall s' \in State.\ [\![C]\!]\langle s, s' \rangle \Rightarrow [\![Q]\!]_{s'})$$

*Consequently, to verify the total correctness of any Hoare triple* $\{P\}\ C\ \{Q\}$ *it suffices to show the validity of the formula* $(P \Rightarrow \mathrm{wp}(C, Q))$.

Unfortunately, corresponding extensions are not so straight-forward for the calculi of strongest postconditions or commands as relations. In the weakest precondition calculus, a command $C$ that does not terminate normally leads to an unsatisfiable weakest precondition (equivalent to "false") from which, according to Proposition 8.9, we can indeed not conclude the total correctness of any Hoare triple $\{P\}\ C\ \{Q\}$. However, in the calculi of strongest postconditions respectively of commands as relations, we must in such a situation *not* return an unsatisfiable formula, since this formula would imply the total correctness of any such triple (see Propositions 8.5 and 8.6). Alternatively, we could return a trivially satisfied postcondition respectively state relation (equivalent to true), but this makes a command that does not terminate normally indistinguishable from a non-deterministic command that always terminates but allows all possible state transitions; this is hardly a convincing solution.

**Judgement**

$\sqcup \downarrow \sqcup \subseteq Command \times Formula$

**Rules for $C \downarrow P$:**

$$\frac{\models P \Rightarrow [\![T]\!]\checkmark}{\{V := T\} \downarrow P} \qquad \frac{\models C \downarrow \exists V : T.\ P}{\{var\ V : S;\ C\} \downarrow P} \qquad \frac{C_1 \downarrow P \quad C_2 \downarrow sp(C_1, P)}{\{C_1; C_2\} \downarrow P}$$

$$\frac{\models P \Rightarrow [\![F]\!]\checkmark \quad C_1 \downarrow (P \wedge F) \quad C_2 \downarrow (P \wedge \neg F)}{\{if\ F\ then\ C_1\ else\ C_2\} \downarrow P} \qquad \frac{\models P \Rightarrow [\![F]\!]\checkmark \quad C \downarrow (P \wedge F)}{\{if\ F\ then\ C\} \downarrow P}$$

$$\frac{\models P \Rightarrow I \quad \models I \Rightarrow [\![F]\!]\checkmark \wedge T \geq 0 \quad \models sp(C, I \wedge F \wedge T = N) \Rightarrow I \wedge T < N}{\{while\ F\ do\ C\} \downarrow P}$$

**Fig. 8.8** Termination calculus (Downward Version)

**Judgement**

$\sqcup \uparrow \sqcup \subseteq Command \times Formula$

**Rules for $C \uparrow P$:**

$$\frac{}{\{V := T\} \uparrow [\![T]\!]\checkmark} \qquad \frac{\models C \uparrow P}{\{var\ V : S;\ C\} \uparrow (\forall V : T.\ P)} \qquad \frac{C_1 \uparrow P_1 \quad C_2 \uparrow P_2}{\{C_1; C_2\} \uparrow (P_1 \wedge wp(C_1, P_2))}$$

$$\frac{C_1 \uparrow P_1 \quad C_2 \uparrow P_2}{\{if\ F\ then\ C_1\ else\ C_2\} \uparrow ([\![F]\!]\checkmark \wedge if\ F\ then\ P_1\ else\ P_2)}$$

$$\frac{C \uparrow P}{\{if\ F\ then\ C\} \uparrow ([\![F]\!]\checkmark \wedge (F \Rightarrow P))}$$

$$\frac{\models I \Rightarrow [\![F]\!]\checkmark \wedge T \geq 0 \quad \models I \wedge F \wedge T = N \Rightarrow wp(C, I \wedge T < N)}{\{while\ F\ do\ C\} \uparrow I}$$

**Fig. 8.9** Termination calculus (Upward Version)

## Termination Calculi

Therefore, rather than further fiddling with the existing calculi, we will introduce a calculus whose sole purpose is to verify the normal termination of a command (independent of its partial correctness). Actually, we give two variants of such a calculus, one with a judgement $C \downarrow P$ and one with a judgement $C \uparrow P$. Both judgements have however the same interpretation which is stated as follows (Figs. 8.8 and 8.9).

**Proposition 8.10** (Soundness of Termination Calculi) *Let $P$ and $C$ be as in Proposition 8.4. If $C \uparrow P$ or $C \downarrow P$ can be derived, then every execution of $C$ that starts in a state satisfying $P$ terminates normally:*

$$C \downarrow P \Rightarrow \forall s \in State. \; [\![\, P \,]\!]_s \Rightarrow [\![\, C \,]\!]\checkmark \langle s \rangle \wedge \langle\!\langle\, C \,\rangle\!\rangle \langle s \rangle$$
$$C \uparrow P \Rightarrow \forall s \in State. \; [\![\, P \,]\!]_s \Rightarrow [\![\, C \,]\!]\checkmark \langle s \rangle \wedge \langle\!\langle\, C \,\rangle\!\rangle \langle s \rangle$$

While thus both judgements ensure that a precondition $P$ guarantees the normal termination of a command $C$, the rules for their derivation differ substantially:

- The rules for $C \downarrow P$ are designed for application in the "downward" direction: they are able to *check* whether a given prestate condition $P$ is sufficiently strong to ensure the normal termination of $C$.
- The rules for $C \uparrow P$ are designed for application in the "upward" direction: they are able to *construct* a prestate condition $P$ that is sufficiently strong to ensure the normal termination of $C$.

By its assumption of a given precondition $P$, the calculus for $C \downarrow P$ is a natural companion of the strongest postcondition calculus: to verify the total correctness of a Hoare triple $\{P\}\, C\, \{Q\}$, it suffices to verify $C \downarrow P$ and $\mathsf{sp}(P, C) \Rightarrow Q$; in fact, in its rule for command sequences this calculus uses itself a strongest postcondition.

Likewise, by its ability to compute a sufficient precondition $P$, the calculus for $C \uparrow P$ is a natural companion of the weakest precondition calculus: to verify the total correctness of a Hoare triple $\{P\}\, C\, \{Q\}$, it suffices to construct a termination formula $F_t$ with $C \uparrow F_t$ and then to verify $P \Rightarrow F_t \wedge \mathsf{wp}(C, Q)$; in fact, in the rule for command sequences this calculus uses itself a weakest precondition. However, in contrast to the extended definition of weakest preconditions given in Fig. 8.7, this calculus delivers in the rule for a loop command only the invariant $I$, i.e., the requirement that $I$ holds in the prestate of the loop; the constraints that $I$ is an invariant and ensures termination is are formulated in the premises of the rule as verification conditions that have to be established separately.

The calculus of commands as relations can make use of either of these calculi: to show $\{P\}\, C\, \{Q\}$: it suffices to derive, e.g., judgements $C \uparrow F_t$ and $C : [\, F \,]^{ls}$ and to verify $(P \Rightarrow F_t)$ and $(P' \Rightarrow (F \Rightarrow Q''))$ (with $P'$ and $Q''$ defined as in Definition 8.3).

## 8.5   Loop Invariants and Termination Measures

The previous sections have introduced various calculi for verifying the correctness of programs (actually program fragments, i.e., commands), but they have said little about the actual practice of program verification. In this section, we want to focus more on this topic with special emphasis on the problem of developing suitable loop invariants and termination measures.

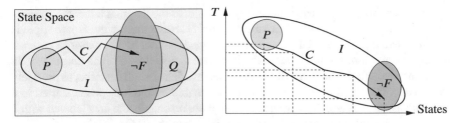

**Fig. 8.10** Visualization of loop invariant $i$ and termination measure $T$

Fundamental results of computability theory (in particular the undecidability of the halting problem shown by Alan Turing already in 1936 and more generally Rice's theorem shown by Henry Gordon Rice in 1953) tell us that there cannot exist any decision procedure that is able to automatically decide the partial correctness or the termination of every given computer program. Nevertheless in the last decades there has been more and more progress in computer-aided verification that allows to automatically or semi-automatically verify many concrete programs. However, for verifying the correctness of programs with loops, all these systems (as all the calculi presented in the previous sections) require loop invariants and termination measures from some external source. While the automated generation of this "meta-information" from given programs and specifications is an active area of research, also these efforts can only hope for partial success; currently the results are limited to programs of moderate complexity. In practice, it is today still the human verifier that has to come up with suitable loop invariants and termination measures. Actually, for most real programs, the verification of termination is less of an issue (the termination measures are comparatively simple to find); the practical source of difficulty is the development of suitable loop invariants.

In this section we discuss on the basis of a number of examples (which gradually grow in complexity) the reasoning processes ("lines of thought") that typically lead us to suitable invariants and termination measures; from these examples we may derive general "strategies" or "heuristics".

**The Role of Invariants in Verifications**

We precede our elaboration of concrete verification examples with an illustration of the general role that an invariant $I$ plays in the verification of a Hoare triple $\{P\}$ while $F$ do $C\,\{Q\}$. For this consider the left diagram depicted in Fig. 8.10 where the gray box depicts the state space in which the whole program operates; every point in this space describes one possible combination of variable values. The zigzag arrow depicts the sequence of states arising from a particular execution of the loop; every segment in this arrow denotes one state transition performed by the execution of the loop body $C$.

This execution starts in the area labeled by $P$, i.e., in a state in which the precondition of the Hoare triple holds; it terminates in the area labeled by $\neg F$, i.e., in the first state in which the loop condition does not hold any more. The Hoare triple is correct if the area labeled by $\neg F$ is completely enclosed by the area labeled by $Q$, i.e., if every state in which the loop can terminate satisfies the postcondition.

However, this condition does generally not hold: as the diagram indicates, the area labeled with $\neg F$ exceeds the area labeled with $Q$. Nevertheless the Hoare triple is still satisfied, if those states that are satisfied by $\neg F$ but not by $Q$ *cannot be reached* by any execution of the loop that starts in a state satisfied by $P$. The purpose of an invariant is exactly to show that this is the case: invariant $I$ shall describe, as precisely as possible, the set of those states that are reachable by the loop when started in a state satisfying $P$ (since the states satisfying $P$ are reachable, the area labeled with $I$ necessarily includes the area labeled with $P$). Then the Hoare triple is (partially) correct if the intersection of the set of these states with those states that satisfy $\neg F$ is contained in the set of those states satisfying $Q$. Therefore it is important to make the formula $I$ as strong as possible (i.e., the set of states denoted by $I$ as small as possible): the more precise the characterization of the set of reachable states is, the more likely it is that the verification succeeds.

The invariant plays a similar role in verifying the termination of the loop: consider the right diagram in Fig. 8.10 where the horizontal axis depicts the sequence of states in an execution of a loop and the vertical axis depicts the value assigned by the termination measure $T$ to that state. This mapping of states to values (in a well-founded domain) must ensure the following constraint: for every state that is reachable by the execution of the loop but that is not a terminal state (i.e., the state is contained in the region labeled with $I$ but not in the region labeled with $\neg F$), an execution of the loop body $C$ in that state leads to another state to which $T$ assigns a value that is smaller than the value it has assigned to the previous one; thus the execution must lead strictly "downwards" along the vertical axis. Only states that are outside the region labeled with $I$ (i.e., states that are not reachable from $P$ by the execution of the loop), are exempted from this constraint.

When performing a verification, we should remember this core purpose of an invariant as a description of the set of "reachable states" and should strive to make this description as precise as possible.

**Example: Squaring**

As our first example (which we will elaborate in somewhat more detail than the subsequent ones) we consider the following command:

```
s := 0; k := 1; i := 0;
while i < n do
{
  s := s+k;
  k := k+2;
  i := i+1
}
```

This command operates program variables $n$, $s$, $k$, and $i$; we assume that the values of these variables are natural numbers. Our goal is to verify the following Hoare triple:

$$\{n = \text{oldn}\} \ \text{s:=0; k:=1; i:=0; while i<n do} \ \dots \ \{n = \text{oldn} \wedge s = n^2\}$$

In other words, the command is expected to compute in output variable $s$ the square of input variable $n$ (which by the computation remains itself unchanged). It is easy to see that the initial part of $C$ (before the loop) yields the following Hoare triple:

$$\{n = \mathsf{oldn}\}\ \mathtt{s:=0;\ k:=1;\ i:=0}\ \{n = \mathsf{oldn} \wedge s = 0 \wedge k = 1 \wedge i = 0\}$$

Therefore it remains to derive the following triple:

$$\{n = \mathsf{oldn} \wedge s = 0 \wedge k = 1 \wedge i = 0\}\ \mathtt{while\ i\ <\ n}\ \ldots\ \{n = \mathsf{oldn} \wedge s = n^2\}$$

To verify this triple, we have to come up with a suitable loop invariant $I$ and termination measure $T$. To get some intuition into the behavior of the loop we trace the values of the program variables before/after every loop iteration for $n = \mathsf{oldn} = 5$:

| $s$ | $k$ | $i$ |
|-----|-----|-----|
| 0 | 1 | 0 |
| 1 | 3 | 1 |
| 4 | 5 | 2 |
| 9 | 7 | 3 |
| 16 | 9 | 4 |
| 25 | 11 | 5 |

Since $I$ is a loop invariant, it must be a condition that describes the relationship between the values of the program variables in *every* row of this table; furthermore, it should be a condition that is as strong as possible, i.e., it should describe the relationship between the program variables as *accurately* as possible. This principle will guide our subsequent elaboration of the invariant.

Since the loop does not change $n = \mathsf{oldn}$, we have not listed the value of $n$ in above table. Nevertheless, $I$ must record its value explicitly:

$$I :\Leftrightarrow n = \mathsf{oldn} \wedge \ldots$$

Now, considering the trace of variable $i$ across the iterations of the loop, we notice that it takes every value from 0 up to $n$:

$$I :\Leftrightarrow n = \mathsf{oldn} \wedge 0 \leq i \wedge i \leq n \wedge \ldots$$

Here we note that $i \leq n$ is weaker than the loop condition $i < n$ which is *necessarily* the case: a loop invariant must be also true after the last iteration of the loop, i.e., when the loop condition does *not* hold any more.

As for variable $k$, it is easy to see that its value is always odd. Moreover, a closer inspection reveals that actually its value directly depends on the value of $i$:

$$I :\Leftrightarrow n = \mathsf{oldn} \wedge 0 \leq i \wedge i \leq n \wedge k = 2 \cdot i + 1 \wedge \ldots$$

Finally, the value of variable $s$ can be clearly computed from $i$:

$$I :\Leftrightarrow n = \mathsf{oldn} \wedge 0 \leq i \wedge i \leq n \wedge k = 2 \cdot i + 1 \wedge s = i^2$$

Having no more idea about further properties of the program variables, we validate above proposal for $I$ by quickly checking whether it indeed holds in every possible initial state of the loop, i.e., whether it is implied by the loop's precondition. Indeed it is not difficult to to see that $I$ is satisfied for $n = \mathsf{oldn}$, $s = 0$, $k = 1$, and $i = 0$.

Furthermore, we notice that $I$ is a generalized version of the postcondition: like the postcondition it contains the condition $n = \mathsf{oldn}$ but it also has a condition $s = i^2$ that is true for every $i$ with $0 \leq i \leq n$, in particular also for the final value $n$ of $i$. From this, we are able to conclude that the postcondition $s = n^2$ holds. In general, it is typically true that the invariant of a loop contains conditions that are similar to the postcondition of the loop, because ultimately from the invariant (and the negation of the loop condition) the postcondition must be concluded.

After this first validation of the invariant (a more thorough verification will follow below), we turn our attention to the termination measure $T$, a value that needs to become smaller in every iteration. However, we only have program variables increasing in every iteration, in particular $i$ is such a variable. Therefore we consider the negation of $i$ as a first candidate for $T$:

$$T := -i$$

Across the loop iterations this yields the following values of $T$:

| $i$ | $T$ |
|---|---|
| 0 | 0 |
| 1 | −1 |
| 2 | −2 |
| 3 | −3 |
| 4 | −4 |
| 5 | −5 |

While $T$ therefore gets indeed smaller in every iteration, it also becomes clearly negative with minimum $-5 = -n$. Thus to ensure the non-negativity of $n$ we augment $T$ by a "correction" $n$:

$$T := -i + n = n - i$$

This correction indeed yields a suitable sequence of values:

| $i$ | $T$ |
|---|---|
| 0 | 5 |
| 1 | 4 |
| 2 | 3 |
| 3 | 2 |
| 4 | 1 |
| 5 | 0 |

Thus we may attempt our verification with the following definitions:

$$I :\Leftrightarrow n = \mathsf{oldn} \wedge 0 \leq i \wedge i \leq n \wedge k = 2 \cdot i + 1 \wedge s = i^2$$
$$T := n - i$$

For this it suffices to show the following verification conditions

$$n = \mathsf{oldn} \wedge s = 0 \wedge k = 1 \wedge i = 0 \Rightarrow I \tag{A}$$
$$I \Rightarrow T \geq 0 \tag{B}$$
$$I \wedge \neg(i < n) \Rightarrow n = \mathsf{oldn} \wedge s = n^2 \tag{D}$$

and to derive the following Hoare triple:

$$\{I \wedge T = N\}\ \texttt{s:=s+k; k:=k+2; i:=i+1}\ \{I \wedge T < N\}$$

By weakest precondition reasoning, this Hoare triple can be reduced to the the following verification condition:

$$I \wedge T = N \Rightarrow$$
$$I[i + 1/i][k + 2/k][s + k/s] \wedge$$
$$T[i + 1/i][k + 2/k][s + k/s] < N \tag{C}$$

The first three verification conditions can be easily checked: (A) immediately follows from the equalities in the prerequisite of the implication. From $I$, we have $i \leq n$ and thus (B). From $I$ and $\neg(i < n)$ we have $s = i^2$ and $i = n$ and thus (D).

It remains to verify (C):

$$n = \mathsf{oldn} \wedge 0 \leq i \wedge i \leq n \wedge k = 2 \cdot i + 1 \wedge s = i^2 \wedge T = N \Rightarrow$$
$$n = \mathsf{oldn} \wedge 0 \leq i + 1 \wedge i + 1 \leq n \wedge$$
$$k + 2 = 2 \cdot (i + 1) + 1 \wedge s + k = (i + 1)^2 \wedge$$
$$n - (i + 1) < N \tag{C}$$

This condition can be checked by simple substitution and arithmetic reasoning. In particular, we can derive the core condition by the following sequence of equalities:

$$s + k = i^2 + (2 \cdot (i + 1) + 1 - 2) = i^2 + 2 \cdot i + 1 = (i + 1)^2$$

This completes (our sketch of) the verification.

**The Development of Invariants**

Before dealing with further examples, we summarize above exposition to some general recommendations for the practical development of invariants that are suitable to "carry" the verification of a loop command:

- **Investigate variable traces:** choose concrete input values and derive from these traces of the values of the program variables, starting with the values immediately before the start of the loop and ending with values immediately after its termination. Investigate these traces to determine the general relationships between the values of the program variables in all iterations of the loop (i.e, in all rows of the trace). Do not forget to consider also those variables that are not changed by the loop.
- **Consider the loop condition:** a weakened variant of the loop condition (but not the loop condition itself) is typically a part of the invariant. This condition must not only hold before every subsequent loop iteration but also after the last one, i.e., when the loop condition does not hold any more.
- **Consider the precondition:** the invariant must hold before the *first* loop iteration, i.e., it must be implied by the precondition of the loop. Check that this is indeed the case before proceeding any further with the other parts of the verification (it is easy to find subtle errors in the invariant already by performing this check).
- **Consider the postcondition:** upon termination of the loop, the loop invariant must imply the postcondition; thus the same "kinds" of formulas that appear in the postcondition must also appear in the invariant. Check that the loop invariant together with the negation of the loop condition indeed implies the postcondition, before proceeding any further with the other parts of the verification (there is no point in proceeding if the invariant does not achieve its core goal).

As the last (and typically most difficult) part of a verification, we have to check that the invariant is indeed preserved by every iteration of the loop. If this fails, it may essentially due to two reasons:

- the invariant is too strong: it claims too much about the poststate of an iteration;
- the invariant is too weak: it assumes too little about the prestate of an iteration.

More often than not a (proposed) invariant suffers from the second problem, potentially even from both problems: it assumes too little and claims too much. Here ingenuity is needed to determine step by step the source of a problem and to fix it by appropriately strengthening and/or weakening the invariant. Since after every change to the invariant the verification has to be repeated, the verification of programs with loops is typically an *iterative* process that requires patience and diligence.

**Example: Searching**

As our next example, we consider the following program fragment:

```
i := 0; r := -1;
while i < n ∧ r = -1 do
{
  if a[i] = x
    then r := i
    else i := i+1
}
```

This command performs "linear search" in the first $n$ positions of an array $a$ for an element $x$. In detail, the command shall satisfy the Hoare triple

$$\{P\}\ \texttt{i := 0; r := -1; while i < n} \land \texttt{r = -1 do} \ldots \{Q\}$$

where precondition $P$ and postcondition $Q$ are defined as follows:

$P :\Leftrightarrow 0 \le n \land n \le |a| \land a = \mathsf{olda} \land n = \mathsf{oldn} \land x = \mathsf{oldx}$

$Q :\Leftrightarrow P \land$
    if $\forall k \in \mathbb{Z}.\ 0 \le k \land k < n \Rightarrow a(k) \ne x$
       then $r = -1$
       else $0 \le r \land r < n \land a(r) = x \land \forall k \in \mathbb{Z}.\ 0 \le k \land k < r \Rightarrow a(k) \ne x$

In other words, if $x$ does not occur in the first $n$ positions of $a$, then $r$ is set to $-1$; otherwise $r$ is set to the smallest position where $x$ occurs. The program leaves the input values unchanged and thus preserves the precondition.

By strongest postcondition reasoning, it suffices to derive the following triple:

$$\{P \land i = 0 \land r = -1\}\ \texttt{while i < n} \land \texttt{r = -1 do} \ldots \{Q\}$$

To get ideas for a loop invariant and termination measure that are suitable to verify this triple, we choose example inputs $a := [2, 3, 5, 7, 11]$, $n = 5$, and $x = 7$; from these we get the following trace for the variables modified by the command:

| $i$ | $r$ |
|---|---|
| 0 | $-1$ |
| 1 | $-1$ |
| 2 | $-1$ |
| 3 | $-1$ |
| 3 | 3 |

As we can see, in every row of this trace either $i$ or $r$ is changed. Every but the last iteration increments $i$ by 1 and leaves $r$ as $-1$; only the last loop iteration sets $r$ to $i$ and leaves $i$ unchanged.

However, since this trace only covers the case of a successful search, we also choose inputs $a := [2, 3, 5, 7, 11]$, $n = 5$, and $x = 13$; from these we derive the following second trace that describes a failed search:

| $i$ | $r$ |
|---|---|
| 0 | $-1$ |
| 1 | $-1$ |
| 2 | $-1$ |
| 3 | $-1$ |
| 4 | $-1$ |
| 5 | $-1$ |

Here ultimately $i$ reaches $n = 5$ and $r$ always remains $-1$.

Since the program leaves the input variables unchanged, we start (as also suggested by the postcondition $Q$) with the following approximation of invariant $I$:

$$I :\Leftrightarrow P \land \ldots$$

Since $i$ runs (in both traces) only over the indices of $a$ up to maximum $n$, we can immediately refine this definition as follows:

$$I :\Leftrightarrow P \land 0 \le i \land i \le n \land \ldots$$

However, the further characterization of $i$ and $r$ requires deeper investigations. As a starting point, we notice that $r$ is always $-1$ except in the last state in the first trace, i.e., the last transition from state $i = 3$ and $r = -1$ is distinctly different from the other ones; therefore we must investigate, why we have proceeded from that state to $i = 3$ and $r = 3$ rather than to $i = 4$ and $r = -1$.

Clearly the critical point is that for $i = 3$ we have $a(i) = x$; therefore the transition to $i = 4$ must be prevented by a condition that implies $a(i - 1) \ne x$. We may be attempted to add this formula itself to the invariant, but should remember that $Q$ demands for the output variable $r$ the condition ($\forall k \in \mathbb{Z}. 0 \le k \land k < r \Rightarrow a(k) \ne x$); since after the initialization of $r$ the value of this variable is only changed by the assignment r := i, this stronger condition (that implies $a(i - 1) \ne x$) must also hold for $i$. Therefore we extend the invariant correspondingly:

$$I :\Leftrightarrow P \land 0 \le i \land i \le n \land (\forall k \in \mathbb{Z}. 0 \le k \land k < i \Rightarrow a(k) \ne x) \land \ldots$$

As for an appropriate characterization of the variable $r$, clearly $r = -1$ holds in all states except when $r = i$ and $a(i) = x$. Therefore we attempt the following invariant (we will see later, however, that it is still not fully adequate):

$$I :\Leftrightarrow P \land 0 \le i \land i \le n \land (\forall k \in \mathbb{Z}. 0 \le k \land k < i \Rightarrow a(k) \ne x) \land$$
$$(r = -1 \lor (r = i \land a(i) = x))$$

As for the termination measure $T$, we notice that a failed search always increases $i$ up to an upper bound $n$, thus we might try $T := n - i$. However, in a successful search this term is only decreased as long as $r$ is $-1$; it is not decreased in the last iteration of the first run by which $r$ becomes $i$. Anyway, we can easily capture this behavior by the following definition that utilizes a conditional term:

$$T := \text{if } r = -1 \text{ then } 0 \text{ else } n - i$$

This gives in the successful search rise to the following trace of $T$:

| $i$ | $r$ | $T$ |
|---|---|---|
| 0 | −1 | 5 |
| 1 | −1 | 4 |
| 2 | −1 | 3 |
| 3 | −1 | 2 |
| 3 | 3 | 0 |

Here the last value $T = 0$ demands the immediate termination of the loop.
  Now, using above definitions of $I$ and $T$, the verification of the triple

$$\{P \wedge i = 0 \wedge r = -1\} \texttt{ while i < n } \wedge \texttt{ r = -1 do} \ldots \{Q\}$$

demands the validity of the verification conditions

$$P \wedge i = 0 \wedge r = -1 \Rightarrow I \tag{A}$$
$$I \Rightarrow T \geq 0 \tag{B}$$
$$I \wedge \neg(i < n \wedge r = -1) \Rightarrow Q \tag{D}$$

and the derivation of the following triple:

$$\{I \wedge i < n \wedge r = -1 \wedge T = N\} \texttt{ if a[i] = x} \ldots \{I \wedge T < N\}$$

It is easy to check the validity of conditions (A) and (B). However, one branch of the proof of condition (D) requires to show, from the assumptions $i \leq n$ and $r = i$, the conclusion $r < n$; this does generally not hold, because the case $i = n$ is possible. The culprit is our definition of $I$ which has to be strengthened as follows:

$$I :\Leftrightarrow P \wedge 0 \leq i \wedge i \leq n \wedge (\forall k \in \mathbb{Z}. \, 0 \leq k \wedge k < i \Rightarrow a(k) \neq x) \wedge$$
$$(r = -1 \vee (r = i \wedge \underline{i < n} \wedge a(i) = x))$$

In our previous definition, we have overlooked that the situation $r = i$ can be only established in the body of the loop when the loop condition $i < n$ still holds. With this strengthened definition, all the conditions (A), (B), and (D) are valid.
  Furthermore, the derivation of above Hoare triple can be reduced to the validity of the following conditions:

$$I \wedge i < n \wedge r = -1 \wedge T = N \Rightarrow 0 \leq i \wedge i < |a| \tag{C1}$$
$$I \wedge i < n \wedge r = -1 \wedge T = N \wedge a(i) = x \Rightarrow I[i/r] \wedge T[i/r] < N \tag{C2}$$
$$I \wedge i < n \wedge r = -1 \wedge T = N \wedge \neg(a(i) = x) \Rightarrow I[i + 1/i] \wedge T[i + 1/i] < N \tag{C3}$$

Here the first condition arises from showing the validity of the well-definedness condition $[\![ a[i] ]\!]\checkmark \Leftrightarrow 0 \le i \land i < |a|$; the validity of all three conditions can be shown by simple substitutions and arithmetic reasoning.

## Example: Partitioning

Our next example deals with the "partitioning" part of the *Quicksort* algorithm:

```
i := m; j := n-1;
while i ≤ j do
{
  if a[i] ≤ x then
    i := i+1
  else if x ≤ a[j] then
    j := j-1
  else
  { var e:int; e := a[i]; a[i] := a[j]; a[j] := e }
}
```

This program fragment rearranges those elements of integer array $a$ that are in the index range $m \ldots n - 1$ into two parts. The first part only contains values less than equal some given pivot value $x$ while the second part only contains values greater than equal $x$. The final value of variable $i$ indicates the boundary between these parts. We will later show an example of the operation of this code and focus now on its formal specification.

In detail, we expect this code fragment to satisfy the Hoare triple

$$\{P\}\, \text{i := m; j := n-1; while i < j do } \{\ldots\}\, \{Q\}$$

with precondition $P$ and postcondition $Q$ defined as follows:

$$
\begin{aligned}
P :&\Leftrightarrow 0 \le m \land m \le n \land n \le |a| \land \\
&\quad a = \text{olda} \land m = \text{oldm} \land n = \text{oldn} \land x = \text{oldx} \\
Q :&\Leftrightarrow P \land \text{permuted}\langle a, \text{olda}\rangle \land m \le i \land i \le n \land \\
&\quad (\forall k \in \mathbb{Z}.\, m \le k \land k < i \Rightarrow a(k) \le x) \land \\
&\quad (\forall k \in \mathbb{Z}.\, i \le k \land k < n \Rightarrow a(k) \ge x)
\end{aligned}
$$

The postcondition embeds the precondition, which ensures that the values of the input variables are not changed. The predicate *permuted*$\langle a, \text{olda}\rangle$ ensures that $a$ is a *permutation* of olda, i.e, that the array has after the execution of the command the same elements that it had before:

$permuted \subseteq \mathbb{Z}^* \times \mathbb{Z}^*$

$permuted\langle a, b\rangle :\Leftrightarrow$

   let $n = |a|$ in $n = |b| \wedge$

   $\exists p \in \mathbb{Z}^n.$

      $\left(\forall k \in \mathbb{Z}. 0 \le k \wedge k < n \Rightarrow 0 \le p(k) \wedge p(k) < n\right) \wedge$

      $\left(\forall k_1 \in \mathbb{Z}, k_2 \in \mathbb{Z}. 0 \le k_1 \wedge k_1 < k_2 \wedge k_2 < n \Rightarrow p(k_1) \ne p(k_2)\right) \wedge$

      $\left(\forall k \in \mathbb{Z}. 0 \le k \wedge k < n \Rightarrow a(k) = b(p(k))\right)$

Here $p$ denotes a mapping of every index $i$ of $a$ to that index $p(i)$ of $b$ where $b$ holds the element that $a$ holds at $i$. For instance, $a = [2, 3, 5]$ is a permutation of $b = [5, 2, 3]$ for $p = [1, 2, 0]$ (i.e., $p(0) = 1$, $p(1) = 2$, and $p(2) = 0$), because $a(0) = 2 = b(1)$, $a(1) = 3 = b(2)$, and $a(2) = 5 = b(0)$. The mapping is bijective because it $p$ is not allowed to map two indices of $a$ to the same index in $b$; therefore we have indeed a one-to-one correspondence between the elements of $a$ and the elements of $b$.

By strongest postcondition reasoning, it clearly suffices to verify the following Hoare triple:

$$\{P \wedge i = m \wedge j = n - 1\} \text{ while } \text{i} < \text{j} \text{ do } \{\ldots\} \{Q\}$$

To determine a suitable invariant $I$ and termination term $T$ for this verification, we investigate the following trace of a sample program run for the input values $a = [2, 6, 7, 4, 1, 3, 5]$, $m = 0$, $n = 7$, and $x = 4$:

| $i$ | $j$ | $a$ | |
|---|---|---|---|
| 0 | 6 | $[2, 6, 7, 4, 1, 3, 5]$ | |
| 1 | 6 | $[2, 6, 7, 4, 1, 3, 5]$ | |
| 1 | 5 | $[2, \underline{6}, 7, 4, 1, 3, 5]$ | |
| 1 | 5 | $[2, \mathbf{3}, 7, 4, 1, \mathbf{6}, 5]$ | $\leftarrow$ |
| 2 | 5 | $[2, 3, 7, 4, 1, 6, 5]$ | |
| 2 | 4 | $[2, 3, \underline{7, 4, 1}, 6, 5]$ | |
| 2 | 4 | $[2, 3, \mathbf{1}, 4, \mathbf{7}, 6, 5]$ | $\leftarrow$ |
| 3 | 4 | $[2, 3, 1, \underline{4, 7}, 6, 5]$ | |
| 4 | 4 | $[2, 3, 1, 4, \underline{7}, 6, 5]$ | |
| 4 | 3 | $[2, 3, 1, 4, 7, 6, 5]$ | |

Here the underlined part of $a$ indicates the index range $i \ldots j$, the arrows mark those arrays where the array has been changed; the modified elements are typeset in bold style. We notice that in every iteration of the loop either $i$ is incremented or $j$ is decremented (in either case the index range shrinks) or the elements $a(i)$ and $a(j)$ have been swapped.

Considering postcondition $Q$, our first sketch of invariant $I$ is naturally as follows:

$$I :\Leftrightarrow P \wedge permuted\langle a, \mathsf{olda}\rangle \wedge \ldots$$

Thus $I$ states that precondition $P$ must be preserved and every iteration $a$ must remain a permutation of the original array.

Also naturally the postcondition formula $m \leq i \wedge i \leq n$ lets us consider the range conditions for variables $i$ and $j$. Since $i$ is initialized with $m$ and only incremented, we clearly have $m \leq i$; likewise, since $j$ is initialized with $n - 1$ and only decremented, we have $j \leq n - 1$. Since the increments take place in steps of one, the loop condition $i \leq j$ gives us $i \leq j + 1$. Thus we have our next approximation of the invariant:

$$I :\Leftrightarrow P \wedge permuted\langle a, \mathsf{olda}\rangle \wedge m \leq i \wedge i \leq j + 1 \wedge j \leq n - 1 \wedge \ldots$$

Here we may note that $i \leq j + 1$ and $j \leq n - 1$ ensures the desired postcondition $i \leq n$. However, we should also note that the stronger range conditions $i < n$ and $0 < j$ do generally *not* hold:

- If $a$ only holds values greater than $x$, we increment $i$ up to $i = n$.
- If $a$ only holds values less than $x$ and $m = 0$, we decrement $j$ down to $j = -1$.

As for the second condition, indeed the inequalities $0 \leq m$, $m \leq i$, and $i \leq j + 1$ (which is equivalent to $i - 1 \leq j$) only ensure $0 \leq i$ and thus $-1 \leq j$. Therefore any attempt to add $i < n$ and $0 < j$ to the invariant will lead to a failure in the proof that the invariant is preserved by the assignment statements that increment $i$ respectively decrement $j$ in the body of the loop.

Further hints for the refinement of the invariant are given by the two postcondition formulas $(\forall k \in \mathbb{Z}. \ m \leq k \wedge k < i \Rightarrow a(k) \leq x)$ and $(\forall k \in \mathbb{Z}. \ i \leq k \wedge k < n \Rightarrow a(k) \geq x)$ which must hold for the final value of variable $i$. Actually, as the trace indicates, the first formula holds through all iterations, leading us to our next version of the invariant:

$$I :\Leftrightarrow P \wedge permuted\langle a, \mathsf{olda}\rangle \wedge m \leq i \wedge i \leq j + 1 \wedge j \leq n - 1 \wedge$$
$$(\forall k \in \mathbb{Z}. \ m \leq k \wedge k < i \Rightarrow a(k) \leq x) \wedge \ldots$$

The second formula, however, only holds after the final iteration of the loop, i.e., for $i = j + 1$, which leads to $(\forall k \in \mathbb{Z}. \ j + 1 \leq k \wedge k < n \Rightarrow a(k) \geq x)$ or equivalently to the slightly simpler formula $(\forall k \in \mathbb{Z}. \ j < k \wedge k < n \Rightarrow a(k) \geq x)$. Indeed the trace confirms that this formula holds through all loop iterations. Thus we get the next version of the invariant:

$$I :\Leftrightarrow P \wedge permuted\langle a, \mathsf{olda}\rangle \wedge m \leq i \wedge i \leq j + 1 \wedge j \leq n - 1 \wedge$$
$$(\forall k \in \mathbb{Z}.\ m \leq k \wedge k < i \Rightarrow a(k) \leq x) \wedge$$
$$(\forall k \in \mathbb{Z}.\ j < k \wedge k < n \Rightarrow a(k) \geq x)$$

Since there is no further information apparent in the example trace, we will attempt the verification with this formulation.

As for the termination measure $T$, we notice that in most loop iterations the size of the index range $i \ldots j$ (i.e., the value of the term $j - i + 1$) shrinks by one of the first two branches of the nested conditional statements; this leads us to our first (not yet correct) attempt:

$$T := j - i + 1$$

However, those iterations where the third branch is executed leave the value of $i$ and $j$ unchanged, i.e., they do not decrease the value of $T$. To also account also for these iterations, we have to determine what measure is decreased by the third branch. Since the only effect of this branch is to swap the values of the array elements $a[i]$ and $a[j]$, the measure has to relate to the content of this array; it must be bigger before the swap than afterwards. Furthermore, since the swap only affects the order of the elements in the array, the measure has to refer to this order.

Here we notice that the swap is only performed if the first branch condition $a[i] \leq x$ does not hold, and also the second branch condition $a[j] \geq x$ does not hold, where $i$ and $j$ are values satisfying the loop condition $i \leq j$. In other words, a swap is only performed if the loop encounters values $i$ and $j$ for which the condition $(i \leq j \wedge a[i] > x \wedge a[j] < x)$ does hold. After the swap, however, this condition clearly does not hold any more, i.e., we have one pair $\langle i, j \rangle$ of indices less for which the conditions hold. In total, thus the number of all pairs of such values $i$ and $j$ in the index range $m \ldots n - 1$ has been decreased. Since either the size of the iteration range or this number is changed, we may take the *sum* of these two values as the measure and write the following definition:

$$T := (j - i + 1) +$$
$$(\#i \in \mathbb{Z}, j \in \mathbb{Z}.\ m \leq i \wedge i \leq j \wedge j \leq n - 1 \wedge a[i] > x \wedge a[j] < x)$$

Here the quantified term $(\#x.\ F) := |\{x \mid F\}|$ denotes the number of all values for variable $x$ (respectively a combination of such variables) that satisfies formula $F$ (provided that $F$ is only satisfied by a finite set of such values).

For the already shown example with $x = 4$, the value of the termination measure changes as follows:

| $i$ | $j$ | $a$ | $j - i + 1$ | # | $T$ |
|---|---|---|---|---|---|
| 0 | 6 | $[2, 6, 7, 4, 1, 3, 5]$ | 7 | 4 | 11 |
| 1 | 6 | $[2, \underline{6}, 7, 4, 1, 3, 5]$ | 6 | 4 | 10 |
| 1 | 5 | $[2, \underline{6}, 7, 4, 1, 3, 5]$ | 5 | 4 | 9 |
| 1 | 5 | $[2, \mathbf{3}, 7, 4, 1, \mathbf{6}, 5]$ | 5 | 1 | 6 |
| 2 | 5 | $[2, 3, \underline{7}, 4, 1, 6, 5]$ | 4 | 1 | 5 |
| 2 | 4 | $[2, 3, \underline{7}, 4, 1, 6, 5]$ | 3 | 1 | 4 |
| 2 | 4 | $[2, 3, \mathbf{1}, 4, \mathbf{7}, 6, 5]$ | 3 | 0 | 3 |
| 3 | 4 | $[2, 3, 1, \underline{4}, \underline{7}, 6, 5]$ | 2 | 0 | 2 |
| 4 | 4 | $[2, 3, 1, 4, \underline{7}, 6, 5]$ | 1 | 0 | 1 |
| 4 | 3 | $[2, 3, 1, 4, 7, 6, 5]$ | 0 | 0 | 0 |

Here the column marked as "#" indicates the number of pairs for which the "swap condition" holds: for $a = [2, 6, 7, 4, 1, 3, 5]$ we have $a(1) = 6 > 4$ but $a(4) = 1 < 4$ and $a(5) = 3 < 4$ as well as $a(2) = 7 > 4$ but $a(4) = 1 < 4$ and $a(5) = 3 < 4$, thus four pairs of indices satisfying this condition. After the first swap, we have $a(2) = 7 > 4$ but $a(4) = 1 < 4$, thus only only one such pair; after the second swap, we have no more such pair any more. We see that in the two cases where the index range does not shrink, the number of pairs of values satisfying the satisfying the swap condition does; thus the termination measure decreases in any case.

Now, to verify the Hoare triple

$$\{P \wedge i = m \wedge j = n - 1\} \, \texttt{while i < j do} \, \{\ldots\} \, \{Q\}$$

it is necessarily to show the validity of the verification conditions

$$P \wedge i = m \wedge j = n - 1 \Rightarrow I \tag{A}$$
$$I \Rightarrow T \geq 0 \tag{B}$$
$$I \wedge \neg(i < j) \Rightarrow Q \tag{D}$$

which are easy to check. Furthermore, we have to derive the Hoare triple

$$\{I \wedge i \leq j \wedge T = N\} \, C \, \{I \wedge T < N\}$$

where $C$ denotes the body of the loop. By weakest precondition reasoning this reduces to verifying the following verification conditions:

$$I \wedge i < n \wedge T = N \wedge a(i) \leq x \Rightarrow$$
$$I[i + 1/i] \wedge T[i + 1/i] < N \tag{C1}$$
$$I \wedge i < n \wedge T = N \wedge \neg(a(i) \leq x) \wedge x \leq a(j) \Rightarrow$$
$$I[j + 1/j] \wedge T[j + 1/j] < N \tag{C2}$$
$$I \wedge i \leq j \wedge T = N \wedge \neg(a(i) \leq x) \wedge \neg(x \leq a(j)) \Rightarrow$$
$$I[a[i \mapsto a(j)][j \mapsto a(i)]/a] \wedge T[a[i \mapsto a(j)][j \mapsto a(i)]/a] < N \tag{C3}$$

Here in condition (C3) the formula $I[a[i \mapsto a(j)][j \mapsto a(i)]/a]$ respectively the term $T[a[i \mapsto a(j)][j \mapsto a(i)]/a]$ are logically equivalent to the more complex formula $I'$ respectively $T'$ that are actually derived by weakest precondition calculation:

$$I' :\Leftrightarrow \forall e \in \mathbb{Z}.\ I[a[j \mapsto e]/a][a[i \mapsto a(j)]/a][a(i)/e]$$
$$T' :\Leftrightarrow \forall e \in \mathbb{Z}.\ T[a[j \mapsto e]/a][a[i \mapsto a(j)]/a][a(i)/e]$$

Conditions (C1) and (C2) are relatively straight-forward to show; the main challenge is posed by condition (C3) which requires reasoning about array updates. For example, we have to show for arbitrary $k \in \mathbb{Z}$ with $m \leq k \wedge k < i$ the following goal:

$$a[i \mapsto a(j)][j \mapsto a(i)](k) \leq x$$

Since in the context of this proof we have the assumptions $k < i$ and $i < j$, the array axioms reduce this goal to

$$a(k) \leq x$$

which follows from the formula $(\forall k \in \mathbb{Z}.\ m \leq k \wedge k < i \Rightarrow a(k) \leq x)$ contained in assumption $I$. We skip over the further details of this verification.

**Example: Sorting**

Our final example deals with the classical problem of sorting an integer array which we solve by the well-known *insertion sort* algorithm:

```
var i:int; i := 1;
while i < a.length do
{
  var x:int; x := a[i];
  var j:int; j := i-1;
  while j ≥ 0 ∧ a[j] > x do
  {
    a[j+1] := a[j];
    j := j-1
  }
  a[j+1] := x;
  i := i+1
}
```

We expect this code fragment to satisfy the Hoare triple

$$\{P_1\}\ \texttt{var i:int; i:=1; while ...}\ \{Q_1\}$$

for the following definitions of precondition $P_1$ and postcondition $Q_1$:

$$P_1 :\Leftrightarrow a = \mathsf{olda}$$
$$Q_1 :\Leftrightarrow \mathit{permuted}\langle a, \mathsf{olda}\rangle \wedge \mathit{sorted}\langle a, |a|\rangle$$

We have already introduced the predicate *permuted* in the previous example; the predicate *sorted* is defined as follows:

$$sorted \subseteq \mathbb{Z}^* \times \mathbb{Z}^*$$

$$sorted\langle a, n\rangle :\Leftrightarrow \forall i \in \mathbb{Z}.\ 0 \leq i \wedge i < n \Rightarrow a(i) \leq a(i+1)$$

Thus the code shall construct a permutation of the original array contents that is sorted in ascending order.

To verify this claim, it suffices to verify the following triple:

$$\{P_1 \wedge i = 1\}\ \texttt{while i < a.length do}\ \{\ldots\}\ \{Q_1\}$$

To investigate the behavior of the code, we construct for $a = [5, 0, 2, 3, 1, 4]$ a variable trace that depicts the values of $i$ and $a$ after every iteration of the main (outer) loop:

| $i$ | $a$ |
|---|---|
| 1 | $[\underline{5}, 0, 2, 3, 1, 4]$ |
| 2 | $[\underline{0, 5}, 2, 3, 1, 4]$ |
| 3 | $[\underline{0, 2, 5}, 3, 1, 4]$ |
| 4 | $[\underline{0, 2, 3, 5}, 1, 4]$ |
| 5 | $[\underline{0, 1, 2, 3, 5}, 4]$ |
| 6 | $[\underline{0, 1, 2, 3, 4, 5}]$ |

Here the underlined part denotes the index range of the array less than $i$ that has already been sorted. Clearly at every iteration of the loop, $a$ is a permutation of the original array; we therefore start with the following approximation of the invariant $I_1$ of the outer loop:

$$I_1 :\Leftrightarrow permuted\langle a, \mathsf{olda}\rangle \wedge \ldots$$

Variable $i$ starts with value 1 and is only incremented; thus clearly $1 \leq i$ always holds. From the trace, it is also tempting to assume $i \leq |a|$, but this is actually not true for $|a| = 0$. We thus extend our invariant a bit more carefully as follows:

$$I_1 :\Leftrightarrow permuted\langle a, \mathsf{olda}\rangle \wedge 1 \leq i \wedge (|a| \geq 1 \Rightarrow i \leq |a|) \wedge \ldots$$

From the trace it is clear that the part of the array left to $i$ is sorted while the part extending from $i$ to right has not yet been modified. These considerations give us our final definition of $I_1$:

$$I_1 :\Leftrightarrow permuted\langle a, \mathsf{olda}\rangle \wedge 1 \leq i \wedge (|a| \geq 1 \Rightarrow i \leq |a|) \wedge$$
$$sorted\langle a, i\rangle \wedge equals(a, \mathsf{olda}, i, |a| - 1)$$

Here predicate *equals* is defined as follows:

$$equals \subseteq \mathbb{Z}^* \times \mathbb{Z}^* \times \mathbb{Z} \times \mathbb{Z}$$
$$equals\langle a, b, m, n\rangle :\Leftrightarrow \forall k \in \mathbb{Z}.\ m \le k \land k \le n \Rightarrow a(k) = b(k)$$

Thus $equals\langle a, b, m, n\rangle$ states that arrays $a$ and $b$ are equal in index range $[m, n]$.

Since $i$ is incremented in every iteration of the outer loop but does not exceed the length of $a$, its termination measure $T_1$ is straight-forward:

$$T_1 := |a| - i$$

Now, to verify the correctness of

$$\{P_1 \land i = 1\}\ \texttt{while i < a.length do }\{\ldots\}\ \{Q_1\}$$

it is first necessary to to show the conditions

$$a = \mathsf{olda} \land i = 1 \Rightarrow I_1 \tag{A}$$
$$I_1 \Rightarrow T_1 \ge 0 \tag{B}$$
$$I_1 \land \neg(i < |a|) \Rightarrow permuted\langle a, \mathsf{olda}\rangle \land sorted\langle a, |a|\rangle \tag{D}$$

which is rather straight-forward. However, we also have to derive the Hoare triple

$$\{I_1 \land i < |a| \land T_1 = N_1\}\ \texttt{var x:int; } \ldots \texttt{; i:=i+1}\ \{I_1 \land T_1 < N_1\}$$

which, by strongest postcondition and weakest precondition reasoning, boils down to the verification of the condition

$$I_1 \land i < |a| \land T_1 = N_1 \land i < |a| \Rightarrow 0 \le i \land i < |a|$$

(which arises from the well-definedness formulas of the terms a[i] and a[j+1] evaluated in the body of the outer loop and is easy to show) and to the derivation of the following triple:

$$\{P_2\}\ \texttt{while j} \ge \texttt{0} \land \texttt{a[j] > x do }\{\ldots\}\ \{Q_2\}$$

This triple represents the verification of the inner loop for precondition $P_2$ and postcondition $Q_2$ defined as follows:

$$P_2 :\Leftrightarrow I_1 \land i < |a| \land T_1 = N_1 \land x = a(i) \land j = i - 1$$
$$Q_2 :\Leftrightarrow I_1[i + 1/i][a[j + 1 \mapsto x]/a] \land T_1[i + 1/i][a[j + 1 \mapsto x]/a] < N_1$$

It is this verification of the inner loop which poses the real challenge. Actually we will verify the Hoare triple

$$\{P_2 \land a = \mathsf{olda2}\}\ \texttt{while j} \ge \texttt{0} \land \texttt{a[j] > x do }\{\ldots\}\ \{Q_2\}$$

where olda2 is a new constant that represents the state of the array before the execution of the inner loop; this constant will become instrumental in the subsequent verification. Nevertheless, since this constant does not appear in the rest of the triple, the original triple is valid if and only if the new one is.

To better understand the behavior of the inner loop, we consider the situation $i = 4$ and $a = [0, 2, 3, 5, 1, 4]$ derived from above trace for the outer loop. With these initial values for the body of the outer loop, we start the execution of the inner loop with $x = a(i) = 1$ and $j = i - 1 = 3$. This gives rise to the following variable trace:

| $j$ | $a$ |
|---|---|
| 3 | $[0, 2, 3, 5, 1, 4]$ |
| 2 | $[0, 2, 3, 5, \underline{5}, 4]$ |
| 1 | $[0, 2, 3, \underline{3}, 5, 4]$ |
| 0 | $[0, 2, \underline{2}, 3, 5, 4]$ |

Here in every row the underlined value represents the array element that has been updated by the corresponding loop iteration. The loop terminates either with $j = -1$ or when $a(j)$ is not greater than $x$ (as in above trace where $a(0) = 0 \not> x = 1$).

For our initial attempt to the invariant $I_2$ of the inner loop, we harvest the precondition $P_2$ (and thus of the invariant $I_1$ of the other loop) for all information that is preserved during all iterations of the inner loop:

$$I_2 :\Leftrightarrow 1 \leq i \wedge (|a| \geq 1 \Rightarrow i \leq |a|) \wedge i < |a| \wedge T_1 = N_1 \wedge \ldots$$

Next we extend this invariant by the apparent knowledge about the newly introduced variables $x$ and $j$:

$$I_2 :\Leftrightarrow 1 \leq i \wedge (|a| \geq 1 \Rightarrow i \leq |a|) \wedge i < |a| \wedge T_1 = N_1 \wedge$$
$$x = \text{olda}(i) \wedge -1 \leq j \wedge j \leq i - 1 \wedge \ldots$$

Here it should be noted that the condition $x = a(i)$ only holds in the initial state of the loop, while $x = \text{olda}(i)$ generally holds.

However, this formulation does not describe the range of $j$ accurately enough, because it always allows the lower bound $j = -1$. Thus we have to consider why in above trace we have already stopped with $j = 0$ and $a = [0, 2, \underline{2}, 3, 5, 4]$ rather than, by performing one more loop iteration, with $j = -1$ and $a = [0, \underline{0}, 2, 3, 5, 4]$. The difference clearly is that by the earlier termination we have only moved values larger than $x = 1$ such that we have values greater than $x$ in all positions greater equal $j + 1 = 1$ and less than $i = 4$ (all the underlined positions); this condition would be violated by one more iteration. We thus get the following more refined invariant:

$$I_2 :\Leftrightarrow 1 \leq i \wedge (|a| \geq 1 \Rightarrow i \leq |a|) \wedge i < |a| \wedge T_1 = N_1 \wedge$$
$$x = \text{olda}(i) \wedge -1 \leq j \wedge j \leq i - 1 \wedge$$
$$\left(\forall k \in \mathbb{Z}.\ j + 1 \leq k \wedge k < i \Rightarrow a(k) > x\right) \wedge \ldots$$

Now we turn our attention to the actual operation of the inner loop and how it changes the content of array $a$. Here we have to clearly understand the roles of the constants olda and olda2 and the program variable $a$ such that we are able to describe their relationships adequately:

- Constant olda denotes the initial content of $a$, in particular the content of $a$ immediately before the outer loop started its execution.
- Constant olda2 denotes the content of $a$ immediately before the inner loop started its execution.
- Variable $a$ denotes the current content of $a$, i.e., the content of $a$ before/after an arbitrary iteration of the inner loop.

We are now going to describe in turn the relationship between olda and olda2 and the relationship between olda2 and $a$.

The relationship between olda and olda2 can be deduced from invariant $I_1$ where variable $a$ denoted the state of the array before/after every iteration of the outer loop which thus also covers the value of olda. We thus get the following conditions:

$$permuted\langle\text{olda2, olda}\rangle \wedge sorted\langle\text{olda2}, i\rangle \wedge equals\langle\text{olda2, olda}, i, |\text{olda2}| - 1\rangle$$

In other words, olda2 is a permutation of olda (and thus has the same length), is sorted up to position $i$, and is from position $i$ on identical to olda.

For the relationship between olda2 and $a$, we have to consult the variable trace above. Clearly $a$ is not always a permutation of olda2, but it has the same length, i.e., we have $|a| = |\text{olda2}|$. Furthermore, $a$ and olda2 have the same elements up to position $j + 1$ and from position $i + 1$ on, i.e., we have $equals\langle a, \text{olda2}, 0, j + 1\rangle$ and $equals\langle a, \text{olda2}, i + 1, |a|\rangle$. Furthermore, we see that after position $j + 1$ and before position $i + 1$ the elements of $a$ are shifted one position to the right compared to the elements of olda2, i.e, we have $\left(\forall k \in \mathbb{Z}. \ j + 1 < k \wedge k < i + 1 \Rightarrow a(k) = \text{olda2}(k - 1)\right)$. Altogether this gives us the following conditions:

$$|a| = |\text{olda2}| \wedge equals\langle a, \text{olda2}, 0, j + 1\rangle \wedge equals\langle a, \text{olda2}, i + 1, |a|\rangle \wedge$$
$$\left(\forall k \in \mathbb{Z}. \ j + 1 < k \wedge k < i + 1 \Rightarrow a(k) = \text{olda2}(k - 1)\right)$$

Considering the relationships above, we get the next version of the invariant:

$$
\begin{aligned}
I_2 :\Leftrightarrow \ & 1 \leq i \wedge (|a| \geq 1 \Rightarrow i \leq |a|) \wedge i < |a| \wedge T_1 = N_1 \wedge \\
& x = \text{olda}(i) \wedge -1 \leq j \wedge j \leq i - 1 \wedge \\
& \left(\forall k \in \mathbb{Z}. \ j + 1 \leq k \wedge k < i \Rightarrow a(k) > x\right) \wedge \\
& permuted\langle\text{olda2, olda}\rangle \wedge sorted\langle\text{olda2}, i\rangle \wedge equals\langle\text{olda2, olda}, i, |\text{olda2}| - 1\rangle \\
& |a| = |\text{olda2}| \wedge equals\langle a, \text{olda2}, 0, j + 1\rangle \wedge equals\langle a, \text{olda2}, i + 1, |a|\rangle \wedge \\
& \left(\forall k \in \mathbb{Z}. \ j + 1 < k \wedge k < i + 1 \Rightarrow a(k) = \text{olda2}(k - 1)\right)
\end{aligned}
$$

Since nothing seems to be left to say, we continue with the natural termination measure $T_2$ for the inner loop:

$$T_2 := j + 1$$

Here apparently $j + 1$ rather than $j$ has to be chosen, since the minimum $j = -1$ is possible.

Now to verify the triple

$$\{P_2 \wedge a = \text{olda2}\} \, \texttt{while j} \geq \texttt{0} \wedge \texttt{a[j]} > \texttt{x do \{\ldots\}} \{Q_2\}$$

we have to show the validity of the verification conditions

$$P_2 \wedge a = \text{olda2} \Rightarrow I_2$$
$$I_2 \Rightarrow T_2 \geq 0$$
$$I_2 \wedge \neg(j \geq 0 \wedge a(j) > x) \Rightarrow Q_2$$

and to derive the triple:

$$\{I_2 \wedge j \geq 0 \wedge a(j) > x \wedge T_2 = N_2\} \, \texttt{a[j+1]:=} \ldots \{I_2 \wedge T_2 < N_2\}$$

For this, we have to show the condition

$$I_2 \wedge j \geq 0 \wedge a(j) > x \wedge T_2 = N_2 \Rightarrow$$
$$0 \leq j \wedge j < |a| \wedge 0 \leq j + 1 \wedge j + 1 < |a| \wedge$$
$$I_2[j - 1/j][a[j + 1 \mapsto a(j)]/a] \wedge$$
$$T_2[j - 1/j][a[j + 1 \mapsto a(j)]/a] < N_2$$

which includes the well-definedness of the array accesses, the preservation of the invariant, and the decrease of the termination measure. A detailed verification of these conditions is possible but tedious by manual efforts; here clearly the support of automatic reasoners or semi-automatic reasoning tools is desired.

Above example demonstrated that the reasoning about programs with doubly nested loops proceeds in a "top-down" fashion: first the invariant of the outer loop is elaborated and partially verified; in particular it is shown that the given precondition ensures the invariant at the entry of that outer loop and that the exit from that loop ensures the desired postcondition. Only afterwards we dive into the body of the outer loop to ensure that the stated invariant is preserved by execution of the inner loop. For developing the invariant of the inner loop, all the information contained in the outer invariant has to be considered.

Actually to make programs simpler to understand and to verify, it is much better, rather than to embed into a loop body another loop, to call a *procedure* which encapsulates the loop and thus hides the details of the corresponding computation; then we only need to be concerned with the verification of singly nested loops. Thus calling a (not yet defined) procedure in a loop body can be considered as an abstract computation that is assumed to satisfy a certain expectation; the implementation of this procedure can be considered as a refinement of the abstraction which has to ensure that this expectation is indeed met. In the following sections, we will consider these topics of refinement, procedures, and programs with procedures in more detail.

## 8.6   The Refinement of Commands

In Sect. 8.7, we will move beyond the domain of plain commands to that of procedures. Procedures represent abstractions of commands, thus the caller of a procedure
only has to know *what* effect is achieved by the call but not *how* it is achieved; actually the same effect may be achieved by different commands. The specification of a
desired effect itself can be considered as an "abstract" command that is *refined* by the
implementation to a "concrete" command. Before actually discussing procedures,
we therefore turn our attention to the topic of refinement.

**The Notion of Refinement**

Intuitively, a command $C_1$ refines another command $C_2$, if any occurrence of $C_2$
in a program can be replaced by $C_1$ without without affecting the outcome of the
program. Our goal is to elaborate this intuition into a precise formalization.

---

**Definition 8.4** (*Command Refinement*)   Let $C_1, C_2 \in Command$ be commands. We define the relation $C_1 \sqsubseteq C_2$ (read: "$C_1$ refines $C_2$") as follows:

$$\_ \sqsubseteq \_ \subseteq Command \times Command$$
$$C_1 \sqsubseteq C_2 :\Leftrightarrow$$
$$\forall s, s' \in State.$$
$$([\![ C_2 ]\!]\checkmark \langle s \rangle \Rightarrow [\![ C_1 ]\!]\checkmark \langle s \rangle) \wedge (\langle\!\langle C_2 \rangle\!\rangle \langle s \rangle \Rightarrow \langle\!\langle C_1 \rangle\!\rangle \langle s \rangle) \wedge$$
$$([\![ C_2 ]\!]\checkmark \langle s \rangle \wedge \langle\!\langle C_2 \rangle\!\rangle \langle s \rangle \wedge [\![ C_1 ]\!]\langle s, s' \rangle \Rightarrow [\![ C_2 ]\!]\langle s, s' \rangle)$$

In other words, $C_1$ refines $C_2$, if $C_1$ terminates normally on every input state
$s$ on which $C_2$ terminates normally, and then may result in an output state $s'$
only, if also an execution of $C_2$ on $s$ may result in $s'$.

---

It is not difficult to see that the refinement relation is reflexive (every command
refines itself) and transitive (if $C_1$ refines $C_2$ and $C_2$ refines $C_3$, then $C_1$ refines $C_3$)
but not necessarily symmetric (if one command refines another one, the reverse need
not be the case).

**Example 8.29** Consider the following commands $C_1$ and $C_2$ operating on an integer
variable $x$:

$$C_1 := \texttt{x:=0}$$
$$C_2 := \texttt{while x≠0 do x:=x-1}$$

Then $C_1$ refines $C_2$: like $C_2$, $C_1$ terminates for every prestate value $x \geq 0$ with the
poststate value $x = 0$ (leaving all other variables unchanged). However, $C_2$ does not

refine $C_1$: while $C_1$ terminates for every $x$, $C_2$ does not terminate for $x < 0$. Thus we may substitute every occurrence of $C_2$ in a program by $C_1$ but not vice versa. $\square$

Above notion of refinement is designed to satisfy the following demand.

**Proposition 8.11** (Refinement Preserves Total Correctness) *For all commands $C_1, C_2 \in Command$ and formulas $P, Q \in Formula$, we have the following result:*

$$C_1 \sqsubseteq C_2 \wedge \{P\} C_2 \{Q\} \Rightarrow \{P\} C_1 \{Q\}$$

*In other words, if $C_1$ refines $C_2$ and $C_2$ is totally correct with respect to a given specification, then also $C_1$ is totally correct with respect to that specification.*

**Proof** Let $C_1, C_2$ be commands and $P, Q$ be formulas with $C_1 \sqsubseteq C_2$ and $\{P\} C_2 \{Q\}$. We show $\{P\} C_1 \{Q\}$. Take $s \in State$ with $[\![ P ]\!]_s$. From $\{P\} C_2 \{Q\}$ and Proposition 8.8, we have $[\![ C_2 ]\!] \checkmark \langle s \rangle$ and $\langle\!\langle C_2 \rangle\!\rangle \langle s \rangle$ and thus, from $C_1 \sqsubseteq C_2$, also $[\![ C_1 ]\!] \checkmark \langle s \rangle$ and $\langle\!\langle C_1 \rangle\!\rangle \langle s \rangle$. Now take $s' \in State$ with $[\![ C_1 ]\!] \langle s, s' \rangle$. Since we have $[\![ C_2 ]\!] \checkmark \langle s \rangle$, $\langle\!\langle C_2 \rangle\!\rangle \langle s \rangle$, and $[\![ C_1 ]\!] \langle s, s' \rangle$, we have, by $C_1 \sqsubseteq C_2$, also $[\![ C_2 ]\!] \langle s, s' \rangle$. $\square$

### Specification Commands

For highlighting the core purpose of refinement, we extend our language of commands introduced in Sect. 7.1 by a new command that satisfies a specification with given precondition, postcondition, and frame.

**Definition 8.5** (*Specification Command: Syntax*) Consider the language of commands introduced by the grammar given in Definition 7.1. We extended this language by adding the following syntactic option for a *specification command*:

$$C ::= \dots \mid [\, F_1 \rightsquigarrow F_2 \,]^{Vs}$$

Here $[\, F_1 \rightsquigarrow F_2 \,]^{Vs}$ denotes a *specification command* with *input condition* (or *precondition*) $F_1$, *output condition* (or *postcondition*) $F_2$ and *frame* $Vs$, where $F_1, F_2 \in Formula$ and $Vs \in \mathsf{Set}(Variable)$.

Correspondingly we extend the type system in Fig. 7.1 by the following rule:

$$\frac{\Sigma, Vt \vdash F_1^0 : \text{formula} \quad \Sigma, Vt \vdash F_2^0 : \text{formula}}{\Sigma, Vt \vdash [F_1 \rightsquigarrow F_2]^{Vs} : \text{command}}$$
$$\forall V \in Vs. \; \exists S \in Sort. \; \langle V, S \rangle \in Vt$$

Here, as in Definition 8.3, $F^0$ denotes a duplicate of formula $F$ where every free variable occurrence old_$V$ and every free variable occurrence new_$V$ has been replaced by the reference $V$.

In a nutshell, a specification command $[F_1 \rightsquigarrow F_2]^{Vs}$ "reifies" a specification with precondition $F_1$, postcondition $F_2$, and input/output variables $Vs$ to a command, i.e., it considers a specification as a command satisfying that specification. Formula $F_1$ now represents the condition that ensures that the execution of the command terminates normally in a state that satisfies the condition $F_2$. As in the calculus of "commands as relations" introduced in Sect. 8.3, we use in $F_1$ and $F_2$ the syntax old_$V$ to refer to the value of program variable $V$ in the prestate of the command and new_$V$ to refer to its value in the poststate (we just write $V$ if $V$ is not in $Vs$, i.e., old_$V$ = new_$V$). The variable set $Vs$ represents the frame of the command; every variable not in this frame remains unchanged by the execution of the command.

**Example 8.30** Consider the following specification command operating on integer variables $x$ and $y$:

$$[0 \le y \wedge y \le \text{old\_}x \rightsquigarrow \text{new\_}x = \text{old\_}x - y]^{\{x\}}$$

The command may be executed in a state where $y$ has a value from 0 to $x$; then it decrements $x$ by $y$ and leaves every other program variable unchanged.          □

We are now going to formalize our intuitive understanding of specification commands by its denotational semantics in relational style (see Sect. 7.2).

**Definition 8.6** (*Specification Command: Semantics*)  We extend the relational semantics, well-definedness condition, respectively termination condition given in Figs. 7.3, 7.4, respectively 7.5 as follows:

$$[\![ [F_1 \rightsquigarrow F_2]^{Vs} ]\!] \langle s, s' \rangle :\Leftrightarrow$$
$$[\![ F_1^0 ]\!] \checkmark \langle s \rangle \wedge [\![ F_1^0 ]\!]_s \wedge [\![ F_2^0 ]\!] \checkmark \langle s \rangle \wedge [\![ F_2 ]\!]_{s,s'} \wedge$$
$$\forall V \in Variable. \; V \notin Vs \Rightarrow s(V) = s'(V)$$
$$[\![ [F_1 \rightsquigarrow F_2]^{Vs} ]\!] \checkmark \langle s \rangle :\Leftrightarrow [\![ F_1^0 ]\!] \checkmark \langle s \rangle \wedge [\![ F_2^0 ]\!] \checkmark \langle s \rangle$$
$$\langle\!\langle [F_1 \rightsquigarrow F_2]^{Vs} \rangle\!\rangle \langle s \rangle :\Leftrightarrow [\![ F_1^0 ]\!]_s$$

Here, as in Proposition 8.6, $[\![\, F\, ]\!]_{s,s'}$ denotes a variant of the semantics of formula $F$ where every occurrence old_$V$ is evaluated as an occurrence $V$ in state $s$ and every occurrence of new_$V$ is evaluated as an occurrence $V$ in state $s'$.

Any occurrence of the "abstract" specification command in a program represents a placeholder for a "concrete" command that implements this specification. By using such a placeholder we may perform the verification in a *modular* way by splitting the verification of a program into two layers. In the higher layer we verify the correctness of the program with the abstract command without (yet) worrying how the specification can be actually implemented by a concrete command. In the lower layer, we refine the specification command by a concrete command; due to Proposition 8.11 the correctness of the program is preserved, if we replace the abstract command by the concrete one. We are now going to discuss these two steps in more detail.

**Reasoning about Specification Commands**

As for the higher verification layer, we need to reason about the total correctness of specification commands; for this, we extend the Hoare calculus correspondingly.

---

**Proposition 8.12** (Hoare Rule for Specification Commands) *We extend the rules of the Hoare calculus depicted in Fig. 8.1 by the following rule:*

$$Variable = \{V_1, \ldots, V_n\} \quad Variable \setminus Vs = \{U_1, \ldots, U_m\}$$
$$\models P \Rightarrow [\![\, F_1\, ]\!]\checkmark \wedge [\![\, F_2\, ]\!]\checkmark \wedge F_1^0$$
$$\models \big(\exists \mathsf{old\_}V_1 \colon S_1, \ldots, \mathsf{old\_}V_n \colon S_n.\ P[\mathsf{old\_}V_1/V_1] \ldots [\mathsf{old\_}V_n/V_n] \wedge$$
$$F_2[\mathsf{old\_}V_1/V_1] \ldots [\mathsf{old\_}V_n/V_n][V_1/\mathsf{new\_}V_1] \ldots [V_n/\mathsf{new\_}V_n] \wedge$$
$$U_1 = \mathsf{old\_}U_1 \wedge \ldots \wedge U_m = \mathsf{old\_}U_m\big) \Rightarrow Q$$

---

$$\{P\}\, [\, F_1 \rightsquigarrow F_2\, ]^{Vs}\, \{Q\}$$

*Here $S_1, \ldots, S_n$ represent the sorts of program variables $V_1, \ldots, V_n$. Also, as in Sect. 8.4, $[\![\, F\, ]\!]\checkmark$ denotes a formula that represents the well-definedness condition of formula $F$.*
*Also extended by this rule, the Hoare calculus is totally sound in the sense of Proposition 8.8.*

---

In a nutshell, to show the triple $\{P\}\, [\, F_1 \rightsquigarrow F_2\, ]^{Vs}\, \{Q\}$, it suffices to show that the given precondition $P$ implies the precondition $F_1$ of the command and that the desired postcondition $Q$ is implied by the knowledge that $P$ held in the prestate, that the poststate satisfies the postcondition $F_2$ of the command, and that the variables not in the frame have not changed; here the prestate value of every program variable $V$ is represented by the existentially quantified variable old_$V$.

We illustrate the application of this rule by the following example.

**Example 8.31**  Consider the following program fragment $C$:

```
x := 0;
[ ∃k ∈ ℤ. 0 ≤ k ∧ k < |a| ∧ a(k) = x ⤳ 0 ≤ new_r ∧ new_r < |a| ∧ a(new_r) = x ]{r} ;
a[r] := x+1
```

We would like to ensure $\{P\}\, C\, \{Q\}$ with precondition $P$ and postcondition $Q$ defined as follows:

$$
\begin{aligned}
P :\Leftrightarrow\ &\exists k \in \mathbb{Z}.\, 0 \leq k \wedge k < |a| \wedge a(k) = 0 \wedge \\
&(\forall k' \in \mathbb{Z}.\, 0 \leq k' \wedge k' < |a| \wedge a(k') = 0 \Rightarrow k' = k) \\
Q :\Leftrightarrow\ &\forall k \in \mathbb{Z}.\, 0 \leq k \wedge k < |a| \Rightarrow a(k) \neq 0
\end{aligned}
$$

In other words, assuming that before the execution of this command array $a$ has exactly one occurrence of element 0, it has no more such occurrence afterwards. The specification command embedded into $C$ states that, provided that $a$ holds element $x$ at some position, the command sets variable $r$ to such a position (and leaves every other variable unchanged).

To verify this Hoare triple, by weakest precondition and strongest postcondition reasoning it suffices to verify $\{P \wedge x = 0\}\ [\ldots \rightsquigarrow \ldots]^{\{r\}}\ \{Q[a[r \mapsto x+1]/a]\}$. By the Hoare rule for specification commands, it thus suffices to show the following verification conditions:

$$
\begin{aligned}
&P \wedge x = 0 \Rightarrow (\exists k \in \mathbb{Z}.\, 0 \leq k \wedge k < |a| \wedge a(k) = x) \\
&(\exists \text{old\_}a \in \mathbb{Z}^*,\, \text{old\_}x \in \mathbb{Z},\, \text{old\_}r \in \mathbb{Z}. \\
&\qquad P[\text{old\_}a/a][\text{old\_}x/x][\text{old\_}r/r] \wedge \text{old\_}x = 0 \wedge \\
&\qquad 0 \leq r \wedge r < |\text{old\_}a| \wedge \text{old\_}a(r) = \text{old\_}x \wedge \\
&\qquad a = \text{old\_}a \wedge x = \text{old\_}x) \Rightarrow \\
&\qquad Q[a[r \mapsto x+1]/a]
\end{aligned}
$$

Here the first condition is self-evident and the second one can be simplified to the following one:

$$
P \wedge x = 0 \wedge 0 \leq r \wedge r < |a| \wedge a(r) = x \Rightarrow Q[a[r \mapsto x+1]/a]
$$

So, from the original precondition, the additional knowledge $x = 0$ established from the command before the specification statement, and the knowledge established by the specification statement, we have to ensure that the desired postcondition holds; this requires some reasoning about array updates but is not particularly difficult.  □

## The Refinement of Specification Commands

We now turn to the lower layer of verification. Here we actually need not immediately refine a specification command to a concrete command. Rather we may perform a refinement in multiple sublayers by first refining the (abstract) specification command by another (a bit less abstract) specification command. This sort of refinement is governed by the following correctness rule.

---

**Proposition 8.13** (Refinement of Specification Commands) *Let* $[\,P_1 \rightsquigarrow Q_1\,]^{Vs_1}$ *and* $[\,P_2 \rightsquigarrow Q_2\,]^{Vs_2}$ *be specification commands with formulas* $P_1$, $P_2$, $Q_1$, $Q_2$, *and variable sets* $Vs_1$, $Vs_2$. *Then we have the following result:*

$$(\llbracket P_2 \rrbracket \checkmark \Rightarrow \llbracket P_1 \rrbracket \checkmark) \wedge (\llbracket Q_2 \rrbracket \checkmark \Rightarrow \llbracket Q_1 \rrbracket \checkmark) \wedge$$
$$(P_2 \Rightarrow P_1) \wedge (Q_1 \Rightarrow Q_2) \wedge (Vs_1 \subseteq Vs_2) \Rightarrow$$
$$[\,P_1 \rightsquigarrow Q_1\,]^{Vs_1} \sqsubseteq [\,P_2 \rightsquigarrow Q_2\,]^{Vs_2}$$

*Thus the first specification command refines the second one, if the first command has weaker (or equivalent) well-definedness conditions, a weaker (or equivalent) precondition, a stronger (or equivalent) postcondition, and a smaller (or equal) frame than the second command.*

---

**Proof** Take commands $C_1 := [\,P_1 \rightsquigarrow Q_1\,]^{Vs_1}$ and $C_2 := [\,P_2 \rightsquigarrow Q_2\,]^{Vs_2}$ with $(\llbracket P_2 \rrbracket \checkmark \Rightarrow \llbracket P_1 \rrbracket \checkmark)$, $(\llbracket Q_2 \rrbracket \checkmark \Rightarrow \llbracket Q_1 \rrbracket \checkmark)$, $(P_2 \Rightarrow P_1)$, $(Q_1 \Rightarrow Q_2)$, and $(Vs_1 \subseteq Vs_2)$. To show $C_1 \sqsubseteq C_2$, we have to show for arbitrary $s, s' \in State$ the following conditions:

- $\llbracket C_2 \rrbracket \checkmark \langle s \rangle \Rightarrow \llbracket C_1 \rrbracket \checkmark \langle s \rangle$: We assume $\llbracket C_2 \rrbracket \checkmark \langle s \rangle$, i.e., $\llbracket P_2^0 \rrbracket \checkmark \langle s \rangle$ and $\llbracket Q_2^0 \rrbracket \checkmark \langle s \rangle$, and show $\llbracket C_1 \rrbracket \checkmark \langle s \rangle$, i.e., $\llbracket P_1^0 \rrbracket \checkmark \langle s \rangle$ and $\llbracket Q_1^0 \rrbracket \checkmark \langle s \rangle$. From $\llbracket P_2^0 \rrbracket \checkmark \langle s \rangle$ and $(\llbracket P_2 \rrbracket \checkmark \Rightarrow \llbracket P_1 \rrbracket \checkmark)$, we know $\llbracket P_1^0 \rrbracket \checkmark \langle s \rangle$; from $\llbracket Q_2^0 \rrbracket \checkmark \langle s \rangle$ and $(\llbracket Q_2 \rrbracket \checkmark \Rightarrow \llbracket Q_1 \rrbracket \checkmark)$, we know $\llbracket Q_1^0 \rrbracket \checkmark \langle s \rangle$.
- $\langle\!\langle C_2 \rangle\!\rangle \langle s \rangle \Rightarrow \langle\!\langle C_1 \rangle\!\rangle \langle s \rangle$: We assume $\langle\!\langle C_2 \rangle\!\rangle \langle s \rangle$, i.e., $\llbracket P_2 \rrbracket_s$, and show $\langle\!\langle C_1 \rangle\!\rangle \langle s \rangle$, i.e., $\llbracket P_1 \rrbracket_s$, which follows from $\llbracket P_2 \rrbracket_s$ and $(P_2 \Rightarrow P_1)$.
- $\llbracket C_2 \rrbracket \checkmark \langle s \rangle \wedge \langle\!\langle C_2 \rangle\!\rangle \langle s \rangle \wedge \llbracket C_1 \rrbracket \langle s, s' \rangle \Rightarrow \llbracket C_2 \rrbracket \langle s, s' \rangle$: We assume $\llbracket C_2 \rrbracket \checkmark \langle s \rangle$, $\langle\!\langle C_2 \rangle\!\rangle \langle s \rangle$, and $\llbracket C_1 \rrbracket \langle s, s' \rangle$, i.e., $\llbracket P_2^0 \rrbracket \checkmark \langle s \rangle$ and $\llbracket Q_2^0 \rrbracket \checkmark \langle s \rangle$, and $\llbracket P_2 \rrbracket_s$, and $\llbracket Q_1 \rrbracket_{s,s'}$. Since we have $\llbracket P_2^0 \rrbracket \checkmark \langle s \rangle$, $\llbracket Q_2^0 \rrbracket \checkmark \langle s \rangle$, and $\llbracket P_2 \rrbracket_s$, to show $\llbracket C_2 \rrbracket \langle s, s' \rangle$, it suffices to show $\llbracket Q_2 \rrbracket_{s,s'}$ which follows from $\llbracket Q_1 \rrbracket_{s,s'}$ and $(Q_1 \Rightarrow Q_2)$.

This completes the proof.                                                            □

As above proposition states, when refining specification commands, we may strengthen postconditions, but we only may *weaken* preconditions and well-definedness conditions.

**Example 8.32** Consider the specification statement $[\, P_1 \rightsquigarrow Q_1 \,]^{\{r\}}$ introduced in Example 8.31 where precondition $P_1$ and postcondition $Q_2$ were defined as follows:

$$P_1 :\Leftrightarrow \exists k \in \mathbb{Z}.\ 0 \le k \wedge k < |a| \wedge a(k) = x$$
$$Q_1 :\Leftrightarrow 0 \le \mathsf{new\_}r \wedge \mathsf{new\_}r < |a| \wedge a(\mathsf{new\_}r) = x$$

This command is refined by the specification command $[\, P_2 \rightsquigarrow Q_2 \,]^{\{r\}}$ where precondition $P_2$ and postcondition $Q_2$ are defined as follows:

$$P_2 :\Leftrightarrow \mathsf{true}$$

$Q_2 :\Leftrightarrow$ if $\forall k \in \mathbb{Z}.\ 0 \le k \wedge k < |a| \Rightarrow a(k) \ne x$

$\qquad$ then $\mathsf{new\_}r = -1$

$\qquad$ else $0 \le \mathsf{new\_}r \wedge \mathsf{new\_}r < |a| \wedge a(\mathsf{new\_}r) = x\ \wedge$

$\qquad\qquad \forall k \in \mathbb{Z}.\ 0 \le k \wedge k < \mathsf{new\_}r \Rightarrow a(k) \ne x$

This command has the same frame, a weaker precondition (which does not demand that element $x$ occurs in array $a$, and a stronger postcondition (which also states that the result is $r = -1$, if $x$ does not occur in $a$, and that it denotes the *smallest* occurrence of $x$, otherwise. $\qquad\square$

### The Implementation of Specification Commands

Ultimately we have to refine a specification command by a concrete command, i.e., to constructively *implement* the command; this implementation is governed by the following correctness rule.

---

**Proposition 8.14** (Implementing Specification Commands) *Consider command $C \in Command$, formulas $P, Q \in Formula$, variable set $Vs \in \mathsf{Set}(Variable)$, and variables $V_1, \ldots, V_n, U_1, \ldots, U_m \in Variable$ such that $Variable = \{V_1, \ldots, V_n\}$ and $Variable \setminus Vs = \{U_1, \ldots, U_m\}$. Then we have*

$$C \sqsubseteq [\, P \rightsquigarrow Q \,]^{Vs}$$

*if we can derive the Hoare triple $\{P'\}\, C\, \{Q'\}$ with precondition $P'$ and postcondition $Q'$ defined as follows:*

$P' :\Leftrightarrow P^0 \wedge V_1 = \mathsf{old\_}V_1 \wedge \ldots V_n = \mathsf{old\_}V_n$

$Q' :\Leftrightarrow Q[V_1/\mathsf{new\_}V_1]\ldots[V_n/\mathsf{new\_}V_n] \wedge U_1 = \mathsf{old\_}U_1 \wedge \ldots \wedge U_m = \mathsf{old\_}U_m$

---

By this rule, the refinement of a specification command is ultimately reduced to the verification of a Hoare triple.

**Example 8.33** Consider the specification command $[\, P_2 \leadsto Q_2 \,]^{\{r\}}$ introduced in Example 8.32 where precondition $P_2$ and postcondition $Q_2$ were defined as follows:

$$P_2 :\Leftrightarrow \text{true}$$

$$Q_2 :\Leftrightarrow \text{if } \forall k \in \mathbb{Z}.\ 0 \leq k \wedge k < |a| \Rightarrow a(k) \neq x$$
$$\text{then new\_}r = -1$$
$$\text{else } 0 \leq \text{new\_}r \wedge \text{new\_}r < |a| \wedge a(\text{new\_}r) = x \wedge$$
$$\forall k \in \mathbb{Z}.\ 0 \leq k \wedge k < \text{new\_}r \Rightarrow a(k) \neq x$$

We claim that this command is refined by the command $C$ defined as follows:

```
var i:int;
i := 0; r := -1;
while i < a.length ∧ r = -1 do
{
  if a[i] = x
     then r := i
     else i := i+1
}
```

To verify this claim, we have to derive the Hoare triple $\{P_3\}\ C\ \{Q_3\}$ with precondition $P_3$ and postcondition $Q_3$ defined as follows:

$$P_3 :\Leftrightarrow a = \text{old\_}a \wedge x = \text{old\_}x \wedge r = \text{old\_}r$$

$$Q_3 :\Leftrightarrow \big(\text{if } \forall k \in \mathbb{Z}.\ 0 \leq k \wedge k < |a| \Rightarrow a(k) \neq x$$
$$\text{then } r = -1$$
$$\text{else } 0 \leq r \wedge r < |a| \wedge a(r) = x \wedge$$
$$\forall k \in \mathbb{Z}.\ 0 \leq k \wedge k < r \Rightarrow a(k) \neq x\big) \wedge$$
$$a = \text{old\_}a \wedge x = \text{old\_}x$$

A similar verification was performed in Sect. 8.5.                                    □

The modular style of verification presented so far dealt with programs that embedded abstract (specification) commands which were gradually refined to less abstract commands and ultimately replaced by concrete (executable) commands. Also in practical software engineering, the iterative approach of "stepwise refinement" gradually transforms an abstract problem to concrete code. In real programming languages, however, the different layers of abstractions are provided by *procedures* where a higher-level (more abstract) procedure calls lower-level (more concrete) procedures. We will now investigate how the concepts presented so far can be applied also to these program constructs.

## 8.7     Reasoning About Procedures

We will now extend our considerations to programming languages with procedures. In Sect. 7.5, we have introduced *procedure declarations* of form

$$\text{procedure } I(X_1; \text{ ref } X_2) \ C$$

where a command $C$ was abstracted to a procedure of name $I$ with value parameters $X_1$ and reference parameters $X_2$. Correspondingly we have introduced as a new kind of commands *procedure calls* of form

$$\text{call } I(Ts; Vs)$$

by which the execution of $C$ is triggered in a context where the value parameters denote the values of the argument terms $Ts$ and the reference parameters denote the addresses of the argument variables $Vs$. The goal of this section is to elaborate a reasoning calculus that deals with procedure declarations and procedure calls.

**Monolithic versus Modular Reasoning**

One approach to deal with procedures is to consider them as "macros" that are textually expanded by procedure calls to corresponding commands. For instance, given a procedure

```
procedure div(a:Nat,b:Nat; ref q:Nat,r:Nat) {
  q := 0;
  while a ≥ b do { q := q+1; a := a-b };
  r := a;
}
```

we might expand a call

```
call div(u*v,u+v;x,y)
```

to a command

```
{
  var a:Nat; a := u*v;
  var b:Nat; b := u+v;
  x := 0;
  while a ≥ b do { x := x+1; a := a-b };
  y := a
}
```

Here the value parameters are replaced by corresponding local variables that receive their initial values from the argument terms and that are forgotten when the enclosing block is left. All occurrences of the variable parameters, however, have to be syntactically replaced by the corresponding argument variables. If this syntactic replacement is appropriately performed (in particular we have to consider those local variables in a procedure whose names conflict with those of variables occurring in the arguments of the procedure call), the command resulting from the macro expansion indeed behaves indistinguishable from the actual procedure. Since thus procedure calls can be removed from a program, we do not consider the semantics of procedure calls any more when reasoning about the correctness of programs.

One disadvantage of this *monolithic* approach to reasoning about procedures is that it is not applicable to procedures that call themselves recursively. The main reason, however, why we will *not* pursue this route further is that the approach does not scale: if a procedure is called in multiple places, every call leads to another verification of an instance of the procedure body with a lot of redundancies (since all the instances are essentially the same); even more, for programs with nested procedure calls, the complexity of verification grows exponentially. Most important, every change in the implementation of a procedure invalidates previous verifications of all programs that call this procedure.

The approach that we will indeed investigate further for the verification of programs with procedures enables *modular* reasoning, as presented in Sect. 8.6. Every procedure will be equipped with a *contract* (specification) such that program verification can be decomposed into two parts:

- A verification that the implementation of a procedure satisfies its contract.
- The verification of the program calling the procedure under the assumption that the procedure satisfies its contract.

The verification of a program with procedure calls is thus decoupled from the details of the implementation of the procedure: if the implementation of a procedure changes, only the first kind of verification has to be repeated; as long as the contract of a procedure does not change, the verifications of all programs calling the procedure remain valid. Furthermore, the complexity of the verifications of a program with procedure calls only depends on the complexity of the procedure's contract, not on the size of its implementation. Contract-based modular reasoning thus avoids the redundancies of the monolithic approach; its complexity is linear in the number of procedure calls and thus also scales to programs with many and deeply nested calls.

**Procedure Declarations with Contracts**

To support modular reasoning, we will now assume that every procedure declaration has the following form:

procedure $I(X_1;$ ref $X_2)$ requires $F_1$ ensures $F_2$ assignable $Vs$ $\{$ $C$ $\}$

The declaration is equipped with a *contract* which consists of precondition $F_1$, postcondition $F_2$, and frame $Vs$ with formulas $F_1, F_2 \in Formula$ and variable set $Vs \in \mathsf{Set}(Variable)$. The precondition $F_1$ is intended to constrain the prestate of the procedure call; correspondingly the postcondition $F_2$ constrains its poststate. The frame $Vs$ may only contain the procedure's reference parameters or global variables visible to the procedure; only these variables may have in the prestate of the procedure call values that are different from their values in the poststate of the call.

For every variable $V$ in $Vs$, we use in $F_2$ the syntax old_$V$ to refer to the value of $V$ in the prestate of the procedure call and new_$V$ to its value in the poststate; for every other variable, a plain reference $V$ always denotes the value of the variable in the prestate. This is even true, if $V$ is a value parameter that is changed in the body of the procedure: the contract describes the view of the *caller* who only knows about the initial value of a value parameter; to her any local modifications to this parameter are invisible and therefore irrelevant.

**Example 8.34** Consider the following contract-annotated declaration of a procedure which operates on natural number variables:

```
procedure div(a:Nat,b:Nat; ref q:Nat,r:Nat)
  requires b ≠ 0
  ensures  a = b·new_q+new_r ∧ new_r < b
  assignable q, r
{
  q := 0;
  while a ≥ b do { q := q+1; a := a-b };
  r := a;
}
```

This contract states that the procedure may be called with $b \neq 0$ and only changes reference parameters $q$ and $r$ by setting them to the quotient respectively remainder of the division of $a$ by $b$. Although the procedure changes the value of parameter $a$ during the computation, it satisfies the contract, because the contract refers by the reference $a$ to the initial value of the variable.                                      □

### The Verification of Procedure Contracts

In the following we describe how procedure contracts can be verified and how the result of such a verification is to be semantically interpreted.

---

**Proposition 8.15** (Verification of Procedure Declarations) *Consider a given well-typed program with a contract-annotated procedure declaration*

    procedure $I(X_1;$ ref $X_2)$ requires $F_1$ ensures $F_2$ assignable Vs { $C$ }

*where    $I \in$ Identifier,    $X_1, X_2 \in$ Parameters,    $F_1, F_2 \in$ Formula,*
*$Vs \in$ Set(Variable),   and   $C \in$ Command.   Let   $Ps \in$ Set(Variable)*
*be   the   set   of   all   variables   in   $X_1$.   Furthermore,   let*
*$V_1, \ldots, V_n, P_1, P_2, \ldots, P_p, U_1, \ldots, U_m \in$ Variable be such that Variable $=$*
*$\{V_1, \ldots, V_n\}$, $Ps = \{P_1, \ldots, P_p\}$, and Variable\\$(Ps \cup Vs) = \{U_1, \ldots, U_m\}$.*
   *Assume that we can derive for this declaration the Hoare triple $\{P\}\, C\, \{Q\}$ with precondition P and postcondition Q defined as follows:*

$$P :\Leftrightarrow F_1^0 \wedge V_1 = \text{old\_}V_1 \wedge \ldots V_n = \text{old\_}V_n$$
$$Q :\Leftrightarrow F_2[\text{old\_}P_1/P_1] \ldots [\text{old\_}P_p/P_1][V_1/\text{new\_}V_1] \ldots [V_n/\text{new\_}V_n] \wedge$$
$$U_1 = \text{old\_}U_1 \wedge \ldots \wedge U_m = \text{old\_}U_m$$

*Then every call of this procedure in the program satisfies the declared contract:*

$$\forall e \in Environment, a \in Address, s, s' \in State.\ [\![\, \text{call}\ I^{Ss_1, Ss_2}(Ts\, ; Vs)\, ]\!]_a^e \langle s, s' \rangle \Rightarrow$$

$\forall vs \in Value^*, n \in \mathbb{N}, as_1, as_2 \in Address^*, s_1 \in State, P \in ProcSem.$

$\forall e_p, e_c Environment, var_1, var_2 \in VarEnv, ass_1, ass_2 \in Assignment.$

$[\![ Ts ]\!]^{ass(e,s)} \langle vs \rangle \wedge n = length(vs) \wedge n = length(as_1) \wedge$

$(\forall i \in \mathbb{N}_n. \ as_1(i) = a + i) \wedge as_2 = [\![ Vs ]\!]^e \wedge$

$(\forall i \in \mathbb{N}_{a+n}. \ s_1(i) = \text{if } i < a \text{ then } s(i) \text{ else } vs(i - a)) \wedge$

$\langle \langle I, Ss_1, Ss_2 \rangle, \langle P, e_p \rangle \rangle \in e.proc \wedge$

$[\![ X_1 ]\!]^{as_1} \langle e_p.var, var_1 \rangle \wedge [\![ X_2 ]\!]^{as_2} \langle var_1, var_2 \rangle \wedge e_c = \langle var{:}var_2, proc{:}e_p.proc \rangle \wedge$

$ass_1 = ass(e_c, s_1) \wedge ass_2 = ass(e_c, s') \wedge$


$\big( \forall V_1 \in Variable, a_1 \in Address. \ \langle V_1, a_1 \rangle \in var_1 \Rightarrow a_1 < a \wedge$

$\forall V_2 \in Variable, a_2 \in Address. \ \langle V_2, a_2 \rangle \in var_2 \wedge V_1 \neq V_2 \Rightarrow a_1 \neq a_2 \big) \Rightarrow$

$\big( [\![ F_1^0 ]\!]_{ass_1} \Rightarrow [\![ F_2 ]\!]_{ass_1, ass_2} \wedge \forall V \in Variable \backslash Vs. \ ass_1(V) = ass_2(V) \big)$

*Thus, provided that all variables denote different store addresses below the current top of stack, if the procedure body is executed in a state $s_1$ that satisfies the precondition $F_1$, it terminates in a state $s'$ that satisfies the postcondition $F_2$ but leaves the values of the variables not in frame Vs unchanged.*

Above proposition demands the verification of a Hoare triple which resembles the one given in Proposition 8.14; the major difference is that every reference to a parameter $P_i$ is replaced by old_$P_i$, because the contract refers by $P_i$ to the initial value of the parameter. Furthermore the postcondition is only extended by a condition $U_j = $ old_$U_j$, if $U_j$ is not a value parameter, because the procedure is actually allowed to modify every value parameter during its execution.

**Example 8.35** We consider the procedure for natural number division declared in Example 8.34 with body $C$ defined as follows:

```
q := 0;
while a ≥ b do { q := q+1; a := a-b };
r := a;
```

To verify its contract it suffices to derive the Hoare triple $\{P\} \ C \ \{Q\}$ with precondition $P$ and postcondition $Q$ defined as follows:

$$P :\Leftrightarrow b \neq 0 \wedge a = \text{old\_}a \wedge b = \text{old\_}b \wedge q = \text{old\_}q \wedge r = \text{old\_}r$$

$$Q :\Leftrightarrow \text{old\_}a = \text{old\_}b \cdot q + r \wedge r < \text{old\_}b$$

Here the postcondition $Q$ refers by old_$a$ and old_$b$ to to the old values of $a$ and $b$; indeed $C$ changes the value of $a$; correspondingly $Q$ does not contain conditions $a = $ old_$a$ or $b = $ old_$b$.  ☐

In the semantic interpretation given in Proposition 8.15, the formalization in small print mimics the semantics of procedure calls given in Fig. 7.20; it sets up the context (environment and store) in which the procedure body is executed and correspondingly the contract formulas $F_1$ and $F_2$ are evaluated. In the condition $\langle\langle I, Ss_1, Ss_2\rangle, \langle P, e_p\rangle\rangle \in e.\text{proc}$ we deviate from that semantics by assuming that the procedure environment $e.\text{proc}$ contains for procedure $I$ not only the state relation $P$ (which remains unused in the proposition) but also the environment $e_p$ at the point of the procedure declaration. It is this environment from which the environment $e_c$ is derived in which the command is executed and the formulas are evaluated.

The formalization contains a side constraint that demands that all variables denote different addresses below the current top of stack. Without this constraint, different variables might represent *aliases* to the same store location; thus an assignment to one variable might also affect the value of another variable, which would destroy the soundness of the Hoare calculus. Actually, this constraint may be violated (only) by the reference parameters of a procedure: if we pass as a reference argument a global variable, the procedure may refer to the same store address either via the name of the global variable or the name of the reference parameter; the same problem arises if we pass the same variable as different reference arguments.

In our simple language, it would be possible to devise a small extension to the type-checker to reject programs with aliases. However, in more expressive programming languages, in particular in languages that support dynamic memory allocation and pointers, the possibility of variable aliases represents a major problem to verification (which can be tackled by more powerful techniques for static analysis or by special logics that are beyond the scope of this text).

**Reasoning about Procedure Calls**

Proposition 8.15 is actually incomplete: it does not yet tell us how to reason about the call of a contract-annotated procedure. We are now going to mend this deficiency.

---

**Proposition 8.16** (Hoare Calculus with Procedure Calls) *Consider a given well-typed program with contract-based procedure declarations as described in Proposition 8.15 where procedure $I$ is declared as follows:*
   procedure $I(X_1; \text{ref } X_2)$ requires $F_1$ ensures $F_2$ assignable Vs { C }
*Assume that for this procedure the Hoare triple denoted in Proposition 8.15 can be derived and extend the Hoare calculus depicted in Fig. 8.1 by the following rule:*

$$Variable = \{V_1, \ldots, V_n\} \quad Variable \backslash Vs = \{U_1, \ldots, U_m\}$$
$$\models P \Rightarrow [\![ F_1' ]\!]\checkmark \wedge [\![ F_2' ]\!]\checkmark \wedge (F_1^0)'$$
$$\frac{\models \big(\exists \text{old\_}V_1 \colon S_1, \ldots, \text{old\_}V_n \colon S_n.\ P[\text{old\_}V_1/V_1]\ldots[\text{old\_}V_n/V_n] \wedge \\ F_2[\text{old\_}V_1/V_1]\ldots[\text{old\_}V_n/V_n][V_1/\text{new\_}V_1]\ldots[V_n/\text{new\_}V_n]'' \wedge \\ U_1 = \text{old\_}U_1 \wedge \ldots \wedge U_m = \text{old\_}U_m\big) \Rightarrow Q}{\{P\}\ \texttt{call}\ I(Ts;Rs)\ \{Q\}}$$

*Here, $S_1, \ldots, S_n$ denote the sorts of variables $V_1, \ldots, V_n$. Furthermore, for every formula $F \in Formula$, we denote by $F'$ the formula $F[\overline{Ts}/\overline{X_1}][\overline{Rs}/\overline{X_2}]$ and by $F''$ the formula $F[\overline{Ts}/\text{old\_}\overline{X_1}][\overline{Rs}/\text{old\_}\overline{X_2}]$ where*

- *$F[\overline{Ts}/\overline{Vs}]$ denotes the simultaneous substitution $F[T_1/V_1, \ldots, T_n/V_n]$ for term sequence $Ts = [T_1, \ldots, T_n]$ and value sequence $Vs = [V_1, \ldots, V_n]$;*
- *old\_$Vs$ denotes [old\_$V_1, \ldots,$ old\_$V_n$] for variable sequence $Vs = [V_1, \ldots, V_n]$;*
- *$\overline{X_1}, \overline{X_2}, \overline{Rs} \in Variable^*$ denote the sequences of variables listed in $X_1$, $X_2$, and Rs, respectively, and $\overline{Ts} \in Term^*$ denotes the sequence of terms listed in Ts.*

*Then this extension preserves the total soundness of the Hoare calculus with respect to the verification of the given program.*

Above rule (whose soundness follows from the semantics of procedure calls and Proposition 8.15) closely mimics the Hoare rule for specification commands given in Proposition 8.12; the only difference is the tedious substitution of the concrete arguments for the formal parameters in all contract formulas. Please note that this substitution has to be performed simultaneously for all parameters such that e.g., $[\![ F[T_1/V_1, T_2/V_2] ]\!]_s = [\![ F ]\!]_{s'}$ with $s' = s[V_1 \mapsto [\![ T_1 ]\!]_s][V_2 \mapsto [\![ T_2 ]\!]_s]$; we omit a more detailed formalization.

**Example 8.36** We consider the procedure for natural number division declared in Example 8.34 with the following header/contract:

```
procedure div(a:Nat,b:Nat; ref q:Nat,r:Nat)
   requires b ≠ 0
   ensures  a = b·new_q+new_r ∧ new_r < b
   assignable q, r
```

Our goal is to derive the Hoare triple

$$\{even\langle n \rangle\} \text{ call } \text{div}(n,2,x,y) \{2 \cdot x = n \wedge y = 0\}$$

with $even\langle n \rangle :\Leftrightarrow \exists m \in \mathbb{N}.\ 2 \cdot m = n$. For this, it suffices to verify these two conditions:

$$even\langle a \rangle \Rightarrow 2 \neq 0$$
$$\big(\exists \text{old\_}n \in \mathbb{N}, \text{old\_}x \in \mathbb{N}, \text{old\_}y \in \mathbb{N}.$$
$$even\langle \text{old\_}n \rangle \wedge \text{old\_}n = 2 \cdot x + y \wedge y < 2 \wedge n = \text{old\_}n\big) \Rightarrow$$
$$2 \cdot x = n \wedge y = 0$$

Here the first condition is trivial and the second one can be simplified to

$$even\langle n \rangle \wedge n = 2 \cdot x + y \wedge y < 2 \Rightarrow 2 \cdot x = n \wedge y = 0$$

which is not difficult to show by some arithmetic reasoning.                    □

### The Termination of Recursive Procedures

The techniques shown so far can be also applied to the verification of the partial correctness and non-abortion of programs with procedures that call themselves recursively. However, they are not yet sufficient to ensure their termination: a program might encounter an infinite sequence of recursive procedure invocations where every invocation leads to another one.

Section 8.4 already demonstrated how we can verify that a loop does not run into an infinite sequence of iterations, namely by the introduction of a *termination measure*, a term whose value is increased by every iteration but does not become negative. We may apply this idea also to recursive procedures, by extending the contract of every such procedure with such a measure:

```
procedure I(X₁; ref X₂) ... decreases T { C }
```

This termination measure may only refer to the parameters of the procedure (and to global variables); in every recursive procedure invocation its value must be decreased but not become negative.

**Example 8.37** Consider the following procedure which recursively computes the factorial $r := n!$ of a natural number $n$:

```
procedure fact(n:Nat, ref r:Nat)
   ...
   decreases n
{
   if n = 0 then
     r := 1
   else
     { call fact(n-1,r); r := n·r }
}
```

For this procedure, the termination measure is $n$ itself: for $n = 0$, the execution of the program does not cause a recursive invocation; for $n > 0$, the procedure calls itself recursively with $n - 1$.                                          □

In general, multiple procedures may call themselves mutually recursively. To ensure the proper termination of such programs, a corresponding verification calculus has to determine whether the call of an arbitrary procedure may lead to a recursive invocation of the current one; if yes, the contracts of both procedures must be equipped with a termination measure that is decreased by the respective procedure calls. We content ourselves with this short sketch and omit a more detailed formalization.

**The Verification of Programs**

We have now all the building blocks to verify a complete program in the language presented in Sect. 7.5:

```
Ds; program I(X) requires F₁ ensures F₂ assignable Vs { C }
```

For this purpose, we just need to annotate (in addition to the procedures declared in $Ds$) the program with precondition $F_1$ and postcondition $F_2$ that may only refer to the program parameters $X$, which represent the only interface to the environment; the frame $Vs$ lists those parameters whose values may change by the execution of the program. The verification is then governed by the following rule.

---

**Proposition 8.17** (Verification of Programs) *Consider a well-typed program*

```
Ds; program I(X) requires F₁ ensures F₂ assignable Vs { C }
```

*where $Ds \in Declarations$ (with all procedures in $Ds$ contract-annotated as described in Proposition 8.15), $I \in Identifier$, $X \in Parameters$, $F_1, F_2 \in Formula$, and $C \in Command$. Let $Ps \in \mathsf{Set}(Variable)$ be the set of all variables in $X$. Furthermore, let $V_1, \ldots, V_n, U_1, \ldots, U_m \in Variable$ be such that $Ps = \{V_1, \ldots, V_n\}$ and $Ps \setminus Vs = \{U_1, \ldots, U_m\}$.*
*Assume that we can derive for the program body the Hoare triple $\{P\}\, C \,\{Q\}$ with precondition $P$ and postcondition $Q$ defined as follows:*

$$P :\Leftrightarrow F_1^0 \wedge V_1 = \mathsf{old\_}V_1 \wedge \ldots V_n = \mathsf{old\_}V_n$$
$$Q :\Leftrightarrow F_2[\mathsf{old\_}P_1/P_1]\ldots[\mathsf{old\_}P_p/P_1][V_1/\mathsf{new\_}V_1]\ldots[V_n/\mathsf{new\_}V_n] \wedge$$
$$U_1 = \mathsf{old\_}U_1 \wedge \ldots \wedge U_m = \mathsf{old\_}U_m$$

*Then the program satisfies its declared contract:*

$$\forall vs, vs' \in Value^*.\ [\![\, Ds\,;\, \mathsf{program}\ I\,(X)\ C\,]\!]\langle vs, vs' \rangle \Rightarrow$$
$$\forall n \in \mathbb{N}, a, a' \in Assignment.$$
$$n = |\overline{X}| \wedge n = |vs| \wedge n = |vs'| \wedge$$
$$a = \{\langle V, v \rangle \mid \exists i \in \mathbb{N}_n.\ V = \overline{X}(i) \wedge v = vs(i)\} \wedge$$
$$a' = \{\langle V, v' \rangle \mid \exists i \in \mathbb{N}_n.\ V = \overline{X}(i) \wedge v' = vs'(i)\} \Rightarrow$$
$$\left([\![\, F_1^0\,]\!]_a \Rightarrow [\![\, F_2\,]\!]_{a,a'} \wedge \forall V \in Ps \setminus Vs.\ a(V) = a'(V)\right)$$

*Here $\overline{X}$ denotes the sequence of all parameters listed in $X$.*
   *Thus, provided that in the prestate of the program execution, the values vs of the parameters satisfy the input condition $F_1$, their values vs' in the poststate satisfy the output condition $F_2$, while the program leaves the values of the parameters not in the frame Vs unchanged.*

---

The Hoare triple listed in above proposition is essentially the same one as that in Proposition 8.15 except for the different role of value parameters there. The semantic constraint in above proposition formalizes in its last line very succinctly the intuitive meaning of a contract consisting of precondition, postcondition, and frame.

**Example 8.38** Consider the following program with parameters $a$ and $b$ denoting natural numbers:

```
procedure div(a:Nat,b:Nat; ref q:Nat,r:Nat) ... ;
procedure divide(a:Nat, b:Nat)
  requires b ≠ 0
  ensures ∃r:Nat. old_a = b·new_a+r ∧ r < b
  assignable a
{
  var r:Nat;
  call div(a, b, a, r)
}
```

The program fulfills its contract to set $a$ to the truncated quotient of $a$ divided by $b$.                                                                                         □

## Exercises

Download from the following URL:
https://www.risc.jku.at/people/schreine/TP/exercises/ex-programs.pdf.

## Further Reading

The verification of computer programs has its roots in the classical works of McCarthy [111], Floyd [49], Hoare [73,74], Manna [107], and Dijkstra [41,42]; these publications represent still a good read. In particular, Hoare's seminal work [73] serves as an easily understandable introduction to the Hoare calculus.

Various textbooks contain introductions to formal methods and program verification, mostly focusing on the Hoare calculus and weakest precondition reasoning, often embedded in more general discussions on logic and formal reasoning. Introductions to formal methods in German language are Nebel's book [127] and Kleuker's book [88]; also in German are the introductions to mathematics and logic for computer scientists by Berghammer [14], Schenke [159], and Buchberger and Lichtenberger [29]. Concise introductions to formal methods can be found in O'Regan's books [137] and [138], in Nielson and Nielson's book [129], while Laski's book [100] present's an integrated approach to software verification and analysis; Huth's book [81] on logic for computer science also briefly discusses verification.

More in-depth discussions on program verification and program logic's can be found in Apt's, de Boer's and Olderog's reference textbook [7], Bradley's and Manna's book [26], and Cousot's handbook chapter [39]. The practice of program verification also in industrial contexts is discussed in Drechler's book [44], in Hinchey's, Bowen's and Olderog's book [72], in Alagić's book [5], in Kundu's, Lerner's and and Gupta's book [95] and in Ray's book [149]; state of the art and new directions are discussed in the book [24] of Boca, Bowen, and Siddiqi. A thorough discussion of the topic of loop invariants with various examples of their derivation can be found Furia's, Meyer's, and Velder's paper [53].

The soundness of axiomatic semantics with respect to the denotational semantics of a programming languages is discussed in various textbooks on program semantics, in particular the book [133] by Nipkow, Paulson, and others, Nielson's and Nielson's book [131], Mitchell's book [119], Winskel's book [180], Schmidt's book [160], as well as Gumm's paper [62].

The topics "commands as relations", "specifications as commands", and "program refinement" have a long history as documented by the works of Back [8], Hoare and Jifeng [78], Hehner [68,69], Morgan [121]; see also Boute's paper [25] and the proceedings [27] of Broy and Steinbrüggen. Various texts on program verification, analysis, and semantics also discuss the issue of refinement, such as the book [133] by Nipkow, Paulson, and others, Nielson's and Nielson's book [131], Nielson's book [130], and Bradley's and Manna's book [26]. Refinement is at the core of various specification and modeling languages mentioned below.

Numerous real programming languages such as Java, C#, or C have been equipped with *behavioral interface specification languages* which allow to attach formal contracts to procedures, for a general overview see the paper [66] by Hatcliff, Leavens, and others. For example, the book [3] by Ahrendt, Beckert, and others presents the system KeY for the verification of Java programs whose contracts are specified in the Java Modeling Language (JML). Various language-independent modeling languages allow to describe the behavior of software in a more abstract way, e.g. the Vienna Development Method (VDM) described in Jones's book [84], the language Z described in Potter's, Sinclair's, and Till's book [147], the language B whose latest incarnation Event-B is described in Abrial's book [2], and Abstract State Machines (ASM) described in Börger's and Stärk's book [31]. For general overviews on specification languages see Bjørner's and Henson's book [22] and Alagar's and Periyasamy's book [4].

## Program Reasoning in RISCAL and the RISC ProgramExplorer

There are numerous software systems that support the verification of programs in various languages, either by checking their executions in finite models or by proving verification conditions. We will present two such systems:

- *RISCAL* [169, 170] is a specification language and associated software system for modeling algorithms and specifying their behavior in first-order logic. The type system of RISCAL ensures that the domains of all programs and formulas are finite; this allows a model checker to fully automatically check in small models all possible program executions with respect to their contracts and other annotations (thus ensuring that invariants are not too strong) but also the validity of verification conditions (thus ensuring that invariants are not too weak). Thus errors may be detected quickly before attempting a proof-based verification for models of arbitrary size.
- The *RISC ProgramExplorer* [167, 168] is a system for the verification of programs written in a subset of Java where the contracts of Java methods are specified in first-order logic. The system translates the bodies of procedures to state relations which are appropriately simplified and presented to the user such that she may inspect the relations with respect to their adequacy to satisfy the specified contracts. If no apparent errors are detected, she may discharge the verification conditions generated from the state relations with the help of the previously presented RISC ProofNavigator [165, 166] as a semi-automatic proof assistant.

The specifications used in the following presentations can be downloaded from the URLs

https://www.risc.jku.at/people/schreine/TP/software/prog/prog.txt
https://www.risc.jku.at/people/schreine/TP/software/prog/Prog.java

and loaded by executing from the command line the following commands:

```
RISCAL prog.txt &
ProgramExplorer & (in an empty directory containing Prog.java)
```

### RISCAL

RISCAL (RISC Algorithm Language) is a software system with a graphical user interface; when started from the command line with the name prog.txt of a text file as an argument, the system opens the window depicted in Fig. 8.11. This window displays the contents of the file in an editor area to the left and various control elements and an output area to the right.

The file contains the model of a variant of the linear search algorithm that was specified in Sect. 8.1 and verified in Sect. 8.5. This model starts with the

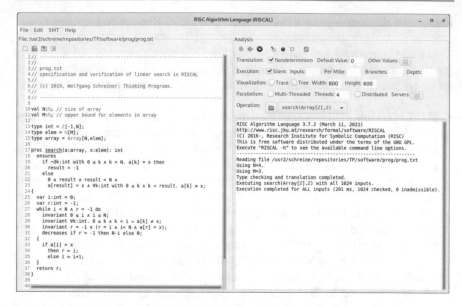

**Fig. 8.11** The RISCAL graphical user interface

declaration of various constants and types that describe the domain in which
the algorithm operates:

```
val N:ℕ; val M:ℕ;
type int = ℤ[-1,N];
type elem = ℕ[M];
type array = Array[N,elem];
```

Here the constant $N$ denotes the length of the arrays we consider; their elements are natural numbers up to the maximum value $M$. Furthermore *int*,
*elem* and *array* denote the types of array indices (including $-1$ and $N$), of
array elements, and of the arrays themselves, respectively. The specification
may contain Unicode symbols such as $\mathbb{N}$ and $\mathbb{Z}$; these symbols may be generated by typing in the editor Nat respectively Int and then pressing the key
shortcut <Ctrl>+#.

The remainder of the file contains the declaration of a procedure annotated
with its contract:

```
proc search(a:array, x:elem): int
  ensures
    if ¬∃k:int with 0 ≤ k ∧ k < N. a[k] = x then
      result = -1
    else
      0 ≤ result ∧ result < N ∧
      a[result] = x ∧ ∀k:int with 0 ≤ k ∧ k < result. a[k] ≠ x;
{ ... }
```

This contract specifies no precondition (which is equivalent to the precondition "true") but by the clause ensures a postcondition; in the postcondition,

the special name *result* refers to the return value of the procedure. Thus, if the given argument array $a$ does not contain at any position the element $x$, the procedure returns the value $-1$; otherwise it returns the smallest index at which $x$ occurs in $a$.

The body of the procedure marked as "Ě" is given below:

```
var i:int = 0;
var r:int = -1;
while i < N ∧ r = -1 do
{
  if a[i] = x
     then r := i;
     else i := i+1;
}
return r;
```

This body represents essentially the command given in Sect. 8.5. To check the correctness of the procedure, we open via the "Other Values" button ▤ a menu that allows us to give values to the model constants, e.g., $N := 4$ and $M := 3$. We then select in the "Operation" panel the operation search and press the "Start Execution" button ➡. This lets the procedure run for *all* possible values of its input parameters, which produces the following output:

```
Executing search(Array[ℤ],ℤ) with all 1024 inputs.
Execution completed for ALL inputs (70 ms, 1024 checked, 0 inadmissible).
```

In the course of these executions the postcondition was checked for every result value of the procedure. For instance, if we change the test a[i] = x erroneously to a[i] ≠ x, the checker produces the following error message:

```
ERROR in execution of search([0,0,0,0],0): evaluation of
    ensures if ¬(∃k:int with (0 ≤ k) ∧ (k < N). (a[k] = x)) then ...
at line 18 in file prog.txt:
    postcondition is violated by result -1 for application search(...)
```

To further validate the correctness of the procedure, we annotate its core loop with an invariant and a termination measure:

```
while i < N ∧ r = -1 do
    invariant 0 ≤ i ∧ i ≤ N;
    invariant ∀k:int. 0 ≤ k ∧ k < i ⇒ a[k] ≠ x;
    invariant r = -1 ∨ (r = i ∧ i< N ∧ a[r] = x);
    decreases if r = -1 then N-i else 0;
  { ... }
```

Here the invariant is represented by the conjunction of the formulas specified in the multiple `invariant` clauses; the termination measure is specified by the term in the `decreases` clause. With these annotations, every execution of the procedure now also checks whether the invariant holds before and after every loop iteration and whether every loop iteration decreases the termination measure but does not make it negative. Thus, if we introduce the same error as above, we get now the following message:

**Fig. 8.12** Performing the validation and verification tasks

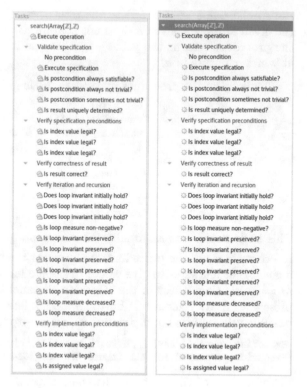

```
ERROR in execution of search([0,0,0,0],0): evaluation of
    invariant ∀k:int. (((0 ≤ k) ∧ (k < i)) ⇒ (a[k] ≠ x));
at line 29 in file prog.txt:
    invariant is violated
```

Likewise, if we introduce by a clause decreases N-i a wrong termination measure, we get the following message:

```
ERROR in execution of search([0,0,0,0],0): evaluation of
    decreases N-i;
at line 31 in file prog.txt:
    variant value 4 is not less than old value 4
```

By such checks we may validate the loop annotations, in particular we may ensure that an invariant is not too strong, i.e., that it is not violated by any iteration of the loop. However, this does not yet rule out that the invariant is too weak, i.e., that it would not be able to carry a proof-based verification.

To further increase our confidence in the adequacy of the program and its specification, we press the "Show/Hide Tasks" button 🗔, which opens an additional panel depicted in the left part of Fig. 8.12. This panel lists a number of tasks that are still "open" (the red color indicates that the tasks have not yet been performed). Apart from "Execute Operation" (which we have already demonstrated above), there are essentially two sets of tasks, those for validating a specification, and those for verifying the procedure.

The tasks listed under "Validate Specification" allow us to investigate the adequacy of the procedure contract (i.e., to ensure that it really expresses our intentions). For instance, the task "Execute Specification" generates from the contract the implicitly specified function

```
fun _search_0_Spec(a:array, x:elem): int = choose result:int with
   if ... then result = (-1) else ... ;
```

If we execute this operation by a double-click (with the options "Nondeterminism" selected and "Silent" not selected), the system prints out all pairs of input/output values admitted by the specification:

```
Executing search(Array[Z],Z) with all 1024 inputs.
Run 0 of deterministic function search([0,0,0,0],0):
Result (0 ms): 0
Run 1 of deterministic function search([1,0,0,0],0):
Result (0 ms): 1
...
Run 5 of deterministic function search([1,1,0,0],0):
Result (0 ms): 2
...
Run 1023 of deterministic function search([3,3,3,3],3):
Result (0 ms): 0
Execution completed for ALL inputs (6757 ms, 1024 checked, ...).
```

Likewise, the task "Is result uniquely determined?" generates a theorem

```
theorem _search_0_PostUnique(a:array, x:elem) ⇔
   ∀result:int with ...  . (∀_result:int with ... . (result = _result));
```

whose validity we may check to ensure that the specification does not allow for one input multiple outputs:

```
Executing _search_0_PostUnique(Array[Z],Z) with all 1024 inputs.
Execution completed for ALL inputs (82 ms, 1024 checked, 0 inadmissible).
```

The majority of the tasks, however, deals with the validity of the verification conditions generated from the program and its annotations. For this purpose, RISCAL implements a variant of the weakest precondition calculus introduced in Sect. 8.3; this variant does not result in a single monolithic verification condition but in a lot of small conditions; if a particular condition is not valid, the source of the error thus becomes easier to find. The condition "Is result correct?" is the core condition: it checks whether the precondition of the procedure implies the weakest precondition calculated from the body of the procedure. The tasks listed under "Verify implementation precondition" are side conditions that check whether all operations in the procedure are well-defined, i.e., whether they are only applied to legal arguments.

The conditions listed under "Verify iteration and recursion" deal with the verification of loops and recursive procedures: in our example, they check whether the various invariants hold in the initial state of the loop and whether every branch of the loop body preserves their validity and decreases the termination measure. By single-clicking a task, the relevant part of the code is high-lighted, by double-clicking its validity is checked. For example, if we single-click the first task labeled "Is loop invariant preserved?", the following piece of code is highlighted:

```
var i:int = 0;
var r:int = -1;
while i < N ∧ r = -1 do
  invariant 0 ≤ i ∧ i ≤ N;
  invariant ∀k:int. 0 ≤ k ∧ k < i ⇒ a[k] ≠ x;
  invariant r = -1 ∨ (r = i ∧ i< N ∧ a[r] = x);
  decreases if r = -1 then N-i else 0;
{
  if a[i] = x
    then r = i;
    else i = i+1;
}
return r;
```

The verification task thus checks the preservation of the first invariant in the first branch of the loop body. If we double-click the task, the validity of the condition is checked:

```
Executing _search_0_LoopOp4(Array[ℤ],ℤ) with all 1024 inputs.
Execution completed for ALL inputs (164 ms, 1024 checked, ...).
```

If we select by a right-click the top-level task folder a menu pops up with an entry "Execute all tasks". By selecting this entry, all tasks are closed in a few seconds and thus turn blue as indicated in the right part of Fig. 8.12.

By the fully automatic checking of some small models, RISCAL allows to quickly *falsify* programs, in particular also to find errors and inadequacies in their contracts and their formal annotations. This is much easier than finding the source of errors from failed proof attempts, which may be due to errors in the models, but (more often than not) also due to inadequacies in the proof strategy respectively proof automation. However, to ultimately *verify* a program, i.e., to show that is is correct in all models of arbitrary size, indeed proof-based verification is required.

### RISC ProgramExplorer

When the RISC ProgramExplorer is started, a window pops up that displays in the left panel a list of all Java files/classes in the current working directory. By double-clicking the class Prog the file Prog.java is loaded and semantically processed as displayed in Fig. 8.13. In the center of the window we find an editor with the contents of the file where all formal annotations are "folded" away; by clicking on a marker "+" in the left bar, the corresponding annotation is shown in blue color.

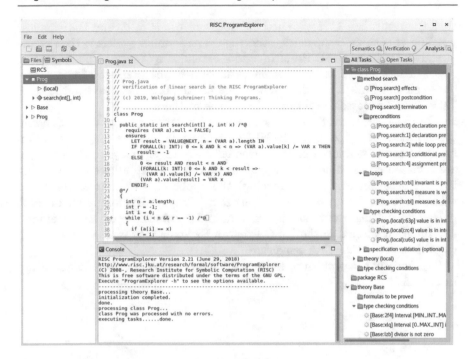

**Fig. 8.13** The RISC programexplorer graphical user interface

Class `Prog` contains a static Java method `search` that implements the linear search algorithm specified in Sect. 8.1 and verified in Sect. 8.5:

```
public static int search(int[] a, int x) /*@
  requires (VAR a).null = FALSE;
  ensures
    LET result = VALUE@NEXT, n = (VAR a).length IN
    IF FORALL(k: INT): 0 <= k AND k < n =>
        (VAR a).value[k] /= VAR x THEN
      result = -1
    ELSE
        0 <= result AND result < n AND
        (FORALL(k: INT): 0 <= k AND k < result =>
          (VAR a).value[k] /= VAR x) AND
        (VAR a).value[result] = VAR x
    ENDIF;
  @*/
  { ... }
```

Here the special comment markers /*@ ... @*/ embed the formal contract of the procedure. Its clause `requires` states as a precondition that parameter array *a* must not be *null*; its clause `ensures` states as a postcondition that the result is $-1$, if parameter $x$ does not occur in $a$, and the smallest position of such an occurrence, otherwise. The formula syntax is that of the already introduced RISC ProofNavigator extended by the special terms (OLD $a$) to denote the value of program variable $a$ in the prestate of the procedure call and (VAR $a$) to denote its value in the poststate; likewise the term VALUE@NEXT denotes the return value of the method.

The RISC ProgramExplorer maps all Java data types to corresponding types in the language of the RISC ProofNavigator. As the major deviation from the actual semantics of Java, heap entities such as arrays or class objects are represented as values rather than as pointers, which allows (by ignoring the complication of variable aliasing) the application of simpler reasoning rules. However, the type checker of the RISC ProgramExplorer controls that indeed no aliasing occurs in the program and rejects the program otherwise; any program verified in the RISC ProgramExplorer thus is also correct with respect to the actual Java semantics.

The body of the procedure embeds the Java version of the command given in Sect. 8.5:

```
int n = a.length;
int r = -1;
int i = 0;
while (i < n && r == -1)
/*@ ... @*/
{
  if (a[i] == x)
    r = i;
  else
    i = i+1;
}
return r;
```

The while loop annotation /*@ ... @*/ equips the loop with the following invariant and termination measure:

```
invariant (VAR a).null = FALSE AND VAR n = (VAR a).length
    AND 0 <= VAR i AND VAR i <= VAR n
    AND (FORALL(k: INT): 0 <= k AND k < VAR i =>
        (VAR a).value[k] /= VAR x)
    AND (VAR r = -1 OR (VAR r = VAR i AND VAR i < VAR n AND
        (VAR a).value[VAR r] = VAR x));
decreases IF VAR r = -1 THEN VAR n - VAR i ELSE 0 ENDIF;
```

The RISC ProgramExplorer implements a version of the *commands as relations* semantics introduced in Sect. 8.3. A right-click on the symbol "search" in the left panel opens a menu whose entry "Show Semantics" displays the

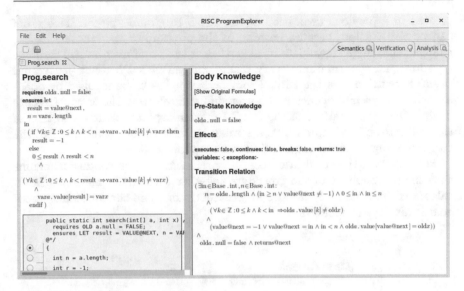

**Fig. 8.14** The semantics view of a method

view shown in Fig. 8.14. Its core is the following state relation calculated from the method body and its invariant annotation:

$$(\exists \text{in} \in \text{Base} . \text{int}, n \in \text{Base} . \text{int}:$$
$$n = \text{old}a . \text{length} \wedge (\text{in} \geq n \vee \text{value@next} \neq -1) \wedge 0 \leq \text{in} \wedge \text{in} \leq n$$
$$\wedge$$
$$(\forall k \in \mathbb{Z} : 0 \leq k \wedge k < \text{in} \Rightarrow \text{old}a . \text{value}[k] \neq \text{old}x)$$
$$\wedge$$
$$(\text{value@next} = -1$$
$$\vee$$
$$\text{value@next} = \text{in} \wedge \text{in} < n \wedge \text{old}a . \text{value}[\text{value@next}] = \text{old}x))$$
$$\wedge$$
$$\text{old}a . \text{null} = \text{false} \wedge \text{returns@next}$$

In a nutshell, this relation states that there exists some value *in* (the final value of the program variable *i*) such that element *x* does not occur in array *a* at any position less than *in*. Furthermore, if the method's return value *value@next* is −1, then *in* equals the length *n* of *a*, i.e., *x* does not occur in *a* at all; otherwise the return value is *in*, which is then an index in *a* at which *x* occurs. Thus, provided that the invariant annotation is indeed correct, the transition relation essentially confirms our expectations about the method. In the same way every command of the method body can be selected from the active labels in the left panel for the inspection of its state relation and termination condition.

After these semantic investigations we turn our attention to the verification tasks displayed in the right panel of Fig. 8.13. Every task displayed in color blue was already (automatically) closed by application of the RISC ProofNavigator. Every task displayed in purple has a proof that was already created in a previous invocation of the RISC ProgramExplorer. This proof can be still trusted, because nothing in the program or its annotations has changed that might invalidate it; nevertheless the proof may be still rerun as an additional confirmation. Every task displayed in red has not yet a valid proof. For such a task the RISC ProofNavigator has to be invoked by opening with a right mouse click a pop-up menu and selecting the entry "Execute Task". Then the interface of the RISC ProgramExplorer switches to that of the RISC ProofNavigator where a semi-automatic (interactive) proof of the corresponding verification condition can be performed.

For instance, the verification condition labeled as "postcondition" claims that the precondition of the procedure together with the transition relation calculated from its body implies its postcondition. The corresponding proof goal is displayed as follows:

3x2  $a_{\text{old}} . \text{null} = \text{false}$

$\wedge$

$(\exists \text{in} \in \text{int}, n \in \text{int} :$

$n = a_{\text{old}} . \text{length} \wedge (\text{value\_}(\text{now\_})+1=0 \Rightarrow \text{in} \geq n) \wedge 0 \leq \text{in}$

$\wedge$

$\text{in} \leq n$

$\wedge$

$(\forall k \in \mathbb{Z} : x_{\text{old}} = a_{\text{old}} . \text{value}[k] \Rightarrow 0 > k \vee k \geq \text{in})$

$\wedge$

$(\text{value\_}(\text{now\_})+1=0$

$\vee$

$\text{in} = \text{value\_}(\text{now\_}) \wedge \text{in} < n$

$\wedge$

$x_{\text{old}} = a_{\text{old}} . \text{value}[\text{value\_}(\text{now\_})]))$

$\wedge$

$a_{\text{old}} . \text{null} = \text{false} \wedge \text{returns\_}(\text{now\_})$

$\Rightarrow$

if $\forall k \in \mathbb{Z} : x_{\text{old}} = a_{\text{old}} . \text{value}[k] \Rightarrow 0 > k \vee k \geq a_{\text{old}} . \text{length}$ then

$\text{value\_}(\text{now\_})+1=0$

else

$0 \leq \text{value\_}(\text{now\_}) \wedge \text{value\_}(\text{now\_}) < a_{\text{old}} . \text{length}$

$\wedge$

$(\forall k \in \mathbb{Z} : x_{\text{old}} = a_{\text{old}} . \text{value}[k] \Rightarrow 0 > k \vee k \geq \text{value\_}(\text{now\_}))$

$\wedge$

$x_{\text{old}} = a_{\text{old}} . \text{value}[\text{value\_}(\text{now\_})]$

endif

We may perform this proof by an application of the scatter command bound to the button ⬇ and a repeated application of auto bound to ♻. This yields the proof tree shown in the left part of Fig. 8.15. In the whole verification of method search, the most complicated proof is that of the condition "invariant is preserved", which involves the proof commands decompose and split (to simplify the proof structure), followed by scatter and auto; the corresponding proof tree is shown in the right part of the figure.

The RISC ProgramExplorer has been mainly developed for didactic purposes but it has also been applied to several non-trivial verifications. The main advantage of the system is the transparency of the generation of the verification conditions (from user-inspectable state relations generated from

**Fig. 8.15** Proof trees of the verification

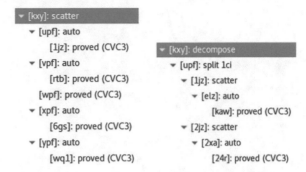

the commands) as well as the smooth integration of the RISC ProofNavigator with its clear separation between the automation of mechanical tasks (based on SMT solvers) and the interactive direction of the human for creative tasks (such as decomposing the proof structure and instantiating quantified formulas). However, there are much more powerful verification systems for much more comprehensive subsets of real programming languages such as the KeY verifier for a "realistic" part of Java [3].

# Concurrent Systems

**9**

> *Was mir an deinem System am besten gefällt? Es ist so*
> *unverständlich wie die Welt. (What do I like most about your*
> *system? It is as incomprehensible as the world.)*
> —Franz Grillparzer

The previous chapters have modeled computer programs mainly as "black boxes" that accept some input and produce some output; the internal operation of a program to achieve this behavior was considered as "irrelevant" to the external observer and thus deliberately hidden. While this view is indeed adequate to model sequential program executions, it fails if we wish to consider *concurrent systems*, i.e., systems where multiple components (cores, processors, computers) execute multiple activities (threads, processes) in parallel. The various components may interact with each other by synchronization respectively communication; such systems of components that react to influences from other components are also called *reactive systems*.

In this chapter, we will consider the formal modeling and reasoning about such systems. This presentation is based on system models expressed in the language of first-order logic as well as on property specifications expressed in linear temporal logic, an extension of first-order logic that is able to talk not only about a single state or a fixed number of states but about arbitrary sequences of such states (to keep our presentation compact, we will not discuss the alternative "branching-time logic" where states are organized in tree-like structures).

The first part of this chapter focuses on the formal modeling of concurrent systems. Section 9.1 lays the groundwork by introducing the core concept of *labeled transition systems* that represent the formal semantics of concurrent systems; these models can be described by formulas in first-order logic. Sections 9.2 and 9.3 continue by presenting two languages of *shared systems* (systems that consist of closely coupled threads interacting via shared variables in a single store) respectively *dis-*

*tributed systems* (systems composed of loosely coupled processes that interact via well-defined interfaces by exchanging messages); we formally define a denotational semantics of both languages as labeled transition systems.

Further on, we turn our attention to the formal specification and verification of properties of labeled transition systems. Section 9.4 introduces the language of *linear temporal logic* that is adequate to express many relevant system properties; we also provide a semantic characterization of these properties as *safety* properties, *liveness* properties, or properties that are mixtures of both aspects. In Sect. 9.5, we consider in more detail the verification of *invariance*, a particular class of safety properties that is also fundamental to reasoning about other properties; the notion of invariance can be considered as a generalization of the partial correctness of sequential programs. Likewise, we discuss in Sect. 9.6 the verification of *response*, a particular class of liveness properties that can be considered as a generalization of the notion of the termination of sequential programs. Finally Sect. 9.7 concludes by discussing the formal *refinement* of systems as the basis of a systematic methodology for concurrent system development.

## 9.1    Labeled Transition Systems

### Systems, States, Actions, and Runs

For modeling and analyzing concurrent systems, it is not only the two states at the start and at the end of the execution that matter; we rather have to consider also all intermediate states. In other words, we have to deal with complete *system runs* of the following kind:

$$s_0 \xrightarrow{l_0} s_1 \xrightarrow{l_1} s_2 \xrightarrow{l_2} \ldots$$

Such a run embeds the sequence $[s_0, s_1, s_2, \ldots]$ of those states that arise in an execution of the system; as in Chap. 7, each state is represented by a mapping of variables to values. Every "step" $s \xrightarrow{l} s'$ of the run represents the execution of an *action* with label $l$ that performs a *transition* from one intermediate state $s$, the *prestate* of the transition, to another intermediate state $s'$, the transition's *poststate*. In contrast to sequential computer programs, systems may intentionally not terminate, thus the state sequence may not only be finite but also infinite.

A system run exposes all those states that are "visible" among the components of the system; however, a transition $s \xrightarrow{l} s'$ may also involve some "hidden" intermediate states occurring between prestate $s$ and poststate $s'$, provided that these intermediate states do not affect the interactions among the components. For example, consider the execution of the following system *XY1* (whose notation will be formally introduced in Sect. 9.2):

```
shared XY1 {
  var x:nat, y:nat;
  init { x := 0; y := 0 }
  action inc { x := x+1; y := y+x }
}
```

The states of this system are pairs of natural numbers $x \in \mathbb{N}$ and $y \in \mathbb{N}$. Starting with the initial state $[x \mapsto 0, y \mapsto 0]$, the values are repeatedly incremented by an action *inc*, which gives rise to the following system run:

$$[x \mapsto 0, y \mapsto 0] \overset{inc}{\to} [x \mapsto 1, y \mapsto 1] \overset{inc}{\to} [x \mapsto 2, y \mapsto 3] \overset{inc}{\to} \dots$$

In each step of this run both $x$ and $y$ are simultaneously incremented, "hiding" the action's intermediate state where $x$ has been already incremented but $y$ has not yet changed. Each action thus represents an "atomic" unit of execution whose effects are *simultaneously* exposed to the external observer. The "granularity" of a system run may thus be adapted by combining multiple commands into a single action respectively by separating commands into multiple actions.

Above example system contains a single action, which naturally models a system with a single component. The execution of such a system, starting in some given initial state, clearly yields a predetermined sequence of successor states; we call such a system *deterministic*. However, if a system consists of multiple components that may execute concurrently with each other, we can model such a system only by multiple actions, at least one per component. Then in a given system state it can be generally not predetermined which of the actions is executed next, which gives rise to multiple possible system runs with different sequences of states; we call such a system *nondeterministic*. As an example, consider the following system *XY2*:

```
shared XY2 {
  var x:nat, y:nat;
  init { x := 0; y := 0 }
  action incX { x := x+1 }
  action incY { y := y+x }
}
```

This system has two actions *incX* and *incY* incrementing $x$ and $y$, respectively. In every system state it is possible that each of these actions is executed; thus there are always two possible successor states. All possible executions of the system can be represented by a labeled directed graph as follows:

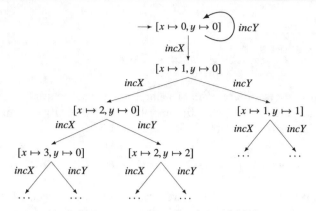

Every node in this graph represents a *state* of the system; every edge represents a *transition* from a prestate (denoted by the source of the edge) to a poststate (denoted by the target of the edge); the edge label denotes the *action* that performs the transition. We can draw (a finite portion of) the graph by drawing, starting with a node for every initial system state (depicted by nodes with incoming arrows without sources and labels), an edge for every transition, which (potentially) leads to new states for which new graph nodes are created from which again new edges arise. The graph is not necessarily a tree, because in general the same poststate may be reached from different prestates; the graph may even contain cycles that describe how from a certain state a number of transitions lead to the same state again.

Every path in the graph starting from an initial state (i.e., a node representing such a state) describes a potential *run* of the system. Every depicted graph node represents a *reachable* state, i.e., a state that may arise in some system run. If a run leads to a node with no outgoing edge (we will see later that this is possible), the run is finite. A run is infinite, if it passes through infinitely many different states or if it infinitely often returns to the same state again.

### Labeled Transition Systems

We are now going to formalize the concepts we have introduced above, in particular a mathematical model of "systems".

**Definition 9.1** (*Labeled Transition Systems*) Given a set $L$ of elements which we call *labels* and a set $S$ of elements which we call *states*, we define the set

$$LTS^{L,S} \subseteq \text{space:Set}(S) \times \text{init:Set}(S) \times \text{next:Set}(L \times S \times S)$$

of *labeled transition systems* over $L$ and $S$ as follows:

$$LTS^{L,S} := \{\langle \text{space:}S_0, \text{init:}I, \text{next:}R \rangle \mid S_0 \subseteq S \wedge I \subseteq S_0 \wedge R \subseteq L \times S_0 \times S_0\}$$

Thus a labeled transition system (*LTS*) *lts* over $L$ and $S$ consists of a *state space* $S_0 := lts.\text{space} \subseteq S$, an *initial state condition* $I := lts.\text{init} \subseteq S_0$, and a *(labeled) transition relation* $R := lts.\text{next} \subseteq L \times S_0 \times S_0$.

This definition differentiates between the set $S$ of system states and the state space $S_0$ of a concrete LTS; this difference is only due to technical reasons in the subsequent formal definitions and may be mainly ignored.

Given a fixed LTS with label set $L$, state space $S_0$, and transition relation $R$, we write $s \xrightarrow{l} s'$ to indicate the transition $R\langle l, s, s' \rangle$ where $l \in L$ is the *label* of the *action* performing the transition, $s \in S_0$ is its *prestate*, and $s' \in S_0$ is its *poststate*.

A LTS gives rise to system runs of the following kind.

**Definition 9.2** (*System Runs*) Given a set $S$, we define the set

$$Run^S := S^* \cup S^\omega$$

as the set of *system runs* (short: *runs*) over $S$. Thus a run $r \in Run^S$ is a finite or infinite sequence of values from $S$.

Now we are ready to define the denotational semantics of a LTS as a set of system runs (respectively as a predicate that holds for any such run).

**Definition 9.3** (*Runs of a LTS*) Given a set $L$ of labels, a set $S$ of states, and a labeled transition system $lts \in LTS^S$ over $L$ and $S$, we define the set $[\![ lts ]\!] \subseteq Run^S$ of *runs* of *lts* and the corresponding set $lruns(lts) \subseteq Run^{l:L \times s:S}$ of *labeled runs* in the following way:

$[\![ . ]\!] : LTS^{L,S} \to \text{Set}(Run^S)$
$[\![ lts ]\!] \langle r \rangle :\Leftrightarrow$
$\quad \exists l \in lruns(lts). \, domain(r) = domain(l) \wedge \forall i \in domain(r). \, r(i) = l(i).s$

$lruns: LTS^{L,S} \to \text{Set}(Run^{l:L \times s:S})$
$lruns(lts)\langle r \rangle :\Leftrightarrow$
$\quad \text{let } S_0 = lts.\text{space}, I = lts.\text{init}, R = lts.\text{next in}$
$\quad r \in Run^{l:L \times s:S_0} \wedge$
$\quad \text{if } r \in (l{:}L \times s{:}S_0)^\omega \text{ then}$
$\quad\quad I\langle r(0).s \rangle \wedge (\forall i \in \mathbb{N}. \, R\langle r(i).\text{l}, r(i).s, r(i+1).s \rangle)$
$\quad \text{else}$

let $n = |r|$ in
if $n > 0$ then
$\quad I \langle r(0).s \rangle \land (\forall i \in \mathbb{N}. \, i + 1 < n \Rightarrow R \langle r(i).l, r(i).s, r(i+1).s \rangle) \land$
$\quad (\neg \exists l \in L, s' \in S_0. \, R \langle l, r(n-1).l, s' \rangle)$
else
$\quad \neg \exists s \in S_0. \, I \langle s \rangle$

Thus a run $r$ of a labeled transition system is a sequence of states over the state space of the system that satisfies one of the following conditions:

- $r$ is infinite and there exists an infinite sequence of system transitions

$$r(0) \xrightarrow{l_0} r(1) \xrightarrow{l_1} r(2) \xrightarrow{l_2} \ldots$$

whose initial state $r(0)$ satisfies the initial state condition of the system and every state $r(i)$ is related to $r(i+1)$ by a transition $r(i) \xrightarrow{l_i} r(i+1)$ (for some label $l_i$);
- $r$ has length $n > 0$ and there exists a sequence of $n-1$ transitions

$$r(0) \xrightarrow{l_0} r(1) \xrightarrow{l_1} r(2) \xrightarrow{l_2} \ldots \xrightarrow{l_{n-2}} r(n-1)$$

whose initial state $r(0)$ satisfies the initial state condition and every state $r(i)$ is related to $r(i+1)$ by a transition $r(i) \xrightarrow{l_i} r(i+1)$ (for some label $l_i$), except for the last state $r(n-1)$ for which there does not exist any transition that relates this state to some poststate;
- $r$ has length 0 and no state satisfies the initial state condition.

The labeled version

$$r(0).s \xrightarrow{r(0).l} r(1).s \xrightarrow{r(1).l} r(2).s \xrightarrow{r(2).l} \ldots$$

of a run records in the component $.l$ of its elements the labels of the state transitions and the states themselves in the component $.s$.

**Defining Labeled Transition Systems**

While there are many different languages to define concurrent systems (we will introduce in Sects. 9.2 and 9.3 also some notations), there is a generic "pure logic" approach to define any concurrent system as a LTS by defining its state space, initial state condition, and labeled transition relation.

**Example 9.1** Consider the system that has been previously informally described:

```
shared XY2 {
  var x:nat, y:nat;
  init { x := 0; y := 0 }
  action incX { x := x+1 }
  action incY { y := y+x }
}
```

The behavior of this system is formally defined by the LTS

$$XY2 := \langle \text{space:}S, \text{init:}I, \text{next:}R \rangle$$

with the following components:

$$S := \text{x:}\mathbb{N} \times \text{y:}\mathbb{N}$$
$$I\langle s \rangle :\Leftrightarrow s.x = 0 \wedge s.y = 0$$
$$R\langle l, s, s' \rangle :\Leftrightarrow$$
$$(l = incX \wedge s'.x = s.x + 1 \wedge s'.y = s.y) \vee$$
$$(l = incY \wedge s'.x = s.x \wedge s'.y = s.y + s.x)$$

However, we can also apply a more compact "pattern matching" notation to define the components of *XY2* as follows:

$$S := \mathbb{N} \times \mathbb{N}$$
$$I\langle x, y \rangle :\Leftrightarrow x = 0 \wedge y = 0$$
$$R\langle l, \langle x, y \rangle, \langle x', y' \rangle \rangle :\Leftrightarrow$$
$$(l = incX \wedge x' = x + 1 \wedge y' = y) \vee$$
$$(l = incY \wedge x' = x \wedge y' = y + x)$$

We might also use auxiliary predicates for the individual actions of the system:

$$S := \mathbb{N} \times \mathbb{N}$$
$$I\langle x, y \rangle :\Leftrightarrow x = 0 \wedge y = 0$$
$$R\langle l, \langle x, y \rangle, \langle x', y' \rangle \rangle :\Leftrightarrow incX\langle l, \langle x, y \rangle, \langle x', y' \rangle \rangle \vee incY\langle l, \langle x, y \rangle, \langle x', y' \rangle \rangle$$
$$incX\langle l, \langle x, y \rangle, \langle x', y' \rangle \rangle :\Leftrightarrow l = incX \wedge x' = x + 1 \wedge y' = y$$
$$incY\langle l, \langle x, y \rangle, \langle x', y' \rangle \rangle :\Leftrightarrow l = incY \wedge x' = x \wedge y' = y + x$$

The later two forms syntactically closely resembles the original system description. Please note the crucial points of the formal definitions:

- The state space $S$ is a product of the state spaces of system variables.
- The initial state condition $I \subseteq S$ is defined by a formula on the system variables that describes the effect of the system initialization command on these variables.

- The labeled transition relation $R \subseteq L \times S \times S$ is described by a formula on the action label, the prestate values of the system variables, and the poststate values of the variables. This formula is

  - a *disjunction* of formulas each of which describes one action of the system by
  - a *conjunction* of formulas that describe how the action relates the values of the system variables in the prestate to their values in the poststate.

  The disjunction thus describes in logical form the fact that in every state *some* of the actions can be performed, while the conjunctions describe *all* the effects of the corresponding actions. Please note that these effect descriptions must also describe which variables remain *unchanged* since otherwise their values in the poststate may be arbitrary.                                                          □

As above example demonstrates, we may thus describe in a concise logical form the behavior of arbitrary systems independently of particular concrete programming/system description languages.

**Example 9.2**  Consider a concurrent system of two components: one component (the "producer") repeatedly sets a variable $x$ to arbitrary natural numbers that it generates; the other component (the "consumer") repeatedly reads from $x$ these values and sums them up in a variable $y$. However, since the producer and the consumer operate asynchronously, the problem arises that the producer might write into $x$ the next value before the consumer has read the previous one; conversely, the consumer might read from $x$ the same value again before the producer has written a new one. Therefore the producer and the consumer use a third variable $z$ to synchronize their interaction: the producer sets $z$ to the value 1 to indicate that a new value of $x$ is available; the consumer sets $z$ to the value 0 to indicate that it has consumed that value. The system may thus exhibit for example the following run with states $\langle x, y, z \rangle \in \mathbb{N}^3$ and actions $P$ ("produce") and $C$ ("consume"):

$$\langle 0, 0, 0 \rangle \xrightarrow{P} \langle 2, 0, 1 \rangle \xrightarrow{C} \langle 2, 2, 0 \rangle \xrightarrow{P} \langle 3, 2, 1 \rangle \xrightarrow{C} \langle 3, 5, 0 \rangle \xrightarrow{P} \langle 3, 5, 1 \rangle \xrightarrow{C} \langle 3, 8, 0 \rangle \xrightarrow{P} \dots$$

We model this system by a LTS

$$PC := \langle \text{space} : S, \text{init} : I, \text{next} : R \rangle$$

with the following components:

$$S := \mathbb{N} \times \mathbb{N} \times \mathbb{N}_2$$
$$I \langle x, y, z \rangle :\Leftrightarrow x = 0 \wedge y = 0 \wedge z = 0$$
$$R \langle l, \langle x, y, z \rangle, \langle x', y', z' \rangle \rangle :\Leftrightarrow$$
$$(l = P \wedge z = 0 \wedge x' \in \mathbb{N} \wedge y' = y \wedge z' = 1) \vee$$
$$(l = C \wedge z = 1 \wedge y' = y + x \wedge z' = 0)$$

Here the description of the various actions has the following particularities:

- The condition $x' \in \mathbb{N}$ (which is actually redundant and can be dropped) does not uniquely define the value of $x$ in the poststate; indeed this intentionally describes that the producer may set $x$ to an arbitrary natural number.
- Each action formula has a *guard conditions* $z = 0$ respectively $z = 1$ which constrains the *prestate* of the transition: thus the corresponding action is only *enabled* in a state where the guard condition is satisfied; if this condition is not satisfied, the action is blocked.

In above system the guard conditions ensure that *either* the producer *or* the consumer performs an action; thus every execution of the system indeed perpetually interleaves the actions of both components.                                                                   □

As shown above, guard conditions may block actions. Thus, if a system shall run permanently, in every reachable state at least one guard condition must be satisfied, i.e., at least one action must be enabled. If all actions are blocked, the system is in an (intentional) *termination* state or in an (unintentional) *deadlock* state.

**Example 9.3** We consider a variant of the system described in Example 9.2, a system where a producer generates a sequence of natural numbers in a variable $x$ and a consumer adds up these numbers in a variable $y$. Now we do not allow both components to write to the same variable; thus the producer and the consumer synchronize their activities by two variables $p$ and $q$ which may be only written by the producer respectively the consumer. The system can be described by the following pseudo-code where two concurrent components $P$ and $C$ repeatedly execute two commands in an infinite loop:

```
var x, y:=0, p:=0, p0:=1, c:=0, c0:=1

P: loop
     1: x,p,p0 := any,p0,1-p0
     2: wait c = 1-p0
   ||
C: loop
     1: wait p = c0
     2: y,c,c0 := y+x,c0,1-c0
```

Thus the producer repeatedly generates a new value for $x$ and flips the value of $p$ from 1 to 0 and back, remembering in variable $p_0$ the next value of $p$. Likewise the consumer repeatedly consumes $x$ and flips the value of $c$ from 0 to 1 and back, remembering in variable $c_0$ the next value of $c$. Both the producer and the consumer wait for the new values of $c$ respectively $p$ to be set by the respective other process before producing respectively consuming the next value.

Both components run their code in loops whose bodies consist of two consecutive statements, one statement performing the updates of the variables, the other statement waiting for the permission of the other process to proceed. To adequately model the components, we therefore have to consider the values of their "program counters" in addition to the values of their variables.

We model this system by a LTS

$$PC := \langle \text{space:} S, \text{init:} I, \text{next:} R \rangle$$

with the following components:

$$S := \{1, 2\}^2 \times \mathbb{N}^2 \times (\mathbb{N}_2)^4$$

$$I \langle a, b, x, y, p, p_0, c, c_0 \rangle :\Leftrightarrow$$
$$\quad a = 1 \wedge b = 1 \wedge y = 0 \wedge p = 0 \wedge p_0 = 1 \wedge c = 0 \wedge c_0 = 1$$

$$R \langle l, \langle a, b, x, y, p, p_0, c, c_0 \rangle, \langle a', b', x', y', p', p_0', c', c_0' \rangle \rangle :\Leftrightarrow$$
$$\quad (P \langle l, \langle a, x, p, p_0, c \rangle, \langle a', x', p', p_0' \rangle \rangle \wedge b' = b \wedge y' = y \wedge c' = c \wedge c_0' = c_0) \vee$$
$$\quad (C \langle l, \langle b, y, c, c_0, p \rangle, \langle b', y', c', c_0' \rangle \rangle \wedge a' = a \wedge x' = x \wedge p' = p \wedge p_0' = p_0)$$

$$P \langle l, \langle a, x, p, p_0, c \rangle, \langle a', x', p', p_0' \rangle \rangle :\Leftrightarrow$$
$$\quad (l = P_1 \wedge a = 1 \wedge a' = 2 \wedge x' \in \mathbb{N} \wedge p' = p_0 \wedge p_0' = 1 - p_0) \vee$$
$$\quad (l = P_2 \wedge a = 2 \wedge c = 1 - p_0 \wedge a' = 1 \wedge x' = x \wedge p' = p \wedge p_0' = p_0)$$

$$C \langle l, \langle b, y, c, c_0, p \rangle, \langle b', y', c', c_0' \rangle \rangle :\Leftrightarrow$$
$$\quad (l = C_1 \wedge b = 1 \wedge p = c_0 \wedge b' = 2 \wedge y' = y \wedge c' = c \wedge c_0' = c_0) \vee$$
$$\quad (l = C_2 \wedge b = 2 \wedge b' = 1 \wedge y' = y + x \wedge c' = c_0 \wedge c_0' = 1 - c_0)$$

The state space $S$ is the product of the state spaces of the program counters of both processes, called $a$ and $b$, of the two natural number variables $x$ and $y$, and of the four bit variables $p$, $p_0$, $c$, $c_0$. The initial state condition $I$ sets all variables to their initial values except for variable $x$ which is left arbitrary; thus the system has infinitely many initial states.

The transition relation $R$ is described with the help of two auxiliary relations $P$ and $C$ that describe the behavior of the producer respectively the consumer. These auxiliary relations receive as their first argument the tuple of the prestate values of all variables that are read by the corresponding component and as their second argument the tuple of the poststate values of all variables that are written by the component; all other variables are determined by $R$ to remain unchanged.

Each auxiliary relation describes by a disjunction the effect of both commands in the body of the while loop as the basic actions of the system; in total the system thus has four actions labeled $P_1$, $P_2$, $C_1$, $C_2$. The corresponding conjunctions constrain by a guard condition the program counter to the position of the respective command and describe as an additional effect the respective program counter's new value.

Actually, the effect of the actions $P_1$ and $C_2$ could (by using additional values for the program counters) be broken into finer grained actions each of which performs only a single assignment; this would still preserve the correctness of the value transfer protocol. □

Above example demonstrates the basic principle of how by the use of program counters the effects of arbitrary commands can be modeled as multiple actions. While we refrain from describing a complete translation of programming language commands to actions, later sections will describe the translation of high level system descriptions into labeled transition systems.

**The Composition of Systems**

A labeled transition system can represent a *component* of a larger system that consists of multiple such components. In the following, we describe how the LTS of the composed system can be constructed from the labeled transition systems of the individual components. For example, let us consider the composition of two systems $A$ and $B$ defined as follows:

$$A := \langle \mathsf{space}{:}\mathbb{N}_3, \mathsf{init}{:}(\pi x. \; x = 0), \mathsf{next}{:}(\pi l, x, x'. \; l = a \wedge x' = x + 1 \bmod 3) \rangle$$

$$B := \langle \mathsf{space}{:}\mathbb{N}_2, \mathsf{init}{:}(\pi y. \; \mathsf{true}), \mathsf{next}{:}(\pi l, y, y'. \; l = b \wedge y' = y + 1 \bmod 2) \rangle$$

System $A$ operates in the state space $\mathbb{N}_3 = \{0, 1, 2\}$ by, starting with value $x = 0$, repeatedly performing a transition $a$ that increments $x$ to $x + 1 \bmod 3$:

$$A: \; 0 \xrightarrow{a} 1 \xrightarrow{a} 2 \xrightarrow{a} 0 \xrightarrow{a} 1 \xrightarrow{a} 2 \xrightarrow{a} 0 \xrightarrow{a} \dots$$

Likewise $B$ operates in the state space $\mathbb{N}_2 = \{0, 1\}$ by, starting with either value $y = 0$ or value $y = 1$, repeatedly performing a transition $b$ that increments $y$ to $y + 1 \bmod 2$ (thus effectively flipping the value from 0 to 1 and back):

$$B: \; 0 \xrightarrow{b} 1 \xrightarrow{b} 0 \xrightarrow{b} 1 \xrightarrow{b} 0 \xrightarrow{b} \dots$$

$$1 \xrightarrow{b} 0 \xrightarrow{b} 1 \xrightarrow{b} 0 \xrightarrow{b} 1 \xrightarrow{b} \dots$$

Our goal is to construct from these components the LTS

$$AB := \langle \mathsf{space}{:}S, \mathsf{init}{:}I, \mathsf{next}{:}R \rangle$$

that describes their concurrent composition:

- **State Space**: Since a state of the composed system consists of the states of its components, the state space is the *product* of the component state spaces:

$$S := A.\mathsf{space} \times B.\mathsf{space} = \mathbb{N}_3 \times \mathbb{N}_2$$

Each state of $S$ is thus a tuple $\langle x, y \rangle$ of values $x \in \mathbb{N}_3$ and $y \in \mathbb{N}_2$. Consequently, since the components operate in finite state spaces, the size of the state space is the product of the sizes of the individual state spaces:

$$|S| = |A.\mathsf{space}| \cdot |B.\mathsf{space}| = |\mathbb{N}_3| \cdot |\mathbb{N}_2| = 3 \cdot 2 = 6$$

- **Initial State Condition**: The composed system is in its initial state, if both components are in their initial states. The initial state condition is thus the *conjunction* of the initial state conditions of the components:

$$I\langle x, y\rangle :\Leftrightarrow A\text{.init}\langle x\rangle \wedge B\text{.init}\langle y\rangle \Leftrightarrow x = 0 \wedge \text{true} \Leftrightarrow x = 0$$

  Thus, if all components have finitely many initial states, the number of initial states of the composed system is the product of the number of the initial states of the components (the example system has $2 = 1 \cdot 2$ initial states).
- **Next State Relation**: As for the question when the composed system makes a transition from one state to the next, we may choose among several models:
  - **Synchronous Execution**: In this model, the system makes a step if *all* of its components make a step. The transition relation is thus defined by a *conjunction* of the transition relations of the individual components:

$$R\langle l, \langle x, y\rangle, \langle x', y'\rangle\rangle :\Leftrightarrow l = ab \wedge A\text{.next}\langle a, x, x'\rangle \wedge B\text{.next}\langle b, y, y'\rangle$$
$$\Leftrightarrow l = ab \wedge a = a \wedge x' = x + 1 \bmod 3 \wedge$$
$$b = b \wedge y' = y + 1 \bmod 2$$
$$\Leftrightarrow l = ab \wedge x' = x + 1 \bmod 3 \wedge y' = y + 1 \bmod 2$$

  The transition of the composed system receives a new label $l = ab$; its effect subsumes the effect of the transitions $a$ respectively $b$ of the components. The two possible runs of this system are the following ones:

$$\langle 0, 0\rangle \xrightarrow{ab} \langle 1, 1\rangle \xrightarrow{ab} \langle 2, 0\rangle \xrightarrow{ab} \langle 0, 1\rangle \xrightarrow{ab} \langle 1, 0\rangle \cdots$$
$$\langle 0, 1\rangle \xrightarrow{ab} \langle 1, 0\rangle \xrightarrow{ab} \langle 2, 1\rangle \xrightarrow{ab} \langle 0, 0\rangle \xrightarrow{ab} \langle 1, 1\rangle \cdots$$

  In each step these runs modify the values of both state components.
  - **Asynchronous Execution**: In this model, the system makes a step if *at least one* of its components makes a step. For a system with two components $A$ and $B$, the transition relation is thus defined by *disjunction* of three conjunctions corresponding to the three possibilities that component $A$ makes a step, component $B$ makes a step, or both components make a step:

$$R\langle l, \langle x, y\rangle, \langle x', y'\rangle\rangle :\Leftrightarrow (A\text{.next}\langle l, x, x'\rangle \wedge y' = y) \vee$$
$$(x' = x \wedge B\text{.next}\langle l, y, y\rangle) \vee$$
$$(l = ab \wedge A\text{.next}\langle a, x, x'\rangle \wedge B\text{.next}\langle b, y, y'\rangle)$$
$$\Leftrightarrow (l = a \wedge x' = x + 1 \bmod 3 \wedge y' = y) \vee$$
$$(l = b \wedge x' = x \wedge y' = y + 1 \bmod 2) \vee$$
$$(l = ab \wedge x' = x + 1 \bmod 3 \wedge y' = y + 1 \bmod 2)$$

  Each conjunction combines the effects of those component transitions that participate in the system transition; if a component does not make a step, an equality indicates that the value of that component remains unchanged. A possible run of this system is the following one:

$$\langle 0, 0 \rangle \xrightarrow{a} \langle 1, 0 \rangle \xrightarrow{ab} \langle 2, 1 \rangle \xrightarrow{b} \langle 2, 0 \rangle \xrightarrow{b} \langle 2, 1 \rangle \cdots$$

In each step this run modifies either one of the components of the system state or both simultaneously.

- **Interleaving Execution**: In this model, the system makes a step if *exactly one* of its components makes a step. In a system with two components, the transition relation is thus a *disjunction* of two conjunctions:

$$R\langle l, \langle x, y \rangle, \langle x', y' \rangle \rangle :\Leftrightarrow (A.\text{next}\langle l, x, x' \rangle \wedge y' = y) \vee$$
$$(x' = x \wedge B.\text{next}\langle l, y, y \rangle)$$
$$\Leftrightarrow (l = a \wedge x' = x + 1 \bmod 3 \wedge y' = y) \vee$$
$$(l = b \wedge x' = x \wedge y' = y + 1 \bmod 2)$$

The transition formula thus contains all the conjunctions of the formula of the asynchronous execution model except those where more than one component makes a step. A possible run of this system is the following one:

$$\langle 0, 0 \rangle \xrightarrow{a} \langle 1, 0 \rangle \xrightarrow{a} \langle 2, 0 \rangle \xrightarrow{b} \langle 2, 1 \rangle \xrightarrow{a} \langle 0, 1 \rangle \cdots$$

In each step this run (as does every run of the system) modifies exactly one component of the system state.

The execution possibilities of the various models are visualized in Fig. 9.1 by "grid-like" directed graphs with six points that denote the state space of the example system; the coordinates of the points represent the values of the component states. The initial states of the system are depicted by the incoming arrows without sources; every other arrow denotes a transition from one state to the next one. In the synchronous model, every transition $ab$ changes the state of both components such that a run proceeds along the "diagonals" of the grid. In the interleaving model, every transition $a$ or $b$ changes the state of one component, leading to executions along the "axes" of the grid. The asynchronous model encompasses all the transitions of the other two models.

Synchronous composition is the model that is easiest to understand. However, it is only applicable if the components of a system are closely coupled by a synchronization mechanism that ensures that all components operate in "lock-step". This is typically only the case for concurrent hardware systems where the execution of each step of all components is triggered by the signal of a central clock.

Concurrent software components typically operate with their own individual clocks, which makes the asynchronous execution model more suitable. This model, however, has the disadvantage that the transition relation of a system composed of $n$ components requires a conjunction of $2^n - 1$ clauses, corresponding to all possible

**Fig. 9.1** Models of composition

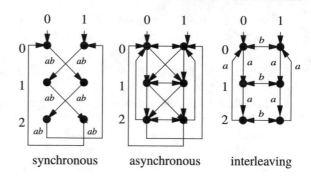

synchronous          asynchronous          interleaving

simultaneous executions of the individual components (excluding only the case that no component makes a transition). This considerably increases the complexity of reasoning about such a system.

We therefore choose as the basis for our subsequent modeling of concurrent software systems the *interleaving model* where the composition of $n$ components only requires a disjunction with $n$ clauses, one for each component. Such a model is not only manageable but it also describes the actual system accurately enough: in most contexts the simultaneous execution of multiple components (which is ignored in the interleaving model) is indistinguishable from the subsequent execution of these components in an arbitrary order (which the model adequately handles).

The following definition formalizes the interleaving execution model for a system with an arbitrary number of components.

---

**Definition 9.4** (*Interleaving Execution*) Let $n \in \mathbb{N}$, $L$ be a label set, and $S$ be a state set. Then the *composition* of $n$ labeled transition systems $C \in \left(LTS^{L,S}\right)^n$ with *interleaving execution* is a labeled transition system $\langle \text{space:}S_0, \text{init:}I, \text{next:}R \rangle$ whose constituents $S_0 \in \text{Set}(S)^n$, $I \in \text{Set}(S_0)$, and $R \in \text{Set}(L \times S_0 \times S_0)$ are defined as follows:

$S_0 := \lambda i \in \mathbb{N}_n.\, C(i).\text{space}$

$I\langle x \rangle :\Leftrightarrow \forall i \in \mathbb{N}_n.\, C(i).\text{init}\langle x(i) \rangle$

$R\langle l, x, x' \rangle :\Leftrightarrow \exists i \in \mathbb{N}_n.\, C(i).\text{next}\langle l, x(i), x'(i) \rangle \wedge \forall j \in \mathbb{N}_n.\, i \neq j \Rightarrow x(j) = x'(j)$

---

Thus in the interleaving execution of a system composed from $n$ components there are in general $n$ choices which component $i$ makes the next step; only the state of this component is changed, while the state of every component $j \neq i$ remains unchanged.

In above formalization the state space of the composed system is constructed as the product of the disjoint state spaces of the individual components; consequently every component operates on its own portion of the state space without any effect on the other components or being affected by them. In practice, however, we would like the individual components to interact with each other; this may be achieved by generalizing the model to allow also shared portions of the system state that can be read and written by multiple components. Furthermore, if the number of components is not fixed but there are arbitrarily many instances of a particular component template, we may describe this by a model of the component that receives as an additional parameter the identity of the instance. We refrain from formalizing these ideas and rather illustrate them by the following example.

**Example 9.4** Consider the following system with one server and $N$ clients:

```
Client(i):
  loop
    0: sendServer(i);
    1: waitServer();
    ... // critical region
    2: sendServer(i);

Server:
  var given := N, waiting := {}
  loop
    var i := receiveClient();
    if i = given then
      if waiting = {} then
        given := N
      else
        choose given in waiting;
        waiting := waiting\{given};
        sendClient(given)
    else if given = N then
      given := i;
      sendClient(given)
    else
      waiting := waiting U {i}
```

Every client forever runs in a loop that sends a message with the identity $i \in \mathbb{N}_N$ of the client to the server and then waits for an answer from the server that allows it to enter the critical region. When the client leaves the critical region it informs the server about this by another message.

The purpose of the server is to ensure that only one client at a time enters the critical region. For this it maintains in variable *given* the identify of the client that was given access to the critical region; if there is no such client, *given* has value $N$. Furthermore, the server records in set *waiting* the identities of the clients that have asked for permission to enter the critical region but not yet received an answer. The server forever processes arriving messages as follows:

- If the message is from the client that was previously given access to the critical region, this message indicates that the client has left the region. If there is no client waiting for access, *given* is set to $N$; otherwise, it is set to the value of any of the waiting clients. The server removes this client from the set and sends to it the permission to enter the critical region.
- Otherwise the message is from a client that asks for access to the critical region. If *given* is $N$, no other client is in the critical region, so the server sets *given* to the identity of the requesting client and sends the client the permission to enter the critical region. Otherwise, the identity of the client is recorded in the set *waiting*.

We model this system by a LTS with the state space $S$ defined as follows:

$$S := \mathbb{N} \times \text{Set}(\mathbb{N}) \times \mathbb{N}^N \times \text{Set}(\mathbb{N}) \times \text{Set}(\mathbb{N})$$

Here every state $\langle given, waiting, pc, req, ans \rangle \in S$ is interpreted as follows:

- *given* and *waiting* represent the server variables.
- *pc(i)* represents the program counter of client $i$.
- *req* and *ans* represent the messages pending in the network i.e., the set of messages from the clients to the server respectively the set of answers from the server.

The system is described by a composition of one server component and one client component (representing the execution of the $N$ clients) where the server component operates on the portion $\langle given, waiting \rangle$ of the state space and the client component operates on the portion $pc$. Thus the state space is the product of the state spaces of the two components extended by the portion $\langle req, ans \rangle$ which represents the network that both components use in common for communication with each other.

We thus model the initial state condition $I$ of the system by a conjunction of conditions of the server component, the client component, and the network component:

$$
\begin{aligned}
&I\langle given, waiting, pc, req, ans \rangle :\Leftrightarrow \\
&(given = N \wedge waiting = \varnothing) \wedge \\
&(\forall i \in \mathbb{N}_N.\, pc(i) = 0) \wedge \\
&(req = \varnothing \wedge ans = \varnothing)
\end{aligned}
$$

The transition relation $R$ of the system is (according to the interleaving model of execution) defined by a disjunction of the transition relation $RS$ of the server and the transition relation $RC$ of the client:

$$
\begin{aligned}
&R\langle l, \langle given, waiting, pc, req, ans \rangle, \langle given', waiting', pc', req', ans' \rangle \rangle :\Leftrightarrow \\
&\quad (RS\langle l, \langle given, waiting, req, ans \rangle, \langle given', waiting', req', ans' \rangle \rangle \wedge \\
&\quad\quad pc' = pc) \vee \\
&\quad (\exists i \in \mathbb{N}_N.\, RC\langle l, i, \langle pc, req, ans \rangle, \langle pc', req', ans' \rangle \rangle \wedge \\
&\quad\quad given' = given \wedge waiting' = waiting)
\end{aligned}
$$

Here the server relation $RS$ operates on the server portion and the network portion of the state space. The client relation $RC$ is parameterized over the identity $i \in \mathbb{N}_N$ of the client performing the transition; it operates on the corresponding client portion and on the network portion. The existential quantification of variable $i$ effectively yields a disjunction of the $N$ transitions of the respective clients.

In detail we define the client relation as follows:

$$RC\langle l, i, \langle pc, req, ans \rangle, \langle pc', req', ans' \rangle \rangle :\Leftrightarrow$$
$$\big((l = C0_i \wedge pc(i) = 0 \wedge pc'(i) = 1 \wedge$$
$$\neg(i \in req) \wedge req' = req \cup \{i\} \wedge ans' = ans) \vee$$
$$(l = C1_i \wedge pc(i) = 1 \wedge pc'(i) = 2 \wedge i \in ans \wedge$$
$$req' = req \wedge ans' = ans \backslash \{i\}) \vee$$
$$(l = C2_i \wedge pc(i) = 2 \wedge pc'(i) = 0 \wedge$$
$$\neg(i \in req) \wedge req' = req \cup \{i\} \wedge ans' = ans)\big) \wedge$$
$$(\forall j \in \mathbb{N}_N. \; j \neq i \Rightarrow pc'(j) = pc(j))$$

A client thus can perform one of three possible transitions depending on the value of its program counter.

Finally we define the server relation as follows:

$$RS\langle l, \langle given, waiting, req, ans \rangle, \langle given', waiting', req', ans' \rangle \rangle :\Leftrightarrow$$
$$\exists i \in req.$$
$$req' = req \backslash \{i\} \wedge$$
$$\big((l = S0_i \wedge i = given \wedge waiting = \varnothing \wedge$$
$$given' = N \wedge waiting' = waiting \wedge ans' = ans) \vee$$
$$(l = S1_i \wedge i = given \wedge \exists given \in waiting. \; given' = given \wedge$$
$$waiting' = waiting \backslash \{given\} \wedge$$
$$\neg(given \in ans) \wedge ans' = ans \cup \{given\}) \vee$$
$$(l = S2_i \wedge i \neq given \wedge given = N \wedge$$
$$given' = i \wedge waiting' = waiting \wedge$$
$$\neg(given' \in ans) \wedge ans' = ans \cup \{given'\}) \vee$$
$$(l = S3_i \wedge i \neq given \wedge given \neq N \wedge$$
$$waiting' = waiting \cup \{i\} \wedge given' = given \wedge ans' = ans)\big)$$

This relation is only enabled if there is a message in set $req$ which is removed; the server then can perform one of four transitions depending on the value of the message received and the values of its variables.    □

Above example sketches how we may model a *distributed* system where components interact with each other by exchanging messages. In Sect. 9.3 we will introduce a language for such systems and define their semantics as labeled transition systems.

## 9.2    Modeling Shared Systems

Section 9.1 has demonstrated that we can model concurrent systems by labeled transition systems whose essential components (the initial state condition and the transition relation) can be described by logical formulas. This description is generic, i.e., it does not require any special notation for system modeling but relies purely on the language of logic and is thus universally applicable. Nevertheless, in reality systems are mostly defined by programming languages respectively program-like modeling notations. In this section, we introduce one such notation for modeling *shared (memory) systems*, i.e., systems where concurrent activities (threads) operate in the same state; they interact with each other by reading and writing variables in that state. The denotational semantics of this language is defined by a mapping to a labeled transition system. The language is a bit more abstract than typical concurrent programming languages such that the translation from a shared system to an LTS is not very complicated; nevertheless this translation may serve as a starting point for giving a semantics also to lower-level concurrent languages.

### Syntax and Type System

We begin by introducing the abstract syntax of the modeling language.

**Definition 9.5** (*Shared Systems*)   A *shared system* is a phrase $Sy \in System$ which is formed according to the following grammar:

$Sy \in System,\ As \in Actions$
$X \in Parameters,\ I \in Identifier,\ C \in Command,\ F \in Formula$

$Sy ::=$ shared $I$ { var $X$; init $C$; $As$ }
$As ::= \_ \mid As$ action $I(X)$ with $F$ { $C$ }

Here *Actions* denotes the syntactic domain of *actions*. The syntactic domains *Parameters* of *parameters*, *Identifier* of *identifiers*, *Command* of *commands*, and *Formula* of *formulas* are formed as in Definition 7.1.

Figure 9.2 describes the type system of the language. In a nutshell it states that command $C$ and actions $As$ are type-checked in the environment $Vt$ set up by the variable declaration $X$ giving rise to the set $Is$ of action names. Each action has a distinct name $I$; its formula $F$ and command $C$ are checked in an environment $Vt_0$ derived from $Vt$ by the elaboration of the action parameters $X$.

If an action has empty parameter list $X$, we subsequently drop in the concrete syntax the parameter list () from its definition. Likewise, if an action has a trivial formula $F \Leftrightarrow$ true, we drop the clause with $F$.

Before formally defining the semantics of this language, we explain it informally by a couple of examples.

---

**Rules for** $\Sigma \vdash Sy$: **shared**($I$):

$$\frac{\Sigma \vdash X: \text{parameters}(Vt, Ss) \quad \Sigma, Vt \vdash C: \text{command} \quad \Sigma, Vt \vdash As: \text{actions}(Is)}{\Sigma \vdash \text{shared } I \ \{ \text{var } X; \text{ init } C; As \ \}: \text{shared}(I)}$$

**Rules for** $\Sigma, Vt \vdash As$: **actions**($Is$):

$$\frac{}{\Sigma, Vt \vdash \sqcup: \text{actions}(Is)} \quad Is = \varnothing$$

$$\frac{\begin{array}{c} \Sigma, Vt \vdash As: \text{actions}(Is_0) \quad \neg(I \in Is_0) \quad Is = Is_0 \cup \{I\} \\ \Sigma \vdash X: \text{parameters}(Vt_1, Ss_1) \quad Vt_0 = Vt \leftarrow Vt_1 \\ \Sigma, Vt_0 \vdash F: \text{formula} \quad \Sigma, Vt_0 \vdash C: \text{command} \end{array}}{\Sigma, Vt \vdash As \ \text{action } I(X) \ \text{with } F \ \{ C \ \}: \text{actions}(Is)}$$

**Fig. 9.2** Type checking shared systems

## Examples

An example for a shared system in this language was already given in Example 9.1 of the previous section:

```
shared XY2 {
  var x:nat, y:nat;
  init { x := 0; y := 0 }
  action incX { x := x+1 }
  action incY { y := y+x }
}
```

The state of the system *XY2* is denoted by the value space of the variables $x$ and $y$. The initial state of the system is derived by executing in an arbitrary state the command $C$ that may set these variables to their initial values (if a variable is not set, its value remains arbitrary). The system proceeds from one state to the next by the interleaved execution of the actions *incX* and *incY*; the execution of such an action (i.e., of the command associated to the action) is *atomic*, i.e., the execution must terminate before the execution of another action can begin.

In above example, the actions have no formulas attached to them; this implies that in *every* state all actions are enabled such that each one may be nondeterministically chosen for execution. However, an action may be also annotated by a `with` clause with a guard formula that constrains the states in which the execution of the action may take place.

**Example 9.5** Consider the following system:

```
shared IncDec {
  var x:nat3;
  init { }
  action dec with x > 0 { x := x-1 }
  action inc with x < 3 { x := x+1 }
}
```

In this system the guard formula $x > 0$ constrains the state in which the execution of the action *dec* is enabled; only if the condition is true, $x$ may be decreased the action. Correspondingly the guard formula $x < 3$ constrains the state in which the execution of the action *inc* is enabled; only if the condition is true, $x$ may be increased

the action. Starting with an arbitrary value $x \in \mathbb{N}_3 = \{0, 1, 2\}$ (since the `init` clause contains an empty command sequence) the system proceeds as follows: If $x = 0$, then only the action *inc* is enabled and thus chosen for execution, which sets $x$ to 1. If $x = 1$, then both actions are enabled and one is nondeterministically chosen: if *inc* is executed, $x$ is set to 2; if *dec* is executed, $x$ is set to 0. If $x = 2$, then only the action *dec* is enabled and thus chosen for execution, which sets $x$ to 1.

Thus the system operates in the finite space $\mathbb{N}_3$ for the values of $x$. This state space and the transitions enabled in each state can be visualized as follows:

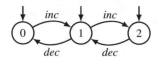

A possible run of this system is the following one:

$$1 \xrightarrow{inc} 2 \xrightarrow{dec} 1 \xrightarrow{inc} 2 \xrightarrow{dec} 1 \xrightarrow{dec} 0 \xrightarrow{inc} 1 \xrightarrow{inc} 2 \xrightarrow{dec} 1 \cdots$$

This model describes essentially a shared system with two concurrent threads that compete with each other for execution. In some states only one the execution of one thread is possible, then the next execution step is deterministic. In other states, however, the execution of multiple threads is enabled, the next execution step is chosen nondeterministically.                                                                    □

A shared system with multiple threads may be thus modeled in our language by at least one action per thread. Since actions are atomic, it may be necessary to decompose a programming language thread into multiple actions to adequately describe its effects.

**Example 9.6** Consider the following pseudo-code for a multithreaded program already given in Example 9.3:

```
var x, y:=0, p:=0, p0:=1, c:=0, c0:=1

P: loop
   1: x,p,p0 := any,p0,1-p0
   2: wait c = 1-p0
||
C: loop
   1: wait p = c0
   2: y,c,c0 := y+x,c0,1-c0
```

This pseudo-program can be described as the following shared system:

```
shared PC {
  var x:nat, y:nat, p:bit, p0:bit, c:bit, c0:bit, a:nat, b:nat;
  init { y:=0; p:=0; p0:=1; c:=0; c0:=1; a:=1; b:=1 }
  action P1 with a = 1 { var any:nat; x:=any; p:=p0; p0:=1-p0; a:=2 }
  action P2 with a = 2 and c = 1-p0 { a:=1 }
  action Q1 with b = 1 and p = c0 { b:=2 }
  action Q2 with b = 2 { y:=y+x; c:=c0; c0:=1-c0; b:=1 }
}
```

Here the behavior of the loops in both threads is described with the help of two
"program counters" $a$ and $b$ that cycle among the two positions 1 and 2 of the loops.
The effect of the command at each position is described by a corresponding action
that is atomically executed (as mentioned in Example 9.3, the effect of actions *P1*
and *Q2* could by the use of additional values for the program counters be also
broken into a sequence of finer grained actions each of which performs only a single
assignment).                                                                    □

Since our language requires the declaration of a fixed set of actions, it seems that
we can only model systems with a fixed number of threads. However, the following
example demonstrates that we may use parameterized actions to also model systems
with an arbitrarily large (even dynamically changing) set of threads.

**Example 9.7**  Consider the following system:

```
shared Threads {
  var ids:idset, x:natmap;
  init { ids:=∅; x:=[ ] }
  action create(i:id) with ¬(i ∈ ids) { ids:=ids∪{i}; x:=x[i ↦ 0] }
  action inc(i:id) with i ∈ ids { x[i]:=x[i]+1 }
  action dec(i:id) with i ∈ ids ∧ x[i] > 0 { x[i]:=x[i]-1 }
}
```

The state of this system consists of two entities:

- a set *ids* of elements that we consider as the identifiers of active threads;
- a map $x$ of the identifiers of active threads to natural numbers; every active thread $i$
  is in charge of maintaining its value $x[i]$.                                 □

The `init` clause initializes *ids* to the empty set and $x$ to the empty map; thus initially
there are no active threads and no values maintained.

In every state, the action *create* may be chosen with an arbitrary thread identifier $i$
that satisfies the guard condition $¬(i ∈ ids)$, i.e., with the identifier of a thread that is
not yet active. The execution of the action "activates" thread $i$ by adding its identifier
to *ids* and by initializing the value $x[i]$ maintained by the thread to 0. The system
thus has $N = |ids|$ active threads.

In every state the action *inc* may be chosen with the identifier $i$ of one of the $N$
active threads, i.e., an identifier that satisfies the guard condition $i ∈ ids$; by the
execution of the action thread $i$ increments $x[i]$. Similarly, the action *dec* may be
chosen with the identifier $i$ of an active thread that maintains a positive value $x[i]$,

$[\![\,.\,]\!]$: *System* → *LTS*$^{Label, State}$

$[\![$ shared $I$ { var $X$; init $C$; $As$ } $]\!]$ :=
   let $S_0 = space(X)$ in
   let $I = \pi s \in S.\ \exists s_0 \in S_0.\ [\![\,C\,]\!] \langle s_0, s \rangle$ in
   let $R = [\![\,As\,]\!]$ in
   $\langle$ space:$S_0$, init:$I$, next:$R \rangle$

$[\![\,.\,]\!]$: *Actions* → **Set**(*Label* × *State* × *State*)

$[\![\,\sqcup\,]\!]\langle l, s, s' \rangle :\Leftrightarrow$ false
$[\![\,As\ \text{action}\ I(X)\ \text{with}\ F\ \{\ C\ \}\,]\!]\langle l, s, s' \rangle :\Leftrightarrow$
   $[\![\,As\,]\!]\langle l, s, s' \rangle \vee$
   $(\exists vs \in choose(X).\ l = \langle I, vs \rangle \wedge \text{let}\ s_0 = write(X, vs, s)\ \text{in}\ [\![\,F\,]\!]_{s_0} \wedge [\![\,C\,]\!]\langle s_0, s' \rangle)$

**Auxiliary Definitions**

   *Label* := *Identifier* × *Value*$^*$

   *space*: *Parameter* → **Set**(*State*)
   *space*($\sqcup$) := *State*
   *space*($X, V: S$) := $\{ s \in space(X) \mid s(V) \in A(S) \}$

   *choose*: *Parameter* → **Set**(*Value*$^*$)
   *choose*($\sqcup$) := ∅
   *choose*($X, V: S$) := $\{ vs \circ [v] \mid vs \in choose(X) \wedge v \in A(S) \}$

**Fig. 9.3** Shared systems as labeled transition systems

i.e., an identifier that satisfies the guard condition $i \in ids \wedge x[i] > 0$; by the execution of the action the thread decrements that value.

In general the execution of a system is nondeterministic in the sense that in each state any of the actions whose guard condition is satisfied by the state can be chosen for execution. Moreover, the use of action parameters extends this nondeterminism to choosing, in addition to an action, also values for the parameters of the action. As above example demonstrates, this has the same effect as creating multiple instances of the action, one for each value of the parameter that satisfies the guard condition. By interpreting the parameter values as thread identifiers and by using as state variables arrays indexed by these identifiers, we may thus simulate the execution of an arbitrary number of threads, each with its own set of variable values.

**Denotational Semantics**

Figure 9.3 defines the denotational semantics of a shared memory system by mapping it to labeled transition system whose labels are pairs of identifiers and value sequences and whose states are mappings of variables to values. The state space $S_0$ of the LTS is derived from the list $X$ of variable declarations by a function application $space(X)$; this application returns the set of those states where, for every declaration $V: S$ in $X$, variable $V$ has a value from the interpretation $A(S)$ of sort $S$ (as in Chap. 7 we assume some fixed $\Sigma$-algebra $A$ as given). The initial state condition $I$ of the LTS permits every state $s$ that can be derived from an arbitrary state $s_0$ by the execution of the initialization command $C$.

The next-state relation $R$ of the LTS is derived from the actions $As$ of the system by constructing a disjunction that consists of one condition for every action $I$:

- First, this condition states that the label $l$ of the transition consists of the action identifier $I$ and the nondeterministically chosen values $vs$ of the action parameters $X$ (the auxiliary function $choose(X)$ determines the set of all possible choices).
- Second, the condition states that the following two conditions are satisfied, for an intermediate state $s_0$ that is identical to the prestate $s$ of the transition except that the parameters $X$ hold the chosen values $vs$:

  - The guard formula $F$ holds in $s_0$.
  - The action body $C$ transforms $s_0$ to the poststate $s'$ of the transition.

The disjunction of these conditions thus describes the nondeterministic choice of an action; the existentially quantified state $s_0$ denotes the result of the nondeterministic choice of values for the parameters of the action.

**Example 9.8** Consider the following shared system:

```
shared XY3 {
  var x:nat, y:nat;
  init { x := 0 }
  action incX { x := x+1 }
  action incY with x > 0 { y := y+x }
}
```

This system starts in an initial state $[x \mapsto 0, \ldots]$ where variable $x$ is mapped to 0 and $y$ is mapped to an arbitrary natural number. Both actions have no parameter clause which corresponds in the abstract syntax to empty parameter lists. Action $incX$ has no `with` clause which corresponds in the abstract syntax to the guard condition true. This action is thus enabled in every state; if selected for execution, it increments $x$ by one. Action $incY$ is only enabled in a state when $x$ is not zero, i.e., when action $incX$ has been executed at least once; if selected for execution, it increments $y$ by $x$.

The denotational semantics of this system is the LTS $\langle \text{space:}(S), \text{init:}I, \text{next:}R \rangle$ whose components are defined as described below:

- **State Space**:

$$S := space(\text{x:nat},\text{y:nat}) = \{s \in State \mid s(x) \in \mathbb{N} \wedge s(y) \in \mathbb{N}\}$$

Thus every state in the state space maps both $x$ and $y$ to natural numbers (assuming the interpretation $A(nat) = \mathbb{N}$). If we interpret states not as mappings of variable names to values but as tuples of the values of the declared variables, we may also define the state space as follows:

$$S := \mathbb{N} \times \mathbb{N}$$

- **Initial State Condition**:

$$I \langle s \rangle :\Leftrightarrow \exists s_0 \in S. \ [\![ x := 0 ]\!] \langle s_0, s \rangle$$
$$\Leftrightarrow \exists s_0 \in S. \ s = s_0[x \mapsto 0]$$
$$\Leftrightarrow s(x) = 0$$

Thus state $s$ is initial if it maps variable $x$ to 0. If we interpret states not as mappings of variable names to values but as tuples of the values of the declared variables, we may thus define the initial state condition as follows:

$$I \langle x, y \rangle :\Leftrightarrow x = 0$$

- **Next State Relation**: Since the actions have no parameters, we use in the following definition plain action identifiers as transition labels.

$$R \langle l, s, s' \rangle :\Leftrightarrow [\![ \texttt{action incX} \ldots \texttt{action incY} \ldots ]\!] \langle l, s, s' \rangle$$
$$\Leftrightarrow (l = incX \wedge [\![ \textsf{true} ]\!]_s \wedge [\![ x := x + 1 ]\!] \langle s, s' \rangle) \vee$$
$$(l = incY \wedge [\![ x > 0 ]\!]_s \wedge [\![ y := y + x ]\!] \langle s, s' \rangle)$$
$$\Leftrightarrow (l = incX \wedge s' = s[x \mapsto s(x) + 1]) \vee$$
$$(l = incY \wedge s(x) > 0 \wedge s' = s[y \mapsto s(y) + s(x)])$$
$$\Leftrightarrow (l = incX \wedge s'(x) = s(x) + 1 \wedge$$
$$\forall I \in Identifier. \ I \neq x \Rightarrow s'(I) = s(I)) \vee$$
$$(l = incY \wedge s(x) > 0 \wedge s'(y) = s(y) + s(x) \wedge$$
$$\forall I \in Identifier. \ I \neq y \Rightarrow s'(I) = s(I))$$
$$\Rightarrow (l = incX \wedge s'(x) = s(x) + 1 \wedge s'(y) = s(y)) \vee$$
$$(l = incY \wedge s(x) > 0 \wedge s'(y) = s(y) + s(x) \wedge s'(x) = s(x))$$

In the last step above we have derived a formula that describes only the values of variables $x$ and $y$ (and ignores the values of all variables not declared in the system). If we interpret states as tuples of these variable values, we may thus define the next-state relation as follows:

$$R \langle l, \langle x, y \rangle, \langle x', y' \rangle \rangle :\Leftrightarrow (l = incX \wedge x' = x + 1 \wedge y' = y) \vee$$
$$(l = incY \wedge x > 0 \wedge y' = y + x \wedge x' = x) \qquad \square$$

Above example demonstrates the general pattern that allows us to "manually" translate a system description to the LTS underlying the system, choosing as the state space the product of the variable domains and defining the initial state condition and the next-state relation by formulas on these domains. The following example generalizes this pattern to actions with parameters.

**Example 9.9** Consider the system introduced in Example 9.7:

```
shared Threads {
    var ids:idset, x:natmap;
    init { ids:=∅; x:=[ ] }
    action create(i:id) with ¬(i∈ids) { ids:=ids∪{i}; x:=x[i ↦ 0] }
    action inc(i:id) with i∈ids { x[i]:=x[i]+1 }
    action dec(i:id) with i∈ids ∧ x[i] > 0 { x[i]:=x[i]-1 }
}
```

The denotational semantics of this system is the LTS $\langle \text{space}:S, \text{init}:I, \text{next}:R\rangle$ whose components are defined as follows:

$$S := \{s \in State \mid s(ids) \in \text{Set}(\mathbb{N}) \wedge s(x) \in (\mathbb{N} \to_{\perp} \mathbb{N})\}$$

The initialization command gives rise to the following initial state condition:

$$I\langle s\rangle :\Leftrightarrow s(ids) = \varnothing \wedge s(x) = [\,]$$

The next-state relation is defined as a disjunction of the state relations of the various system actions:

$$R\langle l, s, s'\rangle :\Leftrightarrow Create\langle l, s, s'\rangle \vee Inc\langle l, s, s'\rangle \vee Dec\langle l, s, s'\rangle$$

These action relations are defined as follows:

$$
\begin{aligned}
Create\langle l, s, s'\rangle :\Leftrightarrow{} & \exists v_i \in \mathbb{N}.\, l = create(v_i) \wedge \\
& \text{let } s_0 = s[i \mapsto v_i] \text{ in} \\
& [\![ \neg(i \in ids)\,]\!]_{s_0} \wedge [\![\,ids:=ids \cup \{i\}\,;\, x:=x[i \mapsto 0]\,]\!]\langle s_0, s'\rangle \\
\Leftrightarrow{} & \exists v_i \in \mathbb{N}.\, l = create(v_i) \wedge \\
& \neg(v_i \in s(ids)) \wedge \\
& s' = s[ids \mapsto s(ids) \cup \{v_i\}][x \mapsto s(x)[v_i \mapsto 0]]
\end{aligned}
$$

$$
\begin{aligned}
Inc\langle l, s, s'\rangle :\Leftrightarrow{} & \exists v_i \in \mathbb{N}.\, l = inc(v_i) \wedge \\
& \text{let } s_0 = s[i \mapsto v_i] \text{ in } [\![\, i \in ids \,]\!]_{s_0} \wedge [\![\, x[i]:=x[i] + 1 \,]\!]\langle s_0, s'\rangle \\
\Leftrightarrow{} & \exists v_i \in \mathbb{N}.\, l = inc(v_i) \wedge \\
& v_i \in s(ids) \wedge s' = s[x \mapsto s(x)[v_i \mapsto s(x)(v_i) + 1]]
\end{aligned}
$$

$$
\begin{aligned}
Dec\langle l, s, s'\rangle :\Leftrightarrow{} & \exists v_i \in \mathbb{N}.\, l = dec(v_i) \wedge \\
& \text{let } s_0 = s[i \mapsto v_i] \text{ in} \\
& [\![\, i \in ids \wedge x[i] > 0 \,]\!]_{s_0} \wedge [\![\, x[i]:=x[i] - 1 \,]\!]\langle s_0, s'\rangle \\
\Leftrightarrow{} & \exists v_i \in \mathbb{N}.\, l = dec(v_i) \wedge \\
& v_i \in s(ids) \wedge s(x)(v_i) > 0 \wedge \\
& s' = s[x \mapsto s(x)[v_i \mapsto s(x)(v_i) - 1]]
\end{aligned}
$$

If we interpret states as tuples of the values of the declared variables, we can simplify these definitions to the following ones:

$$S := \mathsf{Set}(\mathbb{N}) \times (\mathbb{N} \to_\perp \mathbb{N})$$
$$I\langle ids, x\rangle :\Leftrightarrow ids = \varnothing \wedge x = [\,]$$
$$R\langle l, s, s'\rangle :\Leftrightarrow Create\langle l, s, s'\rangle \vee Inc\langle l, s, s'\rangle \vee Dec\langle l, s, s'\rangle$$
$$Create\langle l, \langle ids, x\rangle, \langle ids', x'\rangle\rangle :\Leftrightarrow$$
$$\quad \exists v_i \in \mathbb{N}.\; l = create(v_i) \wedge \neg(v_i \in ids) \wedge ids' = ids \cup \{v_i\} \wedge x' = x[v_i \mapsto 0]$$
$$Inc\langle l, \langle ids, x\rangle, \langle ids', x'\rangle\rangle :\Leftrightarrow$$
$$\quad \exists v_i \in \mathbb{N}.\; l = inc(v_i) \wedge v_i \in ids \wedge x' = x[v_i \mapsto x(v_i) + 1] \wedge ids' = ids$$
$$Dec\langle l, \langle ids, x\rangle, \langle ids', x'\rangle\rangle :\Leftrightarrow$$
$$\quad \exists v_i \in \mathbb{N}.\; l = dec(v_i) \wedge v_i \in ids \wedge x(v_i) > 0 \wedge x' = x[v_i \mapsto x(v_i)-1] \wedge ids' = ids$$

By the brevity of these definitions, we may also inline the definitions of the action relations into that of the overall system relation:

$$R\langle l, \langle ids, x\rangle, \langle ids', x'\rangle\rangle :\Leftrightarrow$$
$$\quad (\exists v_i \in \mathbb{N}.\; l = create(v_i) \wedge \neg(v_i \in ids) \wedge ids' = ids \cup \{v_i\} \wedge x' = x[v_i \mapsto 0]) \vee$$
$$\quad (\exists v_i \in \mathbb{N}.\; l = inc(v_i) \wedge v_i \in ids \wedge x' = x[v_i \mapsto x(v_i) + 1] \wedge ids' = ids) \vee$$
$$\quad (\exists v_i \in \mathbb{N}.\; l = dec(v_i) \wedge v_i \in ids \wedge x(v_i) > 0 \wedge x' = x[v_i \mapsto x(v_i)-1] \wedge ids' = ids)$$

This form very clearly demonstrates the nondeterministic nature of the system that chooses both an action (indicated by the disjunction) and the parameters of the action (indicated by the existential quantification).                                            □

## 9.3   Modeling Distributed Systems

In the previous section, we have modeled systems where multiple activities are closely coupled by reading and writing the same shared state; a prominent example of such systems are processes with concurrent threads operating in a single address space. In this section, we turn our attention towards *distributed (memory) systems* which consist of multiple components that also operate concurrently; however, each component encapsulates its own state in the sense that only this component may read and write this state. Interaction between components is only possible via well-defined interfaces that allow one component to trigger an action in another component by sending to that component a message (which may carry some values as payload). Examples of such systems are e.g. client/server applications communicating over the Internet via network sockets.

In more detail, we will model systems as the one depicted in Fig. 9.4. This system consists of two components labeled $A$ and $B$ each of which encapsulates a separate state; e.g., component $A$ has a state with variables $a, b$, while component $B$ has a

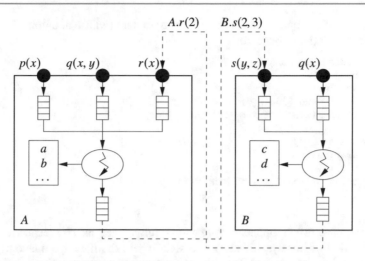

**Fig. 9.4** A component-based distributed system

state with variables $c, d$. Each component provides an externally visible interface that consists of multiple actions each of which has a name and some parameters. For instance, component $A$ is equipped with parameterized action $p(x)$ with name $p$ and parameter $x$, as well as actions $q(x, y)$ and $r(y)$. Likewise, component $B$ has an interface with actions $s(y, z)$ and $q(x)$.

Each component sequentially executes a sequence of its actions. This execution may read and write the state of the component, but it may also perform a send command; e.g., an action in component $A$ may perform the command send $B.s(2, 3)$. This command generates a message $s(2, 3)$ that is placed into an output queue of messages from where a communication subsystem eventually transfers it to component $B$; there the message it stored in an input queue associated to the action $s$ of the component. The execution of the send command does not block the action: its execution immediately continues with the next command.

The execution of every action in a component is only enabled, if there exists a message in the input queue associated to the action. The component nondeterministically selects a message from the front of any queue, removes the message from the queue, and then performs the corresponding action, with the values in the message substituted for the action parameters. The execution of this action may update the local state of the component and lead to the generation of new messages (to any component including the current one). When the execution of an action terminates, the component nondeterministically selects another message that triggers the execution of another action. If all queues of a component are empty, the execution of the component is blocked until a message from another component arrives.

The execution of the system is initialized by executing an initialization command in every component that may generate some initial messages. The execution of the system terminates if no action is executed and all queues are empty. The system may, however, also run forever generating infinitely many messages.

As a simple example consider the following system (whose notation will be subsequently formally introduced):

```
distributed PingPong {
  component Ping {
    var x:nat;
    init { send Ping.r(0) }
    action r(i:nat) { x := i+1; send Pong.s(x) }
  }
  component Pong {
    var y:nat;
    init { }
    action s(i:nat) { send Ping.r(i); y := i }
  }
}
```

This system has two components *Ping* and *Pong* with actions $r(i)$ respectively $s(i)$. The execution of this system starts in a state where the message queue of action *Ping.r* contains a single message $r(0)$ and the message queue of action *Pong.s* is empty; thus only the execution of *Ping.r* is enabled. The execution of this action sets the local variable $x$ of *Ping* to 1 and sends the message $s(1)$ to *Pong*. After that the queue of *Ping.r* is empty; thus only the execution of *Pong.s* is enabled. The execution of this action sends the message $r(1)$ back to *Ping* and then sets the local variable $y$ of *Pong* to 1. This infinitely repeating behavior leads to the following run:

$$\langle x{:}?, y{:}?\rangle \xrightarrow{Ping.r(0)} \langle x{:}1, y{:}?\rangle \xrightarrow{Pong.s(1)} \langle x{:}1, y{:}1\rangle$$
$$\xrightarrow{Ping.r(1)} \langle x{:}2, y{:}1\rangle \xrightarrow{Pong.s(2)} \langle x{:}2, y{:}2\rangle$$
$$\xrightarrow{Ping.r(2)} \dots$$

Above presentation describes a variant of the *actor model*, a mathematical model of systems that communicate by message passing. Alternatively, the execution of a command send $o.m(\dots)$ can be also considered as an *asynchronous* call of a method $m$ in an object $o$, i.e., a method call that is actually executed at an arbitrary later time, does not block the caller of the procedure, and does not immediately return a result; it is thus a natural extension of object-oriented programming model to a concurrent environment.

**Syntax and Type System**

Now we formally define our language of distributed systems.

> **Definition 9.6** (*Distributed Systems*)  A *distributed system* is a phrase $Sy \in System$ which is formed according to the following grammar:
>
> $Sy \in System, \ Cs \in Components, \ As \in Actions$
> $I \in Identifier, \ N \in Numeral, \ X \in Parameters$

$C \in Command, \; F \in Formula, \; Ts \in Terms$

$Sy ::= \texttt{distributed } I \; \{ \; Cs \; \}$
$Cs ::= \; \llcorner \; | \; Cs \; \texttt{component } I[N] \; \{ \; \texttt{var } X; \; \texttt{init } C; \; As \; \}$
$As ::= \; \llcorner \; | \; As \; \texttt{action } I(X) \, \texttt{with } F \; \{ \; C \; \}$
$C ::= \; \ldots \; | \; \texttt{send } I_1[T].I_2(Ts)$

Here *Components* denotes the syntactic domain of *components* while *Actions* denotes the syntactic domain of *actions*. The syntactic domain *Parameters* of *parameters*, *Identifier* of *identifiers*, *Command* of *commands*, and *Formula* of *formulas* are formed as in Definition 7.1 except that the domain *Command* is extended by a "send" command as indicated above. The domain *Numeral* consists of *numerals*, i.e., numerical literals that denote natural numbers.

In essence, every component resembles a shared system as introduced in Definition 9.5 except that for every action $\texttt{action } I(X)$ the values of the parameters $X$ are externally determined by the component that sends the message which triggers the execution of the action; in a shared system, these values are internally chosen nondeterministically by the component performing the action. A declaration $\texttt{component } I[N]$ indicates that there are actually $N$ instances $I[0], \ldots, I[N-1]$ of the component; as shown in the introductory example, we will drop the qualification $[N]$, if $N = 1$. Furthermore the domain of commands is extended by the command $\texttt{send } I_1[T].I_2(Ts)$ that sends a message to that instance of component $I_1$ that is denoted by $T$ (a term denoting a natural number term); this triggers in that component instance the execution of action $I_2$ with the values of the terms $Ts$. We will subsequently drop the qualification $[T]$, if $T = 0$, i.e., if the message is directed to the first instance.

**Example 9.10** Consider the following distributed system with one server component and $N$ instances of a client component (here $N$ denotes an arbitrary natural number):

```
distributed ClientServer {
  component Server {
    var client:id;
    init { client := N }
    action request(i:id) with client = N
    { client := i; send Client[i].enter() }
    action return(i:id)
    { client := N }
  }
  component Client[N] {
    var req:bool, use:bool;
    init { req := 0; use:= 0; send Client[this].ask() } }
    action ask()
    { req := 1; send Server.request(this) }
    action enter()
    { req := 0; use := 1; send Client[this].exit() }
```

```
        action exit()
        { use := 0; send Server.return(this); send Client[this].ask() }
    }
}
```

This system models access to a single shared resource. The clients repeatedly request from the server the permission to enter a critical region where they use the resource. The main task of the server is to ensure that no two clients enter this region at the same time (*mutual exclusion*).

In more detail, each client $i \in \{0, \ldots, N-1\}$ performs the following actions (in the client component, the variable `this` represents this client number $i$):

- Action *ask*() sends message *request*($i$) to the server, thus asking for permission to enter the critical region; initially this action is enabled for every client, i.e., they compete for the permission to enter the critical region.
- Action *enter*() is only enabled by the server (see below). It allows client $i$ to enter the critical region; the client subsequently enables action *exit*($i$).
- Action *exit*() leaves the critical region informing the server by a message *return*($i$); the client subsequently enables action *ask*() again.

Thus each client $i$ executes perpetually a sequence of actions *Client*[$i$].*ask*() $\rightarrow$ *Client*[$i$].*enter*() $\rightarrow$ *Client*[$i$].*exit*() $\rightarrow$ *Client*[$i$].*ask*() $\rightarrow \cdots$ of requesting permission to enter the critical region, entering the critical region, and leaving it again; the actions of the different clients are interleaved in the execution of the whole system. Furthermore, a local variable *req* $\in \{0, 1\}$ records whether the client has a pending request (the request has been sent but no answer has yet been received); local variable *use* $\in \{0, 1\}$ records whether the client currently is in the critical region.

On the other hand, the server performs the following actions:

- By action *request*($i$) the server accepts the request of client $i$ to enter the critical region; the server then sends the response *give*() to the client.
- By action *return*($i$) the server receives the message of client $i$ that indicates that the client has left the critical region.

The actions of the server are thus triggered only by the clients. Additionally, however, the server maintains a local variable *client* that holds the identity of the client $i \in \{0, \ldots, N-1\}$ that is currently allowed to enter the critical region. Action *request*($i$) is only enabled if *client* $= N$ (which indicates that no client holds the resource) and sets by its execution *client* to $i$. Action *return*($i$) resets *client* to $N$.

Assuming $N = 2$, above system thus allows an action sequence

$$Client[1].ask() \rightarrow Client[0].ask() \rightarrow Server.request(1)$$
$$\rightarrow Client[1].enter() \rightarrow Client[1].exit() \rightarrow Server.return(1)$$
$$\rightarrow Client[0].enter() \rightarrow \cdots$$

**Rules for** $\Sigma \vdash Sy$: **distributed**$(I)$**:**

$$\frac{\Sigma \vdash Cs:\text{ctyping}(Ct) \quad \Sigma, Ct \vdash Cs:\text{components}(Is)}{\Sigma \vdash \text{distributed } I \text{ \{ } Cs \text{ \}}:\text{distributed}(I)}$$

**Rules for** $\Sigma \vdash Cs$: **ctyping**$(Ct)$**:**

$$\frac{Ct = \varnothing}{\Sigma \vdash {}_{\sqcup}:\text{ctyping}(Ct)} \qquad \frac{\Sigma \vdash Cs:\text{ctyping}(Ct_0) \quad \Sigma \vdash As:\text{atyping}(At) \quad Ct = Ct_0[I \mapsto At]}{\Sigma \vdash Cs \text{ component } I[N] \text{ \{ var } X;\text{ init } C;\, As \text{ \}}:\text{ctyping}(Ct)}$$

**Rules for** $\Sigma \vdash As$: **atyping**$(At)$**:**

$$\frac{At = \varnothing}{\Sigma \vdash {}_{\sqcup}:\text{atyping}(At)} \qquad \frac{\Sigma \vdash As:\text{atyping}(At_0) \quad \Sigma \vdash X:\text{parameters}(Vt, Ss) \quad At = At_0[I \mapsto Ss]}{\Sigma \vdash As \text{ action } I(X) \text{ with } F \text{ \{ } C \text{ \}}:\text{atyping}(At)}$$

**Rules for** $\Sigma, Ct \vdash Cs$: **components**$(Is)$**:**

$$\frac{Is = \varnothing}{\Sigma, Ct \vdash {}_{\sqcup}:\text{components}(Is)}$$

$$\frac{\begin{array}{c}\Sigma, Ct \vdash Cs:\text{components}(Is_0) \quad \neg(I \in Is_0) \quad Is = Is_0 \cup \{I\} \\ \Sigma \vdash X:\text{parameters}(Vt, Ss) \quad Vt_0 = Vt[\text{this} \mapsto \text{nat}] \\ \Sigma, Vt_0, Ct \vdash C:\text{command} \quad \Sigma, Vt_0, Ct \vdash As:\text{actions}(Is_1)\end{array}}{\Sigma, Ct \vdash Cs \text{ component } I[N] \text{ \{ var } X;\text{ init } C;\, As \text{ \}}:\text{components}(Is)}$$

**Rules for** $\Sigma, Vt, Ct \vdash As$: **actions**$(Is)$**:**

$$\frac{Is = \varnothing}{\Sigma, Vt, Ct \vdash {}_{\sqcup}:\text{actions}(Is)}$$

$$\frac{\begin{array}{c}\Sigma, Vt, Ct \vdash As:\text{actions}(Is_0) \quad \neg(I \in Is_0) \quad Is = Is_0 \cup \{I\} \\ \Sigma \vdash X:\text{parameters}(Vt_1, Ss_1) \quad Vt_0 = Vt \leftarrow Vt_1 \\ \Sigma, Vt_0 \vdash F:\text{formula} \quad \Sigma, Vt_0, Ct \vdash C:\text{command}\end{array}}{\Sigma, Vt, Ct \vdash As \text{ action } I(X) \text{ with } F \text{ \{ } C \text{ \}}:\text{actions}(Is)}$$

**Rules for** $\Sigma, Vt, Ct \vdash C$: **command:**

$$\ldots \quad \frac{\Sigma, Vt \vdash T:\text{term(nat)} \quad \Sigma, Vt \vdash Ts:\text{terms}(Ss) \quad \langle I_1, At \rangle \in Ct \quad \langle I_2, Ss \rangle \in At}{\Sigma, Vt, Ct \vdash \text{send } I_1[T].I_2(Ts):\text{command}}$$

**Fig. 9.5** Type checking distributed systems

where first client 1 enters the critical region and, after it has left, client 0 does. However, the system does not allow a sequence

$$Client[1].ask() \rightarrow Client[0].ask() \rightarrow Server.request(1)$$
$$\rightarrow Client[1].enter() \rightarrow Server.request(0) \rightarrow Client[0].enter() \rightarrow \cdots$$

where the server allows both clients to enter the critical region at the same time.  □

Figure 9.5 gives the type system of the language; the judgements make use of an *action typing At* and *component typing Ct* in the following sense.

**Definition 9.7** (*Action and Component Typing*) An *action typing* is a partial mapping of (action) identifiers to sequences of sorts while a *component typing* is a partial mapping of (component) identifiers to action typings:

$$ActionTyping := Identifier \rightarrow_\perp \Sigma.s^*$$
$$ComponentTyping := Identifier \rightarrow_\perp ActionTyping$$

In a nutshell, the type system determines in a first step by the judgement $\Sigma \vdash Cs$: ctyping($Ct$) a component typing $Ct$ which in turn determines the names of all components and the names and signatures of all actions in these components. In a second step this information is used in the judgement $\Sigma, Ct \vdash Cs$: components($Is$) to determine the well-formedness of the system. Furthermore, the type systems for commands is extended by a rule for the judgement $\Sigma, Vt, Ct \vdash$ send $I_1[T].I_2(Ts)$: command which uses the information in $Ct$ to determine whether the send command indeed refers to an existing action with the appropriate number and types of arguments.

Within a component, the special variable this denotes the index of the component instance for which the initialization command respectively action is executed. This variable is given sort nat; in the following, we assume that a system declaration is evaluated in a $\Sigma$-algebra that maps nat to $\mathbb{N}$.

**Denotational Semantics**

Figure 9.6 defines the denotational semantics of a distributed system by a mapping to a labeled transition system; this definition is based on a couple of auxiliary definitions given in Fig. 9.7. The mapping $\llbracket$ distributed $I$ { $Cs$ } $\rrbracket$ is defined in two steps:

1. Every component in $Cs$ is translated to an LTS, which yields a mapping $lts$ of component identifiers to labeled transition systems.
2. The labeled transition systems of the components are composed to a LTS $\langle$space:$S_0$, init:$I$, next:$R\rangle$ of the whole system.

For the composition, we use the *interleaving model* of composition, i.e., the systems makes a step if exactly one component makes a step.

The translation of a component yields a value of type $LTS^{Message,StateC}$, i.e.:

- Every transition is labeled with the message $m \in Message$ that triggers the action performing the transition. This message is a quadruple of an identifier $m$.comp denoting the name of the component that has received the message, a natural number $m$.id denoting the specific instance of the component, an identifier $m$.act denoting the action in that instances that processes the message, and the argument values $m$.arg for the parameters of the action.

$\llbracket\,.\,\rrbracket$: *System* → *LTS*$^{Message,StateS}$

$\llbracket$ distributed $I$ { $Cs$ } $\rrbracket$ :=
  let $lts = \llbracket Cs \rrbracket$ in
  let $ids = domain(lts)$ in
  let $S_0 = \{s \in StateS \mid domain(s) = ids \land$
            $\forall I \in ids.\ domain(s(I)) = domain(lts(I)) \land$
            $\forall i \in domain(lts(I)).\ s(I)(i) \in lts(I)(i).\text{space}\}$ in
  let $I_0 = \pi s \in S_0.\ \exists s_0 \in S_0.$
    $(\forall I \in ids,\ i \in domain(lts(I)).\ lts(I)(i).\text{init}\langle s_0(I)(i)\rangle) \land send\langle s_0, s\rangle$ in
  let $R_0 = \pi m \in Message,\ s \in S_0,\ s' \in S_0.$
    $\exists I_r \in ids,\ i_r \in domain(lts(I_r)),\ s_0 \in lts(I_r)(i_r).\text{space},\ s_1 \in lts(I_r)(i_r).\text{space}.$
    $receive\langle m, I_r, i_r, s, s_0\rangle \land lts(I_r)(i_r).\text{next}\langle m, s_0, s_1\rangle \land$
    $send\langle s[I_r \mapsto s(I_r)[i_r \mapsto s_1]], s'\rangle$ in
  $\langle\text{space}: S_0, \text{init}: I_0, \text{next}: R_0\rangle$

$\llbracket\,.\,\rrbracket$: *Components* → $\big(Identifier \to_{\perp} \mathbb{N} \to_{\perp} LTS^{Message,StateC}\big)$

$\llbracket \sqcup \rrbracket := \varnothing$
$\llbracket Cs$ component $I[N]$ { var $X$; init $C$; $As$ } $\rrbracket$ :=
  let $S_0 = spaceC(X)$ in
  let $I_0 = \lambda i \in \mathbb{N}_{\llbracket N \rrbracket}.\ \pi s \in S_0.\ \exists s_0 \in S_0.$
    $s_0.\text{in} = \lambda I_a \in names(As).\ [\ ] \land s_0.\text{out} = [\ ] \land s_0.\text{state}(\text{this}) = i \land \llbracket C \rrbracket\langle s_0, s\rangle$ in
  let $R_0 = \lambda i \in \mathbb{N}_{\llbracket N \rrbracket}.\ \pi m \in Message,\ s \in S_0,\ s' \in S_0.$
    $\llbracket As \rrbracket\langle m, \langle s.\text{state}[\text{this} \mapsto i], s.\text{in}, s.\text{out}\rangle, s'\rangle$ in
  $\llbracket Cs \rrbracket[I \mapsto \lambda i \in \mathbb{N}_{\llbracket N \rrbracket}.\ \langle\text{space}: S_0, \text{init}: I_0(i), \text{next}: R_0(i)\rangle]$

$\llbracket\,.\,\rrbracket$: *Actions* → **Set**(*Message* × *StateC* × *StateC*)

$\llbracket \sqcup \rrbracket\langle m, s, s'\rangle :\Leftrightarrow$ false
$\llbracket As$ action $I(X)$ with $F$ { $C$ } $\rrbracket\langle m, s, s'\rangle :\Leftrightarrow$
  $\llbracket As \rrbracket\langle m, s, s'\rangle \lor$
  $(I = m.\text{act} \land \exists s_0 \in StateC.\ \llbracket X \rrbracket\langle m.\text{args}, s, s_0\rangle \land \llbracket F \rrbracket\langle s_0.\text{state}\rangle \land \llbracket C \rrbracket\langle s_0, s'\rangle)$

$\llbracket\,.\,\rrbracket$: *Parameters* → **Set**(*Value*$^{*}$ × *StateC* × *StateC*)

$\llbracket \sqcup \rrbracket\langle vs, s, s'\rangle :\Leftrightarrow vs = [\ ] \land s' = s$
$\llbracket X, V : S \rrbracket\langle vs, s, s'\rangle :\Leftrightarrow$
  $\exists vs_0 \in Value^{*},\ v \in Value,\ s_0 \in StateC.$
  $vs = vs_0 \circ [v] \land v \in A(S) \land \llbracket X \rrbracket\langle vs_0, s, s_0\rangle \land$
  $s'.\text{state} = s_0.\text{state}[V \mapsto v] \land s'.\text{in} = s.\text{in} \land s'.\text{out} = s.\text{out}$

**Fig. 9.6** Distributed systems as labeled transition systems

- Every state $s \in StateC$ extends the plain state $s.\text{state} \in State$ (which captures the values of the local variables of the component) by two additional entities $s.\text{in}$ and $s.\text{out}$. The entity $s.\text{in}$ maps the names of the actions of the component to sequences of messages; this represents the input queues of the component. The entity $s.\text{out}$ represents the component's single output queue.

In the following, we describe the translation of the components before we discuss the composition of the results in more detail.

The translation $\llbracket Cs \rrbracket$ of the components is a corresponding generalization of the translation of a shared system presented in Fig. 9.3 with the help of the func-

**Auxiliary Definitions**

$Message := $ comp:$Identifier \times$ id:$\mathbb{N} \times$ act:$Identifier \times$ args:$Value^*$
$StateC := $ state:$State \times$ in:$(Identifier \rightarrow_\perp Message^*) \times$ out:$Message^*$
$StateS := Identifier \rightarrow_\perp \mathbb{N} \rightarrow_\perp StateC$
$StateRelationC := Set(StateC \times StateC)$

$names: Actions \rightarrow Set(Identifier)$
$names(_\sqcup) := \varnothing$
$names(As \ \texttt{action} \ I(X) \ \texttt{with} \ F \ \{ \ C \ \}) := names(As) \cup \{I\}$

$send \subseteq StateS \times StateS$
$send\langle s, s'\rangle :\Leftrightarrow$
  $\exists t \in StateS^*, n \in \mathbb{N}.$
  $n = |t| \wedge t(0) = s \wedge t(n-1) = s' \wedge (\forall i \in \mathbb{N}_{n-1}. \ transfer\langle t(i), t(i+1)\rangle) \wedge$
  $(\forall I \in domain(s'), i \in domain(s'(I))). \ s'(I)(i).\text{out} = [\ ])$

$transfer \subseteq StateS \times StateS$
$transfer\langle s, s'\rangle :\Leftrightarrow$
  $domain(s') = domain(s) \wedge$
  $\exists m \in Message, I_s \in domain(s), i_s \in domain(s(I_s)).$
  $\exists I_r \in domain(s), i_r \in domain(s(I_r)), I_a \in Identifier.$
  $s(I_s)(i_s).\text{out} = [m] \circ s'(I_s)(i_s).\text{out} \wedge s'(I_s)(i_s).\text{in} = s(I_s)(i_s).\text{in} \wedge$
  $I_r = m.\text{comp} \wedge i_r = m.\text{id} \wedge I_a = m.\text{act} \wedge$
  $s'(I_r)(i_r).\text{in} = s(I_r)(i_r).\text{in}[I_a \mapsto s(I_r)(i_r).\text{in}(I_a) \circ [m]] \wedge$
  $s'(I_r)(i_r).\text{out} = s(I_r)(i_r).\text{out} \wedge$
  $(\forall I \in domain(s). \ domain(s'(I)) = domain(s(I)) \wedge$
    $\forall i \in domain(s(I)). \ s'(I)(i).\text{state} = s(I)(i).\text{state} \wedge$
    $(i \neq i_s \Rightarrow s'(I_s)(i).\text{in} = s(I_s)(i).\text{in} \wedge s'(I_s)(i).\text{out} = s(I_s)(i).\text{out}) \wedge$
    $(i \neq i_r \Rightarrow s'(I_r)(i).\text{in} = s(I_r)(i).\text{in} \wedge s'(I_r)(i).\text{out} = s(I_r)(i).\text{out}))$

$receive \subseteq Message \times Identifier \times \mathbb{N} \times StateS \times StateC$
$receive\langle m, I_r, i_r, s, s_0\rangle :\Leftrightarrow$
  $\exists I_a \in Identifier.$
  $I_r = m.\text{comp} \wedge i_r = m.\text{id} \wedge I_a = m.\text{act} \wedge$
  $s(I_r)(i_r).\text{in}(I_a) = [m] \circ s_0.\text{in}(I_a) \wedge s_0.\text{state} = s(I_r)(i_r).\text{state} \wedge s_0.\text{out} = s(I_r)(i_r).\text{out} \wedge$
  $(\forall I \in domain(s_0.\text{in}). \ I \neq I_a \Rightarrow s_0.\text{in}(I) = s(I_r)(i_r).\text{in}(I))$

$spaceC: Parameter \rightarrow Set(StateC)$
$spaceC(_\sqcup) := StateC$
$spaceC(X, V:S) := \{s \in spaceC(X) \mid s.\text{state}(V) \in A(S)\}$

**Fig. 9.7** Distributed systems as LTS (auxiliary definitions)

tion *spaceC*; this function generalizes the corresponding functions of the shared system translation to deal with states of type *StateC* rather than *State* (there is no need of a corresponding generalization of the function *chooseC* because the values of the parameters $X$ are determined by the message $m$ rather than by an internal choice). One difference is that in every instance of a component the execution of the initialization command starts in a state with empty input queues, empty output queue, and the variable this set to the number of the instance; likewise the processing of every action starts in a state with this correspondingly set (thus overriding any changes to the variable that a previous transition might have made). As another difference, in the translation of an action `action` $I(X)$, from the prestate $s$ of the

$[\![ . ]\!]$ : *Commands* → *StateRelationC*

$[\![ V := T ]\!] \langle s, s' \rangle :\Leftrightarrow$
  $[\![ T ]\!] \checkmark \langle s.\text{state} \rangle \wedge s'.\text{state} = s.\text{state}[V \mapsto [\![ T ]\!]_{s.\text{state}}] \wedge$
  $s'.\text{in} = s.\text{in} \wedge s'.\text{out} = s.\text{out}$

$\ldots$

$[\![ \text{send } I_1[T].I_2(Ts) ]\!] \langle s, s' \rangle :\Leftrightarrow$
  $[\![ T ]\!] \checkmark \langle s.\text{state} \rangle \wedge [\![ Ts ]\!] \checkmark \langle s.\text{state} \rangle \wedge$
  $s'.\text{out} = s.\text{out} \circ [\langle \text{comp}:I_1, \text{id}:[\![ T ]\!]_{s.\text{state}}, \text{act}:I_2, \text{args}:[\![ Ts ]\!]_{s.\text{state}} \rangle] \wedge$
  $s'.\text{state} = s.\text{state} \wedge s'.\text{in} = s.\text{in}$

**Fig. 9.8** Distributed systems as LTS (command semantics)

transition processing a message $m$ an additional intermediate state $s_0$ is generated that maps the parameters $X$ to the message values $m$.args; it is from this state $s_0$ that the execution of the action body generates the poststate $s'$.

Furthermore, the transition relation $[\![ C ]\!] \langle s, s' \rangle$ of every command $C$ has to be extended to deal with the additional state components $s$.in and $s$.out respectively $s'$.in and $s'$.in. Figure 9.8 sketches this adaptation for the assignment $V := T$ that leaves these components unchanged. The same is true for all other atomic commands of the core language while the compound commands "pass through" any updates of these components. Only the send command send $I_1[T].I_2(Ts)$ modifies any of these components: this command adds to the output queue $s$.out a new message with component name $I_1$, action name $I_2$, and the values of the terms $Ts$ as arguments.

As already stated, $[\![ \text{distributed } I \{ Cs \} ]\!]$ first derives the translation of the various components $Cs$ as a mapping $lts = [\![ Cs ]\!]$ of the component names *comp* to labeled transition systems and then combines these to a single LTS $\langle \text{space}:S_0, \text{init}:I, \text{next}:R \rangle$ for the whole system. This LTS is of type $LTS^{Message, StateS}$, i.e.:

- Every transition label is a message $m \in Message$ that indicates that the instance $m$.id of the component $m$.comp executes the action $m$.act with arguments $m$.arg.
- Every system state $s \in StateS$ is a mapping of component identifiers and instance numbers to component states; one may consider this as a product of the states of all component instances.

In the following, we describe the entities of this LTS.

- **State Space**: The state space $S_0$ consists of those system states that map every component identifier and legal instance number of that component to a state of the corresponding component instance. Thus we can consider the state space of the system essentially as the product of the state spaces of the component instances.
- **Initial State Condition**: The initial state condition $I$ holds on system state $s$ if for every instance $i$ of the component with name $I$ the initial state condition $lts(I)(i)$.init holds on some component state $s_0$ with empty input and output queues. From this intermediate state $s_0$, the actual initial state $s(I)(i)$ of the component

is derived by sending all the messages in $s(I)(i)$.out to their respective receivers as described by the auxiliary predicate *send* (therefore the initial state $s(I)(i)$.out of the output queue is always empty).

- **Next State Relation**: The next-state relation $R$ describes the transition with label $m$ from prestate $s$ to poststate $s'$ as a process that consists of three steps:

  1. The message $m$ is removed from the head of the queue of some instance $i_r$ of some component $I_r$; this yields from the prestate $s$ of the transition the intermediate state $s_0$ of the component instance as described by the auxiliary predicate *receive*.
  2. The next-state relation $lts(I_r)(i_r)$.next of component $I_r$ determines from $m$ and $s_0$ some poststate $s_1$ of the instance $i_r$ and thus also the updated poststate $s[I_r \mapsto s(I_r)[i_r \mapsto s_1]]$ of the whole system (the states of all other instances are not affected by this update).
  3. All messages in the output buffer of the component instance are transferred to their respective receivers as described by the auxiliary predicate *send* yielding the poststate $s'$ of the system (thus the poststate $s'(I)$.out of the component's output queue is empty).

The auxiliary predicates mentioned above abstract from the following details:

- Predicate $send\langle s, s' \rangle$ describes the transition from a system state $s$ with messages in the output buffers of some components to a state $s'$ where all messages have been delivered to their respective receivers. This is performed in a finite sequence $t$ of intermediate steps where in every step one message is transferred as described by the predicate application $transfer\langle t(i), t(i+1) \rangle$. The sequence terminates if the output queues of all components are empty.
- Predicate $transfer\langle s, s' \rangle$ describes the transition from a system state $s$ to a system state $s'$ resulting from the transfer of some message $m$ from some sender component $I_s$ to the input queue of some action $I_a$ in some receiver component $I_r$; this transfer leaves the input and output queues of all other components and the variable states of all components unchanged.
- Predicate $receive\langle m, I_r, i_r, s, s_0 \rangle$ describes how, by removing message $m$ from the input queue associated to some action of instance $i_r$ of component $I_r$ in a system with state $s$, the state $s_0$ of that instance is derived.

Since in the initial state of the system the output queues of all components are empty and this property is preserved by every transition of the system, actually in *every* state of the system the output queues of all components are empty and therefore could be omitted. However, these queues are used in the invisible intermediate states used in predicates *send* and *transfer*; we have retained the queues also in the visible states to simplify the definitions.

**Example 9.11**  Consider the system introduced at the beginning of this section:

```
distributed PingPong {
  component Ping {
    var x:nat;
    init { send Ping.r(0) }
    action r(i:nat) { x := i+1; send Pong.s(x) }
  }
  component Pong {
    var y:nat;
    init { }
    action s(i:nat) { send Ping.r(i); y := i }
  }
}
```

In this system there exists exactly one instance of every component, thus we drop the qualification with instance numbers as well in the system description as in the subsequently derived labeled transition system.

According to the denotational semantics, we can derive the labeled transition systems of the individual components as follows (for simplicity, we identify a component state with the value of its single variable and use a single input queue rather than a mapping of the single action to the queue; we also perform some simplifications on the formulas):

$Ping := \langle \text{space:} S_1, \text{init:} I_1, \text{next:} R_1 \rangle$

$S_1 := \text{state:} \mathbb{N} \times \text{in:} Message^* \times \text{out:} Message^*$

$I_1 \langle \text{state:} x, \text{in:} in, \text{out:} out \rangle :\Leftrightarrow in = [\,] \wedge out = [\langle \text{comp:} Ping, \text{id:} 0, \text{act:} r, \text{arg:} [0] \rangle]$

$R_1 \langle m, \langle \text{state:} x, \text{in:} in, \text{out:} out \rangle, \langle \text{state:} x', \text{in:} in', \text{out:} out' \rangle \rangle :\Leftrightarrow$

   $\exists i \in \mathbb{N}. \ m = \langle \text{comp:} Ping, \text{id:} 0, \text{act:} r, \text{arg:} [i] \rangle \wedge$

      $x' = x + i \wedge in' = in \wedge out' = out \circ [\langle \text{comp:} Pong, \text{id:} 0, \text{act:} s, \text{arg:} [x'] \rangle]$

$Pong := \langle \text{space:} S_2, \text{init:} I_2, \text{next:} R_2 \rangle$

$S_2 := \text{state:} \mathbb{N} \times \text{in:} Message^* \times \text{out:} Message^*$

$I_2 \langle \text{state:} y, \text{in:} in, \text{out:} out \rangle :\Leftrightarrow in = [\,] \wedge out = [\,]$

$R_2 \langle m, \langle \text{state:} y, \text{in:} in, \text{out:} out \rangle, \langle \text{state:} y', \text{in:} in', \text{out:} out' \rangle \rangle :\Leftrightarrow$

   $\exists i \in Message. \ m = \langle \text{comp:} Pong, \text{id:} 0, \text{act:} s, \text{arg:} [i] \rangle \wedge$

      $y' = i \wedge in' = in \wedge out' = out \circ [\langle \text{comp:} Ping, \text{id:} 0, \text{act:} r, \text{arg:} [i] \rangle]$

The composition of the labeled transition systems yields the LTS given below. For simplicity we model the state of the system as a product of the component state spaces (rather than a mapping of component identifiers and instance numbers to spaces):

$PingPong := \langle \text{space:} S, \text{init:} I, \text{next:} R \rangle$

$S := S_1 \times S_2$

$I \langle s_1, s_2 \rangle :\Leftrightarrow$

$$s_1.\text{in} = [\langle \text{comp:}Ping, \text{id:0}, \text{act:}r, \text{arg:}[0]\rangle] \wedge s_2.\text{in} = [\ ] \wedge s_1.\text{out} = [\ ] \wedge s_2.\text{out} = [\ ]$$

$$R\langle m, \langle s_1, s_2 \rangle, \langle s_1', s_2' \rangle \rangle :\Leftrightarrow$$

$$(\exists i \in \mathbb{N}.\ m = \langle \text{comp:}Ping, \text{id:0}, \text{act:}r, \text{arg:}[i]\rangle \wedge s_1.\text{in} = [m] \circ s_1'.\text{in} \wedge$$

$$x' = x + i \wedge s_2'.\text{in} = s_2.\text{in} \circ [\langle \text{comp:}Pong, \text{id:0}, \text{act:}s, \text{arg:}[x']\rangle] \wedge$$

$$s_1.\text{out} = [\ ] \wedge s_2.\text{out} = [\ ]) \vee$$

$$(\exists i \in \mathbb{N}.\ m = \langle \text{comp:}Pong, \text{id:0}, \text{act:}s, \text{arg:}[i]\rangle \wedge s_2.\text{in} = [m] \circ s_2'.\text{in} \wedge$$

$$y' = i \wedge s_1'.\text{in} = s_1.\text{in} \circ [\langle \text{comp:}Ping, \text{id:0}, \text{act:}r, \text{arg:}[i]\rangle] \wedge$$

$$s_1.\text{out} = [\ ] \wedge s_2.\text{out} = [\ ])$$

The system can then perform the following transitions (we ignore the empty output buffers and use the labels in1 and in2 to indicate the values of the input buffers of both components):

$$\langle \text{x:?}, \text{y:?}, \text{in1:}[\langle \text{comp:}Ping, \text{id:0}, \text{act:}r, \text{arg:}[0]\rangle], \text{in2:}[\ ]\rangle \xrightarrow{Ping[0].r(0)}$$

$$\langle \text{x:1}, \text{y:?}, \text{in1:}[], \text{in2:}[\langle \text{comp:}Pong, \text{id:0}, \text{act:}s, \text{arg:}[1]\rangle]\rangle \xrightarrow{Pong[0].s(1)}$$

$$\langle \text{x:1}, \text{y:1}, \text{in1:}[\langle \text{comp:}Ping, \text{id:0}, \text{act:}r, \text{arg:}[1]\rangle], \text{in2:}[\ ]\rangle \xrightarrow{Ping[0].r(1)}$$

$$\langle \text{x:2}, \text{y:1}, \text{in1:}[], \text{in2:}[\langle \text{comp:}Pong, \text{id:0}, \text{act:}s, \text{arg:}[2]\rangle]\rangle \xrightarrow{Pong[0].s(2)}$$

$$\langle \text{x:2}, \text{y:2}, \text{in1:}[\langle \text{comp:}Ping, \text{id:0}, \text{act:}r, \text{arg:}[2]\rangle], \text{in2:}[\ ]\rangle \xrightarrow{Ping[0].r(2)} \dots$$

The system thus starts in a state with a message in the input buffer of the first component; the execution of the system proceeds as indicated in the example at the beginning of this section. □

As demonstrated in above example, for the investigation of an distributed system, it is typically more convenient to use (rather than the LTS that mechanically results from the formal semantics) a manually crafted LTS that retains the essential characteristics of the original but simplifies some technical details. In the following, we will freely make use of such simplifications to make the presentation easier to understand.

## 9.4    Specifying System Properties

In the previous sections we have discussed the formal *modeling* of concurrent systems as labeled transition systems, i.e., the precise description of their actual behaviors. Now we are turning our attention to the formal *specification* of such systems, i.e., to the precise description of properties that these behaviors shall satisfy. For this purpose we need to talk and reason not only about a fixed number of states (such as the pre- and post-state of a program) but about arbitrary sequences of such states (such as the complete run of a system). However, while this is perfectly possible within the framework of classical first-order logic (which actually leads to the most expressive specification framework), the resulting formulas become cumbersome: to denote a particular value of a system variable $x$ in some state of a run, the corresponding variable reference has to be indexed as $x(i)$ by the index $i$ of that state; consequently formulas have to be quantified over state indices to describe the evolution of the variable values in a run.

For these reasons, for the specification of concurrent systems typically some variant of the more appropriate *temporal logic* is preferred. This logical framework copes without state indices by the use of *temporal operators* that implicitly determine the state in which a reference $x$ is to be evaluated. In the following we will discuss the particular variant of *linear temporal logic* where formulas are interpreted over the linear sequences of states that arise from system runs.

### Linear Temporal Logic

We begin with the syntax and type system of linear temporal logic (LTL).

---

**Definition 9.8** (*Linear Temporal Logic*)   A formula in *linear temporal logic* (an *LTL-formula*) is a phrase $F \in LTL$ formed according to the following grammar:

$F \in LTL$
$F ::= \ldots \mid I(Ts) \mid \ldots$ (atomic formulas)
$\quad \mid \ldots \mid F_1 \wedge F_2 \mid \ldots$ (propositional combinations)
$\quad \mid \ldots \mid \forall X\!:\! S.\ F \mid \ldots$ (quantified formulas)
$\quad \mid \bigcirc F \mid \Box F \mid \Diamond F \mid F_1 \cup F_2 \mid F_1 \mathsf{W} F_2 \mid F_1 \mathsf{R} F_2 \mid F_1 \mathsf{M} F_2$

This domain includes all the atomic formulas, propositional combinations, and quantified formulas introduced in Definition 6.1 respectively Definition 2.1 and extends them by the use of the *temporal operators* $\bigcirc, \Box, \Diamond, \mathsf{U}, \mathsf{W}, \mathsf{R}, \mathsf{M}$.

---

We read $\bigcirc F$ as "next time $F$", $\Box F$ as "always $F$", $\Diamond F$ as "eventually $F$", $F_1 \cup F_2$ as "$F_1$ (strong) until $F_2$", $F_1 \mathsf{W} F_2$ as "$F_1$ weak until $F_2$", $F_1 \mathsf{R} F_2$ as "$F_1$ (weakly) releases $F_2$", $F_1 \mathsf{M} F_2$ as "$F_1$ strongly releases $F_2$".

Given a signature $\Sigma$, an LTL-formula $F$ is *well-typed* with respect to variable typing $Vs$ if we can derive a judgement $\Sigma, Vs \vdash F : \mathsf{ltl}$ according to the rules of Fig. 9.9 (which generalize the rules for the judgement $\Sigma, Vs \vdash F : \mathsf{formula}$ of Fig. 6.1). In the remainder of this section, we assume a fixed signature $\Sigma$.

**Example 9.12**   We give some examples of LTL-formulas and their readings respectively informal interpretations:

- $\Box(x = 0 \Rightarrow y = 0)$: "it is always the case that, if $x$ is zero, then also $y$ is zero" (thus it can never be the case that simultaneously $x$ is zero and $y$ is one).

**Rules for** $\Sigma, Vs \vdash L$: ltl:

$$\ldots \text{(rules corresponding to those of Figure 6.1)}$$

$$\frac{\Sigma, Vs \vdash F : \text{ltl}}{\Sigma, Vs \vdash \bigcirc F : \text{ltl}} \qquad \frac{\Sigma, Vs \vdash F : \text{ltl}}{\Sigma, Vs \vdash \Box F : \text{ltl}} \qquad \frac{\Sigma, Vs \vdash F : \text{ltl}}{\Sigma, Vs \vdash \Diamond F : \text{ltl}}$$

$$\frac{\Sigma, Vs \vdash F_1 : \text{ltl} \quad \Sigma, Vs \vdash F_2 : \text{ltl}}{\Sigma, Vs \vdash F_1 \ \mathsf{U} \ F_2 : \text{ltl}} \qquad \frac{\Sigma, Vs \vdash F_1 : \text{ltl} \quad \Sigma, Vs \vdash F_2 : \text{ltl}}{\Sigma, Vs \vdash F_1 \ \mathsf{W} \ F_2 : \text{ltl}}$$

$$\frac{\Sigma, Vs \vdash F_1 : \text{ltl} \quad \Sigma, Vs \vdash F_2 : \text{ltl}}{\Sigma, Vs \vdash F_1 \ \mathsf{R} \ F_2 : \text{ltl}} \qquad \frac{\Sigma, Vs \vdash F_1 : \text{ltl} \quad \Sigma, Vs \vdash F_2 : \text{ltl}}{\Sigma, Vs \vdash F_1 \ \mathsf{M} \ F_2 : \text{ltl}}$$

**Fig. 9.9** Type-checking of LTL-formulas

- $(x = 0 \ \mathsf{U} \ y = 0)$: "$x$ remains zero, until $y$ becomes zero" (thus $y$ must become zero, before $x$ can become non-zero).
- $\forall i \in \mathbb{N}. \ \Diamond(x = i)$: " for every natural number $i$, $x$ will eventually become $i$" (thus for every natural number the value of $x$ will at some time be that number; possibly $x$ will be the same number multiple times).
- $\Box \Diamond(x = 0)$: "it is always the case that eventually $x$ will become zero" (thus $x$ will become infinitely often zero).
- $\forall i \in \mathbb{N}. \ \Box(x = i \Rightarrow \Diamond(x > i))$: "for every natural number $i$ it is always the case that, if $x$ is $i$, then $x$ will eventually become greater than $i$" (thus $x$ will be incremented infinitely often). $\qquad \Box$

Above examples just give some preliminary intuition about the meanings of the formulas; their formal semantics will be introduced below.

**The Semantics of LTL**

In linear temporal logic, formulas are evaluated over sequences of states such as the runs of concurrent systems; thus we need some auxiliary notions about system runs.

**Definition 9.9** (*Adequacy of a Run*) Let $r \in Run^{State}$ be a run and let $Vt$ be a variable typing. Then $r$ is *adequate* with respect to $Vt$, if for every position $i \in domain(r)$, every identifier $I \in Identifier$, and every sort $S \in Sort$ with $\langle I, S \rangle \in Vt$, we have $r(i)(I) \in A(S)$; thus in every state of the run the variable values are adequate with respect to $Vt$.

**Definition 9.10** (*Update of a Run*) Let $r \in Run^{State}$ be a run, $I \in Identifier$ an identifier, $v \in Value$ a value. We define the *updated run* $r[I \mapsto v] \in Run^{State}$ as follows:

$$r[I \mapsto v](i) := r(i)[I \mapsto v]$$

Thus at every position $i \in domain(r)$ the updated run has for variable $I$ value $v$ and for every other variable the value that $r$ holds at $i$ for that variable.

**Definition 9.11** (*Rest of a Run*) Let $r \in Run^{State}$ be a run and $n \in \mathbb{N}$. We define the *rest* $r_{\geq n}$ of $r$ at $n$ as follows:

$$r_{\geq n} := \{ \langle i - n, e \rangle \mid \langle i, e \rangle \in r \wedge i \geq n \}$$

Consequently $r_{\geq n} \in Run^{State}$ is that run that holds the same states as $r$ except that the first $n$ states have been removed.

We are now going to define the semantics of an LTL-formula on the basis of Definition 6.12 which has introduced the truth value $[\![\, F \,]\!]_s^A$ of a first-order formula $F$ in a state $s$ with respect to a $\Sigma$-algebra $A$; however, we will from now on assume a fixed $A$ and drop the corresponding superscript.

**Definition 9.12** (*Semantics of LTL*) Let $F$ be an LTL-formula that is well-typed with respect to some variable typing $Vt$. Let $r \in State^\omega$ be an infinite run that is adequate with respect to $Vt$. We define the semantics $[\![\, F \,]\!]^r$ of $F$ with respect to $r$ as the following relation:

$$[\![\, I(Ts) \,]\!]^r :\Leftrightarrow [\![\, I(Ts) \,]\!]_{r(0)} = \text{true}$$
$$[\![\, F_1 \wedge F_2 \,]\!]^r :\Leftrightarrow [\![\, F_1 \,]\!]^r \wedge [\![\, F_2 \,]\!]^r$$
$$[\![\, \forall I : S.\ F \,]\!]^r :\Leftrightarrow \forall v \in A(S).\ [\![\, F \,]\!]^{r[I \mapsto v]}$$

$$\cdots$$

$$[\![\, \bigcirc F \,]\!]^r :\Leftrightarrow [\![\, F \,]\!]^{r_{\geq 1}}$$
$$[\![\, \square F \,]\!]^r :\Leftrightarrow \forall i \in \mathbb{N}.\ [\![\, F \,]\!]^{r_{\geq i}}$$
$$[\![\, \lozenge F \,]\!]^r :\Leftrightarrow \exists i \in \mathbb{N}.\ [\![\, F \,]\!]^{r_{\geq i}}$$
$$[\![\, F_1\ \mathsf{U}\ F_2 \,]\!]^r :\Leftrightarrow \exists i \in \mathbb{N}.\ [\![\, F_2 \,]\!]^{r_{\geq i}} \wedge \forall j \in \mathbb{N}_i.\ [\![\, F_1 \,]\!]^{r_{\geq j}}$$
$$[\![\, F_1\ \mathsf{W}\ F_2 \,]\!]^r :\Leftrightarrow (\forall i \in \mathbb{N}.\ [\![\, F_1 \,]\!]^{r_{\geq i}}) \vee (\exists i \in \mathbb{N}.\ [\![\, F_2 \,]\!]^{r_{\geq i}} \wedge \forall j \in \mathbb{N}_i.\ [\![\, F_1 \,]\!]^{r_{\geq j}})$$
$$[\![\, F_1\ \mathsf{R}\ F_2 \,]\!]^r :\Leftrightarrow (\forall i \in \mathbb{N}.\ [\![\, F_2 \,]\!]^{r_{\geq i}}) \vee (\exists i \in \mathbb{N}.\ [\![\, F_1 \,]\!]^{r_{\geq i}} \wedge \forall j \in \mathbb{N}_{i+1}.\ [\![\, F_2 \,]\!]^{r_{\geq j}})$$
$$[\![\, F_1\ \mathsf{M}\ F_2 \,]\!]^r :\Leftrightarrow \exists i \in \mathbb{N}.\ [\![\, F_1 \,]\!]^{r_{\geq i}} \wedge \forall j \in \mathbb{N}_{i+1}.\ [\![\, F_2 \,]\!]^{r_{\geq j}}$$

Please note that the semantics of an LTL-formula is here defined over *infinite* runs only (we will discuss later how to deal also with finite system runs).

**Fig. 9.10** Semantics of
LTL-formulas

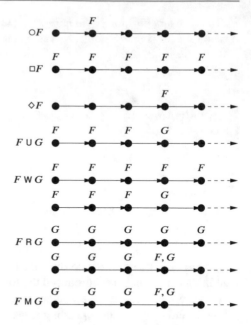

This definition generalizes the semantics of formulas given in Definition 2.16 respectively Definition 6.12. The meaning of the various constructions inherited from first-order logic is a natural generalization of the original semantics:

- Atomic formula $I(Ts)$ holds on run $r$ if it holds on the first state $r(0)$ of the run.
- Propositional combination $F_1 \wedge F_2$ holds on $r$ if both $F_1$ and $F_2$ hold on $r$; the meanings of the other propositional combinations are analogously defined.
- Quantified formula $\forall I: S.\ F$ holds on $r$ if $F$ holds, for every value $v$ in the domain of sort $S$, on the run that is identical to $r$ except that it holds in every state value $v$ for quantified variable $I$. The meanings of the other quantified formulas are analogously defined.

The meaning of the temporal formulas can be understood from the diagrams given in Fig. 9.10 (in the following explanations we say "$F$ holds at a position" to indicate "$F$ holds on the rest of the run starting at that position"):

- **Next Time**: $\bigcirc F$ ("next time $F$") holds if $F$ holds at the *next* position of the run.
- **Always**: $\Box F$ ("always $F$") holds if $F$ holds at *every* position of the run.
- **Eventually**: $\Diamond F$ ("eventually $F$") holds if $F$ holds at *some* position of the run.
- **(Strong) Until**: $F \cup G$ ("$F$ until $G$") holds if $G$ holds at *some* position $i$ of the run and $F$ holds at *every* position $j < i$ before.
- **Weak Until**: $F$ W $G$ holds if $F$ holds always (then $G$ may never hold) or if $G$ holds at *some* position $i$ of the run and $F$ holds at *every* position $j < i$ before.
- **(Weak) Release**: $F$ R $G$ holds if $G$ holds always (then $F$ may never hold) or if $F$ holds at *some* position $i$ and $G$ holds at *every* position $j \le i$ up to $i$ (*including $i$*).

- **Strong Release**: $F$ M $G$ holds if $F$ holds at *some* position $i$ and $G$ holds at *every* position $j \leq i$ up to $i$ (*including* $i$).

The temporal operators of LTL-formulas can be arbitrarily nested, which allows the description of complex system properties.

**Example 9.13** Consider the LTL-formula

$$\Box(F \Rightarrow \Diamond G)$$

to be read as "it is always the case that, if $F$ holds, then eventually also $G$ holds". This formula is *satisfied* by the run

$$\xrightarrow{} \begin{smallmatrix} 1 \\ \end{smallmatrix} \begin{smallmatrix} F \\ \neg G \end{smallmatrix} \xrightarrow{} \begin{smallmatrix} 2 \\ \end{smallmatrix} \begin{smallmatrix} \neg F \\ \neg G \end{smallmatrix} \xrightarrow{} \begin{smallmatrix} 3 \\ \end{smallmatrix} \begin{smallmatrix} \neg F \\ G \end{smallmatrix} \xrightarrow{} \begin{smallmatrix} 4 \\ \end{smallmatrix} \begin{smallmatrix} F \\ \neg G \end{smallmatrix} \xrightarrow{} \begin{smallmatrix} 5 \\ \end{smallmatrix} \begin{smallmatrix} \neg F \\ G \end{smallmatrix} \xrightarrow{} \begin{smallmatrix} 6 \\ \end{smallmatrix} \begin{smallmatrix} F \\ G \end{smallmatrix} \xrightarrow{} \begin{smallmatrix} 7 \\ \end{smallmatrix} \cdots$$

where only states $1, 4, 6$ satisfy $F$ and only states $3, 5, 6$ satisfy $G$: for every occurrence of a state satisfying $F$ there is eventually (*at this time or later*) a state that satisfies $G$. Actually, for the same reason this formula is also satisfied by the run

$$\xrightarrow{} \begin{smallmatrix} 1 \\ \end{smallmatrix} \begin{smallmatrix} F \\ \neg G \end{smallmatrix} \xrightarrow{} \begin{smallmatrix} 2 \\ \end{smallmatrix} \begin{smallmatrix} \neg F \\ \neg G \end{smallmatrix} \xrightarrow{} \begin{smallmatrix} 3 \\ \end{smallmatrix} \begin{smallmatrix} \neg F \\ \neg G \end{smallmatrix} \xrightarrow{} \begin{smallmatrix} 4 \\ \end{smallmatrix} \begin{smallmatrix} F \\ \neg G \end{smallmatrix} \xrightarrow{} \begin{smallmatrix} 5 \\ \end{smallmatrix} \begin{smallmatrix} \neg F \\ \neg G \end{smallmatrix} \xrightarrow{} \begin{smallmatrix} 6 \\ \end{smallmatrix} \begin{smallmatrix} F \\ G \end{smallmatrix} \xrightarrow{} \begin{smallmatrix} 7 \\ \end{smallmatrix} \cdots$$

where again states $1, 4, 6$ satisfy $F$ but only state $6$ satisfies $G$. However, the formula is *violated* by the run

$$\xrightarrow{} \begin{smallmatrix} 1 \\ \end{smallmatrix} \begin{smallmatrix} \neg F \\ G \end{smallmatrix} \xrightarrow{} \begin{smallmatrix} 2 \\ \end{smallmatrix} \begin{smallmatrix} \neg F \\ G \end{smallmatrix} \xrightarrow{} \begin{smallmatrix} 3 \\ \end{smallmatrix} \begin{smallmatrix} \neg F \\ G \end{smallmatrix} \xrightarrow{} \begin{smallmatrix} 4 \\ \end{smallmatrix} \begin{smallmatrix} F \\ \neg G \end{smallmatrix} \xrightarrow{} \begin{smallmatrix} 5 \\ \end{smallmatrix} \begin{smallmatrix} \neg F \\ \neg G \end{smallmatrix} \xrightarrow{} \begin{smallmatrix} 6 \\ \end{smallmatrix} \begin{smallmatrix} \neg F \\ \neg G \end{smallmatrix} \xrightarrow{} \begin{smallmatrix} 7 \\ \end{smallmatrix} \cdots$$

where all states that satisfy $G$ (states $1, 2, 3$) *precede* the only state $4$ that satisfies $F$. □

While LTL-formulas can be read as natural language statements, the naive translation of natural language statements into LTL formulas can be treachery, as the following example demonstrates.

**Example 9.14** Consider the following informal statement:

If $F$ holds, then $G$ must hold before $H$ can hold.

A first attempt to a formalization of this statement might lead to the pattern:

$$F \Rightarrow \cdots$$

However this formula is satisfied by every sequence of shape

$$\neg F \rightarrow F \rightarrow F \rightarrow F \rightarrow \cdots$$

irrespective of the truth values of $G$ and $H$. This is because in above formula $F$ occurs out of the scope of any temporal operator and thus only applies to the first state of the sequence. Actually, the real intention of above sentence can be formulated more precisely as follows:

*Always*, if $F$ holds, then $G$ must hold before $H$ can hold.

This leads to the following correct pattern:

$$\Box(F \Rightarrow \ldots)$$

As for the rest of the formula, the phrase "$G$ ...before $H$" might seem to suggest something like the following formula:

$$\Box(F \Rightarrow (G \cup H))$$

However, this formula is not correct: it prohibits the sequence

$$
\begin{array}{cccc}
F & \neg F & \neg F & \neg F \\
\neg G \rightarrow & G \rightarrow & \neg G \rightarrow & \neg G \rightarrow \ldots \\
\neg H & \neg H & \neg H & H
\end{array}
$$

where $G$ indeed holds before $H$ holds. This is because $G \cup H$ demands that $G$ holds *permanently* until $H$ holds, which is apparently stronger than the informal statement requires. This sequence is actually characterized by the fact that, starting with the state in which $F$ holds, $H$ does *not* hold up to the first time $G$ holds; after that $G$ and $H$ may or may not hold at any time. So we write down

$$\Box(F \Rightarrow (\neg H \cup G))$$

However, this formula makes use of the *strong until* operator which requires that, if $F$ is satisfied, eventually also $G$ is satisfied, which is not required by the informal statement: we also want to allow the sequence

$$
\begin{array}{cccc}
F & \neg F & \neg F & \neg F \\
\neg G \rightarrow & \neg G \rightarrow & \neg G \rightarrow & \neg G \rightarrow \ldots \\
\neg H & \neg H & \neg H & \neg H
\end{array}
$$

where $G$ and $H$ are never satisfied. Thus we weaken our formulation to

$$\Box(F \Rightarrow (\neg H \text{ W } G))$$

which makes use of the "weak until" operator. However, this formulation also allows $H$ to become true in the very *same* state in which $G$ becomes true. If we indeed insist on $H$ to become only true *after* $G$ has become true, we may use the "weak release" operator to write

$$\Box(F \Rightarrow (G \text{ R} \neg H))$$

which says that $G$ "releases" the "obligation" $\neg H$, i.e., that only a state that satisfies $G$ allows a *later* state to satisfy $H$.                                            □

## Properties of LTL

The various notions related to the semantics of first-order formulas can be generalized to LTL, in particular the notions *validity*, *logical consequence*, and *logical equivalence*.

**Definition 9.13** (*Validitiy*) Let $F$ be an LTL-formulas that is well-typed with respect to variable typing $Vt$. Then $F$ is *valid*, written as $\models F$, if for every infinite run $r \in State^{\omega}$ that is adequate with respect to $Vt$ we have $[\![\, F \,]\!]^r$.

**Definition 9.14** (*Logical Consequence and Equivalence*) Let $F_1$ and $F_2$ be two LTL-formulas that are well-typed with respect to variable typing $Vt$. Then $F_2$ is a *logical consequence* of $F_1$, written as $F_1 \models F_2$, if for every infinite run $r \in State^{\omega}$ that is adequate with respect to $Vt$ we have the following: if $[\![\, F_1 \,]\!]^r$ holds then $[\![\, F_2 \,]\!]^r$ holds. Furthermore $F_1$ and $F_2$ are *logically equivalent*, written as $F_1 \equiv F_2$, if we have $F_1 \models F_2$ and $F_2 \models F_1$, i.e. $[\![\, F_1 \,]\!] = [\![\, F_2 \,]\!]$.

Indeed the temporal LTL-formulas satisfy a large variety of logical equivalences, some of which are stated below.

**Definition 9.15** (*LTL Equivalence*) For all LTL-formulas $F$, $G$, $H$ the following equivalences hold.

- **Extraction**: These laws demonstrate how the truth value of a temporal formula can be derived from the value of a subformula in the first state of a run and from the value of the whole formula in the rest of the run:

$$\Box F \equiv F \wedge \bigcirc(\Box F)$$
$$\Diamond F \equiv F \vee \bigcirc(\Diamond F)$$
$$F \ U \ G \equiv G \vee (F \wedge \bigcirc(F \ U \ G))$$

- **Reduction**: These equivalences show how all temporal operators (except for the "next time" operator $\bigcirc$) can be expressed by the "until" operator U:

$$\Box F \equiv \neg(\text{true U} \ \neg F)$$
$$\Diamond F \equiv \text{true U} \ F$$
$$F \ W \ G \equiv (\Box F) \vee (F \ U \ G)$$
$$F \ R \ G \equiv (\Box G) \vee (G \ U \ (F \wedge G))$$
$$F \ M \ G \equiv G \ U \ (F \wedge G)$$

- **Negation**: These laws demonstrate how to "push" negation into the context of a temporal operator, demonstrating a "De Morgan"-like law for "always" and "eventually":

$$\neg(\bigcirc F) \equiv \bigcirc(\neg F)$$
$$\neg(\Box F) \equiv \Diamond(\neg F)$$
$$\neg(\Diamond F) \equiv \Box(\neg F)$$
$$\neg(F \ U \ G) \equiv (\Box\neg G) \vee ((\neg G) \ U \ (\neg F \wedge \neg G))$$

- **Distributivity**: These laws demonstrate how to "distribute" conjunction respectively disjunction over the temporal operators:

$$(\Box F) \wedge (\Box G) \equiv \Box(F \wedge G)$$
$$(\Diamond F) \vee (\Diamond G) \equiv \Diamond(F \vee G)$$
$$(F \wedge G) \ U \ H \equiv (F \ U \ H) \wedge (G \ U \ H)$$
$$F \ U \ (G \vee H) \equiv (F \ U \ G) \vee (F \ U \ H)$$
$$\Box\Diamond(F \vee G) \equiv (\Box\Diamond F) \vee (\Box\Diamond G)$$
$$\Diamond\Box(F \wedge G) \equiv (\Diamond\Box F) \wedge (\Diamond\Box G)$$

The "reduction" equivalences show the reduction of most temporal operators by the "until" operator U; in fact, there are many more alternative reductions of one temporal operator to another one.

Among the listed equivalences, the one for the negation of the "until" operator may be the most unintuitive one; in the following we prove its validity in detail.

**Proof** To prove

$$\neg(F \cup G) \equiv (\Box\neg G) \wedge ((\neg G) \cup (\neg F \wedge \neg G))$$

we take arbitrary run $r \in State^\omega$ and show

$$[\![ \neg(F \cup G) ]\!]^r \Leftrightarrow [\![ (\Box\neg G) \wedge ((\neg G) \cup (\neg F \wedge \neg G)) ]\!]^r.$$

We prove the two directions of this equivalence.

- First we assume $[\![ \neg(F \cup G) ]\!]^r$, i.e.:

$$\forall i \in \mathbb{N}. \; [\![ G ]\!]^{r \geq i} \Rightarrow \exists j \in \mathbb{N}_i. \; \neg [\![ F ]\!]^{r \geq j} \tag{1}$$

and show $[\![ (\Box\neg G) \vee ((\neg G) \cup (\neg F \wedge \neg G)) ]\!]^r$, i.e.:

$$(\forall i \in \mathbb{N}. \; \neg [\![ G ]\!]^{r \geq i}) \vee$$
$$\left(\exists i \in \mathbb{N}. \; (\neg [\![ F ]\!]^{r \geq i}) \wedge (\neg [\![ G ]\!]^{r \geq i}) \wedge \forall j \in \mathbb{N}_i. \; (\neg [\![ G ]\!]^{r \geq j})\right) \tag{F}$$

To show (a), we assume (2) $\exists i \in \mathbb{N}. \; [\![ G ]\!]^{r \geq i}$ and show

$$\exists i \in \mathbb{N}. \; (\neg [\![ F ]\!]^{r \geq i}) \wedge (\neg [\![ G ]\!]^{r \geq i}) \wedge \forall j \in \mathbb{N}_i. \; (\neg [\![ G ]\!]^{r \geq j}) \tag{b}$$

From (2), we have some minimal $i \in \mathbb{N}$ with (3) $[\![ G ]\!]^{r \geq i}$, i.e., (4) $\forall j \in \mathbb{N}_i. \neg [\![ G ]\!]^{r \geq j}$. From (1) and (3) we have some (5) $j \in \mathbb{N}_i$ with (6) $\neg [\![ F ]\!]^{r \geq j}$ from (4) and (5) we also have (7) $\neg [\![ G ]\!]^{r \geq j}$. We now show (b) for $i := j$: from (6) and (7), it suffices to show $\forall k \in \mathbb{N}_j. \; (\neg [\![ G ]\!]^{r \geq k})$ this follows from (4) and (5).

- Second we assume $[\![ (\Box\neg G) \vee ((\neg G) \cup (\neg F \wedge \neg G)) ]\!]^r$, i.e.:

$$(\forall i \in \mathbb{N}. \; \neg [\![ G ]\!]^{r \geq i}) \vee$$
$$\left(\exists i \in \mathbb{N}. \; (\neg [\![ F ]\!]^{r \geq i}) \wedge (\neg [\![ G ]\!]^{r \geq i}) \wedge \forall j \in \mathbb{N}_i. \; (\neg [\![ G ]\!]^{r \geq j})\right) \tag{1}$$

and show $[\![ \neg(F \cup G) ]\!]^r$, i.e.:

$$\forall i \in \mathbb{N}. \; [\![ G ]\!]^{r \geq i} \Rightarrow \exists j \in \mathbb{N}_i. \; \neg [\![ F ]\!]^{r \geq j} \tag{a}$$

From (1), we have two cases. Case $(\forall i \in \mathbb{N}. \; \neg [\![ G ]\!]^{r \geq i})$ immediately implies (a). The other case gives us some $i_0 \in \mathbb{N}$ with (2) $\neg [\![ F ]\!]^{r \geq i_0}$, (3) $\neg [\![ G ]\!]^{r \geq i_0}$ (3), and (4) $\forall j \in \mathbb{N}_{i_0}. \; (\neg [\![ G ]\!]^{r \geq j})$. To show (a), we take $i_1 \in \mathbb{N}$ with (5) $[\![ G ]\!]^{r \geq i_1}$ and show

$$\exists j \in \mathbb{N}_{i_1}. \; \neg [\![ F ]\!]^{r \geq j} \tag{b}$$

From (4) and (5) we know $i_1 \geq i_0$. From (3) and (5) we know $i_1 \neq i_0$ and thus $i_1 > i_0$, i.e., (6) $i_0 \in \mathbb{N}_{i_1}$. From (2) and (6), we have (b) for $j := i_0$. $\qquad\Box$

There are many more equivalences, but the o nes given above typically suffice to transform formulas into alternative forms that are, e.g., easier to understand. Such formulas often arise from the specification of concurrent systems.

**Specifying Labeled Transition Systems in LTL**

While the semantics of LTL-formulas is only defined for infinite runs, the runs of an labeled transition system can be finite or infinite. To bridge this gap, we introduce the following concept.

---

**Definition 9.16** (*Infinite Extension of a Run*) We define the *infinite extension* $r^\omega \in State^\omega$ of a run $r \in Run^{State}$ as follows:

$$r^\omega := \text{if } r \in State^\omega \text{ then } r$$
$$\text{else let } n = |r| \text{ in if } n = 0 \text{ then anyof } State^\omega$$
$$\text{else } r \cup \{\langle i, r(n-1)\rangle \mid i \in \mathbb{N} \wedge i \geq n\}$$

Thus the infinite extension $r^\omega = [r(0), \ldots, r(n-1), r(n-1), r(n-1), \ldots]$ of a finite run $r = [r(0), \ldots, r(n-1)]$ consists of the states of $r$ followed by an infinite repetition of the last state $r(n-1)$ of $r$; the infinite extension of an empty run is arbitrary.

---

We are now in the position to formalize the specification of a labeled transition system by an LTL-formula.

---

**Definition 9.17** (*LTS Satisfying an LTL-Formula*) Let $F$ be an LTL-formula that is well-typed with respect to some variable typing $Vt$. Let $lts \in LTS^{L,State}$ be a labeled transition system for some label set $L$ such that every run $r \in [\![ lts ]\!]$ of $lts$ is adequate with respect to $Vt$. We define the relation $lts \models F$ (read: "$lts$ *satisfies* $F$") as follows:

$$lts \models F :\Leftrightarrow \forall r \in [\![ lts ]\!]. \, [\![ F ]\!]^{r^\omega}$$

Consequently, $lts$ satisfies $F$, if every (infinitely extended) run $r^\omega$ of $lts$ satisfies $F$.

---

Thus we may specify also the expected behavior of a (shared or distributed) system $Sy$ by an LTL-formula $F$: the system satisfies the specification if $[\![ Sy ]\!] \models F$, i.e., its denotational semantics $[\![ Sy ]\!]$ (a labeled transition system) satisfies $F$. In Sects. 9.5 and 9.6 we will discuss the verification of such specifications.

## Patterns of LTL

While generally the temporal operators of LTL can be nested to arbitrary depth, in the specification of concurrent systems LTL-formulas have rarely more than two nested temporal operators. In fact, many LTL specifications match typical *patterns* such as the ones shown below (here $F$, $G$, $H$ denote formulas that do *not* involve temporal operators; their truth value therefore does only depend on the *current* state):

- **Invariance**: $\Box F$
  This formula states that $F$ is an *invariant* of the system, i.e., the system only encounters states that satisfy $F$:

$$F \to F \to F \to F \to F \to \cdots$$

  Invariants play in the reasoning about concurrent systems the same crucial role that loop invariants do for verifying the partial correctness of programs; Sect. 9.5 will discuss this in greater detail.

- **Guarantee**: $\Diamond F$
  This formula states that $F$ is a *guarantee* of the system, i.e., the system will eventually encounter at least one state that satisfies $F$, e.g.:

$$\neg F \to \neg F \to \neg F \to \neg F \to F \to \cdots$$

  The guarantee of a system generalizes the termination property of a computer program; Sect. 9.6 will discuss the verification of such formulas.

- **Recurrence**: $\Box \Diamond F$
  This formula states the *recurrence* of $F$, i.e., every state of the system either itself satisfies $F$ or is followed later by another state that satisfies $F$; this can only be the case if the system encounters *infinitely many* states that satisfy $F$:

$$\neg F \to \neg F \to F \to \neg F \to F \to \neg F \to \cdots$$

  This property still allows every state satisfying $F$ to be be followed later by another state that does not satisfy $F$ (which however must be followed by another state that satisfies $F$ again); every occurrence of a state satisfying $F$ may thus indicate some progress of the system.

- **Stability**: $\Diamond \Box F$
  This formula states that $F$ becomes *stable*, i.e., that the system eventually reaches a state from which on every state satisfies $F$:

$$\neg F \to \neg F \to F \to F \to F \to \cdots$$

  This property still allows that the system encounters a finite number of states that violate $F$; however there must be some point from which on no more violations occur. The stability of $F$ implies the recurrence of $F$ (but not the other way round).

- **Response**: $\Box(F \Rightarrow \Diamond G)$

  This formula states that every state that triggers some *request* (indicated by the satisfaction of formula $F$) is eventually followed by a state that receives a *response* (indicated by the satisfaction of formula $G$):

$$\begin{array}{c} F \\ \neg G \end{array} \rightarrow \begin{array}{c} \neg F \\ \neg G \end{array} \rightarrow \begin{array}{c} \neg F \\ G \end{array} \rightarrow \begin{array}{c} F \\ \neg G \end{array} \rightarrow \begin{array}{c} F \\ \neg G \end{array} \rightarrow \begin{array}{c} \neg F \\ G \end{array} \rightarrow \cdots$$

  This property allows that the response $G$ occurs at the same time as the request $F$ (if this is not desired, the formula $\bigcirc G$ should be used rather than $G$); it also does not ensure a one-to-one relationship between requests and responses (multiple requests may be followed by a single response).

- **Precedence**: $\Box(F \Rightarrow H \cup G)$

  This formula generalizes the response pattern in that it also demands that every request $F$ is followed by a response $G$; additionally, however, it demands that, as long as the response is not given, the intermediate property $H$ holds:

$$\begin{array}{c} F \\ H \\ \neg G \end{array} \rightarrow \begin{array}{c} \neg F \\ H \\ \neg G \end{array} \rightarrow \begin{array}{c} \neg F \\ H \\ \neg G \end{array} \rightarrow \begin{array}{c} \neg F \\ \neg H \\ G \end{array} \rightarrow \begin{array}{c} \neg F \\ \neg H \\ \neg G \end{array} \rightarrow \begin{array}{c} F \\ H \\ G \end{array} \rightarrow$$

  Again this property allows response $G$ to occur at the same time as request $F$; it also does not demand that $H$ becomes false when $G$ holds.

**Example 9.15** We reconsider the system described in Example 9.10:

```
distributed MutEx {
  component Server {
    var client:id;
    init { client := N }
    action request(i:id) with client = N
    { client := i; send Client[i].enter() }
    action return(i:id)
    { client := N }
  }
  component Client[N] {
    var req:bool, use:bool;
    init { req := 0; use:= 0; send Client[this].ask() } }
    action ask()
    { req:=1; send Server.request(this) }
    action enter()
    { req := 0; use := 1; send Client[this].exit() }
    action exit()
    { use := 0; send Server.return(this); send Client[this].ask() }
  }
}
```

Recall that the main task of the server is to ensure that no two of the $N$ clients enter the critical region at the same time (*mutual exclusion*). Another core requirement is that every client requesting access to the resource shall eventually receive the resource (*no starvation*). The state of the system consists of the values of the variables *client*, *req* and *use* (in addition to messages pending in the input queues of the actions).

For formulating the system requirements as LTL-formulas, let us assume that variable *client* refers to the value of the corresponding variable in the server component and that, for $i \in \mathbb{N}_N$, $req(i)$ and $use(i)$ refer to the values of the corresponding program variables in instance $i$ of the client component (we will see later in Example 9.17 how in our semantic model of distributed systems these references actually look like). The core requirements can be thus specified as follows.

- **Mutual Exclusion**: at no time two different clients enter the critical region.

$$\forall i_1 \in \mathbb{N}_N, i_2 \in \mathbb{N}_N. \ \Box(use(i_1) = 1 \wedge use(i_2) = 1 \Rightarrow i_1 = i_2)$$

There are also various other logically equivalent formulations, e.g.:

$$\neg \exists i_1 \in \mathbb{N}_N, i_2 \in \mathbb{N}_N. \ \Diamond(use(i_1) = 1 \wedge use(i_2) = 1 \wedge i_1 \neq i_2)$$

- **No Starvation**: every request is eventually answered.

$$\forall i \in \mathbb{N}_N. \ \Box(req(i) = 1 \Rightarrow \Diamond(use(i) = 1))$$

In addition, we may also express various other properties on the expected overall behavior of the system:

- **Clients Make Permanent Progress**: every client infinitely often enters the critical region and leaves it again.

$$\forall i \in \mathbb{N}_N. \ \Box\Diamond(use(i) = 1) \wedge \Box\Diamond(use(i) = 0)$$

- **Server Gives Access**: the server does not permanently prevent all clients from entering the critical region.

$$\exists i \in \mathbb{N}_N. \ \Diamond(client = i)$$

- **Server View is Correct**: only the client registered by the server can enter the critical region.

$$\forall i \in \mathbb{N}_N. \ \Box(use(i) = 1 \Rightarrow client = i)$$

- **Server Remembers Client**: the server maintains the registration of a client until it is informed that the client has left the critical region.

$$\forall i \in \mathbb{N}_N. \ \Box(client = i \Rightarrow (client = i \ \mathsf{W} \ client = N))$$

These properties are not logically independent, e.g. the property "Server View is Correct" implies "Mutual Exclusion". The formulas are mainly variations of the previously indicated patterns.                                                             □

### Safety and Liveness

There is a fundamental difference between a system property defined by an LTL formula $\Box F$ and a system property defined by a formula $\Diamond F$ (in the following, we assume that $F$ is a formula without temporal operators). A system run violates $\Box F$ ("always $F$") if $F$ is violated by some state of the run; there is no way that the run may continue to "undo" that violation. On the other hand, $\Diamond F$ ("eventually $F$") is not violated by a particular state of the run; there is always a way that the run may continue to make the property "eventually" true. Thus $\Box F$ demands the *safety* of a system ("something bad must never happen") while $\Diamond F$ demands its *liveness* ("something good must eventually happen"). Consequently a violation of the safety by an infinite system run can be observed just from the finite past of the run up to the current state; however, the liveness of the run depends on its infinite future.

We are now going to formalize these concepts of "safety" and "liveness" on the basis of the following auxiliary notion.

---

**Definition 9.18** (*Prefix of a Run*) Let $r \in Run^{State}$ be a run and let $n \in \mathbb{N}$ be a natural number. We define the *prefix* $r_{<n}$ of $r$ with length $n$ as follows:

$$r_{<n} := \{\langle i, e \rangle \mid \langle i, e \rangle \in r \wedge i < n\}$$

Consequently $r_{<n} \in Run^{State}$ is that run that contains the first $n$ states of $r$ (only $m$ states if $r$ is finite with length $m < n$).

---

**Definition 9.19** (*Safety and Liveness*)

• $S \subseteq \mathrm{Set}(State^\omega)$ is a *safety property* if every run $r$ violating $S$ has a prefix that cannot be extended to a run satisfying $S$:

   $safety\langle S \rangle :\Leftrightarrow \forall r \in State^\omega.\ \neg S\langle r \rangle \Rightarrow \exists i \in \mathbb{N}.\ \forall s \in State^\omega.\ \neg S\langle r_{<i} \circ s \rangle$

• $L \subseteq \mathrm{Set}(State^\omega)$ is a *liveness property* if every prefix of an arbitrary run can be extended to a run satisfying $L$:

   $liveness\langle L \rangle :\Leftrightarrow \forall r \in State^\omega, i \in \mathbb{N}.\ \exists s \in State^\omega.\ L\langle r_{<i} \circ s \rangle$

Consequently, a property $P$ is *not* a safety property if there is some run that violates $P$, although every prefix of the run can be extended to satisfy $P$. Likewise, $P$ is *not* a liveness property if there is some prefix that cannot be extended to satisfy $P$.

The prototypical example of a safety property is the denotation of the LTL-formula $\Box F$; a corresponding example for a liveness property is the denotation of the LTL-formula $\Diamond F$ (again we assume that $F$ does not contain temporal operators). However, as the following example demonstrates, not every system property is a safety property or a liveness property.

**Example 9.16** Take the property $P$ denoted by the LTL-formula $(\Box F) \wedge (\Diamond G)$:

- $P$ is *not* a safety property: Consider the run

$$\begin{matrix} F \\ \neg G \end{matrix} \rightarrow \begin{matrix} F \\ \neg G \end{matrix} \rightarrow \begin{matrix} F \\ \neg G \end{matrix} \rightarrow \begin{matrix} F \\ \neg G \end{matrix} \rightarrow \cdots$$

where $G$ never (but $F$ always) holds. This run violates $P$, although every prefix

$$\begin{matrix} F \\ \neg G \end{matrix} \rightarrow \begin{matrix} F \\ \neg G \end{matrix} \rightarrow \cdots \rightarrow \begin{matrix} F \\ \neg G \end{matrix}$$

can be extended to a run

$$\begin{matrix} F \\ \neg G \end{matrix} \rightarrow \begin{matrix} F \\ \neg G \end{matrix} \rightarrow \cdots \rightarrow \begin{matrix} F \\ \neg G \end{matrix} \rightarrow \begin{matrix} F \\ G \end{matrix} \rightarrow \cdots$$

that satisfies $P$.
- $P$ is also *not* a liveness property: Consider the prefix

$$\begin{matrix} \neg F \\ \neg G \end{matrix}$$

whose first state violates $F$. This prefix cannot be extended in any way to a run that satisfies $P$. $\qquad \Box$

Thus the class of system properties is not neatly decomposed into safety properties and liveness properties. The real relevance of the distinction between safety and liveness is revealed by the following theorem (due to Leslie Lamport).

**Theorem 9.1** (Decomposition into Safety and Liveness) *Every system property $P \subseteq \mathsf{Set}(State^\omega)$ is a "conjunction" (i.e., intersection) $S \cap L$ of a safety* *property S and a liveness property L:*

$$\forall P \subseteq \mathsf{Set}(State^{\omega}).$$
$$\exists S \subseteq \mathsf{Set}(State^{\omega}),\ L \subseteq \mathsf{Set}(State^{\omega}).$$
$$safety\langle S \rangle \wedge liveness\langle L \rangle \wedge$$
$$\forall r \in State^{\omega}.\ P\langle r \rangle \Leftrightarrow S\langle r \rangle \wedge L\langle r \rangle$$

If $P$ is itself a safety property, then this decomposition is given by $S = P$ and the trivial liveness property $L = True$ where $True := \mathsf{Set}(State^{\omega})$; if $P$ is a liveness property, then the decomposition is given by $L = P$ and the trivial safety property $S = True$. The general definitions of $S$ and $L$ are given in the following proof.

***Proof*** Take arbitrary $P \subseteq \mathsf{Set}(State^{\omega})$. We define $S$ and $L$ as follows:

$$S\langle r \rangle :\Leftrightarrow P\langle r \rangle \vee \forall i \in \mathbb{N}.\ \exists s \in State^{\omega}.\ P\langle r_{<i} \circ s \rangle$$
$$L\langle r \rangle :\Leftrightarrow P\langle r \rangle \vee \exists i \in \mathbb{N}.\ \forall s \in State^{\omega}.\ \neg P\langle r_{<i} \circ s \rangle$$

Intuitively $S$ extends $P$ by all those runs where all prefixes can be extended to satisfy $P$; thus every run not in $S$ violates $P$ and has some prefix that cannot be extended to satisfy $P$ (which is the definition of a safety property). On the contrary, $L$ extends $P$ by all those runs where there is some prefix that cannot be extended to satisfy $P$; thus every run that is not in $L$ violates $P$ but has only prefixes that can be extended to satisfy $P$ (which is the definition of a liveness property).

As for the formal proof that $S$ and $L$ indeed have the required properties, we note $(\exists i \in \mathbb{N}.\ \forall s \in State^{\omega}.\ \neg P\langle r_{<i} \circ s \rangle) \equiv (\neg \forall i \in \mathbb{N}.\ \exists s \in State^{\omega}.\ P\langle r_{<i} \circ s \rangle)$. Furthermore, it is easy to derive $A \equiv (A \vee B) \wedge (A \vee \neg B)$, for arbitrary formulas $A$ and $B$. Therefore, from the definitions of $S$ and $P$, we have $\forall r \in State^{\omega}.\ P\langle r \rangle \Leftrightarrow S\langle r \rangle \wedge L\langle r \rangle$. Thus it remains to show $safety\langle S \rangle$ and $liveness\langle L \rangle$.

- $safety\langle S \rangle$: We take arbitrary $r \in State^{\omega}$ with (1) $\neg S\langle r \rangle$ and show

$$\exists i \in \mathbb{N}.\ \forall s \in State^{\omega}.\ \neg S\langle r_{<i} \circ s \rangle \tag{a}$$

From (1) and the definition of $S$ we have (2) $\neg P\langle r \rangle$ and some $i \in \mathbb{N}$ with (3) $\forall s \in State^{\omega}.\ \neg P\langle r_{<i} \circ s \rangle$. We now show (a) for $i$: thus we take arbitrary $s \in State^{\omega}$ and show $\neg S\langle r_{<i} \circ s \rangle$, i.e.

$$\neg P\langle r_{<i} \circ s \rangle \tag{b}$$

$$\exists j \in \mathbb{N}.\ \forall t \in State^{\omega}.\ \neg P\langle (r_{<i} \circ s)_{<j} \circ t \rangle \tag{c}$$

From (3), we have (b). We now show (c) for $j := i$: thus we take arbitrary $t \in State^\omega$ and show $\neg P \langle (r_{<i} \circ s)_{<i} \circ t \rangle$. Since $(r_{<i} \circ s)_{<i} = r_{<i}$, this follows from (3).

- *liveness*$\langle L \rangle$: We take arbitrary $r \in State^\omega$ and $i \in \mathbb{N}$ and show

$$\exists s \in State^\omega. \, L \langle r_{<i} \circ s \rangle \qquad \text{(a)}$$

From the definition of $L$, it suffices to show

$$\exists s \in State^\omega. \, P \langle r_{<i} \circ s \rangle \vee \exists j \in \mathbb{N}. \, \forall t \in State^\omega. \, \neg P \langle (r_{<i} \circ s)_{<j} \circ t \rangle$$

In the case of $\exists s \in State^\omega. \, P \langle r_{<i} \circ s \rangle$ we are done. Otherwise we have the case (1) $\forall s \in State^\omega. \, \neg P \langle r_{<i} \circ s \rangle$. To prove (b), we take arbitrary $s \in State^\omega$ and prove

$$\exists j \in \mathbb{N}. \, \forall t \in State^\omega. \, \neg P \langle (r_{<i} \circ s)_{<j} \circ t \rangle \qquad \text{(b)}$$

for $j := i$. Thus we take arbitrary $t \in State^\omega$ and prove $\neg P \langle (r_{<i} \circ s)_{<i} \circ t \rangle$. Since $(r_{<i} \circ s)_{<i} = r_{<i}$, this follows from (2). $\qquad \square$

Reasoning about a system property is simplified if we are able to separate its safety aspect and its liveness aspect into distinct LTL-formulas. Such a separation is not always obvious but may be aided by the following considerations.

---

**Proposition 9.1** (LTL-Formulas and Safety/Liveness)

- *If in LTL-formula F no temporal operator occurs, then it denotes a safety property.*
- *If LTL-formulas F and G denote safety properties, then $F \wedge G$, $F \vee G$, $\forall x. \, F$, $\exists x. \, F$, $\bigcirc F$, $\square F$, F W G, and F R G denote safety properties.*
- *For every LTL-formula F, $\Diamond F$ denotes a liveness property.*
- *If LTL-formulas F and G denote liveness properties, then $F \wedge G$, $F \vee G$, $\forall x. \, F$, $\exists x. \, F$, $\bigcirc F$, $\square F$ denote liveness properties.*
- *Let F and G be LTL-formulas that denote safety properties. Then the property (denoted by) F U G can be decomposed into the safety property F W G and the liveness property $\Diamond G$. Likewise F M G can be decomposed into the safety property F R G and the liveness property $\Diamond F$.*

---

The main usefulness of this proposition lies in the first two rules that determine a class of LTL-formulas that are guaranteed to denote pure safety properties. Essentially these are the formulas constructed from arbitrary first order formulas as building blocks by applying a certain set of "safe" temporal operators (in particular, only the "weak" binary forms); the resulting formulas may be further combined in a restricted way that does not lead to their negation (which by the equivalence $\neg \Box G \equiv \Diamond \neg F$ may turn a safety property into a liveness property, see below). Thus, e.g., the formula pattern $\Box(F \Rightarrow (G \mathrel{W} H))$, which is equivalent to $\Box((\neg F) \vee (G \mathrel{W} H))$, always denotes a safety property (assuming that $F$, $G$, $H$ are formulas without temporal operators).

The third rule identifies the "eventually" operator as the (only) one that introduces a "pure" liveness property from scratch; the fourth rule allows to combine liveness formulas in a restricted form. The last rule decomposes a property expressed by the binary temporal operators "strong until" respectively "strong release" into a safety property expressed by "weak until" respectively "weak release" and a liveness property expressed by "eventually". Consequently we may decompose e.g. the property expressed by $\Box(F \Rightarrow (G \mathrel{U} H))$ into a safety property $\Box(F \Rightarrow (G \mathrel{W} H))$ and a liveness property $\Box(F \Rightarrow \Diamond H)$; however, that this last formula indeed denotes a liveness property is not established by above rules and thus remains to be shown.

There is no simple syntactic criterion to generally decide whether an arbitrary LTL-formula denotes a pure safety property, a pure liveness property, or a conjunction of both; in practice, however, above guidelines are often sufficient. In the following we will discuss techniques for the verification of systems with respect to (special cases of) such safety and liveness properties.

## 9.5    Verifying Invariance

If a system uses only a finite number of variables and every variable has only a finite number of possible values, also the state space of the system is finite; a run of such a system can be infinite only by infinitely often cycling through a finite set of states. For such *finite state systems* an automatic decision procedure, a *model checker*, can determine whether the system satisfies arbitrary LTL-formulas; furthermore, if a formula is violated, the model checker produces a "counterexample", i.e., a run of the system that does not satisfy the formula. However, model checking can be generally not applied to infinite state systems and teaches us actually little about how to think as a human about concurrent systems. In this section and the following one, we will discuss techniques that are applicable for humans to reason about the correctness of arbitrary systems; we will, however, restrict ourselves to the verification of LTL-formulas of special forms only.

The goal of this section is to verify that a formula $F$ is an *invariant* of a system, i.e., that the system satisfies the LTL-formula $\Box F$. Here we assume that $F$ is a first-order formula that does not contain any temporal operator and therefore can be evaluated on a single state; consequently $\Box F$ denotes a safety property of the system. As a special case, we may formulate the *partial correctness* of a computation: the formula $\Box(E \Rightarrow Q_{x,y})$ indicates that, always if the computation has ended (as indicated by some "end condition" $E$), the result $y$ of the computation is correct with respect to input $x$, i.e., it satisfies some postcondition $Q_{x,y}$.

We start with a basic property of formulas of form $\Box F$.

**Proposition 9.2** (Propagation of Invariance) *Let F and G be LTL-formulas that are adequate with respect to some variable typing Vt. Then we have the following logical consequence:*

$$(\Box F) \wedge (F \Rightarrow G) \models (\Box G)$$

Consequently, to show $\Box G$ it suffices to show $\Box F$ for some stronger property $F$, i.e., a property $F$ that implies $G$. In this sense, the validity of $\Box F$, once established, can be transferred to the validity of $\Box G$ and from this to another property $\Box H$ with $G \Rightarrow H$. However, such a sequence of "validity transfers" needs to be started with a property $\Box F$ whose validity has to be established by other means. These means are provided by the following theorem.

**Theorem 9.2** (Verifying Invariance) *Let lts = ⟨space:S, init:I, next:R⟩ be a labeled transition system with some label set L, state space S, initial state condition I, and transition relation R.*

*Let condition $F \subseteq S$ on S be an* inductive invariant *of system lts, i.e., let F satisfy the following two conditions:*

1. *Every initial state of the system satisfies F:*

$$\forall s \in S. \ I\langle s \rangle \Rightarrow F\langle s \rangle$$

2. *If a state s satisfies F and the system can make a transition from s to a state s', then also s' satisfies F:*

$$\forall l \in L, s \in S, s' \in S. \ F\langle s \rangle \wedge R\langle l, s, s' \rangle \Rightarrow F\langle s' \rangle$$

*Then in every run of lts every state satisfies F:*

$$\forall r \in [\![ lts ]\!]. \ [\![ \Box F ]\!]^{r^{\omega}}$$

This theorem in essence states that we may verify $\Box F$ by the principle of *induction*: the induction base is represented by the initial state set $I$ of the system, while the induction step is performed by the application of the next state relation $R$.

**Proof** We informally sketch the correctness of this proof rule. Consider an arbitrary run $r$ of the system. To show that $r^{\omega}$ satisfies $\Box F$, it suffices to show that $F$ is satisfied by every state $r(i)$ with $i \in domain(r)$. We prove this by induction on $i$. By the first condition of the rule, the initial state $r(0)$ satisfies $F$, which establishes the induction base. Now assume as the induction assumption that $r(i)$ satisfies $F$ and that $i + 1 \in domain(r)$. Then the second part of the rule ensures that also $r(i + 1)$ satisfies $F$, which establishes the induction step.                                         □

However, if our goal is to prove $\Box F$, it is often the case that, although $F$ is indeed an invariant of the system, it is not *inductive*, i.e., is too weak to adequately serve as an induction assumption in the second condition of the proof rule. In such situations, we have to come up with a stronger invariant $G$, i.e., a property $G$ that implies $F$ and that is inductive, such that the proof of $\Box G$ succeeds. Then, from the validity of $\Box G$ and the fact that $G$ implies $F$ we also have the validity of $\Box F$. *System invariants* play for the verification of the safety of concurrent systems the same role that *loop invariant* play for the verification of the partial correctness of loops (see Sect. 8.5). It is therefore of crucial importance to characterize concurrent systems by invariants that are as strong as possible, i.e., that characterize the set of the reachable states of the system as accurately as possible.

**Example 9.17** Consider the following pseudo-code of a program with shared variable $s$ and two concurrent threads $A$ and $B$ that forever iterate three commands:

```
var s:=1
A: loop          || B: loop
      0: lock(s)          0: lock(s)
      1: ...              1: ...
      2: unlock(s)        2: unlock(s)
```

The command $lock(s)$ blocks a thread if the value of the "semaphore" $s$ is 0; however, if the value of $s$ is 1, it sets $s$ to 0 and lets the thread proceed. Thus it shall be ensured that both threads mutually exclusive the commands marked as "...".

In our system language, this shared system can be described as follows:

```
shared Semaphore {
  var a:nat, b:nat, s:nat;
  init { a:=0; b:=0; s:=1 }
  action lockA with a=0 ∧ s=1 { a:=1; s:=0 }
  action criticalA with a=1 { a:=2 }
  action unlockA with a=2 { a:=0; s:=1 }
  action lockB with b=0 ∧ s=1 { b:=1; s:=0 }
  action criticalB with b=1 { b:=2 }
  action unlockB with b=2 { b:=0; s:=1 }
}
```

Here the values of the variables $a$ and $b$ represent the "program counters" of the threads which periodically cycle among the values 0, 1, and 2. The two sets of three actions each model the commands in the bodies of the respective loops. Each action is only enabled if the program counter has a particular value; the execution of the action updates this value. The actions *lockA* and *lockB* are special in that they also are only enabled if the semaphore variable $s$ has value 1; they update this value to 0 together with the program counter.

This definition gives rise to the labeled transition system $\langle space\!:\!S, init\!:\!I, next\!:\!R \rangle$ whose components are defined as follows:

$$S := \mathbb{N} \times \mathbb{N} \times \mathbb{N}$$

$$I\langle a, b, s \rangle :\Leftrightarrow a = 0 \wedge b = 0 \wedge s = 1$$

$$R\langle l, \langle a, b, s \rangle, \langle a', b', s' \rangle \rangle :\Leftrightarrow$$
$$(l = lockA \wedge a = 0 \wedge s = 1 \wedge a' = 1 \wedge s' = 0 \wedge b' = b) \vee$$
$$(l = criticalA \wedge a = 1 \wedge a' = 2 \wedge b' = b \wedge s' = s) \vee$$
$$(l = unlockA \wedge a = 2 \wedge a' = 0 \wedge s' = 1 \wedge b' = b) \vee$$
$$(l = lockB \wedge b = 0 \wedge s = 1 \wedge b' = 1 \wedge s' = 0 \wedge a' = a) \vee$$
$$(l = criticalB \wedge b = 1 \wedge b' = 2 \wedge a' = a \wedge s' = s) \vee$$
$$(l = unlockB \wedge b = 2 \wedge b' = 0 \wedge s' = 1 \wedge a' = a)$$

Our goal is to verify the invariance of a formula $F$ (i.e., the validity of the LTL-formula $\Box F$) whose semantics is denoted by the following relation:

$$F\langle a, b, s \rangle :\Leftrightarrow \neg(a = 1 \wedge b = 1)$$

For this, we first have to prove the "induction base"

$$I\langle a, b, s \rangle \Rightarrow F\langle a, b, s \rangle$$

which clearly holds. Then we have to prove the "induction step"

$$F\langle a, b, s \rangle \wedge R\langle l, \langle a, b, s \rangle, \langle a', b', s' \rangle \rangle \Rightarrow F\langle a', b', s' \rangle$$

According to the definition of $R$, this boils down to proving six formulas, one for each transition:

- $\neg(a = 1 \wedge b = 1) \wedge (l = lockA \wedge a = 0 \wedge s = 1 \wedge a' = 1 \wedge s' = 0 \wedge b' = b)$
  $\Rightarrow \neg(a' = 1 \wedge b' = 1)$
- $\neg(a = 1 \wedge b = 1) \wedge (l = criticalA \wedge a = 1 \wedge a' = 2 \wedge b' = b \wedge s' = s)$
  $\Rightarrow \neg(a' = 1 \wedge b' = 1)$
- $\ldots$

However, already the first formula is *not* valid. The "induction assumption" $\neg(a = 1 \wedge b = 1)$ is too weak to let us from the transition formula $(l = lockA \wedge a = 0 \wedge s = 1 \wedge a' = 1 \wedge s' = 0 \wedge b' = b)$ conclude the conclusion $\neg(a' = 1 \wedge b' = 1)$: since the case $a = 0 \wedge b = 1$ is not prohibited, the transition condition $a' = 1 \wedge b' = b$ might indeed lead to the situation $a' = 1 \wedge b' = 1$.

Therefore we have to come up with a stronger system invariant $G$ that describes the state space of the labeled transition system much more accurately:

$$G\langle a, b, s \rangle :\Leftrightarrow$$
$$(a = 0 \lor a = 1 \lor a = 2) \land (b = 0 \lor b = 1 \lor c = 2) \land (s = 0 \lor s = 1) \land$$
$$(a = 1 \lor a = 2 \Rightarrow s = 0 \land b \neq 1 \land b \neq 2) \land$$
$$(b = 1 \lor b = 2 \Rightarrow s = 0 \land a \neq 1 \land a \neq 2) \land$$
$$(s = 0 \Rightarrow a = 1 \lor a = 2 \lor b = 1 \lor b = 2)$$

Apart from restricting the ranges of the variables, this condition ensures that, if process $A$ is in the critical region, the semaphore is locked and process $B$ is not in the critical region, and dually for the process $B$. Furthermore it is claimed that if the semaphore is locked, it is because one of the processes is in the critical region.

It is easy to show that $G$ is indeed stronger than $F$, i.e.:

$$G\langle a, b, s \rangle \Rightarrow F\langle a, b, s \rangle$$

For this, assume $G\langle a, b, s \rangle$ and $\neg G\langle a, b, s \rangle$, i.e., $a = 1$ and $b = 1$. However, from $a = 1$ the fourth clause in the conjunction of $G\langle a, b, s \rangle$ implies the contradiction $b \neq 1$.

Thus, to show that $F$ is an invariant of the system, it suffices to show that $G$ is an invariant. Again the "induction base"

$$I\langle a, b, s \rangle \Rightarrow G\langle a, b, s \rangle$$

clearly holds. For showing the "induction step"

$$G\langle a, b, s \rangle \land R\langle l, \langle a, b, s \rangle, \langle a', b', s' \rangle\rangle \Rightarrow G\langle a', b', s' \rangle$$

we have to follow the following six conditions:

- $G\langle a, b, s \rangle \land (l = lockA \land a = 0 \land s = 1 \land a' = 1 \land s' = 0 \land b' = b) \Rightarrow G\langle a', b', s' \rangle$
- $G\langle a, b, s \rangle \land (l = criticalA \land a = 1 \land a' = 2 \land b' = b \land s' = s) \Rightarrow G\langle a', b', s' \rangle$
- $G\langle a, b, s \rangle \land (l = unlockA \land a = 2 \land a' = 0 \land s' = 1 \land b' = b) \Rightarrow G\langle a', b', s' \rangle$
- $G\langle a, b, s \rangle \land (l = lockB \land b = 0 \land s = 1 \land b' = 1 \land s' = 1 \land a' = a) \Rightarrow G\langle a', b', s' \rangle$
- $G\langle a, b, s \rangle \land (l = criticalB \land b = 1 \land b' = 2 \land a' = a \land s' = s) \Rightarrow G\langle a', b', s' \rangle$
- $G\langle a, b, s \rangle \land (l = unlockB \land b = 2 \land b' = 0 \land s' = 1 \land a' = a) \Rightarrow G\langle a', b', s' \rangle$

Now we can indeed show that all conditions are valid. For instance, for the proof of the first condition we assume

$$G\langle a, b, s \rangle \tag{1}$$
$$(l = lockA \land a = 0 \land s = 1 \land a' = 1 \land s' = 0 \land b' = b) \tag{2}$$

and show $G\langle a', b', s'\rangle$, i.e.

$$(a' = 0 \vee a' = 1 \vee a' = 2) \wedge (b' = 0 \vee b' = 1 \vee c' = 2) \wedge (s' = 0' \vee s' = 1) \tag{a}$$
$$(a' = 1 \vee a' = 2 \Rightarrow s' = 0 \wedge b' \neq 1 \wedge b' \neq 2) \tag{b}$$
$$(b' = 1 \vee b' = 2 \Rightarrow s' = 0 \wedge a' \neq 1 \wedge a' \neq 2) \tag{c}$$
$$(s' = 0 \Rightarrow a' = 1 \vee a' = 2 \vee b' = 1 \vee b' = 2) \tag{d}$$

From (1) we know $(a = 0 \vee a = 1 \vee a = 2) \wedge (b = 0 \vee b = 1 \vee c = 2) \wedge (s = 0 \vee s = 1)$; from (2) we know $a' = 1 \wedge s' = 0 \wedge b' = b$ and thus (a). To prove (b), we assume $a' = 1 \vee a' = 2$ and show $s' = 0 \wedge b' \neq 1 \wedge b' \neq 2$. From (2) we know $s' = 0$. Also from (2) we know $b' = b$, thus it remains to show $b \neq 1 \wedge b \neq 2$. Since (2) implies $s = 1$, this follows from the fifth clause in the conjunction of (1). Since the proof of (c) proceeds analogously to the proof of (b), it remains to prove (d). From (2), we know $a' = 1$ and are therefore done.

The proofs of the other conditions proceed in a similar way. $\qquad\square$

**Example 9.18** Consider again the system introduced in Example 9.4 with a server and $N$ clients. We have modeled this system by a LTS with the following state space $S$, initial state condition $I$, and next state relation $R$:

$$S := \mathbb{N} \times \mathsf{Set}(\mathbb{N}) \times (\mathbb{N}_N \rightarrow \mathbb{N}) \times \mathsf{Set}(\mathbb{N}) \times \mathsf{Set}(\mathbb{N})$$

$I\langle given, waiting, pc, req, ans\rangle :\Leftrightarrow$

$\quad (given = N \wedge waiting = \varnothing) \wedge$

$\quad (\forall i \in \mathbb{N}_N. pc(i) = 0) \wedge$

$\quad (req = \varnothing \wedge ans = \varnothing)$

$R\langle l, \langle given, waiting, pc, req, ans\rangle, \langle given', waiting', pc', req', ans'\rangle\rangle :\Leftrightarrow$

$\quad (RS\langle l, \langle given, waiting, req, ans\rangle, \langle given', waiting', req', ans'\rangle\rangle \wedge$

$\quad\quad pc' = pc) \vee$

$\quad (\exists i \in \mathbb{N}_N. RC\langle l, i, \langle pc, req, ans\rangle, \langle pc', req', ans'\rangle\rangle \wedge$

$\quad\quad given' = given \wedge waiting' = waiting)$

$RC\langle l, i, \langle pc, req, ans\rangle, \langle pc', req', ans'\rangle\rangle :\Leftrightarrow$

$\quad ((l = C0_i \wedge pc(i) = 0 \wedge pc'(i) = 1 \wedge$

$\quad\quad \neg(i \in req) \wedge req' = req \cup \{i\} \wedge ans' = ans) \vee$

$\quad (l = C1_i \wedge pc(i) = 1 \wedge pc'(i) = 2 \wedge i \in ans \wedge$

$\quad\quad req' = req \wedge ans' = ans \setminus \{i\}) \vee$

$\quad (l = C2_i \wedge pc(i) = 2 \wedge pc'(i) = 0 \wedge$

$\quad\quad \neg(i \in req) \wedge req' = req \cup \{i\} \wedge ans' = ans)) \wedge$

$\quad (\forall j \in \mathbb{N}_N. j \neq i \Rightarrow pc'(i) = pc(i))$

$RS\langle l, \langle given, waiting, req, ans\rangle, \langle given', waiting', req', ans'\rangle\rangle :\Leftrightarrow$

$\quad \exists i \in req.$

$req' = req \setminus \{i\} \wedge$

$\big((l = S0_i \wedge i = given \wedge waiting = \varnothing \wedge$

  $given' = N \wedge waiting' = waiting \wedge ans' = ans) \vee$

$(l = S1_i \wedge i = given \wedge \exists given \in waiting.\ given' = given \wedge$

  $waiting' = waiting \setminus \{given\} \wedge$

  $\neg(given \in ans) \wedge ans' = ans \cup \{given\}) \vee$

$(l = S2_i \wedge i \neq given \wedge given = N \wedge$

  $given' = i \wedge waiting' = waiting \wedge$

  $\neg(given' \in ans) \wedge ans' = ans \cup \{given'\}) \vee$

$(l = S3_i \wedge i \neq given \wedge given \neq N \wedge$

  $waiting' = waiting \cup \{i\} \wedge given' = given \wedge ans' = ans)\big)$

Our goal is to verify that the system maintains the *mutual exclusion* property

$$\Box MutEx\langle client, request, return, c\rangle$$

with the predicate *MutEx*⟨*client, request, return, c*⟩ defined as follows:

$$MutEx\langle given, waiting, pc, req, ans\rangle :\Leftrightarrow$$
$$\forall i_1 \in \mathbb{N}_N, i_2 \in \mathbb{N}_N.\ pc(i_1) = 2 \wedge pc(i_2) = 2 \Rightarrow i_1 = i_2$$

For this we define the following system invariant:

$Invariant\langle given, waiting, pc, req, ans\rangle :\Leftrightarrow$

  $\forall i \in \mathbb{N}_N.$

   $(i = given \Rightarrow$

    $(pc(i) = 0 \wedge i \in req) \vee$

    $(pc(i) = 1 \wedge i \in ans) \vee$

    $(pc(i) = 2 \wedge \neg(i \in req) \wedge \neg(i \in ans))) \wedge$       (a)

   $(i \in waiting \Rightarrow i \neq given \wedge pc(i) = 1 \wedge \neg(i \in req) \wedge \neg(i \in ans)) \wedge$   (b)

   $(i \in req \Rightarrow \neg(i \in ans)) \wedge$       (c)

   $(i \in ans \Rightarrow given = i) \wedge$       (d)

   $(pc(i) = 0 \Rightarrow \neg(i \in ans) \wedge (i \in req \Rightarrow i = given)) \wedge$       (e)

   $(pc(i) = 1 \Rightarrow i \in req \vee i \in waiting \vee i \in ans) \wedge$       (a)

   $(pc(i) = 2 \Rightarrow i = given)$       (g)

This invariant constrains the system state by the following conditions on every client $i$:

- Condition (a) states the information we have if the server has given the client permission to enter the critical region: the client may be at position 0, with the message returning the permission still pending in the network; it may be at position 1 with the answer to the request still in the network; or it may be at position 2 with neither a request nor an answer pending.
- Condition (b) states the information we have if the server has recorded the client in set *waiting*: the client has not been given the resource, it is still waiting at position $i$, and no message from/to that client is pending.
- Condition (c) states that there cannot be simultaneously a request from the client and an answer to the client pending in the network.
- Condition (d) states that if an answer to client $i$ is pending in the network, then the server has recorded $i$ in *given*.
- Condition (e) states the information we have if the client is at position 0: there is no answer to that client pending, and, if a message from the client is pending, this message returns the permission but has not yet been processed by the server.
- Condition (e) states the information we have if the client is at position 1: its request for permission to enter the critical region is still pending in the network or the server has recorded the request in set *waiting* or the server has already answered it.
- Condition (e) states the information we have if the client is at position 2: the server has recorded the permission of client $i$ in *given*.

This invariant implies the mutual exclusion property, i.e.:

$$Invariant\langle given, waiting, pc, req, ans\rangle \Rightarrow MutEx\langle given, waiting, pc, req, ans\rangle$$

To show this, assume that the invariant holds and take arbitrary $i_1, i_2 \in \mathbb{N}_N$ with $pc(i_1) = 2 \wedge pc(i_2) = 2$. From condition (g) of the invariant we have $pc(i_1) = given$ and $pc(i_2) = given$ and thus $i_1 = i_2$.

So it suffices to show $Invariant\langle given, waiting, pc, req, ans\rangle$. It is easy to verify

$$I\langle given, waiting, pc, req, ans\rangle \Rightarrow Invariant\langle given, waiting, pc, req, ans\rangle$$

i.e., that the invariant holds in the initial state of the system.

The real challenge is to prove that the invariant is preserved by every transition. We only demonstrate this for the client transition labeled as $C2_i$, i.e., we sketch for arbitrary $i \in \mathbb{N}_N$ the proof of the following property:

$$\begin{aligned}
&Invariant\langle given, waiting, pc, req, ans\rangle \wedge \\
&RC\langle l, i, \langle C2_i, req, ans\rangle, \langle pc', req', ans'\rangle\rangle \wedge \\
&(\forall j \in \mathbb{N}_N.\ j \neq i \Rightarrow pc'(i) = pc(i)) \wedge \\
&given' = given \wedge waiting' = waiting \Rightarrow \\
&\quad Invariant\langle given', waiting', pc', req', ans'\rangle
\end{aligned}$$

So we assume

$$Invariant\langle given, waiting, pc, req, ans\rangle \tag{1}$$
$$pc(i) = 2 \tag{2}$$
$$pc'(i) = 0 \tag{3}$$
$$\neg(i \in req) \tag{4}$$
$$req' = req \cup \{i\} \tag{5}$$
$$ans' = ans \tag{6}$$
$$(\forall j \in \mathbb{N}_N.\ j \neq i \Rightarrow pc'(i) = pc(i)) \tag{7}$$
$$given' = given \tag{8}$$
$$waiting' = waiting \tag{9}$$

and show

$$Invariant\langle given', waiting', pc', req', ans'\rangle$$

Take arbitrary $i_0 \in \mathbb{N}_N$. From (7) and (1), the body of the invariant holds for $i_0 \neq i$. We thus consider only the case $i_0 = i$:

- To show (a), assume $i = given'$. From (3) and (5), we have $pc'(i) = 0$ and $i \in req'$.
- To show (b), we assume $i \in waiting'$ and show a contradiction. From (9), we have $i \in waiting$, which contradicts (2), (1.b), and (1.g).
- To show (c), we assume $i \in req'$ and show $\neg(i \in ans')$; this follows from (2), (1.g), and (1.a), and (6).
- To show (d), we assume $i \in ans'$ and show $given' = i$; this follows from (6), (1.d), and (8).
- We show (e). From (2), (1.g), (1.a), and (6), we have $\neg(i \in ans')$. Furthermore, from (2), (1.g), and (8), we have $i = given'$.
- From (3), we have (f) and (g).

This completes the proof sketch that transition $C2_i$ preserves the invariant.   □

**Example 9.19** Consider again the distributed system introduced in Example 9.10:

```
distributed ClientServer  {
  component Server {
    var client:id;
    init { client := N }
    action request(i:id) with client = N
    { client := i; send Client[i].enter() }
    action return(i:id)
    { client := N }
  }
  component Client[N] {
```

```
    var req:bool, use:bool;
    init { req := 0; use:= 0; send Client[this].ask() } }
    action ask()
    { req:=1; send Server.request(this) }
    action enter()
    { req := 0; use := 1; send Client[this].exit() }
    action exit()
    { use := 0; send Server.return(this); send Client[this].ask() }
  }
}
```

We are now going to sketch the invariant-based proof that every execution of the system satisfies the *mutual exclusion* property specified in Example 9.15. As a starting point we describe the semantics of the system by an LTS with the state space $S$ defined as follows:

$$S := \mathbb{N} \times \mathbb{N}^* \times \mathbb{N}^* \times (\mathbb{N}_N \rightarrow C)$$
$$C := \mathbb{N}_2 \times \mathbb{N}_2 \times \mathbb{N}^* \times \mathbb{N}^* \times \mathbb{N}^*$$

A state $\langle client, request, return, c \rangle \in S$ is interpreted as follows:

- *client* denotes the value of the corresponding variable in the server component.
- *request* and *return* denote the message queues of the corresponding actions.
- $c(i)$ denotes a state $\langle req, use, ask, enter, exit \rangle$ of client $i$ interpreted as follows:
    - *req* and *use* denote the values of the corresponding client variables.
    - *ask*, *enter*, and *exit* denote the message queues of the corresponding actions.

Then the initial state condition $I$ of the system is defined as follows:

$I \langle client, request, return, c \rangle :\Leftrightarrow$
  $client = N \wedge |request| = 0 \wedge |return| = 0 \wedge$
  $\forall this \in \mathbb{N}_N, req \in \mathbb{N}_2, use \in \mathbb{N}_2, ask \in \mathbb{N}^*, enter \in \mathbb{N}^*, exit \in \mathbb{N}^*.$
    $c(this) = \langle req, use, ask, enter, exit \rangle \Rightarrow$
      $req = 0 \wedge use = 0 \wedge ask = [this] \wedge \wedge |enter| = 0 \wedge |exit| = 0$

The transition relation of the system is a disjunction of the transition relations of the actions of all components. For instance, the transition relation for the server action *request* is as follows:

$Request \langle \langle client, request, return, c \rangle, \langle client', request', return', c' \rangle \rangle :\Leftrightarrow$
  $\exists i \in \mathbb{N}_N, req \in \mathbb{N}_2, use \in \mathbb{N}_2, ask \in \mathbb{N}^*, enter \in \mathbb{N}^*, exit \in \mathbb{N}^*.$
    $request = [i] \circ request' \wedge client = N \wedge client' = i \wedge return' = return \wedge$
    $c(i) = \langle req, use, ask, enter, exit \rangle \wedge c'(i) = \langle req, use, ask \circ [0], enter, exit \rangle \wedge$
    $(\forall j \in \mathbb{N}_N.\ j \neq i \Rightarrow c'(j) = c(j))$

Likewise, the transition relation for the client action *exit* is as follows:

$Exit(\langle client, request, return, c \rangle, \langle client', request', return', c' \rangle) :\Leftrightarrow$

$\exists this \in \mathbb{N}_N, req \in \mathbb{N}_2, use \in \mathbb{N}_2, ask \in \mathbb{N}^*, enter \in \mathbb{N}^*, exit \in \mathbb{N}^*.$

$\exists use' \in \mathbb{N}_2, ask' \in \mathbb{N}^*, exit' \in \mathbb{N}^*.$

$c(this) = \langle req, use, ask, enter, exit \rangle \wedge$

$c'(this) = \langle req, use', ask', enter, exit' \rangle \wedge exit = [0] \circ exit' \wedge$

$use' = 0 \wedge return' = return \circ [this] \wedge ask' = ask \circ [this] \wedge$

$(\forall j \in \mathbb{N}_N. j \neq this \Rightarrow c'(j) = c(j)) \wedge client' = client \wedge request' = request$

The formalization of the transition relations of the other actions proceeds analogously.

Based on this model, we define the *mutual exclusion* property of this system by the LTL-formula

$$\Box MutEx \langle client, request, return, c \rangle$$

with the predicate $MutEx \langle client, request, return, c \rangle$ defined as follows:

$MutEx \langle client, request, return, c \rangle :\Leftrightarrow$

$\forall i_1 \in \mathbb{N}_N, i_2 \in \mathbb{N}_N.$

let $use_1 = c(i_2).2, use_2 = c(i_1).2$ in $use_1 = 1 \wedge use_2 = 1 \Rightarrow i_1 = i_2$

For the verification of this property (and any other safety property of the system), we propose the following invariant:

$Invariant \langle client, request, return, c \rangle :\Leftrightarrow$

$\forall i \in \mathbb{N}_N, req \in \mathbb{N}_2, use \in \mathbb{N}_2, ask \in \mathbb{N}^*, enter \in \mathbb{N}^*, exit \in \mathbb{N}^*.$

$c(i) = \langle req, use, ask, enter, exit \rangle \Rightarrow$

$|ask| + |enter| + |exit| + num(request, i) = 1 \wedge$                       (a)

$num(return, i) \leq 1 \wedge$                                                    (b)

$(num(return, i) = 1 \Rightarrow |enter| = 0 \wedge |exit| = 0) \wedge$         (c)

$(req = 1 \Leftrightarrow num(request, i) = 1 \vee |enter| = 1) \wedge$         (d)

$(use = 1 \Leftrightarrow |exit| = 1) \wedge$                                          (e)

$(client = i \Leftrightarrow |enter| = 1 \vee |exit| = 1 \vee num(return, i) = 1)$     (a)

This invariant utilizes the auxiliary function $num(q, i)$ which denotes the number of messages from client $i$ in the queue $q$:

$$num(q, i) := \#k \in \mathbb{N}. k < |q| \wedge q(k) = i$$

The invariant constrains every client $i$ by the following conditions:

- Condition (a) states that exactly one of the following is true: there is a message in the queue of one of the actions of the client (i.e., the corresponding action is enabled) or there is a message of client $i$ in the queue of the server action *request*. Furthermore, the condition implies that there cannot be more one message in the client queues and not more than one message of client $i$ in the *request* queue.
- Condition (b) states that there cannot be more than one message of client $i$ in the queue of the server action *return*. This condition is separated from Condition (a), because there can be simultaneously a message of client $i$ in the return queue and a message in the queue of action *ask* of that client.
- However, condition (c) states that if there is a message of client $i$ in the *return* queue, then the actions *enter* and *exit* of the client are not enabled (i.e., the client is neither in the critical region nor wants to enter it).
- Condition (d) states that the variable *req* of client $i$ is set to one if and only if a request of that client is pending in the *request* queue of the server or there is a response of the server in the *enter* queue of the client (which enables the client to enter the critical region).
- Condition (e) states that the variable *use* of the client is set to one if and only if the client is in the critical region, i.e., the action *exit* is enabled that allows the client to leave the region.
- Condition (f) states that the variable *client* in the server is set to $i$ if and only if the client is ready to enter the critical region, is ready to exit the critical region, or there is a message of that client pending in the *return* queue of the server.

The invariant implies the mutual exclusion property, i.e.:

$$Invariant\langle client, request, return, c\rangle \Rightarrow MutEx\langle client, request, return, c\rangle$$

To show this, assume that the invariant holds and that the values of the variable *use* of client $i_1$ and $i_2$ are both 1. Then invariant condition (e) implies that we have in both clients $|exit| = 1$ and thus, by condition (f), for the server variable *client* both $client = i_1$ and $client = i_2$; this implies $i_1 = i_2$.

Thus it remains to show that $Invariant\langle client, request, return, c\rangle$ is indeed an invariant of the system. It is easy to show

$$I\langle client, request, return, c\rangle \Rightarrow Invariant\langle client, request, return, c'\rangle$$

i.e., that the invariant holds in the initial state of the system. The real challenge is to verify that the invariant is preserved by every transition. We only demonstrate this for the client transition *exit*, i.e., we sketch the proof of the following property:

$$Invariant\langle client, request, return, c\rangle \wedge$$
$$Exit\langle\langle client, request, return, c\rangle, \langle client', request', return', c'\rangle\rangle \Rightarrow$$
$$Invariant\langle client', request', return', c'\rangle$$

So we assume

$$Invariant\langle client, request, return, c\rangle \tag{1}$$

$$Exit\langle\langle client, request, return, c\rangle , \langle client', request', return', c'\rangle\rangle \tag{2}$$

From (2) we have some $this \in \mathbb{N}_N, req \in \mathbb{N}_2, use \in \mathbb{N}_2, ask \in \mathbb{N}^*, enter \in \mathbb{N}^*, exit \in \mathbb{N}^*, use' \in \mathbb{N}_2, ask' \in \mathbb{N}^*, exit' \in \mathbb{N}^*$ with

$$c(this) = \langle req, use, ask, enter, exit\rangle \tag{3}$$

$$c'(this) = \langle req, use', ask', enter, exit'\rangle \tag{4}$$

$$exit = [0] \circ exit' \tag{5}$$

$$use' = 0 \tag{6}$$

$$return' = return \circ [this] \tag{7}$$

$$ask' = ask \circ [this] \tag{8}$$

$$\forall j \in \mathbb{N}_N. \ j \neq this \Rightarrow c'(j) = c(j) \tag{9}$$

$$client' = client \tag{10}$$

$$request' = request \tag{11}$$

We have to show

$$Invariant\langle client', request', return', c'\rangle(a)$$

We take arbitrary $i \in \mathbb{N}_N, req' \in \mathbb{N}_2, use' \in \mathbb{N}_2, ask' \in \mathbb{N}^*, enter' \in \mathbb{N}^*, exit' \in \mathbb{N}^*$ with

$$c'(i) = \langle req', use', ask', enter', exit'\rangle \tag{12}$$

which from (4) implies $req' = req$ and $enter' = enter$; the other primed variables are identical to those used in (4). We show

$$|ask'| + |enter'| + |exit'| = 1 + num(request', i) \tag{a}$$

$$num(return', i) \leq 1 \tag{b}$$

$$num(return', i) = 1 \Rightarrow |enter'| = 0 \land |exit'| = 0 \tag{c}$$

$$req' = 1 \Leftrightarrow num(request', i) = 1 \lor |enter'| = 1 \tag{d}$$

$$use' = 1 \Leftrightarrow |exit'| = 1 \tag{e}$$

$$client' = i \Leftrightarrow |enter'| = 1 \lor |exit'| = 1 \lor num(return', i) = 1 \tag{a}$$

If $i \neq this$, these conditions clearly hold. So we now assume $i = this$ and show each condition in turn:

- From (1.a), (3), (4), (5), and (8) we have (a).
- From (5) and (1.a) we have $|exit| = 1$. Thus we have from (1.c) and (1.b) $num(return, i) = 0$. Thus (7) gives us (b).
- From $|exit| = 1$ and (1.a), we have $|enter| = 0$ from which (4) and (12) give us $|enter'| = 0$. Likewise (3), (4), (5) and (12) give us $|exit'| = 0$ and thus (c).
- From (1.d), (3), (4), and (11), we have (d).
- From (6), (3), (4), and $|exit'| = 0$, we have (e).
- Proof direction $\Rightarrow$: From (1.f), (10), and $|exit| = 1$, we have $num(return, i) = 0$ and from (7) thus $num(return', i) = 1$. Proof direction $\Leftarrow$: From (1.f) and $|exit| = 1$, we have $client = i$.

This completes the proof sketch that transition *exit* preserves the invariant.  □

As above examples demonstrate, proofs that transitions preserve invariants are generally tedious and error-prone. In practice, they are only feasible with sufficient support by (semi)-automatic reasoning tools. Initially some proofs usually fail due to inadequacies of the invariant; its correct formulation has to be elaborated simultaneously with the verification in multiple iterations. Indeed also the invariants in above examples were derived in this way with the help of an automatic checker.

## 9.6    Verifying Response

**Response**

We now turn our attention to the verification of liveness properties denoted by the following kind of LTL-formulas.

---

**Definition 9.20** (*Response Formula*) Let $F$ and $G$ be LTL-formulas that are adequate with respect to some variable typing $Vt$. We define the *response formula* $F \rightsquigarrow G$ (read as "$F$ leads to $G$") as the following syntactic abbreviation of an LTL-Formula:

$$F \rightsquigarrow G := \Box(F \Rightarrow \Diamond G)$$

---

In other words, $F \rightsquigarrow G$ holds if every time when $F$ holds, eventually also $G$ holds; this is clearly a liveness property. This pattern covers also the *termination* guarantee $P_x \rightsquigarrow E$: always if the computation starts with some input $x$ that satisfies precondition $P_x$, it will eventually terminate, as indicated by some "end condition" $E$. Together with the partial correctness guarantee $\Box(E \Rightarrow Q_{x,y})$ formulated in the previous section, this implies the *total correctness* guarantee $P_x \rightsquigarrow (E \wedge Q_{x,y})$: always, if the computation starts with input $x$ in a state that satisfies precondition $P_x$, it eventually terminates with output $y$ that satisfies postcondition $Q_{x,y}$.

**Proposition 9.3** (Propagation of Response) *Let F, G, and H be LTL-formulas that are adequate with respect to some variable typing Vt. Then we have the following logical consequence:*

$$(F \rightsquigarrow G) \wedge (G \rightsquigarrow H) \models (F \rightsquigarrow H)$$

Consequently the response relation is transitive. Thus to show $F \rightsquigarrow H$, it suffices to find some "intermediate" property $G$ for which we can show $F \rightsquigarrow G$ and $G \rightsquigarrow H$. Ultimately, however, we need a direct way of showing the validity of such a formula.

**Verifying Response by a Measure**

We start by introducing an auxiliary concept.

**Definition 9.21** (*Enabledness*)  Let  $lts = \langle \text{space:} S, \text{init:} I, \text{next:} R \rangle$  be a labeled transition system with some label set $L$, state space $S$, initial state condition $I$, and transition relation $R$.

- A transition with label $l \in L$ is *enabled* in a state $s \subseteq S$ if by the execution of the transition the system can step from $s$ to some state $s'$:

$$enabled\langle l, s \rangle^{lts} :\Leftrightarrow \exists s' \in S.\ R\langle l, s, s' \rangle$$

- The system *lts* is *enabled* in $s$, if there is some transition enabled in $s$:

$$enabled\langle s \rangle^{lts} :\Leftrightarrow \exists l \in L.\ enabled\langle l, s \rangle^{lts}$$

We may establish a liveness property in a similar way as we prove the termination of a loop (see Sect. 8.4), by introducing a measure that bounds the number of steps until the desired situation is reached.

**Theorem 9.3** (Verifying Response by a Measure) *Consider the labeled transition system lts = ⟨space:S, init:I, next:R⟩ with some label set L, state space S, initial state condition I, and transition relation R.*
*Let condition F ⊆ S and G ⊆ S be properties on S. Furthermore, let T : S → ℤ be a measure of lts with respect to F and G, i.e., a function from S to ℤ such that in every state that satisfies F but not G, the following two conditions hold:*

1. *The system is enabled.*
2. *Every transition leads to a state $s'$ that satisfies $G$ or satisfies $F$ again; in the later case, the transition decreases the value of $T$ without making it negative.*

*These requirements are formalized as follows:*

$$\forall s \in S.\ F\langle s \rangle \wedge \neg G\langle s \rangle \Rightarrow$$
$$enabled\langle s \rangle^{lts} \wedge$$
$$\left(\forall l \in L, s' \in S.\ R\langle l, s, s' \rangle \Rightarrow \right.$$
$$\left. G\langle s' \rangle \vee (F\langle s' \rangle \wedge 0 \le T(s') \wedge T(s') < T(s)))\right)$$

*Then in every run of lts every state that satisfies $F$ eventually leads to a state that satisfies $G$:*

$$\forall r \in [\![\, lts\, ]\!].\ [\![\, F \rightsquigarrow G\, ]\!]^{r^\omega}$$

Above theorem in essence states a rule how to prove a LTS formula $F \rightsquigarrow G$; this rule is only applicable if $F$ holds as long as $G$ does not hold. Furthermore, the rule requires the help of an upper bound $T$ on the number of transitions that, starting with a state that satisfies $F$ but not $G$, via arbitrarily many states that satisfy $F$ but not $G$, eventually lead to a state that satisfies $G$.

***Proof*** We sketch the proof of the validity of this proof rule. Assume a (possibly infinitely extended) run that satisfies $F$ at some point. We show that the assumption that (from that point on) $G$ never holds leads to a contradiction. From the second condition of the rule, since $G$ never holds, $F$ always holds. Then the first condition establishes that the system is always enabled, which implies that the run consists of infinitely many applications of some transition; the second conditions demands that these transitions infinitely often decrease the value of $T$ but do not make it negative, which contradicts the well-foundedness of the set of natural numbers.            □

Actually, like the rule for the termination of loops, also this rule can be generalized such that the domain of $T$ is any "well-founded" domain $D$, a set with a *well-founded relation* $\_ \prec \_ \subseteq D \times D$ (see Sect. 8.5).

**Example 9.20** Consider following pseudo-code for a system with two threads operating on a shared variable $s$:

```
var s:=0
A: loop          ||  B: loop
     0: wait s=0          0: wait s=1
```

```
1: ...                    1: ...
2: s:=1                   2: s:=0
```

The command `wait` $F$ blocks the execution of the thread unless formula $F$ holds; above system thus maintains mutual exclusion of the regions marked as "1" by letting both threads alternate in their access to the region.

This system can be modeled by the LTS $lts := \langle \text{space:} S, \text{init:} I, \text{next:} R \rangle$ with the following auxiliary definitions (the variables $a$ and $b$ model the values of the program counters of thread $A$ and $B$, respectively):

$$S := \mathbb{N}_3 \times \mathbb{N}_3 \times \mathbb{N}_2$$
$$I \langle a, b, s \rangle :\Leftrightarrow a = 0 \wedge b = 0 \wedge s = 0$$
$$R \langle l, \langle a, b, s \rangle, \langle a', b', s' \rangle \rangle :\Leftrightarrow$$
$$\quad (l = waitA \wedge a = 0 \wedge s = 0 \wedge a' = 1 \wedge s' = s \wedge b' = b) \vee$$
$$\quad (l = criticalA \wedge a = 1 \wedge a' = 2 \wedge b' = b \wedge s' = s) \vee$$
$$\quad (l = exitA \wedge a = 2 \wedge a' = 0 \wedge s' = 1 \wedge b' = b) \vee$$
$$\quad (l = waitB \wedge b = 0 \wedge s = 1 \wedge b' = 1 \wedge s' = s \wedge a' = a) \vee$$
$$\quad (l = criticalB \wedge b = 1 \wedge b' = 2 \wedge a' = a \wedge s' = s) \vee$$
$$\quad (l = exitB \wedge b = 2 \wedge b' = 0 \wedge s' = 0 \wedge a' = a)$$

Our goal is now to verify the response property

$$a = 0 \rightsquigarrow a = 0 \wedge s = 0$$

which states that if thread $A$ requests access to the critical region, it is eventually granted that access. This is the crucial problem in verifying the progress cycle

$$a = 0 \rightsquigarrow a = 0 \wedge s = 0$$
$$a = 0 \wedge s = 0 \rightsquigarrow a = 1$$
$$a = 1 \rightsquigarrow a = 2$$
$$a = 2 \rightsquigarrow a = 0$$

from which Proposition 9.3 implies that thread $A$ loops forever. The verification of the last three properties is simple: in the relevant prestates only one transition is enabled and the execution of this transition immediately leads to the desired poststate.

For verifying the first property, we define the following measure:

$$T(a, b, s) := \text{if } s = 0 \text{ then } 0 \text{ else if } b = 2 \text{ then } 1 \text{ else if } b = 1 \text{ then } 2 \text{ else } 3$$

The core idea of this measure is that it "counts" the number of steps that the execution of thread $B$ needs to reach a state with $s = 0$; this number can be deduced from the value of program counter $b$.

For this measure, it now suffices to prove the requirements stated in Theorem 9.3. For this, we take arbitrary $a \in \mathbb{N}_3, b \in \mathbb{N}_3, s \in \mathbb{N}_2$, for which we assume the following:

$$a = 0 \land \neg(a = 0 \land s = 0) \tag{1}$$

The first requirement is now as follows:

$$enabled^{lts}\langle a, b, s \rangle \tag{a}$$

It is easy to prove this goal: from (1), we have $s = 1$; since $b \in \mathbb{N}_3$, thus one of the transitions *waitB*, *criticalB*, or *exitB* of thread $B$ is enabled.

Now we take arbitrary label $l$ and $a' \in \mathbb{N}_3, b' \in \mathbb{N}_3, s' \in \mathbb{N}_2$, for which we assume:

$$R\langle l, \langle a, b, s \rangle, \langle a', b', s' \rangle \rangle \tag{2}$$
$$\neg(a' = 0 \land s' = 0) \tag{3}$$

We have to show the following goals:

$$a' = 0 \tag{b}$$
$$0 \leq T(a', b', s') \tag{c}$$
$$T(a', b', s') < T(a, b, c) \tag{d}$$

From (1) we have $a = 0$ and $s = 1$; from the definition of $R$, we thus have (b). From the definition of $T$, we immediately have (c).

It remains to show (d). From (1), we know that no transition of thread $A$ is enabled; therefore (2) implies that $l = waitB$, $l = criticalB$, or $l = exitB$:

- If $l = waitB$, we have $s = 1, b = 0, s' = 1$, and $b' = 1$. Thus we have $T(a', b', s') = 2 < 3 = T(a, b, s)$.
- If $l = criticalB$, we have $s = 1$, $b = 1$, $s' = 1$, and $b' = 2$. Thus we have $T(a', b', s') = 1 < 2 = T(a, b, s)$.
- If $l = waitB$, we have $s = 1, b = 2, s' = 0$, and $b' = 0$. Thus we have $T(a', b', s') = 0 < 1 = T(a, b, s)$.

Thus in all three cases we have (d). The fact that in the last two cases we have $s = 1$ is actually a consequence of the system invariant $\Box((s = 0 \Rightarrow b = 0) \land (s = 1 \Rightarrow a = 0))$. The verification of this invariant is easy; we omit the details. $\qquad\Box$

### Verifying Response by Fairness

Not all liveness arguments can be based on explicit measures. For another kind of such arguments, we introduce the following concept.

**Definition 9.22** (*Weak and Strong Fairness*) Let $lts = \langle \text{space}{:}S, \text{init}{:}I, \text{next}{:}R \rangle$ be a labeled transition system over some label set $L$ with state space $S$, initial state condition $I$, and transition relation $R$. Let $r \in Run^S$ be a run of $lts$, i.e., let it satisfy $[\![\,lts\,]\!]\langle r \rangle$. Let $l \in L$ be a transition label.

We define the following two notions of *fairness*:

- Transition $l$ is scheduled by run $r$ with *weak fairness*, written as $WF\langle l, r \rangle^{lts}$ if and only if the following is true: if $l$ is (in the infinite extension $r^\omega$ of $r$) eventually permanently enabled, then $l$ is infinitely often executed:

$$WF\langle l, r \rangle^{lts} :\Leftrightarrow (\exists i \in \mathbb{N}. \, \forall j \in \mathbb{N}. \, j \geq i \Rightarrow enabled\langle l, r^\omega(j) \rangle^{lts}) \Rightarrow$$
$$(\forall i \in \mathbb{N}. \, \exists j \in \mathbb{N}. \, j \geq i \wedge R\langle l, r^\omega(j), r^\omega(j+1) \rangle)$$

- Transition $l$ is scheduled by run $r$ with *strong fairness*, written as $SF\langle l, r \rangle^{lts}$ if and only if the following is true: if $l$ is (in the infinite extension $r^\omega$ of $r$) infinitely often enabled, then $l$ is infinitely often executed:

$$SF\langle l, r \rangle^{lts} :\Leftrightarrow (\forall i \in \mathbb{N}. \, \exists j \in \mathbb{N}. \, j \geq i \wedge enabled\langle l, r^\omega(j) \rangle^{lts}) \Rightarrow$$
$$(\forall i \in \mathbb{N}. \, \exists j \in \mathbb{N}. \, j \geq i \wedge R\langle l, r^\omega(j), r^\omega(j+1) \rangle)$$

To be scheduled with strong fairness, it suffices that a transition is infinitely often enabled. However, to be scheduled with weak fairness, it is necessary that the transition is eventually permanently enabled, which is a stronger requirement (to be eventually permanently enabled implies to be infinitely often enabled but not vice versa). Consequently, strong fairness indeed represents the stronger notion of fairness: if a run schedules a transition with strong fairness, it also schedules it with weak fairness (but not vice versa).

Now another kind of liveness arguments depends on the fact that the desired state is reached by executing a transition which is scheduled with some notion of fairness.

**Theorem 9.4** (Verifying Response by Fairness) *Consider the labeled transition system* $lts = \langle \text{space}{:}S, \text{init}{:}I, \text{next}{:}R \rangle$ *over some label set* $L$ *with state space* $S$, *initial state condition* $I$, *and transition relation* $R$. *Let condition* $F \subseteq S$ *and* $G \subseteq S$ *be properties on* $S$.

- **Weak Fairness**: *Let* $\bar{l} \in L$ *be the label of a "lucky" transition, i.e., a transition such that in every state that satisfies* $F$ *but not* $G$, *the following three conditions hold:*

1. *The execution of transition $\bar{l}$ leads to a state $s'$ that satisfies G.*
2. *Transition $\bar{l}$ is enabled.*
3. *Every transition leads to a state $s'$ that satisfies G or satisfies F again:*

*These requirements are formalized as follows:*

$$\forall s \in S. \ F\langle s \rangle \wedge \neg G\langle s \rangle \Rightarrow$$
$$(\forall s' \in S. \ R\langle \bar{l}, s, s' \rangle \Rightarrow G\langle s' \rangle) \wedge$$
$$enabled\langle \bar{l}, s \rangle^{lts} \wedge$$
$$\big( \forall l \in L, s' \in S. \ R\langle l, s, s' \rangle \Rightarrow G\langle s' \rangle \vee F\langle s' \rangle \big)$$

*Then in every run r of lts that schedules transition $\bar{l}$ with weak fairness every state that satisfies F eventually leads to a state that satisfies G:*

$$\forall r \in [\![ \ lts \ ]\!]. \ WF\langle \bar{l}, r \rangle^{lts} \Rightarrow [\![ \ F \ \rightsquigarrow G \ ]\!]^{r^{\omega}}$$

- **Strong Fairness**: *Let $\bar{l} \in L$ be the label of a "lucky" transition with a measure $T : S \rightarrow \mathbb{Z}$, i.e., a transition such that in every state that satisfies F but not G, the following three conditions hold:*

1. *The execution of transition $\bar{l}$ leads to a state $s'$ that satisfies G.*
2. *The system is enabled.*
3. *Every transition leads to a state $s'$ that satisfies G or satisfies F again; in the later case this transition enables transition $\bar{l}$ or it decreases the value of T without making it negative.*

*These requirements are formalized as follows:*

$$\forall s \in S. \ F\langle s \rangle \wedge \neg G\langle s \rangle \Rightarrow$$
$$(\forall s' \in S. \ R\langle \bar{l}, s, s' \rangle \Rightarrow G\langle s' \rangle) \wedge$$
$$enabled\langle s \rangle^{lts} \wedge$$
$$\big( \forall l \in L, s' \in S. \ R\langle l, s, s' \rangle \Rightarrow$$
$$G(s') \vee (F\langle s' \rangle \wedge (enabled\langle \bar{l}, s \rangle^{lts} \vee (0 \leq T(s') \wedge T(s') < T(s)))) \big)$$

*Then in every run r of lts that schedules transition $\bar{l}$ with strong fairness every state that satisfies F eventually leads to a state that satisfies G:*

$$\forall r \in [\![ \ lts \ ]\!]. \ SF\langle \bar{l}, r \rangle^{lts} \Rightarrow [\![ \ F \ \rightsquigarrow G \ ]\!]^{r^{\omega}}$$

This theorem gives two rules how to prove the LTS formula $F \leadsto G$; these rules require that $F$ holds as long as $G$ does not hold.

The first rule is applicable in a system that schedules (at least) some "lucky" transition $\bar{t}$ with weak fairness; this requires to show that the lucky transition is permanently enabled as long as $G$ does not hold.

***Proof*** We sketch the proof of the validity of the first part of the theorem. Assume a (possibly infinitely extended) run that satisfies $F$ at some point. We show that the assumption that, from that point on, $G$ never holds leads to a contradiction. From the third part of the rule, since $G$ never holds, $F$ always holds and thus the second part of the rule establishes that transition $\bar{l}$ is always enabled. Therefore the weak fairness condition ensures that this transition is eventually executed; by the first part of the rule this establishes $G$, which contradicts the assumption.                □

Analogously, the second rule is applicable in a system that schedules (at least) some "lucky" transition $\bar{t}$ with strong fairness; while this does not require the lucky transition to be permanently enabled as long as $G$ does not hold, it does not allow the transition to be never enabled. The rule therefore demands an upper bound $T$ on the number of transitions that do not enable the lucky transition; please note that this is much weaker than requiring an upper bound on the number of times that the lucky transition is not executed.

***Proof*** We sketch the proof of the validity of the second part of the theorem. Assume a (possibly infinitely extended) run that satisfies $F$ at some point. We show that the assumption that, from that point on, $G$ never holds leads to a contradiction. From the third part of the rule, since $G$ never holds, $F$ always holds. If there are infinitely many occurrences where transition $\bar{l}$ is enabled, the strong fairness condition ensures that the transition is eventually executed; by the first part of the rule this establishes $G$, which contradicts the assumption. However, if there are only finitely many occurrences where transition $\bar{l}$ is enabled, the second part of the rule demands that after the last such occurrence there are still infinitely many transitions following. The last part of the rule demands that these transitions infinitely often decrease the value of the term $T$ without making it negative, which again represents a contradiction.   □

**Example 9.21** Consider the following pseudo-code for a system with two threads that independently increment shared variables $a$ respectively $b$ forever:

```
var a:=0, b:=0
A: loop           || B: loop
      a := a+1           b := b+1
```

This system can be modeled by the following LTS $lts := \langle \text{space}:S, \text{init}:I, \text{next}:R\rangle$:

$$S := \mathbb{N} \times \mathbb{N}$$
$$I\langle a, b\rangle :\Leftrightarrow a = 0 \wedge b = 0$$
$$R\langle l, \langle a, b\rangle, \langle a', b'\rangle\rangle :\Leftrightarrow$$
$$(l = incA \wedge a' = a + 1 \wedge b' = b) \vee (l = incB \wedge b' = b + 1 \wedge a' = a)$$

Our goal is to verify the simple progress property

$$a = 0 \rightsquigarrow a = 1$$

which due to the initial state condition $I$ ensures $\Diamond a = 1$. However, while it seems self-evident that the execution of the system eventually yields the state $a = 1$, this is actually *not* guaranteed: without any further assumptions on the fairness of system execution, it may be the case that only thread $B$ is executed, which forever increments $b$ but leaves $a = 0$ unchanged.

To make the property true, it is critical that eventually the "lucky" transition $incA$ is executed that moves the state $a = 0$ to the state $a = 1$; therefore we assume that every run $r$ schedules $incA$ with weak fairness:

$$WF\langle incA, r\rangle^{lts}$$

The system has clearly only infinite runs, i.e., we have $r = r^\omega$. Therefore, to show that $r$ satisfies $a = 0 \rightsquigarrow a = 1$, according to the first rule of Theorem 9.4 it suffices to take $a \in \mathbb{N}$ and $b \in \mathbb{N}$ with $a = 0$ (which implies $a \neq 1$) and show the following properties for arbitrary $a' \in \mathbb{N}$, $b' \in \mathbb{N}$ and transition label $l$:

$$R\langle incA, \langle a, b\rangle, \langle a', b'\rangle\rangle \Rightarrow a' = 1 \tag{a}$$
$$enabled\langle incA, \langle a, b\rangle\rangle^{lts} \tag{b}$$
$$R\langle l, \langle a, b\rangle, \langle a', b'\rangle\rangle \Rightarrow a' = 1 \vee a' = 0 \tag{c}$$

From $a = 0$ and the definition of $R$, we clearly have (a) and (c). Also from the definition of $R$, $incA$ is always enabled, i.e., we have $enabled\langle incA, \langle a, b\rangle\rangle^{lts} \Leftrightarrow \text{true}$ and thus (b). $\qquad\square$

As above example demonstrates, the assumption of weak fairness suffices if the lucky transition, once enabled, is never disabled by another transition. However, there are also systems where this assumption is not satisfied.

**Example 9.22**  Consider again the following pseudo-code presented in Example 9.17 for a system with two threads operating on a shared semaphore $s$:

```
var s:=1
 A: loop            || B: loop
      0: lock(s)          0: lock(s)
      1: ...              1: ...
      2: unlock(s)        2: unlock(s)
```

This system was modeled by the following LTS $lts := \langle \text{space}:S, \text{init}:I, \text{next}:R \rangle$:

$$S := \mathbb{N} \times \mathbb{N} \times \mathbb{N}$$
$$I \langle a, b, s \rangle :\Leftrightarrow a = 0 \wedge b = 0 \wedge s = 1$$
$$R \langle l, \langle a, b, s \rangle, \langle a', b', s' \rangle \rangle :\Leftrightarrow$$
$$(l = lockA \wedge a = 0 \wedge s = 1 \wedge a' = 1 \wedge s' = 0 \wedge b' = b) \vee$$
$$(l = criticalA \wedge a = 1 \wedge a' = 2 \wedge b' = b \wedge s' = s) \vee$$
$$(l = unlockA \wedge a = 2 \wedge a' = 0 \wedge s' = 1 \wedge b' = b) \vee$$
$$(l = lockB \wedge b = 0 \wedge s = 1 \wedge b' = 1 \wedge s' = 0 \wedge a' = a) \vee$$
$$(l = criticalB \wedge b = 1 \wedge b' = 2 \wedge a' = a \wedge s' = s) \vee$$
$$(l = unlockB \wedge b = 2 \wedge b' = 0 \wedge s' = 1 \wedge a' = a)$$

Our goal is to verify the progress property

$$a = 0 \rightsquigarrow a = 1$$

i.e., if thread $A$ is at the position where it may enter the critical region, it will eventually do so. To make this proposition true, it is necessary to execute the "lucky" transition $lockA$. However, in contrast to Example 9.21, this is not guaranteed if we just assume that every run $r$ schedules this transition with weak fairness: if at one point the enabling condition $a = 0$ and $s = 0$ is satisfied, it may further on be invalidated by thread $B$ whose transition $lockB$ sets $s$ to 0.

Therefore we assume as a stronger requirement that $r$ schedules transition $lockA$ with strong fairness:

$$SF \langle lockA, r \rangle^{lts}$$

We can show that the system has only infinite runs (i.e., it does not deadlock) such that $r = r^\omega$. Therefore, to show that $r$ satisfies $a = 0 \rightsquigarrow a = 1$, according to the second rule of Theorem 9.4 it suffices to take $a \in \mathbb{N}$, $b \in \mathbb{N}$, $s \in \mathbb{N}$ with

$$a = 0 \tag{1}$$

(which implies $a \neq 1$) and show the following properties for arbitrary $a' \in \mathbb{N}, b' \in \mathbb{N}$, $s' \in \mathbb{N}$ and transition label $l$:

$$R\langle lockA, \langle a, b, s \rangle, \langle a', b', s' \rangle \rangle \Rightarrow a' = 1 \qquad \text{(a)}$$

$$enabled\langle a, b, s \rangle^{lts} \qquad \text{(b)}$$

$$R\langle l, \langle a, b \rangle, \langle a', b' \rangle \rangle \Rightarrow a' = 1 \vee$$
$$(a' = 0 \wedge (enabled\langle lockA, \langle a, b, s \rangle \rangle^{lts} \vee$$
$$(0 \leq T(a', b', s') \wedge T(a', b', s') < T(a, b, s)))) \qquad \text{(c)}$$

From the definition of $R$, clearly (a) and (b) hold. For condition (c), we define the following measure:

$$T(a, b, s) := \text{if } s = 1 \text{ then } 0 \text{ else if } b = 2 \text{ then } 1 \text{ else if } b = 1 \text{ then } 2 \text{ else } 3$$

Similar to Example 9.20, the core idea of this measure is that it "counts" the number of steps that the execution of thread $B$ needs to reach a state with $s = 1$, i.e., a state in which the semaphore is not locked. To show (c), we assume

$$R\langle l, \langle a, b \rangle, \langle a', b' \rangle \rangle \qquad \text{(2)}$$
$$a' \neq 1 \qquad \text{(3)}$$

and show

$$a' = 0 \qquad \text{(d)}$$

$$enabled\langle lockA, \langle a, b, s \rangle \rangle^{lts} \vee$$
$$(0 \leq T(a', b', s') \wedge T(a', b', s') < T(a, b, s)) \qquad \text{(e)}$$

From (3), we have (d) by a simple system invariant. To show (e), we assume

$$\neg enabled\langle lockA, \langle a, b, s \rangle \rangle^{lts} \qquad \text{(4)}$$

and show

$$0 \leq T(a', b', s') \qquad \text{(f)}$$
$$T(a', b', s') < T(a, b, s)) \qquad \text{(g)}$$

From the definition of $T$, we clearly have (f). It remains to show (g). From (2) and (4), we know $l \neq lockA$. From (1), the definition of $R$ implies $l \neq critA$ and $l \neq critB$. From (1) and (4), we have $s \neq 1$ (and thus by a simple system invariant $s = 0$) and therefore from the definition of $R$ also $l \neq lockB$. So there remain two cases:

- Case $l = critB$: we have $s = 0$, $b = 1$, $s' = 0$, and $b' = 2$. Thus we have $T(a', b', s') = 1 < 2 = T(a, b, s)$.

- Case $l = unlockB$: we have $s = 0$, $b = 2$, $s' = 1$, and $b' = 0$. Thus we have $T(a', b', s') = 0 < 1 = T(a, b, s)$.

This completes the proof of (g). □

As above examples demonstrate, weak scheduling suffices to adequately allocate processor time to every thread: if the program counter of a thread refers to an action without further precondition, this action is eventually executed. Strong scheduling, however, is needed to guarantee the fair outcome of a "competition" for a resource: if a thread permanently participates in such a competition, it eventually also "wins". It is the responsibility of the implementation to guarantee the adequate fairness of corresponding scheduling mechanisms for processor time respectively resources.

### Fairness as an LTL-Formula

Theorem 9.4 places in the conditions

$$\forall r \in [\![ lts ]\!].\ WF\langle \bar{l}, r \rangle^{lts} \Rightarrow [\![ F \leadsto G ]\!]^{r^\omega}$$
$$\forall r \in [\![ lts ]\!].\ SF\langle \bar{l}, r \rangle^{lts} \Rightarrow [\![ F \leadsto G ]\!]^{r^\omega}$$

the constraints of weak fairness respectively strong fairness on the "meta-level" of the semantics. However, it is also possible to place them on the "object-level" of the LTL-formula that is to be verified. Assume that $E_l$ is a formula in first-order logic that expresses that transition $l$ is enabled and that $X_l$ is a formula in first-order logic that expresses that transition $l$ has been executed, i.e.:

$$[\![ E_l ]\!]_s := \exists s' \in S.\ R\langle l, s, s' \rangle$$
$$[\![ X_l ]\!]_{s'} := \exists s \in S.\ R\langle l, s, s' \rangle$$

We then can define the LTL-formulas $WF_l$ and $SF_l$ that denote the constraints of weak fairness respectively strong fairness with respect to transition $l$ as follows:

$$WF_l := (\Diamond \Box E_l) \Rightarrow (\Box \Diamond X_l)$$
$$SF_l := (\Box \Diamond E_l) \Rightarrow (\Box \Diamond X_l)$$

Here $WF_l$ states that, if transition $l$ is stable (eventually permanently enabled), it is infinitely often executed. Likewise, $SF_l$ states that, if transition $l$ is infinitely often enabled, it is infinitely often executed. Thus the semantics of the formulas matches the definition of the corresponding relations given in Definition 9.22. With the help of these formulas, it becomes now possible to express the fairness constraints as part of the property specifications. Thus above conditions can be written as follows:

$$\forall r \in [\![ lts ]\!].\ [\![ WF_l \Rightarrow (F \leadsto G) ]\!]^{r^\omega}$$
$$\forall r \in [\![ lts ]\!].\ [\![ SF_l \Rightarrow (F \leadsto G) ]\!]^{r^\omega}$$

This is equivalent to stating

$$lts \models (WF_l \Rightarrow (F \rightsquigarrow G))$$
$$lts \models (SF_l \Rightarrow (F \rightsquigarrow G))$$

i.e., to verify that the labeled transition system satisfies the extended specifications.

**Example 9.23** Consider the system verified in Example 9.19. We define

$$SF_{lockA} := (\Box \Diamond a = 0) \Rightarrow (\Box \Diamond a = 1)$$

Then the LTL-formula

$$SF_{lockA} \Rightarrow (a = 0 \rightsquigarrow a = 1)$$

states that thread $A$ infinitely often enters the critical region under the assumption that every run of the system schedules transition *lockA* with strong fairness.   $\Box$

Since we may formulate fairness constraints as part of LTL specifications, model checkers for LTL may cope without special "builtin" support for dealing with fairness.

## 9.7   The Refinement of Systems

We now turn our attention to the *refinement* of a labeled transition system $lts_1$ to a system $lts_2$. Here we can consider $lts_1$ as an *abstract* system (a "model") that describes the desired system behavior without going into low-level details of how this behavior is achieved. On the other hand, $lts_2$ is a (more) *concrete* system (an "implementation" of the model) that describes the missing details of how to achieve this behavior; thus $lts_2$ can be considered as the system's *implementation*.

To develop a notion of refinement that is adequate to the execution of systems, we have to consider two aspects:

- The state spaces of the two systems may be different; in particular, the concrete implementation may extend the set of variables of the abstract model by additional variables that are necessary for the implementation.
- Due to its richer state space, the concrete implementation may have transitions that the abstract model does not have; however, an additional transition must not change any of the variables that appear in the model.

Considering these two aspects, every run of the concrete implementation (in a certain sense) "simulates" some run of the abstract model such that the implementation has "in essence" the same behavior as the model.

**Example 9.24** Consider the labeled transition systems

$$HourClock := \langle \text{space:}S_1, \text{init:}I_1, \text{next:}R_1 \rangle$$
$$MinuteClock := \langle \text{space:}S_2, \text{init:}I_2, \text{next:}R_2 \rangle$$

defined as follows:

$$S_1 := \mathbb{N}_{24}$$
$$I_1\langle h \rangle :\Leftrightarrow h = 0$$
$$R_1\langle l, h, h' \rangle :\Leftrightarrow$$
$$(l = tick \land h < 23 \land h' = h + 1) \lor$$
$$(l = tock \land h = 23 \land h' = 0)$$

$$S_2 := \mathbb{N}_{24} \times \mathbb{N}_{60}$$
$$I_2\langle h, m \rangle :\Leftrightarrow h = 0 \land m = 0$$
$$R_2\langle l, \langle h, m \rangle, \langle h', m' \rangle \rangle :\Leftrightarrow$$
$$(l = tick1 \land m < 59 \land h' = h \land m' = m + 1) \lor$$
$$(l = tick2 \land m = 59 \land h < 23 \land h' = h + 1 \land m' = 0) \lor$$
$$(l = tock \land h = 59 \land h = 23 \land h' = 0 \land m' = 0)$$

These systems describe the execution of two clocks, one clock with only one hand for the hour, one clock with two hands, both for the hour and for the minute.

Then *MinuteClock* refines *HourClock* in the following sense. Take the only run of system *MinuteClock*:

$$\langle h, m \rangle = \langle 0, 0 \rangle \xrightarrow{tick1} \langle h, m \rangle = \langle 0, 1 \rangle \xrightarrow{tick1} \cdots \xrightarrow{tick1} \langle h, m \rangle = \langle 0, 59 \rangle$$
$$\xrightarrow{tick2} \langle h, m \rangle = \langle 1, 0 \rangle \xrightarrow{tick1} \langle h, m \rangle = \langle 1, 1 \rangle \xrightarrow{tick1} \cdots \xrightarrow{tick1} \langle h, m \rangle = \langle 1, 59 \rangle$$
$$\xrightarrow{tick2} \cdots$$

$$\xrightarrow{tick2} \langle h, m \rangle = \langle 23, 0 \rangle \xrightarrow{tick1} \langle h, m \rangle = \langle 23, 1 \rangle \xrightarrow{tick1} \cdots \xrightarrow{tick1} \langle h, m \rangle = \langle 23, 59 \rangle$$
$$\xrightarrow{tock} \langle h, m \rangle = \langle 0, 0 \rangle \xrightarrow{tick1} \cdots$$
$$\xrightarrow{tick2} \cdots$$

This run simulates the following run of *HourClock*:

$$h = 0 \xrightarrow{tick} h = 1 \xrightarrow{tick} \cdots \xrightarrow{tick} h = 23 \xrightarrow{tock} h = 0 \xrightarrow{tick} \cdots$$

This "simulation" maps every state $\langle h, m \rangle$ of *MinuteClock* to the corresponding state $h$ of *HourClock*. The *MinuteClock* transitions labeled with *tick2* and *tock* simulate the corresponding *HourClock* transitions labeled with *tick* and *tock*: they update $h$

in the same way and thus (ignoring the extra state component $m = 0$) yield the same sequence of states. The *MinuteClock* transitions labeled with *tick1* have no counterpart in *HourClock*; however, this does not matter, because they leave the value of $h$ unchanged and are thus "invisible" to *HourClock* (such transitions that do not change the state of the system are also called *stuttering*). □

Having sketched the basic ideas, we are now going to formalize the central notion of this section.

**Definition 9.23** (*System Refinement*) Let $lts_1 \in LTS^{L_1, S_1}$ and $lts_2 \in LTS^{L_2, S_2}$ be labeled transition systems with state spaces $S_1$ respectively $S_2$ and label sets $L_1$ respectively $L_2$. Let $a : S_2 \to S_1$ be an "abstraction" function from the state space of $lts_2$ to the state space of $lts_1$.

We define the relation $lts_2 \sqsubseteq^a lts_1$ (read: "$lts_2$ refines $lts_1$ under $a$") as follows:

$$ \_ \sqsubseteq^{\smile} \_ \subseteq LTS^{L_2, S_2} \times (S_2 \to S_1) \times LTS^{L_1, S_1} $$

$$ lts_2 \sqsubseteq^a lts_1 :\Leftrightarrow $$
$$ (\forall s \in S_2.\ lts_2.\text{init}\langle s \rangle \Rightarrow lts_1.\text{init}\langle a(s) \rangle) \wedge $$
$$ (\forall l_2 \in L_2, s, s' \in S_2.\ \exists l_1 \in L_1. $$
$$ lts_2.\text{next}\langle l_2, s, s' \rangle \Rightarrow lts_1.\text{next}\langle l_1, a(s), a(s') \rangle \vee a(s) = a(s')) $$

In other words, $lts_2$ refines $lts_1$ under $a$ if $a$ maps every initial state $s$ of $lts_2$ to an initial state $a(s)$ of $lts_1$ and every pair $s, s'$ of states with a transition in $lts_2$ to a pair $a(s), a(s')$ of states that have a transition in $lts_1$ or are identical. Furthermore we define the relation $lts_2 \sqsubseteq lts_1$ (read: "$lts_2$ refines $lts_1$") as follows:

$$ \_ \sqsubseteq \_ \subseteq LTS^{L_2, S_2} \times LTS^{L_1, S_1} $$

$$ lts_2 \sqsubseteq lts_1 :\Leftrightarrow \exists a : S_2 \to S_1.\ lts_2 \sqsubseteq^a lts_1 $$

In other words, $lts_2$ refines $lts_1$ if $lts_2$ refines $lts_1$ refines $a$ under some abstraction $a$.

**Example 9.25** We show *MinuteClock* $\sqsubseteq$ *HourClock* for the systems *MinuteClock* and *HourClock* defined in Example 9.24. For this we define $a(h, m) := h$. Then, for arbitrary $h \in \mathbb{N}_{24}$ and $m \in \mathbb{N}_{60}$, we have $I_2\langle h, m \rangle \Leftrightarrow h = 0 \Rightarrow I_1\langle h \rangle \Leftrightarrow I_1\langle a(h, m) \rangle$. Now take arbitrary label $l_2$, $h, h' \in \mathbb{N}_{24}$, and $m, m' \in \mathbb{N}_{60}$. We assume

$$ R_2\langle l_2, \langle h, m \rangle, \langle h', m' \rangle \rangle $$

and show for some label $l_1$

$$R_1 \langle l_1, a(\langle h, m \rangle), a(\langle h', m' \rangle) \rangle \vee a(h, m) = a'(h', m')$$

i.e., according to the definition of $a$,

$$R_1 \langle l_2, h, h' \rangle \vee h = h'$$

From $R_2 \langle l_2, \langle h, m \rangle, \langle h', m' \rangle \rangle$, we have the following cases:

- Case ($l_2 = tick1 \wedge m < 59 \wedge h' = h \wedge m' = m + 1$): we have $h = h'$.
- Case ($l_2 = tick2 \wedge m = 59 \wedge h < 23 \wedge h' = h + 1 \wedge m' = 0$):   we   have $R_1 \langle tick, h, h' \rangle$.
- Case ($l_2 = tock \wedge m = 59 \wedge h = 23 \wedge h' = 0 \wedge m' = 0$): we have $R_1 \langle tock, h, h' \rangle$.

This completes the proof.                                                    □

**Example 9.26** Consider the following pseudo-code for a system with two threads that starting with variable values $a = 0$ and $b = 0$ forever add 2 to the values of the variables; thus these variables can only take even values:

```
var a:=0, b:=0
A: loop          ||  B: loop
      a := a+2            b := b+2
```

This system can be modeled by the LTS $AB := \langle \text{space:} S_1, \text{init:} I_1, \text{next:} R_1 \rangle$ defined as follows:

$$S_1 := \mathbb{N} \times \mathbb{N}$$
$$I_1 \langle a, b \rangle :\Leftrightarrow a = 0 \wedge b = 0$$
$$R_1 \langle l, \langle a, b \rangle, \langle a', b' \rangle \rangle :\Leftrightarrow$$
$$(l = incA \wedge a' = a + 2 \wedge b' = b) \vee (l = incB \wedge b' = b + 2 \wedge a' = a)$$

Now consider the following pseudo-code for a similar system:

```
var a:=0, b:=0, s:=1
A: loop              ||   B: loop
      0: lock(s)               0: lock(s)
      1: c:=a                  1: c:=b
      2: a:=c+1                2: b:=c+1
      3: unlock(s)             3: unlock(s)
      4: a:=a+1                4: b:=b+1
```

This system also consists of two threads incrementing $a$ and $b$. In contrast to the system above, however, every increment operation is split into two transitions that use a shared variable $c$; access to this variable is only allowed in a critical section protected by a semaphore $s$. Furthermore, in the critical section $a$ respectively $b$ are only incremented by 1, another increment by one takes place separately.

This system can be modeled by the following LTS $ABC := \langle \text{space:} S_2, \text{init:} I_2, \text{next:} R_2 \rangle$ where variables $p$ and $q$ represent the program counters of thread $A$ respectively $B$:

$$S_2 := \mathbb{N} \times \mathbb{N} \times \mathbb{N} \times \mathbb{N} \times \mathbb{N} \times \mathbb{N}$$

$$I_2 \langle p, q, a, b, c, s \rangle :\Leftrightarrow p = 0 \land q = 0 \land a = 0 \land b = 0 \land s = 1$$

$$R_2 \langle l, \langle p, q, a, b, c, s \rangle, \langle p', q', a', b', c', s' \rangle \rangle :\Leftrightarrow$$

$(l = lockA \land p = 0 \land p' = 1 \land s = 1 \land s' = 0 \land q' = q \land a' = a \land b' = b \land c' = c) \lor$

$(l = readA \land p = 1 \land p' = 2 \land c' = a \land q' = q \land a' = a \land b' = b \land s' = s) \lor$

$(l = writeA \land p = 2 \land p' = 3 \land a' = c + 1 \land q' = q \land b' = b \land c' = c \land s' = s) \lor$

$(l = unlockA \land p = 3 \land p' = 4 \land s' = 1 \land q' = q \land a' = a \land b' = b \land c' = c) \lor$

$(l = incA \land p = 4 \land p' = 0 \land a' = a + 1 \land q' = q \land b' = b \land c' = c \land s' = s) \lor$

$(l = lockB \land q = 0 \land q' = 1 \land s = 1 \land s' = 0 \land p' = p \land a' = a \land b' = b \land c' = c) \lor$

$(l = readB \land q = 1 \land q' = 2 \land c' = b \land p' = p \land a' = a \land b' = b \land s' = s) \lor$

$(l = writeB \land q = 2 \land q' = 3 \land b' = c + 1 \land p' = p \land a' = a \land c' = c \land s' = s) \lor$

$(l = unlockB \land q = 3 \land q' = 4 \land s' = 1 \land p' = p \land a' = a \land b' = b \land c' = c)$

$(l = incB \land q = 4 \land q' = 0 \land b' = b + 1 \land p' = p \land b' = b \land c' = c \land s' = s) \lor$

Our goal is to prove $ABC \sqsubseteq AB$, i.e., we have to show that the "increment 1" updates performed by $ABC$ simulate the "increment 2" updates performed by $AB$. For this we define the following abstraction $a$ that maps states of $ABC$ to states of $AB$:

$$a(p, q, a, b, c, s) :=$$

let $a_0 = $ if $p \leq 2$ then $a$ else $a - 1$, $b_0 = $ if $q \leq 2$ then $b$ then $b - 1$ in $\langle a_0, b_0 \rangle$

The function thus maps the values of the variables $a$ and $b$ in $ABC$ to those in $AB$ except in those states of $ABC$ where (the simulations of) the "increment 2" updates have not yet been completed; in these states the the original value of $a$ respectively $b$ (before the incomplete update) is used instead.

In the following proof, let $p, q, a, b, c, s, p', q', a' b', c', s', l_2$ be arbitrary values of the respective domains.

We have $I_2 \langle p, q, a, b, c, s \rangle \Rightarrow I_1 \langle a(p, q, a, b, c, s) \rangle$, because $I_2 \langle p, q, a, b, c, s \rangle$ implies $p = 0$, $q = 0$, $a = 0$, and $b = 0$, and thus $a(p, q, a, b, c, s) = \langle a, b \rangle = \langle 0, 0 \rangle$.

We now assume

$$R_2 \langle l_2, \langle p, q, a, b, c, s \rangle, \langle p', q', a', b', c', s' \rangle \rangle$$

and show for some label $l_1$

$$R_1\langle l_1, a(p, q, a, b, c, s), a(p', q', a', b', c', s')\rangle \vee$$
$$a(p, q, a, b, c, s) = a(p', q', a', b', c', s')$$

For this we assume the invariant $\Box Inv\langle p, q, a, b, c, s\rangle$ of $ABC$ (which has to be verified separately) where

$$Inv\langle p, q, a, b, c, s\rangle :\Leftrightarrow$$
$$s \leq 1 \wedge p \leq 4 \wedge q \leq 4 \wedge$$
$$(p \leq 2 \Leftrightarrow 2|a) \wedge (q \leq 2 \Leftrightarrow 2|b) \wedge$$
$$(p = 2 \vee p = 3 \Rightarrow c = a) \wedge (q = 2 \vee q = 3 \Rightarrow c = b) \wedge$$
$$(s = 1 \Leftrightarrow (p = 0 \vee p = 4) \wedge (q = 0 \vee q = 4)) \wedge$$
$$(p = 0 \vee p = 4 \vee q = 0 \vee q = 4)$$

Let $a_0, b_0$ such that $\langle a_0, b_0\rangle = a(p, q, a, b, c, s)$ and $a_0', b_0'$ such that $\langle a_0', b_0'\rangle = a(p', q', a', b', c', s')$ (to show $a(p, q, a, b, c, s) = a(p', q', a', b', c', s')$), it thus suffices to show $a_0 = a_0'$ and $b_0 = b_0'$). From $R_2\langle l_2, \langle p, q, a, b, c, s\rangle, \langle p', q', a', b', c', s'\rangle\rangle$ we have the following cases of transitions of $A$:

- Case $l_2 = lockA$: We have $p = 0$, $p' = 1$, $s = 1$, $q' = q$, $a' = a$, $b' = b$, $c' = c$ and show $a_0 = a_0'$ and $b_0 = b_0'$. From $p = 0$ and $p' = 1$, we have $a_0 = a$ and $a_0' = a'$, from which $a' = a$ gives us $a_0 = a_0'$. From $q' = q$ and $b' = b$, we have $b_0 = b_0'$.
- Case $l_2 = readA$: We have $p = 1$, $p' = 2$, $c' = a$, $q' = q$, $a' = a$, $b' = b$, $s' = s$ and show $a_0 = a_0'$ and $b_0 = b_0'$. From $p = 1$ and $p' = 2$, we have $a_0 = a$ and $a_0' = a'$, from which $a' = a$ gives us $a_0 = a_0'$. From $q' = q$ and $b' = b$, we have $b_0 = b_0'$.
- Case $l_2 = writeA$: We have $p = 2$, $p' = 3$, $a' = c + 1$, $q' = q$, $b' = b$, $c' = c$ and show $a_0 = a_0'$ and $b_0 = b_0'$. From $p = 2$, we have $a_0 = a$ and by the invariant also $c = a$. From $p' = 3$, we have $a_0' = a' - 1 = c = a$ and thus $a_0 = a_0'$. From $q' = q$ and $b' = b$, we have $b_0 = b_0'$.
- Case $l_2 = unlockA$: We have $p = 3$, $p' = 4$, $s' = 1$, $q' = q$, $a' = a$, $b' = b$, $c' = c$ and show $a_0 = a_0'$ and $b_0 = b_0'$. From $p = 2$ and $p' = 3$ we have $a_0 = a - 1$ and $a_0' = a' - 1$; thus $a = a'$ yields $a_0 = a_0'$. From $q' = q$ and $b' = b$, we have $b_0 = b_0'$.
- Case $l_2 = incA$: We have $p = 4$, $p' = 0$, $a' = a + 1$, $q' = q$, $b' = b$ and show $R_2\langle incA, \langle a_0, b_0\rangle, \langle a_0', b_0'\rangle\rangle$, i.e., $a_0' = a_0 + 2$ and $b_0' = b_0$. From $p = 4$, we have $a_0 = a - 1$. From $p' = 0$, we have $a_0' = a' = a + 1$ and thus $a_0' = a_0 + 2$. From $q' = q$ and $b' = b$, we have $b_0' = b_0$.

This proofs of the other five cases of transitions of $B$ proceed analogously.                    $\Box$

The central requirement to any notion of "refinement" is that it preserves the fundamental properties of the entity that is refined. In case of the refinement of systems, this requirement can be formalized as stated below.

**Theorem 9.5**  *Let $lts_1 \in LTS^{L_1, S_1}$ and $lts_2 \in LTS^{L_2, S_2}$ be labeled transition systems with state spaces $S_1$ respectively $S_2$ and label sets $L_1$ respectively $L_2$. Furthermore, let $a: S_2 \to S_1$ be a function from the state space of $lts_2$ to the state space of $lts_1$. Finally, let $F$ be an LTL-formula that denotes a safety property and that does neither contain the "next-time" operator $\bigcirc$ nor the "release" operator R. Then the following holds:*

$$(lts_1 \models F) \wedge (lts_2 \sqsubseteq^a lts_1) \Rightarrow$$
$$\forall r \in [\![\, lts_2 \,]\!].$$
$$\text{let } r_a := \lambda i \in domain(r).\, a(r(i))$$
$$\text{in } [\![\, F \,]\!]^{(r_a)^\omega}$$

*In other words, if $lts_1$ satisfies safety property $F$ and $lts_2$ refines $lts_1$ under $a$, then every run $r$ of $lts_2$ is mapped by $a$ to a run $r_a$ that satisfies $F$.*

Refinement thus preserves every safety property of a specification that can be expressed by an LTL-formula, provided that this formula does not talk about the immediate successor of any state, either in the form $\bigcirc F$ (which states that $F$ holds in the successor of the current state) or in the form $F \text{ R } G$ (which states that $G$ may not hold before the successor of the first state in which $F$ holds). This is because two different states of a run $r$ of the concrete implementation may be mapped by $a$ to the same state in the abstracted run $r_a$; thus, even if in $r_a$ state $s'_a = a(s')$ is the immediate successor of state $s_a = a(s)$, this does not imply that in $r$ state $s'$ is also the immediate successor of state $s$.

**Example 9.27**  Consider the systems *HourClock* and *MinuteClock* introduced in Examples 9.24 and 9.25 where *MinuteClock* refines *HourClock* under the function $a(h, m) := h$. Clearly *HourClock* satisfies the following LTL-formula:

$$\Box(h = 23 \Rightarrow \Diamond(h = 0))$$

This formula is also satisfied by the run $r$

$$\cdots \to \langle h, m \rangle = \langle 23, 0 \rangle \to \langle h, m \rangle = \langle 23, 1 \rangle \to \cdots \to \langle h, m \rangle = \langle 0, 0 \rangle \to \cdots$$

of *MinuteClock*, respectively by its abstraction $r_a$:

$$\cdots \to h = 23 \to h = 23 \to \cdots \to h = 0 \to \cdots$$

However, *HourClock* also satisfies the following stronger formula:

$$\Box(h = 23 \Rightarrow \bigcirc(h = 0))$$

This formula is violated by $r$ respectively $r_a$ where a state with $h = 23$ may be followed by another state with $h = 23$.                                         $\Box$

By a sequence of refinements, an abstract model may be gradually transformed via more and more concrete models until a model is reached that can be considered as an actual "implementation". If the abstract model satisfies a safety property described by an (appropriately restricted) LTL-formula $F$, also the ultimate implementation satisfies this property.

The presented notion of refinement preserves pure safety properties but generally not properties that involve some liveness aspect: while the original system may step through an infinite sequence of states, the refining system might make *no* step at all (if its initial state condition is unsatisfiable), only a *finite* number of steps (if the execution reaches a state in which no transition is enabled), or an infinite number of steps whose states, however, are abstracted to the same state of the original system, i.e., in that system they actually represent *no progress*. To verify that the refining system actually also ensures appropriate progress from the point of view of the original system requires more complex reasoning along the lines of the proof principles presented in Sect. 9.6; this, however, goes beyond the scope of our introductory presentation.

**Exercises**
Download from the following URL:
https://www.risc.jku.at/people/schreine/TP/exercises/ex-systems.pdf .

**Further Reading**
Our presentation has focused on the *logical* approach of modeling and reasoning about concurrent systems, but there are also other branches of concurrency theory.

In the 1960s, Petri developed the concept of (what is now called) *Petri Nets* where a system is modeled by a directed graph whose nodes are either "places" or "transitions". A state of the system is modeled by a distribution of "tokens" to places; the execution of the system is modeled by the flow of tokens across transitions from one place to another according to certain rules that describe when transitions may "fire". The semantics of Petri nets can be described, without resorting to the interleaving model of execution, by a "true concurrency" model based on partially ordered multisets, which is sometimes cited as an advantage of the Petri net model, as is its intuitive visual interpretation. There is vast literature on Petri nets; see, e.g., Peterson's survey paper [143] or Reisig's introductory book [150].

In the 1970s and early 1980s, the *algebraic* approach to the formalization of concurrency was developed. A starting point was Dijkstra's guarded command language [41,42], an imperative language that is in its core nondeterministic and thus allows (via the interleaving model of execution) to describe concurrent shared memory systems. Hoare adopted elements of this language in his imperative language "Communicating Sequential Processes" (CSP) [75], a distributed language where processes communicate by explicit message passing. Variants of the guarded command language respectively CSP are today still used in frameworks for modeling and analyzing concurrent systems such as the Spin model checker [12,80].

CSP influenced the simultaneous development of Milner's "Calculus of Communicating Systems" (CCS) [115], a declarative variant of a message passing language with an operational semantics in the form of a labeled transition system; Park introduced the concept of "bisimulation" to describe the equivalence of processes. Thus an equational calculus of was elaborated, which gave rise to the field of "process calculus" or "process algebra". There are many representatives, in particular an updated version of CCS [116,117,180], a reformulation of CSP [77,154], Bergstra's and Klop's "Algebra of Communicating Processes" (ACP) [15], Hennessy's "Algebraic Theory of Processes" [70], and Milner's $\pi$-Calculus [118], which extends CCS to also address *mobile systems*. Modern presentations can be found in Fokking's books [50,51], the book [60] of Groote and Mousavi, the book [28] of Bruni and Montanari, and the book [105] of Magee and Kramer, which also address corresponding tool support.

However, purely algebraic approaches lack the ability to specify on a high-level system properties and verify these properties; for this some kind of "logic" is required. For CCS, Hennessy and Milner developed in 1980 the "Hennessy-Milner logic", a modal logic that allows to describe in a declarative form sequences of events of a process [71]; see also Winskel's book [180] for a short account. For concurrent programs in structured imperative languages, Owicki and Gries developed already in the 1970s an extension of the Hoare calculus [139,140]. In a nutshell, this "Owicki-Gries method" combines Hoare triples for concurrent components to a Hoare triple for their composition; this requires that the proof outlines for the individual Hoare triples are "interference free", i.e., the execution of one component does not invalidate the correctness proof of another component. This method is also described in the book [7] of Apt and others, in the book [6] of Andrews, and in Schneider's book [163]. It should be noted that the Owicki-Gries method sticks to the classical framework of describing the behavior of a program with respect to its pre- and poststate only.

A completely new direction was opened in 1977 by Pnueli in his seminal paper [145] where he suggested to use a variant of modal logic, namely *temporal logic*, to specify the behavior of whole program executions, by evaluating these formulas over the program's transition system defined by an initial state condition and a transition relation (i.e., to use the systems as the "Kripke structure" over which modal formulas are interpreted). This initiated a completely a new strand of research but also some period of confusion: it took some time to clarify the difference between two branches of temporal logic, *linear temporal logic* whose formulas are evaluated over sequences of states and *branching-time logic* whose formulas are evaluated trees of states; this clarification is essentially due to Lamport [96].

Nowadays, temporal logic is one of the corner stones of concurrent system specification and verification. The books of Manna and Pnueli [108–110] describe the basic principles of the linear temporal logic for the specification and the verification of system properties. The theory is described in numerous resources, e.g., Emerson's handbook chapter [47], Kröger's books [92,93], Fred Schneider's book [163], Klaus

Schneider's book [164], or Kropf's book [91]; shorter accounts are given in Alagar's book [4] and (in German) in Schenke's book [159]. The book [52] of Furia and others considers extensions of temporal logic to model real time specifications.

In the 1980s and 1990s, the area of *model checking* emerged with efficient techniques to fully automatically check finite state systems of non-trivial size with respect to temporal logic properties; these techniques are described in detail in the book [46] of Clarke and others and in the handbook [37] edited by Clarke and others; a more compact account can be found in the book [30] of Bérard and others.

Lamport's "Temporal Logic of Actions" (TLA) [1,97] uses temporal logic, not only to specify system properties, but to model the system itself. The atomic formulas of TLA are interpreted over both the pre-state and the post-state of a transition; thus a formula $S$ of form $I \wedge \Box R$ can describe the execution of a system with initial state condition $I$ and transition relation $R$. To verify that system $S$ satisfies specification $F$ we prove the logical implication $S \Rightarrow F$; to verify that system $S$ refines system $S'$, we prove $S \Rightarrow S'$. Based on TLA, the formal specification language TLA$^+$ was developed with a supporting software system for the modeling and analysis of concurrent systems. TLA$^+$ is extensively described in Lamport's book [98] and in the book chapter [113] of Merz (and will be briefly sketched in the next section). Our own presentation has been inspired by Lamport's work.

# System Analysis in TLA$^+$ and RISCAL

In this section we present two instances of the logical model of the client-server system introduced in Examples 9.4 and 9.18 of the previous chapter; we will verify these instances with the help of model checking software.

The first instance is developed in the *TLA$^+$ toolbox* [99] which is based on Lamport's *Temporal Logic of Actions* [97] and is described in the book [98] and the chapter [113]. This toolbox is an integrated development environment for writing system specifications in the TLA$^+$ modeling language, including the TLC model checker and the TLAPS proof system. The language and system have been used for modeling and analyzing industrial hardware and software systems, e.g., Amazon Web Services (AWS). TLA$^+$ system models are essentially (unlabeled) transition systems as presented in this chapter expressed as linear temporal logic (LTL) formulas constructed from an initial state condition and a transition relation. The TLC model checker can fully automatically verify the correctness of finite instances of such systems with respect to arbitrary correctness properties expressible as LTL formulas. For infinite state systems, the interactive TLPS prover can be used to construct invariant-based proofs of safety properties as described in this chapter, using internal tactics or external SMT solvers or provers (such as Isabelle) as backends. In our presentation, however, we will focus on the TLC checker.

The second instance is developed in the *RISCAL* software [169,170] that was introduced on page xxx. RISCAL also supports the modeling of shared and distributed systems as described in Sects. 9.2 and 9.3. We demonstrate the verification of a safety property of such systems, by checking the property in all reachable states and by checking invariant-based verification conditions.

The specifications used in this chapter can be downloaded from the URLs

  https://www.risc.jku.at/people/schreine/TP/software/systems/System.tla
  https://www.risc.jku.at/people/schreine/TP/software/systems/System.txt

and loaded by executing from the command line the following commands:

```
toolbox System.tla &
RISCAL System.txt &
```

## The TLA$^+$ ToolBox

When starting the toolbox from the command line with the name of the specification file `System.tla` as an argument, a graphical user interface is displayed as shown in Fig. 9.11. This file essentially describes the model of the client-server system described in Example 9.4 as a module `System`:

```
----------------------- MODULE System -----------------------
EXTENDS   Naturals
CONSTANTS N
VARIABLES given, waiting, pc, req, ans
```

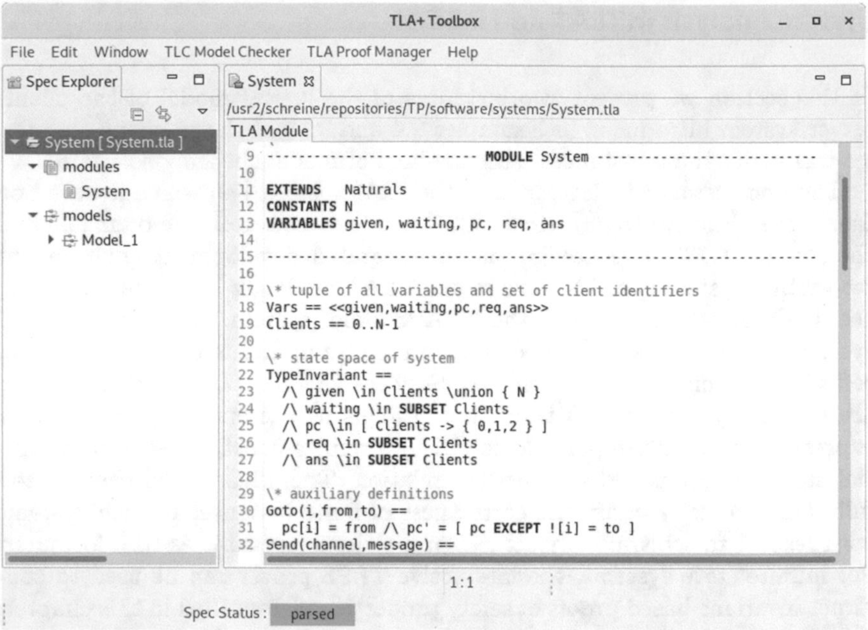

**Fig. 9.11** The TLA$^+$ toolbox

The module starts with the import of another module Naturals that contains some auxiliary operations on the domain of natural numbers; then it introduces a model constant $N$ that denotes the number of clients and it declares those variables whose values represent the state of the system. Now we capture by the following definition the list of system variables under the name *Vars* for later reference:

```
Vars == <<given,waiting,pc,req,ans>>
```

Every such definition introduces an "operator" that describes a named entity (the left side of the definition) by an expression (a term or a formula on the right side of the definition). It is easiest to think of an operator as a syntactic abbreviation where any later occurrence of the operator name is replaced by its defining expression. Operators may also have parameters and then be instantiated by arbitrary argument expressions; they can be considered as the functions respectively predicates of TLA$^+$.

TLA$^+$ has no static type system, but we may define a predicate that describes the expected domain of these variables (the invariance of this property will be later verified with the TLC checker):

```
Clients == 0..N-1
TypeInvariant ==
```

```
/\ given \in Clients \union { N }
/\ waiting \in SUBSET Clients
/\ pc \in [ Clients -> { 0,1,2 } ]
/\ req \in SUBSET Clients
/\ ans \in SUBSET Clients
```

As shown in this definition, conjunctions respectively disjunctions may be expressed not only by the logical infix connectives /\ and \/ but also as "itemized lists" where the connectives serve as labels at the beginnings of the list items; this simplifies the reading of specifications.

The initial state condition of the system is defined as a predicate *I*:

```
I ==
  /\ given = N
  /\ waiting = {}
  /\ pc = [ i \in Clients |-> 0 ]
  /\ req = {}
  /\ ans = {}
```

For a concise definition of the transition relation, we introduce a couple of auxiliary predicates:

```
Goto(i,from,to) ==
  pc[i] = from /\ pc' = [ pc EXCEPT ![i] = to ]
Send(channel,message) ==
  message \notin channel /\ channel' = channel\union{message}
Receive(channel,message) ==
  message \in channel /\ channel' = channel\{message}
```

Any plain variable reference such as *pc* refers here to the value of the system variable in the prestate of a transition while the primed reference *pc'* refers to the value of the variable in the poststate. With the help of these predicates, we may define the transition relation *RC(i)* of client *i* as follows:

```
RC0(i) ==
  \/ Goto(i,0,1) /\ Send(req,i)      /\ ans' = ans
  \/ Goto(i,1,2) /\ Receive(ans,i) /\ req' = req
  \/ Goto(i,2,0) /\ Send(req,i)      /\ ans' = ans
RC(i) == RC0(i) /\ given' = given /\ waiting' = waiting
```

Similarly we describe the transition relation *RC(i)* of the server when processing a message from client *i*:

```
GiveWaiting(g) ==
  g \in waiting /\
```

```
    given' = g /\ waiting' = waiting\{g} /\ Send(ans,given')
  RS0(i) ==
    \/ (i = given /\ waiting = {} /\
        given' = N /\ waiting' = waiting /\ ans' = ans)
    \/ (i = given /\ \E g \in Clients: GiveWaiting(g))
    \/ (i /= given /\ given = N /\
        given' = i /\ waiting' = waiting /\ Send(ans,given'))
    \/ (i /= given /\ given /= N /\
        waiting' = waiting\union{i} /\ given' = given /\ ans' = ans)
  RS(i) == Receive(req,i) /\ RS0(i) /\ pc' = pc
```

TLA$^+$ is based on *unlabeled* transitions; in order to refer to particular transitions (as is required in fairness constraints which will be discussed later) it
is advisable to introduce predicates for subtransitions. In above definition we
use such a predicate GiveWaiting(g) to represent (the core of) the transition
that gives the resource to a client $g$ in the waiting set of the server.

From above definitions we can now describe the transition relation $R$ of
the system as follows:

```
  R == \E i \in Clients: RC(i) \/ RS(i)
```

Here the token \E denotes the existential quantifier ∃ thus any transition of
the system is either the transition *RC(i)* of some client $i$ or the transition
*RS(i)* of the server processing a message from client $i$.

Finally we describe the whole system by the following predicate:

```
  ClientServer == I /\ [][R]_Vars
```

This predicate is defined by an LTL formula that is expressed as a conjunction of the initial state condition *I* and the formula [][*R*]_*Vars*, essentially a
shortcut for the temporal formula [](*R* \/ *Var'* = *Var*); it states that every
step of the system execution is either a transition described by relation $R$ or
a "stuttering step" that leaves the value of the term Var unchanged. Systems
in TLA$^+$ require such stuttering steps in order to allow the refinement of
more abstract systems by more concrete ones (see Sect. 9.7).

Now we define the core safety property expected from the system:

```
  MutEx ==
    \A i \in Clients, j \in Clients: i /= j => ~ (pc[i] = 2 /\ pc[j] = 2)
```

Here the token  A denotes the universal quantifier ∀ thus the property states
that two clients $i$ and $j$ must not be simultaneously in the critical region.

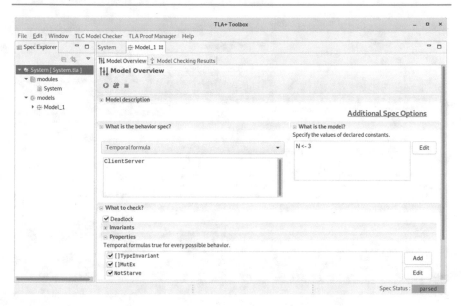

**Fig. 9.12** The TLA$^+$ toolbox (model overview)

To check the invariance of the conditions *TypeInvariant* and *MutEx*, we select in the menu "TLC Model Checker" the entry "New Model" which gives us a new view "Model Overview" depicted in Fig. 9.12. In this view we determine the system to be checked by selecting under the header "What is the behavior spec?" the menu entry "Temporal formula" and enter in the text field our predicate *ClientServer*. Furthermore, we define under the header "What is the model?" for the model constant $N$ the value 3; this creates a finite instance of the system whose state space can be exhaustively checked. Under the header "What to Check?" we enter the LTL properties []*TypeInvariant* and []*MutEx* and then press the green arrow button "Runs TLC on the model". After a second the model check terminates with a new view "Model Checking Results" that reports the success of the check and displays some statistical information; this validates that the system satisfies the defined properties.

We continue by defining the liveness property

```
NotStarve == \A i \in Clients: pc[i] = 1 ~> pc[i] = 2
```

Here a "progress formula" of form F ~> G is a shortcut for [](F => <>G); the property *NotStarve* thus states that every client that requests the resource eventually receives it. However, checking this property yields the error view displayed in Fig. 9.13. This view displays under the header "Error-Trace" a sequence of states, starting with a state satisfying the initial state condition, and continuing with those states each of which is derived from the previous one by an application of the transition relation. In our case, the sequence

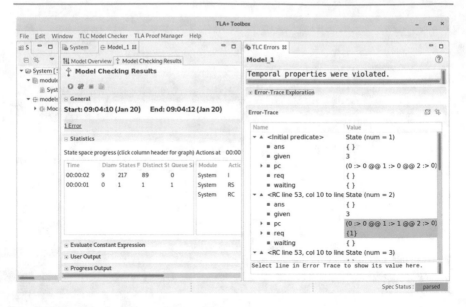

**Fig. 9.13**  The TLA$^+$ toolbox (model checking results)

contains seven distinct states followed by an infinite sequence of stuttering steps; in the last state there is a message in the channel *ans* that is, however, not received by the server, although the server is ready to do so.

The reason for the failure is that our system does not specify any fairness constraints which actually enforce that every system component that is permanently ready to make a step eventually performs this step. We overcome this deficiency by extending our system specification as follows:

```
ClientServer == I /\ [][R]_Vars
    /\ \A i \in Clients: WF_Vars(RC(i)) /\ WF_Vars(RS(i))
```

This definition adds to the system the weak fairness requirements that

- every client *i* that is permanently ready to make a step will eventually make this step,
- every message from client *i* that is permanently waiting to be processed by the server will eventually be processed by the server.

Above formulations demonstrate why we have defined the parameterized predicates *RC(i)* and *RS(i)*: it does not suffice to ensure that *some* client makes a step and the message of *some* client is processed; we must require these properties explicitly for *every* client.

Indeed, if we define the model constant $N=2$, above formulation suffices and the check succeeds. However, if we define $N=3$, the checker again reports an error. The corresponding error trace displays an execution where access to the resource is alternated between two clients; the request of the third client, after having been received by the server, remains forever in its set *waiting*. To also rule out this behavior, we extend our system specification as follows:

```
ClientServer ==
  /\ I /\ [][R]_Vars
  /\ \A i \in Clients: WF_Vars(RC(i)) /\ WF_Vars(RS(i))
  /\ \A g \in Clients:
      SF_Vars(\E i \in Clients: Receive(req,i) /\ i = given /\
        GiveWaiting(g) /\ pc' = pc)
```

Here we require *strong* fairness for every message from a client $g$ in the set *waiting* of the server: if it is infinitely often the case that, by the receipt of a message $i$ of some client that was given the resource (and that thus by the message gives back the resource), it is possible to give the resource to the waiting client $g$, this must be eventually also done. It is really necessary to demand strong (not just weak) fairness, because, if the resource is given to another client, the enabling condition is disabled until the message is returned again. The predicate described in the fairness condition is exactly the enabling condition of the second transition of the server:

```
RS0(i) ==
  \/ ...
  \/ (i = given /\ \E g \in Clients: GiveWaiting(g))
  \/ ...
RS(i) == Receive(req,i) /\ RS0(i) /\ pc' = pc
```

Indeed with this refined definition of the server also for $N=3$ the check of property *NotStarve* succeeds.

## RISCAL

This section shows how to verify in RISCAL invariance conditions for shared and distributed systems as presented in Sects. 9.2 and 9.3.

### Shared Systems

First we are going to model the RISCAL version of the shared system presented in Example 9.4. For this we introduce the number $N$ of clients aund various auxiliary types:

```
val N:N; axiom minN ⇔ N ≥ 1;
type Client = N[N-1]; type Client0 = N[N]; type PC = N[2];
```

The core declaration of the system describes its state space and its initial state as follows:

```
shared system ClientServer1
{
    var given: Client0; var waiting: Set[Client];
    var pc: Array[N,PC]; var req: Set[Client]; var ans: Set[Client];
    init()
    {
      given := N; waiting := ∅[Client];
      pc := Array[N,PC](0); req := ∅[Client]; ans := ∅[Client];
    }
    ...
}
```

This declaration is extended by the transitions of every client *i*:

```
        action cask(i:Client) with pc[i] = 0 ∧ i ∉ req;
        { pc[i] := 1; req := req ∪ {i}; }
        action cget(i:Client) with pc[i] = 1 ∧ i ∈ ans;
        { pc[i] := 2; ans := ans \ {i}; }
        action cret(i:Client) with pc[i] = 2 ∧ i ∉ req;
        { pc[i] := 0; req := req ∪ {i}; }
```

Likewise we add the transitions of the server:

```
action sget(i:Client) with i ∈ req ∧ given = N ∧ i ∉ ans;
{ req := req \ {i}; given := i; ans := ans ∪ {i}; }
action swait(i:Client) with i ∈ req ∧ given ≠ N ∧ given ≠ i;
{ req := req \ {i}; waiting := waiting ∪ {i}; }
action sret1(i:Client) with i ∈ req ∧ given = i ∧ waiting = ∅[Client];
{ req := req \ {i}; given := N; }
action sret2(i:Client,j:Client)
with i ∈ req ∧ given = i ∧ j ∈ waiting ∧ j ∉ ans;
{ req := req \ {i}; given := j;
   waiting = waiting \ {j}; ans := ans ∪ {j}; }
```

Furthermore, we annotate the system with the invariant that shall be satisfied by every reachable state of the system, i.e., the mutual exclusion property:

```
invariant ¬∃i1:Client,i2:Client with i1 < i2. pc[i1] = 2 ∧ pc[i2] = 2;
```

Now checking the system with $N=4$ quickly demonstrates its correctness:

```
Executing system ClientServer1.
340 system states found with search depth 213.
Execution completed (46 ms).
```

However, if we invalidate the correctness of the system by commenting out the command given := i in server action *sget* we get the following result:

```
Executing system ClientServer1.
ERROR in execution of system ClientServer1: evaluation of
  invariant ¬(∃i1:Client, i2:Client with i1 < i2. ...);
at line 28 in file System.txt:
  invariant is violated
The system run leading to this error:
  0:[given:4,waiting:{},pc:[0,0,0,0],req:{},ans:{}]->cask(0)->
  1:[given:4,waiting:{},pc:[1,0,0,0],req:{0},ans:{}]->cask(1)->
  2:[given:4,waiting:{},pc:[1,1,0,0],req:{0,1},ans:{}]->sget(0)->
  3:[given:4,waiting:{},pc:[1,1,0,0],req:{1},ans:{0}]->cget(0)->
  4:[given:4,waiting:{},pc:[2,1,0,0],req:{1},ans:{}]->sget(1)->
  5:[given:4,waiting:{},pc:[2,1,0,0],req:{},ans:{1}]->cget(1)->
  6:[given:4,waiting:{},pc:[2,2,0,0],req:{},ans:{}]
ERROR encountered in execution.
```

This message produces a counterexample run, i.e., a run that leads to a state violating the invariant. The system may be further annotated with more invariants (see Example 9.18):

```
invariant ∀i:Client with i = given.
  (pc[i] = 0 ∧ i ∈ req) ∨ (pc[i] = 1 ∧ i ∈ ans) ∨
  (pc[i] = 2 ∧ i ∉ req ∧ i ∉ ans);
invariant ∀i:Client with i ∈ waiting.
  i ≠ given ∧ pc[i] = 1 ∧ i ∉ req ∧ i ∉ ans;
invariant ∀i: Client with i ∈ req. i ∉ ans;
invariant ∀i: Client with i ∈ ans. given = i;
invariant ∀i: Client with pc[i] = 0. i ∉ ans ∧ (i ∈ req ⇒ i = given);
invariant ∀i: Client with pc[i] = 1. i ∈ req ∨ i ∈ waiting ∨ i ∈ ans;
invariant ∀i: Client with pc[i] = 2. i = given;
```

Also these invariants can be quickly checked:

```
Executing system ClientServer1.
340 system states found with search depth 213.
Execution completed (39 ms).
```

Actually, these invariants are inductive and imply mutual exclusion; we will check later that they can be also used to prove the correctness of the system.

**Distributed Systems**

Now we demonstrate the RISCAL version of the distributed system presented in Example 9.10. For this we introduce (in addition to the entities defined at the beginning of this section) the size $B$ of the message buffers associated the actions of the server:

$$val\ B:ℕ;\ axiom\ minB ⇔ B ≥ 1;$$

The actual system declaration then looks as follows:

```
distributed system ClientServer2
{
  component Server
  {
    var client: Client0;
    init() { client := N; }
    action[B] request(i:Client) with client = N;
    { client := i; send Client[i].enter(); }
    action[B] giveback(i:Client) { client := N; }
  }
  component Client[N]
  {
    var req: ℕ[1]; var use: ℕ[1];
    init() { req := 0; use := 0; send Client[this].ask(); }
    action ask() { req := 1; send Server.request(this); }
    action enter() { req := 0; use := 1; send Client[this].exit(); }
    action exit()
    { use := 0; send Server.giveback(this); send Client[this].ask(); }
  }
}
```

For this system, we claim the following mutual exclusion property:

```
invariant ∀i1:Client, i2:Client.
  (Client[i1].use = 1 ∧ Client[i2].use = 1 ⇒ i1 = i2);
```

Here a term such as *Client[i].use* denotes the value of the state variable *use* in instance $i$ of component *Client*. A check of this property for $N=B=4$ demonstrates the correctness of this system:

```
Executing system ClientServer2.
453 system states found with search depth 140.
Execution completed (22 ms).
```

However, if we invalidate the protocol by commenting out the assignment `client := i` in server action *request*, the system displays the following counterexample leading to a state violating mutual exclusion:

```
Executing system ClientServer2.
ERROR in execution of system ClientServer2: evaluation of
  invariant ∀i1:Client, i2:Client. ...);
at line 107 in file System.txt:
  invariant is violated
The system run leading to this error:
  0:[Server:[4,...],Client:[...]]->Client[0].ask()->
  1:[Server:[4,...],Client:[...]]->Server.request(0)->
  2:[Server:[4,...],Client:[...]]->Client[1].ask()->
  3:[Server:[4,...],Client:[...]]->Server.request(1)->
  4:[Server:[4,...],Client:[...]]->Client[0].enter()->
  5:[Server:[4,...],Client:[...]]->Client[1].enter()->
  6:[Server:[4,...],Client:[...]]
ERROR encountered in execution.
```

The system may be also annotated with more invariants (see Exmple 9.19):

```
invariant ∀i:Id with i < N.
  (Client[i].ask_number + Client[i].enter_number + Client[i].exit_number
  + number(Server.request, Server.request_number, i) = 1);
invariant ∀i:Id with i < N.
  number(Server.giveback, Server.giveback_number, i) ≤ 1;
invariant ∀i:Id with i < N.
  number(Server.giveback, Server.giveback_number, i) = 1 ⇒
  Client[i].enter_number = 0 ∧ Client[i].exit_number = 0;
invariant ∀i:Id with i < N.
  (Client[i].req = 1 ⇔
    number(Server.request, Server.request_number, i) = 1 ∨
    Client[i].enter_number = 1);
invariant ∀i:Id with i < N.
  (Client[i].use = 1 ⇔ Client[i].exit_number = 1);
invariant ∀i:Id with i < N.
  (Server.client = i ⇔
    Client[i].enter_number = 1 ∨ Client[i].exit_number = 1 ∨
    number(Server.giveback, Server.giveback_number, i) = 1);
```

Here, e.g., *Server.request* denotes the message buffer associated to the corresponding action and *Server.request_number* denotes the number of elements in that buffer. The invariants use the following function that returns the number of the first *bn* messages in buffer *b* whose values equal *i*:

```
fun number(b:Array[B,Record[i:Client]],bn:ℕ[B],i:Client):ℕ[B] =
  #k:ℕ[B] with k < bn ∧ b[k].i = i;
```

Also the correctness of these invariants can be quickly checked:

```
Executing system ClientServer2.
453 system states found with search depth 140.
Execution completed (69 ms).
```

### Invariant-Based Verification

Above examples have illustrated the checking of system executions with respect to safety properties; now we briefly sketch a "manual" approach to the invariant-based verification of such properties (an "automated" approach will be given at the end). We start by defining the state space of the system by the introduction of a type *State*:

```
type State = Record[
  given: Client0, waiting: Set[Client], pc: Map[Client,PC],
  req: Set[Client], ans: Set[Client] ];
```

The initial state condition of the system can be then defined as follows:

```
pred I(s:State) ⇔
  s.given = N ∧ s.waiting = ∅[Client] ∧ s.pc = Map[Client,PC](0) ∧
  s.req = ∅[Client] ∧ s.ans = ∅[Client];
```

To speed up the subsequent verification, we will not directly use the system's transition relation $R$ but introduce a type *Action* of system actions with the following property:

```
∀s:State,s0:State. R(s,s0) ⇔
    ∃a:Action. enabled(a,s) ∧ s0 = execute(a,s);
```

Here the predicate *enabled* expresses the enabling condition of an action and the function *execute* denotes its effect (see below). In detail, the type *Action* comprises the three client actions and the four server actions:

```
rectype(0) Action =
  Request(Client) | Get(Client) | Return(Client) |
  Return1(Client) | Return2(Client,Client) |
  Request1(Client) | Request2(Client);
```

Each action is parameterized by the identity $i$ of the client that performs the action respectively that is the sender of the message processed by the server; the server action *Return2* additionally receives the identity $g$ of the client that is taken from the set *waiting* and given the resource. Then we define the enabling condition of an action as follows:

```
pred enabled(a:Action,s:State) ⇔
  match a with {
    ...
    Return2(i:Client,g:Client) ->
      i ∈ s.req ∧ i = s.given ∧ g ∈ s.waiting ∧ ¬(g ∈ s.ans);
    ...
  };
```

The effect of the execution of an action to a state is then described as follows:

```
fun execute(a:Action,s:State):State =
  match a with {
    ...
    Return2(i:Client,g:Client) ->
      s with .req = s.req\{i} with .given = g
        with .waiting = s.waiting\{g} with .ans = s.ans∪{g};
    ...
  };
```

Now we define the mutual exclusion property and the system invariant (as the conjunction of the conditions shown in the shared system above):

```
pred MutEx(s:State) ⇔
  ∀i:Client,j:Client with i≠j. ¬(s.pc[i] = 2 ∧ s.pc[j] = 2);
pred Invariant(s:State) ⇔ ∀i:Client. ...
```

We can then define the verification conditions as follows:

```
theorem MutEx0(s:State) ⇔ Invariant(s) ⇒ MutEx(s);
theorem MutEx1(s:State) ⇔ I(s) ⇒ Invariant(s);
theorem MutEx2(s:State) ⇔ Invariant(s) ⇒
  ∀a:Action with enabled(a,s). let s0 = execute(a,s) in Invariant(s0);
```

All conditions can be quickly checked for $N=3$:

```
Executing MutEx0(..) with all 55296 inputs.
Execution completed for ALL inputs (216 ms, 55296 checked, ...).
Executing MutEx1(...) with all 55296 inputs.
Execution completed for ALL inputs (127 ms, 55296 checked, ...).
Executing MutEx2(...) with all 55296 inputs.
Execution completed for ALL inputs (223 ms, 55296 checked, ...).
```

Actually, all the effort above has not really been necessary: RISCAL itself implements a verification condition generator that automatically generates the necessary conditions: if we simply press the "Show/Hide Tasks" button and then select in the popup menu of "system ClientServer1" the menu entry "Execute All Tasks" or alternatively "Apply SMT Solver to All Theorems", the system quickly confirms that the stated invariants are indeed inductive:

```
The SMT solver Yices started execution.
Theorem _ClientServer1_0_initPre_cverify_0 is valid.
...
Theorem _ClientServer1_0_actionPre_6_cverify_7 is valid.
Total time: 736 ms, translation: 116 ms, decision: 596 ms.
```

We may thus use RISCAL to validate whether invariants are strong enough for a proof-based verification of a system with an arbitrary number of clients.

# References

1. M. Abadi, L. Lamport, Conjoining specifications. Trans. Program. Lang. Syst. **17**(3), 507–535 (1995). https://doi.org/10.1145/203095.201069
2. J.-R. Abrial, *Modeling in Event-B –, System and Software Design* (Cambridge University Press, Cambridge, 2010). https://doi.org/10.1017/CBO9781139195881
3. W. Ahrendt, B. Beckert, R. Bubel, R. Hähnle, P.H. Schmitt, M. Ulbrich, (eds.), *Deductive Software Verification — The KeY Book — From Theory to Practice*. Lecture Notes in Computer Science, vol. 10001 (Springer International Publishing, Cham, 2016). https://doi.org/10.1007/978-3-319-49812-6
4. V.S. Alagar, K. Periyasamy, *Specification of Software Systems*. Texts in Computer Science, 2nd edn. (Springer, London, 2011). https://doi.org/10.1007/978-0-85729-277-3
5. S. Alagić, Software Engineering: Specification, Implementation, Verification (Springer International Publishing. Cham (2017). https://doi.org/10.1007/978-3-319-61518-9
6. G.R. Andrews, *Concurrent Programming – Theory and Practice* (Addison-Wesley, Menlo Park, 1991). https://www.pearson.com/us/higher-education/program/Andrews-Concurrent-Programming-Principles-and-Practice/PGM73364.html
7. K.R. Apt, F.S. de Boer, E.-R. Olderog, *Verification of Sequential and Concurrent Programs*. Texts in Computer Science, 3rd edn. (Springer, London, 2009). https://doi.org/10.1007/978-1-84882-745-5
8. R.-J. Back, J. Wright, *Refinement Calculus – A Systematic Introduction. Texts in Computer Science* (Springer, New York, 1998). https://doi.org/10.1007/978-1-4612-1674-2
9. T. Ball, E. Bounimova, V. Levin, R. Kumar, J. Lichtenberg, The static driver verifier research platform, in *Computer Aided Verification, 22nd International Conference, CAV 2010, Edinburgh, UK, July 15–19, 2010*. Lecture Notes in Computer Science, ed, by T. Touili, B. Cook, P. Jackson, vol. 6174 (Springer, Berlin, 2010), pp. 119–122. https://doi.org/10.1007/978-3-642-14295-6_11
10. C. Barrett, R. Sebastiani, S.A. Seshia, C. Tinel, Satisfiability Modulo theories, in *In [21]*, vol. chapter 26 (IOS Press, Amsterdam, 2009), pp. 825–885. https://doi.org/10.3233/978-1-58603-929-5-825
11. C. Barrett, C. Tinelli, *CVC3. Department of Computer Science* (New York University, New York, 2015). https://www.cs.nyu.edu/acsys/cvc3
12. M. Ben-Ari, Principles of the Spin Model Checker (Springer. London (2008). https://doi.org/10.1007/978-1-84628-770-1
13. M. Ben-Ari, *Mathematical Logic for Computer Science*, 3rd edn. (Springer, London, 2012). https://doi.org/10.1007/978-1-4471-4129-7

14. R. Berghammer, Mathematik für die Informatik – Grundlegende Begriffe, Strukturen und ihre Anwendung, 2nd edn. (Springer Vieweg, Wiesbaden, 2017). https://doi.org/10.1007/978-3-658-16712-7

15. J. A. Bergstra, A. Ponse, S.A. Smolka, (eds.), *Handbook of Process Algebra* (Elsevier Science, New York, 2001). https://www.elsevier.com/books/handbook-of-process-algebra/bergstra/978-0-444-82830-9

16. D. Beyer, CPAchecker – the configurable software-verification platform (2021). https://cpachecker.sosy-lab.org/

17. D. Beyer, T.A. Henzinger, R. Jhala, R. Majumdar, The software model checker BLAST: applications to software engineering. Int. J. Softw. Tools Technol. Transf. **9**(5), 505–525 (2007). https://doi.org/10.1007/s10009-007-0044-z

18. D. Beyer, M. Erkan Keremoglu, CPAchecker: a tool for configurable software verification, in *Computer Aided Verification, 23rd International Conference, CAV 2011, Snowbird, UT, USA, July 14–20, 2011.* Lecture Notes in Computer Science, ed. by G. Gopalakrishnan, S. Qadeer, vol. 6806 (Springer, Berlin, 2011), pp. 184–190. https://doi.org/10.1007/978-3-642-22110-1_16

19. M. Bidoit, P.D. Mosses, *CASL User Manual – Introduction to Using the Common Algebraic Specification Language.* Lecture Notes in Computer Science, vol. 2900 (Springer, Berlin, 2004). https://doi.org/10.1007/b11968 and http://www.informatik.uni-bremen.de/cofi/CASL-UM.pdf

20. A. Biere, Bounded model checking, in *[21]*, vol. Chapter 14 (IOS Press, Amsterdam, 2009), pp. 457–481. https://doi.org/10.3233/978-1-58603-929-5-457

21. A. Biere, M. Heule, H. van Maaren, T. Walsh, (eds.), *Handbook of Satisfiability.* Frontiers in Artificial Intelligence and Applications, vol. 185, 2nd edn. (IOS Press, Amsterdam, 2009). https://ebooks.iospress.nl/volume/handbook-of-satisfiability

22. D. Bjørner, M. Henson, (eds.), *Logics of Specification Languages* (Springer, Berlin, 2008). https://doi.org/10.1007/978-3-540-74107-7

23. D. Bjørner, K. Havelund, 40 years of formal methods, in *FM 2014: Formal Methods, 19th International Symposium, Singapore, May 12–16, 2014.* Lecture Notes in Computer Science, vol. 8442, ed. by C. Jones, P. Pihlajasaari, J. Sun (Springer International Publishing, Cham, 2014), pp. 42–61. https://doi.org/10.1007/978-3-319-06410-9_4

24. P.P. Boca, J.P. Bowen, J.I. Siddiqi, (eds.), *Formal Methods: State of the Art and New Directions* (Springer, London, 2010). https://doi.org/10.1007/978-1-84882-736-3

25. R.T. Boute, Calculational semantics: deriving programming theories from equations by functional predicate calculus. ACM Trans. Program. Lang. Syst. **28**(4), 747–793 (2006). https://doi.org/10.1145/1146809.1146814

26. A.R. Bradley, Z. Manna, The Calculus of Computation - Decision Procedures with Applications to Verification (Springer. Berlin (2007). https://doi.org/10.1007/978-3-540-74113-8

27. M. Broy, R. Steinbrüggen, (eds.), *Calculational System Design.* NATO Science Series, vol. 173 (IOS Press, Amsterdam, 2000). http://www.iospress.nl/book/calculational-system-design

28. R. Bruni, U. Montanari, *Models of Computation.* Texts in Theoretical Computer Science (Springer, Berlin, 2017). https://doi.org/10.1007/978-3-319-42900-7

29. B. Buchberger, F. Lichtenberger, Mathematik für Informatiker I - Die Methode der Mathematik, 2nd edn. (Springer, Berlin, 1981). https://doi.org/10.1007/978-3-642-68351-0 and http://www.risc.jku.at/publications/download/risc_2230/mathematik_informatiker_bookmarks.pdf

30. B. Bérard, M. Bidoit, A. Finkel, F. Laroussinie, A. Petit, L. Petrucci, P. Schnoebelen, Systems and Software Verification - Model-Checking Techniques and Tools (Springer. Berlin (2001). https://doi.org/10.1007/978-3-662-04558-9

31. E. Börger, R. Stärk, Abstract State Machines - A Method for High-Level System Design and Analysis (Springer. Berlin (2003). https://doi.org/10.1007/978-3-642-18216-7

32. CakeML. CakeML — A Verified Implementation of ML (2021). https://cakeml.org

33. L. Cardelli, P. Wegner, On understanding types, data abstraction, and polymorphism. ACM Comput. Surv. **17**(4), 471–522 (1985). https://doi.org/10.1145/6041.6042 and http://lucacardelli.name/papers/onunderstanding.a4.pdf

34. P. Chalin, J.R. Kiniry, G.T. Leavens, E. Poll, Beyond assertions: advanced specification and verification with JML and ESC/Java2, in *Formal Methods for Components and Objects 4th International Symposium, FMCO 2005, Amsterdam, The Netherlands, November 1–4, 2005*, volume 4111 of *Lecture Notes in Computer Science*, vol. 4111, ed. by F.S. de Boer, M.M. Bonsangue, S. Graf, and W.-P. de Roever (Springer, Berlin, 2006), pp. 342–363 https://doi.org/10.1007/11804192_16

35. M. Christakis, N. Polikarpova, P. Sridhar Duggirala, P. Schrammel, (eds.), *Software Verification: 12th International Conference, VSTTE 2020, and 13th International Workshop, NSV, Los Angeles, CA, USA, July 20–21, 2020*. Lecture Notes in Computer Science, vol. 12549 (Springer, Cham, 2020). https://doi.org/10.1007/978-3-030-63618-0

36. E.M. Clarke, O. Grumberg, D.E. Long, Model checking and abstraction. ACM Trans. Program. Lang. Syst. **16**(5), 1512–1542 (1994). https://doi.org/10.1145/186025.186051

37. E.M. Clarke, T.A. Henzinger, H. Veith, R. Bloem, (eds.), *Handbook of Model Checking* (Springer International Publishing, Cham, 2018). https://doi.org/10.1007/978-3-319-10575-8

38. A. Colmerauer, P. Roussel, The birth of prolog. SIGPLAN Notices **28**(3), 37–52 (1993). https://doi.org/10.1145/155360.155362

39. P. Cousot, Methods and logics for proving programs, in *[177]*, vol. Chapter 15 (Elsevier and MIT Press, Amsterdam, 1990), pp. 841–993

40. R. Diaconescu, *A Methodological Guide to the CafeOBJ Logic, in [22]* (Springer, Berlin, 2008), pp. 153–240

41. E.W. Dijkstra, Guarded commands, nondeterminacy and formal derivation of programs. Commun. ACM **18**(8), 453–457 (1975). https://doi.org/10.1145/360933.360975

42. E.W. Dijkstra, *A Discipline of Programming*. Series in Automatic Computation (Prentice Hall, Englewood Cliffs, 1976). https://books.google.at/books/about/A_Discipline_of_Programming.html?id=nZpQAAAAMAAJ

43. G. Dowek, J.-J. Lévy, *Introduction to the Theory of Programming Languages*. Undergraduate Topics in Computer Science (Springer, London, 2011). https://doi.org/10.1007/978-0-85729-076-2

44. R. Drechsler, (ed.), *Formal System Verification – State-of-the-Art and Future Trends* (Springer International Publishing, Cham, 2018). https://doi.org/10.1007/978-3-319-57685-5

45. H.-D. Ebbinghaus, J. Flum, W. Thomas, Mathematical Logic (Springer. New York (1992). https://doi.org/10.1007/978-1-4757-2355-7

46. E.M. Clarke, Jr., O. Grumberg, D. Kroening, D. Peled, H. Veith, *Model Checking*, 2nd edn. (MIT Press, Cambridge, 2018). https://mitpress.mit.edu/books/model-checking-second-edition

47. E. ALlen Emerson, Temporal and modal logic, in *[177], Chapter 16* (Elsevier and MIT Press, Amsterdam, 1990), pp. 995–1072

48. M. Fernández, *Programming Languages and Operational Semantics – A Concise Overview*. Undergraduate Topics in Computer Science (Springer, London, 2014). https://doi.org/10.1007/978-1-4471-6368-8

49. R.W. Floyd, Assigning meanings to programs, in *Mathematical Aspects of Computer Science*. Proceedings of Symposia in Applied Mathematics, vol. 19, ed. by J.T. Schwartz (American Mathematical Society, Providence, 1967), pp. 19–32. https://doi.org/10.1090/psapm/019/0235771

50. W. Fokkink, *Introduction to Process Algebra*. Texts in Computer Science (Springer, Berlin, 2000). https://doi.org/10.1007/978-3-662-04293-9

51. W. Fokkink, *Modelling Distributed Systems*. Texts in Computer Science (Springer, Berlin, 2007). https://doi.org/10.1007/978-3-540-73938-8

52. C.A. Furia, D. Mandrioli, A. Morzenti, M. Rossi, *Modeling Time in Computing*. Monographs in Theoretical Computer Science (Springer, Berlin, 2012). https://doi.org/10.1007/978-3-642-32332-4

53. C.A. Furia, B. Meyer, S. Velder, Loop invariants: analysis, classification, and examples. ACM Comput. Surv. **46**(3), 34:1–34:51 (2014). https://doi.org/10.1145/2506375

54. K. Futatsugi, A.T. Nakagawa, T. Tamai, (eds.), *CAFE: An Industrial-Strength Algebraic Formal Method* (Elsevier, Amsterdam, 2000). https://www.elsevier.com/books/cafe-an-industrial-strength-algebraic-formal-method/futatsugi/978-0-444-50556-9

55. K. Futatsugi et al., *CafeOBJ* (Japan Advanced Institute of Science and Technology (JAIST), Nomi, 2015). https://cafeobj.org

56. J.H. Gallier, *Logic for Computer Science – Foundations of Automatic Theorem Proving* (Harper and Row, New York, 1986). http://www.cis.upenn.edu/~jean/gbooks/logic.html and http://www.researchgate.net/publication/31634432_Logic_for_computer_science__foundations_of_automatic_theorem_proving__J.H._Gallier

57. M. Gleirscher, D. Marmsoler, Formal methods in dependable systems engineering: a survey of professionals from Europe and North America. Empir. Softw. Eng. **25**(6), 4473–4546 (2020). https://doi.org/10.1007/s10664-020-09836-5

58. J.A. Goguen, G. Malcom, (eds.), *Software Engineering with OBJ — Algebraic Specification in Action*. Advances in Formal Methods, vol. 2 (Springer, New York, 2000). https://doi.org/10.1007/978-1-4757-6541-0

59. M.J. Gordon, A.J. Milner, C.P. Wadsworth, *Edinburgh LCF – A Mechanised Logic of Computation*. Lecture Notes in Computer Science, vol. 78 (Springer, Berlin, 1979). https://doi.org/10.1007/3-540-09724-4

60. J.F. Groote, M.R. Mousavi, *Modeling and Analysis of Communicating Systems* (MIT Press, Cambridge, 2014). https://mitpress.mit.edu/books/modeling-and-analysis-communicating-systems

61. Theorema Working Group, *The Theorema System. Research Institute for Symbolic Computation (RISC)* (Johannes Kepler University, Linz, 2014). https://www.risc.jku.at/research/theorema/software

62. H. Peter Gumm, Generating algebraic laws from imperative programs. Theor. Comput. Sci. **217**(2), 385–405 (1999). https://doi.org/10.1016/S0304-3975(98)00278-3

63. C.A. Gunter, *Semantics of Programming Languages – Structures and Techniques* (MIT Press, Cambridge, 1992). https://mitpress.mit.edu/books/semantics-programming-languages

64. T.C. Hales, The Jordan curve theorem, formally and informally. Am. Math. Mon. **114**(10), 882–894 (2007). https://doi.org/10.1080/00029890.2007.11920481

65. J. Harrison, *Handbook of Practical Logic and Automated Reasoning* (Cambridge University Press, Cambridge, 2009). http://www.cambridge.org/az/academic/subjects/computer-science/programming-languages-and-applied-logic/handbook-practical-logic-and-automated-reasoning

66. J. Hatcliff, G.T. Leavens, K. Rustan M. Leino, P. Müller, M. Parkinson, Behavioral interface specification languages. ACM Comput. Surv. **44**(3), 16:1–16:58 (2012). https://doi.org/10.1145/2187671.2187678

67. C. Hawblitzel, J. Howell, M. Kapritsos, J.R. Lorch, B. Parno, M.L. Roberts, S. Setty, B. Zill, IronFleet: proving safety and liveness of practical distributed systems. Commun. ACM **60**(7), 83–92 (2017). https://doi.org/10.1145/3068608

68. E.C.R. Hehner, *A Practical Theory of Programming*. Texts and Monographs in Computer Science (Springer, New York, 1993). https://doi.org/10.1007/978-1-4419-8596-5 and http://www.cs.toronto.edu/~hehner/aPToP

69. E.C.R. Hehner, Specifications, programs, and total correctness. Sci. Comput. Program. **34**(3), 191–205 (1999). https://doi.org/10.1016/S0167-6423(98)00027-6

70. M. Hennessy, *Algebraic Theory of Processes* (MIT Press, Cambridge, 1988). https://mitpress.mit.edu/books/algebraic-theory-processes

71. M. Hennessy, R. Milner, On observing nondeterminism and concurrency, in *ICALP 1980, Automata, Languages and Programming, Noordwijkerhout, the Netherlands July 14–18*. Lec-

ture Notes in Computer Science, vol. 85, ed. by J. de Bakker, J. van Leeuwen (Springer, Berlin, 1980), pp. 299–309. https://doi.org/10.1007/3-540-10003-2_79

72. M.G. Hinchey, J.P. Bowen, E.-R. Olderog, (eds.), *Provably Correct Systems*. NASA Monographs in Systems and Software Engineering (Springer International Publishing, Cham, 2017). https://doi.org/10.1007/978-3-319-48628-4

73. C.A.R. Hoare, An axiomatic basis for computer programming. Commun. ACM **12**(10), 576–580 (1969). https://doi.org/10.1145/363235.363259

74. C.A.R. Hoare, Proof of correctness of data representations. Acta Inf. **1**(4), 271–281 (1972). https://doi.org/10.1007/BF00289507

75. C.A.R. Hoare, Communicating sequential processes. Commun. ACM **21**(8), 666–677 (1978). https://doi.org/10.1145/359576.359585

76. C.A.R. Hoare, Programs are predicates. Philos. Trans. R. Soc. Lond. **312**(1522), 475–489 (1984). https://www.jstor.org/stable/37446

77. C.A.R. Hoare, *Communicating Sequential Processes* (Prentice-Hall, Upper Saddle River, 1985). http://www.usingcsp.com/

78. C.A.R. Hoare, H. Jifeng, *Unifying Theories of Programming* (Prentice Hall International, Upper Saddle River, 1998). http://www.unifyingtheories.org/

79. HOL Interactive Theorem Prover (2021). https://hol-theorem-prover.org

80. G.J. Holzmann, *The Spin Model Checker – Primer and Reference Manual* (Addison-Wesley, Boston, 2003). https://www.pearson.com/us/higher-education/program/Holzmann-SPIN-Model-Checker-The-Primer-and-Reference-Manual/PGM2819142.html

81. M. Huth, M. Ryan, *Logic in Computer Science – Modelling and Reasoning about Systems*, 2nd edn. (Cambridge University Press, Cambridge, 2004). http://www.cambridge.org/az/academic/subjects/computer-science/programming-languages-and-applied-logic/logic-computer-science-modelling-and-reasoning-about-systems-2nd-edition

82. INRIA. CompCert (2021). https://compcert.org

83. B. Jacobs, J. Rutten, An introduction to (Co)algebra and (Co)induction, in *Advanced Topics in Bisimulation and Coinduction*. Cambridge Tracts in Theoretical Computer Science, chapter 2, vol. 52, ed. by D. Sangiori, J. Rutten (Cambridge University Press, Cambridge, 2011), pp. 38–99. https://doi.org/10.1017/CBO9780511792588.003 and http://www.cwi.nl/~janr/papers/files-of-papers/2011_Jacobs_Rutten_new.pdf and http://citeseerx.ist.psu.edu/viewdoc/summary?doi=10.1.1.37.1418 (1997 version)

84. C.B. Jones, *Systematic Software Development Using VDM*, 2nd edn. (Prentice-Hall, Upper Saddle River, 1990). https://www.researchgate.net/publication/239702274_Systematic_Software_Development_Using_VDM_2nd_edition

85. G. Kahn, Natural semantics, in *STACS 87: 4th Annual Symposium on Theoretical Aspects of Computer Science, Passau, Germany, February 19–21, 1987*. Lecture Notes in Computer Science, vol. 1987, ed. by F.J. Brandenburg, G. Vidal-Naquet, M. Wirsing (Springer, Berlin, 1987), pp. 22–39. https://doi.org/10.1007/BFb0039592

86. G. Klein, J. Andronick, M. Fernandez, I. Kuz, T. Murray, G. Heiser, Formally verified software in the real world. Commun. ACM **61**(10), 68–77 (2018). https://doi.org/10.1145/3230627

87. G. Klein, K. Elphinstone, G. Heiser, J. Andronick, D. Cock, P. Derrin, D. Elkaduwe, K. Engelhardt, R. Kolanski, M. Norrish, T. Sewell, H. Tuch, S. Winwood, seL4: formal verification of an OS kernel, in *SOSP '09: ACM SIGOPS 22nd Symposium on Operating Systems Principles, Big Sky, Montana, USA* (ACM, New York, 2009), pp. 207–220. https://doi.org/10.1145/1629575.1629596 (October)

88. S. Kleuker, Formale Modelle der Softwareentwicklung - Model-Checking, Verifikation, Analyse und Simulation (Vieweg+Teubner. Wiesbaden (2009). https://doi.org/10.1007/978-3-8348-9595-0

89. D. Kozen, A. Silva, Practical Coinduction. Technical report, Computing and Information Science (Cornell University, Ithaca, 2012). http://hdl.handle.net/1813/30510

90. S.G. Krantz, Handbook of Logic and Proof Techniques for Computer Science (Birkhäuser. Boston (2002). https://doi.org/10.1007/978-1-4612-0115-1

91. T. Kropf, Introduction to Formal Hardware Verification (Springer. Berlin (1999). https://doi.org/10.1007/978-3-662-03809-3

92. F. Kröger, Temporal logic of programs, in EATCS Monographs in Theoretical Computer Science (Springer. Berlin (1987). https://doi.org/10.1007/978-3-642-71549-5

93. F. Kröger, S. Merz, *Temporal Logic and State Systems*. Texts in Theoretical Computer Science (Springer, Berlin, 2008). https://doi.org/10.1007/978-3-540-68635-4

94. R. Kumar, M.O. Myreen, M. Norrish, S. Owens, CakeML: a verified implementation of ML, in *POPL 2014: 41st ACM SIGPLAN-SIGACT Symposium on Principles of Programming Languages, San Diego, CA, USA, January 22–24, 2014* (ACM, New York, 2014), pp. 179–191. https://doi.org/10.1145/2535838.2535841

95. S. Kundu, S. Lerner, R.K. Gupta, High-Level Verification: Methods and Tools for Verification of System-Level Designs (Springer. New York (2011). https://doi.org/10.1007/978-1-4419-9359-5

96. L. Lamport, "Sometime" is sometimes "Not Never": on the temporal logic of programs, in *POPL '80, 7th ACM SIGPLAN-SIGACT Symposium on Principles of Programming Languages, Las Vegas, Nevada, USA, January 28–30, 1980* (ACM, New York, 1980), pp. 174–185. https://doi.org/10.1145/567446.567463

97. L. Lamport, The temporal logic of actions. Trans. Program. Lang. Syst. **16**(3), 872–923 (1994). https://doi.org/10.1145/177492.177726

98. L. Lamport, *Specifying Systems – The TLA+ Language and Tools for Hardware and Software Engineers* (Addison-Wesley Professional, Boston, 2002). http://research.microsoft.com/en-us/um/people/lamport/tla/book.html

99. L. Lamport, The TLA+ Home Page (2020). https://lamport.azurewebsites.net/tla/tla.html

100. J. Laski, W. Stanley, Software Verification and Analysis - An Integrated, Hands-On Approach (Springer. London (2009). https://doi.org/10.1007/978-1-84882-240-5

101. X. Leroy, A formally verified compiler back-end. J. Autom. Reason. **43**, 363–446 (2009). https://doi.org/10.1007/s10817-009-9155-4

102. W. Li, *Mathematical Logic — Foundations for Information Science*. Progress in Computer Science and Applied Logic, vol. 25, 2nd edn. (Birkhäuser, Basel, 2014). https://doi.org/10.1007/978-3-0348-0862-0

103. J. Loeckx, H.-D. Ehrich, M. Wolf, *Specification of Abstract Data Types* (Wiley & Teubner, Chichester, 1996). https://books.google.at/books/about/Specification_of_Abstract_Data_Types.html?id=L7NQAAAAMAAJ

104. R. Lover, Elementary Logic - For Software Development (Springer. London (2008). https://doi.org/10.1007/978-1-84800-082-7

105. J. Magee, J. Kramer, *Concurrency – State Models & Java Programs*, 2nd edn. (Wiley, Chichester, 2006). https://www.wiley.com/en-us/Concurrency

106. D. Makinson, *Sets, Logic and Maths for Computing*. Undergraduate Topics in Computer Science (Springer, London, 2008). https://doi.org/10.1007/978-1-4471-2500-6

107. Z. Manna, S. Ness, J. Vuillemin, Inductive methods for proving properties of programs. Commun. ACM **16**(8) (1973). https://doi.org/10.1145/355609.362336

108. Z. Manna, A. Pnueli, The Temporal Verification of Reactive Systems - Specification (Springer. New York (1992). https://doi.org/10.1007/978-1-4612-0931-7

109. Z. Manna, A. Pnueli, Temporal Verification of Reactive Systems - Safety (Springer. New York (1995). https://doi.org/10.1007/978-1-4612-4222-2

110. Z. Manna, A. Pnueli, Temporal verification of reactive systems — response, in *Time for Verification — Essays in Memory of Amir Pnueli*. Lecture Notes in Computer Science, ed. by Z. Manna, D.A. Peled (Springer, Berlin, 2010), pp. 279–361. https://doi.org/10.1007/978-3-642-13754-9_13

111. J. McCarthy, A basis for a mathematical theory of computation, in *Computer Programming and Formal Systems*. Studies in Logic and the Foundations of Mathematics, vol. 35, ed. by P. Braffort and D. Hirschberg (1963), pp. 33–70. https://doi.org/10.1016/S0049-237X(08)72018-4and http://www-formal.stanford.edu/jmc/basis.html

112. W. McCune, Solution of the Robbins problem. J. Autom. Reason. **19**(3), 263–276 (1997). https://doi.org/10.1023/A:1005843212881

113. S. Merz, *The Specification Language TLA+, in [22]* (Springer, Berlin, 2008), pp. 401–451

114. Microsoft. Static Driver Verifier (2021). https://docs.microsoft.com/en-us/windows-hardware/drivers/devtest/static-driver-verifier

115. A.R. Milner, *Calculus of Communicating Systems*. Lecture Notes in Computer Science, vol. 92 (Springer, Berlin, 1980). https://doi.org/10.1007/3-540-10235-3

116. R. Milner, *Communication and Concurrency* (Prentice-Hall, Upper Saddle River, 1989). http://catalogue.pearsoned.co.uk/educator/product/Communication-Concurrency/9780131150072.page

117. R. Milner, *Operational and Algebraic Semantics of Concurrent Processes, in [177], Chapter 19* (Elsevier and MIT Press, Amsterdam, 1990), pp. 1201–1242

118. R. Milner, *Communicating and Mobile Systems: The Pi-Calculus* (Cambridge University Press, Cambridge, 1999). https://www.cambridge.org/gb/academic/subjects/computer-science/communications-information-theory-and-security/communicating-and-mobile-systems-pi-calculus

119. J.C. Mitchell, *Foundations for Programming Languages. Foundations of Computing Series* (MIT Press, Cambridge, 1996). https://mitpress.mit.edu/books/foundations-programming-languages

120. Mizar Home Page (2019). http://www.mizar.org/

121. C. Morgan, *Programming from Specifications*, 2nd edn. (Prentice Hall, Hertfordshire, 1994). http://www.cse.unsw.edu.au/~carrollm/ProgrammingFromSpecifications.pdf

122. T. Mossakowski et al., *Hets – the Heterogeneous Tool Set. Research Group Theoretical Computer Science* (Otto von Guericke Universität Magdeburg, 2015). http://theo.cs.uni-magdeburg.de/Research/Hets.html

123. T. Mossakowski, A.E. Haxthausen, D. Sannella, A. Tarlecki, *CASL - the Common Algebraic Specification Language, in [22]* (Springer, Berlin, 2008), pp. 241–298

124. T. Mossakowski, L. Schrüder, M. Roggenbach, Horst-Reichl, Algebraic-coalgebraic specification in CoCASL. J. Logic Algebraic Program. **67**(1–2), 146–197 (2005). https://doi.org/10.1016/j.jlap.2005.09.006

125. P.D. Mosses, CASL for CafeOBJ users, in *[54], Chapter 6* (Elsevier, Amsterdam, 2000), pp. 121–144. https://doi.org/10.1016/B978-044450556-9/50066-6 and http://www.brics.dk/RS/00/51

126. P.D. Mosses, (ed.), *CASL Reference Manual – The Complete Documentation of the Common Algebraic Specification Language*. Lecture Notes in Computer Science, vol. 2960 (Springer, Berlin, 2004). https://doi.org/10.1007/b96103 and http://www.informatik.uni-bremen.de/cofi/CASL-RM.pdf

127. M. Nebel, *Formale Grundlagen der Programmierung*. Studienbücher Informatik (Vieweg+Teubner, Wiesbaden, 2012). https://doi.org/10.1007/978-3-8348-2296-3

128. C. Newcombe, T. Rath, F. Zhang, B. Munteanu, M. Brooker, M. Deardeuff, How Amazon web services uses formal methods. Commun. ACM **58**(4), 66–73 (2015). https://doi.org/10.1145/2699417

129. F. Nielson, H.R. Nielson, *Formal Methods: An Appetizer*. Undergraduate Topics in Computer Science (Springer International Publishing, Cham, 2019). https://doi.org/10.1007/978-3-030-05156-3

130. F. Nielson, H.R. Nielson, C. Hankin, Principles of Program Analysis (Springer. Berlin (1999). https://doi.org/10.1007/978-3-662-03811-6

131. H.R. Nielson, F. Nielson, *Semantics with Applications: An Appetizer*. Undergraduate Topics in Computer Science (Springer, London, 2007). https://doi.org/10.1007/978-1-84628-692-6

132. T. Nipkow, L. Paulson, et al., *Isabelle* (University of Cambridge, UK, and Technische Universität München, 2021). https://isabelle.in.tum.de

133. T. Nipkow, G. Klein, Concrete Semantics - With Isabelle/HOL (Springer. Heidelberg (2014). https://doi.org/10.1007/978-3-319-10542-0

134. The ocaml.org Team. OCaml (2015). https://ocaml.org

135. University of Illinois and University of Iasi. K Framework (2017). http://www.kframework. org

136. OpenJML.org. OpenJML — Does your program do what it is supposed to do? (2021). https:// www.openjml.org

137. G. O'Regan, *Concise Guide to Formal Methods – Theory, Fundamentals and Industry Applications*. Undergraduate Topics in Computer Science (Springer International Publishing, Cham, 2017). https://doi.org/10.1007/978-3-319-64021-1

138. G. O'Regan, *Mathematics in Computing – An Accessible Guide to Historical, Foundational and Application Contexts*. Undergraduate Topics in Computer Science (Springer International Publishing, Cham, 2020). https://doi.org/10.1007/978-3-030-34209-8

139. S. Owicki, D. Gries, An axiomatic proof technique for parallel programs I. Acta Inform **6**(4), 319–340 (1976). https://doi.org/10.1007/BF00268134

140. S. Owicki, D. Gries, Verifying properties of parallel programs: an axiomatic approach. Commun. ACM **19**(5), 279–285 (1976). https://doi.org/10.1145/360051.360224

141. P. Padawitz. Swinging types = functions + relations + transition systems. Theor. Comput. Sci. **243**, 93–165 (2000). https://doi.org/10.1016/S0304-3975(00)00171-7 and http://fldit-www. cs.uni-dortmund.de/

142. L.C. Paulson, A fixedpoint approach to (co)inductive and (co)datatype definitions, in *Proof, Language, and Interaction — Essays in Honour of Robin Milner*, ed. by G. Plotkin, C.P. Stirling, M. Tofte (MIT Press, Cambridge, 2000), pp. 187–211. http://citeseerx.ist.psu.edu/ viewdoc/summary?doi=10.1.1.146.1835

143. L.J. Peterson, Petri nets. Comput. Surv. **9**(3), 223–252 (1977). https://doi.org/10.1145/356698. 356702

144. G.D. Plotkin, A structural approach to operational semantics. Technical Report DAIMI FN-19, Computer Science Department (Aarhus University, Denmark, Reprinted in the Journal of Logic and Algebraic Programming, vols. 60–61. July-December **2004**, 17–139 (1981). https:// doi.org/10.1016/j.jlap.2004.05.001

145. A. Pnueli, The temporal logic of programs, in *SFCS '77, 18th Annual Symposium on Foundations of Computer Science, Providence, RI, USA, October 31 - November 2* (IEEE Computer Society, Washington, DC, 1977), pp. 46–57. https://doi.org/10.1109/SFCS.1977.32

146. N. Polikarpova, J. Tschannen, C.A. Furia, A fully verified container library, in *FM 2015: Formal Methods, 20th International Symposium, Oslo, Norway, June 24-26, 2015*, volume 10001 of *Lecture Notes in Computer Science*, vol. 10001, ed. by N. Bjørner, F. de Boer (Springer International Publishing, Cham, 2016), pp. 414–434. https://doi.org/10.1007/978-3-319-19249-9_26

147. B. Potter, J. Sinclair, D. Till, *An Introduction to Formal Specification and Z*. International Series in Computer Science, 2nd edn. (Prentice Hall, London, 1996), https://books.google.at/ books/about/An_Introduction_to_Formal_Specification.html?id=KqhQAAAAMAAJ

148. The KeY Project. The KeY to Software Correctness (2021). https://www.key-project.org

149. S. Ray, Scalable Techniques for Formal Verification (Springer. New York (2010). https://doi. org/10.1007/978-1-4419-5998-0

150. W. Reisig, Understanding petri nets - modeling techniques, in Analysis Methods, Case Studies (Springer. Berlin (2013). https://doi.org/10.1007/978-3-642-33278-4

151. J.A. Robinson, A machine-oriented logic based on the resolution principle. J. ACM **12**(1), 23–41 (1965). https://doi.org/10.1145/321250.321253

152. J.A. Robinson, A. Voronkov, (eds.), *Handbook of Automated Reasoning* (North-Holland, Amsterdam, 2001). https://www.sciencedirect.com/book/9780444508133/handbook-of-automated-reasoning

153. M. Roggenbach, L. Schröder, Towards trustworthy specification i: consistency checks, in *Recent Trends in Algebraic Development Techniques: 15th International Workshop, WADT 2001 Joint with the CoFI WG Meeting Genova, Italy, April 1–3, 2001*. Selected Papers, Lecture Notes in Computer Science, vol. 2267, ed. by M. Cerioli G. Reggio (Springer, Berlin, 2002). https://doi.org/10.1007/3-540-45645-7_15

154. A.W. Roscoe, *The Theory and Practice of Concurrency* (Prentice Hall, Upper Saddle River, 1997). http://www.cs.ox.ac.uk/bill.roscoe/publications/68b.pdf
155. K. Rosen, *Discrete Mathematics and Its Applications*, 7th edn. (McGraw-Hill Education, Columbus, 2012). http://highered.mheducation.com/sites/0073383090/information_center_view0/index.html
156. G. Roşu, T. Florin Şerbănuţă, An overview of the K semantic framework. J. Logic Algebraic Program. **79**(6), 397–434 (2010). https://doi.org/10.1016/j.jlap.2010.03.012
157. D. Sangiorgi, On the origins of bisimulation and coinduction. ACM Trans. Program. Lang. Syst. **31**(4), 15:1–15:41 (2009). https://doi.org/10.1145/1516507.1516510
158. D. Sannella, A. Tarlecki, Foundations of Algebraic Specification and Formal Software Development (Springer. Berlin (2012). https://doi.org/10.1007/978-3-642-17336-3
159. M. Schenke, Logikkalküle in der Informatik - Wie wird Logik vom Rechner genutzt? Studienbücher Informatik (Springer Vieweg. Wiesbaden (2013). https://doi.org/10.1007/978-3-8348-2295-6
160. D.A. Schmidt, *Denotational Semantics – A Methodology for Language Development* (Allyn and Bacon, Boston, 1986). http://people.cis.ksu.edu/~schmidt/text/densem.html
161. D.A. Schmidt, *The Structure of Typed Programming Languages* (MIT Press, Cambridge, 1994). https://mitpress.mit.edu/books/structure-typed-programming-languages
162. C. Schneider, The summation package sigma (2021). https://www.risc.jku.at/research/combinat/software/Sigma/
163. F.B. Schneider, *On Concurrent Programming*. Graduate Texts in Theoretical Computer Science (Springer, New York, 1997). https://doi.org/10.1007/978-1-4612-1830-2
164. K. Schneider, *Verification of Reactive Systems - Formal Methods and Algorithms*. Texts in Theoretical Computer Science (Springer, Berlin, 2004). https://doi.org/10.1007/978-3-662-10778-2
165. W. Schreiner, The RISC ProofNavigator: a proving assistant for program verification in the classroom. Formal Aspects Comput. **21**(3), 277–291 (2009). https://doi.org/10.1007/s00165-008-0069-4 and https://www.risc.jku.at/people/schreine/papers/fac2008.pdf
166. W. Schreiner, *The RISC ProofNavigator. Research Institute for Symbolic Computation (RISC)* (Johannes Kepler University, Linz, 2011). https://www.risc.jku.at/research/formal/software/ProofNavigator
167. W. Schreiner, Computer-assisted program reasoning based on a relational semantics of programs. Electron. Proc. Theor. Comput. Sci. (EPTCS) 79, 124–142 (2012). P. Quaresma, R.-J. Back (eds.), *Proceedings of the First Workshop on CTP Components for Educational Software (THedu'11)* (Wrocław, Poland, 2011). https://doi.org/10.4204/EPTCS.79.8 and https://www.risc.jku.at/research/formal/software/ProgramExplorer/papers/THeduPaper-2011.pdf
168. W. Schreiner, *The RISC ProgramExplorer. Research Institute for Symbolic Computation (RISC)* (Johannes Kepler University, Linz, 2015). https://www.risc.jku.at/research/formal/software/ProgramExplorer
169. W. Schreiner, *The RISC Algorithm Language (RISCAL). Research Institute for Symbolic Computation (RISC)* (Johannes Kepler University, Linz, 2019). https://www.risc.jku.at/research/formal/software/RISCAL
170. W. Schreiner, A. Brunhuemer, C. Fürst, Teaching the formalization of mathematical theories and algorithms via the automatic checking of finite models, in Post-Proceedings ThEdu'17, Theorem proving components for Educational software, Gothenburg, Sweden, August 6, , EPTCS, vol. 267, ed. by P. Quaresma. W. Neuper **2018**, 120–139 (2017). https://doi.org/10.4204/EPTCS.267.8
171. D. Scott, C. Strachey, Towards a mathematical semantics for computer languages, in *Proceedings of the Symposium on Computers and Automata*. Microwave Research Institute Symposia Series, ed. by J. Fox, vol. 21 (Polytechnic Institute of Brooklyn Press, New York, 1971), pp. 19–46. Also: Technical Monograph PRG-6, Oxford University Computing Laboratory, Programming Research Group, Oxford, https://www.researchgate.net/publication/237107559_Towards_a_Mathematical_Semantics_for_Computer_Languages
172. seL4 Project. The seL4 Microkernel (2021). https://sel4.systems

173. E.Y. Shapiro, The fifth generation project - a trip report. Commun. ACM **26**(9), 637–641 (1983). https://doi.org/10.1145/358172.358179
174. Stanford Encyclopoedia of Philosophy (2020). https://plato.stanford.edu
175. The Theorema System (2019). https://www.risc.jku.at/research/theorema/software/
176. Vampire (2021). https://vprover.github.io
177. J. van Leeuwen, (ed.), *Handbook of Theoretical Computer Science, Volume B: Formal Models and Semantics* (Elsevier and MIT Press, Amsterdam, 1990). https://mitpress.mit.edu/books/handbook-theoretical-computer-science-volume-b
178. D.J. Velleman, *How To Prove It – A Structured Approach*, 2nd edn. (Cambridge University Press, Cambridge, 2006). http://www.cambridge.org/at/academic/subjects/mathematics/logic-categories-and-sets/how-prove-it-structured-approach-2nd-edition
179. F. Wiedijk, (ed.) *The Seventeen Provers of the World*. Lecture Notes in Artificial Intelligence, vol. 3600 (Springer, Berlin, 2006). https://doi.org/10.1007/11542384
180. G. Winskel, *The Formal Semantics of Programming Languages – An Introduction* (MIT Press, Cambridge, 1994). https://mitpress.mit.edu/books/formal-semantics-programming-languages
181. M. Wirsing, *Algebraic Specification, in [177], Chapter 13* (Elsevier and MIT Press, Amsterdam, 1990), pp. 675–788
182. J. Woodcock, P.G. Larsen, J. Bicarregui, J. Fitzgerald, Formal methods: practice and experience. ACM Comput. Surv. **41**(4), 19 (2009). https://doi.org/10.1145/1592434.1592436
183. Z3 Theorem Prover (2021). https://github.com/Z3Prover
184. T. Florin Şerbănuţă, The K Primer (version 3.3), in *Electronic Notes in Theoretical Computer Science*, vol. 304 (2014), pp. 57–80. https://doi.org/10.1016/j.entcs.2014.05.003

# Index

Printed in the United States
by Baker & Taylor Publisher Services